Recent Developments in Therapeutic Drug Monitoring and Clinical Toxicology

Recent Developments in Therapeutic Drug Monitoring and Clinical Toxicology

edited by

Irving Sunshine

Palo Alto, California

CRC Press is an imprint of the
Taylor & Francis Group, an **informa** business

CRC Press
Taylor & Francis Group
6000 Broken Sound Parkway NW, Suite 300
Boca Raton, FL 33487-2742

First issued in paperback 2019

ISBN-13: 978-0-8247-8586-4 (hbk)
ISBN-13: 978-0-367-40283-9 (pbk)

Visit the Taylor & Francis Web site at
http://www.taylorandfrancis.com

and the CRC Press Web site at
http://www.crcpress.com

Preface

Worldwide attendees assembled in Barcelona, Spain, at the Second International Conference on TDM-Toxicology. During this conference, a series of symposia, workshops and poster sessions were presented. Those involved in these activities were invited to submit manuscripts covering their contributions. This volume is a compilation of the submitted material. As editor, I have made a few necessary editorial corrections. Otherwise, the submitted material represents the authors' attempt to share their respective innovative efforts with their colleagues.

A noteworthy aspect of this conference was its international base. This volume is representative of the scope of this interest. Most of the participants came from western European countries, Japan, and the United States. However, it was gratifying to all that there were a significant number of attendees from the eastern European countries and the Mideast. Hopefully, future meetings will attract a greater and more diverse audience.

The many chapters are grouped into parts. Each part contains related presentations from all aspects of the congress, the symposia, the poster sessions, and the open communications. This arrangement, a relatively arbitrary one, requires the reader to seek a particular title or subject in the Subject Index.

Sandra Beberman and her associates are to be commended for their significant help in producing this volume. A hearty "thank you" is due Marcel Dekker, Inc., for publishing these proceedings.

Irving Sunshine

Contents

Part III Selected High-Performance Liquid Chromatographic Methods

Part I

Therapeutic Monitoring

1

A Case for Therapeutic Drug Monitoring

Ian D. Watson *Fazarkerly Hospital, Liverpool, England*

I. INTRODUCTION

The concept of therapeutic drug monitoring (TDM) may be summarized as follows: If there is a concentration-effect relationship, then the measurement of serum drug concentrations, appropriately sampled, with adequate dosing and clinical history available, allows rational interpretation of results and subsequent dosage adjustment if necessary.

There has been criticism of this concept (1-3). In particular, the relationship between concentration and desired effect has been questioned. Providers of TDM services, while recognizing some of the limitations in the studies supporting TDM, see it in a wider context (4-8); the detection of poor compliance and concentration-associated toxicity are frequently advanced as reasons for measuring serum drug concentrations.

The case for a TDM service based on informed interpretation of serum drug concentrations will be considered and the evidence in support of such a concept presented.

II. PRINCIPLES

TDM is a procedure for assisting a physician in making therapeutic judgments. The questions being asked of a serum drug concentration include the following:

1. Is the patient taking the drug?
2. Is the patient getting enough drug?
3. Is the patient getting too much drug?
4. Is a dose adjustment needed?
5. Is this drug causing the symptoms?

The concept of "too much" and "too little" have led to the idea of a therapeutic range. This enables there to be a subsequent rationale for making a dosage adjustment. This is predictably achievable if pharmacokinetic principles are applied. Too rigid an interpretation is inappropriate: Some individuals respond below the range and some need to be above the range, so a more correct term is target range. Results thereafter are interpreted in the light of clinical information.

Kinetic principles of dosage adjustment still apply if required, but a target range implies a population norm; clinical response must be taken into account. In addition, individual variation in response may require dosage individualization as well as consideration of the modifying influences of drug interactions and disease on pharmacokinetics and pharmacodynamics.

Despite the attractiveness of the rationale of TDM, very few drugs are actually monitored. Spector (1) lists a number of criteria that must be fulfilled before the target concentration strategy can be accepted. The crucial question is: "Is there a concentration-effect relationship?" Both critics (1,2) and proponents (6,7) of TDM agree that there is insufficient data to answer these questions properly. The ideal solution to this is to conduct prospective, randomized, blind studies of sufficient power to confirm or refute the hypothesize for each drug monitored.

How can such studies be conducted? With what can TDM be compared? Some would consider no monitoring to be unethical. Can the costs of such studies be justified, and who is going to meet them? These are a few of the questions that supporters of large-scale trials must answer. We need to know the answers to three questions asked of TDM:

1. Is there a concentration-effect relationship?
2. Is a TDM service used appropriately?
3. Is TDM, used correctly, cost-effective?

Of the studies to answer these questions, particularly the first, many are retrospective in nature. Although there are prospective studies, these are often carried out on small numbers of patients and can be criticized for lack of power.

Below are briefly summarized some studies that strongly suggest that TDM is a successfully tested hypothesis.

III. A CONCENTRATION-EFFECT RELATIONSHIP?

A. Prospective Studies

Digoxin

In 1974 a comparison between two institutions prescribing digoxin, one with and one without concentration-level monitoring, showed that the risk of toxicity was 2.1 times greater if monitoring was not performed (9).

Theophylline

Peak expiratory rates improved more readily in a monitored versus a nonmonitored group, and the former group suffered no fatalities (10). FEV_1 and FVC were statistically significantly improved at theophylline concentrations of 110 μmol/L (20 mg/L) versus 55 μmol/L (10 mg/L) with a shorter duration of therapy for the high-concentration group, yet there was no difference in side-effect rates (11).

Anticonvulsants

Increasing phenytoin concentrations above 40 μmol/L (10 mg/L) resulted in improved seizure control in newly diagnosed epileptics (12). Dramatic improvements, of 98% and 92% decrease in grand mal and partial seizure rates, were found if the plasma concentration of phenytoin or carbamazepine levels were within the target range (13), and in one of the earliest prospective studies there was a decline in mean seizure rate as phenytoin levels increased (14). There is also evidence that there is an improvement in seizure control if serum valproate exceeds around 350 μmol/L (15), although the small numbers in the study do weaken its case.

Aminoglycosides

Studies by Moore et al. indicate that in gram-negative sepsis there is a strong association between clinical response and peak aminoglycoside concentrations of peak aminoglycoside concentrations or peak concentration/MIC ratio (16,17).

Methotrexate

Knowledge of methotrexate serum concentrations predicts toxicity if concentrations are above 1 μmol/L 48 h post-high-dose therapy (18,19).

Lithium

Lithium levels above 1.4 μmol/L result in excessive toxicity with no increased benefit, while below 0.9 μmol/L the benefit rate begins to decline (20).

B. Retrospective Studies

While prospective studies provide some evidence, they do lack power. However, retrospective studies reinforce the findings of the prospective studies.

General

The use of TDM resulted in a decrease in toxic drug reactions based on a meta-analysis of 14 studies with a total of over 3400 subjects (21).

Digoxin

A study with over 5000 subjects showed a nearly fivefold increase in mortality rate, over a background rate of 2%, as the serum digoxin concentration increased above 2.7 nmol/L (22).

Theophylline

Monitoring theophylline levels markedly altered infusion rates in pediatric patients with status asthmaticus and improved outcome (23).

Anticonvulsants

In one of the very earliest studies there was a significant improvement in fit frequency with phenytoin levels above 60 μmol/L (24), while symptoms of phenytoin toxicity have been correlated with phenytoin levels (25). With appropriate use of serum levels, seizure control is significantly improved (26).

Aminoglycosides

The incidence of ototoxicity is associated with duration of high aminoglycoside trough levels (27).

IV. IS A TDM SERVICE USED APPROPRIATELY?

If the premise of TDM is accepted, is it used correctly? A number of studies attest to the fact that many TDM service users misuse the service, as evidenced by a meta-analysis of 16 studies (28). Individual studies indicate high levels of invalid sampling (26,29,30). A prospective randomized crossover study showed that a properly applied TDM service decreased assay numbers and readmission rates (31). Withdrawal of such a TDM service for digoxin resulted in a decrease in the quality of sampling, dosing, and interpretation (32).

V. IS TDM COST-EFFECTIVE?

Vozeh has thoroughly reviewed the cost effectiveness of TDM (7), and there are a number of other studies (33–35). Vozeh concludes his article by noting that there was little direct evidence of cost benefit of TDM and that prospective evaluation was needed, but he went on to note: "There are sufficient data to conclude that the cost-benefit ratio can be improved by performing therapeutic drug monitoring with the appropriate expertise. There is also indirect evidence that for some drugs, when it is applied correctly, therapeutic drug monitoring is a cost-effective procedure."

VI. CONCLUSION

The weight of evidence in favor of TDM for some drugs is very strong. I would submit that TDM has been adequately validated for phenytoin, lithium, theophylline, aminoglycosides, and probably digoxin. There is a reasonable but not conclusive case for carbamazepine, valproate, and phenobarbitone. Reference 8 gives a critique for each of these drugs.

It is instructive to compare the efforts being expended to determine the place of cyclosporine TDM. Despite formidable efforts in the face of a great number of variables, it has still not proved possible to reach a definitive conclusion on the benefit of cyclosporine monitoring to avoid allograft rejection and predict toxicity, a debate which continues. In summarizing the proceedings of the recent Hawks Cay meeting, Holt answers the question, "Is it worth measuring Sandimmun?" by saying: "The best we can hope for is that they [concentrations] assist in the differential diagnosis of clinical events and that they [concentrations] provide, a guide to optimal dosing" (36). The consensus document (37) summarizes the way forward for such monitoring. The position of other monitored drugs is often not dissimilar, though from the evidence of the studies quoted earlier, one can be much more positive about some of them.

TDM need not be restricted to the narrow sense of serum drug concentration measurement. TDM has a teaching role in clinical pharmacology (38). Some work on cyclosporine indicates that the area under the concentration curve may be a more meaningful parameter (39). Inclusion of effect parameters into models could be appropriate. Knowledge of phenotype has been used in the past to determine appropriate dosing, e.g., acetylator status. Susceptibility of tumors to antineoplastic therapy may be predicted from knowing the intracellular glutathione status (40) or even mdr-1 gene expression (41).

Insofar as concentration-effect relationships require further validation, large-scale efforts would be required to definitively prove any of the range of drugs noted above as being worthy of TDM. It is doubtful this will be done unless adequate funding becomes available. Given the studies to date and the weight of daily practice in favor of well-practised TDM, it could be argued that there is no need for such studies on the basis of anticipated cost-benefit ratio. However, perhaps one study of a candidate drug should be conducted to ascertain the viability and practicality of large-scale studies.

The establishment of an international organization for TDM and toxicology could be the ideal vehicle for ascertaining practitioners' feelings on the need for such large-scale studies and, if necessary, assist in organizing them.

REFERENCES

1. R. Spector, G. D. Park, G. F. Johnson, and E. S. Vessell. Therapeutic drug monitoring. *Clin. Pharmacol. Ther., 43*:345–353 (1988).
2. G. T. McInnes. The value of therapeutic drug monitoring to the practising physician—An hypothesis in need of testing. *Br. J. Clin. Pharmacol., 27*:281–284 (1989).
3. This volume.
4. Anonymous. What therapeutic drugs should be monitored. *Lancet, ii*:309–310 (1985).
5. S. Vozeh. Therapeutic drug monitoring. *Clin. Pharmcol. Ther., 48*:713–714 (1988).
6. I. D. Watson and A. H. Thomson. The value of therapeutic drug monitoring to the practising physician—An hypothesis needing sensible application. *Br. J. Clin. Pharmacol., 28*:619–620 (1989).
7. S. Vozeh. Cost-effectiveness of therapeutic drug monitoring. *Clin. Pharmacokinet., 13*:131–140 (1987).
8. M. J. Brodie and J. Feely. Practical clinical pharmacology. Therapeutic drug monitoring and clinical trials. *Br. Med. J., 296*:1110–1114 (1988).
9. J. Koch-Weser, D. W. Duhme, and D. J. Greenblatt. Influence of serum digoxin concentration measurements on frequency of digitoxicity. *Clin. Pharmacol. Ther., 16*:284–287 (1974).
10. S. F. Hurley, L. J. Dziukas, J. J. McNeil, and M. J. Brignell. A randomised controlled clinical trial of pharmacokinetic theophylline dosing. *Am. Rev. Respir. Dis., 134*:1219–1224 (1986).
11. S. Vozeh, G. Kewitz, A. Perruchoud, M. Tschan, C. Kopp, M. Heitz, and F. Follath. Theophylline serum concentration and therapeutic effect in severe acute bronchial obstructions: The optimal use of intravenously administered aminophylline. *Am. Rev. Respir. Dis., 125*:181–184 (1982).
12. E. H. Reynolds, S. D. Shorvon, A. W. Galbraith, D. Chadwick, C. I. Dellaportas, and L. Vydelingum. Phenytoin monotherapy for epilepsy: A long-term prospective study, assisted by serum level monitoring, in previously untreated patients. *Epilepsia, 22*:475–488 (1981).
13. S. D. Shorvon, D. Chadwick, A. W. Galbraith, and E. H. Reynolds. One drug for epilepsy. *Br. Med. J., 1*:474–476 (1978).
14. L. Lund. Anticonvulsant effect of diphenylhydantoin relative to plasma levels. A prospective three-year study in ambulant patients with generalised epileptic seizures. *Arch . Neurol., 31*:289–294 (1974).
15. L. Gram, H. Flachs, A. Wurtz-Jorgensen, J. Parnas, and B. Andersen. Sodium valproate, serum level and clinical effect in epilepsy: A controlled study. *Epilepsia, 20*:303–312 (1979).
16. R. D. Moore, C. R. Smith, and P. S. Leitman. Association of aminoglycoside plasma levels with therapeutic outcome in Gram-negative pneumonia. *Am. J. Med., 77*: 657–662 (1984).
17. R. D. Moore, P. S. Lietman, and C. R. Smith. Clinical response to aminoglycoside therapy: Importance of the ratio of peak concentration to minimal inhibitory concentration. *J. Infec. Dis., 155*:93–99 (1987).
18. R. G. Stoller, K. R. Hande, S. A. Jacobs, S. A. Rosenberg, and B. A. Chabner. Use of plasma pharmacokinetics of predict and prevent methotrexate toxicity. *N. Engl. J. Med. 297*:630–634 (1977).
19. A. Nirenberg, C. Mosende, B. M. Mehta, A. L. Gisolfi, and G. Rosen. High dose methotrexate with citrovorum factor resin. Predictive value of serum methotrexate concentrations and corrective measures to avert toxicity. *Cancer Treatment Rep. 61*:779–783 (1977).
20. R. F. Prien, E. M. Caffey, and C. J. Klett. Relationship between serum lithium level and clinical response in acute mania treated with lithium. *Br. J. Psychiat., 120*: 409–414 (1972).
21. L. D. Ried, J. R. Horn, and D. A. McKenna. Therapeutic drug monitoring, reduces toxic drug reactions: A meta analysis. *Ther. Drug Monitor., 12*:72–78 (1990).

22. G. J. Ordog, S. Benaron, V. Bhasin, J. Wasserberger, and S. Balasubramanium. Serum digoxin levels and mortality in 5,100 patients. *Ann. Emerg. Med., 16*:32–39 (1987).
23. J. Fox, P. Hicks, B. R. Feldman, W. J. Davis, and C. H. Feldman. Theophylline blood levels as a guide to intravenous therapy in children. *Am. J. Dis. Child., 136*:928–930 (1982).
24. F. Buchthal, O. Svensmark, and P. J. Schiller. Clinical and electroencephalographic correlations with serum levels of diphenylhydantoin. *Arch. Neurol., 2*:624–630 (1960).
25. H. Kutt, W. Winters, R. Kokenge, and F. McDowell. Diphenylhydantoin metabolism, blood levels and toxicity. *Arch. Neurol., 11*:642–648 (1964).
26. R. S. Beardsley, J. M. Freeman, and F. A. Appel. Anticonvulsant serum levels are useful only if the physician appropriately uses them. An assessment of the impact of providing serum level data to physicians. *Epilepsia, 24*:330–335 (1983).
27. C. R. Smith, R. D. Moore, and S. A. Lerner. Aminoglycoside ototoxicity—Interaction between duration of therapy and plasma trough levels. *Clin. Pharmacol. Ther., 39*:229 (1986).
28. L. D. Ried, D. A. McKenna, and J. R. Horn. Effect of therapeutic drug monitoring services on the number of serum drug assays ordered for patients: A meta analysis. *Ther. Drug Monitor., 11*:253–263 (1989).
29. H. I. Bussey and E. W. Hoffman. A prospective evaluation of therapeutic drug monitoring. *Ther. Drug Monitor., 5*:245–248 (1983).
30. H. W. Clague, Y. Twum-Barima, and S. G. Carruthers. An audit of requests for therapeutic drug monitoring of digoxin. Problems and pitfalls. *Ther. Drug Monitor., 5*:249–254 (1983).
31. D. S. Wing and H. J. Duff. The impact of a therapeutic drug monitoring programme for phenytoin. *Ther. Drug Monitor., 11*:32–37 (1989).
32. J. K. Michalko and L. Blain. An evaluation of a clinical pharmacokinetic service for serum digoxin levels. *Ther. Drug Monitor., 9*:311–319 (1987).
33. C. J. Destache, S. K. Meyer, M. T. Padomek, and B. G. Ortmeier. Impact of a clinical pharmacokinetic service on patients treated with aminoglycosides for Gram-negative infections. *Drug Intell. Clin. Pharm., 23*:33–38 (1989).
34. K. D. Crist, M. C. Nahata, and J. Ety. Positive impact of a therapeutic drug monitoring programme on total aminoglycoside dose and cost of hospitalisation. *Ther. Drug Monitor., 9*:306–310 (1987).
35. B. J. Kimelblatt, K. Bradbury, L. Chodoff, T. Aggour, and B. Mehl. Cost-benefit analysis of an aminoglycoside monitoring service. *Am. J. Hosp. Pharm., 43*:1205–1209 (1986).
36. D. W. Holt. Conference summary of Hawk's Cay meeting. *Transplant Proc., 22*:1356 (1990).
37. Consensus document: Hawk's Cay meeting on therapeutic drug monitoring of cyclosporine. *Transplant Proc., 22*:1357–1361 (1990).
38. E. P. Brass. The drug analysis laboratory: A resource for teaching clinical pharmacology to students and residents. *Clin. Pharmacol. Ther., 46*:245–249 (1979).
39. J. Grevel, K. L. Napoli, S. Gibbons, and B. D. Kahan. Area-under-the-curve monitoring of cyclosporine therapy: Performance of different assay methods and their target concentrations. *Ther. Drug Monitor., 12*:8–15 (1990).
40. T. C. Hamilton, M. A. Winker, and K. J. Louie. Augmentation of adriamycin, melphalan and cisplatin, cytotoxicity in drug-resistant and sensitive human ovarian carcinoma cell lines by bythionine sulfoximine mediated glutathione depletion. *Biochem. Pharmacol., 34*:2583–2586 (1985).
41. K. Ueda, D. P. Clark, C.-J. Chen, I. B. Roninson, M. M. Gottesman, and I. Pastan. The human multi-drug resistance (mdr-1) gene. *J. Biol. Chem.* 262:505–508 (1987).

2

A Clinical Chemistry Therapeutic Drug Monitoring Service

Leonor Pou and Francesc Campos *Hospital General Vall d'Hebrón, Barcelona, Spain*

I. INTRODUCTION

Therapeutic drug monitoring (TDM) has been shown to be effective, valuable, and necessary in providing optimum patient care (1,2). Clinical pharmacists, clinical pharmacologists, and clinical chemists may be involved in TDM, and consequently many TDM service models have been reported (3,4). Although the most effective model seems to be a multidisciplinary TDM service (5), in some hospitals this is not feasible in routine practice. The purpose of this chapter is to describe TDM in a biochemistry service of a large public hospital and to summarize the results obtained during the 10 years it has been functioning.

II. DESCRIPTION OF THE SERVICE

Vall d'Hebrón is a public institution comprised of four hospitals (General, Trauma, Maternal, and Children's). The drug monitoring unit, housed in the biochemistry service, receives samples from the 930-bed General Hospital and the 375-bed Trauma Hospital as well as from outpatient clinics in the Barcelona area assigned to our sector.

Our TDM service began in 1980. In the first year, only levels of anticonvulsants were determined. The routine assays presently available, and the methods used, are indicated in Table 1. Digoxin is not included in our menu, since it has always been done in the nuclear medicine laboratory.

Besides these routine techniques, specific protocols for monitoring other drugs (trimethoprim, sulfamethoxazol, ciprofloxacin, 5-flucitosine, ganciclovir) are followed.

The laboratory service is available from 8:00 a.m. to 8:00 p.m., Monday through Friday. Drug determination may be done in the emergency laboratory if intoxication is suspected. To monitor therapeutic doses on weekends, only requests for theophylline and those for methotrexate and antibiotics are accepted. In some cases, when drug dosage requires extraction outside the working schedule of the routine laboratory, the sample is accepted and processed the next day.

Table 1 TDM Assays Routinely Available

Drug	Method	Drug	Method
Anticonvulsants		Tricyclic Antidepressants	
Phenobarbital	FPIA	Amytriptyline	EMIT
Phenytoin	FPIA	Nortriptyline	EMIT
Carbamazepine	FPIA	Clomipramine	EMIT
Valproic acid	FPIA	Desmethylclom.	EMIT
Ethosuximide	FPIA	Imipramine	EMIT
		Desipramine	EMIT
Antibiotics			
		Theophylline	FPIA EMIT
Tobramycin	FPIA	Methotrexate	FPIA
Amikacin	FPIA	Lithium	ATOM.ABS.
Gentamicin	FPIA	Cyclosporine	RIA-I125
Vancomycin	FPIA	Pentobarbital	HPLC

A file is prepared with demographic data which refers to dosage for all patients being monitored. In the case of unexpected results or results which are inconsistent with previous available data, the physician is contracted and collection may be repeated if necessary.

Since there is a central pharmacology department for the four hospitals, reports on drug levels are not accompanied by dosage recommendations. However, since February 1990 we have incorporated into our system the Bayesian forecasting programs for predicting aminoglycosides and vancomycin. The reports include the peak and trough levels, the dose simulation graph supplied by the program, recommendations of dose and interval to achieve correct therapeutic levels, and advice concerning when collection may be done.

III. RESULTS

The monthly average of requests in the period between January and July 1990 and the profile of drugs solicited are shown in Fig. 1.

To evaluate the clinician's use of theophylline, a follow-up was done for 2 months on all inpatients who required determinations of this drug. Of the 166 serum levels, 44.0% fell within the therapeutic range, 44.5% had serum values less than 10 mg/L, and 11.5% had potentially toxic concentrations (above 20 mg/L). The changes in dosage as a response to the serum analyses are shown in Fig. 2. In 10% of those patients with serum levels less than 10 mg/L, the dose was decreased (this group included the patients who stopped the treatment). With levels above 20 mg/L, the dose was not increased, although in 24% of patients it was not modified.

Concerning the anticonvulsants, in 1982 only 40% of patients in our hospital were treated on monotherapy; the remaining 60% received more than one anticonvulsant. In 1986 the incidence of monotherapy was 60%, and in 1990 (up to July) it was 65%. In monotherapy we can see an important reduction in the use of phenobarbital and primidone as well as an increase in carbamazepine and valproic acid from 1982 up to

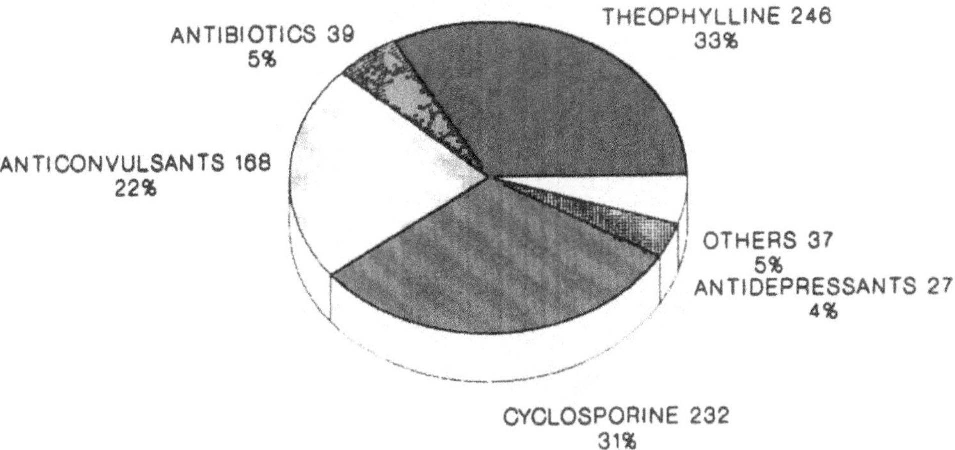

Figure 1 TDM monthly requests. Mean (January–July 1990):736.

the present (Fig. 3). Changes in multidrug therapy may also be observed. Treatments with four anticonvulsants, which in 1982 were 16% of overall multidrug therapy, decreased to 0% later. Similarly, triple therapy decreased from 15% in 1986 to 8.7% in 1990. The association of two anticonvulsants increased, except for phenobarbital and phenytoin, which were used much more frequently in 1982 (Fig. 4).

IV. DISCUSSION

Since TDM is a multidisciplinary subject, different model services have been described. Our TDM unit is a section of the biochemistry service, as is the majority of TDM units in our country.

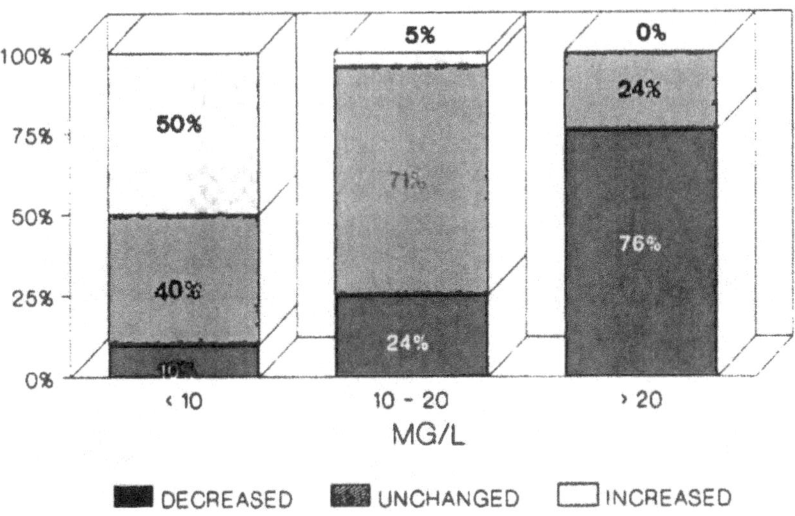

Figure 2 Changes of theophylline dosage according to the serum drug level.

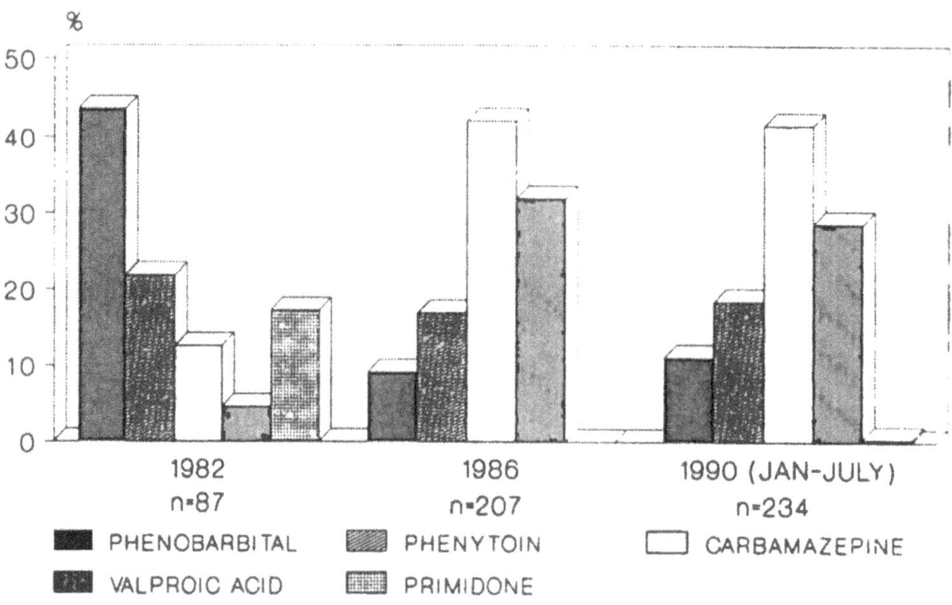

Figure 3 Percentage of anticonvulsant drugs in monotherapy.

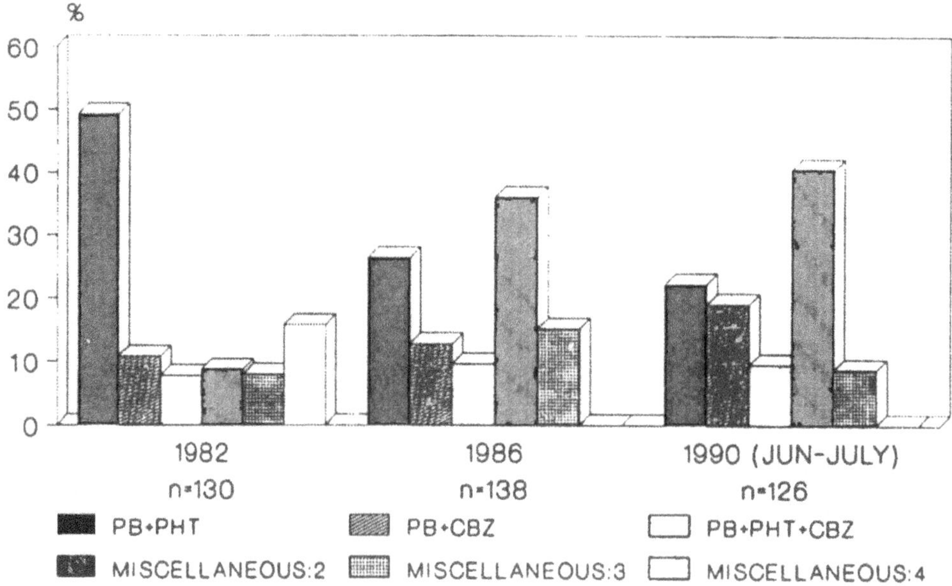

Figure 4 Percentage of anticonvulsants in multidrug therapy. PH = phenobarbital, PHT = phenytoin, CBZ = carbamazepine.

Until now our service has not recommended dose changes based on serum levels, since there is a central pharmacology unit in our institution. However, recently we have incorporated a forecasting program for predicting aminoglycosides and vancomycin. Although it has been functioning for only a short time, its acceptance by clinicians generally has been good, and the overall evaluation is positive. For this reason we are considering including other drugs in this system.

Multicenter studies (6) have shown that more than 80% of epileptic patients were adequately managed on monotherapy. In 1982 only 40% of patients in our hospital were treated with one anticonvulsant. In 1986 the incidence of monotherapy was 60%, and 65% in the period up to the present. Moreover, in multidrug therapy there has been a clear reduction in the number of drugs used. Therefore, we believe that TDM has played an important role in the management of epileptic patients.

V. CONCLUSIONS

TDM is valuable and necessary in providing optimum patient care, especially, if recommendations for changes in therapy based on serum levels are made. Our objectives for the future are to continue the antibiotics program and to extend this service to other drugs.

REFERENCES

1. D. E. Hoffa. Serial pharmacokinetic dosing of aminoglycoside: A community hospital experience. *Ther. Drug Monitor., 11*:574 (1989).
2. R. Spector, G. D. Park, G. F. Johnson, and E. S. Vesell. Therapeutic drug monitoring. *Clin. Pharmacol. Ther., 43*:345 (1988).
3. C. E. Pippenger. Commentary: Therapeutic drug monitoring in the 1990s. *Clin. Chem., 35*:1348 (1989).
4. W. E. Wade and C. Y. McCall. Therapeutic drug monitoring in a community hospital. *Ther. Drug Monitor., 12*:79 (1990).
5. S. Cox and P. D. Walson. Providing effective therapeutic drug monitoring services. *Ther. Drug Monitor., 11*:310 (1989).
6. B. J. Wilder. Treatment considerations in anticonvulsant monotherapy. *Epilepsia, 28 (Suppl. 2)*:S1 (1987).

3

A Successful Experience in Therapeutic Drug Monitoring

Mostafa G. Bigdeli *Tehran University of Medical Sciences, Tehran, Iran*

I. INTRODUCTION

Among few other laboratories offering therapeutic drug monitoring (TDM) services in Iran, our institution was the first to start, in 1984. At that time a young and well-trained neurologist who joined our staff in the university hospital requested antiepileptic drug measurement and encouraged us to set up a specialized section in the laboratory. We started with determination of phenobarbital, PB, and diphenylhydantoin, DPH (phenytoin), blood levels in epileptic patients, employing enzyme-multiplied immunoassay technique (EMIT) and commercially available kits (Syva Co., Palo Alto, California).

At the beginning we were running only a few tests per month, ordered by one specialist. But as the service continued and more clinicians were informed, ordering of these analyses increased so that within 2 years TDM became part of the established repertoire of our clinical laboratory.

In order to assure proper handling of patients and to improve the effectiveness of the program, a questionnaire was designed to collect certain information from each patient (Table 1). The questionnaires were completed by phlebotomists, supervised by attending physicians or nurses. Blood samples for routine antiepileptic drug monitoring were collected 15 to 30 min before the last dose of medication and the sera were separated and stored frozen until the time of analysis. Suspected cases of toxicity were handled "STAT."

II. RESULTS

Figure 1 shows the age distribution of 1826 epileptic patients. The majority of these patients were young, under 20 years of age. Males outnumbered females by the ratio of 5.0 to 3.8. Some of the young male patients had been on active military duty and had returned from the war zone.

Figure 2 demonstrates the pattern of growth of the TDM laboratory. The relatively sharp rise of the first part of the graph is due to a large number of requests from physicians asking for blood drug measurements for patients who had been under medication for 20 years without being monitored. Gradually, as the drug dosage

Table 1 Required Information from Each Patient for TDM

Patient's age, weight, and sex
Kind of therapy (mono- or poly-)
Length of therapy
All other drugs patient is receiving
Reason for drug measurement:
Seizures uncontrolled
Suspected toxicity
Optimization of drug level
Clinical status of patient (renal and hepatic function, etc.)
Time of blood sampling relative to last dose

could be adjusted by monitoring blood concentration, the upward trend slowed down in later years.

Table 2 summarizes the data collected during the last 4 years or our experience. At the beginning of the program a significant percentage of results were in a nonoptimal range (optimal ranges of PB and DPH were considered to be 15 to 40 μg/mL and 5 to 20 μg/mL, respectively). About half of the treated patients had therapeutic blood concentrations and the other half were either high or low. But as the service continued, the proportion increased from 54% to 75% for PB and from 49% to 69% for DPH.

The program not only shifted the majority of patients from the nonoptimized group to the optimized population, but also improved the distribution of therapeutic concentration to a more limited deviation (Table 3).

The program was also useful in reducing the number of cases with combination therapy. Table 4 demonstrates that the number of patients controlled with either PB or DPH alone increased as the dosage of drug was adjusted by the TDM program. Conversely, the number of patients under polytherapy (PB and DPH together or each combined with other drugs) declined gradually.

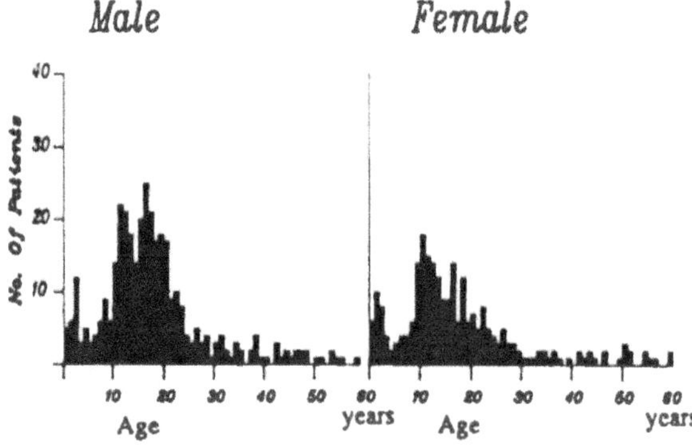

Figure 1 Age distribution for TDM patients.

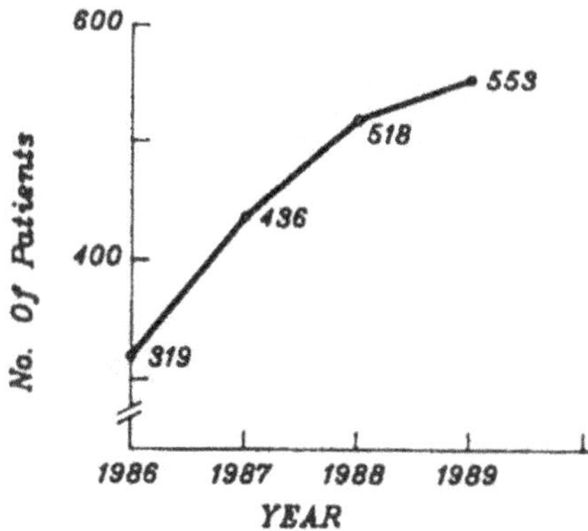

Figure 2 The growth of TDM laboratories.

III. CONCLUSION

We consider our program to be successful because: (1) yearly growth of interested physicians led to increased testing of patients; (2) the number of nonoptimum values, both high and low, decreased significantly, and the proportion of patients in the therapeutic range increased from 54% and 49% to 75% and 69% for PB and DPH, respectively; (3) the program could guide the physicians to try a single appropriate anticonvulsant before a second or third agent was added, so that the number of patients controlled with monotherapy increased; and (4) the program resulted in a shift of therapeutic values after 4 years.

Table 2 Summary of Data Collected During 4 Years of Experience

			Blood Drug Concentration					
			Phenobarbital			Phenytoin		
Year	Total no. of Tests		Low	Ther.	High	Low	Ther.	High
1986	319	(n)	82	123	23	29	45	17
		(%)	36	54	10	32	49	19
1987	436	(n)	46	199	28	44	101	18
		(%)	17	73	10	27	62	11
1988	518	(n)	45	229	23	53	146	22
		(%)	15	77	8	27	62	11
1989	553	(n)	46	238	33	47	165	24
		(%)	15	75	10	20	69	11

Table 3 Average Blood Drug Concentration of Patients Referred to the Laboratory in Different Years

Year	PB (μg/mL)	DPH (μg/mL)
1986	13.1 ± 20.4	10.4 ± 12.2
1987	17.3 ± 16.5	12.2 ± 10.3
1988	18.0 ± 18.1	13.8 ± 10.3
1989	17.7 ± 16.0	13.3 ± 9.8

Since the patients were referred to the TDM laboratory from various clinics, we were unable to control some of the extralaboratory factors such as correct interpretation of data by clinicians and observe the appropriate response of patients. Both of these factors could help us learn more about the effectiveness of the program.

Table 4 Number of Patients Demonstrating Optimum Blood Drug Concentration

	1986	1987	1988	1989
Polytherapy (PH plus or DPH plus)	41.6%	34.5%	31.0%	32.5%
Monotherapy:				
PB alone	52.2%	60.5%	68.3%	66.8%
DPH alone	55.1%	55.9%	60.0%	58.7%

4

The Impact of an Active Laboratory Approach on Therapeutic Drug Monitoring

Sander A. Vinks, Daan J. Touw, and Els Fockens
The Hague Hospitals Central Pharmacy, The Hague, The Netherlands

I. INTRODUCTION

During the last decade, therapeutic drug monitoring (TDM) has expanded enormously. Today almost every laboratory can offer some form of drug monitoring. However, too often the numbers generated by immunoassays are reported to the clinician without proper interpretation and advice on dosing. One of the major problems is the apparent lack of communication between the laboratory staff and the clinicians.

In our experience, an active approach (consultation service) toward the clinician is essential for the success of a TDM service and is the only way to improve dosing strategies based on blood levels.

In the Netherlands, TDM is performed predominantly by the hospital pharmacist, and, in most institutions, the TDM and clinical toxicology laboratory is part of the hospital pharmacy. In The Hague the situation is different from most other hospital pharmacies in the country. The establishment of a central pharmacy for all the local hospitals and 14 nursing homes resulted in the centralization of many services, including the laboratory for drug monitoring and clinical toxicology.

Some 10 years ago, a central-pharmacy-based TDM/pharmacokinetics service for the participating hospitals was started. Included in this program were 2650 hospital beds, 900 psychiatric beds, and 2100 nursing home beds (total of 5660 beds). Today some 12,000 serum levels are determined annually, along with 350 toxicology cases. The success of a TDM/pharmacokinetics service is determined by several factors, including speed of the assay, a 24-h- and 7-days-a-week service, and a clinically relevant and rapid reporting system. However, in our view, the most important cornerstones are a good working relation between the physicians and the laboratory staff, most effectively established by daily and direct consultation by telephone and visits to wards and including feedback presentations at meetings; followed by a report interpreting the measured drug level. Preferably, advice on dosing, computer calculated, is sent with the results.

In order to get an insight on the impact of our service on drug dosing in the different hospitals, we have studied the effect of the serum determinations of theophylline and the aminoglycosides on their respective dosing schedules.

Theophylline was chosen because there is a good concentration-effect relationship, and not too complicated, predominantly linear, kinetics. Our laboratory therefore does not calculate drug dosages on a routine basis but suffices in most instances to offer general advice.

The group of aminoglycosides, on the other hand, is representive for the typical group of drugs for which computer calculation with feedback of one or more drug concentrations after dosage changes, often over a short period of time, is appropriate.

II. PATIENTS AND METHODS

A. Theophylline

The theophylline data from five general hospitals (2490 beds) from patients receiving intravenous theophylline were retrospectively studied from 1986 to 1989. The total number of patients receiving intravenous theophylline could be approximated by dividing the total amount of theophylline used per hospital by the average i.v. dose per patient. A 2-month survey on different wards resulted in a calculated average i.v. dose per patient of 6 g. The survey learned that an average daily dose of 600 mg during 10 consecutive days is a realistic approximation. The international defined daily dose (DDD) is 600 mg/day. In order to get a better insight on the various dosing strategies and whether or not, based on the blood level, the dosage was changed, we subsequently looked at those patients where, in the course of their i.v. treatment, more than one theophylline level was determined.

B. Aminoglycosides

In 1984 our laboratory began computerized pharmacokinetic dosing of aminoglycosides. For the initial dosing a simple user-friendly nomogram was developed by means of a computer calculation (Table 1). Dosing intervals were based on elimination rate constants calculated from creatinine clearance. Creatinine clearance was calculated from the serum creatinine (1). In the nomogram the volume of distribution has fixed at 0.25 L/kg. The initial loading dose was 2 mg/kg. This was followed by 1.7 mg/kg every 12 h. Levels were ordered after the second (or third) dose. Individual pharmacokinetic parameters are calculated by means of a one-compartment computer model according to the peak-trough principle (2), trough 30 min before the next dose and peak 30 min after the end of a 30-min infusion. Every calculated advice on dosing was reported to the physician by phone; a printed report including a concentration-versus-time profile was sent by internal mail.

In order to evaluate the impact of the service, pharmacokinetic dosing strategies based on blood levels from 1984 were compared with those in 1988. Results from 350 patients from five general hospitals were evaluated. Theophylline and aminoglycoside (gentamicine, tobramycine) serum levels were assayed by Abbott's TDx. Among data collected on all patients were demographic data: age, gender, height, weight, and calculated creatinine clearance. Sampling times were taken from the request form.

Table 1 Aminoglycoside Nomogram

Serum Creatinine	<90	120	150	200	250 μmol/L
Age (years)					
40	8	12	12	12	12
50	8–12	12	12	12	12–24
60	12	12	12	12–24	12–24
70	12	12	12	24	24
80	12	12	12	24	24

III. RESULTS

A. Theophylline i.v., 1986 to 1989: The Simple Approach

Figure 1 shows the data of patients on intravenous theophylline from 1986 to 1989. Data are grouped by year and divided into subtherapeutic (<7.5 mg/L), therapeutic (8 to 20 mg/L), toxic (>20 mg/L), and potentially lethal (>40 mg/L).

In 1986 the "therapeutic" score was 55%, subtherapeutic was 33%, and toxic was 12%. Of this group, five patients were given a dose leading to a potentially lethal theophylline concentration (>40 to 50 mg/L). In the following years no dramatic changes occurred. In 1989 the number of patients with a drug concentration within the therapeutic window increased to 66%. The total number of measurements also increased by 54%, from 511 in 1986 to 788 in 1989. The number of serum levels outside the therapeutic range was still large for this intravenous therapy (33%).

To substantiate our feeling that a large number of patients on i.v. theophylline are not monitored, we calculated the theoretical number of patients receiving i.v.

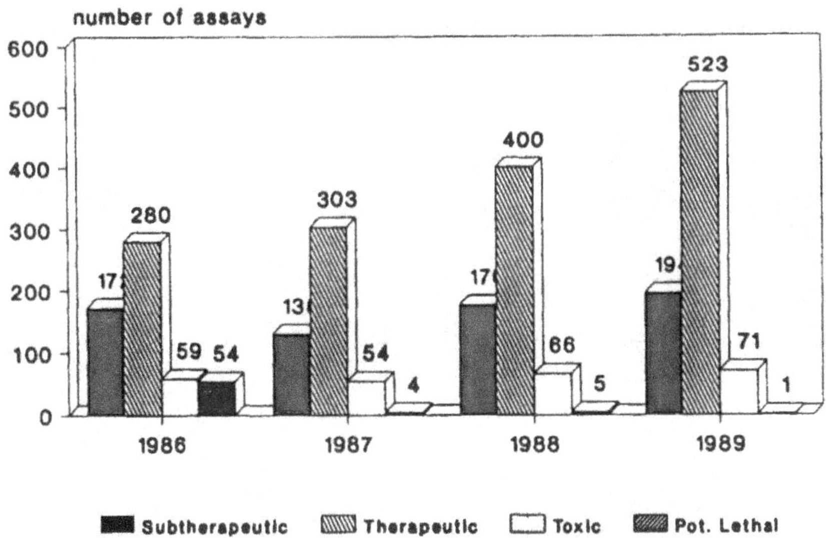

Figure 1 Number of theophylline assays, 1986 to 1989. Subtherapeutic, <7.5 mg/L; therapeutic, 8 to 20 mg/L; toxic, >20 mg/L; potentially lethal, >40 mg/L.

theophylline. Figure 2 shows the calculated number of patients who received intravenous theophylline therapy together with the number of patients in whom theophylline actually was monitored. Data are grouped per hospital and according to the size of the hospital. The percentage of monitored patients varies from 35% to 61%; on average, 46% were monitored.

B. Theophylline Follow-up Levels (n = 94)

Figure 3 shows that surprisingly few patients are monitored more than once in the course of therapy. Only 94 patients (15%) out of 623 were monitored. The initial concentration range varied form 0 mg/L to 45 mg/L, and this range was still fairly large after the second measurement. A few patients were monitored more than twice, but the third (n = 29), the fourth (n = 10), and in some instances the fifth measurement (n = 3) did not result in all theophylline levels being within the target range. Half the levels still were lower than the desired concentration.

C. Aminoglycosides Monitoring: The Computer Approach

The educational effect of the monitoring in combination with our guidelines is shown in Figure 4. A major shift in initial dosing has occurred. In 1984 most patients (90%) were treated with 80 mg b.i.d. A loading dose was seldomly administered. Patients with renal failure were inappropriately dosed: 40 mg t.i.d., sometimes 80 mg b.i.d. In 1988 the initial dose was increased a great deal, and most patients received a loading dose. The interval was prolonged in most cases (54%) to 12 h, and the dosage was

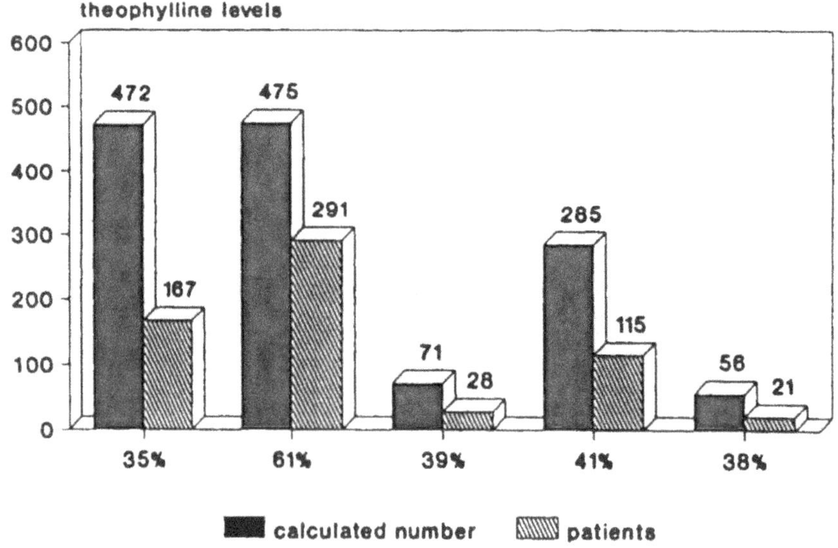

Figure 2 Calculated number of patients receiving intravenous theophylline versus the actual number of monitored patients. Data are grouped per hospital. The sizes of the hospitals were 861, 606, 415, 370, and 237 beds, respectively. Percentages of actual monitored patients are given on the axis.

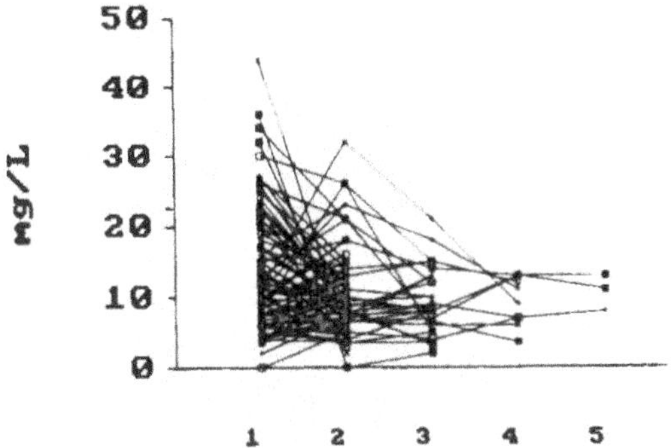

Figure 3 Theophylline follow-up determinations in 1989. Data are presented for 94 patients. The numbers on the axis correspond to the number of follow-up measurements in time.

increased to 120 mg or more. Figure 5 shows the initial dose versus the dose based on blood levels in 1988. In 26% of the patients, a dose of 120 to 140 mg b.i.d. was calculated on the basis of individual pharmacokinetic parameters. In many patients (38%) a higher dosage was needed for adequate therapy. Once-daily dosing was required in 21% of the patients; 80 mg t.i.d. seldom gives the desired drug levels.

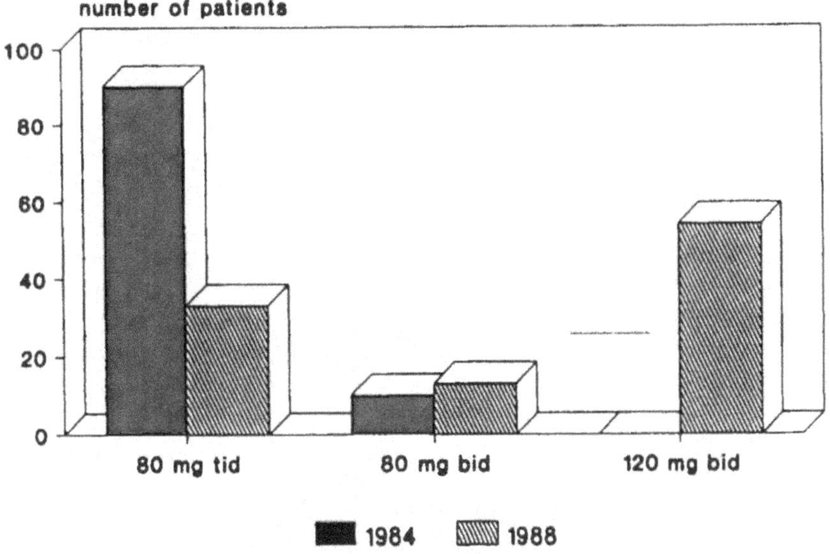

Figure 4 Aminoglycoside dosing in 1984 and 1988. Each bar represents the number of patients at a given initial dose of 80 mg t.i.d., 80 b.i.d., or 120 mg b.i.d. n = 350.

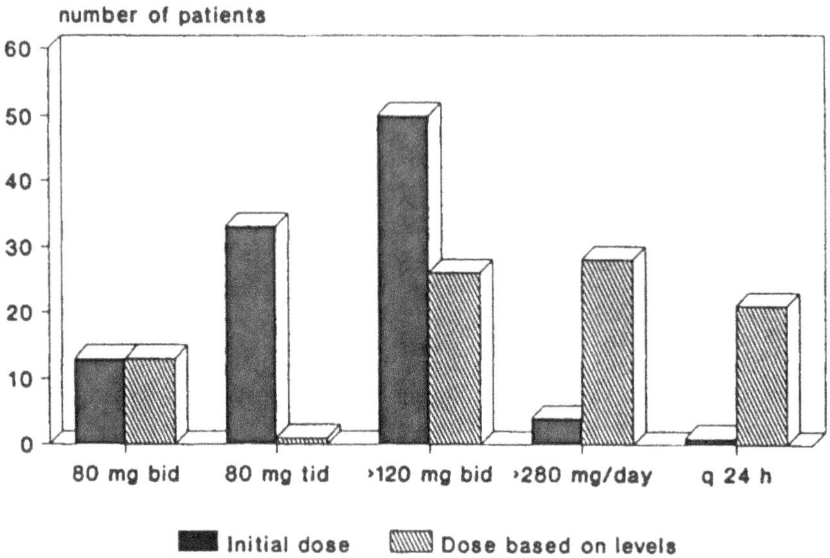

Figure 5 Aminoglycoside dosing in 1988. Each bar represents the number of patients at a given dose. The closed bars give the initial dose, whereas the open bars correspond to the dose given after drug monitoring.

D. Netilmycine Peak/Trough Levels in Cystic Fibrosis

Figure 6 shows the reached peak and trough levels in eight cystic fibrosis (CF) patients. Initial levels show large variation, and follow-up measurements are necessary because of changing kinetics. The volume of distribution in these patients is subject to large variation over short periods of time (range: 0.17 to 0.42 L/kg).

IV. DISCUSSION

The theophylline study was inspired by one of the few studies on this subject (3). This study showed that dosing by "rule of thumb" and clinical impression does not give satisfactory results. Pharmacokinetic dosing advice after the first assay resulted in a major shift of the number of theophylline levels toward the target range. Initially, 29% of the samples were within the therapeutic range; but after pharmacokinetic consultation, 70% of the follow-up samples became "therapeutic." Our data (Figure 1) shows a comparable pattern, although the initial number of patients within the desired concentration range in our study is larger (55%).

From Figure 2 it is obvious that not all patients are included in a drug monitoring program. The total number of patients that receive i.v. theophylline therapy was calculated to be 1360. Only 45% of this group was sampled once, and although the total number of serum levels has increased over the years, it is still rather low: 788 for 2490 beds. In only 94 of all patients (7%) was a follow-up level determined during i.v. therapy. The reason for a follow-up level was not always clear: Therapeutic as well as initial levels outside the desired range result in follow-up measurements (Figure 3).

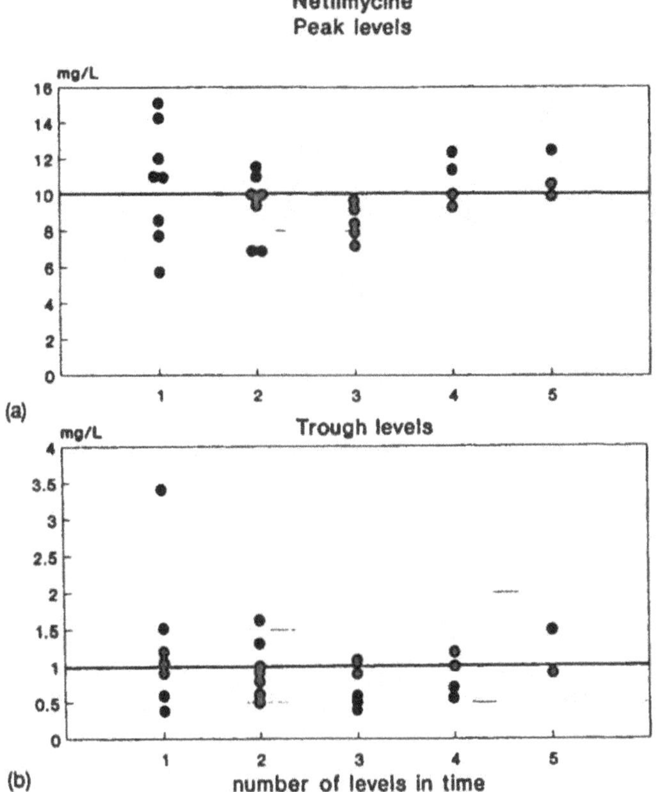

Figure 6 Netilmycine peak (a) and trough (b) levels in eight cystic fibrosis patients. Data are presented for five consecutive measurements. Measurement 2, 3, 4, and 5 give the obtained netilmycine serum levels after pharmacokinetic consultation.

There is a slight preference for nontherapeutic initial concentrations, but in 43% of the cases the first serum level was in the therapeutic range.

By looking at the prescribed dosages on the request forms, it became apparent that changing an intravenous dose not lead automatically to a check of the theophylline concentration. More important, a nontherapeutic initial serum concentration does not, in many instances, result in an obvious dosage change. These findings are now being used to confront the prescribing physicians with their inefficient monitoring strategy. This is probably the only way to dispel the general belief that theophylline monitoring is simple and well under control. In all hospitals, and unfortunately also in teaching hospitals, very little attention is paid to TDM during the residents' education. Workshops and teaching by laboratory staff therefore are a "must." In these workshops cost effectiveness of TDM should be discussed, and physicians should be educated that drug monitoring is preventive medicine.

The calculation of drug dosages based on blood levels is especially extensive for the aminoglycosides. The relative importance which is directed toward good antimicrobial therapy has helped to get the peak/trough principle across, together with some general pharmacokinetic principles. From Figures 4 and 5 it can be concluded that active support from the laboratory has led to better aminoglycoside therapy.

The data from the cystic fibrosis patients show clearly the need for intensive feedback (Figure 6). In the CF patients the initial dose is based on a priori pharmacokinetic parameters. Target values are a peak of 10 mg/L and a trough of 1 mg/L. Thus high-dose therapy is started with initial doses of 180 mg t.i.d. or more. After the second dose, drug levels are requested. Dose adjustment is made on the basis of the serum levels with a computer-generated optimal dose. This advice is given by phone to the physician in charge. After every dosage change, verification of the achieved concentrations is mandated.

It is concluded that therapeutic drug monitoring can be an exciting activity in which the laboratory has to play an important role. Only an active consultation service will lead to the desired drug levels.

V. SUMMARY

The impact of a central clinical pharmacokinetics service for five hospitals (2490 beds) on theophylline and aminoglycoside dosing was evaluated. The "simple approach" which is being used for theophylline does not result in a good response. Analysis of the theophylline data in 1989 showed that only in 46% of all the patients on i.v. therapy was a drug level determined. Of the measured theophylline levels in 1989, 66% were within the therapeutic range, 25% were subtherapeutic (< 8 mg/L), and 9% were toxic (> 20 mg/L). Only 94 samples (12%) were sent in for a follow-up measurement. Of this group, 52 (55%) were therapeutic, 34 (36%) were subtherapeutic, and 8 patients (9%) had levels in the toxic range. A third and fourth measurement did not improve the results. For aminoglycoside antibiotics a "computer approach" is being utilized. In the majority of patients the initial dose is calculated from a nomogram based on population parameters. Pharmacokinetic parameters are calculated after the second or third dose on the basis of peak and trough levels. Every consultation is handled by a hospital pharmacist by phone. The standard report comprises calculated parameters and advice on dose. In 1988, drug level measurements in 350 patients resulted in doses of 120 to 240 mg b.i.d. in 80% of the patients. In 20% a 24-h interval was advised. In 1989, all patients receiving aminoglycoside therapy were monitored, with most levels in the target range after following the first dosing advice. This approach results in a response rate > 95%.

It is concluded that in TDM only an active consultation service leads to the desired drug levels.

REFERENCES

1. D. W. Cockcroft and M. H. Gault. Prediction of creatinine clearance from serum creatinine. *Nephron, 16*:31–41 (1976).
2. R. J. Sawchuk and D. E. Zaske. Pharmacokinetics of regimens with multiple intravenous infusions: Gentamicin in burn patients. *J. Pharmacokin. Biopharm.*, 4:183–195 (1976).
3. B. Whiting, A. W. Kelman, S. M. Bryson, F. H. M. Derkx, A. H. Thomson, G. H. Fotheringham, and S. J. Joel. Clinical pharmacokinetics: A comprehensive system for therapeutic drug monitoring and prescribing. *Br. Med. J., 288*:541–545 (1984).

5

Assessment of the Introduction of a Therapeutic Handbook in a Teaching Hospital

Raymond G. Morris, Rudolf Zacest, Richard J. Leeson, Nunzia C. Saccoia, Sotirios Mangafakis, Chryss Kassapidis, and Linda K. Fergusson *The Queen Elizabeth Hospital, Woodville, South Australia, Australia*

I. INTRODUCTION

The prescribing habits of clinical staff in teaching hospitals is guided by many variables, including formal education (both undergraduate and postgraduate), informal education (including seminars, commercial sources, etc.), experience (including personal and peer), as well as various forms of prescribing guideline documents. In the present study we have considered the impact of the introduction of a "pocket-sized" Therapeutic Handbook (referred to hereafter as the Handbook) oriented specifically toward drugs available in this institution. This Handbook was prepared as a collaborative venture between the Clinical Pharmacology and Pharmacy Departments under the auspices of the Drugs Committee of this hospital. It contains both prescribing information (including pharmacokinetic details) and a formulary for this hospital's pharmacy stocks. The Handbook introduction was timed to coincide with the commencement of a new batch of junior medical staff at the beginning of their preregistration training year.

If used appropriately, the potential exists for an improved quality of prescribing, particularly from less experienced medical staff, by selection of appropriate drug(s), dose(s) and dosage schedules at the outset of introducing an agent. The obvious goal of the Handbook is to increase the frequency of desired therapeutic responses and minimize the incidence of drug toxicity.

II. METHODS

A. Therapeutic Drug Monitoring Laboratory

The effect of the introduction of this Handbook has been reviewed by comparing the distribution of assay results for 12 monthly periods prior to, as well as following, the release of the Handbook with respect to our advertised therapeutic ranges for five of the most frequently requested drugs. These drugs (and number of results included in years 1 and 2 of the study) were digoxin (3156, 3159), theophylline (2400, 2345), phenytoin (1289, 1446), carbamazepine (1272, 1350), and valproic acid (727, 715). Our therapeutic drug monitoring (TDM) laboratory currently performs a total of approximately 12,000 tests per year. Cyclosporin-A was excluded from the study, despite being the fourth most frequently requested assay, as the laboratory had not determined a therapeutic range for this drug at the time.

Assay results were divided on a monthly basis into "below," "within," and "above" the therapeutic range for the two 12-month periods and compared statistically using a chi-square test. The 5% probability level was assigned as the index of significance.

B. Questionnaire Survey

These same junior medical staff (n = 29) who commenced duty at the start of the second year of the study were surveyed by telephone by a clerical officer (in preference to a pharmacologist) to answer the following questionnaire. The purpose of this survey was to seek direct feedback as to the general acceptance of the Handbook. The following questions were asked:

Question 1. How frequently do you use the Handbook?
Within this question the interns were asked to indicate whether they used the Handbook daily, more than once per week, or less than once per week for both their first month of internship and 3 to 6 months later.
Question 2. Is the Handbook sufficiently portable? (Y/N)
Question 3. Has it influenced your prescribing? (Y/N)
Question 4. Does it influence your ordering of drug assays? (Y/N)
Question 5. Does it offer any advantage over other more general publications? (Y/N)
If "no," then which other publication(s) are used regularly?
Question 6. Do you consider that the number of drugs available for general prescribing should be restricted in this hospital? (Y/N)
Question 7. General comments sought.

Twenty-five of the 29 interns completed interviews with respect to the above, and results were pooled.

RESULTS

Figure 1 presents the results of the TDM laboratory for the five most frequently requested drug assays (excluding cyclosporin-A; see above). These data are expressed as the percentage of results which were within the advertised therapeutic range. Obvious differences are apparent in the frequency at which the therapeutic range is achieved for the five different drugs, being highest for digoxin (mean approximately 90%) and lowest for carbamazepine (mean approximately 35%). However, when comparing the two study years for each drug, only minor significant differences were noted, an increase ($p < 0.02$) from 88% to 90% for digoxin assays and a reduction ($p < 0.01$) from 36% to 31% for carbamazepine. The phenytoin data appeared to show a positive improvement in the percentage of therapeutic results over the first 6 months of the second year, but this difference was not sustained in the second half of this

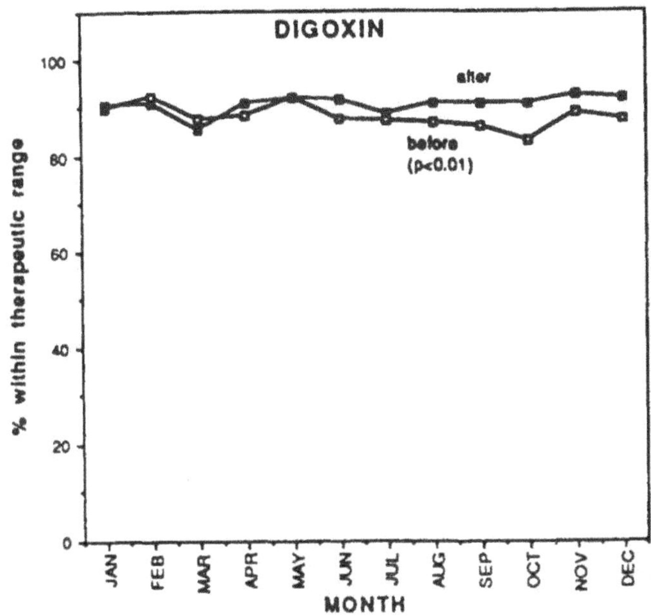

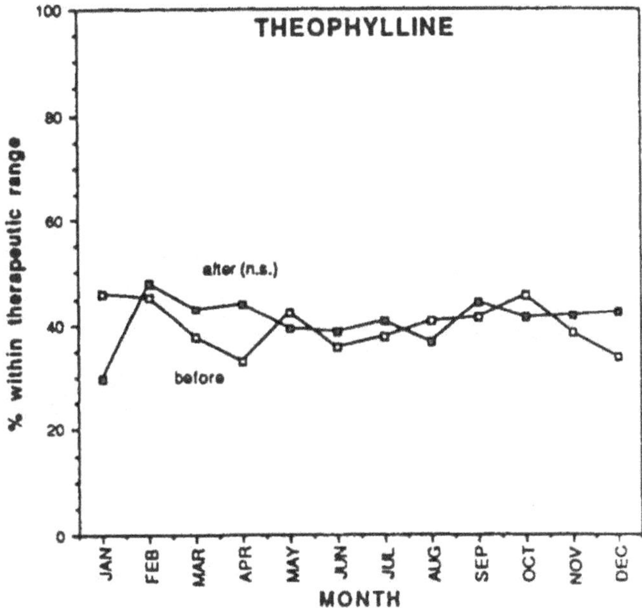

Figure 1 Percentages of assay results from the TDM laboratory which were within the recommended therapeutic range for 12 monthly periods prior to (open squares) and following (closed squares) the introduction of a Therapeutic Handbook in a teaching hospital.

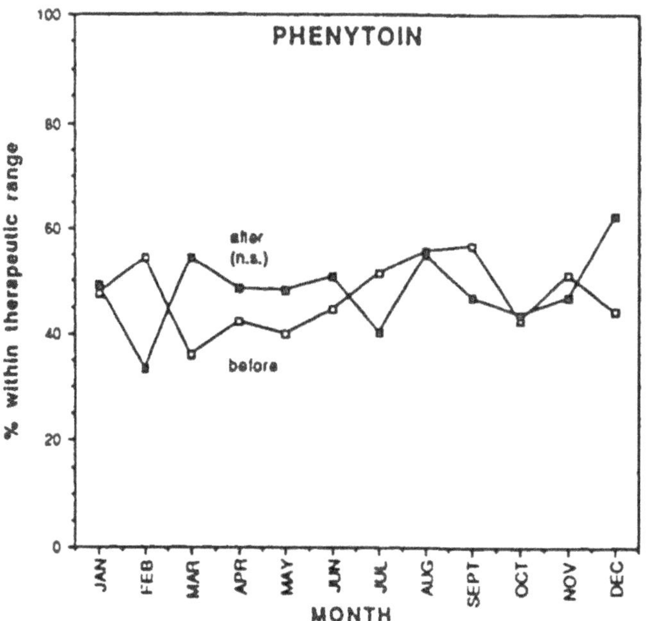

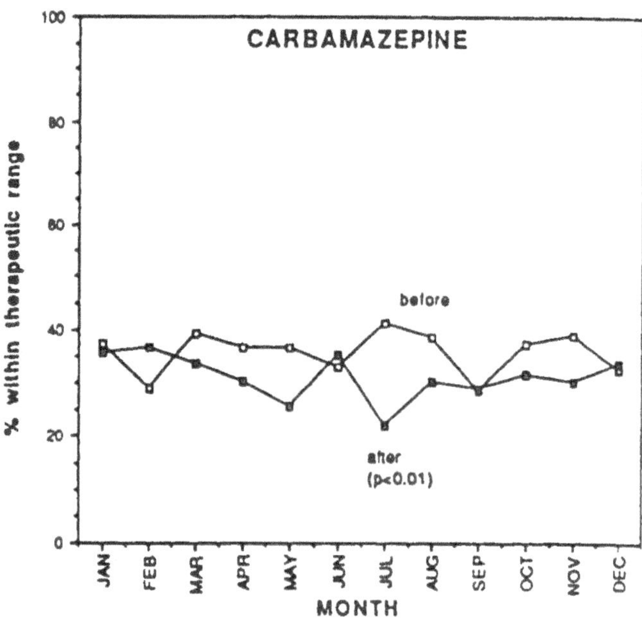

Figure 1 (continued)

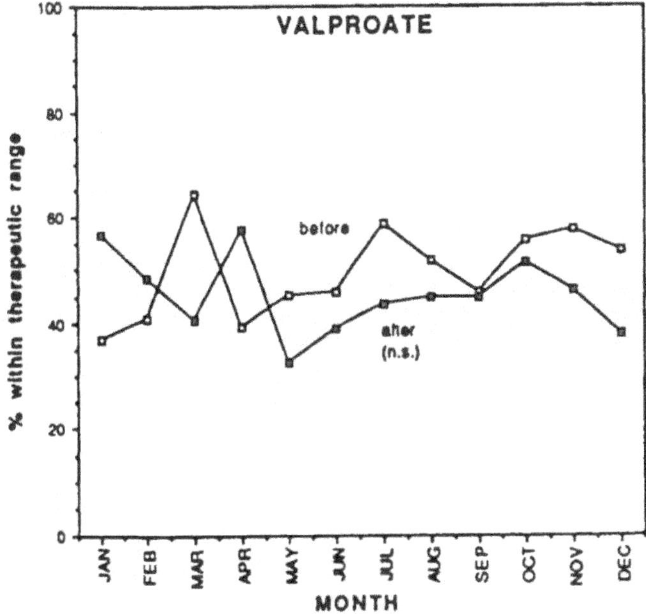

Figure 1 (continued)

second year. Hence there appeared no strong evidence of net positive effect of the Handbook in increasing the frequency of assay results achieving the therapeutic range.

The results for the questionnaire (as itemized above) were as follows from 25 of the 29 (86%) interns.

Question 1. How often the Handbook was used.

	Daily	>Once/Week	<Once/Week
First month of internship	40%	44%	16%
At 3 months	12%	44%	44%

Question 2. Forty-four percent considered it appropriately sized, while 56% considered the Handbook too large.

Question 3. Seventy-two percent accepted that it had positively affected their prescribing.

Question 4. Ninety-two percent indicated that it had not influenced the ordering of drug assays.

Question 5. Seventy-two percent indicated that it did have advantages over other publications.

Question 6. Fifty-six percent of those questioned considered that the hospital should not restrict access to available drugs for prescribing, 32% considered that access should be limited, and 12% did not have an opinion.

Question 7. Twenty percent of responders volunteered that they would appreciate the use of trade names for drugs in addition to generic names. Thirty-two percent indicated that they would appreciate more prescribing information, thereby indicating that they were using only the Formulary section and were not aware of the monograph section of the Handbook, which provided these details.

Four of the 18 respondents who indicated that they were using other publications said that they were using a major publication used throughout this country (1) for dosage guidance.

IV. DISCUSSION

The results of the drug assay survey failed to demonstrate that the introduction of the Therapeutic Handbook had any major impact on the prescribing habits of clinical staff as assessed from the perspective of the TDM laboratory. Only a minor significant ($p < 0.01$) increase (from 88% to 90%) in the percentage of digoxin assays achieving the therapeutic range was demonstrated, and indeed significantly ($p < 0.01$) fewer carbamazepine results were within the recommended range (reduced from 36% to 31%) following the Handbook's release. A limitation of this data is that specimens received for analysis by the TDM laboratory are not necessarily representative of the population of patients prescribed the drug in this hospital. In all probability such data may be biased toward those patients who have not shown an obvious clinical result of the therapy, or those with symptoms which may include drug-related toxicity. There was no evidence to suggest that the Handbook had influenced the ordering of drug assays as suggested both by replies from 92% of the intern staff questioned and by the number of requests for assays received by the TDM laboratory which were essentially unchanged in the second year.

Some possible explanations and controlled variables for these results include the following.

1. The clinical staff were not the same in the two years of the study, although they had all received the same undergraduate course. Hence the possibility exists that junior medical personnel differences may have affected the validity of the comparison. It was hoped that by selecting categorical data, rather than absolute comparisons, some of the potential difference in this area might be accommodated.
2. The clinical staff were not using the Handbook effectively. There was some evidence provided by the answers to questions that 32% asked to have dosing guidelines included, suggesting that they were using only the Formulary section and were not aware of the prescribing details also provided. This further suggests that greater education and reinforcement of the use of the Handbook was required.
3. Such dosing guidelines based on broad population statistics for particular drugs (possibly generated in a different ethnic group) may have less relevance when dealing with the individual patient, who possibly has a spectrum of clinical disorders which may not be stable from day to day and who may be receiving multiple therapies.
4. As mentioned above, the specimens received by the TDM laboratory are probably biased toward the patients who are not showing the desired response or those experiencing drug-related toxicity, rather than being representative of the population of patients prescribed the agent at this hospital.
5. The clinical staff were using other sources of information than the Handbook for prescribing. The questionnaire suggested that, although 92% reported that it had influenced their prescribing, 28% thought it did not offer any additional information

to the other generally available source widely used in this country (1). Hence, a large number of drug assay requesters were using other sources of information as a guide to prescribing and so this group could potentially dilute any impact of the Handbook which might otherwise have been detected in the TDM laboratory.

6. The recommended therapeutic range advertised by the laboratory may not have been the only guide to therapeutic management. Indeed, this department strongly recommends as part of its teaching program that such ranges be used as a guideline in conjunction with the clinically measured physiological endpoints for a therapeutic outcome. Hence the potential exists for a significant percentage of patients prescribed these drugs to be adequately treated with plasma drug concentrations below or above the recommended ranges. In addition, the recommended ranges are commonly modified depending on concomitant therapy. As an example, it is common for neurologists to use half the plasma carbamazepine concentration range when it is used in conjunction with other antiepileptic drugs. The latter observation may at least partially explain the very low number (mean of only approximately 34%) of carbamazepine assay results which were within the therapeutic range.
7. The TDM laboratory data as presented would also not distinguish plasma drug concentrations which may have been modified within the therapeutic range but still resulted in a positive result for the patient, for example, a patient with a drug concentration at the low end of the recommended range who may not have shown an adequate therapeutic outcome but who with dosage increase (possibly as suggested by the Handbook) may have achieved a positive therapeutic result with a plasma concentration at the upper end of the range. Conversely, a toxicity at the top of the therapeutic range may have been reduced with a plasma concentration at the lower end of the range.
8. The clinical training received as an undergraduate may have already optimized the pharmacological expertise to the point that the observations made in the present study represent the normal scatter that could reasonably be expected with any population of patients. This optimistic view is considered unlikely.

There has been a proliferation of therapeutic handbooks and/or formularies at teaching hospitals over many years, at least as early as that produced at the Linkoping District Hospital in Sweden (2). However, there is a paucity of published data on the impact of such publications on improving patient care and the cost effectiveness of such initiatives. Such publications should ideally be extensions of educational and regulatory activities of the hospital. The results of our survey, which indicate that little overall impact occurred, suggest that other facilitatory measures may have to be considered in maximizing the potential contribution of such institutional publications. Better promotion and continued reinforcement may have resulted in increased evidence of utilization of the handbook.

In conclusion, although the results were somewhat disappointing in that they failed to demonstrate a positive impact of the introduction of the "pocket-sized" Therapeutic Handbook on the percentage of patients achieving plasma drug concentrations within the recommended therapeutic ranges, the parallel questionnaire survey of clinical staff indicated its general acceptance as a guide to dosing, source of drugs available within the hospital pharmacy, limited pharmacokinetic data, etc. Other information provided in the Handbook was therefore also considered of value as an adjunct to appropriate therapeutic management of our patients.

ACKNOWLEDGMENT

The authors express their sincere appreciation to The Queen Elizabeth Hospital's Drugs Committee for supporting this study, particularly the Chairman of the Handbook Working Party, Mr. Graham Walker (Chief Pharmacist).

REFERENCES

1. *The MIMS Medical Reference System*. IMS Publishing, Division of Intercontinental Medical Statistics (Australia), Crow's Nest, NSW.
2. R. Berlin and F. Sjoqvist. *Pharmaceutical Agents*. From the District Hospital in Linkoping. AB Ostgotatryck-Linkoping, 1972.

6

A Multidisciplinary Approach to Therapeutic Drug Monitoring Services

Shareen Cox *Children's Hospital, Columbus, Ohio*

Therapeutic drug monitoring (TDM) has been shown to improve patient care when samples are collected properly, measured accurately, and reported and interpreted appropriately. At our institution, a multidisciplinary team consisting of medical technologists, clinical pharmacists, and physician pharmacologists work together to provide an efficient and effective TDM service (1).

Medical technologists are responsible for providing reliable, accurate analytical results and for notifying physicians of "critical" values. In order to fulfill these responsibilities, all results must be interpreted. This requires accurate sample collection and dosing information as well as knowledge of the pharmacokinetics and pharmacology of the drugs monitored.

Our TDM service analyzes and interprets more than 20,000 drug concentrations per year. Of these, approximately 14,000 (70%) are interpreted solely by specially trained medical technologists. Through these interpretations, recommendations concerning monitoring or changes in therapy are suggested and then reviewed by clinical pharmacists. Medical supervision is provided by a physician who is also a pharmacologist. The effectiveness of this team approach is illustrated by the examples below.

In November 1988, 755 theophylline assays were done. Follow-up data is available on 699. Medical technologists made 216 (31%) therapeutic recommendations, and 164 (75%) were followed. The remaining 25% were ignored. Suggestions as to when to recheck concentrations at specific times other than those already scheduled were made on 84 (12%), and 63 (75%) of these recommendations were followed by the attending physician.

Table 1 shows the number and percent (%) of theophylline (THEO) concentrations below (<TR), within (TR), and above (>TR) the therapeutic range as reviewed for November 1988. Further, Table 2 shows the steady-state (SS) plus non-steady state (All) carbamazepine (CBZ) and phenobarbital (PB) concentrations as reviewed from 1984 through 1989. These data show a greater and increased percentage of concentrations within the "therapeutic" range when compared to other published data (2,3,4).

The advantages of this multidisciplinary approach are that if sample collection times are documented, samples are processed correctly, then accurate dosing

Table 1 Theophylline Concentrations

	<TR	TR	>TR
All	327 (43%)	409 (54%)	19 (2%)[a]
Inpatients	218 (37%)	362 (61%)	11 (2%)[b]
Asthma (10 to 20 μg/mL)	152 (29%)	362 (69%)	11 (2%)[b]
Apnea (5 to 10 μg/mL)	20 (30%)	46 (69%)	1 (<1%)

[a]Includes 12 nontherapeutic ingestions.
[b]Includes 9 nontherapeutic ingestions.

Table 2 Steady-State Concentrations 1984–1989

		<TR	TR	>TR
CBZ	SS	52 (18%)	232 (80%)	7 (2%)
	All	1083 (16%)	5325 (77%)	476 (7%)
PB	SS	16 (2%)	631 (94%)	23 (4%)
	All	705 (6%)	10266 (83%)	1410 (11%)

information and appropriate interpretations can be provided. Critical values are caught and changes made while in the "therapeutic" range, and quality assurance activities occur naturally.

Disadvantages include stress created by the supervision of team interactions, unclear professional "territory," ego conflicts, and a need for administrative support. However, the advantages clearly outweigh the disadvantages.

In conclusion, therapeutic drug monitoring is too important to leave to any one professional group. By working together, a greater number of patients can benefit. Many professionals do TDM individually, but a good team can do it better.

BIBLIOGRAPHY

1. S. Cox and P. D. Walson. Providing effective therapeutic drug monitoring services. *Ther. Drug. Monitor.* *11*:310–322 (1989).
2. R. G. D'angio, J. G. Stevenson, T. L. Buford, and J. E. Morgan. Therapeutic drug monitoring: Improved performance through educational intervention. *12*:173–181 (1990).
3. P. N. Houtman, S. K. Hall, A. Green, and G. W. Rylance. Rapid anticonvulsant monitoring in an epilepsy clinic. *Arch. Dis. Children,* *65*:264–268 (1990).
4. M. D. Privitera. Dosing accuracy of antiepileptic drug regimens as determined by serum concentrations in outpatient epilepsy clinic patients. *Ther. Drug Monitor.,* *11*:647–651 (1989).

7

An Audit of a Therapeutic Drug Monitoring Service

I. D. Watson[1], A. Mahoney[2], M. White[2], D. Talwar[2], M. Stewart[2], and J. Elder[2] *[1]Fazakerley Hospital, Liverpool, England and [2]Glasgow Royal Infirmary, Glasgow, Scotland*

I. INTRODUCTION

In 1989 we reviewed the effectiveness of the Clinical Pharmacokinetics Service* at Glasgow Royal Infirmary for 574 digoxin requests and concluded that the service "probably contributed to improved clinical decision making" (1). We have now reviewed a total of 2961 requests, asking the following questions: Was adequate clinical information provided? Did the digoxin concentration confirm or refute the clinical impression? Was the service being sensibly used?

In addition, we were interested in the specific relationship between hypokalaemia and the presence of digoxin in routine care.

II. METHODS

A total of 2961 computerized records, comprising both the clinical information provided on the request form and the results and comments returned, were analyzed.

The request form in use in the department is shown in Figure 1 and seeks information (in addition to standard patient identification) on weight, dosage, time since dose, clinical impression, and concurrent dosage with other drugs. The results form (Figure 2) provides cumulative information on dosage per kilogram, digoxin concentration, and a comment, which may include advice on dosage adjustment.

III. RESULTS

Clinical impressions of the problem were identified from the physician-marked information boxes (Figure 3) on the request form (Figure 1). There was a smaller number of patients suspected of compliance problems, but the proportion above, below, and within our target range for digoxin (1.0–2.5 nmol/L) did not differ.

---

* The Clinical Pharmacokinetics Service is provided jointly by laboratory and pharmacy staff.

GLASGOW ROYAL INFIRMARY DEPARTMENT OF BIOCHEMISTRY LABORATORY No. **D**

SPECIMEN DATE TIME

PLEASE USE ADDRESSOGRAPH LABEL OR BALL POINT PEN
PLEASE COMPLETE ALL SECTIONS OF ID BLOCK

DOCTOR'S NAME (CAPITALS) TEL EXT PAGE No

REASONS FOR REQUEST

Toxicity Abuse Compliance Management Problem

HEIGHT (CM) WEIGHT (kg)

PATHOLOGY

Drugs (include all drugs)	Dose	Dose Frequency	Time since last dose (hrs)

TESTS REQUIRED

Phenytoin	Valproate	Drugs Of Abuse Screen (Urine)
Carbamazepine	Theophylline	Drug Toxicity Screen (Serum)
Phenobarbitone	Digoxin	Drug Toxicity Screen (Urine)
Primidone	Methotrexate	Salicylate/ Paracetamol
Others (Please Specify)		

Please complete this form in full. Further information on the reverse.

Figure 1 Therapeutic Drug Monitoring Service request form.

BIOCHEMISTRY DEPARTMENT GLASGOW ROYAL INFIRMARY **D2**

Sex M 24/ 4/29 GRI Ward 10

Address SMITH

14/ 3/91 1400

Height 167 109654 H

Weight 76

27/ 2/89	1400	6		1.5	6.60	4.3		M
10/ 2/89	1430			1.0	3.2D	5.0		
1/11/90	0940	21		<0.5	2.8D	5.1		M
20/ 2/91	0909	7		1.6	3.3D	4.2		T
26/ 2/91	0900	6		2.0	3.3D	5.5		T
13/ 3/91	1400	6		4.8	3.3D	5.1		T

1/11/90 0940 AN ELEMENT OF NON-COMPLIANCE MAY BE PRESENT
13/ 3/91 1400 AVERAGE WEIGHT USED. PLEASE GIVE ACCURATE WEIGHT.

Note target ranges are ... and are not suitable for publication ISSUE AT 16:50:48 ON 18/ 3/91

Figure 2 Therapeutic Drug Monitoring Service report form.

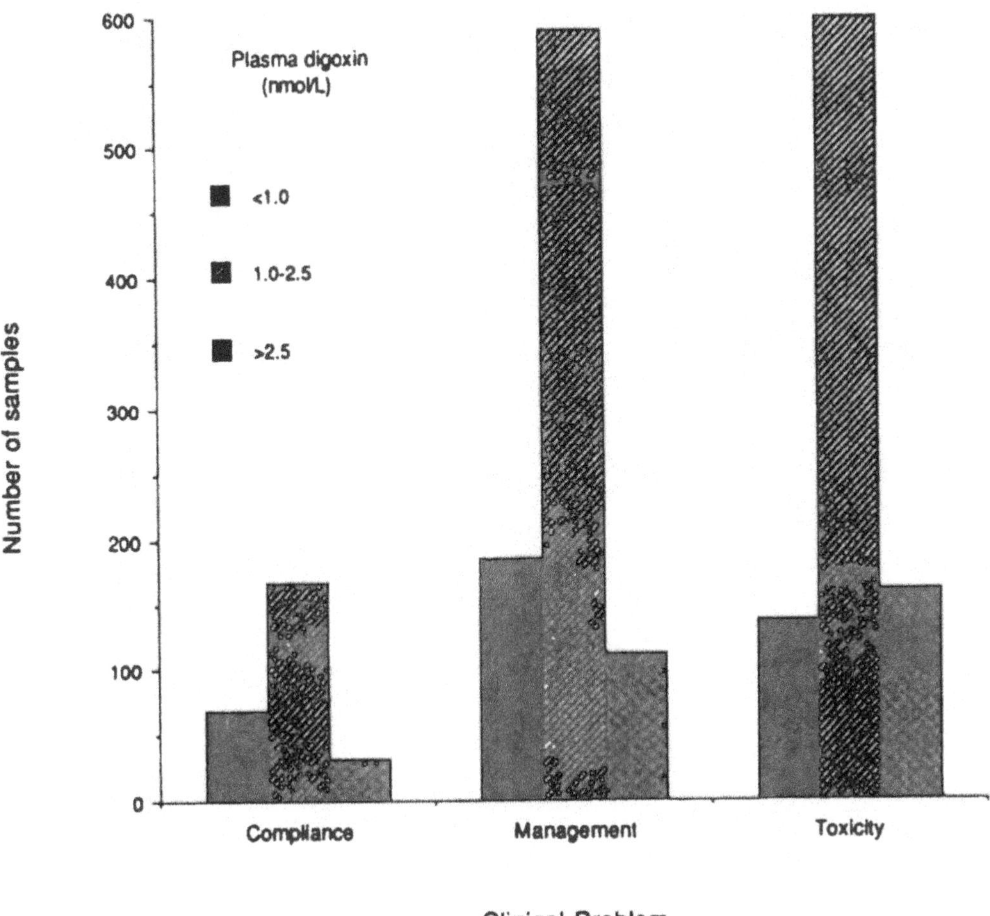

Figure 3 Distribution of digoxin concentrations by clinician-identified problem.

Atrial fibrillation is often treated with higher digoxin concentrations than the norm. As can be seen in Figure 4, the proportion with concentrations above 2.6 nmol/L was less in the group who had atrial fibrillation as part of clinical information than in the group not so identified.

Comments were made on 60.1% of all requests by the reporting team. Follow-up samples were received on 95% of patients on whom comments were made, the majority within 2 months of the first request but some up to a year after (Figure 5).

Hypokalemia was identified in 5% of samples, of which only 22% had queried toxicity; 80.8% had a serum digoxin concentration above 1.0 nmol/L.

IV. DISCUSSION

In 70% of cases, adequate clinical information was provided to allow full analysis. In 20% of cases, the patient was stated to be in atrial fibrillation.

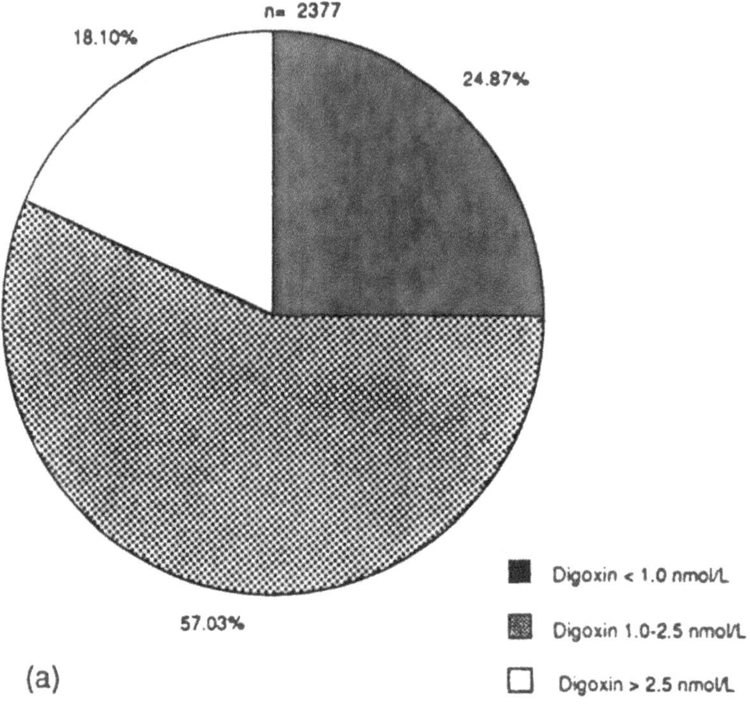

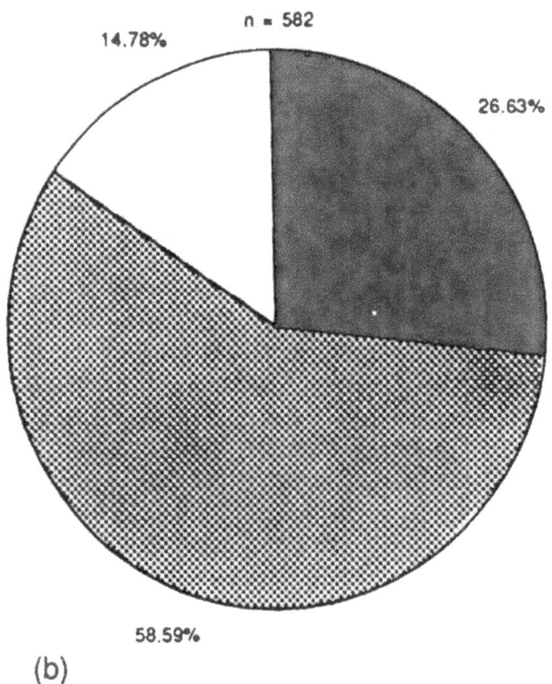

Figure 4 Distribution of digoxin concentration in patients (a) with and (b) without atrial fibrillation.

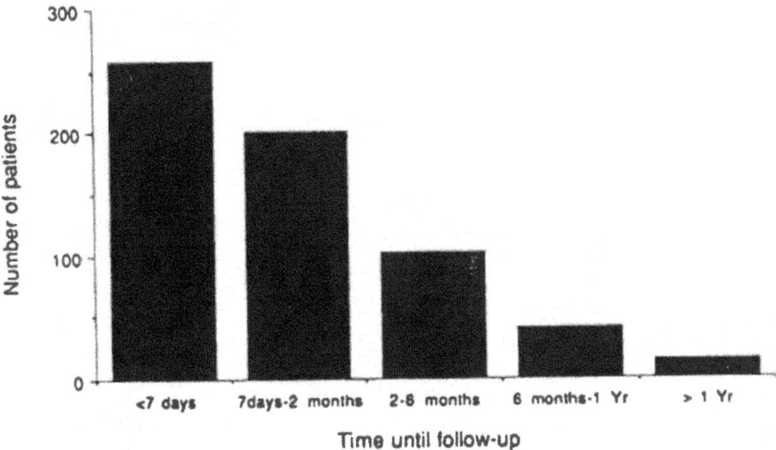

Figure 5 Follow-up of patients on whom comments were made by the Therapeutic Drug Monitoring Service.

The relationship between the clinical impression and the measured digoxin concentration is shown in Figures 2 through 5. The distribution among low, acceptable, and high digoxin concentrations (once inappropriately timed specimens were excluded) showed that there was a similar distribution regardless of the clinical impression (Figure 3).

In all three groups, 63–67% of patients had levels within the range 1 to 2.5 nmol/L.

Of particular interest was the group of 31 cases where the digoxin levels exceeded 2.5 nmol/L and yet faulty compliance was suspected, confirming the fact that the clinical assessment of digoxin status is a difficult procedure.

In patients with atrial fibrillation (AF), there was no higher proportion of subtherapeutic concentrations over those with no AF reported, again indicating that the digoxin concentration is an independent variable in the assessment.

The study of the relationship among digoxin levels, toxicity, and hypokalemia showed that, of 185 patients with high digoxin and low potassium concentrations, only 22% were judged to be clinically toxic.

The effectiveness of the service was judged by the rate of response to comment, as observed by the taking of follow-up samples within a meaningful time limit. The fact that follow-up was performed on 1691 of the 1781 patients on whom a comment was made suggests that the liaison is effective.

We conclude from the results of this study, which is still in the process of more detailed analysis, that measurement of plasma digoxin levels provides information which is not available to the clinician from examination and which is reacted to by further investigation.

REFERENCE

1. I. D. Watson, M. Clark, M. J. Steward, G. Lindsay, and D. Talwar. Organisation of a clinical pharmacokinetics service. *Clin. Chem.*, 35(7):1354–1355 (1989).

8

Free Therapeutic Drug Monitoring

Jerome Barre *Centre Hospitalier Intercommunal, Creteil Cedex, France*

I. INTRODUCTION

It has long been assumed that only free drug is diffusible out of the vascular compartment and capable of eliciting pharmacological action upon reaching sites of action. Accordingly, therapeutic drug monitoring should rely on the measurement of unbound rather than total concentrations of drugs. So far, therapeutic drug monitoring has been based mainly on the assay of total drug. The reasons for this are twofold. First, at the beginning of the therapeutic drug monitoring era, knowledge of the influence of protein binding and of the factors controlling the unbound concentrations of drug was not as much advanced as it is today. Second, the methodology available to separate free drug was not entirely reliable. In addition, this methodology was too cumbersome to be implemented in routine clinical practice. Presently, some ultrafiltration devices which have been developed in the past few years provide suitable techniques for use in the clinical setting. All these advances, in both techniques for determination of free drug concentrations and in the understanding of the pharmacokinetic factors governing unbound concentrations, are leading us to reconsider the value of free drug level monitoring.

II. PHARMACOKINETIC RATIONALE

When monitoring free drug, one should be aware that judgment relying only on the measurement of the unbound fraction (fu) can be misleading. If one is dealing with a low-extraction drug, it can be seen from the following relationships that an increase in fu does not affect the average unbound concentrations at steady state (Cuss,av), whereas it produces a decrease in the average total concentration at steady state (Css,av):

$$\text{Cuss,av} = \frac{F \cdot D}{T \cdot CLu} \qquad (1)$$

$$\text{Css,av} = \frac{F \cdot D}{T \cdot fu \cdot CLu} \qquad (2)$$

where F is the bioavailability factor, D is the dose, T is the dosing interval, and CLu is the unbound clearance.

If, on the other hand, one is dealing with a high-extraction drug, the relationships are different:

$$\text{Cuss,av} = \frac{F \cdot D \cdot fu}{T \cdot Q} \quad (3)$$

$$\text{Css,av} = \frac{F \cdot D}{T \cdot Q} \quad (4)$$

where Q is the blood flow.

The effect of changing fu will be totally different depending on whether the drug is administered intravenously or orally. If the drug is given by the i.v. route, F is equal to 1, and an elevated fu will increase Cuss,av whereas Css,av will not change. When the drug is given orally, a rise in fu producing a slight increase in the extraction ratio can lead to a significant decrease in F. The opposite variations of these two parameters (F and fu) produces little or no change in Cuss,av. However, Css,av will decrease in the same proportion as fu.

The events occurring at peak plasma levels for both low- and high-extraction drugs depend on the extent of the change in unbound volume of distribution cause by the elevation of fu. An increase in fu accounts for a significant decrease in unbound volume of distribution for drugs exhibiting a small total volume of distribution. Accordingly, a transient increase in unbound concentration at peak levels (at steady state) is likely to induce side effects even when the average unbound steady-state concentration remains unchanged.

These theoretical considerations illustrate that a change in fu does not necessarily imply a change in the average unbound concentration at steady state. This is the reason why unbound concentration should preferably be monitored instead of fu.

III. FACTORS INDUCING ALTERATIONS IN UNBOUND FRACTION

One factor inducing alterations in the unbound fraction is concentration-dependent binding (1). Some drugs showing a concentration-dependent binding are listed in Table 1. Some of these drugs are bound predominantly to albumin, whereas the others are bound to alpha-1-glycoprotein. A common characteristic of all these compounds is that their therapeutic molar concentration approaches or even exceeds the physiological molar concentration of the binding protein. This phenomenon accounts for the change in fu within the therapeutic range.

A second factor is drug displacement induced by co-medications, i.e., drugs likely to share the same binding sites as the drug, which is monitored (2) by endogenous compounds such as bilirubin or free fatty acids (3,4) and finally by drug metabolites.

A third factor is binding defect produced by renal impairment. It is now well known that albumin in this pathophysiological condition is carbamylated and is characterized by a change in the amino acid composition (3). Subsequently, the sites do not bind drugs with the same affinity constant as they usually do. In addition, many endogenous compounds which accumulate in blood can compete with drugs for the same sites.

Table 1 Conditions Associated with Altered Albumin Concentration in Plasma

Decrease
Age (neonate, elderly)
Burns
Cancer
Cirrhosis
Enteropathy
Liver abscess
Malnutrition
Nephrotic syndrome
Pregnancy
Renal failure
Surgery
Trauma

Finally, there are many physiological or pathophysiological conditions where albumin or alpha-1-acid glycoprotein levels can be altered (1,5) (Table 2). The effect of dyslipidemia is less well documented (3).

Table 2 Circumstances Associated with Altered Alpha-1-Acid Glycoprotein Concentration in Plasma

Decrease	Increase
Age (neonate)	Age (elderly)
Severe cirrhosis	AMT[a]
Nephrotic syndrome	Cancer
Oral contraceptive	Celiac disease
Pregnancy	CAPD[b] treatment
	CBZ and PB[c] treatment
	Crohn's disease
	Infections
	Inflammatory states
	Myocardial infarction
	Obesity
	Renal failure
	Rheumatoid arthritis
	Stress
	Surgery
	Trauma

[a]AMT = amitriptyline.
[b]CAPD = continuous ambulatory peritoneal dialysis.
[c]CBZ = carbamazepine; PB = phenobarbital.

IV. METHODS FOR PROTEIN BINDING DETERMINATION

Blood samples should be properly collected to avoid spurious estimation in free drug determinations. The main sources of errors are well identified: use of heparin in vivo, stimulating the release of nonesterified free fatty acids which behave as drug displacers, and plasticizers in stoppers of blood collection tubes displacing drugs bound to the alpha-1-acid glycoprotein (2). The effects of thawing, freezing, and storage conditions are poorly documented.

Ultrafiltration is often used for determining free concentration in routine clinical practice (1,2,5). Free drug is forced through a semipermeable membrane by centrifuging the plasma sample. The procedure is rapid and easy to implement. The major limitation is a possible nonspecific adsorption of the drug onto the filter membrane, which leads to an underestimation of the free drug concentration. Accordingly, a check of nonspecific adsorption should always be performed before using an ultrafiltration device. Drug concentration in the ultrafiltrate of a protein drug solution may be smaller than in the nonfiltered solution even when no nonspecific adsorption takes place. This phenomenon, which has been reported as the sieve effect, is caused by the relation between the diameters of the pores and of the drug particles and by the fact that the diameters of the pores of a filter membrane vary within a certain range (6). The sieve effect is all the more important as the molecular weight rises.

Ultracentrifugation can also be used (1,2,7). The advantages of this method lie in the absence of noninterfering membrane effects and also in the simplicity of the method. However, several limitations outweigh the merits. Owing to the low sample capacity of the rotor, only a few samples can be handled per run. In addition, when a drug is bound to low- and very-low-density lipoproteins which float on the surface, the estimation of free drug concentration in the supernatant can be overestimated. Free drug sedimentation can occur with a high-molecular-weight compound. Nonspecific adsorption onto the walls of the tube can take place in the aqueous supernatant phase. Finally, this method is rather expensive and time-consuming.

Equilibrium dialysis is the classical method often used for determining the number of binding sites and the affinity constant of a drug-protein complex (1,2). This method can be used to validate the other two methods provided that the drug is not subjected to nonspecific adsorption either to the membrane or to the cell walls of the dialysis chambers (8,9). The main drawbacks are the following: the Donnan effect; osmotic effects due to the protein, producing a volume shift; and finally, a dilution effect resulting in the passage of free drug from the rententate to the dialysate and causing an underestimation of free concentration. The latter effect is pronounced for drugs showing concentration-dependent binding (8). For further details on the merits and the drawbacks of each method, the reader is referred to a more extensive review on methodology (6).

Saliva sampling has been suggested as an alternative method for obtaining an estimate of unbound concentration (10,11). Salivary measurement can be of value provided that the saliva-to-plasma concentration ratio remains constant between patients, and within the therapeutic range. This ratio should not vary with saliva flow rate and pH. In addition, salivary samples should be collected and handled carefully. Binding of drug to salivary mucoproteins or debris can generate spuriously elevated concentrations of drug. Accordingly, proper centrifugation is needed. Currently, phenytoin and carbamazepine seem to be viable candidates for such measurements, since saliva concentrations are well correlated with plasma unbound concentrations (10,11).

V. CURRENT STATUS OF FREE THERAPEUTIC DRUG MONITORING

Drug candidates for free therapeutic drug monitoring should exhibit the following characteristics:

At least 70% of the drug should be extensively bound to plasma proteins.

The unbound fraction should vary for any of the reasons mentioned above, thus causing a dissociation between free and total levels (i.e., the free concentration is not a constant fraction of total levels).

The drug should reside in a small volume of distribution. Then a small change in the unbound concentration is likely to elicit a change in the pharmacological effect.

The drug should have a narrow therapeutic index.

A clear relationship between free concentration and pharmacodynamic or therapeutic effect should have been established.

Currently, drugs meeting these criteria are few. Only anticonvulsants (valproic acid, phenytoin, carbamazepine) and antiarrythmic drugs (lidocaine and disopyramide) are potential candidates.

The reasons why free drug monitoring has received little attention so far are manifold. One of the main reasons is that no relationship between free drug concentration and therapeutic effect has been established. Accordingly, no clear therapeutic range is currently available (1). There are scarce data coming from investigations performed on small population samples or from case reports, which can sometimes be very informative. For phenytoin, the few available data seem to indicate that signs of toxicity are likely to occur with free levels exceeding 2.5 mg/L, but this point remains to be confirmed (1,7). The range of free drug concentrations is even more uncertain, if not unknown, for carbamazepine, valproic acid, and the antiarrhythmic drugs. So far, all the studies which were performed to attempt to compare the utility of free versus total drug have not shown a clear-cut advantage for free level monitoring over total drug concentration.

However, there are very specific situations where full understanding of a discrepancy between total levels and an unusual response can be gained with the measurement of unbound concentration.

A good example of such a situation is a decrease in CLu and a rise in fu of phenytoin caused by co-administration of valproic acid, which may behave both as a metabolic inhibitor and as a displacer (12,13). As a result, the free phenytoin concentration will increase due to the decrease in CLu, whereas the total concentration will decrease. This example illustrates the dissociation between total and free phenytoin concentrations.

In chronic renal failure of nephrotic syndrome, fu of phenytoin is elevated whereas CLu is not affected. These are other conditions where the physicians may be misled by an abnormally low total concentration of phenytoin although the unbound level will be normal. Only monitoring of free concentrations will allow correct prediction of clinical signs and therefore rational management of the case.

VI. CONCLUSIONS

This short review shows that many reasons warrant free therapeutic drug monitoring. Recent data (14,15), showing a close correlation between free pK_a levels and brain concentrations of phenytoin, substantiate the utility of free drug measurements.

Several methods for separation of free drug, especially ultrafiltration devices, are now available on the market and can provide reliable determinations in the clinical laboratory setting. However, it should be emphasized that a separation method, whatever it is, should not be implemented without careful checking of its reliability. Otherwise the analytical variability may increase and generate confusing results.

The establishment of free concentration versus therapeutic and/or toxic effects relationships is needed. This will help in defining free therapeutic ranges which are currently lacking. In that respect, prospective studies likely to provide such information are difficult to design properly for a number of ethical and technical reasons. The publication of case reports should be encouraged, since these can help offer insight into this complex problem. At this point, use of free levels should not be advocated in routine measurements (except in some specific situations such as those described in the previous section) until the results of investigations in this area clearly show the utility and the cost-effectiveness of free therapeutic drug monitoring.

REFERENCES

1. J. Barre, F. Didey, F. Delion, and J. P. Tillement. Problems in therapeutic drug monitoring: Free drug level monitoring. *Ther. Drug. Monitor.*, *10*:133 (1988).
2. C. K. Svensson, M. N. Woodruff, and D. Lalka. Influence of protein binding and use of unbound (free) drug concentrations. In *Applied Pharmacokinetics* (W. E. Evans, J. J. Schentag, and W. J. Jusko, eds.), Applied Therapeutics, Spokane, WA, p. 187 (1986).
3. R. Zini, P. Riant, J. Barre, and J. P. Tillement. Disease-induced variations in plasma protein levels. Implications for drug dosage regimens (Part I). *Clin. Pharmacokinet.*, 2:147 (1990).
4. R. Zini, P. Riant, J. Barre, and J. P. Tillement. Disease-induced variations in plasma protein levels. Implications for drug dosage regimens (Part II). *Clin. Pharmacokinet.*, 3:218 (1990).
5. C. K. Svensson, M. N. Woodruff, J. G. Baxter, and D. Lalka. Free drug concentration monitoring in clinical practice. Rationale and current status. *Clin. Pharmacokin.*, *11*:450 (1986).
6. H. Kurz, H. Trunk, and B. Weitz. Evaluation of methods to determine protein binding of drugs. *Arzneim. Forsch.*, *27*:1373 (1977).
7. M. Oellerich. Influence of protein binding commentary. In *Applied Pharmacokinetics* (W. E. Evans, J. J. Schentag, and W. J. Jusko, eds.), Applied Therapeutics, Spokane, WA, p. 220 (1986).
8. J. Barre, J. M. Chamouard, G. Houin, and J. P. Tillement. Equilibrium dialysis, ultrafiltration and ultracentrifugation compared for determining the plasma protein binding characteristics of valproic acid. *Clin. Chem.*, *31*:60 (1985).
9. R. H. Levy, P. N. Friel, I. Johno, L. M. Linthicum, L. Colin, K. Koch, V. A. Raisys, A. J. Wilensky, and N. R. Temkin. Filtration for free drug level monitoring: Carbamazepine and valproic acid. *Ther. Drug Monitor.*, *6*:67 (1984).
10. M. Danhof and D. O. Breimer. Therapeutic drug monitoring in saliva. *Ther. Drug Monitor.*, 3:39 (1978).
11. J. C. Mucklow. The use of saliva in therapeutic drug monitoring. *Ther. Drug Monitor.*, 4:229 (1982).
12. J. Bruni, J. M. Gallo, C. S. Lee, R. J. Perchalsky, and B. J. Wilder. Interactions of valproic acid with phenytoin. *Neurology*, *30*:1233 (1980).
13. E. Perucca, S. Hebdige, G. M. Frigo, G. Gatti, and A. Crema. Interaction between phenytoin and valproic acid: Plasma protein binding and metabolic effects. *Clin. Pharmacol. Ther.*, *28*:779 (1980).

14. P. N. Friel, G. A. Ojemann, R. L. Rapport, R. H. Levy, and G. Van Belle, Human brain phenytoin: Correlation with unbound and total serum concentrations. *Epilepsy Res.*, 3:82 (1989).
15. B. Rambeck, T. May, and R. Schnabel. Distribution of phenytoin in different regions of the brain and in the serum: Analysis of autopsy specimens from 23 epileptic patients. Proceedings of 2nd International Congress of Therapeutic Drug Monitoring and Toxicology. *Quim. Clin.*, 9:326 (1990).

9

Population Data in Therapeutic Drug Monitoring

G. P. Mould *St. Luke's Hospital, Guildford, Surrey, England*

It is now widely accepted that dose individualization and interpretation of drug concentration results during therapeutic drug monitoring (TDM), taking into account patient variability, must be made. Population pharmacokinetics describes this variability in terms of a number of factors which may be called fixed or random effects. Fixed effects are the population average values of pharmacokinetic parameters and are a function of patient demographic characteristics, underlying disease, and pharmacological considerations. The random effects quantify the residual due to variability among patients. Both effects need to be considered in order to advise clinicians on initiating suitable dosage regimens and on individualizing future dose regimens following drug concentration measurements.

Data for population parameter estimations are obtained mostly from single-dose patient pharmacokinetic studies. Nevertheless, at steady state following multiple doses, useful information can also be obtained from samples taken at known times during a dosing interval. Provided that sufficient data are collected, these may be analyzed for linear correlations (1) or by nonlinear models, such as NONMEM (2).

The use of population data can take a number of forms. First, intuitive dosing can be done based on previous experience, but a number of studies have shown that underdosing is common (3). Another approach is to use data generated from patients taking a single dose, measuring the concentration at a fixed time point afterward, and correlating that concentration with a subsequent concentration taken during steady-state conditions, e.g., lithium (4).

A third approach is to use nomograms based on population data; many types of nomograms have been produced. Recently, six different gentamicin nomograms were compared and dose predictions based on their recommendations produced initial gentamicin peak levels which were subtherapeutic in 50% of the patients (5).

Most use of population data is during the interpretation of analytically generated patient data during TDM. The method most widely used by practitioners is to compare predicted and measured concentrations through a simple kinetic formula (6). More sophisticated techniques use Bayesian feedback mechanisms, which rely on revising initial population data in the light of one or more subsequent blood measurements (7). The complexity of the required mathematics necessitates the use of computers (8). It therefore takes time to generate recommendations.

In spite of the availability of population data and techniques to use it, in practice the TDM practitioner relies more on intuitive judgment and pharmacokinetic experience to make recommendations, and most patient cases are probably solved in this way. It is therefore essential that experience gained in this manner is communicated to other practitioners. The final outcome is that better patient care will result.

REFERENCES

1. L. B. Sheiner, B. Rosenberg, and V. V. Marathe. Estimation of population characteristics of pharmacokinetic parameters from routine clinical data. *J. Pharmacokin. Biopharm.*, 5:445–479 (1977).
2. L. S. Beal and L. B. Sheiner. NONMEM (User's Guide) Parts I and VI. Technical Report, Division of Clinical Pharmacology, University of California, San Francisco (1979).
3. J. L. Browne, C. S. Huffman, and R. N. Golden. A comparison of pharmacokinetic versus empirical lithium dosing techniques. *Ther. Drug. Monitor.*, 11:149–154 (1989).
4. M. V. Nelson. Comparison of three lithium dosing methods in 950 "subjects" by computer simulation. *Ther. Drug Monitor.*, 10:269–274 (1988).
5. A. H. Thomson, K. C. Campbell, and A. W. Kelman. Evaluation of six gentamicin nomograms using a Bayesian parameter estimation program. *Ther. Drug Monitor.*, 12:258–263 (1990).
6. M. E. Winter. *Basic Clinical Pharmacokinetics*. Applied Pharmacokinetics, San Francisco (1989).
7. L. B. Sheiner, H. Halkin, C. Peck, B. Rosenberg, and K. Melmon. Improved computer assisted digoxin therapy. *Ann. Intern. Med.*, 82:619–627 (1975).
8. B. Whiting, A. W. Kelman, and S. W. Bryson. OPT: A package of computer programs for computer optimisation in clinical pharmacokinetics. *Br. J. Clin. Pharmacol.*, 14:247–256 (1982).

10

Therapeutic Drug Monitoring in Pregnancy

Gideon Koren *The Hospital for Sick Children, University of Toronto, Toronto, Ontario, Canada*

During pregnancy, therapeutic drug monitoring (TDM) has different meaning than before or after gestation, as any consideration of maternal health must incorporate potential effects on the unborn baby. In this communication I wish to discuss therapeutic drug monitoring in three major categories:

1. TDM of therapy given to maintain maternal health
2. TDM of fetal therapy given through the maternal circulation
3. TDM aimed at assessing potential toxicity of xenobiotics to which the mother was exposed

I. TDM OF THERAPY GIVEN TO MAINTAIN MATERNAL HEALTH

Pregnancy is associated with a variety of physiological changes which may affect the natural course of diseases, the way the body handles drugs, or both. This text will focus on possible alterations in drug therapy that may have to take place in pregnancy to deal with pharmacokinetic and pathophysiological changes. In general, two principal groups of changes characterize pregnancy with respect to drug disposition.

A. Pregnancy-Induced Pharmacokinetic Changes

There is a gradual increase in renal function during pregnancy. This will result in an augmented elimination rate of agents which are excreted by the kidney (e.g., ampicillin, gentamicin, amikacin, digoxin) (1–5). Distribution volume may be altered during pregnancy because of increases of 50% in blood (plasma) volume and 30% in cardiac output (6–8). A parallel of 50% increment in renal blood flow and a substantial increase of uterine blood flow to 600 to 700 mL/min also take place. During pregnancy there is a mean increase of 8 L in body water, 60% of which is distributed to the placenta, fetus, and amniotic fluid, and 40% is distributed to maternal tissues (6,9–11). Consequently, a decrease in the serum concentrations of many drugs has been documented. It is very likely that lower serum concentrations in pregnancy will be noted, especially with drugs having a relatively small volume of distribution which corresponds to water components. During pregnancy there is a well-documented fall in the protein binding of drugs, partially because of the decrease in serum albumin concentrations (12). Consequently, the unbound (free) drug is free to move into various tissue compartments. Of potential clinical importance, the protein binding of several antiepileptic drugs, including phenytoin, diazepam, and valproic acid, has been shown to decrease significantly toward the last trimester of pregnancy (13).

Changes in hepatic elimination patterns during pregnancy are less consistent. Hepatic blood flow appears to be unchanged during pregnancy (1). However, there is evidence that the elimination rate of clindamycin, which is metabolized by the liver, is increased during pregnancy (3), suggesting a possible increase in hepatic clearance. It is also possible that the faster elimination of trimetoprim/sulfamethoxazole observed during pregnancy is due to higher liver clearance, although increased renal clearance may be the major determinant of this change (14).

The decrease in drug protein binding may account for higher clearance rates of drugs in pregnancy, as it is the free fraction which is accessible to the metabolizing systems. Studies in epileptic women have shown an increase in the clearance of phenytoin during pregnancy, accounting for lower serum concentrations (13). However, because it is the free drug which is pharmacologically active, lower total levels should not be interpreted as lower free concentrations.

As shown in Table 1, the above-mentioned changes in the pharmacokinetics of many drugs administered during pregnancy may result in a decrease in serum concentrations when compared to levels measured in nonpregnant women. Thus, the standard dose schedule may result in lower concentrations in pregnancy and, as discussed later, these changes may have implications in treating the pregnant woman.

Almost all drugs have been shown to cross the placenta and to appear in measurable concentrations in the fetal blood. Several determinants govern the movement of drugs across the placental barrier and determine the maternal/fetal concentration ratio. In general, this ratio is different from (in most cases less than) unity (Table 2).

Table 1 Selected Drugs Having Lower Serum Concentrations During Pregnancy and Relevant Pharmacokinetic Changes (3,4,13,15–21)

Drug	Elimination $T_{1/2}$	V_d	Clearance	Protein Binding
Amikacin	↓			
Ampicillin	↓	↑	↑	↓
Cephalosporines	↓	↓		
Erythromycin				
Gentamicin	→			
Kanamycin	→			
Methicillin	↑			
Nitrofurantoin				
Oxacillin		↑	↑	
Phenobarbitol		↑	↑	↓
Phenytoin			↑	↓

Source: From G. Koren (ed.), *Maternal-Fetal Toxicology, A Clinician's Guide*, Marcel Dekker, New York (1990).

Table 2 Fetomaternal Concentration Ratio of Antibiotics (3,22–32)

Drug	Fetomaternal Ratio
Ampicillin	0.38–0.87
Cephalosporines	0.13–1.0
Clindamycin	0.4–0.5
Dicloxacillin	0.07–0.27
Gentamicillin	0.21–1.0
Methicillin	0.83–1.43
Penicillin G	0.06–0.7

Source: From G. Koren (ed.), *Maternal-Fetal Toxicology, A Clinician's Guide*, Marcel Dekker, New York (1990).

During pregnancy there is a gradual decrease in the concentration of maternal albumin and an increase in the concentration of fetal albumin. Consequently, at different times in pregnancy, different fetal/maternal ratios of albumin occur. At term, it appears that fetal albumin concentrations are equal or even higher than in the mother. The degree of protein binding of a drug is an important determination of its movement across the placenta. The least protein-bound drugs (e.g., digoxin, ampicillin, 20%) reach higher concentrations in the fetus and in the amniotic fluid. Drugs with high protein binding (e.g., dicloxacillin, 96%) achieve higher maternal and lower fetal concentrations, as only the free fraction of the drug crosses the placental barrier. However, as additional determinants other than protein binding may play an important role in placental passage, drugs such as sulfisoxazole reach therapeutic concentrations in the fetus in spite of their high protein binding.

Fetal blood pH is slightly lower than that of mothers. The pH gradient may influence the movement of drugs according to their pK_a; weak bases, with a pK_a value close to the blood pH, will be mainly non-ionized and consequently will easily cross the placenta, as non-ionized molecules penetrate biological membranes more quickly than do ionized molecules. However, after crossing the placenta and making contact with the more acidic fetal blood, these molecules are ionized, as the fetal pH is less close to their pK_a. This results in an apparent fall in fetal concentrations of non-ionized drugs and to a subsequent concentration gradient, leading to net movement from maternal to fetal systems (33,34). This mechanism is commonly referred to as "ion trapping." In contrast to weak bases, ion trapping induced from fetal to maternal circulations is likely to occur with weak acids.

Other important determinants of drug transport across the placenta are water/lipid solubility, molecular weight, and the surface available for diffusion. A good example of the different effects on placental passage is the higher concentrations of trimetoprim in the fetus than in the mother, corresponding to its relatively low protein binding (42% to 46%), pK_a of 6.6, and poor water solubility. Sulfamethoxazole, on the other hand, has good water solubility at physiological pH, and therefore has more difficulties in crossing the lipid placental barrier. Both drugs appear to reach the amniotic fluid, and their concentration ratio there closely corresponds to that in the fetal serum (36).

Table 3 Therapeutic Serum Concentrations of Drugs Commonly Monitored in Clinical Practice, and Documented Changes in Pregnancy

Drug	Therapeutic Concentration		Documented Changes in Drug Concentration During Pregnancy
	Metric	SI[a]	
Amikacin	Peak <20–30 μg/mL Trough <5–10 μg/mL		↓
Amiodarone	<2.5 μg/mL		
Carbamazepine	4–12 μg/mL	17–51 μM	↓
Chloramphenicol	<25 μg/mL		
Cyclosporine	<200–250 ng/mL		
Digoxin	0.5–2 ng/mL	0.7–2.6 nM	↓
Disopyramide	3–5 μg/mL	9–15 μM	
Ethosuximide	40–100 μg/mL	285–710 μM	↓
Gentamicin	Trough <2.3 μg/mL Peak <8–10 μg/mL		↓
Lidocaine	1.5–5.0 μg/mL	6–21 μM	
Lithium	0.8–1.0 mM		↓
Methotrexate	<5 μM 24 h after high dose		
Phenobarbitol	15–40 μg/mL	65–172 μM	↓
Phenytoin	10–20 μg/mL	39–79 μM	↓
Primidone	5–12 μg/mL	23–55 μM	
Procainamide	4–10 μg/mL	15–37 μM	
Quinidine	2.3–5.0 μg/mL	7–15 μM	
Theophylline	10–20 μg/mL	55–110 μM	
Tobramycin	Trough <2–3 μg/mL Peak <8–10 μg/mL		
Valproic acid	50–100 μg/mL	347–693 μM	
Vancomycin	Trough <10 μg/mL Peak <45 μg/mL		

[a]M = 1 mol/L.
Source: From G. Koren (ed.), *Maternal-Fetal Toxicology, A Clinician's Guide*, Marcel Dekker, New York (1990).

Following repeated doses of drugs, concentrations measured in the fetus appear to be higher than those following a single dose (37). Repeated high-bolus injections of ampicillin or gentamicin yield higher concentrations in fetal serum and amniotic fluids than a similar pattern of transplacental transfer, and after a single intravenous dose a peak umbilical concentration is achieved within 30 to 60 min.

B. Fetal Drug Disposition

Drug metabolism has been documented by the fetal liver as early as 7 to 8 weeks of pregnancy. Virtually all enzymatic processes, including phase 1 (oxidation, dehydrogenation, reduction, hydrolysis, etc.) and phase 2 (glucuronidation, methylation, acetylation, etc.) have been documented in the fetal liver (39). However, the degree of activity is very low in most cases when compared to the adult liver. Similarly, drug-metabolizing activity has been demonstrated in human placental tissue. In summary, the placentofetal unit contributes only marginally to the total elimination capacity of drugs by the maternal body. As pregnancy progresses, higher amounts of antimicrobials are excreted into the amniotic fluids through fetal urine. This process depends on maturation of the fetal kidney.

In general, metabolites are more polar than their parent compounds and are therefore less likely to cross the placental barrier. As a result, metabolites may accumulate in various tissues of the fetus or may be recovered from the amniotic fluid. One recent study has shown that thiamphenicol achieves higher concentrations in the umbilical vein than in the umbilical artery, reflecting some degree of extraction of the drug by the fetus, probably by renal excretion (32).

C. Clinical Implications

As reflected in the previous discussion, it appears that a variety of drugs achieve lower serum concentrations during pregnancy. For agents which maintain good correlation between serum concentrations and pharmacological effects, this may mean that patients may be at a higher risk of suboptimal therapy during pregnancy.

1. *Anticonvulsants*

As discussed earlier, there is good evidence that the serum concentrations of phenytoin, phenobarbital (40), ethosuximide (41), and carbamazepine (42) decrease as pregnancy progresses. Some authors observed an increase in the frequency of seizures as phenytoin clearance increased in pregnancy, and as plasma levels fell (40). A variety of reasons have been forwarded to explain this fall in drug concentrations in pregnancy:

1. Increase in extracellular fluid and tissue volume, leading to increase in distribution volume of the drug.
2. Decreased plasma protein binding, leading to more free drug available for biotransformation. However, higher free drug concentration may secure antiepileptic effect even in the presence of lower total drug level, as it is the free drug that reaches the brain.
3. Folic acid, given to the pregnant woman, may increase liver metabolism of phenytoin (41).
4. Increase in glomerular filtration (GFR), leading to faster clearance rates of drugs eliminated by the kidney.
5. Enhanced hepatic metabolism of drugs.

The question whether epilepsy is worsened during pregnancy is of extreme importance, but current evidence is inconclusive: Schmidt reviewed 2162 pregnancies and could not detect a clear pattern, as some 23% reported improvement, 53% no change, and 24% worsened (43). However, other studies suggest that epilepsy with at least one seizure per month is likely to worsen in pregnancy (44).

Changes in seizure frequency in pregnancy may stem from fluid and sodium retention, hyperventilation, a rise in estrogen levels, emotional and psychological problems, and of course the tendency of drug levels to fall (13). Although presently no research has addressed the contribution of each factor, there are well-documented cases to prove the importance of adequate serum concentrations during pregnancy (13,41).

Caring for the pregnant epileptic patient must, therefore, incorporate careful monitoring of serum concentrations and appropriate adjustment of the antiepileptic dose. Measurement of free phenytoin and carbamazepine may yield more accurate estimate of their pharmacokinetics. After pregnancy, most women will need lower dosages, and failure to adjust their schedule may lead to drug toxicity.

2. *Lithium*

The antidepressant lithium is eliminated almost entirely by the kidney. The pregnancy-induced increase in the GFR is consistent, therefore, with lower serum concentrations of lithium reported sporadically during pregnancy (45). Because in many cases lithium exerts its pharmacological effects at nearly toxic levels, the drop in serum concentrations may lead to suboptimal therapy.

After birth, with the return of GFR to its prepregnancy values, patients may need reduction of their doses.

Some lithium is reabsorbed by the renal tubule, and this process is in competition with sodium reabsorption. Pregnant patients with toxemia are kept on a restricted sodium intake, and therefore they may experience higher levels of lithium owing to higher renal reabsorption of the cation.

3. *Digoxin*

Similar to lithium, digoxin is eliminated in humans mainly by renal excretion, and therefore is expected to maintain lower steady-state concentrations in pregnancy. When measured at term, digoxin serum concentrations in five pregnant women were found to be almost twofold lower than 1 month later (5).

4. *Ampicillin*

Since ampicillin is one of the most widely used antibiotics, knowledge of its pharmacokinetics in pregnancy may yield valuable information about the dose requirement during gestation. Philipson studied the disposition of ampicillin once during and again after pregnancy, and found the plasma concentration to be 50% lower during gestation owing to both the larger distribution volume and faster clearance rate (2). Similar to digoxin and lithium, ampicillin is eliminated mainly through the kidney, and the twofold decrease in its levels is consistent with that described for digoxin.

5. *Other Drugs*

Similar observations have been documented with cephalosporins, clindamycin, erythromycin, kanamycin, amikacin, tobramycin, nitrofurantoin, and sulfamethoxazole-trimethoprim (37,38). In all instances, the lower serum concentrations during pregnancy could be attributed to pharmacokinetic changes.

II. TDM OF FETAL THERAPY GIVEN THROUGH THE MATERNAL CIRCULATION

During the last decades there have been increasing attempts to treat pharmacology babies at risk through the maternal circulation. As most drugs cross the placenta and when time allows, achieve comparable fetal concentrations to maternal levels, it is conceivable that more agents will be used in the future for this end. Use of corticosteroids given to the mother to enhance fetal lung maturation, and administration of antimicrobial agents to treat fetal infections such as syphilis or toxoplasma, may serve as such examples. In at least two situations, mothers are administered potentially toxic drugs, which are normally measured routinely, to treat the baby. Digoxin has been given to treat fetal dysrythmias as well as hydrops fetalis (46). Most investigators agree that to achieve a favorable effect in the fetus, almost toxic maternal concentrations have to be maintained. Using in-situ placental perfusion, we have recently shown that it takes 4 to 5 h for equilibration between maternal and fetal digoxin (47). This figure closely resembles the time needed in vivo to achieve equilibrium (48).

Similar to digoxin, other antiarrhythmic drugs have been effective in reversing dangerous fetal tachyarrhythmias.

During the last decade several well-controlled studies have shown that maternal treatment with phenobarbital prevents serious intracranial bleeding in high-risk preterm infants. The mothers are given phenobarbital to maintain low therapeutic levels, which are not associated with increased maternal risk (49).

With increased diagnostic accuracy in pregnancy, it is likely that more and more fetal conditions which can be managed pharmacologically will be identified and treated through the maternal circulation. TDM is likely to be used more to secure maternal safety as well as to establish a therapeutic window.

III. ASSESSING POTENTIAL TOXICITY OF XENOBIOTICS

Since the thalidomide disaster, an increasing number of xenobiotics have been identified which may adversely affect the fetus. Because in many cases the toxicity is dose (concentration)-dependent, there is an intensive search for a biological marker which will reflect the systemic burden of xenobiotics. During the last decade, in several areas, TDM has achieved impressive accuracy in predicting potential fetal damage: Measuring maternal methyl mercury in hair has a clear dose-response curve; above a certain value, there is high likelihood of fetal neurological damage (50).

Recent use of neonatal hair to identify intrauterine drug exposure to cocaine and heroin is yet another example for a biological marker for significant, otherwise unproven exposure (51). Maternal blood concentration of lead correlate with cognitive development of offspring (52), even at maternal levels that are not associated with adult toxicity.

Carbon monoxide poisoning is a major emergency in pregnancy, as it can cause irreversible fetal asphyxia. Carbon monoxide equilibrates slowly between the mother and fetus; hence, even after maternal carboxyhemoglobin levels return to normal, the mother should be treated with oxygen for several more hours to enhance clearance of residual CO from the fetus (53).

The above are merely several examples where TDM has been proven useful in predicting fetal damage. Many more compounds are likely to follow in the next few years.

IV. SUMMARY

One of the main goals of TDM is to individualize drug therapy. In the context of pregnancy, this concept is even more important, as it may yield a better estimate of fetal well-being in addition to the maternal status. This field is at its very beginning, and I am confident that during the next decade TDM in pregnancy will become even more important in maintaining maternal and fetal health.

ACKNOWLEDGMENTS

This work was supported in part by PSI Toronto, Health & Welfare Canada, MRC, Canada, and the Motherisk Research Fund.

REFERENCES

1. B. Krauer and F. Krauer. Drug kinetics in pregnancy. In *Handbook of Clinical Pharmacokinetics* (M. Gibaldi and L. Prescott, eds.), Section II, ADIS, New York, pp. 1–17 (1983).
2. A. Philipson. Pharmacokinetics of ampicillin during pregnancy. *J. Infect. Dis., 136*:370–376 (1977).
3. A. J. Weinstein, R. S. Gibbs, and M. Gallagher. Placental transfer of clindamycin and gentamicin in term pregnancy. *Am. J. Obstet. Gynecol., 124*:688–691 (1976).
4. B. Bernard, M. Abate, P. F. Thielen, H. Attar, C. A. Ballard, and P. F. Wehnle. Maternal fetal pharmacological activity of amikacin. *J. Infect. Dis., 135*:925–932 (1977).
5. M. E. Rogers, J. T. Willerson, A. Goldblatt, et al. Serum digoxin concentrations in the human fetus, neonate and infant. *N. Engl. J. Med., 287*:1010–1013 (1972).
6. F. E. Hytten and T. Leitch. *The Physiology of Pregnancy*. Blackwell, Oxford (1971).
7. B. B. K. Pizani, D. M. Campbell, and T. McGillivray. Plasma volume in normal pregnancy. *J. Obstet. Gynecol., 80*:884–887 (1973).
8. W. A. W. Wlaters and Y. Lengling. Blood volume and haemodynamics in pregnancy. In *Obstetrics and Gynecology*, Vol. 2, Saunders, London, pp. 301–302 (1975).
9. J. M. Davidson and F. E. Hytten. Glomerular filtration during and after pregnancy. *J. Obstet. Gynecol., 81*:588–595 (1974).
10. I. M. Young. The placenta: Blood flow and transfer. In *Modern Trends in Physiology*, Butterworth, London, pp. 214–244 (1972).
11. M. G. Kerr. Cardiovascular dynamics in pregnancy and labour. *Br. Med. Bull., 24*:19–24 (1968).
12. P. Rebound, J. Groulade, P. Groslambert, et al. The influence of normal pregnancy and the postpartum state on plasma proteins and lipids. *Am. J. Obstet. Gynecol., 86*:820–828 (1963).
13. E. Perucca and A. Richens. Antiepileptic drugs, pregnancy and the newborn. In *Clinical Pharmacology in Obstetrics* (P. Lewis, ed.), Wright PSG, Briston, pp. 264–387 (1983).
14. O. Ylikorkala, E. Sjostedt, P. A. Jarvinen, R. Tikkanen, and T. Raines. Trimetoprim/sulfonamide combination administered orally and intravaginally in the first trimester of pregnancy: Its absorption into serum and transfer to amniotic fluid. *Acta Obstet. Gynecol. Scand., 52*:229–234 (1973).
15. R. E. Bray, R. W. Boe, and W. L. Johnson. Transfer of ampicillin into fetus and amniotic fluid from maternal plasma in late pregnancy. *Am. J. Obstet. Gynecol., 96*:938–942 (1966).
16. B. Bernard, L. Barton, M. Abate, and C. A. Ballard. Maternal fetal transfer of cefazolin in the first twenty weeks of pregnancy. *J. Infect. Dis., 136*:377–382 (1977).
17. A. Philipson, L. D. Saboth, and D. Charles. Erythromycin and clindamycin absorption and elimination in pregnant women. *Clin. Pharmacol. Ther., 19*:68–77 (1976).
18. R. G. Good and G. Johnson. The placental transfer of kanamycin in late pregnancy. *Obstet. Gynecol., 38*:60–62 (1971).
19. M. A. MacAuley, S. R. Berg, and D. Charles. Placental transfer of methicillin. *Am. J. Obstet. Gynecol., 115*:58–65 (1973).
20. K. Amon, I. Amon, and H. Huller. Verteilung und kinetik von nitrofurantoin in der fruhschwangerschaft. *Int. J. Clin. Pharmacol. Ther. Toxicol., 63*:218–222 (1972).
21. G. Bastert, W. G. Muller, K. H. Wallhauser, and H. Hebauf. Pharmacokinetishe untersuchungen zum ubertriff von antibiotika in das fruchtwasser am enter der schwangerschaft. 3 teili. *Oxacillin Geburtshilfe Perinatol., 179*:346–355 (1975).

22. H. A. Hirsch, E. Dreher, A. Perrochet, and E. Schmid. Transfer of ampicillin to the fetus and amniotic fluid during continuous infusion (steady state) and by repeated single intravenous injections to the mother. *Infection*, 2:207–212 (1974).
23. I. Croft and T. C. Forster. Materno-fetal cephadine transfer in pregnancy. *Antimicrob. Agents Chemother.*, 14:924–926 (1978).
24. W. Barr and R. M. Graham. Placental transmission of cephaloxidine. *J. Obstet. Gynecol.*, 74:739–745 (1947).
25. H. A. Hirsch, S. Herbet, R. Lang, L. Dettli, and A. Goblinger. Transfer of a new cephalosporin antibiotic to the fetus and the amniotic fluid during a continuous infusion (steady state) and single repeated intravenous injections to the mother. *Arzneimsttelforschung* 24:1474–1478 (1974).
26. M. A. MacAuley, M. Abou-Sabe, and D. Charles. Placental transfer of dicloxacillin at term. *Am. J. Obstet. Gynecol.*, 102:1162–1168 (1968).
27. L. Forreres, M. Paz, G. Martin, and M. Gobernado. New studies on placental transfer of fosfomycin. *Chemotherapy, Suppl.* 1:175–179 (1977).
28. O. Daubenfeld, H. Modde, and H. A. Hirsch. Transfer of gentamicin to the fetus and amniotic fluid during a steady state in the mother. *Arch. Gynecol.*, 217:233–240 (1974).
29. H. Yoshioka, T. Monma, and S. Matsudo. Placental transfer of gentamicin. *J. Pediatr.*, 80:121–123 (1972).
30. R. Depp, A. C. Kind, W. M. M. Kirby, and W. L. Johnson. Transplacental passage of methicillin and dicloxacillin into the fetus and amniotic fluid. *Am. J. Obstet. Gynecol.*, 197:1054–1057 (1979).
31. D. Charles. Placental transmission of antibiotics. *J. Obstet. Gynecol.*, 61:790–797 (1954).
32. T. A. Plomp, R. A. A. Moes, and M. Thiery. Placental transfer of thaimphenicol in term pregnancy. *Eur. J. Obstet. Gynecol. Reprod. Biol.*, 7:383–388 (1977).
33. G. T. Tucker, R. N. Boyes, and P. O. Bridenbaugh. Binding of anilide-type local anesthetics in human plasma. II. Implications in vivo, with special reference to transplacental distribution. *Anesthesiology*, 33:304–314 (1970).
34. G. Levy. Salicylate pharmacokinetics in the human neonate: In *Basic and Therapeutic Aspects of Perinatal Pharmacology* (Marselli, Garattini, and Sereni, eds.), Raven Press, pp. 319–330 (1975).
35. J. H. Asling and E. L. Way. Placental transfer of drugs. In *Fundamentals of Drug Metabolism and Drug Disposition* (H. La Du, B. Mandel, Way, Y., eds.). Williams & Wilkins, Baltimore, p. 88 (1972).
36. A. M. Walter and L. Heilmeyer. *Antibiotika—Fibel. 4.* Auflage, Georg Thieme, Stuttgart (1975).
37. A. Philipson. Pharmacokinetics of antibiotics in pregnancy and labour. *Clin. Pharmacokinet.*, 4:297–309 (1979).
38. A. W. Chow and P. J. Jewesson. Pharmacokinetics and safety of antimicrobial agents during pregnancy. *Rev. Infect. Dis.*, 7:287–313 (1985).
39. M. R. Juchau, S. T. Chao, and C. J. Omiecinski. Drug metabolism by the human fetus. In *Handbook of Clinical Pharmacokinetics* (M. Gibaldi and L. Prescott, eds.), Section II, ADIS, New York, pp. 58–78 (1983).
40. M. Dam, K. J. Mygind, and J. Christiansen. Antiepileptic drugs: Plasma clearance during pregnancy. *Epileptology*, 179–183 (Jan. 3, 1976).
41. M. J. Eadie, C. M. Lander, and J. H. Tyrer. Plasma drug level monitoring in pregnancy. In *Handbook of Clinical Pharmacokinetics* (M. Gibaldi and L. Prescott, eds.), Section IV, ADIS, New York, pp. 53–62 (1983).
42. M. Dam, J. Christiansen, O. Munck, and K. I. Mygind. Antiepileptic drugs: Metabolism in pregnancy. *Clin. Pharmacokinet.*, 4:53–62 (1979).
43. D. Schmidt. The effect of pregnancy on the natural history of epilepsy. In *Epilepsy, Pregnancy and the Child* (D. Janz, L. Bossi, M. Dam, et al., eds.), Raven Press, New York, pp. 3–14 (1981).
44. A. H. Knight and E. G. Rhind. Epilepsy and pregnancy: A study of 153 pregnancies in 59 patients. *Epilepsia*, 16:99–110 (1975).
45. M. Schou, A. Amidsen, and D. R. Steenstrup. Lithium and pregnancy. II: Hazards to women given lithium during pregnancy and delivery. *Br. Med. J.*, 2:137–138 (1973).

46. J. T. Harrigan, J. Kango, and E. Lisalo. Successful treatment of fetal congestive heart failure secondary to tachycardia. *N. Engl. J. Med.*, *304*:1527–1529 (1981).
47. L. O. Derewlany, S. J. Leeder, R. Kumar, I. C. Radde, B. Knie, and G. Koren. The transport of digoxin across the perfused human placental lobule. *J. Pharmacol. Exp. Ther.*, in press.
48. L. Padeletti, M. C. Porciani, and G. Scimone. Placental transfer of digoxin in man. *J. Clin. Pharmacol. Biopharm.*, *17*:82–83 (1979).
49. S. Donn, D. Raoff, and G. Goldstein. Prevention of intraventricular haemorrhage in preterm infants by phenobarbitone. *Lancet*, 2:215 (1981).
50. D. O. Marsho, T. W. Clarkson, C. Cox, G. J. Meyers, L. Amin Zaki, and Altikritis. Fetal methylmercury poisoning: Relationship between concentration in single strands of maternal hair and child effects. *Arch. Neurol.*, *44*:1017–1022 (1987).
51. K. Graham, G. Koren, J. Klein, and J. Schneiderman. Detecting gestational exposure to cocaine by hair analysis. *J. Am. Med. Assoc.*, *262*:3328–3330 (1989).
52. D. Bellinger, et al. Longitudinal analysis of prenatal and postnatal lead exposure and early cognitive development. *N. Engl. J. Med.*, *316*:1037–1043 (1987).
53. G. Koren, et al. Fetal outcome following carbon monoxide poisoning in pregnancy. Submitted.

11

Issues in Therapeutic Drug Monitoring of Digoxin in Children

Gideon Koren *The Hospital for Sick Children, Toronto, Ontario, Canada*

Two hundred years after its introduction into clinical use, digoxin is still one of the most commonly prescribed drug in clinical medicine. Despite continued controversy on digoxin efficacy, it appears that cardiac glycosides will be in use in the visible future. The therapeutic drug monitoring (TDM) of digoxin, like that of any other drug, stems from two cardinal assumptions:

1. There is a predictable correlation between serum concentrations and clinical or toxic effects of the drug, better than their correlation with the dose.
2. The essay used for determination of the drug is sensitive, specific, and accurate.

Neither of these conditions is fully met in the case of digoxin. While there is some correlation between heart rate and drug concentrations, there is no well-established correlation between serum concentrations and the inotropic effect. As expected, there is more likelihood of toxicity with increased serum concentrations above 2 ng/mL, but even at 5 ng/mL only 50% of pediatric patients exhibit toxicity (1); conversely, patients may experience toxicity even at concentrations below 2 ng/mL, and myocardial disease, hypocalcemia, and hypomagnesemia all increase the likelihood of adverse effects (2).

Of all aspects of the digoxin assay, the issue of specificity is the most pressing. During the last decade it has become apparent that commercially available assays for digoxin, while being sufficiently sensitive and reproducible, also measure endogenous compounds, collectively referred to as endogenous digoxinlike compounds (EDLS) (3).

Sera of newborn infants, pregnant women, and patients with hypertension, renal failure, or hepatic failure give positive digoxin immunoassays (2) when these individuals have not been exposed to the cardiac glycoside. Spiking of "true" digoxin to these samples results in additive levels (2). Hence, we hypothesized that the clinician would have to deal with these spuriously high levels as if they represent potential toxicity and discontinue or decrease digoxin schedule. To prove this hypothesis we analyzed all samples of children in our hospital who had steady-state trough concentrations of digoxin equal or higher than 3 nM. To separate true digoxin from EDLS, a newly devised HPLC method was combined with the FPIA assay (4).

In most samples, true digoxin was lower than the total results reported in the clinical laboratory using the Abbot FPIA method (4.1 ± 1.2 nM versus 3.2 ± 0.9 nM, $p < 0.01$). Of importance, in 5 out of 17 cases, true digoxin was within the therapeutic range; however, the drug was discontinued by the clinicians due to the "toxic levels," despite lack of clinical toxicity in the patients. By using the HPLC method, which is digoxin-specific, insight into "potentially toxic" levels of digoxin is possible. Our results confirm the clinical impression of many pediatricians that toxicity of digoxin is hardly ever seen in children, even at levels of 3 and 4 nM. As important, it is probable that many such children are undertreated due to the false guidance of the immunoassay's RDM result. Is it possible that this is one of the reasons for the difficulties in proving the clinical efficacy of the drug? While awaiting an improved assay for digoxin, the clinician should probably combine serum concentrations obtained by immunoassays with clinical judgment before cutting digoxin dose in a child who is free of adverse effects.

Because of the multiple pharmacokinetic interactions of digoxin with commonly co-administered drugs (e.g., quinidine, verapamil, propafenone, amiodarone) (5,6), TDM is very important when these drugs are used. Similarly, because most of the body load of the glycoside is excreted by the kidneys, it is important to measure levels of the glycoside whenever there is evidence of impaired or changing renal function.

TDM of digoxin is an important tool to identify the noncompliant patient. Whenever trough concentrations are very low or undetectable, despite an adequate dose, such a possibility should be ruled out by pill count, or by the use of the novel MEMS device, where time of each opening of the pill container is recorded by a computerized system.

In summary, TDM of digoxin is far from ideal. Clinicians should be aware of its limitations when using cardiac glycosides in pediatric patients.

ACKNOWLEDGMENT

This work was supported by grant MA 8544 of MRC, Canada.

REFERENCES

1. E. Koren and P. Parker. Interpretation of excessive serum concentrations of digoxin in children. *Am. J. Cardiol.*, 55:1210–1214 (1985).
2. G. Koren and S. Soldin. Cardiac glycosides. *Lab. Med. N. Am.*, 7:587–606 (1987).
3. G. Koren, D. Farine, D. Maresky, J. Taylor, J. Heyes, S. Soldin, and S. M. MacLeod. Significance of the endogenous digoxin like substance in infants and mothers. *Clin. Pharmacol. Ther.*, 36:759–764 (1984).
4. J. Stone, Y. Bentur, E. Zalstein, S. Soldin, E. Geisbrecht, and G. Koren. Effects of endogenous digoxin like substances on the interpretation of high concentrations of digoxin in children. *J. Pediatr.*, 117:321–325 (1990).
5. G. Koren. Interaction between digoxin and commonly coadministered drugs in children. *Pediatrics*, 75:1032–1037 (1985).
6. E. Zalstein, S. M. Bryson, R. Freedom and G. Koren. Interaction between digoxin and propafenone in children. *J. Pediatr.*, 116:310–312 (1990).

12

Practical Aspects of Neonatal Therapeutic Drug Monitoring

Philip D. Walson *The Ohio State University Children's Hospital, Columbus, Ohio*

I. INTRODUCTION

The indications, acceptance, and increasing use of neonatal therapeutic drug monitoring (TDM) have been described (1–5). Certain trends, including medico-legal implications and quality assurance requirements, promise to continue or even accelerate this use (1). Neonatal TDM is useful but complicated. In order to maximize the effectiveness of neonatal TDM, a number of practical problems must be recognized and solved. These problems involve clinical realities as well as analytical and pharmacokinetic/dynamic considerations.

II. CLINICAL REALITIES

Neonatal therapy involves diagnostic uncertainty. Antibiotics, for example, are often started before a diagnosis is confirmed, because the risk of waiting for confirmation of sepsis far exceeds the risk of unnecessary treatment. It is difficult to select an appropriate "therapeutic range" for a drug which may not even be indicated.

In addition, both the developmental status and the clinical condition of the patients, which in large part determine the pharmacokinetic behavior of a drug, are constantly changing (6). Sick neonates may never be at "steady state." The variables which control drug absorption, distribution, and clearance are constantly changing during the course of treatment.

The signs, especially neurological or subjective signs, of efficacy or toxicity are also often difficult or impossible to detect or measure in the newborn. This is a special problem when the sign of toxicity are similar to the signs of inadequate therapy. Septic newborns receiving aminoglycosides, for example, may have renal dysfunction from either too little drug to control sepsis or from excessive doses.

Dosing uncertainty is also created by uncertain or changing weight, measurement errors, spillage, vomiting, lack of accurate dosing times, rapid or changing GI transit times, alterations in dosage formulations (e.g., grinding, cutting, etc.), as well as route, rate, and site-specific problems.

Proper TDM interpretation requires accurate dosing information, which is especially difficult to obtain in neonates. Even patient weight can be either very inaccurate or change significantly from dose to dose. Timing of sample collection with respect to dosing is especially critical in TDM, yet the actual time of drug administration can be very difficult to ascertain in neonates (6). There can be hours delay between the time of addition of a drug to an i.v. line and the actual time that the drug begins to be administered because of problems with low, variable i.v. flow rates and variable drug density and flow characteristics (6).

Drug administration problems have, in the experience of our service, been responsible for a number of otherwise difficult-to-understand assay results, such as "peaks" that were lower than "troughs." As much as 25% of ordered chloramphenicol doses are discarded when i.v. lines are changed daily (7), and many transported neonates arrive with undetectable aminoglycoside concentrations because their first aminoglycoside dose is still in their i.v. line (8). This is especially dangerous for septic neonates and is made worse if i.v. lines containing potentially life-saving drugs are discarded and replaced with fresh lines, which is standard nursing practice in some institutions. Problems with drug administration exist for other routes as well.

Commercially available formulations are often inadequate for newborns, either because of problems with the size of the dose used, problems with "inert" ingredients, difficulty in accurate measurement of dosages, or in use of preparations which are not designed for use or proven to work in neonates (e.g., prolonged-release preparations).

Lack of appropriate formulations are a cause of some of the potentially life-threatening 10-fold dosing errors which are not uncommon (9). It is simply impossible to administer accurately volumes less than 0.1 mL (5 mg/kg of commercially available phenobarbital is 0.077 mL!), and the "dead space" in syringes (up to 0.2 to 0.3 mL), if "flushed" with i.v. fluids, may result in lethal two- to threefold dosing errors for some dangerous drugs (e.g., morphine or aminophylline).

Rectal suppositories can produce erratic dosing due to stools. Prolonged-release preparations are seldom if ever indicated in neonates, since clearance rather than absorption is rate-limiting. Even if clearance were rapid enough to justify a prolonged-release preparation, a preparation designed to release drug over 12 to 24 h will not produce reliable or acceptable absorption in neonates, who have rapid GI transit. These will also simply not work if ground up or cut up.

Unfortunately, the margin for error for either underdosing or overdosing is much smaller in critically ill neonates, yet both the ability to calculate a safe and effective dosage regimen and the ability to administer the selected dosage accurately are difficult in neonates.

All of these practical clinical problems make it imperative that dosage regimens be designed by people who are knowledgeable in the practical and pharmaceutical realities of a neonatal unit, and that the TDM results used to calculate these regimens may be interpreted by individuals trained and experienced in the clinical, analytical, and pharmacokinetic realities of neonatal TDM.

III. ANALYTICAL CONSIDERATIONS

Analytical problems involve sample collection, timing, handling, dosage information, and assay limitations.

There are real limitations on the ability to obtain adequate sample volumes or gain access to skin, veins, or arteries without risk of excessive blood loss, seriously interrupting important therapies, or increasing the risk of infection. The method, skill, and site (10,11) used to collect, transport, separate, or store samples can all influence results.

Many collection methods are used, but venipuncture and skin puncture are the most common. Modern analytical techniques and blood-loss considerations make skin puncture the preferred method of collection. However, this must be done correctly. Knowledge and skill are required to avoid excessive blood loss, skin trauma leading to infection (including osteomyelitis), artifactual interference by hemolyzed blood or dilution by tissue fluids, as well as contamination by the monitored drug or other interfering substances, including maternal drugs.

Skin contamination with the administered drug is a special problem with drugs administered by aerosol (e.g., aminoglycosides) (12), but must also be considered in patients where gastric or oral fluids come in contact with the skin surface used to collect samples. Neonates suck their fingers (or even heels at times) and often vomit, which can have major implications for the accuracy of results.

Contamination is also a major problem if venous samples are obtained from parenteral infusion sites. Both residual administered drug as well as parenteral components (e.g., heparin) can interfere with assay results. A site used to administer digoxin cannot ever be used to draw a digoxin sample, since nanogram quantities are measured while microgram amounts are administered. Collection of a blood sample proximal to drug administration can also be important (e.g., from the antecubital space of a baby receiving the drug in a hand vein, or from the jugular vein of an infant receiving the drug in a scalp vein).

Clinical and dosing information is critical, yet clinical realities (see above) often make data collection difficult. A complete drug history and time of sampling with respect to dosing are especially critical. For example, digoxin samples drawn within 6–8 h of either an oral or even i.v. dose should not be assayed, since they reflect drug distribution and not therapeutic effects.

Other important dosing information, including how long the drug has been given, whether loading doses were used, time of dosing, route, rate, site, and formulation, must all be accurately collected by knowledgeable staff. Accurate recording of sample collection time, site, and method are critical. Unfortunately, some of the staff who collect samples (including medical students, house officers, physicians, or nurses) have not received adequate training in sample collection or documentation. Unfortunately, all too often the time a drug is scheduled to be given is accepted as the time of drug administration.

Sample handling is also important. The transport, separation, and storage of samples can have major effects on assay accuracy for both stable and unstable drugs. Aminoglycoside samples must be placed on ice, then spun down and frozen or analyzed immediately in order to obtain accurate aminoglycoside concentrations in patients receiving penicillins or cephalosporins (2). Sample handling is also a problem for drugs that distribute unevenly between RBCs and plasma (e.g., theophylline or amiodarone) (13), and for measurements of unbound concentrations which can be altered by temperature (14). Plasticizers in collection containers can cause altered binding or assay interference (15). Finally, unstable drugs can degrade, and metabolites or stereoisomers may significantly alter assay results.

Assay technology has improved greatly, but good results require more than "easy" assays. Good equipment and reagents must be obtained and maintained.

Finally, the staff performing the assays must have skill, training, and dedication to produce consistently reliable results and deal with the analytical problems which alter interpretation.

One analytical consideration is volume limitations. Small neonatal samples limit both the choice of assay methodology and the possibility of confirmation testing. Other considerations include endogenous substances (e.g., digoxinlike immunoreactive substance, bilirubin, or lipids) (16), as well as exogenous chemicals (e.g., maternal medications, skin contamination, co-administered interfering drugs), which can also cause significant analytical problems.

Age-specific differences in metabolic rates or pathways must also be considered. For example, both the production of caffeine from theophylline as well as the clearance of both maternal caffeine and theophylline metabolized to caffeine can alter the interpretation of both caffeine and theophylline results (17).

Age-dependent clearance of interfering metabolites can also be a problem. Delayed renal clearance of a phenobarbital metabolite resulted in significant inaccuracy of commercial immunoassays in neonates and other patients with decreased renal function. This inaccuracy was not evident in "normal" adult populations but became obvious as soon as the assay was used in neonates. This example, as well as the well-known problems with digoxin assays in neonates, argue for the routine premarket testing of assays in neonatal populations prior to routine use of assays. Testing in nonneonatal populations will not always predict reliability in neonates.

IV. PHARMACOKINETIC/DYNAMICS

TDM requires accurate results plus interpretation. The interpretation of results is altered by pharmacokinetic and pharmacodynamic factors. A "therapeutic" level in one patient can in fact be a "panic value" in another merely because of the time of collection in relation to dosing.

A phenobarbital level of 10 mg/L, for example, can be "therapeutic" but can also be useless to predict future concentrations if it is drawn immediately after a loading dose. It can be mean that a patient has been noncompliant or can indicate that future concentrations will be toxic, depending on when it was drawn in relation to dosing.

In addition, in neonates the pharmacokinetic factors that affect interpretation are both different from those in older patients and constantly changing during therapy.

Both clinical conditions and developmental status can change drastically during the course of a single hospitalization. Gastric emptying, GI transit, and drug absorption all change rapidly postnatally, as do tissue perfusion, body composition, and protein binding. These changes, as well as changes in metabolic pathways or rates, renal function, and therapeutic ranges must all be considered.

Pharmacodynamic differences are especially important in neonates. Very little data exist on neonatal "therapeutic ranges," and the conditions treated can be different from those seen in adults. For example, antibiotic ranges probably differ due to differences in immune response as well as renal and metabolic clearance. Neonatal seizures almost certainly require different anticonvulsant concentrations even if differences in protein binding are taken into account. The concentrations of phenobarbital required to treat hyperbilirubinemia or prevent intraventricular hemorrhage are not the same as those required to control adult seizures. Broncodilators have different therapeutic and toxic potential in neonates with apnea than in adults or children with asthma. Finally, cardiovascular drugs have different effects in neonates due to differences in baseline diseases, developmental changes in sympathetic and parasympathetic innervation, or even the lack of upright posture.

V. SUMMARY

Neonatal TDM requires the ability to recognize and deal with the problems created by the clinical, analytical, and pharmacological realities of a modern neonatal ICU. The pharmaceutics of the drugs used, drug administration techniques, sample collection, storage and analysis, and the pharmacokinetics, pharmacodynamics, and clinical pharmacology of neonatal drugs must all be considered to provide maximally effective neonatal TDM. In the author's opinion, a multidisciplinary team approach, involving medical technologists, clinical chemists, pharmacists, nurses, and physicians is required in order to do this in a cost-effective way.

REFERENCES

1. G. P. Giacoia. The future of neonatal therapeutic drug monitoring. *Ther. Drug Monitor., 12*:311–315 (1990).
2. P. D. Walson, R. Edwards, and S. Cox. Neonatal therapeutic drug monitoring—Its clinical relevance in neonatal therapy: An update (F. F. Rubatelli and B. Granati, eds.), Elsevier Biomedical Division, pp. 289–296 (1986).
3. P. Gal. Therapeutic drug monitoring in neonates: Problems and issues. *Drug Intell. Clin. Pharmacol., 22*:317 (1988).
4. J. Shipe and D. A. Herold. Drug monitoring in the neonate. *Ann. Clin. Lab. Sci., 12*:296 (1982).
5. J. V. Aranda, T. Turmen, and C. Cote-Boileau. Drug monitoring in the perinatal patient: Uses and abuses. *Ther. Drug Monitor.,* 2:39 (1980).
6. R. J. Roberts. *Drug Therapy in Infants: Pharmacologic Principles and Clinic Experience.* Saunders, Philadelphia (1984).
7. M. C. Nahata. Delayed delivery of antibiotics by retrograde intravenous infusion. *Am. J. Hosp. Pharm., 43*:2237–2239 (1986).
8. R. C. Edwards. Techniques used for the administration of IV antibiotics to newborns in community hospitals. (Poster presentation) American Society of Hospital Pharmacists, Mid-Year Clinical Meeting, New Orleans (1985).
9. M. J. Rieder, D. Goldstein, H. Zinman, and G. Koren. Tenfold errors in drug dosage (letter). *Can. Med. Assoc. J., 139*:12–13 (1988).
10. W. L. Chiou. The phenomenon and rationale of marked dependence of drug concentration on blood sampling site. Implications in pharmacokinetics, pharmacodynamics, toxicology and therapeutics (Part I). *Clin. Pharmacokinet., 17*:175–199 (1989).
11. W. L. Chiou. The phenomenon and rationale of marked dependence of drug concentration on blood sampling site. Implication in pharmacokinetics, pharmacodynamics, toxicology and therapeutics (Part II). *Clin. Pharmacokinet., 17*:275–290 (1989).
12. Y. Bentur, D. Hummel, C. M. Roifman, E. Giesbrecht, and G. Koren. Interpretation of excessive levels of inhaled tobramycin. *Ther. Drug. Monitor., 11*:109–110 (1989).
13. T. J. B. Maling, R. W. L. Siebers, C. D. Burgess, C. Taylor, and G. Purdie. Individual variability of amiodarone distribution in plasma and erythrocytes: Implications for therapeutic monitoring. *Ther. Drug. Monitor., 11*:121–126 (1989).
14. N. Ratnaraj, V. D. Goldberg, and M. Hjelm. Temperature effects in the estimation of free levels of phenytoin, carbamazepine, and phenobarbitone. *Ther. Drug Monitor., 12*:465–472 (1990).

15. J. Leslie, M. Busby, and E. Khazan. Interference from a red top venomjet tube in high-performance liquid chromatographic analysis. *Ther. Drug Monitor., 11:724–725* (1989).
16. D. M. Baer and R. A. Paulson. The effect of hyperlipidemia on therapeutic drug assays. *Ther. Drug Monitor., 9:72–77* (1986).
17. J. V. Aranda, K. Beharry, J. Rex, R. J. Hohannes, and L. Charest-Boule. Caffeine enzyme immunoassay in neonatal and pediatric drug monitoring. *Ther. Drug Monitor., 9:97–103* (1987).

13

Evaluation of Erythropoietin Determination as a Method of Diagnosis and Therapeutic Monitoring

J. Manuel Pena, Carlos Aulesa, J. Luis Alonso, and J. José Ortega *Hospital Infantil Vall d'Hebrón, Barcelona, Spain*

I. INTRODUCTION

Erythropoietin is a glycoprotein which acts essentially as a specific growth factor by stimulating proliferation and final maturation of the red blood cells in bone marrow. It is secreted mainly in the kidneys, and its production is regulated by the relative amount of oxygen available to organs involved in its synthesis; its secretion is increased by hypoxia and decreased by hyperoxia (1–3).

The production of large quantities of pure human recombinant erythropoietin by genetic cloning has permitted its generalized therapeutic use, with encouraging results in the treatment of anemia resulting from chronic kidney failure (4–6). At the same time, analytical methods of erythropoietin determination in serum have greatly improved and can serve as diagnostic and therapeutic aids (7–9).

The methodological evaluation and correlation of two radioisotopic techniques for erythropoietin determination in pediatric samples, together with their application in the diagnosis or indications for biosynthetic hormone therapy, are presented.

II. MATERIAL AND METHODS

Determinations were made from EDTA-plasma which was obtained by cold centrifugation and conserved in aliquots at –30°C for each of the projected assays.

Erythropoietin determination was performed by two methods:

Method A: competitive RIA with recombinant human erythropoietin as tracer and standard (INCSTAR EPO-TRAC 125-I RIA)

Method B: IRMA following a sandwich technique in a step with recombinant erythropoietin and two monoclonal antibodies (125-I EPO COATRIA)

Table 1 Performance Characteristics of EPO Determination

	Method A (INCSTAR)	Method B (COATRIA)
Practicability		
Calibration range (U/L)	10 to 280	5 to 800
Time [min (h)]	180 + (20) + 90	300
Sample volume (μL)	200	200
Material	Counter, centrif.	Counter, shaker
Antigen	Recombinant EPO	Recombinant EPO
Antibodies	Rabbit anti-EPO	Two monoclonals
	Goat anti-rabbit	Anti-EPO
Imprecision and inaccuracy		
Within run (CV%)	2.1	1.9
Between runs (CV%)	9.6	6.3
Inaccuracy controls %	+4.9	+7
	+0.5	+9

RIA parameters		B/T	R/T	ED_{20}	ED_{50}	ED_{80}
Method A	Spline linear	3	31	140	47	17
	Spline log	3	31	160	38	15
	4-place linear	4	30	79	28	10
	4-place log	4	32	80	27	8
Method B	Spline linear	—	—	152	371	597
	Spline log	—	—	153	359	569
	4-place linear	—	—	152	364	589

ED_{20}, ED_{50}, ED_{80}: EPO values (U/L) interpolated from the displacement curve for 20, 50, and 80% of maximum binding.

Basic methodological features are compared and their reliability characteristics are shown in Table 1.

Erythropoietin used was from Boehringer-Mannheim, 5000 IU (Barcelona, Spain) and from Eprex-Cilag AG Inter., 4000 IU (Zug, Switzerland).

III. RESULTS

The correlation between the two methods evaluated is shown in Figure 1.

The validity of the kits (methods A and B) for quantifying the two drugs available on the national pharmaceutical market (Boehringer-Mannheim and Eprex-Cilag) was verified (Figures 2 and 3).

Reference values for the two methods were found in a pediatric population of premature babies with no other pathology, infants, and school children (Table 2). In addition, values in a variety of hematological disorders are listed in Table 3A.

We found that not all patients present with low EPO values in chronic renal insufficiency (CRI) anemia (Table 3B).

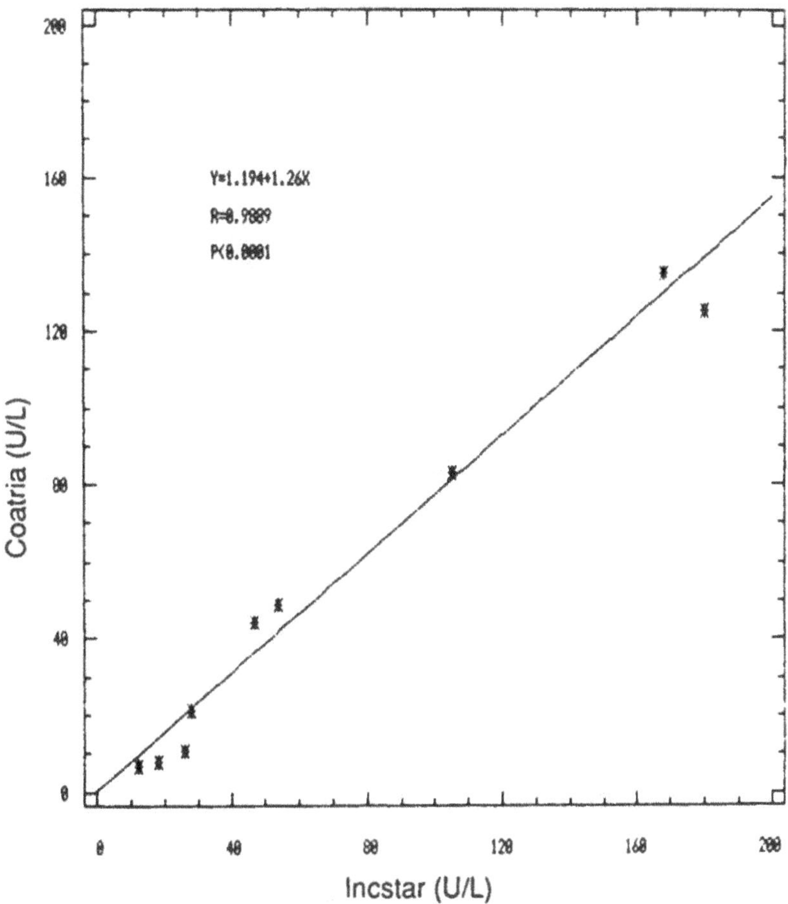

Figure 1 Regression of COATRIA (Method B) on INCSTAR (Method A). EPO (RIA).

IV. DISCUSSION

The comparative study of practicability and reliability of methods A and B shown in Table 1 indicates a wider calibration range and shorter performance time for method B. A correct correlation between both methods is observed.

The reference values found in the pediatric population and also in adults require that methods for EPO determination have a low detection limit if one is to obtain well-defined differentiation between low and normal EPO values (10,11).

Erythropoietin determination is a valuable factor in hematological patients, not only for establishing the physiopathology of erythrocytic disorders, but also for orienting therapeutic indications and verifying the evolutive status of the anemic processes (12).

The finding of high EPO levels in CRI anemia patients suggests that hormone-level determination might help to establish the therapeutic approach that should be followed (13).

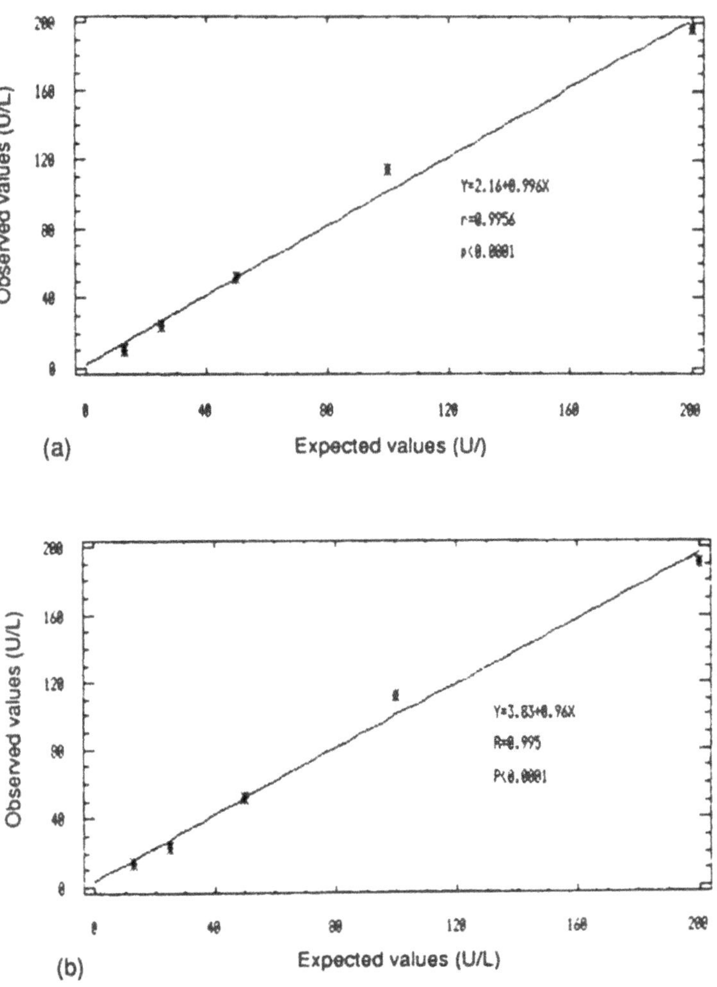

Figure 2 EPO immunoreactivity by INCSTAR (RIA): (a) Eprex-Cilag; (b) Boehringer-Mannheim.

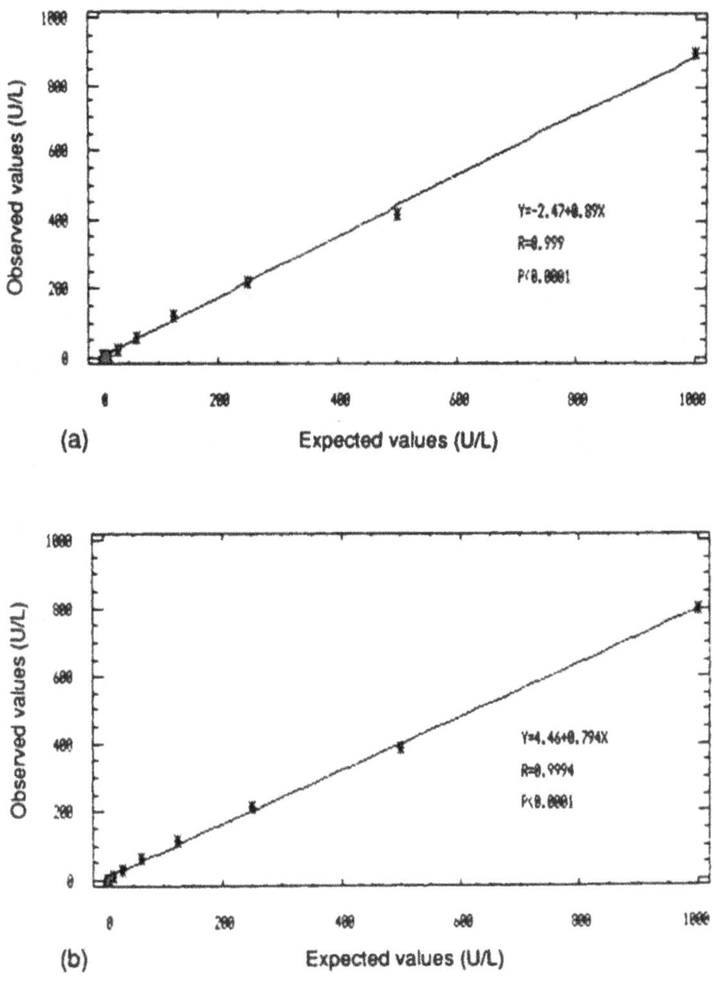

Figure 3 EPO immunoreactivity by COATRIA (RIA): (a) Eprex-Cilag; (b) Boehringer-Mannheim.

Table 2 Reference (Normal) Values

		Premature	Infant	School Child	Adult
Method A:	N	14	16	12	10
	X (U/L)	4.7	11.74	12.36	20.7
	SD	1.1	4.21	4.17	6.41
	Lim. 95%	2.5–6.9	3.3–20.2	4–20.7	7.9–33.5
Method B:	N	6	6	7	6
	X (U/L)	5.8	6.5	7.3	8.5
	SD	0.3	1.6	3.1	1.46
	Lim. 95%	5.2–6.4	3.3–9.7	1.1–13.5	5.6–11.4

Table 3 EPO Levels in Erythrocytic Disorders and CRI Anemia

A: EPO Levels in Erythrocytic Disorders

	N	X (U/L)	SD
Thalassemia major	24	53.43	33.01
Thalassemia major (splenectomized)	31	27.53	27.33
Spherocythemia	28	84.32	20.01
Polycythemia (cardiopathies)	17	14.23	9.24
Blackfan-Diamond active	2	+200	—
Blackfan-Diamond remission	2	15	—

B: EPO Levels in Chronic Renal Insufficiency

	Group I	Group II
N	31	5
X (U/L)	11.03	50.25
SD	5.38	10
Lim. 95%	0.26–21.8	30–70

Thus, EPO determination, besides being helpful for hematologic diagnosis, may establish in which patients biosynthetic erythropoietin therapy would be indicated.

REFERENCES

1. R. D. Christensen. Admistración de fármacos recombinantes de la estimulación de la eritropoyesis como altenativa a la transfusión de hematíes con anemia de la prematuridad. *Pediatrics*, 5:243 (1989).
2. S. C. Cark and R. Kamen. The human hematopoietic colony stimulation factors. *Science*, 236:1229 (1987).

3. K. M. Shanon, G. S. Naylor, and J. C. Torkildson. Circulating erythroid progenitors in the anemia of prematurity. *N. Engl. J. Med., 317*:728 (1987).
4. T. Miyake, C. K. Kung, and E. Goldwasser, Purification of erythropoietin. *J. Biol. Chem.,* 252:5558 (1977).
5. P. Lai, R. Everett, and F. Wnsg. Structural characterization of human erythropoietin. *J. Biol. Chem., 261*:3116 (1986).
6. H. Sasaki, B. Bothner, and A. Dell. Carbohydrate structure of erythropoietin expressed in Chinese hanster of erythropoietin ovary cells by human erythropoietin cDNA. *J. Biol. Chem.,* 262:12059–12076 (1987).
7. J. F. García, J. Sherwood, and E. Goldwasser. Radioimmunoassay of erythropoietin. *Blood Cells, 5*:405 (1979).
8. A. B. Rege, J. Brookins, and J. W. Fisher. A radioimmunoassay for erythropoietin. Serum levels in normal human subjects and patients with hemopoietic disorders. *J. Lab. Clin. Med., 100*:829 (1982).
9. J. B. Sherwood and E. Goldwasser. A radioimmunoassay for erythropoietin. *Blood, 86*:180 (1986).
10. P. R. Dallman. Erythropoietin and the anemia of prematurity. *J. Pediatr., 105*:756 (1986).
11. K. Stoutenborough, J. M. Sutherland, and H. Meineke. Erythropoietin levels in cord blood of control infants and infants with respiratory distress syndrome. *Acta Pediatr. Scand., 58*:121 (1969).
12. J. García, N. Shirley, and L. Hollander. Radioimmunoassay of erythropoietin circulating levels in normal and polycythemic human beings. *J. Lab. Clin. Med., 90*:624 (1982).
13. C. Winearls and O. Desmond. Effect of human erythropoietin derived from recombinant DNA on the anemia of patients maintained by chronic hemodialysis. *Lancet,* 22:1175 (1986).

14

Clinical Validation of a Computerized Method for Therapeutic Monitoring of Aminoglycoside Therapy

Stein Bergan and Oddbjørn Brubakk *Sarpsborg Hospital, Sarpsborg, Norway*

I. INTRODUCTION

During aminoglycoside antibiotic treatment, therapeutic outcome is related to peak serum levels (1,2). Although correlation between toxic effects and trough serum levels is less convincing (3,4), therapeutic monitoring of aminoglycosides usually includes trough as well as peak serum levels. Indications of the potential benefits of clinical pharmacokinetic services versus traditional dosing of aminoglycosides have recently been presented (5).

Wide variability of pharmacokinetic parameters has been demonstrated among patients, even from day to day within the individual patient. Several methods have been suggested to provide adequate interpretation of measured serum concentrations and to ensure proper adjustment of dosage. Such procedures include traditional nomograms, routines for hand-held calculators, and applications on mainframe systems as well as personal computer software. Documentation of the performance of these applications in clinical practice, however, is sparse. We therefore chose to develop our own software, a program called Aminodos. The present study was performed in order to undertake a prospective validation of this program in a relevant patient population.

II. PATIENTS AND METHODS

All hospitalized patients older than 6 months receiving netilmicin were included in this prospective study. Patients were included even in the presence of factors predisposing to unstable pharmacokinetic parameters. It was decided to proceed until 50 patients with more than one pair of evaluable serum netilmicin values were included. Diagnosis, age, sex, body weight, height, serum creatinine, dosing history, and measured serum levels—including timing of serum samples—were recorded. Trough and peak serum levels were measured two or three times per week. The serum samples were analyzed by a fluorescence polarization immunoassay method (TDx,

Abbott Laboratories). No additional blood samples beyond standard practice was drawn. As experience with the computer program was gained, the computed predictions of netilmicin serum levels were used increasingly as support for netilmicin monitoring. It should be noted, however, that in this period any pair of serum levels would be included in the study, regardless of whether or not the computer program had been consulted for dosage recommendations.

Measured serum levels were considered not evaluable and were excluded from the study if any of the following parameters were missing or obviously not properly recorded: (a) timing of any of the two doses immediately preceding a measured peak serum level, or timing of serum sampling for peak measurement; (b) measured peak or trough serum level within a pair; (c) serum creatinine (only when evaluating the first measurement).

When predicting the first pair of serum levels, calculations were based on population-derived kinetic parameters (6). Hence the elimination constant is calculated from creatinine clearance, which in turn is calculated from age, body weight, and serum creatinine (7). The individual volume of distribution is assumed to be proportional to actual body weight, except in obese patients (more than 30% above ideal body weight) in whom "dosing weight" was calculated by reducing excess weight by 60%. Predictions of subsequent serum levels are based exclusively on the most recently reported pair of serum levels and the corresponding dosing history.

The algorithm in the computer program is based on a one-compartment model, and the calculations are performed as described in handbooks (8). However, the individual pharmacokinetic parameters are calculated by iteration rather than by the traditional approximations to steady-state conditions.

The program Aminodos, ver. 3.0, was developed by one of us (S.B.) in Turbo-Pascal, ver. 3.0 (Borland International), and was run on a Zenith 181 portable computer under MS-DOS 3.2. Statistical analysis was performed on the same computer, utilizing Statgraphics ver. 2.6 (STSC, Inc.).

Over 22 months, 79 patients were considered candidates for the study. Among these, seven were excluded: In three cases no serum concentration was obtained, one had only a single sample drawn, and three had from one to three pairs sampled without properly recorded timing. Four patients were admitted and included twice, and one patient was included in the study on four occasions; thus the 72 admissions were represented by 65 patients—41 males and 24 females. From these patients 207 pairs of serum samples were obtained, of which 32 (in 24 patients) were excluded as incomplete according to the protocol. The reasons for these exclusions were timing of sample not adequately recorded (20 pairs), trough or peak level inadequate or missing (3 pairs), dosage not properly recorded (3 pairs), and other reasons or combinations (6 pairs). Thus 175 pairs of serum levels were adequate for analysis in this study. Demographic data for the 72 admissions are summarized in Table 1.

For each evaluable pair of serum levels, the trough and peak levels were predicted by the computer program. The performance of the program was quantified in terms of correlation between predicted trough and peak levels with the corresponding measured levels.

III. RESULTS

In Figures 1 through 4 the measured serum netilmicin concentrations are plotted versus the corresponding concentrations predicted by the computer program. When

Table 1 Demographic Characteristics at Start of Netilmicin Therapy (n = 72 admissions)

	Average (± SD)	Range
Age	59.5 (± 18.5)	14–83 years
Body weight	69.2 (± 16.2)	38–112 kg
Dosing weight[a]	67.0 (± 14.1)	35–104 kg
Serum creatinine	104.4 (± 33.2)	58–247 μmol/L

[a]"Dosing weight" is body weight corrected for obesity as described in the text.

prediction was based on demographic data (Figures 1 and 2), the prediction error (predicted versus measured level) was 1.5 ± 2.2 mg/L and 2.8 ± 3.4 mg/L (mean ± standard deviation) for trough and peak values, respectively. With predictions based on a previously measured pairs of serum levels (Figures 3 and 4), prediction errors were reduced to 0.09 ± 1.1 mg/L and −0.5 ± 2.3 mg/L, respectively. From a practical point of view the results may be summarized as follows: Among the 103 pairs predicted from the immediately preceding measurements, 11 predicted trough levels were more than 1 mg/L too high and 11 predicted peak levels were more than 2 mg/L too low.

Pharmacokinetic parameters are presented in Table 2. Average values at baseline are shown along with the relative change from baseline to the last value during netilmicin therapy. The average minimum and average maximum values of the

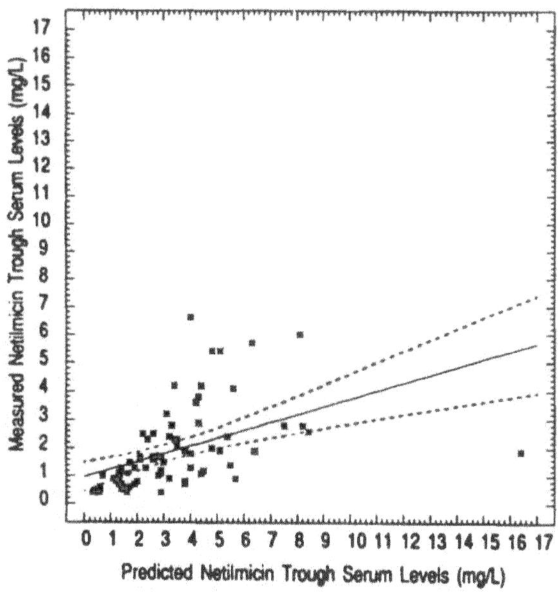

Figure 1 Netilmicin trough serum levels. Predictions based on S-creatinine, etc. (n = 72). Broken lines represent 95% confidence limits. Correlation coefficient = 0.47.

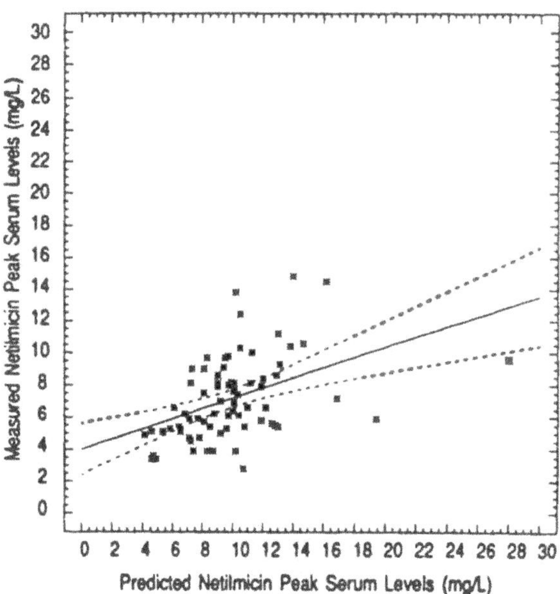

Figure 2 Netilmicin peak serum levels. Predictions based on S-creatinine, etc. (n = 72). Broken lines represent 95% confidence limits. Correlation coefficient = 0.45.

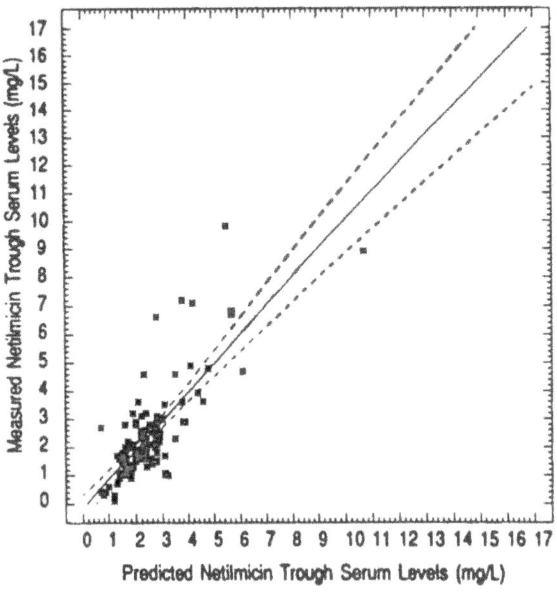

Figure 3 Netilmicin trough serum levels. Predictions based on previous measurement (n = 103). Broken lines represent 95% confidence limits. Correlation coefficient = 0.78.

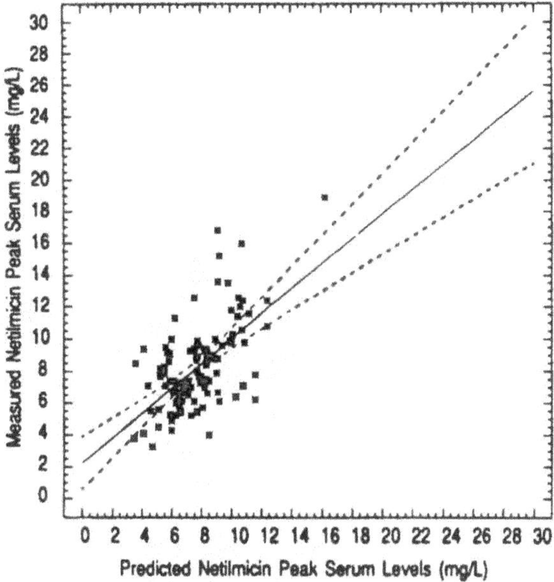

Figure 4 Netilmicin peak serum levels. Predictions based on previous measurement (n = 103). Broken lines represent 95% confidence limits. Correlation coefficient = 0.60.

Table 2 Creatinine Clearance and Pharmacokinetic Parameters in Patients with More than One Pair of Measured Serum Levels (n = 50 admissions)

	Baseline[a]	Relative Increase[b]
Creatinine clearance (mL/min)	65 (± 28)	1.2% (± 21%)
Elimination constant (L/h)	0.14 (± 0.048)	2.3% (± 27%)
Distribution volume (L/kg)	0.38 (± 0.17)	3.9% (± 44%)

[a]Mean values (± SD).
[b]Average relative increase (± SD) from baseline to last value calculated during therapy.
Elimination constant and volume of distribution calculated on the basis of a measured pair of serum levels and the corresponding dosing history.

studied population are also presented. The differences between the averages of baseline and later values are not statistically significant for any of these parameters.

IV. DISCUSSION

Traditional interpretation of observed serum levels based on clinical judgment will often prove unsatisfactory. While adjustment should frequently affect both dosing interval *and* dose, such adjustments of dosage regimen are not easily performed on an empirical basis. Consequently, *dose* is often corrected while dosing interval is left unaltered. In contrast, proper handling of both variables is facilitated by the tested software. The program Aminodos is flexible regarding input of variable dosing intervals, time of sampling, and infusion time. When testing the software, an important aspect would be operation outside steady-state conditions. In the present study, non-steady-state conditions were in part provided by serum sampling soon after adjusting dosage. Even in cases with unadjusted dosage, however, non-steady-state conditions were in fact obtained by quite irregular dosing intervals—e.g., an 11 h interval between evening and morning dose in "8-hourly" regimens. Accordingly, it is not relevant to classify these observations into subgroups of steady-state versus non-steady-state conditions.

Wide variations in pharmacokinetic parameters were experienced among the patients in our study. The range of distribution volume for aminoglycosides is commonly referred to as 0.15 to 0.25 L/kg (9). In the present population the average was 0.38 L/kg. Since "dosing weight" and ordinary weight were quite similar, the calculation of distribution volume per kilogram of "dosing weight" rather than actual body weight did not contribute significantly to this high value. Indeed, distribution volumes higher than normal have even been reported by others (10–12). The mean elimination constant was close to values reported by others (12), but with a rather wide range. As a consequence of such variations, the performance of new dosing methods should preferably be validated in patients included prospectively and on an "intention-to-treat" basis. In our study, patients were included without regard to conditions such as unstable renal function or fluctuations in body fluids.

The rather imprecise prediction of the first pair of serum levels, when based solely on patient characteristics and reported population averages, is not surprising. This is due to the large variability in population-derived parameters. In the population studied, the precision might be improved by some adjustment of the proportional factors relating netilmicin clearance to creatinine clearance and volume of distribution to "dosing weight." The huge variations in dose requirements among individuals would remain, however, merely reflecting the rationale for individual monitoring of aminoglycoside therapy.

With predictions based on immediately preceding serum levels, however, precision was greatly improved. Still, considerable fluctuations in pharmacokinetic parameters were observed during therapy in a few patients. This is reflected by the standard deviations reported in Table 2.

A major point with this kind of computer-assisted therapeutic monitoring of aminoglycosides is to avoid the uncertainty related to steady-state approximations. Minimizing toxic effects is important, but in clinical practice subtherapeutic peak levels may be a more frequent problem. When utilizing this software, our recommendation is to start therapy with relatively aggressive dosage, measure serum levels by the second dose, and adjust dosage accordingly as assisted by the computer program.

This may prevent treatment failures resulting from subtherapeutic peak levels as well as adverse effects related to persistence of overly high trough levels. Subsequent monitoring will of course still be necessary.

It is concluded that in the present patient population the tested software has proved sufficiently precise for practical use in individual monitoring of netilmicin during clinical routines.

ACKNOWLEDGMENTS

We thank the staff at the involved departments of Sarpsborg Hospital for their cooperation during this study. The Departments of Microbiology and Clinical Chemistry at the Fredrikstad Hospital are acknowledged for performing the serum-level assays.

Financial support for the computer equipment has been provided by Schering-Plough, Eli Lilly, Roussel, and the Rôche grant for hospital pharmacy in Norway (in cooperation with the Association of Norwegian Hospital Pharmacists).

REFERENCES

1. R. D. Moore, C. R. Smith, and P. S. Lietman. Association of aminoglycoside plasma levels with therapeutic outcome in Gram-negative pneumonia. *Am. J. Med., 77*:657–662 (1984).
2. R. D. Moore, C. R. Smith, and P. S. Lietman. The association of aminoglycoside plasma levels with mortality in Gram-negative bacteremia. *J. Infect. Dis., 149*:443–448 (1984).
3. R. D. Moore, C. R. Smith, and P. S. Lietman. Risk factors for the development of auditory toxicity in patients receiving aminoglycosides. *J. Infect. Dis., 153*:23–30 (1984).
4. R. D. Moore, C. R. Smith, J. J. Lipsky, E. D. Mellits, and P. S. Lietman. Risk factors for nephrotoxicity in patients treated with aminoglycosides. *Ann. Intern. Med., 100*:352–357 (1984).
5. C. J. Destache, S. K. Meyer, M. T. Padomek, and B. G. Ortmeier. Impact of a clinical pharmacokinetic service on patients treated with aminoglycosides for Gram-negative infections. *Drug Intell. Clin. Pharm., 23*:33–38 (1989).
6. J. S. Kaka and E. C. Buchanan. Aminoglycoside pharmacokinetics on a microcomputer. *Drug Intell. Clin. Pharm., 17*:33–38 (1983).
7. D. W. Cockroft and M. H. Gault. Prediction of creatinine clearance from serum creatinine. *Nephron., 16*:31–41 (1976).
8. W. A. Ritschel (ed.). *Handbook of Basic Pharmacokinetics . . . Including Clinical Applications.* Hamilton, IL: Drug Intelligence Publications (1986).
9. D. Zaske. Aminoglycosides. In *Applied Pharmacokinetics: Principles of Therapeutic Drug Monitoring,* (W. E. Evans, J. Schentag, and W. J. Jusko, eds.), Applied Therapeutics, Spokane, WA, pp. 331–381 (1986).
10. M. M. Tointon, M. L. Job, T. T. Peltier, J. E. Murphy, and E. S. Ward. Alterations in aminoglycoside volume of distribution in patients below ideal body weight. *Clin. Pharm., 6*:160–162 (1987).
11. D. E. Zaske, P. Irvine, L. M. Strand, R. J. Strate, R. J. Cipolle, and J. Rotschafer. Wide interpatient variations in gentamicin dose requirements for geriatric patients. *J. Am. Med. Assoc., 248*:3122–3126 (1982).
12. D. W. Fuhs, H. J. Mann, C. A. M. Kubajak, and F. B. Cerra. Intrapatient variation of aminoglycoside pharmacokinetics in critically ill surgery patients. *Clin. Pharm., 7*:207–213 (1988).

15

Evaluation of a New Software Package for the Individualization of Amikacin Dosing in Intensive-Care-Unit Patients

B. Lacarelle *Centre de Pharmacologie Clinique et d'Evaluation Therapeutique, Marseille, France* **D. Guelon** *Hôpital St Jacques, Clermont-Ferrand, France* **G. Voloury** *Hôpital Dupuytren, Limoges, France* **O. Pourrat** *Reanimation Polyvalente, Niort, France* **M. Pinsard** *Hôpital J. Bernard, Poitiers, France* **M. C. Andro** *Laboratoires Bristol, Paris, France* **A. Durand** *Centre de Pharmacologie Clinique et d'Evaluation Thérapeutique, Marseille, France*

I. INTRODUCTION

Amikacin is an effective antibiotic in the treatment of severe infections, particularly those due to Gram-negative bacilli. Unfortunately, this aminoglycoside has a low therapeutic index and exhibits a wide inter- and intraindividual variability in its pharmacokinetic parameters (1). On the other hand, it is well known that amikacin serum levels are in good correlation with both efficacy and toxicity. Therapeutic drug monitoring improves the outcome and reduces the adverse effects of amikacin, particularly in intensive-care-unit (ICU) patients. Many aminoglycoside dosing methods described in the literature (2–4) can be applied to amikacin. Among these, Bayesian methods are considered the most attractive (5,6). However, there is a need for efficient and user-friendly software to allow a more general use of these methods.

In that context, Abbott Laboratories has developed a new computer software program (Abbott Pharmacokinetic System) which allows individualization of dosage for various drugs including amikacin. The purpose of this study was to determine the performance of this new software in the individualization of amikacin dosage in ICU patients.

II. PATIENTS AND METHODS

A. Patients

The 23 patients (4 women, 19 men) involved in this multicenter study were admitted to four different ICUs. Mean (± SD) age was 55 (± 10) years. Mean weight (± SD) and serum creatinine level were 72 ± 10 kg and 97 ± 63 ± μmol/L, respectively.

B. General Methodology

After a standard infusion dose (7.5 mg/kg), amikacin serum levels were measured, using a TDX, at the following times after the end of infusion: 0, 0.5, 1, 2, 4, 6, 8, and 11.5 h. The "true" pharmacokinetic parameters of each patient were then estimated by using a classical weighted least-square estimation method and the APIS software (7). These parameters were compared with Bayesian estimates provided by the Abbott Pharmacokinetic System using only two drug serum concentrations (t = 0.5 and 11.5 h). In addition, during this study, dosing recommendations were obtained using the Abbott Pharmacokinetic System. These recommendations were checked by comparison of predicted and measured amikacin levels on day 3.

When necessary, a new recommendation was obtained on day 3 and a new comparison (predicted versus measured) was performed on day 5.

C. Software Package

1. *APIS Program*

The APIS program has been described previously (7). It runs on an IBM PS/2 and allows classical weighted least-square estimation of pharmacokinetic parameters.

2. *Abbott Pharmacokinetic System*

The Abbott Pharmacokinetic System is based on the Bayesian estimation principle. In our study this program was run on an IBM PS/2 equipped with an Intel 8087 co-processor. The pharmacokinetic model used assumes that the aminoglycosides can be described by a one-compartment model and that total clearance consists of both renal (Clr) and nonrenal clearance (Clnr), as depicted in the following equation (8):

$$\mathrm{Cl} = \mathrm{Clr} + \mathrm{Clnr}$$

Clr can be related to creatinine clearance (Clcr) as follows:

$$\mathrm{Clr} = m \times \mathrm{Clcr}$$

Thus clearance can be represented mathematically by the following equation:

$$\mathrm{Cl} = m \times \mathrm{Clcr} + \mathrm{Clnr}$$

Clcr was estimated by using serum creatinine and demographic data (height, weight, age, and sex) (9).

By use of a Bayesian estimation method, this software provides the most probable estimated of m (aminoglycoside clearance/creatinine clearance), Clnr, and Vd. Given these parameters, one can choose the optimal dosing schedule and predict peak and trough serum amikacin concentrations.

The estimated population characteristics for amikacin are shown in the following table:

	Mean	SD
Slope (m)	0.8150	0.3260
Clnr (mL/min/kg)	0.0417	0.0104
Vd (L/kg)	0.350	0.111

D. Assessment of the Abbott Pharmacokinetic System Performance

Estimated pharmacokinetic parameters provided by the two methods (weighted least-square and Bayesian estimation) were statistically compared using the suggestions of Sheiner and Beal (10).

The predictive performances (bias and precision) were assessed by comparison of predicted and measured amikacin serum levels on days 3 and 5.

Bias (the tendency for serum concentrations to be above or below the predicted concentration) was defined as the predicted amikacin serum level minus the observed level. Bias was computed separately for peak and trough concentrations.

III. RESULTS AND DISCUSSION

As shown in Table 1, the pharmacokinetic results obtained with the two estimation methods did not differ significantly. The computation of mean error (bias) revealed that the Abbott Pharmacokinetic System provided unbiased pharmacokinetic parameters.

The predictive performances of the Abbott Pharmacokinetic System with regard to peak and trough amikacin concentrations are presented in Table 2.

Table 1 Precision and Bias for Peak and Trough Amikacin Concentrations (μg/mL)[a]

	Precision (RMSE)	Bias (ME)
Vd (L/kg)	0.09 (0.04; 0.12)	–0.01 (–0.05; 0.04) NS
$T_{1/2}$ (h)	0.83 (–0.63; 1.33)	0.29 (–0.07; 0.06) NS
cl (mL/h/kg)	33.6 (17.8; 44.1)	–12.1 (–26.7; 2.60) NS

[a]Values are point-estimated (95% confidence interval). RMSE, root-mean-square prediction error; ME, mean prediction error; NS, nonsignificant bias.

Table 2 Precision and Bias for Peak and Trough Amikacin Concentrations (μg/mL)[a]

	Day 3		Day 5	
	Peak Level	Trough Level	Peak Level	Trough Level
Precision (RMSE)	5.48 (4.09; 6.59)	2.10 (-0.75; 3.06)	5.01 (2.73; 6.53)	1.60 (0.74; 2.15)
Bias (ME)	-3.70 (S) (-5.59; -1.82)	-1.08 (S) (-1.94; -0.22)	-1.39 (-3.70; 0.91)	-0.73 (-1.44; 0.03)

[a]Values are point-estimated (95% confidence interval). RMSE, root-mean-square prediction error; ME, mean prediction error; S, statistically significant bias ($p < 0.05$).

On day 3 a significant bias was found between predicted and measured amikacin concentrations. It appears that the pharmacokinetic parameters obtained after the first dose do not reflect those observed after multiple doses. By contrast, no significant bias was found between predicted and measured amikacin concentrations on day 5.

The daily doses recommended on day 3 by the Abbott Pharmacokinetic System were highly variable among individuals (1.84 to 28.1 mg/kg/day). The observed concentrations on day 5 were within the desired therapeutic range (i.e., 1 to 5 μg/mL for trough level and 15 to 30 μg/mL for peak level) for 70% of the patients. This percentage was only 35% on day 1. From a practical point of view, the Abbott Pharmacokinetic System is a very user-friendly software.

In conclusion, this software appears to be suitable for individualization of amikacin dosage in ICU patients. It is, however, advisable to wait for the third day of treatment before establishing the first dosing recommendation.

REFERENCES

1. B. Lacarelle, C. Granthil, P. Crevat-Pisano, J. C. Manelli, G. Francois and J. P. Francois. Inter- and intraindividual variability of amikacin pharmacokinetic parameters in intensive care unit patients. *J. Pharm. Clin.*, *6*:109–116 (1987).
2. R. E. Cutler and B. M. Orme. Correlation of serum creatinine concentration and kanamycine half-life. Therapeutic implications. *J. Am. Med. Assoc.*, *209*:539–542 (1969).
3. F. A. Sarubbi and J. H. Hull. Amikacin serum concentrations predictions of levels and dosage guidelines. *Ann. Intern. Med.*, *89*:612–618 (1978).
4. R. J. Sawchuck, D. E. Zaske, R. J. Cipolle, W. A. Wargin, and R. G. Strate. Kinetic model for gentamicin dosing with the use of individual patient parameters. *Clin. Pharmacol. Ther.*, *21*:362–369 (1977).
5. M. E. Burton, D. C. Brater, Q. S. Chen, R. B. Day, P. J. Huber, and M. R. Vasko. A Bayesian feedback method of aminoglycoside dosing. *Clin. Pharmacol. Ther.*, *37*:349–357 (1985).
6. B. Lacarelle, C. Granthil, J. C. Manelli, N. Bruder, G. Francois, and J. P. Cano. Evaluation of a Bayesian method of amikacin dosing in intensive care unit patients with normal or impaired renal function. *Ther. Drug Monitor.*, *9*:154–160 (1987).
7. A. Iliadis. APIS: A computer program for clinical pharmacokinetics. *J. Pharm. Clin.*, *4*:573–577 (1985).

8. C. C. Peck and J. H. Rodman. Analysis of clinical pharmacokinetic data for individualizing drug dosage regimens. In *Applied Pharmacokinetics, Principles of Therapeutic Drug Monitoring* (W. E. Evans, J. J. Schentag, and W. Jusko, eds.), Applied Therapeutics, San Francisco, pp. 55–82 (1986).
9. D. W. Cockkroft and M. H. Gault. Prediction of creatinine clearance from serum creatinine. *Nephron, 16*:31–41 (1976).
10. L. B. Sheiner and S. L. Beal. Some suggestions for measuring predictive performances. *J. Pharmacokinet. Biopharm., 4*:503–512 (1981).

16

Evaluation of the Abbott Base Pharmacokinetic System for the Individualization of Amikacin, Vancomycin, and Theophylline Dosing

B. Lacarelle, F. Marre, D. Masset, P. Pisano, T. Blanc-Gauthier, N. Fenina, C. Le Guellec, and A. Durand
Laboratoire de Pharmacocinétique et de Toxicocinétique, Marseille, France

I. INTRODUCTION

It is now well known that for several drugs the use of "standard" doses to effect desired clinical response is unreliable if many if not most patients. This fact has led to the development of many methods to optimize drug dosing, including use of nomograms and pharmacokinetics.

The antibiotics amikacin and vancomycin have a narrow therapeutic index, serum concentrations correlate with efficacy (1) and toxicity (2), and a wide interpatient variability in their pharmacokinetics exists (3,4). During theophylline therapy it is recommended that one monitor drug serum levels in smoking patients, during disease states, or when concurrent medications may alter theophylline pharmacokinetics. Therefore, drug dosing methods appear to be very necessary for amikacin, vancomycin, and theophylline.

Since the work by Sheiner et al. (5,8) the reliability of Bayesian estimation methods has been proved, and these methods have been proposed for the optimization of various drug therapies including amikacin, vancomycin, and theophylline. Thus, Abbott Laboratories has developed a new computer software package (Abbott Base Pharmacokinetic System) which allows designing dosage regimens of the above-mentioned drugs.

The aim of the present study was to validate this new pharmacokinetic system in routine clinical practice for amikacin, vancomycin, and theophylline.

Table 1 Patient Personal Data

	Amikacin	Vancomycin	Theophylline
n	50 ICU patients	10 ICU patients	10 outpatients
Sex	40 males, 10 females	5 males, 5 females	7 males, 3 females
Mean age (range)	57 years (18–88)	45 years (19–76)	69 years (58–85)
Main pathology	Infectious diseases		Congestive heart failure, chronic asthma
Dosage form	i.v., 30 min	i.v., 60 min	Theophyllilne SR (Armophylline)
Initial daily dose	14.3 ± 6 mg/kg	28.6 ± 6.7 mg/kg	412 ± 126 mg

II. MATERIALS AND METHODS

A. Patients

Patients' personal data are summarized in Table 1.

B. Software

The population parameters used in the software were determined from clinical studies. Since these pharmacokinetic parameters are related to demographic data or patient status, patient information requirements are

Sex, height, weight, age
Dosing history
Serum levels data
Creatinine serum level (for antiinfectious drugs)
Disease factors and concurrent medications (for theophylline)

The Cockroft-Gault formula was used for creatinine clearance estimation (6). A one-compartment model was used for amikacin and theophylline and a two-compartment model for vancomycin. The software ran on an IBM PS/2 or a compatible unit (more than 512 KB).

III. GENERAL METHODOLOGY

By using this software one or more times during the treatment, dosing recommendations were obtained in real time for all the patients. The predictive performances (bias and precision) were assessed by comparing predicted drug concentration with those measured after dosage individualization (7). In the case of aminoglycosides and vancomycin, precision and bias were computed separately for peak and trough concentrations.

IV. ANALYTICAL METHOD

The drug levels were measured using fluorescence polarization immunoassay (TDX, Abbott). Blood samples were drawn 5 min before (trough) and 30 min after (peak) administrations of the antiinfectious drugs and immediately before theophylline administration.

V. RESULTS AND DISCUSSION

Results showed that recommended daily doses were highly variable among individuals for amikacin (250 to 2400 mg), vancomycin (300 to 2800 mg), and theophylline (100 to 600 mg).

Precision and bias (μg/mL) values are summarized for amikacin peak and trough levels (Table 2A), for vancomycin peak and trough levels (Table 2B), and for theophylline concentrations (Table 2C).

For amikacin and vancomycin the differences between predicted and observed concentrations were not statistically significant for both peak and trough levels. For theophylline, a significant bias was found. The mean error, however, was slight (1.1 μg/mL) and had no clinical consequence.

The mean (± SD) daily dose for amikacin (16.3 ± 6.8 mg/kg) and vancomycin (32.8 ± 10.5 mg/kg) were slightly higher than mean initial doses. However, daily dose was highly increased or reduced in many patients. Finally, the difference between initial and recommended daily doses was highly significant (paired t-test; $p < 0.01$).

The mean individualized daily dose of theophylline (682 ± 279 mg) was higher than the initial one (412 ± 126 mg).

Table 2 Precision and Bias for the Prediction of the Different Drug Concentrations (μg/mL)[a]

	A. Amikacin	
	Peak Level	**Trough Level**
Precision (RMSE)	4.80	2.59
Bias (ME)	0.22 (−9.41; 9.85) NS	0.20 (−4.99; 5.39) NS
	B. Vancomycin	
	Peak Level	**Trough Level**
Precision (RMSE)	5.91	3.52
Bias (ME)	0 (−2.34; 2.33) NS	0.40 (−0.95; 1.76) NS
	C. Theophylline	
Precision (RMSE)	2.61	
Bias (ME)	−1.06 (−2.04; −0.08) S	

[a]Values are point-estimated (95% confidence intervals). RMSE, root-mean-square prediction error; ME, mean prediction error; S, statistically significant bias ($p < 0.05$); NS, nonsignificant bias.

VI. CONCLUSION

The Abbott Base Pharmacokinetic System appears to be accurate for the prediction of drug doses necessary to achieve effective and nontoxic serum concentrations in patients.

This method is compatible with standard therapeutic drug monitoring, since, for a given patient, the usual times of sampling (peak and trough levels) are sufficient for the estimation of pharmacokinetic parameters and the determination of optimum dosage.

However, complementary studies with large population patients are needed for theophylline and vancomycin.

REFERENCES

1. R. D. Moore, C. R. Smith, and P. S. Lietman. The association of aminoglycoside plasma levels with Gram-negative bacteremia. *J. Infect. Dis., 149*:443–448 (1984).
2. J. J. Schentag, M. E. Plant, F. E. Cena, P. B. Wells, P. Walezak, and R. J. Buckley. Aminoglycoside nephrotoxicity in critical ill surgical patients. *J. Surg. Res., 26*:270–279 (1979).
3. H. Yasuhara, S. Kobayashi, K. Sakamoto, et al. Pharmacokinetics of amikacin and cephalosporin in bedridden elderly patients. *J. Clin. Pharmacol.,* 22:403–409 (1982).
4. L. A. Bauer, R. A. Blouin, W. D. Griffin, K. E. Record, and R. M. Bell. Amikacin pharmacokinetics in morbidly obese patients. *Am. J. Hosp. Pharm., 37*:519–522 (1980).
5. L. B. Sheiner, S. Beal, B. Rosenberg, and V. V. Marathe. Forecasting individual pharmacokinetics. *Clin. Pharmacol. Ther.,* 26:294–305 (1979).
6. D. W. Cockroft and M. H. Gault. Prediction of creatinine clearance from serum creatinine. *Nephron, 15*:31–41 (1976).
7. L. B. Sheiner and S. L. Beal. Some suggestions for measuring predictive performances. *J. Pharmacokinet. Biopharm.,* 9:503–512 (1981).

17

The Impact of the Cardiac Arrhythmia Suppression Trial (CAST) on Antiarrhythmic and Cardioactive Drug Monitoring

David W. Holt and Atholl Johnston *St. George's Hospital Medical School, London, England*

I. INTRODUCTION

The object of this chapter is to summarize some of the factors relating to the monitoring of antiarrhythmic and cardioactive drugs in the light of the results of the Cardiac Arrhythmia Suppression Trial (CAST).

As a consequence of the findings of the CAST study, we are at a turning point in attitudes toward the administration of antiarrhythmic drugs (1,2). The results are having an effect throughout the field of cardiology with respect to trial design and criteria for the collection of data, both before and after registration of a new drug. In particular, it has called into question the use of surrogate endpoints as markers of a drug's ability to reduce the incidence of sudden cardiac death (3).

For some time, a mainstay of antiarrhythmic drug therapy has been their use as prophylactic agents following acute myocardial infarction. The rationale for such use has been the contention that, by suppressing ventricular premature beats (VPBs), the incidence of sudden death in this setting was reduced. CAST set out to establish this in a randomized placebo controlled trial.

II. CAST

Patients were recruited into CAST if they had asymptomatic or mildly symptomatic ventricular arrhythmia following acute myocardial infarction. In an open-label phase of the study they received one of three antiarrhythmic drugs—encainide, flecainide, or moricizine—all members of the Class I group in the Vaughan-Williams classification, the first two being members of the subclass 1c. If the patients responded to the drug treatment, they were randomized to either the effective therapy or placebo.

A total of 1727 patients were randomized for a mean follow-up period of 10 months before the preliminary results were assessed and the trial stopped (4). The results showed an excess of deaths in the treatment group associated with the two Class 1c agents. Deaths were 4.5% in the treated group, compared with 1.2% in the placebo group.

A number of reasons can be put forward for these unexpected findings. They include the fact that many patients at risk of sudden cardiac death were excluded by the design of the study. Many high-risk patients died during the initial open-label titration. As a result, patients at least risk of dying were entered into the study, and this could have contributed to the very low mortality in the placebo group. Indeed, compared with historical controls, even the results in the treatment group were good. However, notwithstanding these considerations, the results were shocking and have provoked considerable controversy.

The implications of the findings were that the drugs were not only effective in the suppression of VPBs but could also provoke serious arrhythmias, leading to arrhythmic death. The concept of arrhythmogenicity is not new, nor is it confined to compounds of Class 1c. Recent analyses of previous studies have shown a similar finding for quinidine (5). However, the findings of CAST have had effects in several areas. They have caused cardiologists to question the routine use of antiarrhythmic agents in this setting. They have provoked the pharmaceutical industry to search for new compounds which can be shown to be both antiarrhythmic and free from significant arrhythmogenic potential. Finally, they have led drug regulatory authorities to reassess the criteria for testing these and similar drugs, and to impose more stringent guidelines on the data which must be collected on drug pharmacokinetics/pharmacodynamics.

The results of CAST should also be seen in the context of the therapeutic alternatives to antiarrhythmic drugs which are now available to clinicians. There have been major advances in surgery, including the surgical excision of aberrant electrical conducting pathways, and the use of percutaneous transluminal coronary angioplasty (PTCA). Mortality following acute myocardial infarction has been reduced by the use of thrombolytic agents, and sudden cardiac death has been reduced by the introduction of implantable defibrillators.

While the use of implantable defibrillators is limited at present (15,000 worldwide), it is anticipated that, in the United States alone, 100,000 will be implanted per year by 1995. Their use has implications for drug trials designed to test the thesis that a drug has the potential to reduce sudden cardiac death, since patients at risk could be entered into a placebo treatment group without ethical problems if these devices were available. In this way, the assessment of the drug would be focused on a hard endpoint, death, rather than the surrogate endpoints used at present. Efficacy at present is often assessed by suppression of VPBs or by the use of programmed electrical stimulation, to provoke arrhythmias in a laboratory setting and monitor the effects of test doses of a chosen agent.

III. ANTIARRHYTHMIC DRUG THERAPY

What antiarrhythmic drugs are left to the clinician, in the light of the doubts surrounding the Class 1c agents? At opposite ends of the spectrum there are two drugs with very different pharmacological effects and pharmacokinetics.

Amiodarone is a highly effective Class III agent, used in the treatment of both ventricular and supraventricular arrhythmias. However, it is far from ideal because of its exceptionally long elimination half-life and profile of adverse effects. Monitoring of the compound has been shown to be effective in some clinical settings (6). In contrast, adenosine is a purine nucleoside which is highly effective for the treatment of paroxysmal supraventricular tachyarrhythmias. It is a very short-acting drug, having a half-life of only a few seconds, and must be given intravenously (7). It is unique in being approved by the FDA but, as yet, not by European regulatory authorities.

With regard to compounds in development, the impact of the CAST study has been to cancel the clinical evaluation of many promising new drugs, especially if they had Class 1c activity. Drug manufacturers are tending to distance themselves from this class of compounds as fast as possible, and it is likely that the CAST findings will result in the accelerated demise of the Vaughan-Williams classification in favor of a system more related to the site(s) of action of a drug in the heart.

Many compounds in development show antiarrhythmic potential, with much of the clinical interest being focused on Class III compounds, in view of the spectacular efficacy of amiodarone. The tendency is toward the development of highly potent drugs, given in relatively low doses, leading to very low plasma concentrations. Some have more than one class of activity as judged by the Vaughan-Williams classification. Two examples will suffice to show the trends in development.

First, UK-68,798 is a highly potent Class III compound (8). Its pharmacokinetic profile is very acceptable, and doses of only 1.5 to 4.5 μg/kg have proved effective. Metabolites of the drug have not been shown to have clinical activity, and plasma concentrations of the parent compound have been shown to relate to a pharmacodynamic response, namely, measurements of the corrected QT interval. Thus, there may be a role for drug-level monitoring as a guide to therapy.

From the same company, UK-52,046 is an example of a compound with more than one class of activity (9). It is both an alpha- and a beta-blocking compound, but with less peripheral alpha-blocking activity than compounds such as prazosin. Interest in this compound stems from a growing recognition that there are many endogenous factors which enhance the potential for arrhythmia. In this case, ischemia alpha receptors are unmasked and have increased sensitivity. A compound combining both alpha- and beta-blocking activity is, therefore, of potential interest, since beta blockers alone have been shown to improve prognosis following acute myocardial infarction (10).

IV. DRUG EVALUATION

What measurements should we be making for the optimal evaluation of new drugs affecting the heart or cardiovascular system?

It is increasingly obvious that drugs must be shown to be both effective and safe in the target groups for which they are intended. A key area in the early clinical evaluation of drugs is, therefore, the identification of risk groups. For many of the compounds under consideration here, this will involve the study of pharmacokinetics of the compound in patients with angina, heart failure, previous myocardial infarction, or hypertension.

After characterizing the basic pharmacokinetic parameters of absorption, distribution, metabolism, and elimination in the populations at risk, there is a growing checklist of subsidiary areas to explore. These include drug interactions, the potential for pharmacological activity or toxicity of metabolites, whether metabolism is genetically linked, chirality, chronopharmacokinetics, and dose ranging. If a strong relationship between plasma concentrations and a clinical parameter, either efficacy or toxicity, can be shown, there may even be a need to establish a population pharmacokinetic model.

One of the antiarrhythmic agents currently under evaluation is propafenone; it serves as a good example of many of the points alluded to above. Although it is a Class 1c compound, it is still of clinical interest, partly because there is considerable clinical experience with its use in Germany and partly because the parent compound has beta-blocking activity. However, other factors could confound its general acceptance. It has stereoisomers, and the plasma clearance of the R and S forms differ (11). It is metabolized by the liver to 5-hydroxy, 5-hydroxy,4-methoxy, and mono-N-dealkylated metabolites. Of these, the 5-hydroxy metabolite has been shown to have Class 1c activity but little beta-blocking activity. The picture is further complicated by the fact that the formation of this metabolite is genetically linked, giving rise to two groups of patients—extensive and poor metabolizers (12).

As a result, evaluation of this and similar antiarrhythymic agents may require chiral separations of the parent compound and, if necessary, metabolites. Following this, an appreciation of the comparative clinical efficacy of metabolites, and the identification of patients who may be at risk of drug accumulation because of their genetic phenotype, can be made.

One factor, which affects the interpretation of pharmacokinetic data but which has received comparatively little attention in the past, is that of dose timing. Langner and Lemmer (13) noted that the maximum plasma concentration (C_{max}) for propranolol was significantly higher following a morning dose (08.00 h) compared with doses taken at three other time points throughout the day (14.00, 20.00, and 02.00 h). Recently, we have noticed a similar chronopharmacokinetic effect for the calcium channel blocker nicardipine (14). Both mean C_{max} and area under the time-concentration curve (AUC) were significantly higher following a morning (08.00-h) dose than after an evening (20.00-h) dose (Figure 1). Interestingly, there was also a chronopharmacodynamic effect. Both systolic and diastolic blood pressures were significantly lower after the morning dose, which produced the highest plasma concentrations of nicardipine.

The matching of pharmacokinetic and pharmacodynamic data is becoming of particular relevance in the evaluation of new compounds. It is often approached by means of the dose ranging study, since drug regulatory authorities are increasingly anxious that investigators establish that initial clinical studies provide adequate data on which to recommend a rationale dosage schedule. The chosen doses should have been shown to be the minimum effective doses consistent with efficacy. This may be relatively easy for some classes of compounds, e.g., beta blockers, for which blood pressure can be monitored, or ACE inhibitors, for which plasma angiotensin-converting enzyme activity can be monitored. But for antiarrhythmic agents the data are often highly variable (15).

In Figure 2 the plasma concentrations of propafenone which were sufficient to produce a 90% reduction in VPBs (EC_{90}) in two patients are shown. These concentrations differ by almost two orders of magnitude. Clearly, dose recommendations can be made only on the basis of the "average" patient, but it is obvious from data such as those illustrated in Figure 2 that factors other than the concentration of an

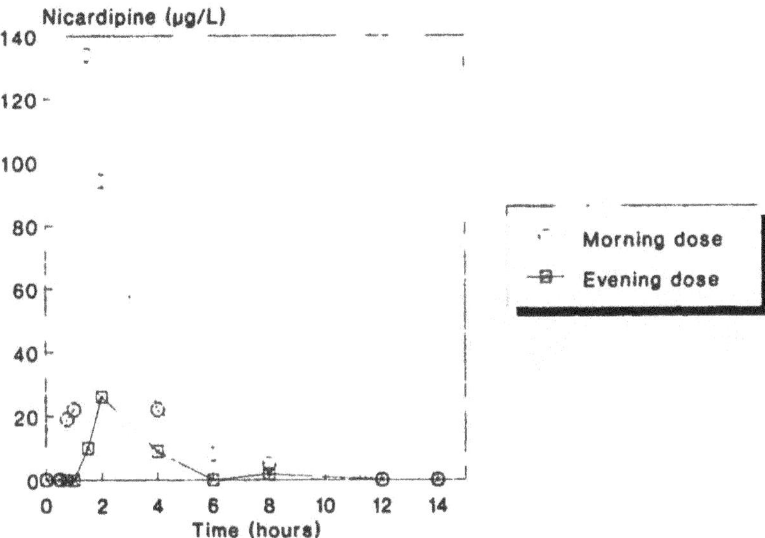

Figure 1 Plasma concentrations of nicardipine in a volunteer subject following a 40-mg dose administererd either in the morning or evening.

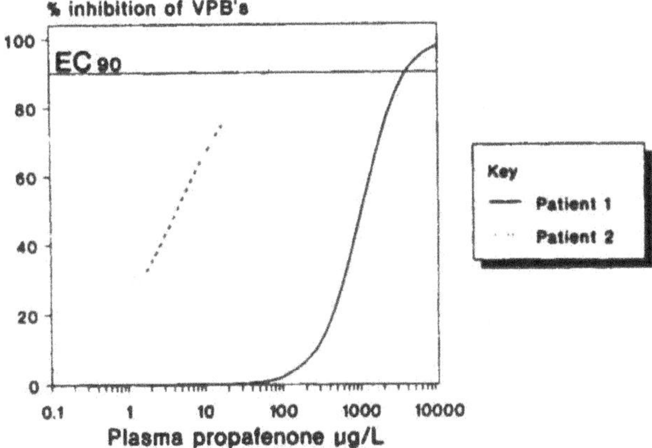

Figure 2 Plasma concentrations of propafenone sufficient to reduce the incidence of ventricular premature beats (VPBs) by 90% compared with control data (EC_{90}) in two patients receiving chronic therapy with the drug.

antiarrhythmic drug in plasma have an impact on the efficacy of the drug. We will devote the last section of this chapter to discussing some of the additional factors which must be considered in order to interpret drug concentrations fully and which point the way toward the measurements we should be making in this field, besides the drug itself.

V. FACTORS AFFECTING INTERPRETATION

There is an increasing awareness that a very broad range of plasma concentrations of antiarrhythmic drugs may be consistent with efficacy. Assessment of the drug concentration alone is too simplistic an approach, since a number of substrates which promote arrhythmia may have an effect on the efficacy of the drug.

In addition, the interpretation of drug concentrations may be influenced by such factors as the type of arrhythmia, its spontaneous variability, other drug therapy, other underlying morbidity, and the age of the patient. To this list should be added the variability of a number of endogenous factors, some of which could be measured at the same time as the drug concentrations.

For instance, the influence of low intra- or extracellular potassium concentrations on the efficacy of such drugs as digoxin is well known, but other aspects of the ionic milieu are also important. There is considerable interest in the role of magnesium in the etiology of arrhythmia, and on those factors which may promote its loss from tissue (16). Recently, in our own laboratory, we have developed a method for the measurement of magnesium in biopsy samples with a dry weight of less than 1 mg. This method is being used to assess the relationship between arrhythmia and low tissue magnesium following therapy with such drugs as diuretics or cyclosporin.

Another factor which could influence the efficacy of antiarrhythmic drugs is blood rheology, by its effect on ischemia. Blood flow is improved by some drugs which increase red cell deformity, such as pentoxifylline or the calcium channel blocker bepridil.

Perhaps one of the most important growth areas in this field is the measurement of markers of free-radical activity. In our own laboratory, interest stems from the investigation of the role of free radicals in the etiology of reperfusion arrhythmias following coronary angioplasty, and in tissue damage following cold ischemia during solid organ transplantation. A number of approaches can be taken to obtain an indirect measurement of free-radical activity, one of the simplest being the measurement of the ratio allantoin/uric acid in plasma (17). An alternative approach is to measure the activities of free-radical-scavenging enzymes in either plasma or red cells. While controlled studies in defined patient populations are still underway, the field is an exciting one, since it offers the prospect of treatment with antioxidants, trace elements, or genetically engineered enzyme preparations if specific defects can be identified.

Finally, the measurement of endothelin may be of value in some settings. This potent peptide vasoconstrictor is produced by endothelial cells. Its production is stimulated by a number of substances, including adrenalin, angiotensin II, thrombin, and tissue plasminogen activator. Thus, it may affect ischemia and influence arrhythmia as a result of other endogenous compounds or drugs. Its effect is long-lasting but can be reversed by calcium channel blockers which are, themselves, frequently used in association with antiarrhythmic agents.

VI. CONCLUSIONS

It can be concluded that drug measurements in the field of cardioactive drugs are of particular value during the initial assessment of new compounds. Basic pharmacokinetic parameters can be determined and dose ranging studies verified. Patients thought to be at particular risk, through such variables as genetically linked metabolism, can be identified and monitored.

There is less convincing evidence that the measurements can be used as a routine guide to therapy, because of the many factors which can influence the initiation of arrhythmia and the response to the drugs. Single drug measurements are unlikely to be of diagnostic value, except as a check on compliance, and therapy should always be optimized in the light of clinical data. To this end, there is a growing interest in the measurement of endogenous compounds that affect the cardiovascular system.

The requirements for analytical methodology are forever increasing sensitivity and specificity, to cope with highly potent compounds and the measurement of their metabolites. In our experience, this has led to a growing use of immunoassay techniques at the expense of high-performance liquid chromatography. For some drugs and their metabolites it may be necessary to make stereoselective measurements to obtain a meaningful relationship between pharmacokinetic and pharmacodynamic data.

The results of the CAST study have altered many attitudes to the use of antiarrhythmic drugs, by removing a pivotal reason for their prescription. The study has had a number of effects on the collection of data for the registration of new antiarrhythmic and cardioactive compounds. It has concentrated the mind, but in the final analysis, has not altered the goals which many of us in this field have been striving toward for some years.

REFERENCES

1. Task Force of the Working Group on Arrhythmias of the European Society of Cardiology. CAST and beyond. Implications of the Cardiac Arrhythmia Suppression Trial. *Eur. Heart. J., 11*:194 (1990).
2. A. B. de Luna, J. Guindo, J. Borja, M. Roman, and C. Madoery. Recasting the approach to the treatment of potentially malignant ventricular arrhythmias after the CAST study. *Cardiovasc. Drugs & Ther., 4*:651 (1990).
3. R. L. Woosley. CAST: Implications for drug development. *Clin. Pharmacol. Ther., 47*:553 (1990).
4. J. T. Bigger. The events surrounding the removal of encainide and flecainide from the Cardiac Arrhythmia Suppression Trial (CAST) and why CAST is continuing with moricizine. *J. Amer. Col. Cardiol., 15*:243 (1990).
5. P. E. Coplen, E. M. Antman, J. A. Berlin, P. Hewitt, and T. C. Chalmers. Prevention of recurrent atrial fibrillation by quinidine. A meta analysis of randomized trials. *Circulation, 80*(Suppl. II):633 (1989).
6. D. W. Holt, G. T. Tucker, P. R. Jackson, and W. J. McKenna. Amiodarone pharmacokinetics. *Br. J. Clin. Pract., 40*:109 (1986).
7. L. Belardinelli and B. B. Lerman. Electrophysiological basis for the use of adenosine in the diagnosis and treatment of cardiac arrhythmias. *Br. Heart J., 63*:3 (1990).
8. T. C. K. Tham, B. A. MacLennan, D. W. G. Harron, P. E. Coates, D. Walker, and H. S. Rasmussen. Pharmacodynamics and pharmacokinetics of the novel Class III antiarrhythmic drug UK-68,798 in man. *Br. J. Clin. Pharmacol., 31*:243–244p (1990).
9. B. Tomlinson, J. C. Renondin, B. R. Graham, and B. N. C. Prichard. Alpha-adrenoceptor antagonism and haemodynamic effects of UK-52,046 compared with prazosin. *Br. J. Clin. Pharmacol., 27*:686P (1989).

10. N. Rehnqvist. Prevention of sudden death by beta-blockade. *Cardiovasc. Drugs & Ther.*, 4:675 (1990).
11. H. K. Kroemer, C. Funch-Brentano, D. J. Silberstein, A. J. J. Wood, M. Eichelbaum, R. L. Woosley, and D. M. Roden. Stereoselective disposition and pharmacological activity of propafenone enantiomers. *Circulation*, 79:1068 (1989).
12. L. A. Siddoway, K. A. Thompson, C. B. McAllister, G. R. Wilkinson, D. M. Roden, and R. L. Woosley. Polymorphism of propafenone metabolism and disposition in man: Clinical and pharmacokinetic consequences. *Circulation*, 75:785 (1987).
13. B. Langner and B. Lemmer. Circadian changes in the pharmacokinetics and cardiovascular effects of oral propranolol in healthy subjects. *Eur. J. Clin. Pharmacol.*, 33:619 (1988).
14. D. W. Holt, R. J. Eastwood, S. Walker, K. K. Hla, and J. A. Henry. Effect of dose time on the pharmacokinetics of nicardipine. *Eur. J. Pharmacol.*, 183:1595 (1990).
15. J. Morganroth. Determination of antiarrhythmic drug efficacy in the treatment of ventricular arrhythmias. *Cardiovasc. Drugs & Ther.*, 4:669 (1990).
16. S. A. Abraham, D. Rosenmann, and M. Kramer. Magnesium in the prevention of lethal arrhythmias in acute myocardial infarction. *Arch. Intern. Med.*, 147:1768 (1986).
17. Grootveld and B. Halliwell. Measurement of allantoin and uric acid in human body fluids. A potential index of free radical reactions in vivo? *Biochem. J.*, 243:803 (1987).

18

Clinical Usefulness of Imipramine and Desipramine Serum-Level Monitoring in Patients Treated for Endogenous Depression

Joanna Szymura-Oleksiak and Andrzej Wasieczko
Medical Academy, Kraków, Poland

I. INTRODUCTION

Therapeutic drug monitoring of imipramine (IMI) enhances the physician's ability to use this agent more rationally, increasing both its safety and efficacy.

Clinical studies have demonstrated that IMI efficacy was achieved in most patients with total serum concentration in the range of 150 to 300 ng/mL [IMI + active metabolite desipramine (DMI)] (1,2).

The aim of the present study was to individualize the dosage regimen of IMI based on the steady-state serum concentration of IMI + DMI measurement and clinical evaluation of patients treated for depression.

II. PATIENTS AND METHOD

Fifteen endogenous unipolar or bipolar depression patients (7 women and 8 men), aged 28 to 56 years, were treated with IMI and, if necessary a benzodiazepine or neuroleptic. All patients were in good physical health and without any history of notable renal or hepatic disease.

Depth of depression was assessed using the 24-item Hamilton Depression Rating Scale (HDRS) according to the following criteria: 0 to 6 scores, without depression; 7 to 17 and 18 to 24 scores, mild and moderate depression, respectively; and 25 scores or more, severe depression (3).

IMI (Imipramine, Polfa) was given orally in two divided doses at 8:00 a.m. and 1 p.m. Minimum steady state of IMI + DMI serum concentration was measured after 3 weeks of therapy or after an adjusting dose of IMI. Blood samples were collected in the morning just before the intake of the next dose. Serum concentration (IMI + DMI) was determined by the FPIA method (Abbott, TDx system).

Based on the knowledge of one steady-state serum concentration of IMI + DMI (C^{ss}_{start}), which was determined from starting dosage (D_{start}), dosage adjustment was performed (D_{adj}). The new steady-state serum concentration was predicted (C^{ss}_{pred}) according to the following equation:

$$C^{ss}_{pred} = D_{adj} \times \frac{C^{ss}_{start}}{D_{start}} \tag{1}$$

The accuracy and precision of prediction method were evaluated on the basis of mean prediction error (ME) and root-mean-square prediction error (RMSE), respectively (4).

III. RESULTS AND DISCUSSION

The patient's demographic data, concomitant treatment, dosage of IMI, and assessment of the depth of depression in the responders and nonresponders group are presented in Tables 1 and 2, respectively. Figure 1 shows the minimum steady-state serum concentration of IMI + DMI determined in each patient after starting and adjusting dosage of IMI.

In the responders group (Table 1), the IMI + DMI steady-state serum concentration and the clinical status of the patient indicated that the starting dose of IMI (a) should be unchanged (four patients with serum concentration within the therapeutic range or higher and without side effects); (b) should be increased (three patients with low serum concentration); or (c) should be decreased (three patients with high concentration and noticeable side effects).

In seven patients out of 10 responders, the steady-state serum concentration of IMI + DMI was between 150 and 300 ng/mL, within the therapeutic range.

Final HDRS in this group indicated remission of depression in seven cases, mild and moderate depression in one and two patients, respectively. It was achieved in no more than 6 weeks.

In the nonresponders group (Table 2), monitoring the serum concentration helped in the earlier decision to change the tricyclic antidepressant (patient RK) or intensified the diagnostic process which then detected additional pathology (patients KZ, AM).

In one patient (MK) monitoring the serum concentration confirmed a suspicion of noncompliance, which caused a temporary change in the way of administration of the drug from oral to intramuscular.

The correlation between the predicted steady-state serum concentration of IMI + DMI using Eq. (1) and measured in seven patients after adjustment of the dosage of IMI is presented in Figure 2.

Predictive performance evaluation is presented in Table 3. A negative value of ME indicates that the prediction method had a tendency to underestimate the steady-state serum concentration of IMI + DMI.

IV. CONCLUSIONS

Serum concentration monitoring of IMI + DMI in the responders group intensified the efficiency of treatment by optimizing the dosage of IMI earlier than would have been possible based only on assessing the clinical effects. In the nonresponders group, it

Table 1 Demographic Data, Concomitant Treatment, and Assessment of the Depth of Depression in the Responders Group

				Daily IMI dose (mg)		HDRS (scores)		
Patient	Age (years)	Sex	Other drugs taken	Start	Adj.	Init.	Final	Comments
J.F.	56	M	Clobazam	100	100	19	6	Response adequate
A.K.	53	M	Levopromazine	150	150	27	16	
J.C.	39	F	Levopromazine	100	100	39	6	
G.P.	27	M	Levopromazine, lithium	150	150	30	6	Starting dose unchanged
J.B.	45	F	Perazine	150	225	53	4	Poor response
Z.N.	42	M	Levopromazine, perazine	150	200	34	6	IMI serum concentration low
M.G.	34	F	Temazepam	150	200	32	5	Dose increased
J.L.	28	M	Perazine	175	125	37	9	Side effects
E.S.	44	M	Levopromazine, propranolol	150	125	38	16	IMI serum concentration high
K.K.	41	F	Levopromazine, lithium, propranolol	150	125	36	2	Dose decreased

Table 2 Demographic Data, Concomitant Treatment, and Assessment of the Depth of Depression in the Nonresponders Group

				Daily IMI dose (mg)		HDRS (scores)		
Patient	Age (years)	Sex	Other drugs taken	Start	Adj.	Init.	Final	Comments
U.Z.	35	F	Oxazepam	150	150	36	36	Schizoaffective depression
R.K.	28	M	Levopromazine	100	125	24	18	Optimal IMI concentration, depression resistant to IMI
K.Z.	47	F	Perazine	125	125	30	20	Depression with organic mental syndrome
A.M.	37	F	Levopromazine, carbamazepine, oxazepam	100	100	42	24	Depression with organic mental syndrome
M.K.	38	M	Levopramazine	150	225	33	25	Noncompliance

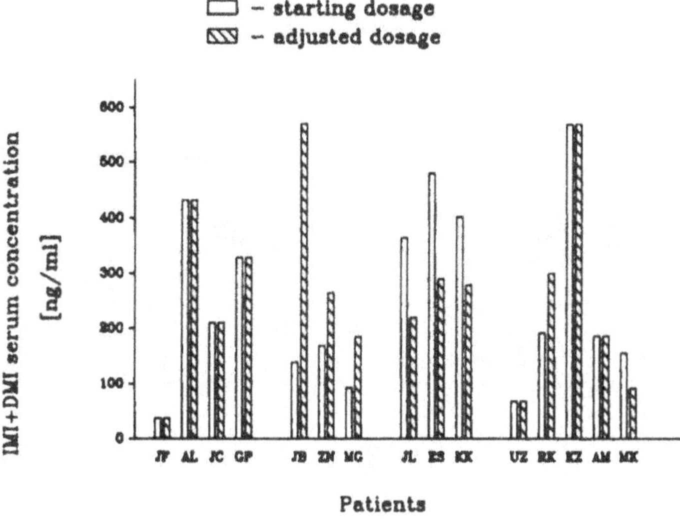

Figure 1 Steady-state serum concentration of IMI + DMI determined after starting and adjustment of IMI dosage.

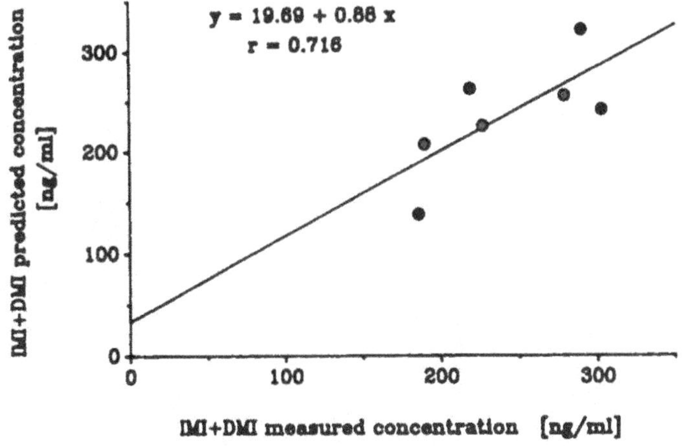

Figure 2 Relationship between measured and predicted IMI + DMI steady-state serum concentration.

Table 3 Predictive Performance Evaluation of IMI + DMI Steady-State Serum Concentration[a]

$C_{meas.}$ ± SD (ng/mL)	$C_{pred.}$ ± SD (ng/mL)	ME	MSE	RMSE
242 ± 48	236 ± 60	−8 (−241; 226)	2 (−7; 10)	40 (0; 102)

[a]Values in parentheses are lower and upper 95% confidence limits.

helped in making an earlier decision to change tricyclic antidepressant and increased the diagnostic accuracy.

REFERENCES

1. Tricyclic antidepressants—Blood level measurements and clinical outcome: An APA Task Force report. *Am. J. Psychiatr.*, 142:2 (1985).
2. S. H. Preskorn, C. Dorey, and G. Jerkovich. Therapeutic drug monitoring of tricyclic antidepressants. *Clin. Chem.*, 34:5 (1988).
3. M. Hamilton. A rating scale for depression. *J. Neurol. Neurosurg. Psychiatr.*, 23:56 (1960).
4. L. Sheiner and S. Beal. Some suggestions for measuring predictive performance. *J. Pharmacokin. Biopharm.*, 9:4 (1981).

19

Light-Induced Racemization of the Diastereoisomeric Pairs of Thioridazine 5-Sulfoxide: Origin of Artifacts in the TDM of Thioridazine

C. B. Eap, A. Souche, L. Koeb, and P. Baumann *Département Universitaire de Psychiatrie Adulte, Prilly-Lausanne, Switzerland*

I. INTRODUCTION

Thioridazine (THD), a phenothiazine drug, is a very commonly prescribed neuroleptic for the treatment of psychoses. After oral adsorption, THD is extensively metabolized by side-chain oxidation leading to mesoridazine (THD 2-SO) and sulforidazine (THD 2-SO_2), by N-demethylation to northioridazine (N-THD), and by ring sulfoxidation to thioridazine 5-sulfoxide (THD 5-SO). THD is believed to act mainly via antidopaminergic functions in the central nervous system, and studies measuring the abilities of THD, mesoridazine, and sulforidazine to displace spiperone from binding sites on striatal homogenates indicate that these two metabolites are the major determinants in THD's antipsychotic and side effects (1–4). On the other hand, THD 5-SO, which is the metabolite found in largest concentrations in serum after chronic THD administration (5), seems to be devoid of antipsychotic effects but contributes to cardiotoxicity (4,6,7) and is more potent than the parent drug (8,9). THD possesses an asymmetrical carbon at position 2 in the piperidyl ring, and oxidation of the ring sulfur atom creates an additional chiral center. THD 5-SO therefore exists in the form of two diastereoisomeric pairs of enantiomers, called fast-eluting THD 5-SO (FE) and slow-eluting THD 5-SO (SE) (10). Until now, all studies in which THD 5-SO concentrations in human and rat serum or urine have been determined (11–16) have not revealed significant differences in the concentrations of the two diastereoisomeric pairs. This suggests a lack of stereoselectivity in their biotransformation and renal clearance in rats and humans. In a series of controlled experiments, we have observed, as reported here, that these isomeric pairs are light-sensitive and racemize. Moreover, data concerning the THD ring sulfoxide concentrations measured with light-protected extraction steps in the plasma of 11 psychiatric patients having received THD for at least 1 week and up to 2 months, with doses varying between 30 and 400 mg per day, are also presented (full results and detailed description of the methods used are given in another paper: Eap et al., submitted to *Therapeutic Drug Monitoring*).

II. MATERIAL AND METHODS

Extraction, chromatographic methods, and detection were performed according to Kilts et al. (17) with small modifications. In particular, in internal standard (pipothiazine) was added before the extraction steps, and the methanol content of the HPLC mobile phase was lowered to 7% to allow better separation between mesoridazine and THD 5-SO FE. For the racemization experiments, an ultraviolet lamp, Mineralight UVSL 25, was used. Urines and heparinized plasmas, after sampling and immediate centrifugation, were stored at –20°C in aluminum-wrapped tubes.

III. RESULTS

Due to the impossibility of obtaining pure isomeric THD 5-SO pairs during their isolation processes, solvent pH, and temperature conditions were checked to avoid racemization. Using the isomeric pairs, diluted 500 ng/mL in the HPLC injection solvent, it was found that this phenomenon was light-induced and roughly proportional both to the quantity of light and to the duration of exposure. No precise quantitative measurements of racemization have been performed because the results were found to be dependent on (a) uncontrolled factors such as weather conditions (even if the samples were not exposed to direct sunlight) (b) the location of the apparatus for extraction (vortex, shaker) or (c) HPLC injection (automatic injector). Nevertheless, it was noticed that racemization was complete within a few minutes when the samples were exposed to direct sunlight; it took several hours when they were stored in a room without direct sunlight but with window blinds open, and several days when the window blinds were closed, using neon as unique source of light. Therefore, it was decided that all manipulations must be performed with aluminum-wrapped tubes, in a room with window blinds closed, using only neon lighting. One must mention that, most probably, photolysis also occurs in addition to racemization: A gradual decrease of the peak heights was observed as a function of time, and no more peaks were detectable at 254 nm after half an hour of exposure to direct sunlight or after 1 h to a UV lamp located about 2 cm from the injection glass vials (in this condition, racemization was complete in less than 1 min).

In Figure 1 are shown kinetic experiments in which both racemization and disappearance of peaks are clearly demonstrated (one can notice appearance of new peaks which might correspond to degradation products). In Table 1 are shown the concentrations of THD 5-SO (FE) and THD 5-SO (SE) in the plasma and urine of 11 thioridazine-treated psychiatric patients. A Wilcoxon test indicates that plasma and urine concentrations of THD 5-SO (FE) are significantly higher than THD 5-SO (SE): $p < 0.05$ and $p < 0.01$, respectively. Upon dividing THD 5-SO (FE) by THD 5-SO (SE) concentrations, values ranging from 0.89 to 1.75 in plasma and from 1.15 to 2.05 in urine are obtained.

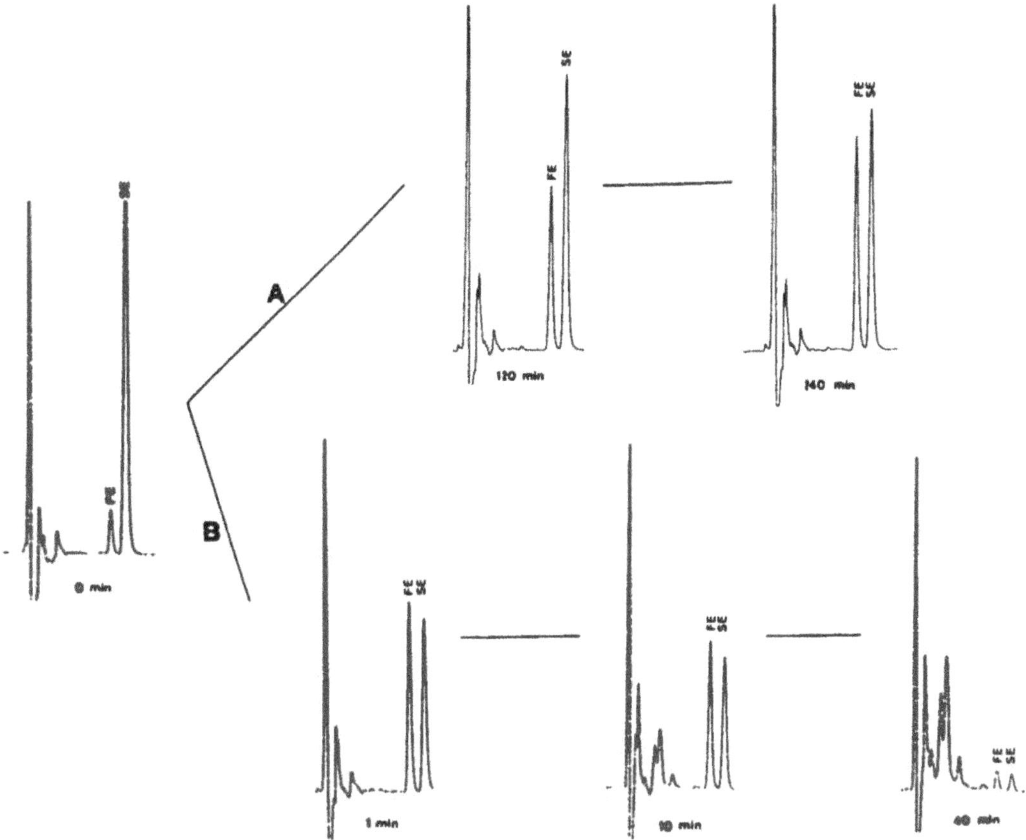

Figure 1 HPLC chromatograms obtained during a kinetic experiment with a glass vial containing slow-eluting (SE) thioridazine 5-sulfoxide diluted in the injection solvent at approximately 5 μg/mL. The vial was stored without any light protection (A) in a room with the window blinds open; (B) at a distance of about 2 cm from a UV lamp. At the indicated time, 100 μL were injected directly into the HPLC (Eap et al., submitted to *Therapeutic Drug Monitoring*). Similar results were obtained with fast-eluting (FE) thioridazine 5-sulfoxide (chromatograms not shown).

IV. DISCUSSION

Our results suggest a potential stereoselectivity in the biotransformation of thioridazine to THD 5-SO. They are in discrepancy with previously published studies which did not find significant differences in the concentrations of the two stereoisomeric pairs in serum and urine from rats and humans (11–16). A possible explanation is that no particular care was taken against light-induced racemization in the former studies.

A potential stereoselectivity in the biotransformation of THD warrants that studies be initiated to check a possible stereoselectivity in the cardiotoxicity of the ring

Table 1 Concentrations (ng/mL) of Thioridazine Ring Sulfoxide Concentrations, as Measured in Plasma (P) and Urine (U) of 11 Psychiatric Patients Medicated with Thioridazine[a]

	Age/Sex	Dose		THD 5-SO (FE)	THD 5-SO (SE)	(FE)/(SE)
1	62/F	30	P	155	96	1.61
			U	772	494	1.56
2	44/F	60	P	161	99	1.63
			U	2294	1121	2.05
3	33/F	90	P	614	692	0.89
			U	3914	3401	1.15
4	42/M	100	P	508	429	1.18
			U	3077	2136	1.44
5	32/M	100	P	323	237	1.36
			U	3965	2281	1.74
6	26/F	100	P	336	268	1.25
			U	1770	1348	1.31
7	37/F	100	P	442	252	1.75
			U	1146	761	1.51
8	20/M	100	P	227	224	1.01
			U	503	424	1.19
9	38/F	125	P	223	187	1.19
			U	2339	1616	1.45
10	19/F	200	P	698	456	1.53
			U	16266	9309	1.75
11	35/F	400	P	650	393	1.65
			U	7092	4069	1.74

[a]Dose: in mg of THD administered per day; M: male; F: female; THD 5-SO; thioridazine 5-sulfoxide; (FE): fast-eluting; (SE): slow-eluting.

sulfoxide stereoisomeric pairs, which are the metabolites found in largest concentration in serum after chronic THD administration (5). Indeed, among all phenothiazine drugs used, abnormalities in the electrocardiogram are more frequent with THD and occur in 50% of all patients treated with this drug (18). THD 5-SO seems to be devoid of antipsychotic effects but contributes to the drug cardiotoxicity (4,6,7) even more potently than the parent compound (8,9); it has already been suggested that the two stereoisomeric pairs may have different pharmacodynamic properties, which may explain discrepant results on the cardiotoxicity of the ring sulfoxide (8).

ACKNOWLEDGMENT

This work was supported in part by the Swiss National Research Foundation (Project No. 32-27579.89).

REFERENCES

1. D. M. Niedzwiecki, R. B. Mailman, and L. X. Cubeddu. Greater potency of mesoridazine and sulforidazine compared with the parent compound, thioridazine, on striatal dopamine autoreceptors. *J. Pharm. Exp. Ther.*, *228*:636–639 (1984).
2. C. D. Kilts, D. I. Knight, R. B. Mailman, E. Widerlöv, and G. R. Breese. Effects of thioridazine and its metabolites on dopaminergic function: Drug metabolism as a determinant of the antidopaminergic actions of thioridazine. *J. Pharm. Exp. Ther.*, *231*:334–342 (1986).
3. D. M. Niedzwiecki, L. X. Cubeddu, and R. B. Mailman. Comparative antidopaminergic properties of thioridazine, mesoridazine and sulforidazine on the corpus striatum. *J. Pharm. Exp. Ther.*, *250*:117–125 (1989).
4. S. G. Dahl. Active metabolites of neuroleptic drugs: Possible contribution to therapeutic and toxic effects. *Ther. Drug Monitor.* *4*:33–40 (1982).
5. E. Martensson, G. Nyberg, R. Axelsson, and K. Serck-Hansen. Quantitative determination of thioridazine and nonconjugated thioridazine metabolites in serum and urine of psychiatric patients. *Curr. Ther. Res.*, *18*:687–700 (1975).
6. L. A. Gottschalk, E. Dinovo, R. Biener, and B. R. Nandi. Plasma concentrations of thioridazine metabolites and ECG abnormalities. *J. Pharmaceut. Sci.*, *67*:155–157 (1978).
7. R. Axelsson and E. Martensson. Side effects of thioridazine and their relationship with the serum concentrations of the drug and its main metabolites. *Curr. Ther. Res.*, *28*:463–489 (1980).
8. P. W. Hale and A. Poklis. Thioridazine 5-sulfoxide cardiotoxicity in the isolated, perfused rat heart. *Tox. Lett.*, *21*:1–8 (1984).
9. A. Heath, C. Svensson, and E. Martensson. Thioridazine toxicity—An experimental cardiovascular study of thioridazine and its major metabolites in overdose. *Vet. Hum. Toxicol.*, *27*:100–105 (1985).
10. E. C. Juenge, C. E. Wells, D. E. Green, I. S. Forrest, and J. N. Shoolery. Synthesis, isolation, and characterization of two stereoisomeric ring sulfoxides of thioridazine. *J. Pharmaceut. Sci.*, *72*:617–621 (1983).
11. P. W. Hale, S. Melethil, and A. Poklis. The disposition and elimination of stereoisomeric pairs of thioridazine 5-sulfoxide in the rat. *Eur. J. Drug Metab. Pharmacokinet.*, *10*:333–341 (1985).
12. P. W. Hale. Thioridazine 5-sulfoxide diastereoisomers in serum and urine from rats and man after chronic thioridazine administration. *J. Anal. Toxicol.*, *9*:197–201 (1985).
13. G. M. Watkins, R. Whelpton, D. G. Buckley, and S. H. Curry. Chromatographic separation of thioridazine sulfoxide and N-oxide diastereoisomers: Identification as metabolites in the rat. *J. Pharm. Pharmacol.*, *38*:506–509 (1986).
14. P. W. Hale and A. Poklis. The biotransformation of thioridazine to thioridazine 5-sulfoxide stereoisomers in phenobarbital induced rats. *Eur. J. Drug Metab. Pharmacokinet.*, *11*:283–289 (1986).
15. A. Poklis, C. E. Wells, and E. C. Juenge. Thioridazine and its metabolites in post mortem blood, including two stereoisomeric ring sulfoxides. *J. Anal. Toxicol.*, *6*:250–252 (1982).
16. C. Svensson, G. Nyberg, M. Soomögi, and E. Martensson. Determination of the serum concentrations of thioridazine and its main metabolites using a solid-phase extraction technique and high-performance liquid chromatography. *J. Chromatogr.*, *529*:229–236 (1990).
17. C. D. Kilts, K. S. Patrick, G. R. Breese, and R. B. Mailman. Simultaneous determination of thioridazine and its S-oxidized and N-demethylated metabolites using high-performance liquid chromatography on radially compressed silica. *J. Chromatogr.*, *231*:377–391 (1982).
18. L. R. Weiss. The cardiotoxicity of neuroleptic and tricyclic antidepressant drugs. In *Cardiac Toxicology, Volume II* (T. Balazs, ed.), CRC Press, pp. 125–143, Boca Raton, FL (1981).

20

Cephalosporins

Milap C. Nahata *The Ohio State University Wexner Institute for Pediatric Research, Children's Hospital, Columbus, Ohio*

I. INTRODUCTION

Cephalosporins are among the most commonly used antibiotics in pediatric and adult patients. These drugs are effective for both prevention and treatment of infections caused by a variety of Gram-positive and Gram-negative bacteria. Nearly 25 cephalosporins have been approved by the U.S. Food and Drug Administration (FDA), and additional agents are available in Europe, Japan, and other parts of the world. These drugs have been grouped under three generations of cephalosporins. It is difficult to assure optimal usage of these drugs for several reasons, including: (a) there are so many agents with similar generic and brand names; (b) the clinically important differences in antimicrobial activity, pharmacokinetics, efficacy, and adverse effects are not clearly known or understood; and (c) there is a tendency to prescribe an expensive cephalosporin in situations where a cheaper alternative antibiotic could be equally effective. This provides an excellent opportunity for a therapeutic drug monitoring (TDM) service to improve patient care.

The purpose of this chapter is to provide a brief review of cephalosporins, and to define the role of TDM in assuring optimal use of these drugs in patients.

II. THREE GENERATIONS

Although the classification of cephalosporins by generation may be somewhat arbitrary, it is based on their antimicrobial activity. In general, first-generation cephalosporins are most active against Gram-positive microorganisms, while third-generation cephalosporins have the greatest activity against Gram-negative bacteria. The FDA-approved drugs, under each generation, are listed in Table 1.

III. ANTIMICROBIAL ACTIVITY

The first-generation parenteral cephalosporins have similar antimicrobial spectrum. These are active against a variety of microorganisms, including *Streptococcus pneumoniae*, *S. pyogenes*, *S. viridans*, *Neisseria gonorrhoea*, and *N. meningidis*. Higher concentrations are required to inhibit Gram-negative bacteria such as *E. coli* and *Proteus mirabilis*. It is important to note that methicillin-resistant staphylococci may be sensitive in vitro to cephalosporins, but the cephalosporins are generally ineffective to

Table 1 Three Generations of Cephalosporins

First Generation	Second Generation	Third Generation
Cefazolin	Cefuroxime	Cefotaxime
Cephalothin	Cefoxitin	Ceftriaxone
Cephapirin	Cefotetan	Ceftizoxime
Cephaloridine	Cefamandole	Ceftazidime
Cephradine	Cefonicid	Cefoperazone
Cephalexin	Ceforanide	Cefmenoxime
Cefadroxil	Cefmetazole	Moxalactam
	Cefaclor	Cefixime
	Cefuroxime axetil	

treat these infections. Oral agents (cephalexin, cefadroxil, cephradine) possess activity against certain Gram-positive (streptococci, staphylococci) and Gram-negative organisms (*E. coli, P. mirabilis, Klebsiella* sp.), but the activity is lower than that of parenteral drugs.

The second-generation parenteral cephalosporins may differ in their activity from one agent to another. For example, cefuroxime and cefamandole have greater activity against *E. coli, Enterobacter* sp., *Klebsiella* sp., *P. mirabilis*, and *Hemophilus influenzae*, while cefoxitin and cefotetan have higher activity against *Bacteroides fragilis*. The oral drugs (e.g., cefaclor and cefuroxime axetil) are active against Gram-negative organisms including *H. influenzae, E. coli, P. mirabilis*, and *K. pneumoniae*.

The third-generation parenteral cephalosporins are generally active against *S. pneumoniae, S. pyogenes, H. influenzae, N. meningitidis, N. gonorrhoea, E. coli, P. mirabilis, K. pneumoniae, Salmonella* sp., and *Shigella* sp. Ceftazidime, however, appears to be most active against *Pseudomonas aeruginosa* and *Acinetobacter*. One oral agent (cefixime) is active against *E. coli, K. Pneumoniae, P. mirabilis, P. vulgaris, H. influenzae, N. gonorrhoea, Providencia, Salmonella* sp., and *Shigella* sp.

IV. PHARMACOKINETICS

The assessment of pharmacokinetics is important in designing optimal dosage regimens of drugs. The parameters, including serum concentrations, cerebrospinal fluid concentrations, urinary concentrations, bioavailability, protein binding, distribution volume, total body clearance, renal clearance, and elimination of half-life, should be determined using accurate, reproducible, and specific analytical methods to characterize the pharmacokinetics of antibiotics in patients of various ages (e.g., premature newborns, full-term newborns, infants, children, adolescents, adults, and elderly) and organ function (e.g., liver or renal disease). These data, however, are not available for many agents, which makes it difficult to compare these drugs. A summary of some pharmacokinetics data is presented in Tables 2 through 5.

As can be seen from these tables, there are some important differences and striking similarities in the pharmacokinetic properties of cephalosporins. In general, the clearance is lowest and distribution volume as well as half-life greatest in the premature newborns compared with older infants, children, and adults. Most of the cephalosporins are eliminated primarily by the kidney, and yet specific dosage

Table 2 Examples of Pharmacokinetic Data of Some First- and Second-Generation Parenteral Cephalosporins

Drug	Clearance (mL/min/kg)	Distribution volume (L/kg)	Urinary excretion (%)	Half-life (h)
Cefazolin	0.85	0.12	88	1.7–2.0
Cephalothin	6.7	0.2–0.9	60	0.3–0.6
Cephapirin[a]		0.23	75	0.4
Cefuroxime	2.2–2.8	0.27–0.65	95	1.2–5.5
Cefoxitin	1.8–6.0	0.16–0.58	80	0.8–3.8
Cefamandole	3.0	0.27–0.45	75	0.8–1.5

[a]Adults only.
Source: Modified from Ref. 1.

Table 3 Examples of Pharmacokinetic Data of Some Third-Generation Parenteral Cephalosporins

Drug	Clearance (mL/min/kg)	Distribution volume (L/kg)	Urinary excretion (%)	Half-life (h)
Cefotaxime	0.8–4.5	0.25–0.60	60	0.8–4.6
Ceftriaxone[a]	0.23–0.83	0.15–0.52	46–70	4.7–17
Ceftazidime	1.0–1.9	0.25–0.54	90	1.8–7.0
Cefoperazone[a]	0.8–1.2	0.17–0.53	25–55	2.1–7.5

[a]Plasma protein binding >90%.
Source: Modified from Ref. 1.

Table 4 Examples of Pharmacokinetic Data of Oral Cephalosporins

Drug	Peak conc. (μg/mL)	Time to peak (h)	Bioavailability	Urinary excretion (%)	Half-life (h)
Cephalexin	9.0–23.4	0.5–3.0	0.9[a]	40–90	0.9–5
Cephradine	9.9–21.3	0.5–1.0	0.9[a]	92[a]	0.7–1
Cefadroxil	10–13.7	1–2	0.94[a]	93[a]	1.3–1.7
Cefaclor	6–13.1	0.5–3	0.9[a]	70[a]	0.6–1
Cefuroxime axetil	3.6–6.2	1.5–2	0.3–0.5[a]	30–40[a]	1.3–1.4
Cefixime	1–4	2.7–4.3	0.4–0.5[a]	20[a]	3–4.5

[a]Adults only.
Source: Modified from Ref. 1.

Table 5 CSF Penetration of Some Cephalosporins

Drug	CSF Penetration (% of serum conc.)[a]
Cefuroxime	14–18
Cefotaxime	20–30
Ceftriaxone	17
Ceftazidime	25
Moxalactam	20–30
Cefixime	6–9

[a]These percentages are less important than the actual CSF concentrations relative to its MIC for pathogens.
Source: Modified from Ref. 1.

guidelines in certain patients (e.g., infants and children) with renal dysfunction are not available. Similarly, bioavailability and certain pharmacokinetic data have not been reported for pediatric and elderly patients. These data should be considered for TDM of cephalosporins.

V. COMPARATIVE EFFICACY

Limited comparative studies have been performed among cephalosporins to demonstrate their relative efficacy. Thus it is difficult to contrast efficacy of cefazolin versus cephalothin, cephalexin versus cefadroxil, cefoxitin versus cefotetan, or cefotaxime versus ceftriaxone. In such cases, the selection of one agent over another is primarily on relative safety, convenience of dosing (once or twice versus three or four times daily), and cost. In other cases, cefuroxime is selected over cefamandole to treat systemic *H. influenzae* infections in pediatric patients because the latter may not penetrate into the CSF adequately. Some comparative studies are available suggesting that ceftriaxone would be a better agent that cefuroxime to treat bacterial meningitis in infants and children. Single cephalosporins (e.g., ceftriaxone and cefotaxime) have been compared with traditional drug combinations (e.g., ampicillin plus chloramphenicol) to show equivalent efficacy in bacterial meningitis in infants and children.

VI. ADVERSE EFFECTS AND DRUG INTERACTIONS

Cephalosporins are among the safest groups of antimicrobials. However, certain adverse effects have been associated with some agents. Hypoprothrombinemia and bleeding has been caused by several third-generation cephalosporins. Moxalactam, cefmenoxime, cefmetazole, cefotetan, cefoperazone, and cefamandole have methyl-5-thiotetrazole as part of their chemical structure, but coagulation defects have been reported most frequently with moxalactam. Coagulopathies are treated with vitamin K. Displacement of bilirubin from albumin binding sites has been caused in vitro by ceftriaxone, cefonicid, cefoperazone, cefotetan, and cefmetazole. Diarrhea has been reported with ceftriaxone and cefoperazone, although this has not required alteration in therapy. Biliary pseudolithiasis may occur in some patients receiving ceftriaxone. Serum sickness has been associated with cefaclor. Finally, alcohol intolerance similar to disulfiram reaction has been caused most commonly by moxalectam. Thus, alcohol or medications containing alcohol should be avoided in patients receiving moxalactam or structurally related cephalosporins.

Although no major drug interactions have occurred with cephalosporins, concurrent administration of aminoglycosides or loop diuretics with cephalothin and cephaloridine may increase the risk of nephrotoxicity. Combinations of cephalosporins may be antagonistic for antibacterial activity.

High concentrations of drugs including cefadroxil, cefaclor, cefamandole, ceforanide, cefoxitin, and cefotaxime may cause false-positive results if the copper-reduction method (Clinitest) is used for testing glycosuria. Glucose-oxidase tests (Diastix and Tes-Tape) can be used without any interaction with any of these drugs. Serum creatinine determinations by Jaffe's method can also give falsely elevated values, particularly when cephalothin and cefoxitin are being used in patients. Blood samples should be obtained at least 2 h after drug administration to patients with normal renal function; this method should not be used in patients with renal dysfunction.

VII. PATIENT CONVENIENCE AND COST

Successful therapy depends on maximizing efficacy; minimizing adverse effects, microbial resistance, or interactions; and offering the most convenient dosage schedule (which may improve compliance) at the lowest cost of therapy. Cefazolin can be given three times daily compared with the four-times-daily regimen of cephalothin. Ceftriaxone is the only third-generation parenteral cephalosporin which can be given once daily to treat infections except bacterial meningitis, which may require twice-daily dosing. Cefixime can also be given once daily to treat certain infections. However, the potential for missing a dose by a patient, who would not be covered for 24 h, should be considered. Antibiotic therapy at home is an attractive alternative to therapy in hospitals for the treatment of certain infections.

The cost of therapy should include costs associated with drug administration sets (e.g., tubing, syringe, pump), and nursing or pharmacy services. Relative total costs should be calculated for a typical course of therapy in order to select the least expensive drug. However, this should be put in proper perspective with relative efficacy, toxicity, and convenience of drug therapy.

VIII. ANTIBIOTIC UTILIZATION

Antibiotics account for one-third to one-half of the drug budget in hospitals. Thus, these drugs are targeted for evaluating overall usage for both prevention and treatment of infections. The literature is replete with reports of inappropriate use of antibiotics. The common problems include:

1. Inappropriate selection (e.g., use of cephalosporin when an alternative could be equally or more effective)
2. Prolonged use (e.g., use of a cephalosporin for prophylaxis for longer than 1 to 2 days)
3. Incorrect prescribing (e.g., use of cefazolin four times rather than three times daily)

The suggestions for improving the use of cephalosporins may include implementation of an antibiotic order form, automatic stop order, selective formulary, limited susceptibility reports, and ongoing utilization review of antibiotic usage and microbial susceptibility pattern. The TDM service can assist in conducting such evaluations.

IX. SUMMARY

Cephalosporins are useful antibiotics for the management of pediatric and adult patients. There are many similarities and clinically important differences within and among the three generations of cephalosporins. Although many of these drugs have been studied to a certain degree, additional studies are needed to determine their pharmacokinetics (particularly in newborns and the elderly), comparative efficacy and safety in all groups of patients. Further, the influence of renal or liver dysfunctions on dosage requirements in pediatric patients is not known for many cephalosporins. Effect of gastrointestinal disease on pharmacokinetics and dosage requirements is also incompletely understood. It would also be important to find alternatives to parenteral cephalosporins; oral drugs could be used more conveniently and safely in both hospitals and at home.

TDM personnel should assure optimal clinical use of cephalosporins. The routine serum concentration determination of cephalosporin is not necessary. However, a joint effort among infectious disease and other physicians, nurses, pharmacists, microbiologists, epidemiologists, i.v. therapists, and TDM laboratory personnel is most likely to achieve the best results. These efforts should be coordinated among different departments and committees such as a pharmacy and therapeutics committee, an infection control committee, and an antibiotic utilization review committee.

REFERENCES

1. J. S. Leeder and R. Gold. Cephalosporins in antimicrobial therapy in infants and children. (G. Koren, C. G. Prober, and R. Gold, eds.), 1988. Marcel Dekker, New York, pp. 173–235 (1988).
2. J. J. Marr, H. J. Moffet, and C. M. Kunin. Guidelines for improving the use of antimicrobial agents in hospitals: A statement by the Infectious Diseases Society of America. *J. Infect. Dis., 157*:869–876 (1988).
3. C. Paap, M. C. Nahata, M. Mentser, and J. Mahan. Pharmacokinetics of cefotaxime and its active metabolite in children with renal dysfunction. *J. Clin. Pharmacol.* (in press).
4. C. Paap and M. C. Nahata. Clinical pharmacokinetics of antibacterial drugs in neonates. *Clin. Pharmacokinet., 19*:280–318 (1990).
5. M. C. Nahata, D. E. Durrell, and W. J. Barson. Pharmacokinetics and cerebrospinal fluid concentrations of cefixime in infants and children with bacterial meningitis. *Pharmacotherapy, 10*:238 (1990).
6. M. C. Nahata, N. Freimer, and M. D. Hilty. Decreased absorption of cefaclor in short bowel syndrome. *Drug. Intell. Clin. Pharm., 17*:201–202 (1983).
7. H. C. Neu. Third generation cephalosporins: Safety profiles after 10 years of clinical use. *J. Clin. Pharmacol., 30*:396–403 (1990).

21

Therapeutic Drug Monitoring for 6-Mercaptopurine: What, When, and How?

Pierre Lafolie *Karolinska Hospital, Stockholm, Sweden*

I. INTRODUCTION: THE WHAT

The purine analog 6-mercaptopurine (6-MP) was synthesized by Hitchings and Elion in the late 1940s (1). The first clinical trial was conducted with an oral regimen, and this trial showed a beneficial effect in the treatment of acute leukemia in children (2). In this report, intravenous administration of 6-MP did not show a better effect than the oral administration. The doses used were obtained from trials in laboratory animals (1). Later the pharmacologic disposition was described (3). It was postulated that the main cytotoxic of effect 6-MP was mediated by negative feedback on the purine de-novo synthesis (4) (Figure 1). With the increased use of 6-MP, more experience was achieved during the next decade. In the 1960s the drug was established as the backbone in orally administrated maintenance therapy of acute lymphoblastic leukemia (ALL) (5,6). *However, it is noteworthy that clinical trials, as they are performed today, have not been used in the development of oral and long-term 6-MP maintenance treatment.* The problems with 6-MP as it is used today are therefore: (a) Is the treatment effective at all? (b) Are there ways to make it more effective? (c) What are those ways?

II. THE ORAL REMISSION MAINTENANCE THERAPY

Until the early 1960s the prognosis in childhood ALL was very poor. Most patients died without achieving a remission. With the introduction of better immunological subgrouping, multidrug combinations, and better supportive care, 90% of patients today achieve a complete remission (7,8). Of these, 40% later relapse (9). The greatest problem in the treatment of ALL today is the prevention of relapse, which most often occurs in the CNS, in the testes, or in the bone marrow (pharmacological sanctuaries) (7). Maintenance therapy, aiming to reduce the relapse risk, includes oral methotrexate (MTX) once weekly and oral 6-MP once daily. It has been presented in many reports (10–16). Oral maintenance treatment with 6-MP and MTX reduces relapses in ALL, and continuous maintenance therapy has been shown to be more effective than intermittent maintenance therapy (17–19).

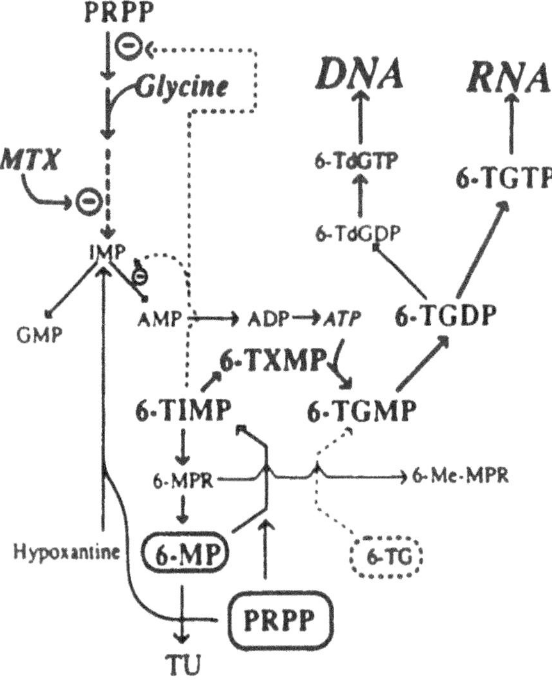

Figure 1 Purine synthesis.

III. CRITICISMS OF ANTINEOPLASTIC ORAL THERAPY

This oral and long-term treatment with a cytostatic drug has been criticized (20,21), comparing the treatment with antimicrobiological chemotherapeutic treatment and claiming that the treatment should be administered intravenously to achieve maximum tolerated dose (22). This has led to ongoing trials with intravenously administered 6-MP (23) and a discussion addressing the question whether or not the maintenance treatment is suboptimal or necessary (24).

IV. OTHER AREAS OF ANTINEOPLASTIC ORAL MAINTENANCE THERAPY

Despite the criticisms mentioned above, oral maintenance treatment with 6-MP has not changed. Positive effects of maintenance treatment with 6-MP and MTX have also been reported in other leukemias, e.g., promyelocytic leukemia (25,26). A possible explanation for the synergistic effect of MTX and 6-MP (Figure 1) has been proposed by Bökkerink et al., as MTX seems to raise the levels of PRPP and thereby the utilization of 6-MP (27,28).

V. LOSS OF 6-MP EFFECT

Loss of therapeutic effect of 6-MP is not easily measured until the relapse is at hand. It has been reported that sulphoxidation might reduce the effect of 6-MP (29). Resistance in children with ALL due to inherited or acquired deficiency in 5'-nucleotidase, leading to low levels of phosphorylated metabolites, has also been proposed (30).

VI. THERAPEUTIC MONITORING OF DRUGS

Therapeutic drug monitoring, as a part of clinical pharmacology, is generally concerned with orally administered drugs. For many orally administered drugs (e.g., digoxin, theophylline) it has been possible to establish correlations between the drug concentration in plasma and the clinical effect. This is true even when the final action on a molecular level is unknown, e.g., for phenytoin. It has been shown that dose adjustments based on measurements in an alternative compartment, such as plasma, instead of awaiting clinical signs of toxicity or underdosing, reduces the cases of therapeutic failure (21–33). *This is certainly true for drugs with pronounced interindividual variation, where a standard dose may produce greatly variable concentrations.*

VII. THERAPEUTIC MONITORING OF ANTINEOPLASTIC DRUGS

Most treatment schedules with anticancer drugs use standardized doses. The doses are often calculated from the patient's body weight or body surface area. This routine does not take interindividual variability in pharmacokinetics into account, e.g., differences in bioavailability after oral dosing or in clearance, which might affect drug effects. Hardly any protocol today includes target plasma concentrations or production of cytotoxic metabolites as rationale for dosing. There are only a few studies that have tried to correlate the use of anticancer drugs to pharmacokinetic data with application to clinical outcome. One exception is a work by Plunkett et al. They show that the intracellular concentration of ara-CTP, the active metabolite of ara-C, may predict the response to treatment of relapsed or primary refractory acute leukemia with high-dose i.v. ara-C (34). Recently it has become obvious that application of the modern principles of clinical pharmacology to oncological and hematological problems makes it possible to individualize and improve treatment in order to reduce side effects and increase therapeutic effectiveness (8,15,20,21,35–44).

VIII. REASONS FOR INTERINDIVIDUAL VARIATION IN 6-MP PHARMACOKINETICS

One reason for relapse may be individual variability in the pharmacokinetics. This interindividual variability in 6-MP pharmacokinetics has been described by different groups (45–47). Defective patient compliance may be another source for this interindividual variation. Failing patient compliance has been suspected as a source of relapse (48). Another source for relapse may be a lack in physician compliance (49), leading to unnecessary dose reductions and undertreatment. We found a 10-fold interindividual variation in peak concentration and AUC. This is seen even after a linear transformation of the results to the scheduled dose of 75 mg/m^2 (45). This indicates that the difference in doses between the patients is not the major reason for the variability. Zimm et al. (47) reported a bioavailability of 5 to 16%, comparing oral and intravenous doses. This low, variable, and, in view of old data, new aspect on 6-MP bioavailability indicates more pronounced first-pass effect than was previously believed. Another reason may be laboratory variation, but in no report has this been so pronounced that it affected the results reported.

IX. 6-MP AND INTAKE OF FOOD

Two studies (50,51) have reported higher AUC and peak concentrations when 6-MP was administered on an empty stomach. They reached these results differently, Riccardi et al. by giving the same dose to all children and Burton et al. by administering different doses to each child. Both measured each patient only once in each state (fasted/after food) and pharmacokinetics after the oral dosing was calculated. We repeated the measurements twice in every patient (52). Among our patients, the interindividual variation was reduced when 6-MP was administered together with food. The reason for this is unknown, but it could be due to an occupation of xanthine oxidase in the gut wall or in the liver by purines in the food.

X. INTRAINDIVIDUAL VARIATION OF 6-MP PHARMACEUTICS

With repeated measurements in the same patient, a pronounced intraindividual variation has been found (52). This kind of variation has also been reported for other drugs (53–55).With regard to 6-MP, only Poplack and Balis (56) seem to have discussed this after repeated drug concentration measurements in one patient. The intraindividual variation may be an important determinant for the clinical outcome in the individual patient. The intraindividual variation was also reduced by administration of 6-MP together with food. After the initial finding, this problem was further studied in more standardized conditions (57). The intraindividual variation appeared to be normally distributed, with a mean C.V. in plasma AUC of 57.9% (Figure 2). Calculating mean values of different courses showed that there were no systemic changes with increasing number of courses. It is concluded that the intraindividual variation is pronounced in some patients and may constitute a background for therapeutic failures.

XI. INDICATION OF A RELATIONSHIP BETWEEN 6-MP PLASMA CONCENTRATIONS AND CLINICAL OUTCOME

The children in our report were studied repeatedly. As AUC takes both time and concentration into account, this parameter was calculated after each course in each child. Due to the intraindividual variation, mean AUC for each child was calculated for the whole maintenance treatment period (58). It was found that five children who relapsed during maintenance therapy had a mean AUC < 270 ng/mL/h and that the outcome was significantly related to each patient's mean plasma AUC ($p < 0.005$). Four patients developed toxicity, expressed as a severe bone marrow depression. They had a mean plasma AUC of > 370 mg/mL/h. They were all restituted after temporary withdrawal of 6-MP. Among the event-free patients were individuals who had values below 270 ng/mL/h, but also above 370 ng/mL/h. The mean AUC in this event-free group was 315 ng/mL/h (Figure 3). Two children relapsed *after* the end of the maintenance treatment period. They showed high concentrations during the first year of treatment, but considerably lower toward the end of the treatment period (Figure 4). Whether these relapses were due to momentarily or persistent low concentrations or other factors cannot be evaluated from our results. A relationship between clinical outcome and plasma concentrations thus seems possible, indicating a beneficial effect of 6-MP in maintenance treatment and the possibility of using plasma determinations of 6-MP in monitoring therapy.

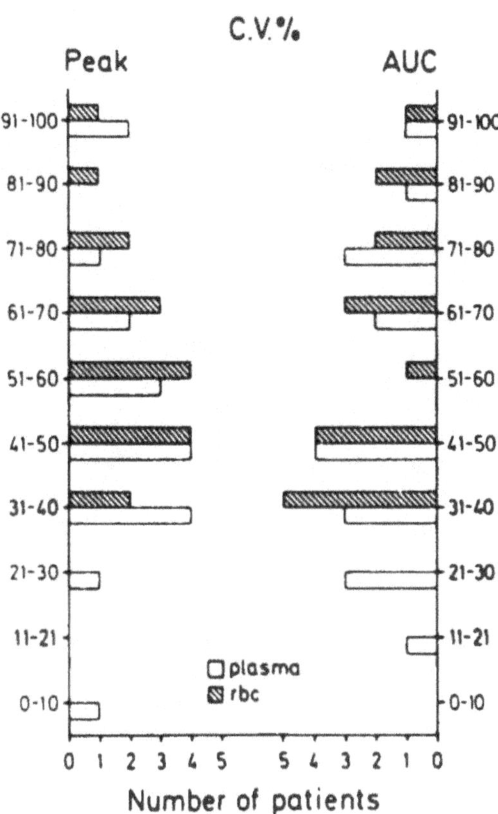

Figure 2 Intraindividual variation of 6-MP pharmacokinetics.

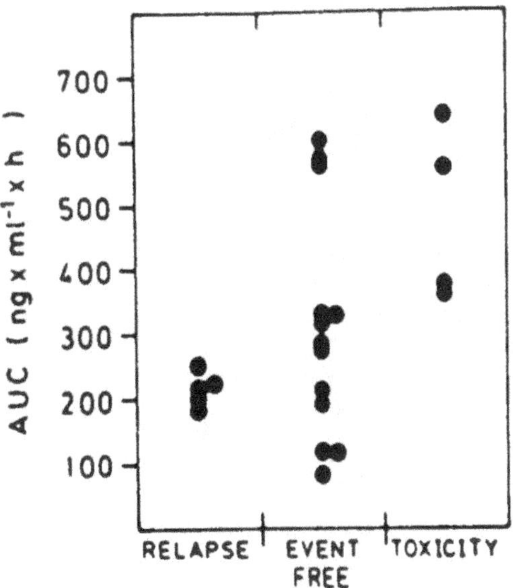

Figure 3 Relationship between the mean plasma $AUC_{0\text{-}4h}$ and the outcome of maintenance therapy.

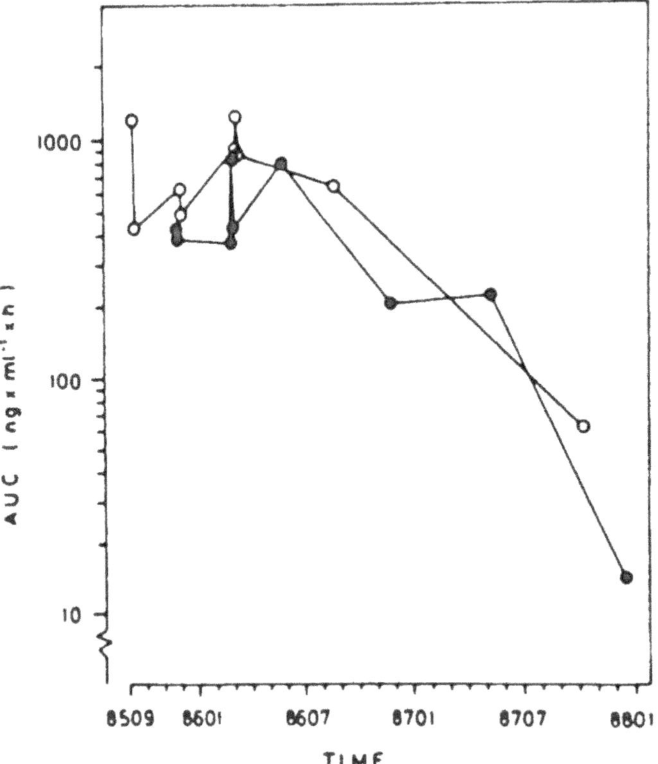

Figure 4 Plasma AUC values at various times during the maintenance period in patients 15 (open circles) and 16 (filled circles), replapsing 6 and 11 months, respectively, after termination of maintenance treatment.

XII. THERAPEUTIC MONITORING OF 6-MP?

Different ways to correlate 6-MP treatment with clinical outcome have been tried. Lennard et al. reported a correlation between neutropenia and concentrations of 6-thioguanosine-nucleotides (6-TGN) in red blood cells (59). Later they studied the relationship between 6-TGN in red blood cells and plasma concentrations of 6-MP, but they could not find any correlation (60). Recently it has been reported that there is a correlation between 6-thioguanosine nucleotides in red blood cells and clinical outcome (61). The relapse rate during the maintenance treatment and the pronounced interindividual variation thus indicate a possibility of individualizing treatment by measuring drug concentrations. However, effective therapeutic levels in vivo are difficult to predict. Different in vitro studies have reported different concentrations to be cytotoxic (27,62,63).

XIII. MAY BLOOD SAMPLING BE PERFORMED MORE EASILY FOR THE PATIENT?

Repeated sampling over several hours is not optimal from a nursing point of view. In 18 patients the concentration in each plasma sample was correlated to the patients'

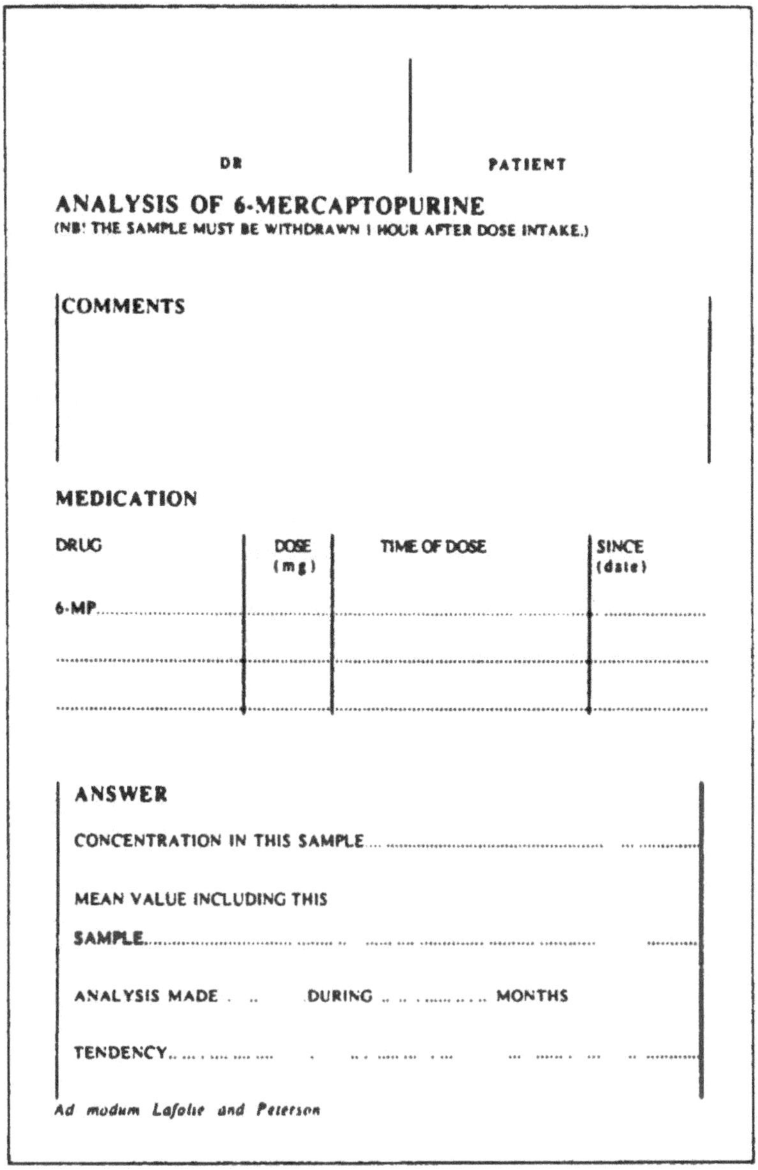

DR | PATIENT

ANALYSIS OF 6-MERCAPTOPURINE
(NB! THE SAMPLE MUST BE WITHDRAWN 1 HOUR AFTER DOSE INTAKE.)

COMMENTS

MEDICATION

DRUG	DOSE (mg)	TIME OF DOSE	SINCE (date)
6-MP			

ANSWER

CONCENTRATION IN THIS SAMPLE ...

MEAN VALUE INCLUDING THIS SAMPLE ...

ANALYSIS MADE ... DURING ... MONTHS

TENDENCY ...

Ad modum Lafolie and Peterson

Figure 5 Sample remittance order.

achieved plasma AUC$_{0-4\ h}$ on the same day. It was found that samples withdrawn at 1.0 h after 6-MP intake on the day before MTX intake correlated to achieved AUC$_{0-4\ h}$ ($r = 0.79$, $p = 0.0002$). If withdrawn at the day of MTX intake, the correlation between the single blood sample and achieved AUC$_{0-4\ h}$ was slightly different ($r = 0.86$, $p = 0.0001$) (64). Single plasma measurements thus may predict plasma AUC$_{0-4\ h}$ in the individual patient. For practical reasons, to shorten time at the outpatient unit, measurements at 1.0 seem appropriate. Measuring 6-MP at 1 h after 6-MP dose thus seems reasonable to predict $AUC_{0-4\ h}$, and use of this may be of value for future studies on the value of therapeutic drug monitoring of 6-MP. By repeated analyses and calculation of mean results, intraindividual variation can be taken into account.

XIV. THE WHEN AND HOW

Our suggestions indicate the need for a remittance order. This order should consider time on treatment, administered dose, time of last dose, time of sampling, actual concentration, cumulative mean concentration, *and tendency*. An example is shown in Figure 5. The single test could be performed when best suited, e.g., every third month. Interpretations of the data should be made with care, especially taking clinical status into account. These recommendations are seen as a background for further trials, to make a prospective study for evaluating the possibility of 6-MP as a future possibility for therapeutic drug monitoring.

REFERENCES

1. G. Elion, G. Hitchings, and H. Vander-Weff. Antagonists of nucleic acid derivatives. VI. Purines. *J. Biol. Chem., 192*:505 (1951).
2. J. Burchenal, M. Murphy, R. Ellison, M. Sykes, T. Tan, et al. Clinical evaluation of a new antimetabolite, 6-mercaptopurine, in the treatment of leukemia and allied diseases. *Blood, 8*:965–999 (1953).
3. H. Skipper. On the mechanism of action of 6-mercaptopurine. *Ann. N.Y. Acad. Sci., 60*:315–321 (1954).
4. L. Hamilton and G. Elion. The fate of 6-mercaptopurine in man. *Ann. N.Y. Acad. Sci., 60*:304–314 (1954).
5. G. Elion, S. Callahan, H. Nathan, S. Bieber, R. Rundles et al. Potentiation by inhibition of drug degradation: 6-Substituted purines and xanthine oxidase. *Biochem. Pharmacol., 12*:85–93 (1963).
6. G. Elion. Biochemistry and pharmacology of purine analogues. *Fed. Proc., 26*:898–903 (1967).
7. A. Mauer. Therapy of acute lymphoblastic leukemia. *Blood, 56*:1–10 (1980).
8. D. Poplack, F. Balis, and S. Zimm. The pharmacology of orally administered chemotherapy. A reappraisal. *Cancer, 58*:473–480 (1986).
9. B. Bell and V. Whitehead. Chemotherapy of childhood acute lymphoblastic leukemia. *Dev. Pharmacol. Ther., 9*:145–179 (1986).
10. F. Balis, S. Jeffries, B. Lange, R. Murphy, K. Doherty, et al. Chronopharmacokinetics of oral methotrexate and 6-mercapto-purine: Is there diurnal variation in the disposition of antileukemic therapy?, *Am. J. Pediatr. Hematol. Oncol., 11*:324–326 (1989).
11. W. Bleyer, J. Drake, and B. Chabner. Neurotoxicity and elevated cerebrospinal-fluid methotrexate concentration in meningeal leukemia. *N. Engl. J. Med., 289*:770–773 (1973).
12. W. Bleyer and R. Dedrick. Clinical pharmacology of intrathecal methotrexate. I. Pharmacokinetics in non-toxic patients after lumbar injection. *Cancer Treat. Rep., 61*:703–708 (1977).
13. W. Bleyer and D. Poplack. Intraventricular versus intralumbar methotrexate for central-nervous-system leukemia: Prolonged remission with the Ommaya reservoir. *Med. Pediatr. Oncol., 6*:207–213 (1979).
14. W. Bleyer. Cancer chemotherapy in infants and children. *Pediatr. Clin. N. Am.,* 32:557–574 (1985).
15. J. Collins. Pharmacologic rationale for regional drug delivery. *J. Clin. Oncol.,* 2:498–504 (1984).
16. D. Covell, P. Narang, and D. Poplack. Kinetic model for disposition of 6-mercaptopurine in monkey plasma and cerebrospinal fluid. *Am. J. Physiol., 248*:147–156 (1985).
17. R. Aur, J. Simone, M. Verzosa, H. Hustu, L. Barker, et al. Childhood acute lymphoblastic leukemia. Study VIII. *Cancer,* 42:2133 (1978).

18. S. Koizumi, T. Fujumoto, T. Takeda, M. Yatabe, J. Utsumi, et al. Comparison of intermittent or continuous methotrexate plus 6-mercaptopurine in regimens for standard-risk acute lymphoblastic leukemia. (JCCLSG-S811). *Cancer, 61*:1292–1300 (1988).
19. D. Lonsdale, E. Gehan, D. Fernbach, M. Sullivan, D. Lane, et al. Interrupted vs. continued maintenance therapy in childhood acute leukemia. *Cancer, 36*:341–352 (1975).
20. E. Gehan. Dose-response relationship in clinical oncology. *Cancer, 54*:1204–1207 (1984).
21. G. Powis. Anticancer drug pharmacodynamics. *Cancer Chemother. Pharmacol., 14*:177–183 (1985).
22. J. Goldie and A. Coldman. The genetic origin of drug resistance in neoplasms: Implications for systemic therapy. *Cancer Res., 44*:3643–3653 (1984).
23. B. Camitta, B. Leventhal, S. Lauer, J. Shuster, S. Adair, et al. Intermediate-dose intravenous methotrexate and mercaptopurine therapy for non-T, non B acute lymphocytic leukemia of childhood: A Pediatric Oncology Group Study, *J. Clin. Oncol., 7*:1539–1544 (1989).
24. L. Lennard. Are children with lymphoblastic leukemia given enough 6-mercaptopurine?, *Lancet, ii*:785–787 (1987).
25. H. Kantarjian, M. Keating, R. Walters, T. Smith, K. McCredie, et al. Role of maintenance chemotherapy in acute promyelocytic leukemia. *Cancer, 59*:1258–1263 (1987).
26. P. Cassileh, D. Harrington, J. Hines, M. Oken, J. Mazza, et al. Maintenance chemotherapy prolongs remission duration in adult acute nonlymphocytic leukemia. *J. Clin. Oncol., 6*:583–487 (1988).
27. J. Bökkerink, M. Bakker, T. Hulsche, R. de Abreu, E. Schretlen, et al. Sequence-, time- and dose-dependent synergism of methotrexate and 6-mercaptopurine in malignant human T-lymphoblasts. *Biochem. Pharmacol., 35*:3549–3555 (1986).
28. J. Bèkkerink, M. Bakker, T. Hulscher, R. de Abreu, and E. Schretlen. Purine de novo synthesis as the basis of synergism of methotrexate and 6-mercaptopurine in human malignant lymphoblasts of different lineages. *Biochem. Pharmacol., 37*:2321–2327 (1988).
29. S. Mitchell, R. Waring, C. Haley, J. Idle, and R. Smith. Genetic aspects on the polymodally distributed sulphoxidation of S-carboxymethyl-L-cysteine in man. *Br. J. Clin. Pharmacol., 18*:507–521 (1984).
30. R. Pieters, D. Huismans, and A. Veerman. Are children with lymphoblastic leukemia resistant to 6-mercaptopurine because of 5'-nucleotidase?, *Lancet, ii*:1471 (1987).
31. J. Koch-Weser, L. Duhme, and D. Greenblatt. Influence of serum digoxin concentration measurements on frequency of digitotoxicity. *Clin. Pharmacol. Ther., 16*:284–287 (1974).
32. W. Taylor and A. Finn. Individualizing drug therapy. Gross, Townsend, Frank, New York, pp. 1–30 (1981).
33. D. Mungall, J. Marshall, D. Penn, et al. Individualizing theophylline therapy: The impact of clinical pharmacokinetics on patients outcome. *Ther. Drug. Monitor., 5*:95–101 (1983).
34. W. Plunkett, S. Iacoboni, E. Estey, I. Danhauser, J. Liliemark, et al. Pharmacologically directed ara-C therapy for refractory leukemia. *Sem. Oncol., 12*:20–30 (1985).
35. F. Balis. Pharmacokinetic drug interactions of commonly used anticancer drugs. *Clin. Pharmacokin., 11*:223–235 (1986).
36. J. Bertino. Blood levels of chemotherapeutic agents and clinical outcome. *J. Clin. Oncol., 5*:996 (1987).
37. W. Brade, E. Freirich, and A. Goldin. Dose-response relationship in clinical oncology. *Cancer, 54*:1226–1228 (1984).
38. L. Brox, L. Birkett, and A. Belch. Clinical pharmacology of oral thioguanine in acute myelogenous leukemia. *Cancer Chemother. Pharmacol., 6*:35–38 (1981).
39. W. Evans, W. Crom, J. Sinkule, G. Yee, G. Stewart, et al. Pharmacokinetics of anticancer drugs in children. *Drug Metabol. Rev., 14*:847–886 (1983).
40. J. Grygiel and D. Raghavan. Clinical pharmacology and cancer therapy: An evolving interface?, *Med. J. Austral., 145*:458–463 (1986).
41. J. Collins, Improving the use of anticancer drugs: Clinical pharmacokinetic approaches. *Isr. J. Med. Sci., 24*:483–487 (1988).
42. J. Collins, D. Zaharko, R. Dedrick, and B. Chabner. Potential roles for preclinical pharmacology in phase 1 clinical trials. *Cancer Treat. Rep., 70*:73–80 (1986).
43. C. Erlichman, R. Donehower, and B. Chabner. The practical benefits of pharmacokinetics in the use of neoplastic agents. *Cancer Chemother. Pharmacol., 4*:139–145 (1980).

44. R. Rosenthal. The impact of pharmacokinetics on cancer chemotherapy. *Vet. Clin. N. Am., 18*:1133–1139 (1988).
45. P. Lafolie, S. Hayder, O. Björk, L. Ahström, J. Liliemark, et al. Large inter-individual variations in the pharmacokinetics of oral 6-mercaptopurine in maintenance therapy of children with acute leukemia and non-Hodgkin lymphoma. *Acta Pediatr. Scand., 75*:797–803 (1986).
46. G. Lönnerholm, B. Lindström, J. Ludvigsson, and U. Myrdal. Plasma and erythrocyte concentrations after oral administration in children. *Pediatr. Hematol. Oncol., 3*:27–35 (1986).
47. S. Zimm, J. Collins, R. Riccardi, D. O'Neill, P. Narang, et al. Variable bioavailability of oral mercaptopurine: Is maintenance chemotherapy in acute lymphoblastic leukemia being optimally delivered?, *N. Engl. J. Med., 308*:1005–1009 (1983).
48. W. Snodgrass, S. Smith, R. Trueworthy, T. Vats, P. Klopovich, et al. Lack of compliance as a factor in leukemia relapse. *Proc. Am. Soc. Clin. Oncol.*, C-794 (1984).
49. M. Peeters, G. Koren, D. Jakubovicz, and A. Zipursky. Physician compliance and relapse rates of acute lymphoblastic leukemia in children. *Clin. Pharmacol. Ther., 43*:228–232 (1984).
50. N. Burton, M. Barnett, G. Aherne, J. Evans, I. Douglas, et al. The effect of food on the oral administration of 6-mercaptopurine. *Cancer Chemother. Pharmacol., 18*:90–91 (1986).
51. R. Riccardi, F. Balis, P. Ferrara, A. Lasorella, D. Poplack, et al. Influence of food intake on bioavailability of oral 6-mercaptopurine in children with acute lymphoblastic leukemia. *Pediatr. Hematol. Oncol., 3*:319–324 (1986).
52. S. Hayder, P. Lafolie, O. Björk, L. Åhström, and C. Peterson. Variability of 6-mercaptopurine pharmacokinetics during oral maintenance therapy of children with acute leukemia. *Med. Oncol. Tumor Pharmacother., 6*:259–265 (1988).
53. A. Grahnén, N. Hammarlund, and T. Lundqvist. Implications of intraindividual variability in bioavailability studies of furosemid. *Eur. J. Clin. Pharmacol., 27*:595–602 (1984).
54. L. Wagner. Intrasubject variation in elimination half-lives of drugs which are appreciably metabolized. *J. Pharmacokin. Biopharm., 1*:165–173 (1973).
55. R. Upton, J. Thiercelin, T. Guentert, S. Wallace, J. Powell, et al. Intraindividual variability in theophylline pharmacokinetics: Statistical verification in 39 of 60 healthy young adults. *J. Pharmacokin. Biopharm., 10*:123–134 (1982).
56. D. Poplack and F. Balis. The influence of route on administration on the pharmacology of antineoplastic drugs: Lessons learned from evaluation of orally administered chemotherapy. *Proceedings in Pediatric Oncology. The Role of Clinical Pharmacology.* SIOP, Trondheim, pp. 103–110 (1988).
57. P. Lafolie, S. Hayder, O. Björk, and C. Peterson. Intraindividual variations in 6-mercaptopurine pharmacokinetics during oral maintenance therapy in children with acute lymphoblastic leukemia. *Eur. J. Clin. Pharmacol.* (in press).
58. S. Hayder, P. Lafolie, O. Björk, and C. Peterson. 6-Mercaptopurine plasma levels in children with acute leukemia: Relation to relapse risk and myelotoxicity. *Ther. Drug. Monitor., 11*:617–622 (1989).
59. L. Lennard, C. Rees, J. Lilleyman, and J. Maddocks. Childhood leukemia: A relationship between intracellular 6-mercaptopurine metabolites and neutropenia. *Br. J. Clin. Pharmacol., 16*:359–363 (1983).
60. L. Lennard, D. Keen, and J. Lilleyman. Oral 6-mercaptopurine in childhood leukemia: Parent drug pharmacokinetics and active metabolite concentrations. *Clin. Pharmacol. Ther., 40*:287–292 (1986).
61. L. Lennard and J. Lilleyman. Variable mercaptopurine metabolism and treatment outcome in childhood lymphoblastic leukemia. *J. Clin. Oncol., 7*:1816–1823 (1989).
62. D. Tidd and A. Paterson. Distinction between inhibition of purine nucleotide synthesis and the delayed cytotoxic reaction of 6-mercaptopurine. *Cancer Res., 34*:733–737.
63. J. Liliemark, B. Petterson, B. Engberg, P. Lafolie, M. Masquelier, et al. On the paradoxically concentration-dependent metabolism of 6-mercaptopurine in WEHI-3b murine leukemia cells. *Cancer Res., 50*:108–112 (1990).
64. P. Lafolie. Clinical pharmacological aspects of 6-mercaptopurine in maintenance therapy of childhood leukemia. Dissertation at Karolinska Institute, Stockholm, Sweden (1990).

22

Selection of Optimal Specifications for Cyclosporine Monitoring: A Study of Variation in Pharmacokinetic Parameters in Response to Dose Changes and Prediction of Adverse Clinical Events Using Four Different Assay Methods

Kimberly L. Napoli, Joachim Grevel, and Barry D. Kahan
University of Texas Health Science Center, Houston, Texas

I. INTRODUCTION

Since the inception of the cyclosporine (CsA) era in the field of solid organ transplantation, validation of CsA monitoring by correlation of levels in bodily fluids to dose and to prediction of clinical events has been a hotly researched subject. Wide intra- and interindividual variations in drug disposition hallmark inconsistent reliability of CsA levels for indication of drug therapy and clinical status. In addition, CsA monitoring strategies are incongruent. Despite the call by the American Association of Clinical Chemistry task force on CsA monitoring (1) and again by the consensus report from the Hawk's Cay Expert Workshop on CsA monitoring (2), numerous variables have precluded standardization of assay methodologies (i.e., use of either whole blood or plasma/serum, use of nonspecific or specific assay method, use of a single trough or multiple samples collected over the entire dosing interval).

This CsA monitoring method evaluation does not merely compare the linear regressions of numerous blood samples analyzed by several assay methods, as has been performed *ad nauseum*. Instead, we examine real patient data as assessed by an educated variety of methods and make valid conclusions on assay performance as well as on choice for a pharmacokinetic parameter, which are practical for adjustment of drug therapy and assessment of clinical status.

This study utilized 274 oral CsA pharmacokinetic profiles (AUC) from 40 renal transplant recipients performed at steady state during 7 to 180 postoperative days (POD) to study the effects of dose changes on trough levels (tr) and average concentration (Cav) (3), each determined by four assay methods: radioimmunoassay using a specific monoclonal antibody on whole blood samples (MSRIA-WB); and three fluorescence polarization immunoassays using a nonspecific polyclonal antibody on whole blood samples (FPIA-WB), a nonspecific polyclonal antibody on serum samples (FPIA-SM), and a specific monoclonal antibody on whole blood samples (FPIA-SP-WB)). Forty-nine pairs of consecutive profiles between which the dose was held constant and 72 pairs of consecutive profiles where the dose was decreased after completion of the first profile were evaluated. Sixteen pairs of consecutive profiles where the dose was increased between collections were insufficient for statistical evaluation. Data on tr and Cav, as determined by the four assay methods, from a subset of 15 patients who followed benign courses over 180 days, were compared to a subset of 12 patients who experienced acute rejection on POD 11 ± 6 for potential prediction and verification of rejection episodes via CsA monitoring. Normalized data [i.e., tr corrected for dose, (tr/dose), and Cav corrected for dose, (Cav/dose)] were also examined for possible correlation to clinical events. Data from patients who experienced accelerated and chronic rejection, acute and chronic nephrotoxicity was insufficient for statistical evaluation.

II. MATERIALS AND METHODS

A. Patient Population

Forty consecutively transplanted adult renal allograft recipients followed an immunosuppressive regimen of CsA/prednisone (4). Clinical decisions involving CsA monitoring were made via MSRIA-WB analysis. Fifteen patients followed benign courses through POD 180, while 12 experienced at least one episode of acute rejection (POD 11 ± 6), with four grafts lost to rejection. During periods of rejection, 2 of 12 subjects were treated with ALG, 10 of 12 with AZA, and 2 of 12 with OKT3. Episodes of acute rejection were verified on the basis of an increase of >0.5 mg/dL in serum creatinine and histologic findings from biopsies. Five additional subjects experienced chronic nephrotoxicity, seven acute nephrotoxicity, three accelerated rejection, and one chronic rejection.

B. Sample Collection

Twelve- or twenty-four-hour AUC studies (0, 2, 4, 6; 10, 13, 24; or 0, 2, 4, 8, 12 h) were performed on POD 7, 14, 30, 60, 90, 120, 150, and 180 during periods of steady-state CsA dosing, where steady state was defined as at least 3-day maintenance on constant dose. Both whole blood (EDTA) and serum (equilibrated at 22°C > 2 h before separation from clot) samples were collected at all time points.

C. Methods of CsA Analysis

1. *MSRIA-WB*

Fresh whole blood samples were assayed as received, according to the manufacturer's protocol (Sandimmun Radioimmunoassay Technical Bulletin) by the Sandoz radioimmunoassay using ^{3}H-tracer and a monoclonal antibody specific for parent compound.

2. *FPIA-WB*

Whole blood samples, stored until analysis at −40°C in polystyrene tubes, were assayed according to Abbott protocol (CsA and Metabolites Whole Blood Assay Information Bulletin) by fluorescence polarization immunoassay using a fluorescein tracer and a polyclonal antibody which cross-reacts with CsA metabolites.

3. *FPIA-SM*

Serum samples stored until analysis at −40°C in polystyrene tubes were assayed according to Abbott protocol (CsA and Metabolites Assay Information Bulletin) by fluorescence polarization immunoassay using a fluorescein tracer and a polyclonal antibody which cross-reacts with CsA metabolites.

4. *FPIA-SP-WB*

Whole blood samples stored until analysis at −40°C in polystyrene tubes were assayed according to Abbott protocol (CsA Monoclonal Whole Blood Assay Information Bulletin) by fluorescence polarization immunoassay using a fluorescein tracer and a monoclonal antibody which cross-reacts minimally with CsA metabolites.

D. Parameters Evaluated from AUC Studies

Dose (mg/24 h), trough (either 12- or 24-h value), Cav (AUC/dosing interval, AUC determined by trapezoidal rule), trough/24-h dose, and Cav/dose were calculated for each AUC profile. Studies where Cav was >700 according to MSRIA-WB were deleted because of the potential for autoinhibition of the cytochrome P450 system by CsA at high concentrations. Percent variations in trough (%V tr) and Cav (%V Cav) between consecutive steady-state AUC studies, where the two studies were between 7 and 60 days apart, were calculated using the following equations:

$$\frac{\text{trAUC2} - \text{trAUC1}}{(\text{trAUC1} + \text{trAUC2})/2} \times 100\%$$

$$\frac{\text{CavAUC2} - \text{CavAUC1}}{(\text{CavAUC1} - \text{CavAUC2})/2} \times 100\%$$

Using consecutive AUC profiles from patients at constant dose, assay methods were evaluated on ability to duplicate tr and Cav within 20% of 0 deviation. In order to determine the minimum decrease in dose that elicits a significant decrease in tr or Cav, %V tr and %V Cav data from patients at constant dose were compared to patients whose CsA dose was decreased after the first steady-state AUC was performed. Parameters from patients who experienced an acute rejection episode were compared to those from patients with uneventful courses according to day of profiling. The Ansari-Bradley method for analysis of unequal dispersions and the Mann-Whitney method for analysis of equal dispersions were used for determination of significant difference ($p < 0.05$) between data sets. An RS1 statistical program was used for all calculations.

III. RESULTS

A. Variation in tr and Cav with No Dose Change

Forty-nine paired tr from consecutive steady-state AUC profiles were compared for %V according to dose level and median tr. Theoretically, the tr value should remain constant at steady state. From plots of %V tr versus dose and median tr, no relation was found between % V tr and dose level or median tr for any assay (i.e., slopes of all linear regressions were near 0, all $r < 0.075$; see Figure 1). Figure 2 illustrates the %V tr observed as a function of percent of total tr pairs for all four assays. With FPIA-SM, only 24% of paired tr fall within the desired range in variation of ±20% deviation from 0. Using other assays, 46% to 50% of paired tr are within 20% of 0, by which, according to this analysis, whole blood assays are clearly favored over the serum-based assay. Ranges of %V tr were −117% to 200% (317%) for MSRIA-WB, −84% to 87% (171%) for FPIA-WB, −83% to 128% (211%) for FPIA-SM, and −62% to 72% (134%) for FPIA-SP-WB. Thus, FPIA-SP-WB exhibits significant technical advantage over the other assays in reproduction of steady-state tr values despite variation in intrapatient drug disposition.

The %V Cav from AUC studies performed at constant dose should also be 0. The %V Cav was plotted versus dose and median Cav and, as with tr, no relation to dose or median Cav was found (see Figure 1). Figure 3 illustrates %V Cav as a function of percent of total Cav pairs. With FPIA-SM, only 31% of paired steady-state Cav fall within ±20% of 0 deviation. FPIA-SP-WB h as the highest percentage (49%) of paired Cav within ±20% of 0. As with tr evaluation, whole blood assays better replicate Cav within the desired range of deviation than the serum-based assay, whereas whole-blood specific and nonspecific assays provide virtually identical results. There appears to be little advantage in using Cav rather than tr, as the percent of studies where %V Cav falls within ±20% of 0 is similar to that observed for tr, except for FPIA-SM where a slight improvement is apparent. Similar ranges of %V Cav were found: −64% to 82% (146%) for MSRIA-WB, −67% to 86% (153%) for FPIA-WB, −89% to 91% (180%) for FPIA-SM, and −58% to 68% (126%) for FPIA-SP-WB. The narrowest ranges of %V for both tr and Cav are achieved with FPIA-SP-WB.

B. Variation in Trough and Cav with Dose Decrease

After 72 steady-state AUC profiles were performed, a physician deemed the resulting Cav "too high," and the CsA dose was decreased. Patients were reequilibrated to steady state and a second AUC profile was performed with the hope that tr and Cav would decrease into a more efficacious clinical range. The resulting tr data (Figure 4) represent the linear relationship found between %V tr and % dose decrease: a 7% dose decrease is reflected by a 10% reduction in tr. The other three assays provided similar relationships. However, regardless of the assay used, at 3% to 5% decrease in dose, the tr *increases* in 67% to 83% of cases rather than decreases; at 6% to 15% decrease in dose, tr decreases in 68% to 82% of cases; while at dose decreases >15%, 17% to 33% of cases still respond via increasing tr.

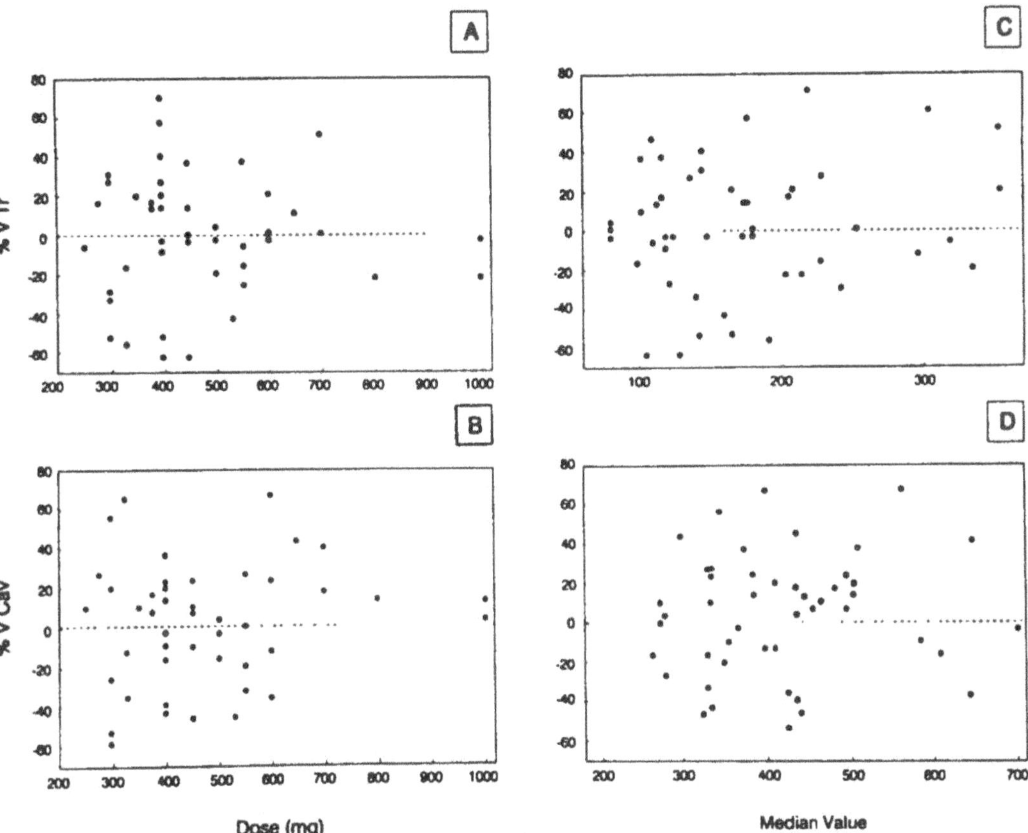

Figure 1 %V tr [y axis for (A) and (C)] and % Cav [y axis for (B) and (D)] for 49 pairs of consecutive steady-state AUC profiles performed at constant CsA dose plotted as a function of dose [x axis for (A) and (B)] and median value (x axis for (C) and (D)]. All tr and Cav determined by FPIA-SP-WB. These plots indicate no relation of %V tr or %V Cav to dose level or to tr or Cav level. Other assays produced similar plots.

Based on variations in tr and Cav expressed in patients at constant dose, %V tr from patients at constant dose was compared to %V tr from patients at four levels of decreases in dose, (a) 3% to 5% (n = 12), (b) 6% to 10% (n = 28), (c) 11% to 15% (n = 18), (d) >15% (n = 15), to determine the effective dose change at which a significant decrease in tr could be predicted. As illustrated in Figure 5, where mean ± SEM of %V tr is plotted as a function of % dose decrease, an increase in mean %V tr is observed with a 3% to 5% decrease in dose. At no level is there a significant difference in tr by MSRIA-WB. At >15% dose decrease, all FPIA assays produce mean tr that are significantly less than the control. Between 6% and 10% but not between 11% and 15% decrease in dose, the mean tr by FPIA-SP-WB is also significantly lower than the control.

The relationship of %V Cav to % decrease in dose is similar to that for %V tr. As shown in Figure 6, a 7% reduction in dose is reflected by a 10% reduction in Cav.

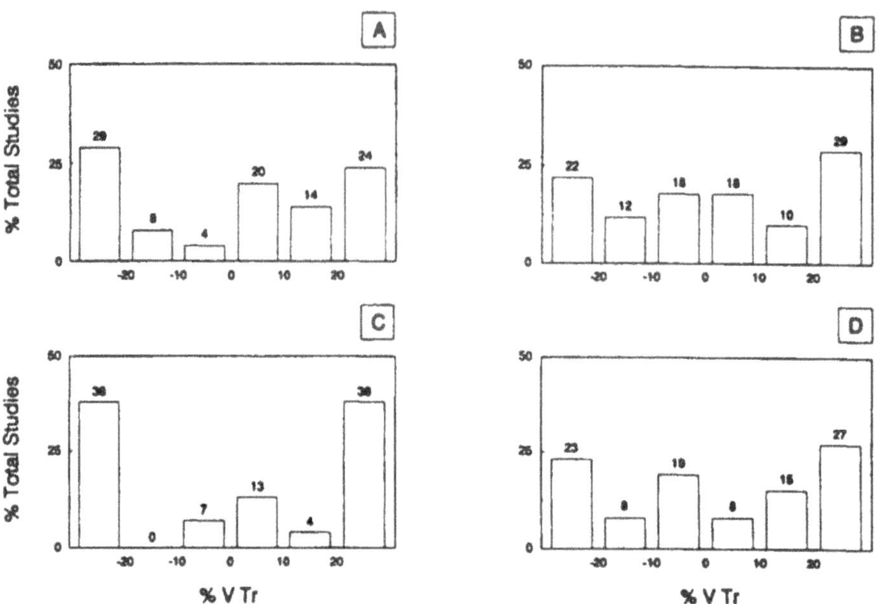

Figure 2 Levels of %V tr (x axis) from 49 pairs of consecutive steady-state AUC profiles at constant CsA dose plotted as a function of percent of total study pairs (y axis) according to assay method: (A) MSRIA-WB; (B) FPIA-WB; (C) FPIA-SM; (D) FPIA-SP-WB. In approximately 46% to 50% of paired studies, tr are ±20% of 0 deviation according to analysis by whole-blood-based assays, whereas, by serum-based assay, only 24% of paired studies are tr ±20% of 0 deviation.

Regardless of assay, at 3% to 5% decrease in dose, rather than decrease, Cav *increases* in 67% to 75% of cases; at 6% to 15% decrease in dose, Cav decreases in 60% to 73% of cases; while at >15% decrease in dose, a small percentage of cases (12% to 22%) still respond with increasing Cav.

Mean %V Cav ± SEM from studies between which the CsA dose was decreased were compared to those from the control group where the dose was held constant (see Figure 7). Using this parameter, both MSRIA-WB and FPIA-SP-WB produced mean Cavs that were significantly less than control at >5% dose decreases, while FPIA-WB and FPIA-SM assays provided mean Cav results that were less than control only at >15% decreases in dose. Improvement in the performance of the MSRIA-WB was due to significant decrease in SEM for Cav as compared to tr, while the SEM for Cav and tr determined by the FPIA assays were similar. The increase in mean %V tr observed between the control and the 3% to 5% decrease in dose level (0.6–2.3 to 17.7–29.9, regardless of assay) was smaller for mean %V Cav (–0.85–5.4 to 7.3–15.3, respective of assay).

C. Utility of CsA Monitoring for Evaluation of Clinical Events

Fifteen of forty renal transplant recipients who followed uneventful postoperative courses had 88 CsA AUC studies performed between POD 7 and 180. Another twelve patients who experienced an acute rejection episode POD 11 ±6 had 65 CsA AUC studies performed between POD 7 and 180. Doses were not significantly different

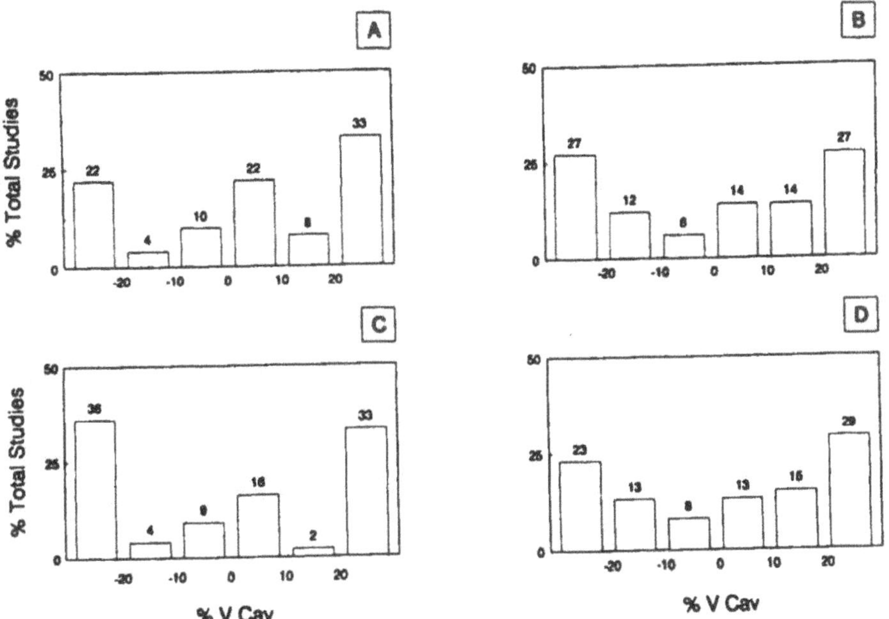

Figure 3 Levels of %V Cav (x axis) from 49 pairs of consecutive steady-state AUC profiles at constant CsA dose plotted as a function of percent of total study pairs (y axis) according to assay method (see Figure 1 for legend). FPIA-SP-WB has the highest percentage of paired studies (49%), where Cav is æ20% of 0 deviation and FPIA-SM has the least (31%).

between the two groups. Comparison of tr and Cav, according to day of study, between benign patients and those who experienced rejection yielded the data as expressed in Table 1. Only POD 14 and 180 were tr as determined by all FPIA, and not MSRIA, from patients who experienced rejection significantly lower than those from patients without rejection. Cav from patients experiencing rejection were lower than Cav from benign patients on POD 14, 150, and 180, but only by FPIA-SM. Tr and Cav data normalized by dose indicated improved utility for evaluation of rejection episodes in that tr/dose from patients who experienced rejection was lower than tr/dose from patients without rejection on POD 7 by all three whole-blood assays, on POD 14 by MSRIA-WB and FPIA-WB, and on POD 180 by all assays. Cav/dose from patients who experienced rejection was less than Cav/dose from benign cases on POD 7 by all assays and POD 180 by all FPIA assays.

IV. DISCUSSION

In spite of wide variations in CsA drug disposition in transplant recipients, the medical community has longed for a method of CsA monitoring that accurately reflects dose changes and that can be used to predict and verify clinical events. Real variation in CsA drug disposition is due in part to decreasing oral clearance with time (6). Seeming variations in drug disposition are due to the crude methods by which

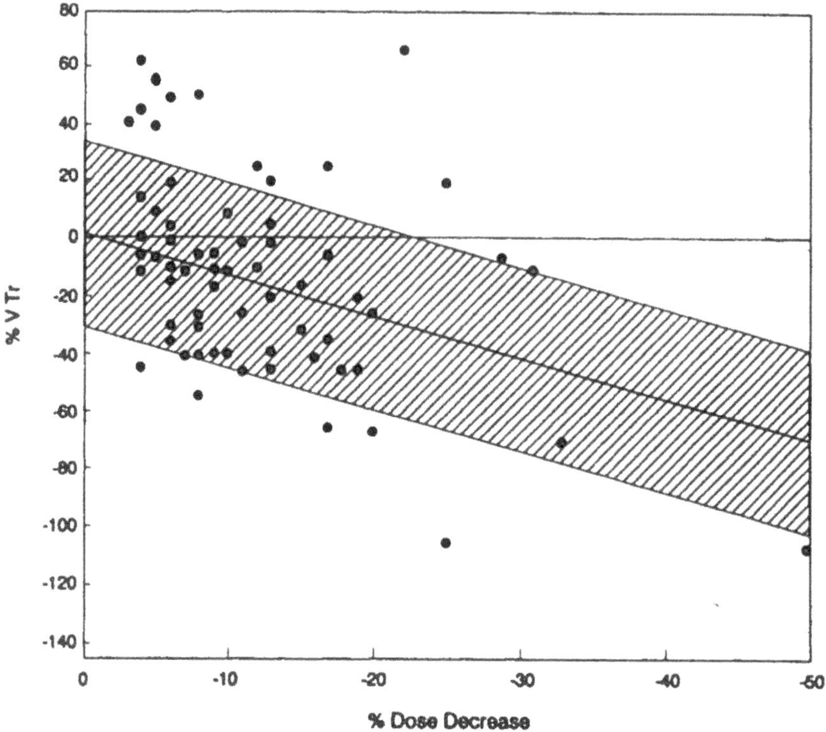

Figure 4 %V (y axis) from 72 pairs of consecutive AUC profiles where CsA dose was decreased between profiles plotted as a function of percent decrease in dose (x axis) according to FPIA-SP-WB. Range of %V tr for FPIA-SP-WB is indicated by dotted lines and one standard deviation by shading. Linear regression analysis yields the following equation: $y = 1.52x + 5.9$, $r = 0.352$. Other assay methods yield similar equations: y (MSRIA-WB) = $1.58x + 8.6$, $r = 0.270$; y (FPIA-WB) = $1.65x + 10.7$, $r = 0.342$; and y (FPIA-SM) = $2.11x + 5.5$, $r = 0.302$.

pharmacokinetic parameters have been determined and to AUC studies performed at nonsteady state. Antibody-mediated methods which quantify not only CsA but various and sundry CsA metabolites, and high-performance liquid chromatographic techniques, which are subject to variable analyte recovery and numerous interferences (6), have precluded well-founded evaluation of CsA disposition. This study examined the capacity of four different assay techniques to follow changes in CsA dose via steady-state tr and Cav. Several aspects of methodology were compared—(a whole blood to serum, (b) specific to nonspecific antibody, (c) narrow to wide assay specifications, (d) steady-state tr to Cav or some other parameter—to select an optimal method for practical use.

Dosing changes do affect tr and Cav. With dose decrease as the most common therapeutic intervention used with CsA, we found our study too limited to permit statistical evaluation of increase in dose to tr and Cav. However, 72 instances where dose was decreased, related to 49 instances where dose was held constant, provided valuable clues to the type of assay which best reflects *actual* changes in tr and Cav with

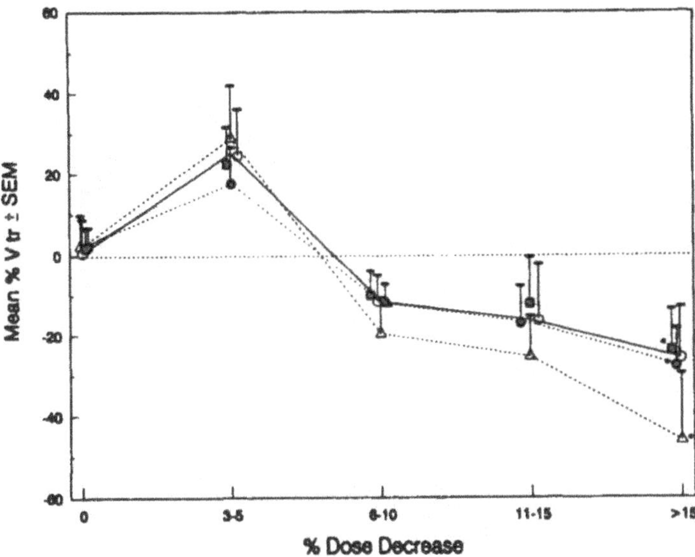

Figure 5 Mean %V ± SEM (y axis) from paired steady-state AUC profiles where CsA dose was decreased 0, 3% to 5%, 6% to 10%, 11% to 15%, and >15% between profiles plotted as a function of percent decrease in dose (x axis) according to assay method: (O) MSRIA-WB; (■) FPIA-WB; (Δ) FPIA-SM; (●) FPIA-SP-WB. Starred symbols (*) indicate significant difference ($p < 0.05$) from control (no dose change).

dose change. From the control steady-state AUC studies performed at constant dose, FPIA-SM proved the least reliable of the assays tested in reproducing both tr and Cav. Given excellent assay specifications (7), technicalities involved in processing serum samples are the causes for error. Current specimens were processed at room temperature. Meanwhile, Humbert et al. (8) have proven that [CsA] in plasma separated from red cells at 37°C, despite varying hematocrits, reflects [CsA] in whole blood, while separation at room temperature, because of varying hematocrit, does not. Neither have Holt et al. (9) found practical advantage to use of serum.

Specimens from liver and heart recipients often have large and fluctuating concentrations of metabolites, which serve to cloud the meaning of results performed according to nonspecific assays. Although specimens from renal recipients, such as those used currently, probably have the least and most consistent quantities of CsA metabolites, cross-reactivity of the polyclonal antisera used in FPIA-SM with metabolites may reduce the efficacy of FPIA-SM. Therefore, because of practical limitations for serum processing and use of a nonspecific polyclonal antisera, the FPIA-SM method is least suitable for following dosing changes through tr or Cav.

Considering the whole-blood assays, the %V Cav range by MSRIA-WB is considerably narrower than the %V tr range, although FPIA-SP-WB provides the narrowest ranges of variation for both steady-state tr and Cav at constant dose. However, by any whole-blood assays studied, only 40% to 50% of steady-state tr and Cav at constant dose can be reproduced within 20% of the original value. Since the same whole-blood samples were analyzed by all three assays, the importance of narrow

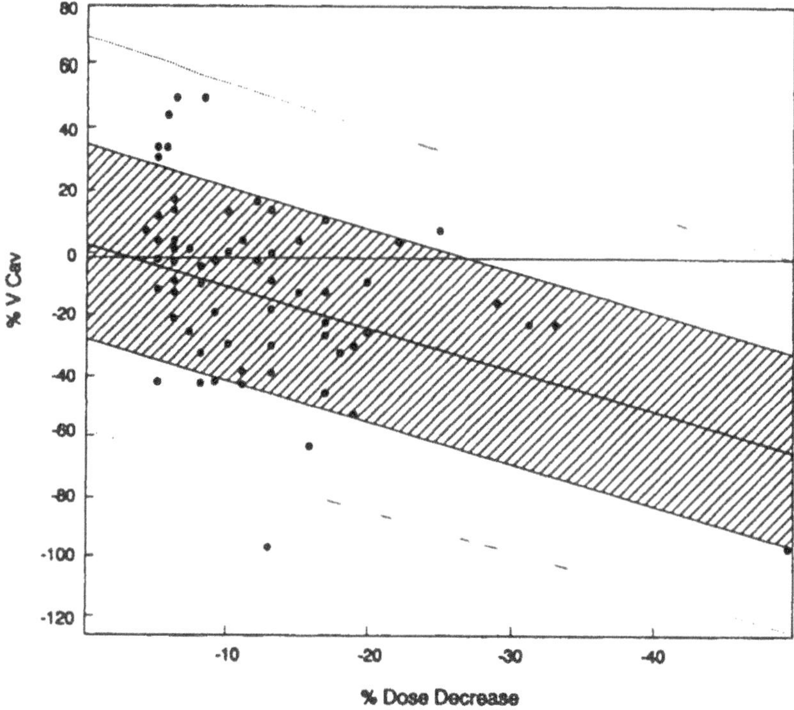

Figure 6 %V Cav from paired steady-state AUC profiles where CsA dose was decreased between profiles (y axis) plotted as a function of percent decrease in dose (x axis) according to FPIA-SP-WB. Range of %V Cav for FPIA-SP-WB is indicated by dotted lines and one standard deviation by shading. Linear regression analysis yields the following equation: $y = 1.46x + 5.1$, $r = 0.415$. Other assays yield similar equations: y (MSRIA-WB) = $1.38x + 4.4$, $r = 0.361$; y (FPIA-WB) = $1.52x + 7.8$, $r = 0.379$; and y (FPIA-SM) = $1.93x + 3.1$, $r = 0.366$.

assay specifications becomes clear through comparison of respective ranges of variation for tr and Cav. While Ball et al. (10) determined a 2% to 7% within-run CV for MSRIA-WB, up to 15% was accepted for this study. An 8% to 13% between-run CV (10) is due to use of a serially diluted standard curve that is prepared daily *in house*, numerous manual steps, and inefficient beta-radiation counting. FPIA-WB and FPIA-SP-WB have 2% to 6% within-run CV (11; unpublished results), while up to 10% was accepted for this study. Low between-run CV of 3% to 6% (11; unpublished results) is due to use of precalibrated standard reagents supplied by the manufacturer and minimal operator intervention.

Therefore, the MSRIA-WB range in %V Cav is narrower than for %V tr because derivation of a parameter (Cav) from several raw measurements serves to smooth error associated with single-sample measurement (e.g., tr). Ranges in %V tr and %V Cav by FPIA-WB and FPIA-SP-WB are similar, regardless of assay, and inherently narrower than for MSRIA-WB because of low within- and between-run CVs. Thus, with FPIA whole-blood assays, measurement of more than a single steady-state tr is unnecessary. With all other FPIA whole-blood assay specifications considered to be

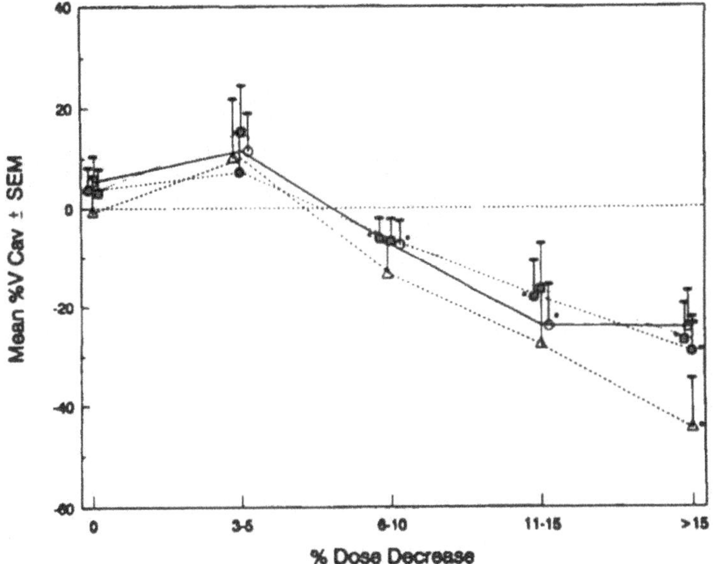

Figure 7 Mean %V Cav ± SEM (y axis) from paired steady-state AUC profiles where CsA dose was decreased 0, 3% to 5%, 6% to 10%, 11% to 15%, and >15% between profiles plotted as a function of percent decrease in dose (x axis) according to assay method (see Figure 3 for legend). Starred symbols (*) indicate significant difference ($p < 0.05$) from control (no dose change).

similar, the wide range in %V Cav for FPIA-WB in relation to FPIA-SP-WB must be cross-reaction of the nonspecific polyclonal antisera used in FPIA-WB with CsA metabolites whose concentrations do not parallel that of CsA.

Current results indicate that tr is equivalent to Cav. It must be remembered, however, that specimens for tr were all collected at steady state at the exact end of the dosing interval. Consequently, it was determined that, via use of any of the four assays tested, a 7% decrease in dose relates proportionally to a 10% decrease in both tr and Cav. Inability of small dose changes to overcome the decrease in oral clearance of CsA observed over time may be evidenced by the sizable increase in mean %V tr observed when dose is decreased only 3% to 5%. Depending on choice of either tr or Cav and assay used, decreases in dose of at least 6% and preferably 15% or more are necessary to elicit a decrease in tr or Cav. The narrower ranges of variation produced by the FPIA-SP-WB method make it the method most likely to reflect real change in tr and Cav in relation to dose. Practical assessment of CsA drug therapy is therefore possible with use of a monitoring technique that has the advantages of low within- and between-run CVs and specificity of antibody for parent compound.

Diagnosis of clinical events in transplant recipients solely on the basis of CsA levels would be a formidable achievement for CsA monitoring techniques, because clinical events are not related solely to CsA level. Nevertheless, use of CsA levels as an aide to prevention and diagnosis of adverse clinical events would be a practical benefit. Previously, tr levels have yielded inconsistent relationships to clinical events (12–16) and often only when the two poles of nephrotoxicity and rejection were

Table 1 CsA Pharmacokinetic Parameters for Diagnosis of Acute Rejection

tr

POD	7	14	30	60	90	120	180
MSRIA							
FPIA-WB		p<0.05					p<0.05
FPIA-SM		p<0.05					p<0.05
FPIA-SP-WB		p=0.005					

tr/dose

POD	7	14	30	60	90	120	180
MSRIA	p=0.49	p<0.01					p<0.01
FPIA-WB	p=0.015	p<0.05					p<0.01
FPIA-SM							p=0.013
FPIA-SP-WB	p=0.040						p=0.005

Cav

POD	7	14	30	60	90	120	180
MSRIA							
FPIA-WB							
FPIA-SM		p=0.028				p=0.004	p=0.009
FPIA-SP-WB							

Cav/dose

POD	7	14	30	60	90	120	180
MSRIA	p=0.023						
FPIA-WB	p<0.05						p=0.018
FPIA-SM	p<0.05					p=0.009	
FPIA-SP-WB	p=0.040		p<0.05				p<0.05

[a]POD where tr, Cav, tr/dose, and Cav/dose from patients who experienced acute rejection were significantly less than patients who had benign courses. Shaded area indicates period of active rejection. Significant p values (≤0.05): < determined by Mann-Whitney method for equal dispersions; = by Ansari-Bradley method for unequal dispersions. Benign cases = 10, 10, 12, 14, 13, 11, 8, 10, regardless of POD. Cases who experienced rejection = 10, 12, 10, 7, 9, 6, 6, 5, regardless of POD.

compared (17,18). Using AUC monitoring, episodes of nephrotoxicity have been more readily discriminated from uneventful courses (3) than have episodes of rejection. Current comparison of tr and Cav, from steady-state AUC profile performed on patients who experienced acute rejection and patients with uneventful courses, yielded suggestive, but by no means conclusive, results. From AUC profiles performed during episodes of rejection, tr, but not Cav, was indicative of rejection in that tr from patients undergoing rejection was significantly lower than in others who were not. Neither tr nor Cav was predictive of events in that data from the group who would later experience rejection was not different from that of benign cases.

Normalization of tr and Cav data by dose was performed to provide closer examination of the potential for parameters determinable by CsA monitoring for prediction and verification of an adverse clinical event. Interestingly, t r/dose from rejecting patients, both before and during rejection, was significantly lower than from patients with benign courses, while Cav/dose was lower only prior to the rejection episode.

If results of the current evaluation can be replicated and amplified using larger patient populations, reasons for early episodes of acute rejection may be clarified. These data suggest that early acute rejection may relate to initial inadequate absorption and high clearance in relation to dose (low Cav/dose and low tr/dose), which may not be detected prior to event by either tr or Cav because of overly wide ranges previously established for efficacious tr and Cav. Significant differences observed in results from POD 180 could be due to insufficient data, as there were only 5 rejecting cases to compare to 10 benign cases on that POD.

While tr, Cav/dose, and especially tr/dose show potential for prediction and verification of acute rejection, there was no clear demarcation between choice of specific or nonspecific assay, and whole-blood or serum-based assay. Perhaps there were insufficient data, or perhaps acute rejection is related to disposition of CsA metabolites as well as parent drug, as suggested by Grevel et al. (19). Expansion of this cohort is clearly necessary in order to better rank the parameters derived from CsA monitoring, to allow a method of choice for diagnosis of clinical events to be determined, and to discern how CsA drug disposition precipitates adverse clinical events.

ACKNOWLEDGMENT

The authors would like to thank Abbott Diagnostics, North Chicago, Illinois, for their generous gifts of cyclosporine assay reagents. The authors would also like to thank Mary Worthington of TDx Research and Development at Abbott Diagnostics for her assistance in coordinating the efforts of these clinical trials with Abbott.

REFERENCES

1. L. Shaw, L. Bowers, L. Demers, et al. Critical issues in cyclosporine monitoring: Report of the Task Force on Cyclosporine Monitoring. *Clin. Chem.*, *33*:1269 (1987).
2. B. Kahan, L. Shaw, D. Holt, et al. Consensus document: Hawk's Cay meeting on therapeutic monitoring of cyclosporine. *Clin. Chem.*, *36*:1510 (1990).
3. J. Grevel, M. Welsh, and B. Kahan. Cyclosporine monitoring in renal transplantation: Area under the curve monitoring is superior to trough-level monitoring. *Ther. Drug Monitor.*, *11*:246 (1989).
4. B. Kahan, M. Lorber, S. Flechner, et al. Comparison of five cyclosporine-prednisone regimens for induction of immunosuppression in cadaveric kidney recipients: A retrospective analysis of 245 cases. *Transplant. Proc.*, *18*:59 (1986).
5. D. Holt, J. Marsden, and A. Johnston. Quality assessment of cyclosporine measurements: Comparison of current methods. *Transplant. Proc.*, *22*:1234 (1990).
6. J. Grevel, K. Napoli, and B. Kahan. Covariables in cyclosporine immunosuppression. *Transplant. Proc.*, *22*:1113 (1990).
7. K. Napoli and B. Kahan. Nonselective measurement of cyclosporine for therapeutic drug monitoring by fluorescence polarization immunoassay with a rabbit polyclonal antibody: I. Evaluation of the serum methodology and comparison with a sheep polyclonal antibody in a ^{3}H-tracer mediated radioimmunoassay. *Transplant. Proc.*, *22*:1175 (1990).

8. H. Humbert, L. Vernillet, M. Cabiac, et al. Influence of different parameters for the monitoring of cyclosporine. *Transplant. Proc.*, 22:1210 (1990).
9. D. Holt, J. Marsden, T. Fashola, and A. Johnston. Performance of the Sandoz radioimmunoassays for cyclosporine. *Transplant. Proc.*, 22:1155 (1990).
10. P. Ball, H. Munzer, H. Keller, et al. Specific ^{3}H radioimmunoassay with a monoclonal antibody for monitoring cyclosporine in blood. *Clin. Chem.*, 34:257 (1988).
11. K. Napoli and B. Kahan. Nonselective measurement of cyclosporine for therapeutic drug monitoring by fluorescence polarization immunoassay with a sheep polyclonal antibody: II. Evaluation of the whole blood methodology and comparison with a ^{3}H-tracer-mediated radioimmunoassay with a sheep polyclonal antibody. *Transplant Proc.*, 22:1181 (1990).
12. B. Kasiske, K. Heim-Duthoy, K. Rao, and W. Awni. The relationship between cyclosporine pharmacokinetic parameters and subsequent acute rejection in renal transplant recipients. *Transplantation*, 46:716 (1988).
13. D. Holt, J. Marsden, A. Johnston, and D. Taube. Cyclosporine monitoring with polyclonal and specific monoclonal antibodies during episodes of renal allograft dysfunction. *Transplant Proc.*, 21:1482 (1989).
14. S. Veremis, M. Maddux, I. Tang, et al. Comparison of cyclosporine (CSA) blood levels and RIA/HPLC ratios among renal transplant patients with normal allograft function, CSA nephrotoxicity or rejection. *Transplant Proc.*, 21:1476 (1989).
15. B. Sommer, D. Sing, M. Henry, et al. Serum cyclosporine kinetic profile: Failure to correlate with nephrotoxicity or rejection episodes following sequential immunotherapy for renal transplantation. *Transplantation*, 45:86 (1988).
16. G. Klintmalm, J. Sawe, O. Ringden, et al. Cyclosporine plasma levels in renal transplant recipients. *Transplantation*, 39:132 (1985).
17. B. Kahan, K. Napoli, M. Welsh, et al. Comparison of the utility of ^{3}H-based specific monoclonal antibody assay on whole blood samples with the fluorescence polarization nonspecific immunoassay on serum samples for diagnosis of adverse events in renal transplant patients. *Transplant Proc.*, 22:1274 (1990).
18. L. Shaw, P. Audet, R. Grossman, et al. Adjustment of cyclosporine dosage in renal transplant patients based on concentration measured specifically in whole blood: Clinical outcome and diagnostic utility. *Transplant Proc.*, 22:1267 (1990).
19. J. Grevel, K. Napoli, M. Welsh, et al. Prediction of acute graft rejection in renal transplantation: The utility of cyclosporine blood concentrations. *Pharm. Res.*, 1991 (in press).

23

Monitoring of Plasma Glucocorticosteroids as an Indicator of Noncompliance in Pediatric Renal Transplant Recipients

James H. McBride, Robert B. Ettenger, Denis O. Rodgerson, Sue S. Park, and Ana F. Reyes *UCLA School of Medicine, Los Angeles, California*

I. INTRODUCTION

In the past decade, patient noncompliance has been identified with increasing frequency as a cause of failure of long-term treatment regimens in tuberculosis, diabetes mellitus, hypertension, and rheumatoid arthritis (1). The adverse impact on the success of renal transplantation of noncompliance to prescribed immunosuppressive therapy (azathioprine and prednisone) was first reported in 1975 (2). Since the advent of cyclosporine (CsA) and prednisone, used in combination in immunosuppression therapies, it has been recognized that the third leading cause of graft failure after allograft rejection and systemic infection is noncompliance; further, noncompliance tends to be a major factor in younger patients (1).

Recently it has been reported that monitoring of steroid treatment by measurement of prednisone, prednisolone, and cortisol can be useful in studying patient compliance in nephrologic disease and postrenal transplantation (3). This study also detailed that compliance tended to be much more of a problem with pediatric rather than adult patients.

The purpose of this study was to establish a rapid liquid chromatographic method for the glucocorticosteroids, prednisone, prednisolone, and cortisol and application of this method to identification of noncompliance in pediatric renal transplant recipients who were receiving both CsA and prednisone. To date there is little information on compliance with these immunosuppressive therapies in pediatric patients.

II. MATERIALS AND METHODS

Plasma samples (n = 68) were obtained from pediatric patients (n = 37) attending the renal transplant outpatient clinic. Patients were identified as noncompliant when their behavior resulted in a rejection episode (minor noncompliance) or graft loss (major noncompliance) as documented during direct interview by members of the transplant team. The diagnosis was suspected when the CsA plasma level was <25 ng/mL as determined both by the Abbott TDx (fluorescence polarization immunoassay) and by high-performance liquid chromatography for the parent drug.

Prednisone (dose 6 to 10 mg/q.d.) and CsA were usually administered daily between 7:00 a.m. and 9:00 a.m. However, on the day of attendance at clinic, blood samples were drawn to obtain trough levels for both steroids and CsA.

A rapid liquid chromatographic method was established for plasma prednisone, prednisolone, and cortisol utilizing an improved sample-extraction technique (4). Briefly, the method was as follows. To 1.0 mL of plasma (blank, standard, or control) add 0.2 mL of internal standard (1 µg/mL methylprednisolone) and vortex for 10 s. Add 1.0 mL of 0.1 mol/L sodium hydroxide and vortex for 10 s before adding the mixture to a 3-mL-capacity Chem Elut column (Analytichem International, Harbor City, California). After 5 min, add 5 mL of methylene chloride and collect the eluate, then add an additional 5 mL of solvent. Combine the eluates and evaporate them to dryness under a stream of nitrogen.

Reconstitute the dried eluates with 200 µL of mobile phase (tetrahydrofuran/water, 25/75 VN) and inject 20 µL onto an Axxi-Chrom, ODS-C_{18} (Cole Scientific, Calabasas, California) reversed-phase column having a particle size of 5µ. Column flow rate was 1.2 mL/min, and typical elution times were 7.4 min for prednisone, 9.1 min for prednisolone, 9.9 min for cortisol and 15.5 min for the internal standard. Detection was at 240 nm with a Shimadzu SPD-6A detector, and the minimum detectable amount of each steroid was 10 ng/mL.

III. RESULTS AND DISCUSSION

Figure 1 shows a typical chromatogram for the glucocorticosteroids detected by this method. From this profile it is evident that there is clear resolution of the steroids in question, and the analysis time permits the processing of four samples per hour.

Linearity studies revealed that the method was completely linear for prednisone, prednisolone, and cortisol up to concentrations of 500 ng/mL. Recovery data is given in Table 1, and intra- and interassay precision are given in Table 2.

Interference studies were conducted with other steroids and a variety of medications commonly encountered in pediatric patients posttransplantation. Table 3 lists the relative retention times of other steroids studied; it is apparent that the method is free from interference from these steroids. Further, commonly administered medications did not interfere in the analysis of the glucocorticosteroids in question, as judged by the relative retention times given in Table 4.

When the method was applied to the study of plasma samples from pediatric renal transplant recipients, it was successful in identifying noncompliance in 5 of 37 patients (13.5%). Table 5 gives three examples of compliance, and it can be seen that total cyclosporine, as measured by Abbott TDx, is within the range (25 to 250 ng/mL) for our group of patients, with the parent drug (CsA) being <25 ng/mL. In these

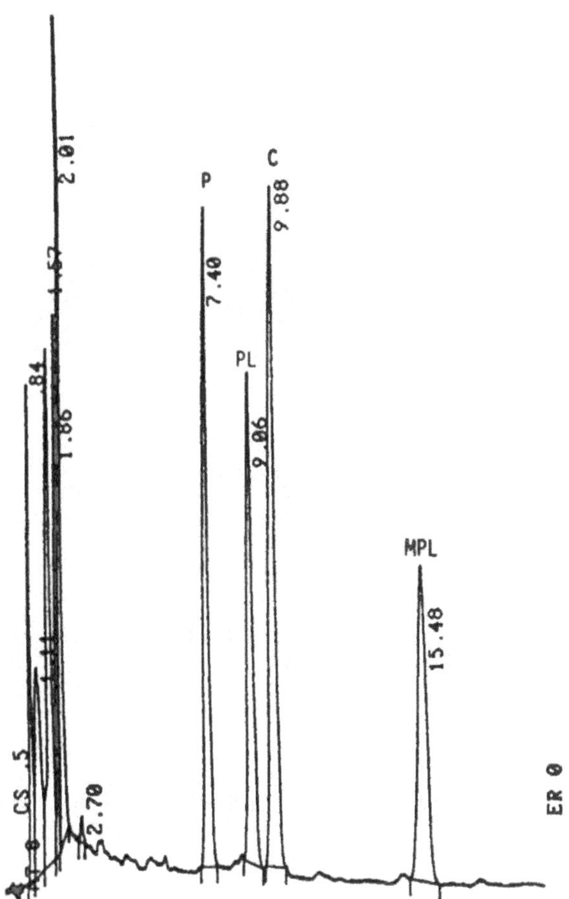

Figure 1 Measurement of 200 ng/mL of prednisone (P), prednisolone (PL), and cortisol (C). Methylprednisolone (MPL) is the internal standard.

patients there is adequate conversion of prednisone to prednisolone, with residual suppression of endogenous cortisol in keeping with what would be expected.

The group of noncompliant patients demonstrated low total cyclosporine concentrations and low (CsA) concentrations (except patient 7) with no detectable prednisone, low concentrations of prednisolone, and adequate reserves of endogenous cortisol (except patient 9). On interview, patient 7 admitted to regular intake of CsA but not prednisone, whereas patient 9 admitted to CsA and prednisone intake immediately before coming to clinic. All five patients in this group had been classified as having experienced rejection episodes (minor compliance) over the course of this study.

In conclusion, we developed a liquid chromatographic method for prednisone, prednisolone, and cortisol which is rapid and can be applied to the analysis of plasma samples on the day of attendance at clinic. The method is successful in identifying noncompliance in pediatric renal transplant recipients and will be used routinely for this purpose. Our results indicated that 13.5% of the patients studied were noncompliant and are very much in agreement with previous studies (1).

Table 1 Recovery Experiments

Amount Added (ng/mL)	Mean Recovery (ng/mL) (n = 10)	Mean Percent Recovery (n = 10)
	Prednisone	
10	12.6	126.0
75	73.3	97.7
150	156.3	104.2
300	303.3	101.1
	Prednisolone	
10	12.3	123.0
75	75.0	100.0
150	150.0	100.0
300	287.0	95.7
	Cortisol	
10	10.8	108.0
75	76.0	101.1
150	149.0	99.3
400	409.0	102.3
500	510.0	102.0

Table 2 Intraassay (TRA) and Interassay (TER) Precision

	Prednisone		Prednisolone		Cortisol	
	(TRA)	(TER)	(TRA)	(TER)	(TRA)	(TER)
			Steroid Conc. (50 ng/mL)			
N	10.0	10.0	10.0	10.0	10.0	10.0
Mean	47.3	49.6	52.7	49.9	55.2	51.8
2 SD	4.0	7.3	4.5	4.7	7.5	8.1
2 CV (%)	8.5	14.6	8.6	9.3	13.5	15.7
			Steroid Conc. (150 ng/mL)			
N	10.0	10.0	10.0	10.0	10.0	10.0
Mean	150.6	153.0	152.2	149.0	156.0	152.3
2 SD	10.7	16.4	11.3	12.2	7.6	16.3
2 CV (%)	7.1	10.7	7.4	8.2	4.8	10.7
			Steroid Conc. (400 ng/mL)			
N	10.0		10.0		—	
Mean	411.0		409.0		—	
2 SD	18.8		36.6		—	
2 CV	4.6		9.0		—	

Table 3 Interference Studies: Other Steroids

Steroids	Conc. (ng/mL)	Retention Time (RT)	Relative Retention Time (RRT)
1. d-Aldosterone	500	3.54	0.28
2. 5α-Cholestan-3-one	10,000	6.88	0.54
3. Cortisol	200	8.14	0.64
4. Cortisone	1,000	7.06	0.55
5. 11-Dehydrocorticosterone	1,000	7.17	0.56
6. Dehydroisoandrosterone	10,000	—	—
7. 11-Ketoeticocholanolone	10,000	—	—
8. Prednisone	20	6.12	0.48
9. Prednisolone	200	7.52	0.59
10. Dexamethasone	1,000	25.15	1.97
11. 11-Deoxy-17-OH corticosterone	1,000	14.92	1.17
12. 6α-Methylprednisolone	200	12.77	1.00
13. 18-OH, Deoxycorticosterone	1,000	13.06	1.02
14. 4-Androsterone-3,17-dione	1,000	18.90	1.48
15. 4-Prednene-11β,17a-diol	1,000	13.24	1.04
16. a-Equilenin	1,000	13.46	1.05

Table 4 Interference Studies: Medications

Medication	Conc.	Retention Time (RT)	Relative Retention Time (RRT)
1. Amoxicillin	1 mg/ml	4.58	0.37
2. Atenolol	1 mg/ml	4.61	0.38
3. Cephalex	10 mg/ml	4.60	0.38
4. Captopril	1 mg/ml	4.61	0.38
5. Clonidine	1 mg/ml	4.58	0.38
6. Ditropan	1 mg/ml	4.57	0.38
7. Hydralazine (Apresoline)	10 mg/ml	9.03	0.74
8. Imuran (Azathioprine)	1 mg/ml	4.60	0.38
9. Lasix	1 mg/ml	4.46	0.37
10. Methylprednisolone	250 ng/ml	12.32	1.00
11. Minoxidil	1 mg/ml	4.68	0.38
12. Normodyne	1 mg/ml	4.56	0.38
13. Procardia (Nifedipine)	1 mg/ml	19.00	1.54
14. Propranolol HCl	1 mg/ml	—	—
15. Ranitidine	1 mg/ml	—	—
16. Prednisone	200 ng/ml	5.92	0.47
17. Rocaltrol	1 mg/ml	—	—
18. Slow Fe	1 mg/ml	4.60	0.38
19. Sulphamethoxazole + Trimethoprim	10 mg/ml	7.10	0.58
20. Vasotex	1 mg/ml	—	—
21. Verapamil HCl	10 mg/ml	7.17	0.58

Table 5 Clinical Results

Patient No.	Cy A (ng/mL)	Total Cy (ng/mL)	Pred. (ng/mL)	Prednis. (ng/mL)	Cortisol (ng/mL)
		Examples of Compliance			
1	<25	114	57	411	84
2	<25	38	20	268	12
3	—	222	38	301	<10
		Examples of Noncompliance			
5	<25	<25	—	25	208
6	<25	<25	—	—	170
7	77	192	—	—	117
8	<25	<25	—	16	46
9	<25	48	—	17	8

It is our intention with future studies to obtain a profile of glucocorticosteroids on patients immediately after renal transplantation and use this as a baseline for noncompliance when assessing those patients later at clinic visits.

REFERENCES

1. R. H. Didlake, K. Dreyfus, R. H. Kerman, C. T. VanBuren, and B. D. Kahan. Patient noncompliance: A major cause of late graft failure in cyclosporine-treated renal transplants. *Transplant Proc.*, *20*(Suppl. 3):63–69 (1988).
2. M. L. Owen, J. G. Maxwell, J. Goodnight, and M. W. Wolcott. Discontinuance of immunosuppression in renal transplant patients. *Arch. Surg.*, *110*:1450–1451 (1975).
3. U. J. Hesse, B. Roth, G. Knuffertz, and T. V. Lilien. Control of patient compliance in the outpatient steroid treatment of nephrologic diseases and renal transplants. European Society for Transplantation, Barcelona, Spain, Abst. 34 (1989).
4. J. T. Stewart, I. L. Honigberg, B. M. Turner, and D. A. Davenport. Improved sample extraction before liquid chromatography of prednisone and prednisolone in human serum. *Clin. Chem.*, *28*:2326–2327 (1982).

24

Therapeutic Drug Monitoring: Role of Alpha$_1$-Acid Glycoprotein (AGP) Binding

P. Baumann and C. B. Eap *Hôpital de Cery, Prilly-Lausanne, Switzerland*

I. INTRODUCTION

α1-Acid glycoprotein (AGP) is an acute-phase reactant protein with a normal plasma concentration ranging between 0.60 and 1.20 mg/mL. Although its exact function in vivo is not clearly known, it may have some immunomodulative properties (1). AGP, albumin (HSA), and lipoproteins represent the most important drug-binding proteins in plasma. While albumin binds mainly acidic drugs, AGP can be the major binding protein for neutral and basic drugs (2). Recently, there was a symposium on state of research about AGP. Therefore, only some aspects related to drug monitoring will be treated here.

II. BIOCHEMISTRY AND GENETICS OF AGP

Much of the present knowledge about the biochemistry of AGP is due to the pioneering work of K. Schmid in the 1950s and 1960s (1). AGP is a glycoprotein with a peptide chain composed of 181 amino acids and characterized by an exceptionally high carbohydrate content of 45% and a great number of sialyl residues, which contribute to its low isoelectric point of 3.4. Two different genes, A and B/BÇé, code for two proteins differing by 22 amino acid substitutions (4). For a better understanding of the drug-binding properties of AGP, it is worthwhile to mention its three microheterogeneities.

The first one heterogeneity, which is not genetically determined and concerns structural differences in the glycan chain, is revealed by affinity by immunoelectrophoresis using Concanavalin A (Con A) (5).

Both the second and third heterogeneities are genetically determined. The former yields "polymorphic forms" which are due to different linkages of the terminal sialic acid residues. By isoelectric focusing, five, six, seven, or eight bands are revealed (6). Amino acid substitutions in the peptide chain yield the genetically determined variants. By classical electrophoresis, a slow- and a fast-migrating band have been demonstrated, yielding the variant forms S-AGP and F-AGP. Modern isoelectrophoretic techniques have shown that S-AGP is composed of ORM2 A and ORM1 S, and F-AGP corresponds to the variant ORM1 F1, encoded by the AGP A gene (4). The variant ORM2 A, encoded by the AGP B/B' gene, differs from ORM1 F1 by 22 amino acid substitutions (7). In a Swiss population, the frequency of the most common phenotypes is 33.7% for ORM1 F1/ORM2 A, 50.5% for ORM1 F1/ORM1 S/ORM2 A, and 15.2% for ORM1 S/ORM2 A (8). Moreover, many other rare variants have been described in the last 5 years (9).

III. FACTORS INFLUENCING AGP PLASMA LEVELS

AGB being a typical "acute-phase reactant," its plasma concentrations may considerably increase in inflammation processes, some forms of cancer, and after surgery (2). Sometimes, a slight but significant increase is observed during some drug treatments, specifically amitriptyline (10) and some antiepileptics (11). Its level is very low in the fetus and in perinatal conditions, but then increases sharply to a more or less constant value during later life (2). Indeed, despite some reports on an increase of AGP in the elderly, it is thought today that this increase is most probably due to some minor and clinically "silent" pathological conditions (12,13). AGP levels have been shown to be sex-independent, but a circadian rhythm has been described, with lowest AGP levels at midnight and highest levels at noon, about 5% to 10% higher than at midnight (14).

IV. DRUG BINDING TO AGP

Many different basic, neutral, but also acidic drugs are bound to AGP, with association constants between 10^4 and 10^7, and a number of binding sites between 0.5 and 2 (2). The fact that basic drugs are preferentially bound explains why very lipophilic psychotropic drugs, such as antidepressants and neuroleptics, are highly bound (15). Chiral drugs show stereospecificity in binding to AGP, which should be considered when the binding of racemates is studied (16,17).

V. GENETICS OF DRUG BINDING

The existence of three microheterogeneities for AGP, two of them being genetically determined, suggests a possible genetic contribution to the interindividual differences in drug binding. This hypothesis has found some support in a study in which the free fraction of amitriptyline was higher in patients whose plasma presented eight bands than in those with only six bands, after isoelectric focusing of the polymorphic forms of AGP (18). However, much more promising were the investigations dealing with the role of the genetically determined variants. In-vitro studies with S- and F-AGP isolated from commercially available AGP demonstrated a preferential binding of many psychotropic drugs to S- rather than to F-AGP (19). An investigation with amitriptyline showed a higher binding capacity of S- than F-AGP (20). More recently, the in-vitro binding of d-, l-, and d,l-methadone was measured in the plasmas of 45 healthy volunteers taking into account the role of other plasma constituents. The results confirmed the role of S-AGP, more precisely of the ORM2 A variant in contrast to the negligible role of ORM1 F1 variant, in the binding of basic drugs (21).

VI. METHODOLOGICAL ASPECTS

When the free fraction of a drug is to be measured in taking account of the role of AGP, the methodological problems are those generally encountered in submitting plasma to equilibrium dialysis of ultrafiltration to separate the free from the bound drug (22). For example, the question arises whether plasma can be deep-frozen and then thawed for practical purposes before the free drug is measured. Indeed, an increase of the methadone-free fraction has been reported in binding experiments performed with thawed compared to freshly isolated plasmas (21). This seems not to be due to AGP, as deep-freezing and thawing have relatively little effect on the binding of drugs to AGP and on its concentration (23), but this manipulation may have some consequences on the behavior of lipoproteins implied in the binding of basic drugs. On the other hand, it is known that a modification of the plasma pH, such as occurs when the plasma is exposed to air, has an influence on the binding of drugs to AGP. These authors also show the importance of limiting dialysis time, as there is a drop of AGP as a function of time, a phenomenon partly explained by osmotic dilution. AGP may be commonly measured by immunodiffusion (Behring-Hoechst), but also by fluorimetry using auramine as a reactant (24). The latter method yields higher AGP levels than the former (25).

In the 1970s, many reports dealt with artefactual results using blood collection tubes containing plasticizers such as tris(butoxyethyl)phosphate (TBEP). This compound has a high affinity for AGP and, as a consequence, displaces bound drugs which are then taken up by erythrocytes. This phenomenon leads to artificially low plasma drug levels. Recent work has shown that some vacutainers are now manufactured without TBEP and are suitable for plasma-level monitoring of antidepressants (26–28).

VII. CLINICAL RELEVANCE OF DRUG BINDING TO AGP

The clinical relevance of monitoring the free fraction has been widely discussed in recent review papers (22,29,30), but the ability to interpret correctly the free drug level is still limited (31). However, an example of the usefulness of monitoring free and total drug concentrations and the AGP concentration will shortly be given (32): In a patient who had suffered a myocardial infarction, AGP, as expected, continued to rise for several days and, during intravenous infusion of lignocaine, total plasma drug concentrations continued to rise despite the expectation that steady-state concentrations should have been reached. Free drug levels remained almost stable during this period, as did, presumably, the drug effect. This may be at least partially explained by an increase of plasma protein binding, which results in a modification of tissue distribution and a reduced free clearance of lignocaine, a "high-clearance" drug. This example demonstrates the danger of misinterpretation of total plasma drug levels. Actually, today, the monitoring of free drug concentrations seems to be clinically relevant only for a few drugs such as disopyramide and lignocaine, which fulfill at least the following criteria related directly to AGP: extensive protein binding; a free fraction likely to vary in clinical conditions; a low volume of distribution (31). On the other hand, presently available methods such as ultrafiltration or equilibrium dialysis have some drawbacks and are not readily available for routine analysis. This leads to the proposal to measure AGP concentrations together with total plasma drug concentrations in order to estimate the free drug concentrations (32). Finally, the pharmacogenetics of binding to AGP should also be considered: The variants differ in their binding properties, and their relative proportion varies among individuals. Using methadone, it has been shown that the decrease in binding observed within the therapeutic range (100 to 800 ng/mL) in one plasma was most probably due to the saturation of the high-affinity binding variant (ORM2 A), which represents only about one-third of this total AGP plasma concentration. It is noteworthy that the relative percent of this variant ranged from 16.9% to 47.8%, as measured in 45 subjects (21).

In conclusion, several factors determine the relative importance of AGO in drug binding in blood. The clinical relevance of monitoring the free fraction is debated. However, for the few drugs for which monitoring seems to be justified, fast and reliable techniques are needed before it can be introduced into daily practice.

REFERENCES

1. K. Schmid. α1-Acid glycoprotein. In *The Plasma Proteins* (F. W. Putnam, ed.), Academic Press, New York, pp. 183–228 (1975).
2. J. M. H. Kremer, J. Wilting, and L. H. M. Janssen. Drug binding to human α1-acid glycoprotein in health and disease. *Pharmacol. Rev.*, *40*:1–47 (1988).
3. P. Baumann, C. B. Eap, W. E. Müller, and J. P. Tillement (eds.). *Alpha1-Acid Glycoprotein. Genetics, Biochemistry, Physiological Functions, and Pharmacology* (Proceedings of a symposium held at Prilly-Lausanne, September 1988). A. R. Liss, New York (1989).
4. L. Tomei, C. B. Eap, P. Baumann, and L. Dente. Use of transgenic mice for the characterization of human α1-acid glycoprotein (orosomucoid) variants. *Hum. Genet.*, *84*:89–91 (1989).
5. J. Agneray. Glycan microheterogeneity forms of α1-acid glycoprotein (AGP): Their identification in biological fluids and variations in their relative proportions in disease states. In *Alpha1-Acid Glycoprotein. Genetics, Biochemistry, Physiological Functions, and Pharmacology* (P. Baumann, C. B. Eap, W. E. Müller, and J. P. Tillement, eds.), A. R. Liss, New York, pp. 47–65 (1989).
6. K. Schmid. Human plasma α1-acid glycoprotein—Biochemical properties, the amino acid sequence and the structure of the carbohydrate moiety, variants and polymorphism. In *Alpha1-Acid Glycoprotein. Genetics, Biochemistry, Physiological Functions, and Pharmacology* (P. Baumann, C. B. Eap, W. E. Müller, and J. P. Tillement, eds.), A. R. Liss, New York, pp. 7–22 (1989).
7. L. Dente, M. G. Pizza, A. Metspalu, and R. Cortese. Structure and expression of the genes coding for human α1-acid glycoprotein. *EMBO J.*, *6*:2289–2296 (1987).
8. C. B. Eap, C. Cuendet, and P. Baumann. Orosomucoid (α1-acid glycoprotein) phenotyping by use of immobilized pH gradients with 8 M urea and immunoblotting. A new variant encountered in a population study. *Hum. Genet.*, *80*:183–185 (1988).
9. C. B. Eap and P. Baumann. The genetic polymorphism of human α1-acid glycoprotein. In: *Alpha1-Acid Glycoprotein. Genetics, Biochemistry, Physiological Functions, and Pharmacology* (P. Baumann, C. B. Eap, W. E. Müller, and J. P. Tillement, eds.), A. R. Liss, New York, pp. 111–125 (1989).
10. P. Baumann, D. Tinguely, and J. Schöpf. Increase of α1-acid glycoprotein after treatment with amitriptyline. *Br. J. Clin. Pharm.*, *14*:102–103 (1982).
11. U. Breyer-Pfaff and M. Brinkschulte. Binding of tricyclic psychoactive drugs in plasma: Contribution of individual proteins and drug interactions. In: *Alpha1-Acid Glycoprotein. Genetics, Biochemistry, Physiological Functions, and Pharmacology* (P. Baumann, C. B. Eap, E. Müller, and J. P. Tillement, eds.), A. R. Liss, New York, pp. 351–361 (1989).
12. S. M. Wallace and R. K. Verbeeck. Plasma protein binding of drugs in the elderly. *Clin. Pharmacokinet.*, *12*:41–72 (1987).
13. D. R. Abernethy and L. Kerzner. Age effects on α1-acid glycoprotein concentration and imipramine plasma protein binding. *J. Am. Geriatr. Soc.*, *32*:705–708 (1984).

14. B. Bruguerolle, C. Arnaud, F. Levi, C. Focan, Y. Touitou, and G. Bouvenot. Physiopathological alterations of α1-acid glycoprotein temporal variations: Implications for chronopharmacology. In *Alpha1-Acid Glycoprotein. Genetics, Biochemistry, Physiological Functions, and Pharmacology* (P. Baumann, C. B. Eap, W. E. Müller, and J. P . Tillement, eds.), A. R. Liss, New York, pp. 199–214 (1989).
15. W. E. Müller. Commentary. A common single binding site for many psychotropic drugs on human α1-acid glycoprotein. Therapeutically relevant observation?, *Pharmacopsychiatr.*, *18*:257–258 (1985).
16. U. K. Walle, T. Walle, S. A. Bai, and L. S. Olanoff. Stereoselective binding of propranolol to human plasma, α1-acid glycoprotein, and albumin. *Clin. Pharmacol. Ther.*, *34*:718–723 (1983).
17. D. B. Campbell. Stereoselectively in clinical pharmacokinetics and drug development. *Eur. J. Drug. Metab. Pharmacokinet.*, *15*:109–125 (1990).
18. D. Tinguely, P. Baumann, M. Conti, M. Jonzier-Perey, and J. Schöpf. Interindividual differences in the binding of antidepressants to plasma proteins: The role of the variants of α1-acid glycoprotein. *Eur. J. Clin. Pharmacol.*, *27*:661–666 (1985).
19. C. B. Eap, C. Cuendet, and P. Baumann. Selectivity in the binding of psychotropic drugs to the variants of α1-acid glycoprotein. *Naum. Schmied. Arch. Pharmacol.*, *337*:220–224 (1988).
20. C. B. Eap, C. Cuendet, and P. Baumann. Binding of amitriptyline to α1-acid glycoprotein and its variants. *J. Pharm. Pharmacol.*, *40*:767–770 (1988).
21. C. B. Eap, C. Cuendet, and P. Baumann. Binding of d-methadone, l-methadone, and dl-methadone to proteins in plasma of healthy volunteers: Role of the variants of α1-acid glycoprotein. *Clin. Pharmacol. Ther.*, *47*:338–346 (1990).
22. C. K. Svensson, M. N. Woodruf, J. G. Baxter, and D. Lalka. Free drug concentration monitoring in clinical practice. Rationale and current status. *Clin. Pharmacokinet.*, *11*:450–469 (1986).
23. D. S. Morse, D. R. Abernethy, and D. J. Greenblatt. Methodologic factors influencing plasma binding of α1-acid glycoprotein-bound and albumin-bound drugs. *Int. J. Clin. Pharmacol. Ther. Toxicol.*, *23*:535–539 (1985).
24. Y. Sugiyama, Y. Suzuki, Y. Sawada, S. Kawasaki, T. Beppu, T. Iga, and M. Hanano. Auramine-O as a fluorescent probe for the binding of basic drugs to human α1-acid glycoprotein (AAG). The development of a simple fluorometric method for the determination of AAG in human serum. *Biochem. Pharmacol.*, *34*:821–829 (1985).
25. H. H. Zhou, A. Adedoyin, and G. R. Wilkinson. Differences in plasma binding of drugs between Caucasians and Chinese subjects. *Clin. Pharmacol. Ther.*, *48*:10–17 (1990).
26. G. Nyberg and E. Martensson. Preparation of serum and plasma samples for determination of tricyclic antidepressants: Effects of blood collection tubes and storage. *Ther. Drug. Monitor.*, *8*:478–482 (1986).
27. J. S. Kennedy, H. Friedman, J. M. Scavone, J. S. Harmatz, R. I. Shader, and D. J. Greenblatt. Effect of blood collection tubes on antidepressant concentrations. *J. Chromatogr.*, *423*:373–375 (1987).
28. P. J. Orsulak, M. Sink, and J. Weed. Blood collection tubes for tricyclic antidepressant drugs: a reevaluation. *Ther. Drug Monitor.*, *6*:444–448 (1984).
29. R. Zini, P. Riant, J. Barré, and J. P. Tillement. Disease-induced variations in plasma protein levels. Implications for drug dosage regimens (Part I). *Clin. Pharmacokinet.*, *19*:147–159 (1990).
30. R. Zini, P. Riant, J. Barré, and J. P. Tillement. Disease-induced variations in plasma protein levels. Implications for drug dosage regimens (Part II). *Clin. Pharmacokinet.*, *19*:218–229 (1990).
31. J. Barré, F. Didey, F. Delion, and J. P. Tillement. Problems in therapeutic drug monitoring: Free drug level monitoring. *Ther. Drug Monitor.*, *10*:133–143 (1988).
32. P. A. Routledge. The plasma protein binding of basic drugs. *Br. J. Clin. Pharmacol.*, *22*:499–506 (1986).

25

Impact of a Clinical Pharmacokinetic Service on Patients Treated with Aminoglycosides: A Cost-Benefit Analysis

Christopher J. Destache, Sharon K. Meyer, Marvin J. Bittner, and Kenneth G. Hermann *Creighton University Schools of Pharmacy and Medicine, Omaha, Nebraska*

I. INTRODUCTION

During the last 25 years, considerable progress has been made in drug therapy (1). The application of pharmacokinetic principles to improve patient care has comprised clinical pharmacokinetics, performed in some cases through a hospital-based clinical pharmacokinetic service (CPS). Aminoglycosides are often monitored using pharmacokinetic principles. These drugs are easily assayed (2). Additionally, these drugs display a wide range of interpatient and intrapatient variability in their disposition (3).

Several studies have assessed the utilization of CPS on patients treated with aminoglycosides. Burn patients who were dosed pharmacokinetically were less likely to die from their injuries (4). However, CPS patients were hospitalized longer. Another study found CPS-monitored patients had considerable savings in a 118-patient study of aminoglycoside use (5) and in another study confined to the use of gentamicin in burn patients (6).

These results may reflect the impact of appropriate therapy rather than a CPS. The authors of another study evaluated the clinical courses of 612 patients with Gram-negative bacteremia (7). Appropriate initial antibiotic therapy was associated with a significant reduction in mortality regardless of the severity of the underlying disease. A 46-patient retrospective cohort study found a 6-day decrease in hospital stay and a 33.4-h decrease in antibiotic therapy when patients were monitored by a CPS (8). We sought to estimate the cost-benefit ratio of using the CPS for patients receiving aminoglycosides for Gram-negative infections.

II. METHODS

Inpatients at Saint Joseph Hospital, more than 18 years old and receiving intramuscular or intravenous aminoglycoside antibiotics for proven or suspected Gram-negative infections, were studied. Patients receiving aminoglycoside therapy for systemic enterococcal infections were also included. From January through June 1988, 200 patients were enrolled. Patients were randomized to either CPS monitoring or to a control group with no CPS monitoring. Randomization assignments were determined by a table of random numbers and placed in sealed envelopes. No CPS monitoring was defined as no intervention by clinical pharmacists.

Among data collected on all patients were demographic data [age, gender, height, weight, and calculated creatinine clearance (9)]. Vital signs were recorded from the start of aminoglycoside therapy until the time of hospital discharge. The length of hospital stay and length of aminoglycoside therapy were recorded. The APACHE II system was used to assess the severity of acute illness (10). Initial steady-state predictions of serum concentrations were calculated based on population data for volume of distribution (Vd) and half-life ($t_{1/2}$) before any levels were drawn and reported (11).

The CPS utilized a one-compartment, computerized, Bayesian model for pharmacokinetic interpretation of serum levels. The target peak therapeutic ranges were based on literature data. The target trough therapeutic ranges are concentrations of < 2.0 μ/mL for gentamicin and tobramycin and 5 to 10 μg/mL for amikacin (2). For every set of serum drug concentrations results, a note in the chart included an interpretation and recommendations for dose optimization.

Analysis of variance (ANOVA) was used to compare length of hospital stay, length of aminoglycoside therapy, age, weight, height, APACHE II score, initial peak and trough concentrations, and number of dosage changes. Multiple regression analyses were used to determine whether changes in hospital stay were due to the CPS. Regressions were performed using length of hospital stay as a continuous variable. The Fisher's exact test compared the presence or absence of aminoglycoside-induced nephrotoxicity. Nephrotoxicity was defined as a >0.5 mg/dL rise in serum creatinine (12). Cost-benefit was performed by comparing the groups:

$$BC = \frac{\Sigma B_t}{\Sigma C_t}$$

where B_t = total benefits for time period t and C_t = total costs for time period t (13). Total benefits were measured by decreased length of hospital stay. Total costs were calculated using a cost-to-charge ratio. The direct costs were calculated by taking the patient's charges of hospitalization (excluding physicians charges) and multiplying a cost-to-charge ratio for the department within the hospital (e.g., nursing, pharmacy, laboratory, radiology) for the time period the patient was part of the study. All direct departmental costs were then added together for that patient to provide the direct cost of hospitalization for each patient. The hospital fiscal expense report for the year ending August 1987 was used as the basis for the calculation of the cost-to-charge ratio during the study period. This method of calculating direct costs is similar to a study that calculated cost of nephrotoxicity (14).

Principle direct costs of the CPS involve salaries and supplies. The variable and fixed costs of monitoring a CPS patient were extrapolated to an annual cost based on 500 patients. Fixed costs were depreciated where appropriate.

III. RESULTS

Two hundred patients were enrolled. One hundred and fifty-nine (79.0%) patients received gentamicin, 22 (10.9%) patients received tobramycin, seven (3.5%) patients had amikacin, and 12 (5.9%) patients had more than one aminoglycoside in succession. Thirty-three (16.4%) of the patients died, and 24 (11.9%) had nephrotoxicity.

The control group consisted of 90 patients. The CPS group consisted of 110 patients; for 35, CPS recommendations were not always followed. These 35 patients were excluded from analysis. Thus, this analysis compares 70 patients in the control group and 75 in the CPS group.

A total of 23% CPS and 24% control patients were admitted to an intensive-care unit. Demographic data and study results are depicted in Table 1. The length of hospital stay was not significantly different between the two groups ($p=0.087$). The mean difference in hospital stay is calculated as 120.21 h/patient.

Peak concentrations in the CPS and control groups were compared. The first peak concentrations were obtained within the first 48 h of therapy. Second peak concentrations were drawn 2 to 7 days into therapy for both groups.

The first peak concentrations were categorized as "adequate" (> 5.0 μg/mL and > 20.0 μ/mL for amikacin), "low" (< 5.0 μg/mL; < 20 μ/mL for amikacin), or not drawn. Of 71 first peak concentrations in the CPS group, 23 (33%) were adequate, 44 (62%) were low, and 4 (6%) were not drawn. Of the 68 first peak concentrations in the control group, 22 (32%) were adequate, 35 (51%) were low, and 12 (17%) were not drawn.

In the CPS group, 22 (50%) of the 44 cases with "low" first peak concentrations achieved an adequate second peak concentration. In four (9%), the second peak was inadequate, and in 18 (42%), a second peak was not drawn. In contrast, 14 (40%) in the control group had a low second peak concentration ($p < 0.05$; Fisher's exact test). Additionally, 11 (31%) of control group patients had an adequate second peak and 10 (29%) did not have a second peak drawn.

Table 1 Study Results

Variable	CPS Group (mean ± SD)	Control Group (mean ± SD)
Age	61.21 ± 22.65	55.75 ± 19.62
Weight (kg)[a]	65.15 ± 19.12	72.41 ± 18.13
APACHE II score	12.13 ± 8.48	10.82 ± 6.88
First measured peak concentration (μg/mL)	5.0 ± 4.5	4.3 ± 3.4
First extrapolated peak concentration (μg/mL)	5.7 ± 4.7	5.1 ± 4.2
Length of hospital stay (h)	322.67 ± 270.28	442.89 ± 536.81
Length of aminoglycoside therapy (h)	109.81 ± 90.31	108.95 ± 79.09
Time for temperature to decrease (h)[a]	50.05 ± 79.38	92.23 ± 122.50
Number of dosage changes[b]	1.14 ± 0.97	0.64 ± 0.83
Direct costs/patient ($)[c]	7,102.56 ± 8,898.19	13,758.64 ± 22,874.31

[a] $p<0.05$.
[b] $p<0.005$.
[c] $p<0.02$.

Table 2 Length of Hospitalization Regressions

Independent Variables	Length-of-Stay Coefficients (95% CI)
Constant	−406.31 (−961.16; 32.63)
Time for temperature to decrease to normal or baseline	1.36 (0.57; 1.51)
Number of levels drawn	45.95 (23.88; 64.85)
Citrobacter sp.	400.37 (−15.87; 587.73)
Pseudomonas sp.	184.98 (180.19; 468.54)
Klebsiella oxytoca	273.27 (−9.95; 469.54)
Parkinsonism	−354.48 (−586.85; 469.54)
Escherichia coli	123.41 (82.99; 325.78)
Asthma	−303.00 (−635.65; −43.70)
No pharmacy involvement	95.30 (−36.87; 190.64)

High trough concentrations were demonstrated after dosage modification in 3 (4.3%) of CPS patients and 8 (14%) in control patients ($p > 0.05$). No difference was demonstrated between the two groups with respect to calculated aminoglycoside $t_{1/2}$ or Vd utilizing a one-compartment model. Six (8%) of CPS patients and 10 (14.4%) of control patients were found to develop nephrotoxicity ($p=0.078$). Direct costs of hospitalization by our method demonstrated significant differences on ANOVA. The total direct costs for the CPS group were $7,102.56 ± 8,898.18 and for the control group $13,758,64 ± 22,874.31 ($p < 0.02$).

Stepwise multiple regression analyses using length of hospital stay were performed. Those patients whose physician accepted <100% of CPS recommendations were excluded from the regression analysis (Table 2). The variables which were determined from the regression model were Gram-negative organisms (*Pseudomonas* sp., *Citrobacter* sp., *Klebsiella oxytoca*, and *Escherichia coli*), number of serum levels drawn, time for temperature to decrease, parkinsonism, asthma, and no pharmacist's involvement in aminoglycoside dosing. The regression model captured 54.2% of the variation in this variable and was significant ($p<0.0001$).

Based on the observed 68% acceptance rate, the total annual saving from the CPS is approximately $2,263,040.00. The mean total costs of providing the service to the CPS group is calculated as $6,363.00 (average of $85.00/patient) for the entire study. Annual costs of providing the service is calculated to be $42,500.00. The annual savings subtracting the annual costs of the service is calculated as $2,220,540.00. The benefit-to-cost ratio based on these figures is 52.25.

IV. DISCUSSION

A CPS was associated with achievement of desired peak aminoglycoside concentrations. Concentrations of > 5.0 μg/mL have been predictive of efficacy (15). The CPS affected patients with sepsis and pneumonia. The average length of hospital stay of CPS patients was 415 h shorter than the controls. For pneumonia, CPS hospital stay averaged 151 h shorter.

Multiple regression analyses demonstrated that the number of serum levels drawn and *Pseudomonas* infections had the highest correlation with length of hospital stay.

Additionally, *E. coli* and the time for temperature to decrease baseline were significantly correlated with length of hospital stay. Parkinsonism and asthma were part of the regression analysis and were correlated with a shorter length of hospital stay. Additionally, no pharmacy intervention in the dosing of aminoglycosides was correlated with a longer length of hospitalization. Of the factors in the regression analyses, only no pharmacy intervention in the dosing of aminoglycosides is controllable. Finally, CPS patients incurred lower direct hospitalization costs. CPS monitoring of all hospitalized adults on aminoglycosides is projected to save ~ $2,220,540.00 annually.

In this study, a cost-to-charge ratio was used to determine the direct costs of hospitalization. Other investigators have used other methods (16,17). The reason for the special cost analyses is that, in most hospital fiscal departments, routine cost accounting omits the calculation of direct costs of hospitalization. Indeed, this study calculated direct costs from each department using the annual expenses for the last fiscal year as the basis for calculating our cost-to-charge ratio.

To our knowledge, this is the first study that used the cost-to-charge ratio, prospectively, for hospitalized patients to determine significant differences in direct hospitalization costs due to a clinical pharmacokinetic service. Additional information on the direct costs of nephrotoxicity has been obtained.

The prospective payment system is emphasizing appropriate utilization of resources. It may, therefore, be appropriate for hospitals to initiate a pharmacokinetic service.

REFERENCES

1. G. Levy. Applied pharmacokinetics: A prospectus. In *Applied Pharmacokinetics*, 2nd ed. (W. E. Evans, J. J. Schentag, and W. J. Jusko, eds.), Applied Therapeutics, Spokane, Washington, pp. i–xxi (1986).
2. D. E. Zaske. Aminoglycosides. In *Applied Pharmacokinetics*, 2nd ed. (W. E. Evans, J. J. Schentag, and W. J. Jusko, eds.), Applied therapeutics, Spokane, Washington, p. 370 (1986).
3. D. E. Zaske, R. J. Cipolle, J. C. Rotschafer, et al. Gentamicin pharmacokinetics in 1,640 patients: Method for control of serum concentrations. *Antimicrob. Agents Chemother.*, *21*:407–411 (1982).
4. J. L. Bootman, A. J. Wertheimer, D. E. Zaske, et al. Individualizing gentamicin dosage regimens in burn patients with Gram-negative septicemia: A cost-benefit analysis. *J. Pharm. Sci.*, *68*:267–272 (1979).
5. B. J. Kimelblatt, K. Bradbury, L. Chodoff, et al. Cost-benefit analysis of an aminoglycoside monitoring service. *Am. J. Hosp. Pharm.*, *43*:1205–1209 (1986).
6. D. E. Zaske, J. L. Bootman, L. B. Solem, et al. Increased burn patient survival with individualized dosages of gentamicin. *Surgery*, *91*:142–149 (1982).
7. B. E. Kreger, D. E. Craven, and W. R. McCabe. Gram-negative bacteremia: Re-evaluation of clinical features and treatment in 612 patients. *Am. J. Med.*, *68*:344–355 (1980).
8. C. J. Destache, S. M. Meyer, B. G. Ortmeier, et al. Impact of a clinical pharmacokinetic service on patients treated with aminoglycosides for Gram-negative infections. *DICP: The Annuals of Pharmacotherapy*, *23*:33–37 (1989).
9. R. W. Jelliffe, and S. M. Jelliffe. A computer program for estimation of creatinine clearance from unstable serum creatinine levels, age, sex, and weight. *Math. Biosci.*, *14*:17–24 (1972).
10. W. A. Knaus, E. A. Draper, D. P. Wagner, et al. APACHE II: A severity of disease classification system. *Crit. Care Med.*, *13*:818–829 (1985).
11. R. W. Jelliffe. Clinical applications for pharmacokinetics and control theory: Planning, monitoring, and adjusting dosage regimens of aminoglycosides, lidocaine, digoxitin, and digoxin. In *Topics of Clinical Pharmacology and Therapeutics*, (R. Maronde, ed.), Springer-Verlag, New York, pp. 26–82 (1986).

12. R. D. Moore, C. R. Smith, and P. S. Leitman. Increased risk of renal dysfunction due to interaction of liver disease and aminoglycosides. *Am. J. Med.*, *80*:1093–1097 (1986).
13. W. E. McGhan, C. R. Rowland, and J. L. Bootman. Cost-benefit and cost-effectiveness: Methodologies for evaluating innovative pharmaceutical services. *Am. J. Hosp. Pharm.*, *36*:368–370 (1979).
14. J. M. Eisenberg, J. Koffer, H. A. Glick, et al. What is the cost of nephrotoxicity associated with aminoglycosides? *Ann. Intern. Med.*, *107*:900–909 (1987).
15. R. D. Moore, C. R. Smith, and P. S. Lietman. The association of aminoglycosides plasma level with mortality in patients with Gram-negative bacteremia. *J. Infect. Dis.*, *194*:443–448 (1984).
16. J. L. Bootman, D. E. Zaske, A. I. Wertheimer, et al. Cost of individualizing aminoglycoside dosage regimen. *Am. J. Hosp. Pharm.*, *36*:368–370 (1979).
17. K. D. Crist, M. C. Nahata, and J. Ety. Positive impact of a therapeutic drug-monitoring program on total aminoglycoside dose and cost of hospitalization. *Ther. Drug Monitor.*, *9*:306–310 (1987).

Part II

Pharmacology

26

Pregnancy in Cyclosporine-Treated Renal Transplant Recipients

Ester Zylber-Katz, Arnon Samueloff, Ethel Ackerman, Olga Peleg, and Michael M. Friedlaender *Hadassah University Hospital, Jerusalem, Israel*

I. INTRODUCTION

Cyclosporine A (CsA), a potent immunosuppressive agent, has in recent years been used increasingly in renal transplantation, resulting in improved allograft survival (1). Among its biological properties, CsA has the ability to suppress the immune response without significant myelotoxicity (2). CsA is also known to have side effects, causing nephrotoxicity, hepatotoxicity, hypertension, hirsutism, tremors, and paresthesias (3). In view of the increase in life expectancy and improved fertility of renal transplant patients, questions have been raised regarding the effect of CsA during pregnancy, labor, and delivery (4). The clinical experience with pregnancy in women treated with CsA is still sporadic (5).

This chapter describes the management of successful pregnancies in three patients receiving CsA after kidney transplantation. Maternal renal function was followed by BUN, serum creatinine levels, creatinine clearance, and protein excretion.

CsA 12-h trough blood-level monitoring, using the original Sandoz radioimmunoassay (RIA) with polyclonal antibodies, was carried out regularly in our patients (6). More recently, two new analytical methods to monitor CsA have been applied. One method, specific for CsA (Sp-CsA), consists of a RIA for CsA with monoclonal antibodies (7), the second measures both CsA and its metabolites (total CsA) by an immunochemical fluorescence polarization method (TDx) (8). Data of CsA measurements in maternal blood and breast milk, as well as in the child's blood, are also presented.

II. SUBJECTS, METHODS, AND RESULTS

A. Patient 1

A 22-year-old nulliparous woman, 47 kg body weight, suffering from familial juvenile medullary cystic kidney disease (juvenile nephronophthisis), received a child's cadaveric renal allograft in January 1987. Figure 1 shows renal function, CsA dose, and blood concentrations. After 6 months, despite early hemiinfarction of the graft, blood creatinine level had stabilized between 150 to 180 µmol/L with creatinine clearance (C_c) between 35 and 60 mL/min. The patient was immunosuppressed with CsA, beginning at 400 mg/day and gradually reduced to 220 mg/day after 18 months. Prednisone 120 mg/day was tapered to 10 mg/day at 6 months. The patient also received azathioprine, 50 mg/day. Prepregnancy total CsA blood levels, as measured by the original Sandoz RIA with polyclonal antibody, ranged between 200 and 500 ng/mL (therapeutic range 250 to 1000 ng/mL). Liver function tests were normal; BUN, uric acid, and cholesterol blood levels were slightly above normal. A salt-losing tendency, presumably from native kidneys, caused fluctuations in clinical hydration status but could be corrected by appropriate diet.

The patient became pregnant 21 months following transplantation. The first trimester was uneventful. In the nineteenth week of pregnancy, renal dysfunction

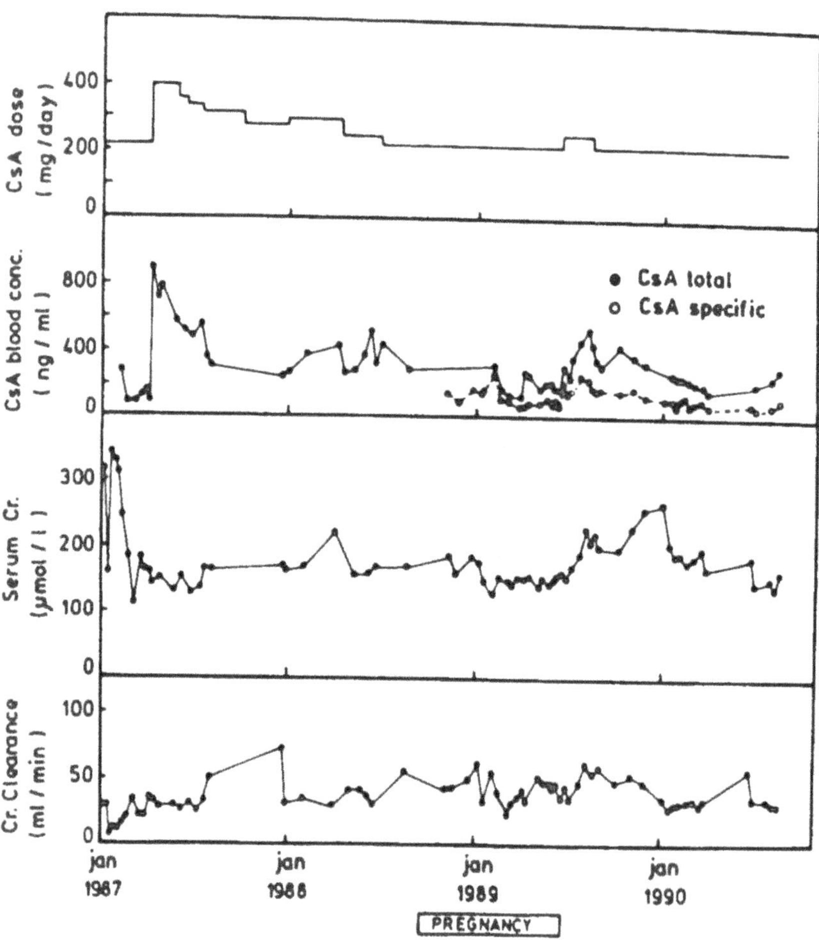

Figure 1 Cyclosporine A (CsA) dose and serial measurements of CsA blood concentration, serum creatinine, and creatinine clearance in patient 1 after transplantation.

caused hospital admission; C_c was 15 mL/min. A urinary tract infection (*Pseudomonas aeroginosa* and *Enterobacter* sp.) was detected and treated with cefsulodin(i.v.), 2 g × 3 per day. After rehydration, renal function rapidly recovered and C_c rose to 40 mL/min. Subsequently, the patient remained hospitalized for continued close monitoring of the mother and the fetus. Blood pressure was stable, between 110/70 and 130/90 mm Hg. CsA dose during pregnancy was 220 mg/day, and regular monitoring of CsA blood levels 12 h after administration, determined by the new radioimmunoassay with specific monoclonal antibodies for CsA (Sp-CsA) (Cyclo Trac SP—whole blood from INCSTAR) (7) ranged between 49 and 268 ng/mL (mean ± SD = 117.2 ± 49, n = 25) (therapeutic range 100 to 300 ng/mL). CsA and metabolites blood levels (total CsA) were measured by fluorescence polarization immunoassay (TDx from Abbott) (8), and ranged from 125 to 308 ng/mL (mean ± SD = 203.0 ± 57.1, n = 12) (therapeutic range 250 to 1000 ng/mL). The mean ratio of total to Sp-CsA concentration was 2.3 ± 0.9, n = 12. Spontaneous labor started at 41 weeks of gestation. The patient was treated with magnesium sulfate infusion and hydralazine, 6.25 mg i.m., for preeclampsia. Hydrocortisone (50 mg × 3), i.v., was also given. A healthy female was born by normal vaginal delivery, weighing 2150 g, with an Apgar score of 8 and 9 at 1 and 5 min, respectively. After delivery, the mother's blood pressure stabilized at 140/90 mm Hg. The dose of CsA was increased to 260 mg/day, prednisone to 20 mg/day, and azathioprine was continued at a dose of 50 mg/day. Transient fever with renal dysfunction (C_c 16 mL/min) occurred at 5 days postpartum. Urinary tract infection with proteus morabilis was diagnosed. Treatment with amoxycillin 500 mg × 3/day caused rapid resolution of the fever, and renal function returned to normal values. Postpartum blood pressure ranged between 110/80 and 120/70 mm Hg.

Maternal blood samples were obtained at 1, 4, 5, 6, and 8 days after delivery for the measurement of CsA levels, and blood samples were obtained from the baby at 24, 48, and 72 h from time of delivery for general biochemistry tests and CsA levels (Table 1). The baby's Sp-CsA level at 24 h after birth was 149 ng/mL, and total CsA concentration 172 ng/mL. The ratio T/Sp was 1.2. At 48 h the Sp-CsA and total CsA concentration obtained were at the lower limit of detection for both assays. CsA was also present in appreciable concentrations in samples of the mother's breast milk obtained on days 5, 6, and 8 postpartum. The patient was advised against breast feeding, and mother and baby were discharged on the tenth day postpartum. On regular follow-up the baby showed normal development with normal weight gain. The mother's renal function remained normal for the next 13 months, and CsA blood levels measured routinely ranged from 61 to 253 ng/mL (mean ± SD = 135.8 ± 53.0, n = 23) for CSA-SP, and from 161 to 537 ng/mL (mean ± SD = 322 ± 102, n = 20) for total CsA.

B. Patient 2

A 37-year-old woman, mother of seven, 95 kg body weight, suffering from adult polycystic kidney disease, received a second cadaveric renal allograft in October 1985, after the first transplant in 1984 failed. Figure 2 shows renal function, CsA dose, and blood levels since the transplantation. Graft function was good and stabilized 1 year posttransplant with serum creatinine level between 98 and 126 μmol/L (mean ± SD = 108.5 ± 8.7, n = 14) and C_c between 55 and 121 mL/min (mean ± SD = 81.1 ± 19.7, n = 13). Posttransplant dose of CsA was 640 mg/day and was reduced to 300 mg/day by October 1986, resulting in total CsA levels of 700 to 800 ng/mL (original Sandoz RIA). The CsA dose was further reduced to 180 mg/day, which resulted in total CsA levels

Table 1 Patient 1, Cyclosporine A Measured in Mother's Blood, Breast Milk, and Infant's Blood

Mother	Days After Delivery	Time After Dose (h)	Cyclosporine Concentration (ng/mL)				B/M Ratio	
			Blood		Milk			
			Total	Specific	Total	Specific	Total	Specific
	1	8	307	204	—	—	—	—
	4	5	496	388	—	—	—	—
	5	3	1069	1036	188	96	5.7	10.8
	6	4	611	336	254	156	2.4	2.2
	8	12	231	147	218	134	1.1	1.1
Infant	**Age (h)**							
	24		172	149				
	48		46	12				
	72		ND	ND				

between 300 and 400 ng/mL. In addition to CsA, the patient received prednisone, 7.5 mg/day, and azathioprine, 50 mg/day. To control hypertension, the patient was treated with atenolol, 100 mg daily; methyldopa, 500 to 1000 mg/day; furosemide (40 mg/week), and nifedipine, 60 mg/day.

The patient was out of the country for about 6 months. On her return in April 1989 she was found to be pregnant (week 27, by ultrasound). The pregnancy continued uneventfully. Serum creatinine was between 84 and 105 μmol/L (mean ± SD = 91.2 ± 6.7, n = 11) and the C_c 73 to 107 mL/min (mean ± SD = 91.1 ± 14.4, n = 8). CsA dosage was 180 mg/day, which resulted in therapeutic blood levels, as measured by the new methods, as follows: total CsA (TDX) ranged from 139 to 442 ng/mL (mean ± SD = 311.0 ± 93.1, n = 9) and Sp-CsA (Cyclo-Trac) from 50 to 157 ng/mL (mean ± SD = 112.5 ± 34.7, n = 10).

At 36 weeks of gestation the patient was hospitalized due to mild preeclampsia; blood pressure was 150/90 mm Hg and proteinuria +1, C_c was 76 mL/min. Biochemical and hematological blood tests were within normal limits. The fetus showed normal growth and development. During hospitalization an episode of bacteriuria (*E. coli*) was treated by amoxycillin, 500 mg × 3 per day and blood pressure was maintained between 120/80 and 150/100 mm Hg. Spontaneous vaginal delivery occurred at 30 weeks of pregnancy, and a healthy male was born, weighing 2610 g, with Apgar scores of 9 and 10 after 1 and 5 min, respectively. On the day of delivery the patient received hydrocortisone (100 mg × 3) i.v., and the postpartum course was unremarkable. Her blood pressure ranged from 120/80 to 140/90 mm Hg and C_c was 80 mL/min. At delivery, simultaneous samples of the mother's peripheral blood and cord blood were obtained for measurement of CsA levels. Additional samples of the baby's blood were obtained 1 and 3 days later for biochemistry tests and CsA levels. Samples of mother's blood and breast milk were obtained simultaneously on the third and fourth day postpartum for measurement of CsA levels and calculation of CsA

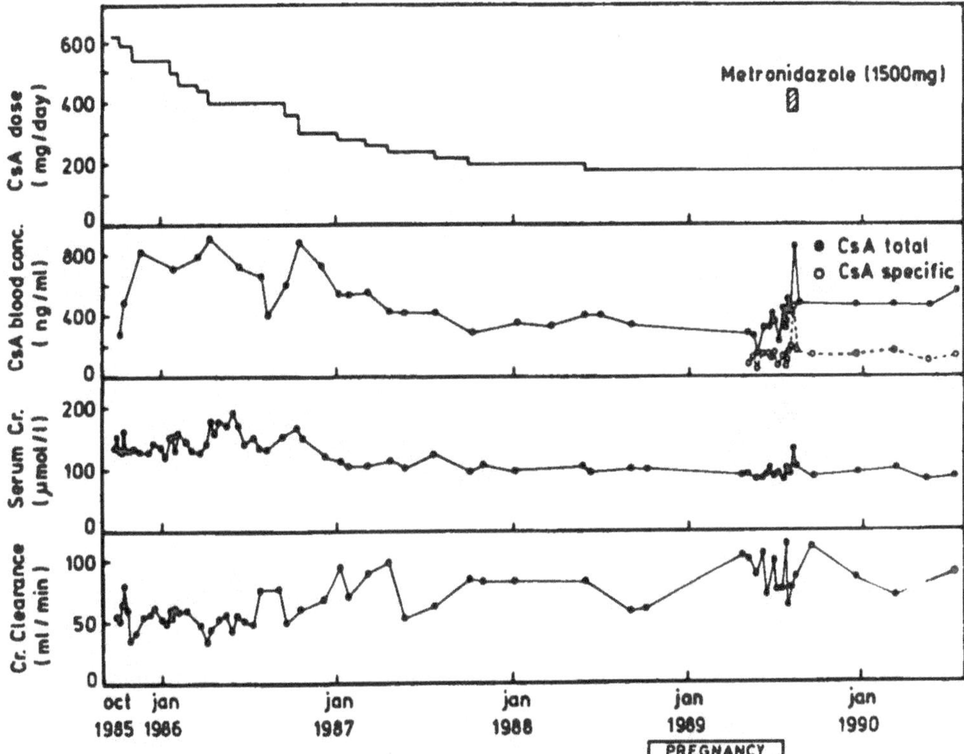

Figure 2 Cyclosporine A (CsA) dose and serial measurements of CsA blood concentration, serum creatinine, and creatinine clearance in patient 2 after transplantation.

distribution ratio between blood and milk (Table 2). Mother and baby were discharged on the fifth postpartum day. However, 3 days later the mother was rehospitalized due to a fever of 40°C caused by an *E. coli* urinary tract infection which did not respond to ampicillin treatment. Subsequently, the patient received cefuroxime (1.5 g × 3 per day) and metronidazole (0.5 g × 3 per day), with rapid resolution of the symptoms. It was noted that during the 6 days of treatment with metronidazole, a dramatic increase in both total and Sp-CsA blood level occurred, which returned to the therapeutic range when treatment with metronidazole was discontinued. A slight increase in serum creatinine and decrease in C_c was simultaneously observed (9). Both mother and infant progressed normally. During the following 12 months the mother was maintained on a CsA dose of 180 mg/day and the total CsA concentrations ranged from 420 to 902 ng/ml (mean ± SD = 537.9 ± 191.3, n = 8) and for specific CsA from 94 to 429 ng/ml (mean ± SD = 198.4 ± 118.1, n = 9).

C. Patient 3

A 32-year-old woman, mother of one, 46 kg body weight, suffering from chronic glomerulonephritis, received a living, nonrelated renal allograft in March 1988, in a

Table 2 Patient 2, Cyclosporine A Measured in Mother's Blood, Breast Milk, and Infant's Blood

Mother	Days After Delivery	Time After Dose (h)	Cyclosporine Concentration (ng/mL) Blood Total	Blood Specific	Milk Total	Milk Specific	B/M Ratio Total	B/M Ratio Specific
	At delivery	12	320	87	—	—	—	—
	3	12	349	106	247	124	1.4	0.9
	4	10	520	172	375	183	1.4	1.0
Infant	**Age (h)**							
	At birth		284	79				
	24		206	40				
	72		ND	ND				

neighboring country. She returned to our clinic in May 1988. Figure 3 shows renal function CsA dose, and blood concentration levels. Renal function was stable during the first 15 months posttransplant, with a serum creatinine level between 83 and 114 μmol/L (mean ± SD = 94.7 ± 8.4, n = 20) and C_c between 20.3 and 50.6 mL/min (mean ± SD = 38.4 ± 9.3, n = 19). Posttransplant CsA dose was 300 mg/day, which was reduced to 160 mg/day after 5 months. The total CsA blood level was between 400 and 866 ng/mL (mean ± SD = 653.3 ± 198.3, n = 7), and the Sp-CsA concentration ranged from 114 to 233 ng/mL (mean ± SD = 188.4 ± 29.7, n = 8). In addition, the patient received prednisone, 25 mg/day, which was tapered to 7.5 mg/day after 1 year, and azathioprine, 50 mg/day. To control blood pressure, the patient received atenolol, 50 mg/day.

The patient became pregnant 15 months after transplant. During pregnancy, mean serum creatinine was 102.6 ± 18.7 μmol/L (n = 20) and mean C_c was 42.2 ± 8.8 mL/min (n = 18). CsA dose was maintained at 160 mg/day, resulting in a mean total CsA level of 232.6 ± 58.9 mg/mL (n = 9) while the mean Sp-CsA level was 100.7 ± 35.7 ng/mL) (n = 16). At 33 weeks of pregnancy the patient was hospitalized due to vaginal bleeding. Ultrasonic examination of the fetus showed a symmetric intrauterine growth retardation (IUGR) with a normal volume of amniotic fluid. At 35 weeks she presented mild preeclampsia, with blood pressure up to 165/100 mm Hg; no proteinuria was present. The ratio of lecithin to sphingomyelin (L/S), determined by amniocentesis at 37 weeks of calculated gestational age, was >2.0. Immediate caesarean section under general anesthesia was performed, resulting in delivery of a 1640-g male infant with Apgar scores of 9 and 10 at 1 and 5 min, respectively, and without any apparent clinical abnormalities. At delivery, simultaneous samples of maternal blood and cord blood were obtained (Table 3). Additional blood samples from the baby were obtained at 19, 44, and 68 h after birth. The elimination half-life of Sp-CsA in the neonate, calculated from the concentration-time data, was 30.2 h. The baby developed normally, and after 20 days his weight was 2040 g. The mother was

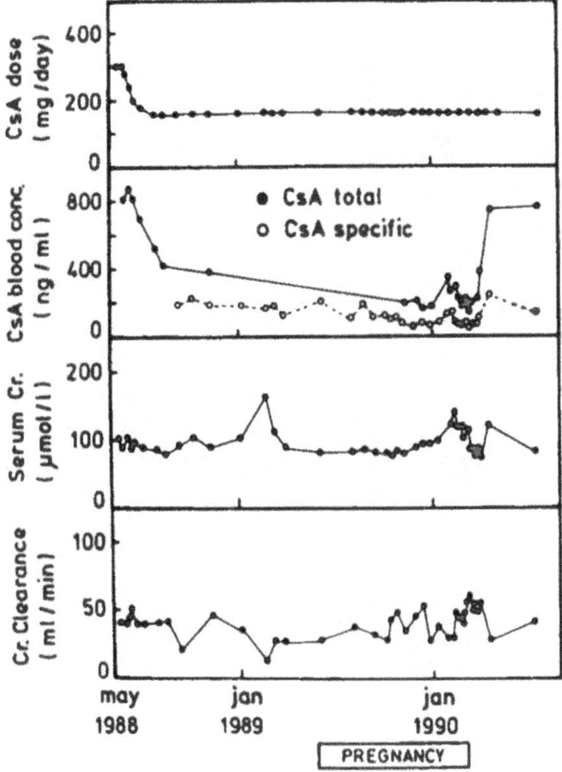

Figure 3 Cyclosporine A (CsA) dose and serial measurements of CsA blood concentration, serum creatinine, and creatinine clearance in patient 3 after transplantation.

Table 3 Patient 3, Cyclosporine A Measured in Mother's Blood and Infant's Blood

			Cyclosporine Conc. (ng/mL)	
			Blood	
Mother	Days After Delivery	Time After Dose (h)	Total	Specific
	At delivery	10	223	81
	1.5	12	147	59
	2.5	12	205	78
	3.5	12	245	108
Infant	Age (h)			
	At birth		154	35
	19		133	29
	44		71	17
	68		34	12

advised to avoid breast feeding and was discharged without change in kidney function. She continues to receive CsA 160 mg/day, and the total CsA blood level ranges between 218 and 775 ng/mL (mean ± SD = 532.8 ± 272.2, n = 4); the specific CsA level ranges between 70 and 233 ng/mL (mean ± SD 151.3 ± 68.9, n = 4).

III. DISCUSSION

With the improvement in both length and quality of life of renal transplant recipients, pregnancy in such patients has become a more frequent occurrence. To date, cases reported in the literature have not shown a significant harmful effect of CsA on fetal development or on the allograft survival (10). The early reports were optimistic about the outcome of pregnancy in patients with well-functioning kidney transplants and treated with CsA (11–23). Although intrauterine growth retardation was observed (13–17), no congenital abnormalities were detected. The possibility of a direct effect of CsA on fetal growth was considered (14), but the influence of risk factors such as impaired renal function and hypertension could not be excluded.

CsA is a lipid-soluble compound, highly protein bound and almost completely eliminated by hepatic metabolism (18). CsA was found to cross the placenta and appeared in the fetal circulation at concentrations similar to those in the mother (11,13). Others did not find detectable levels of CsA in the fetus, probably due to low maternal CsA levels (12). The management of pregnancies in renal transplant patients requires special attention to blood pressure, renal function, and urinary tract infection. In our patients, renal function during pregnancy was stable. The normal increase in glomerular filtration rate (GFR) during pregnancy (19) occurred only in the second case. A mild preeclampsia was observed in all the cases and was treated accordingly. The infants' development was normal, although intrauterine growth retardation was seen in two of the cases. Routine monitoring of the CsA blood levels in the three cases showed no major fluctuations, thus CsA dosage did not require adjustment. After delivery, measurable CsA blood levels were found in the baby, which were undetectable after 2 to 3 days. Breast feeding was avoided, as CsA was also detected in samples of breast milk.

In conclusion, there appears to be no contraindication to the use of CsA in pregnancy of renal transplant recipients, when close monitoring is provided.

REFERENCES

1. N. L. Tilney, E. L. Milford, C. B. Carpenter, J. M. Lazarus, T. B. Strom, and R. L. Kirkman. Long-term results of cyclosporine treatment in renal transplantation. *Transplant. Proc.*, *8*(Suppl. 1):179–185 (1986).
2. A. Laupacis, P. A. Keown, R. A. Ulan, N. McKenzie, and C. R. Stiller. Cyclosporine A: A powerful immunosuppressant. *Can. Med. Assoc. J.*, *126*:1041 (1982).
3. R. J. Ptachcinski, G. J. Burckart, and R. Venkataramanan. Cyclosporine. *Drug Intell. Clin. Pharm.*, *19*:90 (1985).
4. R. Jane Lau and S. R. Scott. Pregnancy following renal transplantation. *Clin. Obstet. Gynecol.*, *28*:339 (1985).
5. I. Cockburn, P. Krupp, and C. Monka. *Present experience of Sandimmun in pregnancy. Transplant. Proc.*, *21*:3730-3732 (1989).
6. E. Zylber-Katz and L. Granit. Cyclosporine blood concentration determined by different assay methods in heart transplant patients. *Ther. Drug Monitor*, *11*:592 (1989).
7. Cyclo-Trac SP—whole blood radioimmunoassay. INCSTAR Corporation, Stillwater, MN, Part #11598 (revised 3/88).

8. TDx cyclosporine and metabolites procedure. Personal communication, Abbott Laboratories, Diagnostic Division (October 1987).
9. E. Zylber-Katz, D. Rubinger, and Y. Berlatzky. Cyclosporine interactions with metronidazole and cimetidine. *Drug Intell. Clin. Pharm.*, 22:504 (1988).
10. D. A. Burrows, T. J. O'Neil, and T. L. Sorrells. Successful twin pregnancy after renal transplant maintained on cyclosporine A immunosuppression. *Obstet. Gynecol.*, 72:459 (1988).
11. G. J. Lewis, C. A. R. Lamont, H. A. Lee, and M. Slapak. Successful pregnancy in a renal transplant recipient taking cyclosporin A. *Br. Med. J.*, *286*:603 (1983).
12. G. Klintmalm, P. Althoff, G. Appleby, and E. Segerbrandt. Renal function in a newborn baby delivered of a renal transplant patient taking cyclosporine. *Transplantation*, *38*:198 (1984).
13. S. M. Flechner, A. R. Katz, A. J. Rogers, Ch. Van Buren, and B. D. Kahan. The presence of cyclosporine in body tissues and fluids during pregnancy. *Am. J. Kidney Dis.*, *5*:60 (1985).
14. M. D. Pickrell, R. Sawers, and J. Michael. Pregnancy after renal transplantation: Severe intrauterine growth retardation during treatment with cyclosporin A. *Br. Med. J.*, *296*:825 (1988).
15. S. Niesert, H. Gunter, and U. Frei. Pregnancy after renal transplantation. *Br. Med. J.*, *296*:1736 (1988).
16. P. F. Williams, I. G. M. Brons, D. B. Evans, R. E. Robinson, and R. Y. Calne. Pregnancy after renal transplantation. *Br. Med. J.*, *296*:1400 (1988).
17. Z. Varghese, S. F. Lui, O. N. Fernando, P. Sweny, and J. F. Moorhead. Pregnancy after renal transplantation. *Br. Med. J.*, *296*:1401 (1988).
18. B. D. Kahan. Cyclosporine. *N. Engl. J. Med.*, *321*:1725 (1989).
19. J. M. Davison. The effect of pregnancy on kidney function in renal allograft recipients. *Kidney Int.*, 2774 (1985).

27

Cyclosporine Neurotoxicity and Hypomagnesemia

A. Ibañez, C. Manzanares, M. Quintanilla, M. C. Muñoz, P. Urruzuno, J. Manzanares, and R. Sanz *Hospital 12 de Octubre Materno-Infantil, Spain*

I. INTRODUCTION

Since the introduction of cyclosporine A (CsA) for the immunosuppression in solid-organ transplantation, the rate of allograft rejection has decreased substantially. However, its use is associated with multiple side effects, such as nephrotoxicity, hypertension, and neurotoxicity. Patients with high levels of CsA, concurrent high-dose methylprednisolone treatment, hypertension, or hypermagnesemia have been associated with central nervous system (CNS) side effects.

In pediatric patients undergoing liver transplantation, we have tried to associate the level of CsA and magnesium (Mg) with CNS side effects such as seizures, tremors, agitation, psychosis, and confusion.

II. MATERIALS AND METHODS

A. Patient Population

We followed the clinical course of five pediatric liver recipients, three males and two females, aged from 2 to 9 years (5.5 ± 2.7), who were treated with CsA during the study period (3 months after transplantation).

B. Samples and Determination

We obtained 80 blood samples to measure magnesium (Mg), calcium (Ca), albumin (Alb), creatinine (Cr), and CsA levels. We also recorded the children's neurological incidences during the posttransplant time.

C. Definitions

Serum magnesium determinations were performed by atomic absorption spectrophotometry (Perkin-Elmer 2280) and cyclosporine A metabolites by a nonspecific polyclonal radioimmunoassay (Sandoz, Inc.). We defined the normal serum Mg concentration as 1.5 to 2.3 mg/dL; hypomagnesemia was a value under 1.5 mg/dL and hypermagnesemia a value over 2.3 mg/d:.

No patient had amphotericin B, aminoglycoside, or diuretic treatment during the time of our study. Each of these is known to cause renal Mg wasting (1–3).

Data were analyzed using Student's t test and comparison of percentages. Data were expressed as mean ± standard deviation (x ± SD). Probability values less than 0.05 were considered to be statistically significant.

III. RESULTS

Serum magnesium concentrations obtained during the study period at the time of neurotoxicity (1.47 ± 0.33 mg/dL) were significantly lower than without neurotoxicity (1.76 ± 0.35 mg/dL) ($p < 0.01$) (Table 1).

We did not find significative differences between CsA level at the onset of neurological symptoms (674.8 ± 319.9 ng/mL) and periods with normal neurological activity (674.9 ± 332.3).

Serum calcium, albumin, and creatinine levels were not significantly different with or without neurotoxic episodes. However, as compared with reference values, albumin levels were lower in both groups.

Magnesium levels of < 1.5 mg/dL (hypomagnesemia) ($p < 0.001$) were associated with 62.5% of neurotoxic episodes and 20% of nontoxic episodes (Figure 1).

We did not find any significative difference in calcium, creatinine, or albumin levels between hypo- and normo- + hypermagnesemic groups (Table 2).

IV. DISCUSSION

The administration of drugs that cause renal Mg wasting (aminoglycosides, digoxin, furosemide, amphotericin B, cyclosporine, etc.) had been associated with the development of hypomagnesemia, and that with the neurotoxicity ascribed to CsA (4). Seizures, tremors, and depression have been reported as major side effects of CsA treatment. At the same time, hypomagnesemia causes a variety of neurological side effects: ataxia, tetany, and symptoms already mentioned for CsA. An association was found between neurotoxicity ascribed to CsA and the development of hypomagnesemia in bone marrow transplant recipients (4). Kone et al. (5) reported that 88% of their 64 bone marrow transplant recipients developed hypomagnesemia, and that incidence was greater in the patients treated with cyclosporine. Renal transplant recipients given CsA had marked hypomagnesemia requiring Mg supplementation

Table 1 Laboratory Values (x ± SD) at Onset of Neurotoxicity (n = 24) and Nontoxicity (n = 56)

	Toxicity	No Toxicity
Magnesium (mg/dL)	1.47 ± 0.33	1.76 ± 0.36[a]
Calcium (mg/dL)	8.88 ± 0.65	8.89 ± 0.86
Creatinine (mg/dL)	0.48 ± 0.15	0.47 ± 0.15
Albumin (g/dL)	3.14 ± 0.22	3.05 ± 0.30
Cyclosporine (ng/mL)	674.8 ± 319.9	674.9 ± 332.3

[a]$p < 0.01$.

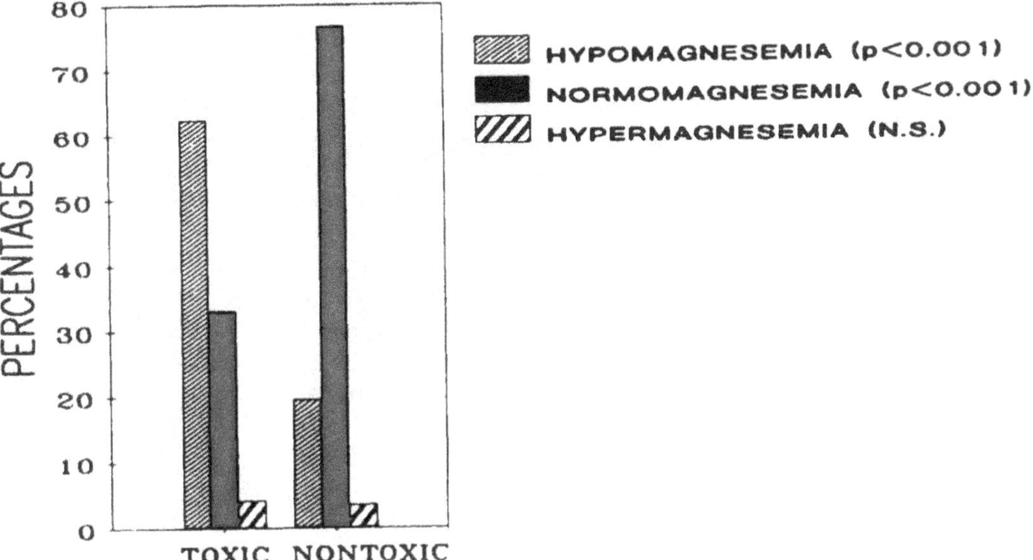

Figure 1 Percentages of hypo-, normo-, and hypermagnesemia during toxic and nontoxic episodes.

(6). Inappropriate urine Mg loss was observed in all patients under CsA treatment for marrow transplantation in the presence of moderate to severe hypomagnesemia (1). Experimental animals treated with CsA showed a significant reduction in serum Mg concentration when compared to both baseline values and to placebo-treated animals (6). Renal Mg wasting caused by impaired tubular reabsorption of this compound is the primary mechanism of the maintenance of reduced serum Mg levels with CsA administration lasting longer than 2 weeks (6).

There is an acute dose-dependent nephrotoxic effect of CsA, principally on cells of the proximal convoluted tubules (7). In addition, CsA appears to promote a shift of Mg into the tissues that may, in part, contribute to the observed reduction in serum Mg concentration (6). Seizures (4) are less common in renal transplant patients (8) than in cardiac (9) or bone marrow transplant recipients (10). A decreased glomerular filtration rate limits the ability of the kidney to remove Mg from the circulation. This agrees with results (11) that did not confirm earlier observations

Table 2 Comparison Between Serum Ca, Cr, and Alb in Our Hypo- and Normo- + Hypermagnesemic Groups

	Hypo-[a]	Normo- + Hyper[b]
Calcium	9.1 ± 0.81	8.8 ± 0.78
Creatinine	0.52 ± 0.13	0.44 ± 0.16
Albumin	3.07 ± 0.27	3.09 ± 0.28

[a]n = 26.
[b]n = 54.

of severe Mg wasting and hypomagnesemia in renal transplant recipients under immunosuppression with CsA.

In contrast with previous studies (4), we did not get significative differences between creatinine levels (as a sign of neurotoxicity) in patients with or without neurotoxicity. There was a higher proportion of hypomagnesemia in the hypocalcemic patients than in the rest of hospitalized patient groups (12). We did not find any difference in calcium levels between the hypomagnesemic and the normo- + hypermagnesemic groups. In that report (12), calcium values were corrected for albumin concentration; our result was serum calcium (not corrected).

Hypomagnesemia or hypermagnesemia often go unrecognized, because symptoms and signs of disturbed Mg levels are not unique nor do they constitute a readily recognizable clinical syndrome (12). In our case, neurotoxicity caused by low Mg levels could have been attributable to CsA side effects, but according to our results could really be a symptom of the loss of circulating Mg.

An association has been established between CsA neurotoxicity and hypomagnesemia in liver pediatric patients. Its prevention during CsA treatment seems likely to reduce the risk of associated neurotoxicity.

V. SUMMARY

Since the introduction of cyclosporine (CsA) for immunosuppression in solid-organ transplantation, the rate of allograft rejection has decreased substantially. However, its use is associated with multiple side effects, such as nephrotoxicity, hypertension, and neurotoxicity, patients with high levels of CsA, concurrent high-dose methylprednisolone treatment, hypertension, or hypomagnesemia have had central nervous system (CNS) side effects with our pediatric transplantation patients we tried to associate the levels of CsA and magnesium with CNS side effects such as seizures, tremors, agitation, psychosis, and confusion. We followed the clinical evolution of five pediatric patients for 3 months after transplantation, and obtained 80 blood samples to measure magnesium (Mg), calcium (Ca), albumin (Alb), and CsA levels. Magnesium levels at the time of neurotoxicity were ($x \pm SD$) 1.47 ± 0.33 mg/dL versus 1.76 ± 0.36 mg/dL without neurotoxicity ($p < 0.01$). Ca, Alb, and CsA levels were not significantly different, but the CsA levels were not specific for CsA, in that they measured CsA and metabolites. We observed that 62.5% of neurotoxic and 20% of nontoxic episodes were associated with Mg levels < 1.5 mg/dL ($p < 0.001$). An association has been established between CsA neurotoxicity and hypomagnesemia. Prevention of hypomagnesemia during CsA treatment seems likely to reduce the risk of associated neurotoxicity.

REFERENCES

1. C. H. June, C. B. Thompson, M. S. Kennedy, J. Nims, and E. D. Thomas. Profound hypomagnesemia and renal magnesium wasting associated with the use of cyclosporine for marrow transplantation. *Transplantation, 39*:620<19624 (1985).
2. C. H. Barton, M. Pahl, N. D. Vaziri, and T. Cesario. Renal magnesium wasting associated with amphotricin B therapy. *Am. J. Med., 77*:471<19674 (1984).
3. R. K. Rude, and F. R. Singer. Magnesium deficiency and excess. *Ann. Rev. Med.,* 32:245 (1981).
4. C. Thompson, J. C. Sullivan, and E. Thomas. Association between cyclosporine neurotoxicity and hypomagnesemia. *Lancet,* 17:1116–1120 (1984).

5. B. C. Kone, A. Whelton, G. Santos, R. Saral, and A. J. Watson. Hypertension and renal dysfunction in bone marrow transplant recipients. *Quart. J. Med.*, *69*:985<19695 (1988).
6. C. H. Barton, M. D. Vaziri, S. Mina-Araghi, S. Crosby, and M. I. Seo. Effects of cyclosporine on magnesium metabolism in rats. *J. Lab. Clin. Med.*, *114*(3):232–236 (1989).
7. R. Devineni, N. McKenzie, J. Duplan, P. Keown, C. Stiller, and A. C. Wallace. Renal effects of cyclosporine: Clinical and experimental observations. *Transplant. Proc.*, *15*(Suppl. 1):2695–2698 (1983).
8. Canadian Multicentre Transplant Study Group. A randomized clinical trial of cyclosporine in cadaveric renal transplantation. *N. Engl. J. Med.*, *309*:809–815 (1983).
9. R. L. Hardesty, B. P. Griffith, and H. T. Bahnson. Experience with cyclosporine in cardiac transplantation. *Transplant. Proc.*, *15*:2553–2558 (1983).
10. D. V. Joss, A. J. Barret, J. R. Kendra, C. F., and S. Desasi. Hypertension and convulsion in children receiving cyclosporine A. *Lancet*, *i*:906 (1982).
11. M. Haag-Weber, P. Schollmeyer, and W. H. Horl. Failure to detect remarkable hypomagnesemia in renal transplant recipients receiving cyclosporine. *Minor Electrolyte Metabl.*, *16*(1):66–68 (1990).
12. E. T. Wong, K. Rude, R. F. Singer, and S. T. Shaw. A high prevalence of hypomagnesemia and hypermagnesemia in hospitalized patients. *Am. J. Clin. Pathol.*, *79*:348–352 (1983).

28

The Role of Lipids and Cyclosporine A in Adverse Side Effects in Pediatric Liver Transplant Patients

C. Manzanares, A. Ibañez, T. Gomez-Izquierdo, S. Larrumbe, A. Erkiaga, P. Urruzuno, J. Manzanares, and R. Sanz *Hospital 12 de Octubre Materno-Infantil, Spain*

I. INTRODUCTION

Cyclosporine A (CsA) is a unique cyclic peptide of fungal origin, with potent immunosuppressive properties but low myelotoxicity. It has had a major impact on organ transplantation, significantly improving 1- and 2-year graft-survival rates and decreasing morbidity in renal, hepatic, heart, heart-lung, and pancreas transplantation (1).

The use of CsA is associated with multiple side effects. Nephrotoxicity and hypertension are most frequently observed; these effects are dose-related and nearly always improve after reduction of the dose of CsA. Less known are the toxic effects of CsA on the central nervous system (confusion, paresthesias, encephalopathy, tremors, and seizures). These effects are more frequently seen in patients with high blood CsA concentrations (2).

CsA is a highly lipophilic drug. In blood, approximately 40% of the drug is taken up by erythrocytes. The binding of the remaining 60% to lipoproteins and albumin is as follows: 57% to HDL, 25% to LDL, 2% to VLDL, and approximately 10% to albumin. As lipoprotein profiles are altered in patients under liver transplant, the extent of CsA binding and the pharmacologic response may have been altered, producing adverse reactions such as nephrotoxicity and neurotoxicity (3).

Our objective was to examine the relationship of cholesterol and triglyceride, the major lipid constituents of lipoproteins, cholesterol-HDL, apolipoprotein A and B, albumin, creatinine, and CsA levels with episodes of neurotoxicity.

II. MATERIAL AND METHODS

A total of 147 venous blood samples from 17 pediatric liver transplant patients (ages between 2 and 13 years), obtained from 1 month to 1 year after transplant, were analyzed, and posttransplant times were recorded. Blood samples, in tubes containing EDTA as anticoagulant, were used to determine CsA levels, measured by nonspecific polyclonal radioimmunoassay (Sandoz, Inc.) (4).

Additionally, blood was collected for serum cholesterol (COL), triglycerides (TG), albumin (ALB), and creatinine (CR) analysis. In 33 samples from five patients, we also measured COL-HDL and apolipoproteins A and B.

Serum COL, COL-HDL, and TG were determined with enzymatic methods, COL-HDL after precipitation of non-HDL lipoproteins (5,6). Apo A and Apo B were analyzed by nephelometry (7).

We retrospectively searched for neurotoxic episodes defined by these clinical criteria: presence of confusion, somnolence, agitation, tremor or seizure not attributable to cause other than CsA. We divided the samples into two groups, those that coincided with neurotoxic episodes and those corresponding to the time of non-neurotoxicity. Data were analyzed using Student's t test. Data were expressed as mean ± standard deviation (x ± SD). Probability values less than 0.05 were considered to be statistically significant.

III. RESULTS

The statistically significant parameters were: COL, toxic (165 ± 52) versus nontoxic (202 ± 109); CsA level (953 ± 522 versus 746 ± 398); COL-HDL (23.5 ± 5 versus 35.7 ± 16.6); and posttransplant time (32 ± 42 versus 61 ± 65) (Table 1). We can observe that COL and COL-HDL are lower when neurotoxicity exists. CsA levels are higher, and neurotoxicity develops earlier when it is compared with the group of patients without neurotoxicity.

IV. DISCUSSION

The results reported in this study show that COL, COL-HDL, and CsA levels, in the group of samples that coincide with neurotoxic episodes, are different from those corresponding to the time when there are no symptoms of toxicity.

There is a significant inverse correlation between the free fraction of CsA and total serum levels of COL, and with COL-HDL (8). With the same objective, studies demonstrated a trend of reduction of unbound fraction with increasing COL concentration, and that, at average lipid values, although the fraction of CsA associated with each lipid remains relatively constant, unbound fraction can change appreciably, with major changes occurring at low lipid concentrations (9).

Our data from samples of patients with neurotoxicity show COL and COL-HDL levels lower than in patients without these symptoms ($p < 0.01$), perhaps because changes in CsA binding produce an altered pharmacological response.

Neurotoxic episodes appear at an average of 31 days after the transplant date. This number of days is lower than the average (61 days) found at the time of no symptoms ($p < 0.01$). Other findings found that the free fraction of CsA was highest immediately after transplantation (9). In our hospital the CsA therapeutic range during the first month after transplant is 500 to 1000 ng/mL, following that it is 200 to 500 ng/mL. This protocol can explain the difference in CsA levels (953 ± 522 versus 756 ± 398), $p < 0.05$, between the two groups.

Table 1 Laboratory Parameters of Neurotoxic and Nontoxic Episodes

	Neurotoxicity	Nonneurotoxicity
	n = 23	n = 124
Post transplant time (days)	32 ± 42	61 ± 65[a]
CsA level (ng/mL)	953 ± 522	746 ± 398[b]
Cholesterol (mg/dL)	165 ± 52	202 ± 109[a]
Triglyceride (mg/dL)	200 ± 158	203 ± 93
Creatinine (mg/dL)	0.6 ± 0.2	0.6 ± 0.3
Albumin (g/dL)	3.3 ± 0.4	3.3 ± 0.5
	n = 5	n = 28
COL-HDL (mg/dL)	23.5 ± 5	35.7 ± 16.6[a]
Apo-A (mg/dL)	71.6 ± 14.4	64.2 ± 46.3
Apo-B (mg/dL)	74.4 ± 19.3	81.3 ± 27.3

[a]$p < 0.01$.
[b]$p < 0.05$.

Several mechanisms possibly relate COL to the CsA-induced neurotoxicity (2). One of them is that CsA could interfere with transport of COL and other lipids into the brain, and low COL levels could magnify this effect.

Since lipoprotein moieties serve as a reservoir for CsA, probably transferring the drug directly into plasma membranes, lipid binding may buffer this effect (10). The toxic effects of CsA on the central nervous system are enhanced if serum COL levels are low. If the total serum COL concentration is low, more CsA than usual may remain unbound and thus able to pass through the blood-brain barrier (11). Our results agree with the conclusions of these authors, but we think we have to continue with the study of lipoproteins because, although there are statistically significant differences with COL-HDL, we have only five samples in the group with neurotoxicity, and we could not calculate COL-LDL values in some samples because the TG value was too high to apply the correct calculations.

V. SUMMARY

Cyclosporine A (CsA) is highly bound, principally to the lipoprotein fractions in human plasma. The binding of CsA is as follows: High-density lipoprotein (HDL) 57%, low-density lipoprotein (LDL) 25%, very-low-density lipoprotein (VLDL) 2%, and approximately 10% is bound to albumin. Changes in the lipids composition may alter the percentage of unbound CsA, producing adverse effects (nephro- and neurotoxicity) or rejection. Our objective was to relate the lipid plasma composition with CsA levels and adverse CsA effects. We measured cholesterol (COL) and triglycerides (TG) as the major lipid constituents of lipoproteins, albumin (ALB), creatinine (CR), and CsA levels in 160 blood samples from 17 pediatric liver transplant patients. These were obtained during time periods between 1 month and 1 year from the transplant date, and retrospectively we searched for neurotoxic episodes. We also measured 30 samples for COL-HDL, Apo A, and Apo B. During neurotoxicity the COL levels were lower than when there was no neurotoxicity ($x \pm SD$) (178 ± 61 versus 198 ± 104

($p < 0.01$), the CsA levels were 919 ± 536 versus 746 ± 398 ($p < 0.05$), and the COL-HDL levels were 23 ± 5 versus 37 ± 18 ($p < 0.01$). If the total serum COL is low, more CsA than usual may remain unbound and thus able to pass through the blood-brain barrier, inducing CNS toxicity.

REFERENCES

1. I. M. Shaw, et al. Critical issues in cyclosporine monitoring: Report of the task force on cyclosporine monitoring. *Clin. Chem.*, 33(7):1269–1288 (1987).
2. P. C. Groen, et al. Central nervous system toxicity after liver transplantation. (The role of cyclosporine and cholesterol.) *N. Engl. J. Med.*, *317*:861-866 (1987).
3. R. J. Ptachcinski, G. J. Burckart, and R. Venkataramanan. Cyclosporine. *Drug Intell. Clin. Pharm.*, *19*:90–100 (1985).
4. P. Donatsch, E. Abish, M. Homberger, et al. A radioimmunoassay to measure cyclosporine A in plasma and serum samples. *Immunoassay*, 2:19–32 (1981).
5. A. W. Wahlefeld. Lipoprotein analysis. In *Methods of Enzymatic Analysis* (H. V. Bergmeyer, ed.). Academic Press, Philadelphia, p. 1831 (1974).
6. C. C. Allain, L. S. Poon, C. S. G. Chang, W. Richmond, and P. C. Fu. Enzymatic determination of total serum cholesterol. *Clin. Chem.*, *20*:470–475 (1974).
7. C. C. Henk, F. Erbe, and P. Flint-Hansen. Immunoephelometric determination of apolipoprotein A in lipoproteinemic serum. *Clin. Chem.*, *29*:120–125 (1983).
8. A. Lindholm and S. Henricsson. Intra- and interindividual variability in the free fraction of cyclosporine in plasma in recipients of renal transplants. *Ther. Drug Monitor.*, *11*:623–630 (1989).
9. B. Legg, S. K. Gupta, and M. Rowland. A model to account for the variation in cyclosporine binding to plasma lipids in transplant patients. *Ther. Drug Monitor.*, *10*:20–27 (1988).
10. B. D. Kahan. Cyclosporine. *N. Engl. J. Med.*, *321*(25):1725–1738 (1989).
12. J. Nemunaitis, H. J. Deeg, and G. C. Yee. High cyclosporine levels after bone marrow transplantation associated with hypertriglyceridemia. *Lancet*, 2:744–745 (1986).

29

In Vitro Immunomodulatory Effect of Doxycycline

M. J. Enguídanos and A. Romar *"La Fé" Hospital, Valencia, Spain*
J. García de Lomas *School of Medicine, University of Valencia, Valencia, Spain*

I. INTRODUCTION

Recent investigations have shown that antimicrobial agents can modify host defense mechanisms. Several of these studies show an immunosuppressive effect of doxycycline, but generally using concentrations of the antibiotic above therapeutic levels (1–3).

In the present study we have evaluated the in vitro effect of doxycycline on two essential lymphocyte functions: proliferation to stimulants and secretion of immunoglobulins. The antibiotic was used at two concentrations, which were similar and above therapeutic levels.

II. MATERIALS AND METHODS

A. Cell Donors

Peripheral blood mononuclear cells (MNC) were obtained from 45 healthy donors.

B. Antibiotic

Doxycycline (Vibravenosa; Pfizer) was evaluated at two final concentrations in cell cultures which were similar to (1 μg/mL) or above (10 μg/mL) therapeutic levels.

C. Cell Culture Conditions for Lymphocyte Proliferation

Antigen PPD (purified protein derivative of tuberculin) was at a final concentration of 10 μg/mL in culture, which was the optimal proliferative dose. Mitogens were used in a ray of doses according to dose-response curves, giving a final concentration for PH (phytohemagglutinin; Sigma):0.2, 1, 5 and 50 μg/mL and for PWM (pokeweed; Sigma): 0.001, 0.1, and 10 μg/mL. Each culture, with a final volume of 200 μL/well and 1×10^5 MNC, was pulsed with 0.5 μCi of (methyl-^{3}H)thymidine (Amesham), harvested, and counted in a beta counter.

D. Cell Culture Conditions for Immunoglobulin Secretion

PWM was used as a stimulant at a final concentration of 1 μg/mL. Each culture, with a final concentration of 1.200 μL/tube and 6×10^5 MNC, was centrifuged after 7 days and the concentration of immunoglobulins (Igs) IgA, IgM, and IgG was determined in the supernatants by enzyme immunoassay.

E. Cell Control Cultures

In each study MNC stimulated and nonstimulated, with and without antibiotic, were included.

F. Calculation of Results

Percentage of stimulation by antibiotic in lymphocyte proliferation was calculated as follows:

$$\left(\frac{\text{c.p.m. cultures with stimulant and with antibiotic}}{\text{c.p.m. cultures with stimulant and without antibiotic}} \times 100\right) - 100$$

Percentage of suppression by antibiotic in lymphocyte proliferation was calculated as follows:

$$100 - \left(\frac{\text{c.p.m. cultures with stimulant and antibiotic}}{\text{c.p.m. cultures with and without antibiotics}} \times 100\right)$$

Percentage of stimulation by the antibiotic in IgA, IgM, and IgG secretion was calculated as follows:

$$\left(\frac{\text{ng of Ig in culture with antibiotic}}{\text{ng of Ig in culture without antibiotic}} \times 100\right) - 100$$

Percentage of suppression by the antibiotic in IgA, IgM, and IgG secretion was calculated as follows:

$$100 - \left(\frac{\text{ng of Ig in culture with antibiotic}}{\text{ng of Ig in culture without antibiotic}} \times 100\right)$$

III. RESULTS

Table 1 shows the plasma levels of doxycycline which are obtained after the in-vivo administration (4).

Figure 1 shows the percentage of stimulation (%) or suppression (%) obtained, in all the cases, after the addition of MNC cultures of two concentrations (1 μg/mL, left; 10 μg/mL, right) of doxycycline. Each percentage is the mean value of individuals values obtained with MNC from 45 donors.

Percentages of each experience has been numbered from 1 to 26 as follows:

Spontaneous proliferation of MNC in control cultures without stimulant and without antibiotics after 3 days (1 and 14) and 6 days (2 and 15)

Table 1 Doxycycline

Dose	Administration	Time (h)	Plasma Levels (μg/mL)
100 mg	Oral	1.5–4	1.5–2.1
200 mg	Oral	1.5–4	2.6–3
100 mg	i.v.	1	2.5
200 mg	i.v.	2	3.6

Proliferation of MNC to PHA: 0.2 μg/mL (3 and 16), 1 μg/mL, (4 and 17), 5 μg/mL (5 and 18), and 50 μg/mL (6 and 19)
Proliferation of MNC to PWM: 0.001 μg/mL (7 and 20), 0.1 μg/mL (8 and 21), and 10 μg/mL (9 and 22)
Proliferation of MNC to PPD: 10 μg/mL (10 and 23)
Secretion of IgA (11 and 24), IgM (12 and 25), and IgG (13 and 26)

The high concentration of doxycycline, above therapeutic levels, clearly shows a suppressive effect, but the low concentration, similar to therapeutic levels, shows a stimulatory effect on lymphocyte proliferation with optimal (5 μg/mL) and supraoptimal (50 μg/mL) doses of PHA (Figure 1, 5 and 6) and to lesser extent on IgM and IgG secretion (Figure 1).

Lymphocyte proliferation stimulated by PWM is suppressed with the high concentration of doxycycline. A suppressive effect is also observed with the low concentration of antibiotic but with some modulation, which is dependent on the mitogen dose and follows the shape of the dose-response curve (Figure 2).

Our results show that concentrations of the antibiotic above therapeutic levels have an immunosuppressive effect, but concentrations which are similar to therapeutic levels have an immunomodulatory effect. These results suggest a double effect of doxycycline on the in-vitro lymphocyte response, which is dependent on the serum concentrations obtained after the in vivo administration.

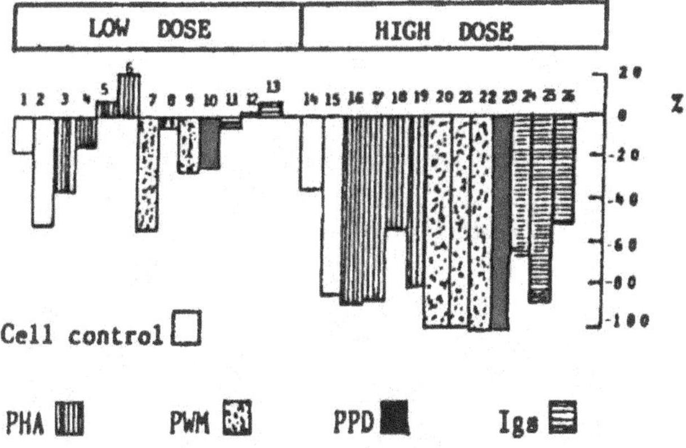

Figure 1 Percentages of stimulation or suppression obtained in all cases after addition to MNC cultures, a low dose (left) or high dose (right) of doxycycline.

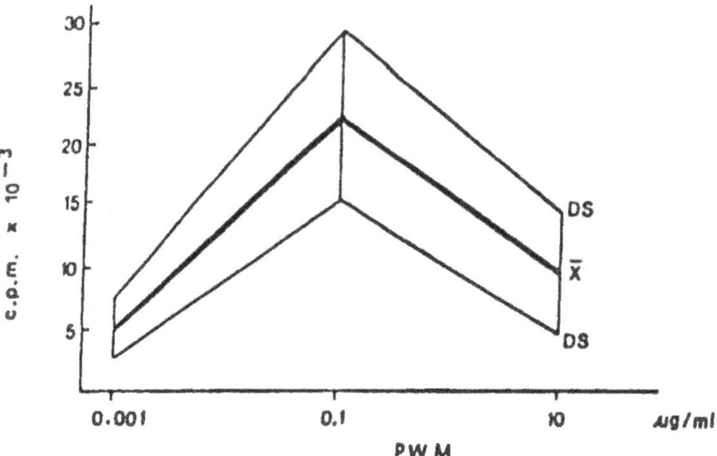

Figure 2 Dose response to PWM of MNC from 20 normal donors.

We conclude that when the effect of an antimicrobial agent on the immune mechanisms is evaluated using an in vitro assay, it is necessary to include as many variables as possible in this assay, in order to obtain the most similar conditions to the in-vivo situation.

IV. SUMMARY

An immunosuppressive effect of doxycycline has been observed in several studies, but generally using concentrations of the antibiotic above therapeutic levels. We have evaluated the in-vitro effect of doxycycline on the proliferation of human lymphocytes stimulated by PPD (purified protein derivative of tuberculin), PHA (phytohemagglutinin), and PWM (pokeweed), and also the effect on the immunoglobulin synthesis of lymphocytes stimulated by PWM. In the experiments we used two concentrations of doxycycline, which were at low (1 µg/mL) and high (10 µg/mL) therapeutic levels.

Our results confirm a suppressive effect of doxycycline at the concentration of 10 µg/mL, but at the concentration of 1 µg/mL we observed a modulatory effect which correlates to the degree of lymphocyte stimulation by PHA and PWM (which is dependent on the mitogen dose) and, indeed, we observed a slight stimulatory effect of IgM and IgG lymphocyte synthesis.

We concluded that in order to investigate the in-vitro effect on immune mechanisms by an antimicrobial agent, it is necessary to use a range of concentrations, which is usually obtained in serum after in-vivo administration.

REFERENCES

1. A. Forsgren, G. Banck, H. Beckman, and A. Bellahsene. Antibiotic-host defense interaction *in vitro* and *in vivo*. *Scand. J. Infect. Dis.*, 24:195–203 (1980).

2. R. C. Potts, H. A. A. Hassan, R. A. Brown, A. McConnachie, J. H. Gibbs, A. J. Robertson, and J. Swason. *In vitro* effects of doxycicline and tetracycline on mitogen stimulated lymphocyte growth. *Clin. Exp. Immunol.*, 53:458–464 (1983).
3. A. Bellahsene and A. Forsgre. Effect of doxycicline on immune response in mice. *Infect. Immun.*, 48:556–559 (1985).
4. American Hospital Formulary Service. *Drug Information, 1987 Ed.* American Society of Hospital Pharmacists, Bethesda, MD (1987).

30

Pharmacology and Toxicology of Quinolones

A. Anadón and M. R. Martínez-Larrañaga *Universidad Complutense, Madrid, Spain*

Quinolone antibacterial agents act on susceptible bacteria by immediately, selectively, and reversibly inhibiting DNA synthesis (10,14,29). The quinolones also inhibit DNA replication, but not DNA repair, in permeabilized cell systems (28,35). The molecular target of the quinolones is bacterial DNA gyrase. The mechanism of action of the quinolone antibacterial agents involves the inhibition (specifically the A subunit) of DNA gyrase (7,24,40).

DNA topoisomerases are a class of enzymes that alter the topology of DNA (38). These enzymes catalyze the reactions called supercoiling, relaxing, knotting, or catenating. Among these enzymes, bacterial topoisomerase II (DNA gyrase) has been purified from various bacterial species such as *Escherichia coli, Pseudomonas aeruginosa, Citrobacter freundii,* and *Bacteroides fragilis* and has been verified to be a major target of quinolones (4,21,31). Inhibition of this enzyme prevents DNA replication and DNA-mediated processes, subsequently leading to cell death (14). Eukaryotic topoisomerase II also has been purified from various sources, including bovine calf thymus. Quinolones such as ciprofloxacin, norfloxacin, and ofloxacin interfere slightly with the function of mammalian topoisomerase II as well. It has been demonstrated that eukaryotic topoisomerase II is much less sensitive to inhibition by quinolones than is bacterial DNA gyrase (19,20).

The molecular mechanism of inhibition of this important DNA-modifying enzyme has been studied extensively. Recently, Shen (34) proposed a cooperative quinolone-DNA binding model for the inhibition of bacterial DNA gyrase. The essential feature of the model is that bound gyrase induces a specific quinolone binding site in the relaxed DNA substrate in the presence of ATP. There is evidence to support the model that drug inhibits the enzyme by binding to a specific site on DNA created by DNA gyrase.

The general structures of two of the most studied classes of quinolones are shown in Figure 1. Nalidixic acid (Figure 2) is the prototype for antibacterial agents of the naphthyridine class, and oxolinic acid (Figure 2) is the prototype of the quinolone class. Molecular modifications of the parent structures have produced compounds with higher potency and antimicrobial spectrum and have greatly enhanced the therapeutic application of quinolones. The qualitative results of structure-activity

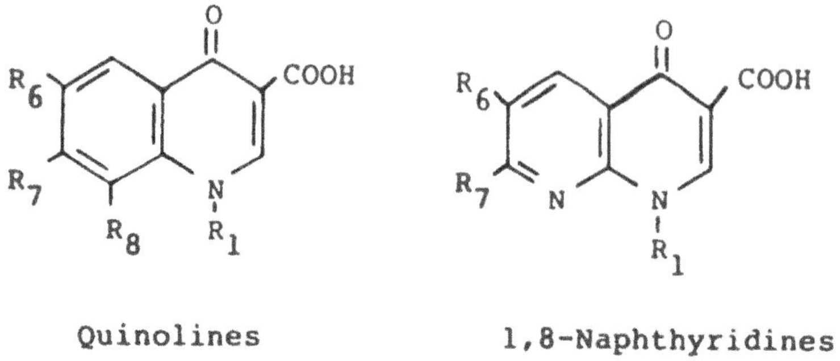

Quinolines 1,8-Naphthyridines

Figure 1 General structures of quinolone antibacterial agents.

studies performed to date can be summarized as follows: maximum in-vitro potency (expressed as MIC_S) as well as in-vivo efficacy occur with a fluorine substituent at C-6 (22) with the concomitant presence of an amino functionality of optimal size at C-7. The most common C-7 substituents of the quinolone molecule are (a) piperazin-1-yl (found in norfloxacin, ciprofloxacin, and enoxacin), (b) 4-methyl-piperazin-1-yl (found in pefloxacin, ofloxacin, difloxacin, fleroxacin, and amifloxacin). Substitution at the N-1 position is important for antibacterial activity (2). Conventional thought had limited the N-1 substituent to groups of similar steric bulk such as ethyl, vinyl, and fluoroethyl. Recent studies (9), however, indicate that a steric bulk factor at N-1 alone cannot account for optimal biologic activity. Modifications of the groups which occupy C-2, C-3 carboxilic acid group, C-5, and C-8 positions have also permitted the development of potent antibacterial agents.

Many studies have yielded results that indicate that modifications of every position in the standard quinolone molecule, with the exception of the C-4 position, have been successful. Nevertheless, the most important advance in structure-activity relationship is the discovery of quinolones with lower side effects, greater water solubility, and improved pharmacokinetics and activity against streptococci and anaerobes.

The antibacterial spectrum of activity of the first-generation quinolones was limited to gram-negative enterobacteriaceae. The new 4-quinolones, such as pefloxacin, ofloxacin, enoxacin, ciprofloxacin, and norfloxacin, have a broader in-vitro

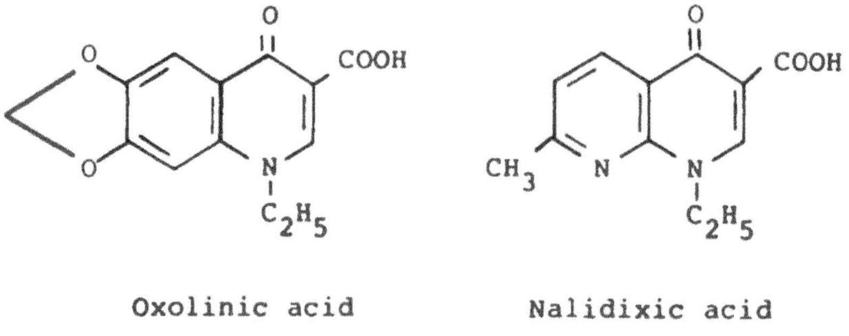

Oxolinic acid Nalidixic acid

Figure 2 Structures of first-generation quinolones.

antibacterial spectrum with a much greater intrinsic activity (reflected in lower MIC_s), including methicillin- and gentamicin-resistant staphylococci, multiresistant nonfermenters, all enterobacteriaceae, *legionella, neisseria* species, *branhamella,* and *hemophilus influenzae.* The activity against other gram-positive organisms (streptococcal groups A, B, C; *streptococcus faecalis, pseudomonas* species), *actinobacter* species, *chlamydia* species, *mycoplasma* species, and gram-positive and gram-negative anaerobes is more variable. The new 4-quinolone compounds have moderate activities against gram-positive microorganisms; anaerobic bacteria are generally resistant.

With the exception of norfloxacin, which is only 30% to 40% bioavailable from oral route, the 4-quinolones are 80% to 100% bioavailable, absorption occurring within 1 to 3 h. Food does not significantly alter C_{max}, AUC, or elimination of half-life, although t_{max} may be increased. The 4-quinolones are widely distributed throughout the body, with volumes of distribution greater than 1.5 L/kg. Protein binding is less than 30% in most cases. Penetration into most tissues is good. With the exception of ofloxacin and lomefloxacin, which are metabolically stable, metabolism of the 4-quinolones occurs primarily at the C-7 position in the piperazinyl ring. Biotransformation is extensive (85%) with pefloxacin, medium (25% to 40%) with ciprofloxacin and enoxacin, and low (20%) with norfloxacin. Elimination half-lives vary between 3 and 5 h (ciprofloxacin) and 8 to 14 h (pefloxacin). Biliary concentrations of the 4-quinolones are 2 to 10 times greater than those in serum or plasma, with several compounds undergoing enterohepatic circulation. There is some evidence that ciprofloxacin, norfloxacin, ofloxacin, and enoxacin have an active renal tubular excretion pathway (27).

The good tissue penetration of the 4-quinolones (12), despite considerable interindividual variation, has been associated with good response rates in a number of types of infection. The clinical uses of these new fluoroquinolones would include the following: (a) urinary tract infections, especially those caused by *P. aeruginosa;* (b) prostatitis; (c) severe bacterial gastroenteritis; (d) pneumonia caused by gram-negative bacilli; (e) selective decolonization of the gastrointestinal tract in granulocytopenic patients; (f) treatment of upper respiratory tract colonization by methicillin-resistant *Staphylococcus aureus;* and (g) gram-negative bacillary osteomyelitis (26).

The published toxicologic profile for the quinolones includes arthropathy, nephropathy, CNS effects, ocular toxicity, impairment of spermatogenesis, metabolic drug interactions, cardiovascular effects, possible mutagenicity, and photosensitivity (8).

Animal toxicology studies have shown that some quinolones damage the articular cartilage of immature animals (15). Various closely related analogs, such as nalidixic, oxolinic, or pipemidic acids or cinoxacin, evoke an arthropathic syndrome in juvenile dogs. Similar effects have been also observed after ofloxacin, pefloxacin, ciprofloxacin, and enoxacin. The pathogenesis of this drug-induced arthropathy remains unexplained. Nevertheless, growing cartilage is a target tissue of quinolone toxicity. As a consequence, quinolones are contraindicated during pregnancy and growth stages.

The risk of quinolones causing ocular, hepatic, and renal effects is low. These adverse effects have been observed in animals only after much greater than therapeutically relevant dosages.

On the other hand, through accumulated clinical experience, attention has been paid to the different CNS-related adverse drug effects after administration of fluoroquinolone carboxilic acid derivatives. Even though the incidence of these neurological effects has been low (5), the most common reactions include headache, dizziness, and restlessness (11,16,23). Seizures and hallucinations, which are rare in patients who receive quinolones, have been observed more frequently in patients who receive quinolones in combination with either theophylline or nonsteroidal antiinflammatory drugs. Although the mechanism by which quinolones exhibit epileptogenic neurotoxicity is unknown, inhibition of γ-aminobutyric acid (GABA) binding to its receptor sites, resulting in central nervous system excitation, might be responsible for such adverse phenomena of quinolones (34,36). Chemically, the new quinolones are characterized in two ways: they have a fluorine atom(s) and either a piperazine or an aminopyrrolidine moiety at the 7 position of the quinolone or naphthyridine ring, where the compounds seem to share a common structure with GABA receptor agonists. The epileptogenic activity of quinolones possibly relates to the GABA-like structures of substituents at their 7 positions, which act as antagonists of GABA receptors (1). Recent results also suggest that clinically observed CNS adverse effects of fluoroquinolones could be mediated at least in part through interactions with the benzodiazepine-$GABA_A$-receptor complex and may be controlled by benzodiazepine agonist administration (37).

The overall rate of adverse reactions associated with the fluoroquinolones is of the order of 4.0% to 8.0%. Reactions have been mild to moderate, leading to drug withdrawal in 1.0% to 6.0% of patients, and have been reversible. However, the new fluoroquinolones should be employed to fill specific therapeutic requirements in areas of infectious disease only partially or poorly filled by currently available agents (16,18).

Drug interactions with the new 4-quinolone antimicrobial agents have now been established. Aluminum-containing antacids and dialysate cause malabsorption of ciprofloxacin (13). Products containing iron and multivitamins with zinc also impair the absorption of ciprofloxacin (30). The concomitant administration of various quinolones and theophylline has been shown to decrease the total body clearance of theophylline in both healthy volunteers and patients and in several instances has resulted in CNS toxicity. Enoxacin is the most potent inhibitor of theophylline metabolism, and it reduces the clearance of theophylline by 42% to 74%; ciprofloxacin and pefloxacin are intermediate (decrease, approximately 30%) (6,32,39). Caffeine, a methylxantine, is similar in structure to theophylline. Interaction between ciprofloxacin and caffeine has also been reported. Ciprofloxacin administered at a dosage of 750 mg every 12 h resulted in a mean decrease in total caffeine clearance of 38% (17). The mechanism of these interactions has been suggested to be inhibition of cytochrome P-450-mediated xanthine derivative metabolism by fluorinated quinolones or their 4-oxo derivatives on the 3 position of the piperazine ring (25,39).

Potentially significant interactions could occur among fluorinated quinolones and other drugs which are metabolized by hepatic cytochrome P-450 enzymes. Ciprofloxacin pretreatment alters pharmacokinetics and metabolism of the classical oxidative substrate antipyrine (3). Ciprofloxacin is capable of inhibiting oxidative drug metabolism. This could be of clinical significance because ciprofloxacin and other new quinolones are usually used in combination with other therapeutic agents.

Even with the limited biochemical, toxicological, and pharmacological knowledge about the quinolones that is available at present, these agents appear to offer certain advantages over many of the existing antimicrobial drugs. However, until further investigations of the effects of long-term administration are available, prudent use of these agents will be important.

REFERENCES

1. K. Akahane, M. Sekiguchi, T. Une, and Y. Osada. Structure-epileptogenic relationship of quinolones with special reference to their interaction with γ-aminobutiric acid receptor sites. *Antimicrob. Agents Chemother.*, *33*:1704–1708.
2. R. Albrecht. Antibacterial activity of quinolonecarboxilic acids. III. Methyl-1-ethyl-4-quinolone-3-carboxylates and some 3-substituted quinolones. *Chim. Ther.*, *8*:45–48 (1973).
3. A. Anadón, M. R. Martinez-Larrañga, M. C. Fernandez, M. J. Diaz, and P. Bringas. Effect of ciprofloxacin on antipyrine pharmacokinetics and metabolism in rats. *Antimicrob. Agents Chemother.*, *34* (in press) (1990).
4. H. Aoyama, K. Sato, T. Fujii, K. Fujimaki, M. Inoue, and S. Mitsuhashi. Purification of *Citrobacter freundii* DNA gyrase and inhibition by quinolones. *Antimicrob. Agents Chemother.*, *32*:104–109 (1988).
5. P. Ball. Adverse reactions and interactions of fluoroquinolones. *Clin. Invest. Med.*, *12*:28–34 (1989).
6. J. Beckmann, W. Elsasser, V. Gundert-Remy, and R. Hertrampf. Enoxacin, a potent inhibitor of theophylline metabolism. *Eur. J. Clin. Pharmacol.*, *33*:227–230 (1987).
7. D. M. Benbrook and R. V. Miller. Effects of norfloxacin on DNA metabolism of *Pseudomonas aeruginosa*. *Antimicrob. Agents Chemother.*, *29*:1–6 (1986).
8. W. Christ, T. Lehnert, and B. Ulbrich. Specific toxicologic aspects of the quinolones. *Rev. Infect. Dis.*, *10*:S141–S146 (1988).
9. D. T. W. Chu and P. B. Fernandes. Structure-activity relationships of the fluoroquinolones. *Antimicrob. Agents Chemother.*, *33*:131–135 (1989).
10. G. C. Crumplin, J. M. Midgley, and J. T. Smith. Mechanism of action of nalidixic acid and its congeners. *Top. Antibiot. Chem.*, *3*:9–38 (1980).
11. P. G. Davey. Overview of drug interactions with the quinolones. *J. Antimicrob. Chemother.*, *22*(Suppl. C):97–107 (1988).
12. D. N. Gerding and I. A. Hitt. Tissue penetration of the new quinolones in humans. *Rev. Infect. Dis.*, *11*:S1046–S1057 (1989).
13. T. A. Golper, A. I. Hartstein, V. H. Morthland, and J. M. Christensen. Effects of antacids and dialysate dwell times on multiple-dose pharmacokinetics of oral ciprofloxacin in patients on continuous ambulatory peritoneal dialysis. *Antimicrob. Agents Chemother.*, *31*:1787–1790 (1987).
14. W. A. Goss, W. H. Deitz, and T. M. Cook. Mechanism of action of nalidixic acid on *Escherichia coli*. II. Inhibition of deoxyribonucleic acid synthesis. *J. Bacteriol.*, *89*:1068–1074 (1965).
15. A. Gough, N. J. Barsoum, L. Mitchell, L. M. E. J. McGuire, and F. A. De La Iglesia. Juvenile canine drug-induced arthropathy: Clinicopathological studies on articular lesions caused by oxolinic and pipemidic acids. *Toxicol. Appl. Pharmacol.*, *51*:177–187 (1979).
16. H. Halkin. Adverse effects of fluoroquinolones. *Rev. Invest. Dis.*, *10*(Suppl. 1):S258–S261 (1988).
17. D. P. Healy, R. E. Polk, L. Kanawati, D. T. Rock, and M. L. Mooney. Interaction between oral ciprofloxacin and caffeine in normal volunteers. *Antimicrob. Agents Chemother.*, *33*: 474–478 (1989).
18. D. C. Hooper and J. S. Wolfson. The fluoroquinolones: Pharmacology, clinical uses and toxicities in humans. *Antimicrob. Agents Chemother.*, *28*:716–721 (1985).
19. K. Hoshino, K. Sato, T. Une, and Y. Osada. Inhibitory effects of quinolones on DNA gyrase of *Escherichia coli* and topoisomerase II of fetal calf thymus. *Antimicrob. Agents Chemother.*, 33:1816–1818 (1989).
20. P. Hussy, G. Maass, B. Tummier, F. Grosse, and U. Schomburg. Effect of 4-quinolones and novobiocin on calf thymus DNA polymerase-α-primase complex, topoisomerases I and II and growth of mammalian lymphoblasts. *Antimicrob. Agents Chemother.*, 29:1073–1978 (1986).

21. Y. Inove, K. Sato, T. Fujii, K. Hirai, M. Inoue, S. Iyobe, and S. Mitsuhashi. Some properties of subunits of DNA gyrase from *Pseudomonas aeruginosa* PAO1 and its nalidixic acid-resistant mutant. *J. Bacteriol., 169*:2322–2325 (1987).
22. H. Koga, A. Itoh, S. Murayama, S. Suzue, and T. Irikura. Structure activity relationship of antibacterial 6,7- and 7,8-disubstituted 1-alkyl-1,4-dihydro-4-oxoquinoline-3-carboxilic acids. *J. Med. Chem., 23*:1358–1363 (1980).
23. J. C. H. Lucet, G. Tilly, J. J. Gres, and H. Piguet. Neurological toxicity related to pefloxacin. *J. Antimicrob. Chemother., 21*:811–812 (1988).
24. J. Morita, K. Watabe, and T. Komano. Mechanism of action of new synthetic nalidixic acid-related antibiotics: Inhibition of DNA gyrase supercoiling catalyzed by DNA gyrase. *Agric. Biol. Chem., 38*:663–668 (1984).
25. M. Nadai, T. Hasegawa, T. Kuzuya, I. Muraoka, K. Takagi, and H. Yoshizumi. Effects of enoxacin on renal and metabolic clearance of theophylline in rats. *Antimicrob. Agents Chemother., 34*:1739–1743 (1990).
26. T. M. Neer. Clinical pharmacologic features of fluoroquinolone antimicrobial drugs. *J. Am. Vet. Med. Assoc., 193*:577–580 (1988).
27. M. Neuman. Clinical pharmacokinetics of the newer antibacterial 4-quinolones. *Clin. Pharmacokinet., 14*:96–121 (1988).
28. A. M. Pedrini, D. Geroldi, A. Siccardi, and A. Falaschi. Studies on the mode of action of nalidixic acid. *Eur. J. Biochem., 25*:359–365 (1972).
29. L. J. Piddock, M. C. Hall, and R. Wise. Mechanism of action of Lomefloxacin. *Antimicrob. Agents Chemother., 34*:1088–1093 (1990).
30. R. E. Polk, D. P. Healy, J. Sahai, L. Dewal, and E. Racht. Effect of ferrous sulfate and multivitamins with zinc on absorption of ciprofloxacin in normal volunteers. *Antimicrob. Agents Chemother., 33*:1841–1844 (1989).
31. K. Sato, Y. Inoue, T. Jufii, H. Aoyama, and S. Mitsuhashi. Antibacterial activity of ofloxacin and its mode of action. *Infection, 14*(Suppl. 4):S226–S230 (1986).
32. J. Schwartz, L. Jauregui, J. Lettieri, and K. Bachmann. Impact of ciprofloxacin on theophylline clearance and steady-state concentrations in serum. *Antimicrob. Agents Chemother., 32*:75–77 (1988).
33. S. Segev, M. Rehavi, and E. Rubinstein. Quinolones, theophylline, and diclofenac interactions with the γ-amino butyric acid receptor. *Antimicrob. Agents Chemother., 32*:1624–1626 (1988).
34. L. L. Shen, L. A. Mitscher, P. N. Sharma, T. J. O'Donnell, D. W. T. Chu, C. S. Cooper, T. Rosen, and A. G. Pernet. Mechanism of inhibition of DNA gyrase by quinolone antibacterials: A cooperative drug-DNA binding model. *Biochemistry, 28*:3886–3894 (1989).
35. W. L. Staudenbauer. Replication of *Escherichia coli* DNA in vitro: Inhibition by oxolinic acid. *Eur. J. Biochem., 62*:491–497 (1976).
36. A. Tsuji, H. Sato, Y. Kume, I. Tamai, E. Okezaki, O. Nagata, and H. Kato. Inhibitory effects of quinolone antibacterial agents on γ-aminobutyric acid binding to receptor sites in rat brain membranes. *Antimicrob. Agents Chemother., 32*:190–194 (1988).
37. E. Unseld, G. Ziegler, A. Gemeinhard, U. Janssen, and U. Klotz. Possible interaction of fluoroquinolones with the benzodiazepine-$GABA_A$-receptor complex. *Br. J. Clin. Pharmacol., 30*:63–70 (1990).
38. J. C. Wang. DNA topoisomerases. *Ann. Rev. Biochem., 54*:665–697 (1985).
39. W. J. A. Wijnands, T. B. Vree, and C. L. A. Van Herwaarden. The influence of quinolone derivatives on theophylline clearance. *Br. J. Clin. Pharmacol., 22*:677–683 (1986).
40. M. M. Zweerink and A. Edison. Inhibition of *Micrococcus luteus* DNA gyrase by norfloxacin and 10 other quinolone carboxilic acids. *Antimicrob. Agents Chemother., 29*:598–601 (1986).

31

Metoclopramide as Inhibitor of Serum Cholinesterase

Manuel Martín Ruiz and Joaquín Herrera Carranza
Universidad de Sevilla, Sevilla, Spain

I. INTRODUCTION

Two related enzymes have the ability to hydrolyze acetylcholine. One is acetylcholinesterase, which is called true cholinesterase, and is found in erythrocytes, lung, spleen, nerve endings, and the gray matter of the brain. This form is responsible for the prompt hydrolysis of acetylcholine, which is released at the nerve endings and mediates transmission of the neural impulse across the synapse. The degradation of acetylcholine is necessary to the depolarization of the nerve. The other form of cholinesterase is acylcholine-acylhydrolase or pseudocholinesterase. This type is found in liver, pancreas, heart, the white matter of the brain, and serum, but its biological role is still unknown. The serum enzyme is the one whose assay is clinically useful (1).

Both enzymes are inhibited by several competitive inhibitors (alkaloids, physostigmine and prostigmine, dibucaine), and both enzymes are irreversibly inhibited by some organic phosphorus compounds. On the other hand, atypical genetic variants of the enzyme are found in the sera of a small fraction of apparently healthy persons. In this respect it is possible to identify genetic variants by assaying both total activity and the extent of inhibition by dibucaine (1).

Metoclopramide, a synthetic substituted benzamide, is a dopamine receptor antagonist, an antiemetic, and a stimulant of upper gastrointestinal (GI) motility. The drug is a derivative of *p*-aminobenzoic acid and is structurally related to procainamide and dibucaine, but lacks local anesthesic and antiarrhythmic properties (2).

In this report, a new inhibitor of serum cholinesterase activity is described. Metoclopramide has been found to produce a reversible and competitive-type inhibition of the enzyme with respect to substrate. This finding suggests a new approach to explain the mechanism of action of the drug.

II. MATERIALS AND METHODS

Serum: Blood extracts and serum specimens were obtained according to usual clinical laboratory methods in order to measure cholinesterase enzyme activities.

Table 1 Inhibition of Serum Cholinesterase by Metoclopramide

Metoclopramide Concentration (μM)	Enzyme activity (mU/mL)	Enzyme activity (%)	Enzyme inhibition (%)
—	4.105	100	0
4.52	3.050	74	26
9.31	2.580	63	37
28.20	1.408	34	66
56.45	852	21	79

The assay of cholinesterase activity, butyrylthiocholine as substrate, was carried out by Boehringer-Mannheim kit, with or without metoclopramide as indicated.

Enzyme activity: Cholinesterase assay activity, with butyrylthiocholine as substrate, was carried out by Boehringer-Mannheim kit, at 25°C.

Drugs and compounds: Metoclopramide was a kind gift from the Laboratorios Delagrange (Madrid, Spain) and dibucaine from Boehringer-Mannheim (Barcelona, Spain). These compounds are soluble in the incubation medium at the concentration described in Section III.

III. RESULTS

The inhibition of serum cholinesterase by metoclopramide is shown in Table 1. The drug produces the inhibition of cholinesterase (80% inhibition of enzyme activity). The inhibition of serum cholinesterase by dibucaine is shown in Table 2 (about 45%). Thus dibucaine is much less effective than metoclopramide is inactivating the enzyme. In order to characterize the inhibition type, Figure 1 shows the competitive inhibition of serum cholinesterase by both metoclopramide and dibucaine. Double-reciprocal plots show a typical competitive inhibition, which increases of the apparent K_m of enzyme for substrate but does not affect V_{max} of the enzyme reaction.

Table 2 Inhibition of Serum Cholinesterase by Dibucaine

Dibucaine Concentration (μM)	Enzyme activity (mU/mL)	Enzyme activity (%)	Enzyme inhibition (%)
—	4.105	100	0
4.28	3.933	96	4
8.60	3.519	86	14
26.30	3.000	73	27
52.60	2.346	57	43

The assay of cholinesterase activity, butyrylthiocholine as substrate, was carried out by Boehringer-Mannheim kit, with or without dibucaine as indicated.

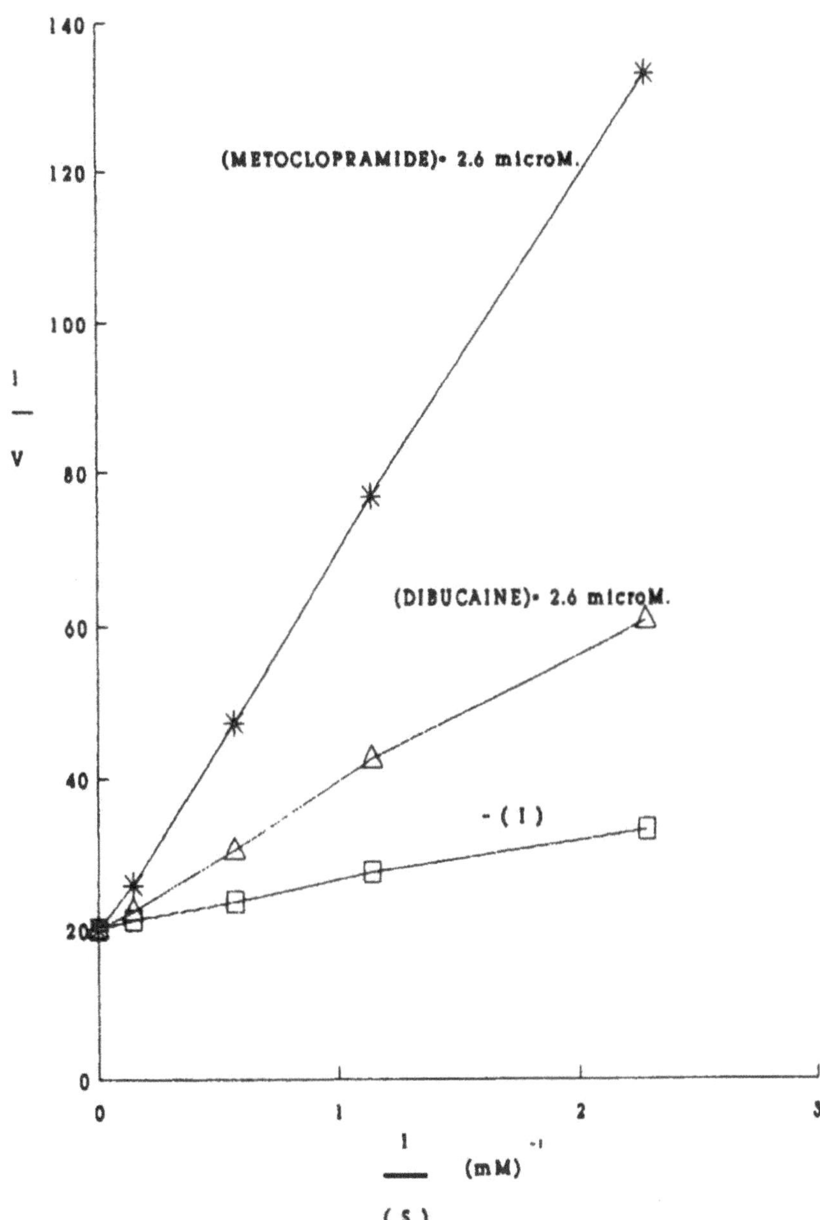

Figure 1 Competitive inhibition of serum cholinesterase by metoclopramide and dibucaine.

IV. DISCUSSION

The pharmacology of metoclopramide is complex, and its mechanism of action is still unclear. The principal pharmacologic effects involve the gastrointestinal tract and the CNS. Metoclopramide is considered a potent dopamine-receptor antagonist; some of the actions of this drug on GI smooth muscle may be mediated via antagonism of dopaminergic neurotransmission (3), but since cholinergic mechanism is responsible for most excitatory motor activity in the GI tract, it suggests that the therapeutic effects of metoclopramide are principally caused by cholinergic-like activity of drug. The effects of drug on GI motility may be mediated via enhancement of cholinergic excitatory processes at the postganglionic neuromuscular junction (4).

On the other hand, the effects of metoclopramide on GI motility are reduced or abolished by anticholinergic drugs (atropine) and potentiated by cholinergic drugs (carbachol and methacholine). These findings suggest that the effects of metoclopramide on GI motility may depend in part on intramural cholinergic neurons of smooth muscle (4).

Moreover, the cholinergic effects of metoclopramide may be due to release of acetylcholine (5). Pinder et al. (6) reported that metoclopramide has no anticholinesterase activity, but we have found that this antiemetic drug produces a reversible inhibition of cholinesterase activity with respect to substrate. In connection with this, double-reciprocal plots show a characteristic competitive inhibition.

Since we have found that metoclopramide produces a reversible and competitive inhibition of cholinesterase activity and some results have indicated that cholinesterase (serum type) is related to intestinal motility (7), these findings suggest a new and perhaps complementary approach to explain the mechanism of action of the drug, in agreement with the hypothesis which proposes that the cholinergic effects of metoclopramide may be due to release of acetylcholine or perhaps the inhibition of cholinesterase by the drug.

REFERENCES

1. S. S. Brown, W. Kalow, W. Pilz, M. Whittaker, and C. L. Woronick. The plasma cholinesterase: A new perspective. *Adv. Clin. Chem.*, 22:1–123 (1981).
2. K. Schulze-Delrieu. Metoclopramide. *Gastroenterology, 77*:768–779 (1979).
3. M. A. Zar, O. O. Ebong, and D. N. Bateman. Effect of metoclopramide in guinea-pig ileum longitudinal muscle: Evidence against dopamine mediation. *Gut*, 23:66–70 (1982).
4. R. A. Harrington, C. W. Hamilton, R. N. Brogden, J. A. Linkewich, J. A. Romankiewicz, and R. C. Heel. Metoclopramide: An update review of its pharmalogical properties and clinical use. *Drugs*, 25:451–494 (1983).
5. A. M. Hay, W. K. Man, and R. F. McCloy. Mechanism of action of metoclopramide: Importance of intrinsic sources of acetylcholine. *Gut, 18*:950–957 (1977).
6. R. M. Pinder, R. N. Brogden, P. R. Sawyer, T. M. Speight, and G. S. Avery. Metoclopramide: A review of its pharmacological properties and clinical use. *Drugs*, 12:81–131 (1976).
7. D. Jamieson. The function of true and pseudocholinesterase in the mammalian ileum. *Biochem. Pharmacol.*, 12:693–703 (1963).

32

Cyclosporine A–Drug Interactions

Alberto M. Castelao *Hospital de Bellvitge, Universidad de Barcelona, Barcelona, Spain*

I. INTRODUCTION

Cyclosporine is a potent immunosuppressive drug with a well-established effect on organ transplantation (1,2).

Because cyclosporine A (CsA) is subject to extensive hepatic metabolism, the possibility of pharmacokinetic drug interactions can be anticipated if drugs which enhance or inhibit its metabolism are co-administered. In addition, there is the possibility of pharmacodynamic interaction if other drugs given with CsA are nephrotoxic, since they may enhance the toxicity of CsA (3–6).

An large number of interactions have been described (Table 1) (7–44). The aim of this study is to show the observed interactions in our transplanted patients.

II. MATERIALS AND METHODS

Three hundred and seventy-seven patients who received a kidney transplant in our unit between March 1984 and March 1990 were treated with a CsA-prednisone (PNS), a CsA-PNS-antilymphocyte globulin (ALG), or a CsA-short prophylactic course of OKT3-PNS protocol (Table 2).

We studied the influence of certain drugs on CsA blood levels (polyclonal RIA in whole blood, n = 300 to 800 ng/mL, or specific monoclonal RIA, n = 100 to 250 ng/mL) in 51 patients (31 men, 20 female, mean age 39 ± 14 years), who presented 60 episodes of variation in CsA levels because of treatment with different drugs.

III. RESULTS

A. Rise in CsA Levels (Synergistic Effects)

We observed a rise in CsA levels in 50 episodes in 44 patients because of the treatment—in 18 patients with erythromycin, in one with ciprofloxacin, in one with norfloxacin, in two with ketoconazole, in 18 with diltiazem, in one with verapamil, and in one with cimetidine. There was a decrease when we stopped cimetidine in three patients, ranitidine in two, and nifedipine in one. We observed a CsA level increase in two patients given norethisterone.

Table 1 CSA Drug Interactions

Well-Substantiated Interactions	
Drugs that increase CsA blood levels:	
Diltiazem	Erythromycin
Verapamil	Josamycin
Nicardipin	Tycarcilline
Ketoconazole, fluconazole, itraconazole	
Choleretics	Tamoxyfen
Chenodeoxicolic ac	Pentazocine
Alcohol	Methoclopramide
Olive oil	Pentoxiphilline
Drugs that decrease CsA blood levels:	
Phenytoin	Sulphamidin-trimethoprim i.v.
Phenobarbitone	Sulfinpirazone
Carbamazepine	Anticoagulants
Rifampin	Metoprolol
Isoniazid	Omeprazole
Drugs that cause increased nephrotoxicity due to an additive effect:	
Amphotericin	Adriamycin
Aminoglycosides	Acyclovir
Mannitol	Ciprofloxacin
Oral cotrimoxazol	Ceftazydime?
Melphalan	Captopril
Digoxin	Metolazone
Suspected Interactions	
Steroids	Nafcillin
H_2 receptor antagonists	Moxalactam?
Thiazide diuretics	Doxycycline?
Sulfinpirazone	Ceftazidime?
Contraceptive pills	Norfloxacin
Androgens, somatostatine	Probenecid
Food	
Nonsteroidal antiinflammatory drugs	
Miscellaneous	
Antracuronium (neuromuscular blocking)	
Chlorpropamide, metronidazol (antabuse effect)	
Lovastatin (myopathy)	
Nifedipine (gingival hyperplasia)	
1,25 dihydroxyvitamin D (immunomodulation)	
Etoposide (immunomodulation)	
Mizoribine (immunomodulation)	
Verapamil (immunomodulation)	
FK-506 (immunomodulation)	
Rapamycin (immunomodulation)	

Table 2 Immunosuppression Protocol

1. CsA + PNS	CsA 5 mg/kbw/day i.v. before RT 3 mg/kbw/day i.v. after RT PNS 0.25 mg/kbw/day
2. CsA + PNS + horse ALG	CsA 3 mg/kbw/day before RT 2.5 mg/kbw/day after RT PNS same doses ALG 15 mg/kbw before RT 10 mg/kbw/alternate days, maximum 6 days
3. CsA + PNS + monoclonal OKT3	CsA same doses PNS same doses OKT3 5 mg i.v. during anesthesia 5 mg/day i.v. 4 days after RT

All the patients received a steady dose of CsA for a minimum of 4 days before starting the treatment. The time between the start of the treatment and the detection of rising levels of CsA was between 3 days (erythromycin) and 23 days (cimetidine).

The influence of erythromycin on CsA levels presented individual variability of interaction. The rise in CsA levels and plasma creatinine and the reduction in the CsA dose needed to maintain normal CsA levels were statistically significant ($p = 0.002$; $p = 0.05$; $p = 0.001$, Wilcoxon test).

Four patients required oral diltiazem (240 mg/day) and one patient oral verapamil (240 mg/day) because of angor pectoris. In 14 others we administered diltiazem to obtain a rise in CsA levels without increasing the CsA dose in order to minimize nephrotoxicity.

We performed a CsA clearance study on two patients, calculated as the intravenous dose divided by the area under the curve (AUC), with and without the questioned drug.

In the first patient, CsA clearance with diltiazem was 1.8 mL/min/kbw, and, without diltiazem, 3.8 mL/m/kbw. The rise in CsA level was significant (before treatment 290 ng/mL, after treatment 1200 ng/mL, $p = 0.009$), but the increase in plasma creatinine was not (before treatment 180 µmol/L, after treatment 158 µmol/L).

In the second patient, CsA clearance with verapamil was 1.12 mL/m/kbw and without verapamil, 2.03 mL/m/kbw. The rise in CsA level was significant (before treatment 460 ng/mL, after treatment 920 ng/mL, $p = 0.05$), but the increase in plasma creatine was not (before treatment 787 µmol/L, after treatment 813 µmol/L).

In one patient treated with 40 mg/day of nifedipine due to arterial hypertension, CsA level decreased from 590 mg/mL to 290 ng/mL when nifedipine was stopped.

Ciprofloxacin, administered to a patient due to a pseudomonas urinary infection, produced an increase in CsA level.

In the case of H2 receptor antagonists, one patient experienced a rise in CsA level after starting cimetidine (600 mg/day orally) and three others presented a decrease after stopping cimetidine in two patients and ranitidine in two patients. The increase in CsA dose (or decrease in the first case) needed to maintain normal CsA levels was 1.1 ± 0.8 mg/kbw/day.

Two of three women treated with oral norethisterone, 10 mg/day for 10 days, due to menorraghia, presented a mild increase in CsA levels (216 ± 156 ng/mL) with no changes in plasma creatinine.

B. Decrease in CsA Levels (Antagonistic Effects)

We observed a decrease in CsA blood levels or persistently low levels in nine patients. Five patients received phenytoin (3) or phenobarbitone (2). The mean CsA level decrease was 209 ± 197 ng/mL, and plasma creatinine diminished 123 ± 56 μmol/L (p n.s.). The patients required an increase of 1.55 ± mg/kg/day of CsA to maintain normal CsA blood levels.

Four patients suffered a strong interaction when rifampin was added to CsA, due to lung tuberculosis (2) or legionella pneumonia (2). CsA levels decreased from 402 ng/mL to 109 ng/mL, requiring an increase in CsA dose of 2.7 ± 0.5 mg/kbw/day to maintain normal blood levels. Plasma creatinine decreased from 272 ± 104 to 236 ± 41 μmol/L.

C. Increased Nephrotoxicity

A synergistic nephrotoxicity was suspected in a patient treated with ciprofloxacin due to a pseudomonas urinary infection. CsA clearance did not change significantly (13 to 15 mL/min/kbw), but plasma creatinine rose. Others (32,33,34) have described potentiation of nephrotoxicity between CsA and ciprofloxacin, as we suspected in our patient.

IV. DISCUSSION

CsA is a potent immunosuppressive drug that is metabolized extensively in the liver by the cytochrome P-450 enzyme system. Drugs that may enhance or inhibit this system could interact with CsA when they are co-administered. Recent studies in rabbit and human mycrosomal enzymes indicate that CsA metabolism is catalyzed by cytochrome P-450 3c in rabbits and HLp and PCN1 in humans, enzymes that are members of the cytochrome P-450 III A gene family.

Many of the interactions are well known, but some others are presently under discussion.

Some drugs act to enhance (as rifampin) or inhibit (as diltiazem) microsomal enzyme activity, modifying CsA clearance. In some other cases (phenytoin) there is a reduction of the CsA intestinal absorption.

Some drugs (erythromycin) could perhaps interact by a double mechanism, inhibiting microsomal activity and increasing intestinal CsA absorption (6).

Within the same pharmacological group of drugs, some can interfere with CsA but some others do not. For example, the calcium channel blockers, diltiazem and verapamil, increase CsA blood levels, but nifedipine does not (11).

Drug interference can produce variations in peak CsA concentration (C_{max}), area under the curve (AUC), or drug half-life, affecting both parent drug or its metabolites (5).

We agree with some authors (45,46) that research on the CsA metabolites may reveal important information on drug interactions.

V. CONCLUSIONS

1. Care should be taken when co-administering drugs that could enhance CsA levels because of the danger of nephrotoxicity and especially with others that decrease CsA levels which may raise the risk of rejection.
2. The interaction intensity presents very marked individual variability.
3. Diltiazem, verapamil, and ketoconazole are drugs that could minimize CsA nephrotoxicity, decreasing CsA doses and reducing the treatment cost, but further studies are needed before this cost-saving approach can be recommended.

REFERENCES

1. R. Y. Calne, K. Rolles, D. J. G. White, S. Thirus, D. B. Evans, R. Henderson, D. R. Hamilton, N. Boowe, P. McMaster, O. Gibby, and R. Williams. Cyclosporin A in clinical organ grafting. *Transplant. Proc., 13*:394–358 (1981).
2. R. Y. Calne, S. Thiru, P. McMaster, G. N. Craddock, D. G. J. White, D. B. Evans, D. C. Dunn, B. D. Pentlow, and K. Rolles. Cyclosporin A in patients receiving renal allografts from cadaver donors. *Lancet*, 2:1323–1327 (1978).
3. G. Maurer, H. R. Loosli, E. Schreier, and B. Keller. Disposition of cyclosporin in several animal species and man. I. Structural elucidation of its metabolites. *Drug Metab. Dispos., 12*:120–126 (1984).
4. B. D. Kahan. Individualization of cyclosporin therapy using pharmacokinetic and pharmacodinamic parameters. *Transplantation, 40*:457–476 (1985).
5. G. C. Yee. Pharmacokinetic interactions between cyclosporine and other drugs. *Transplant. Proc.*, 22:1203–1207 (1990).
6. P. A. Keown, C. R. Stiller, A. L. Laupacis, W. Howson, R. Cloes, M. Stawecki, J. Koegler, G. Carruthers, N. McKenzie, and N. E. Sinclair. The effects and side effects of cyclosporin: Relationship to drug pharmacokinetics. *Transplant. Proc., 14*:659–661 (1982).
7. J. M. Griñó, L. Sabaté, A. M. Castelao, and J. Alsina. Influence of diltiazem on cyclosporin clearance. *Lancet, 1*:1387 (1986).
8. I. Sabaté, J. M. Griñó, A. M. Castelao, and J. Ortolá. Evaluation of cyclosporin-verapamil interaction with observations on parent cyclosporin and metabolites. *Clin. Chem.*, 34:2151 (1988).
9. C. Linholm and S. Henricsson. Verapamil interaction inhibits cyclosporin metabolism. *Lancet, 1*:1262–1263 (1987).
10. B. Bourbigot, J. Guiserix, J. Sairiau, L. Bresolette, J. F. Morin, and J. Cledes. Nicardipine increases cyclosporin blood levels. *Lancet, 1*:1447 (1986).
11. K. L. Tortorice, K. L. Heim-Duthoy, W. M. Awni, K. Wenkateswara Rao, and B. L. Kasiske. The effects of calcium channel blockers on cyclosporine and its metabolites in renal transplant recipients. *Ther. Drug Monitor.*, 12:321–328 (1990).
12. G. R. Morganster, R. Powles, B. Robinson, and T. J. McElwain. Cyclosporine interaction with ketoconazole and melphalan. *Lancet*, 2:1324 (1982).
13. M. Gumbleton, J. E. Brown, G. Hawksworth, and P. H. Whithing. The possible relationship between hepatic drug metabolism and ketoconazole enhancement of cyclosporin nephrotoxicity. *Transplantation, 40*:454–455 (1985).
14. I. T. R. Cockburn and P. Krupp. Sandimmun®. An appraisal of drug interactions. In *Current Therapy in Nephrology* V. Andreucci and A. Dal Canton, (eds.), Kluwer Academic Publishers, Boston, pp. 503–528 (1989).
15. C. T. Veda, M. Lemaire, G. Gsnell, and K. Nussbaumer. Intestinal lymphatic absorption of cyclosporin A following oral administration of an olive oil solution in rats. *B. Opham. and Drug Diop.*, 4:113–124 (1983).
16. J. M. Griñó, I. Sabaté, A. M.Castelao, M. Guardia, D. Serón, and J. Alsina. Erythomycin and cyclosporin. *Ann. Intern. Med., 105*:467–468 (1986).
17. N. K. Wadhwa, Scoeder, E. Flaherty, A. J. Pesce, S. A. Myke, and R. First. The effect of oral metoclopramide on the absorption of cyclosporine. *Transplantation*, 43:211–213 (1987).

18. D. J. Freeman, A. Laupacis, P. A. Keown, C. R. Stiller, and S. G. Carruthers. Evaluation of cyclosporin-phenytoin interaction with observations on cyclosporin metabolites. *Br. J. Clin. Pharmacol., 18*:887–893 (1984).
19. H. Carstensen, N. Jacobsen, and H. Dieperink. Interaction between cyclosporin A and phenobarbitone. *Br. J. Clin. Pharmacol., 21*:550–551 (1986).
20. P. Lele, P. Peterson, S. Yang, B. Jarrell, and J. F. Burkee. Cyclosporin and tegretol, another drug interaction (abstr.). *Kidney Int., 27*:344 (1985).
21. D. L. Modry, E. B. Stinson, P. E. Oyer, S. W. Jamieson, J. C. Baldwin, and N. E. Shumway. Acute rejection and massive cyclosporin requirements in heart transplant recipients treated with rifampin. *Transplantation, 39*:313–314 (1985).
22. E. Langhoff and S. Madsen. Rapid metabolism of cyclosporin and prednisone in kidney transplant patients on tuberculostatic treatment. *Lancet,* 2:1303 (1983).
23. D. K. Jones, M. Hakim, J. Wallwork, T. W. Higenbottam, and D. J. G. White. Serious interaction between cyclosporin A and suphamidine. *Br. Med. .J., 292*:728–729 (1986).
24. M. S. Kennedy, H. J. Deeg, M. Siegel, J. J. Crowley, R. Storb, and E. D. Thomas. Acute renal toxicity with combined use of amphotericin B and cyclosporin after marrow transplantation. *Transplantation, 35*:211–215 (1983).
25. W. M. Bennet. Comparison of cyclosporine nephrotoxicity by aminoglycoside nephrotoxicity. *Clin. Nephrol., 25*(Suppl. 1):s126–s129 (1986).
26. A. Termeer, A. J. Hoitsma, and R. A. P. Koene. Severe nephrotoxicity caused by the combined use of gentamicin and cyclosporin in renal allograft recipients. *Transplantation,* 42:220–221 (1986).
27. F. P. Brunner, M. Hermle, M. J. Mihatsch, and G. Thiel. Mannitol potentiates cyclosporine nephrotoxicity. *Clin. Nephrol., 25*(Suppl. 1):s1130–s1136 (1986).
28. J. F. Thompson, D. H. K. Chalmers, A. G. W. Hunnistett, R. F. M. Wood, and P. J. Morris. Nephrotoxicity of trimethoprim and cotrimoxazol in renal allograft recipients treated with cyclosporine. *Transplantation, 36*:204–206 (1983).
29. R. L. Poules, B. Evans, C. Poole, A. Pedrazzini, M. Crofts, C. Pollard, and G. Hughes. Cyclosporine for the preservation of graft-vs-host disease in 72 patients with acute myeloblastic leukemia in first remission receiving matched sibling bone marrow transplantation. *Transplant. Proc., 15*(Suppl. 1):2624–2627 (1983).
30. P. C. Johnson, K. K. Kumor, M. S. Welsh, J. Woo, and B. D. Kahan. Effects of coadministration of cyclosporine and acyclovir on renal function of renal allograft recipients. *Transplantation, 44*:329–331 (1987).
31. A. M. Castelao, I. Sabaté, J. M. K. Griñó, S. GilVernet, E. Andrés, R. Sabater, and J. Alsina. Cyclosporin A drug interactions. *Transplant. Proc., 20*:66–69 (1988).
32. R. A. Elston and J. Taylor. Possible interaction of ciprofloxacin with cyclosporin. *J. Antimicrob. Chemother., 21*:679–680 (1988).
33. K. K. Tan, A. K. Trull, and S. Shawket. Study of the potential pharmacokinetic interaction between ciprofloxacin and cyclosporinnin man. *Br. J. Clin. Pharmacol., 26*:644–645 (1988).
34. C. K. Avent, D. Krinsky, J. K. Kirklin, R. C. Bourge, and W. D. Figg. Synergistic nephrotoxicity due to ciprofloxacin and cyclosporin. *Am. J. Med., 85*:452–453 (1988).
35. G. Klintmalm and J. Sawe. High dose methylprednisolone increases plasma cyclosporin levels in renal transplant recipients. *Lancet, 1*:731 (1984).
36. R. Giacomelli, V. Filingeri, G. Famularo, A. Calogero, F. Stortoni, S. Nardi, R. Rosati, A. Iacona, S. Sachetti, V. Cervelli, R. Verna, G. Tonietti, and C. U. Csaciani. Cyclosporin A-H_2 receptor antagonists drug interaction in the Sprague-Dawley rat. In *Current Therapy in Nephrology* (V. Andreucci and A. Dal Canton, eds.), Kluwer Academic Publishers, Boston, pp. 529–537 (1989).
37. W. B. Ross, D. Roberts, P. J. A. Griffin, and J. R. Salaman. Cyclosporin interaction with danazol and norethisterone. *Lancet, 1*:330 (1986).
38. E. Broch-Moller and B. Ekelund. Toxicity of cyclosporin during treatment with androgens. *N. Engl. J. Med., 313*:1416 (1986).
39. R. Landgraf, M. M. C. Landgraft-Leurs, J. Nusser, G. Hillebrand, W. D. Illner, D. Abendroth, and W. Land. Effect of somastostatin analogue (SMS 201-995) on cyclosporin levels. *Transplantation, 44*:724–725 (1987).

40. K. P. Harris, D. Jenkins, and J. Walls. Nonsteroidal antiinflammatory drugs and cyclosporine. A potentially serious adverse reaction. *Transplantation, 46*:598–59 (1988).
41. S. A. Veremis, M. S. Maddux, R. Pollack, and M. F. Mozes. Subtherapeutic cyclosporin concentrations during nafcilline therapy. *Transplantation,* 43:913–914 (1987).
42. D. J. Thompson, A. H. Menkis, and F. N. McKenzie. Norfloxacin-cyclosporine interaction. *Transplantation, 46*:312–313 (198x).
43. R. J. Ptachcinski, R. Venkataramanan, J. T. Roenthal, G. J. Burckhart, R. J. Taylor, and T. R. Hakala. The effect of food on cyclosporin absorption. *Transplantation, 40*:174–176 (1985).
44. J. Thomas, C. Matthews, R. Carroll, R. Loreth, and F. Thomas. The immunosuppressive action of FK 506. In vitro induction of allogenic unresponsiveness in human CTL precursors. *Transplantation, 49*:390–396 (1990).
45. R. Morris, B. Meiser, J. Wang, J. Wu, and R. Shorthouse. Rapamycin is a new safe and highly potent and effective means of inducing antigen specific unresponsiveness to allografts. XIII International Congress of the Transplantation Society, San Francisco, Abstr. 223 (1990).
46. G. Burckart, C. P. Wang, R. Venkataramanan, and R. Ptachcinski. Cyclosporine metabolites I. *Transplantation,* 43:932 (1987).
47. T. G. Rosano, M. B. Freed, and N. Lempert. Cyclosporine metabolites II. *Transplantation,* 43:932–933 (1987).

33

Relationship Between Q-T Interval and Serum Concentrations in a Case of Sotalol Poisoning

A. M. Buysse, A. Verstraete, L. Missault, and L. Jordaens
University Hospital, Ghent, Belgium

I. INTRODUCTION

Solatol is a noncardioselective beta-adrenoceptor blocking agent with Class III antiarrhythmic properties. The most important symptoms after overdosage (2.4 g to 8 g) are hypotension, bradycardia, and a prolonged Q-T interval, with the possibility of torsades de pointes (1).

We have developed a rapid HPLC method for the determination of concentrations of sotalol in serum. We measured serum sotalol concentrations and Q-T intervals in a patient after acute sotalol poisoning. The molecular structure of sotalol is shown in Figure 1.

A. Physicochemical Properties

The molecular weight of sotalol is 308.8. Sotalol is an amphoteric compound with pK_a of 8.3 and 9.2 (2). The maximum absorption of its ultraviolet spectrum is at 227 nm.

B. Pharmacokinetic Parameters

The bioavailability of sotalol is 100%. It is not bound to plasma proteins. The distribution is 2 L/kg (1,2). The half-life is 9.2 h. Sotalol is excreted unchanged in the urine.

II. METHODS

Serum sotalol concentrations were determined by HPLC on a Chrompack 10 cm × 3 mm column filled with Hypersil ODS 5 μm (Chrompack, Middelburg, The Netherlands). The mobile phase consisted of 0.116 M phosphate buffer, triethylamine and methanol (89.9:0.1:10). The flow rate was 0.5 mL/min at 55°C. Since sotalol is not protein-bound, sample preparation consisted of ultrafiltration through Amicon Centrifree devices that were centrifuged for 5 min at 1500 g. Twenty microliters of the

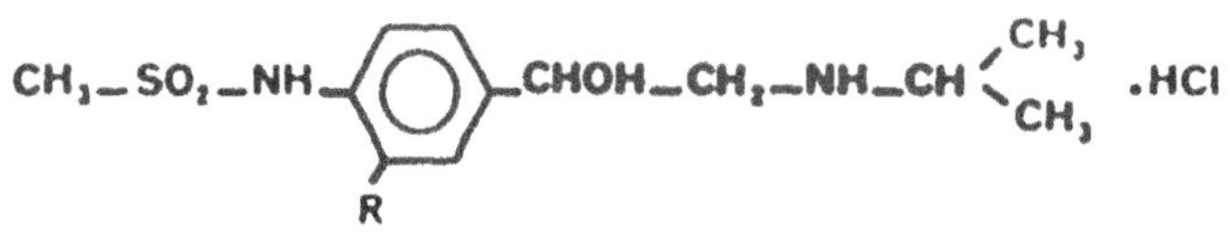

$CH_3-SO_2-NH-C_6H_3(R)-CHOH-CH_2-NH-CH(CH_3)_2 \cdot HCl$

Figure 1 Molecular structure of sotalol (R = H).

ultrafiltrate were injected into the HPLC. The absorbance was measured between 225 and 230 nm with an LKB 2140 rapid spectral detector.

III. RESULTS

The chromatogram obtained from a serum extract is shown in Figure 2. The retention time of sotalol is 2.5 min. The recovery was determined by adding known amounts of sotalol to blank serum samples and was 92.4%. For determination of the precision, six samples containing 1.6 µg/mL were extracted and analyzed on the same day. The coefficient of variation was 2%. The method was found to be linear in the range of 0 to 20 µg/mL. The calibration curve had a correlation coefficient of 0.9997. The calibration curve is shown in Figure 3.

IV. CASE REPORT

A 51-year-old man was admitted to the emergency department after taking 4 g of sotalol in a suicide attempt. Three hours later bradycardia developed, with frequent supraventricular extrasystoles. The QTc and serum sotalol concentrations are shown in Figure 4. The QTc was 0.46 upon admission, increased to 0.61, and returned to normal values in about 20 h. The serum sotalol concentration was 9.5 mg/L upon admission (therapeutic: 1 to 2 mg/L), increased to 23 mg/L 4 h later, and declined subsequently, with a half-life of 8 h.

V. DISCUSSION

Since sotalol is not protein-bound, we used ultrafiltration for sample preparation. The absolute recovery is higher than 90%, and the method has a good precision. In a fatal

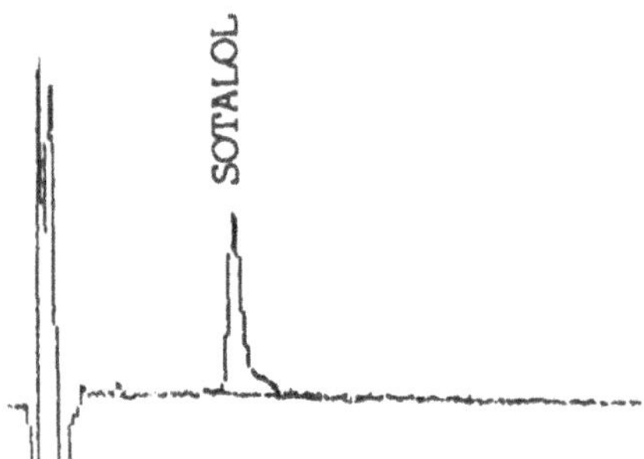

Figure 2 Chromatogram obtained from a serum sample. The retention time of sotalol is 2.5 min.

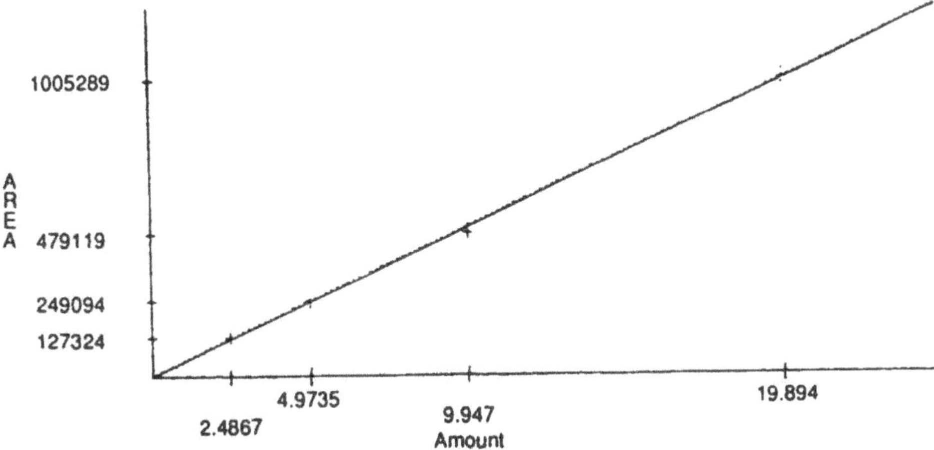

Figure 3 Sotalol calibration curve. The calibrators were made by adding 2.5, 5.0, 10, and 20 mg of sotalol to 1 mL of drug-free serum. No internal standard was used. The correlation coefficient of the calibration curve is 0.9997.

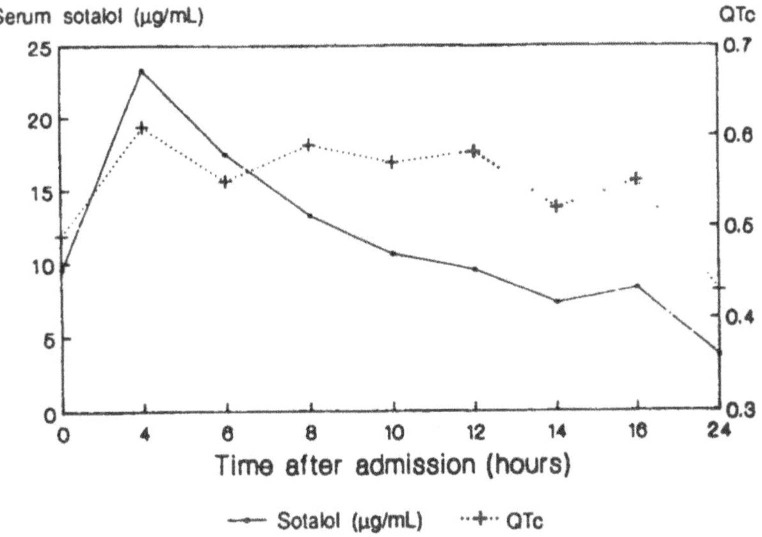

Figure 4 Evolution of serum sotalol concentrations and QTc over time. The time of intake is unknown.

case, a sotalol serum concentration of 65 μg/mL was found (3). The peak serum sotalol concentration in our case is similar to the concentrations found in three cases who survived (36.5, 16, and 7.5 μg/mL, respectively 11, 13 and 22 h after ingestion). In these cases, the maximum QTc were 0.67, 0.68, and 0.70, respectively (4,5). The decrease in serum sotalol shows a relatively good correlation with the QTc, but is more rapid than the decrease in QTc.

REFERENCES

1. Martindale. *The Extra Pharmacopoeia* (1982).
2. W. P. Gluth, F. Sörgel, B. Gluth, et al. Determination of sotalol in human body fluids for pharmacokinetic and toxicokinetic studies using HPLC. *Arzneim-Forsch./Drug Res.*, 3:408–411 (1988).
3. D. Perrot, B. Bui-Xuan, J. Lang, et al. A case of sotalol poisoning with fatal outcome. *J. Toxicol. Clin. Toxicol.*, *26*:389–396 (1988).
4. E. Edvardsson and E. Varnauskas. Clinical course, serum concentrations and elimination rate in a case of massive sotalol intoxication. *Eur. Heart J.*, *8*:544–548 (1987).
5. E. Elonen, P. Neuvonen, L. Tarssanen, et al. Sotalol intoxication with prolonged Q-T interval and severe tachyarrhythmias. *Br. Med. J.*, *5*:1184 (1979).

34

Vancomycin Pharmacokinetics in Hematologic Cancer Patients

Francesc Campos, Leonor Pou, Andrés López, and Lydia García *Hospital General Vall d'Hebron, Barcelona, Spain*

I. INTRODUCTION

Vancomycin is a bactericidal glycopeptide antibiotic which is primarily effective against Gram-positive cocci (1,2). In the last 10 years, vancomycin has been reestablished as an effective antibiotic with limited adverse reactions.

Febrile courses in granulocytopenic patients with hematological malignancies are treated with broad-spectrum antibiotic therapy, generally an association of a β-lactam antibiotic with an aminoglycoside.

The increase in infections with methicillin-resistant coagulase-negative staphylococci has resulted in an increasing use of vancomycin together with an aminoglycoside for febrile episodes in cancer patients (3).

Vancomycin pharmacokinetic parameters were retrospectively calculated in granulocytopenic patients with hematological malignancies, using a two-compartment Bayesian forecasting program.

II. MATERIALS

We have selected 17 courses of therapy in 17 adult granulocytopenic patients (Table 1) with the following characteristics: absolute granulocyte count of less than 1000 μL, hematologic malignancies, aggressive chemotherapy treatment, fever ≥ 38°C, and a presumed (n = 13) or documented infection (n = 4).

Patients selected were treated with vancomycin, amikacin, and ceftazidime. In seven patients amphotericin B was concurrently administered from the seventh day to the end of treatment. The duration of the therapy ranged from 4 to 33 days, and the mean dose of vancomycin was 30.9 mg/kg/day.

Patients received intravenously infused vancomycin for 60 min. Peak vancomycin levels were obtained 30 min after the infusion was completed. Trough levels were obtained just before a new dose was given. All peak and trough levels were obtained after five half-lives had passed. Therapeutic ranges of peak and trough levels were

Table 1 Patient Data

Sex (F/M)	9/8
Age (means ± SD; range)	38.9 ± 17.23 (15–66)
Weight (means ± SD; range)	63.9 ± 13.53 (41–92)
Diagnosis:	
Acute Myeloid Leukemia	13
Other Malignancies	4

between 18 to 30 mg/L and 5 to 10 mg/L, respectively. Amikacin levels were also monitored during treatment.

III. METHODS

Vancomycin serum determinations were done by fluorescence polarization immunoassay (TDx, Abbott Diagnostics, Irving, Texas) with a 8.9% between-day coefficient of variation at a level of 42 mg/L (n = 22).

All patients were tested for renal function before, during, and after treatment. Nephrotoxicity was defined as an increase of ≥0.5 mg/dL in creatinine levels when creatinine was <1.5 mg/dL just before treatment; and ≥1.0 mg/dL when it was >1.5 mg/dL. Patients with a creatinine level of >1.5 mg/dL were considered to have undergone previous kidney malfunction.

Creatinine assays were done with a Hitachi-737 Analyzer (Boehringer, Mannheim). Creatinine clearance (CL_{cr}) was estimated by the method of Cockcroft and Gault (4).

Vancomycin pharmacokinetics in 84 vancomycin serum levels were determined by fitting the data to a two-compartment Bayesian forecast program (Abbott Pharmacokinetic System).

To evaluate pharmacokinetic parameters, available patient data, including age, weight, height, sex, serum creatinine concentrations, vancomycin dosage, and vancomycin serum concentrations at a steady state, were analyzed retrospectively.

Analysis of regression (least-squares method) was done to correlate vancomycin clearance with creatinine clearance and age.

Vancomycin pharmacokinetics from patients before and after amphotericin B co-administration were tested for normal distribution (Kolmogorof-Smirnoff test) and for significant differences with Student's t test.

IV. RESULTS

The mean of vancomycin serum determinations per patient was 3.76 mg/L. Peak and trough levels are described in Table 2. Pharmacokinetic parameters are shown in Table 3. Table 4 depicts the changes in vancomycin pharmacokinetics before and after amphotericin B was added to the therapy.

No patient had any previous kidney disease. Creatinine clearance before and during treatment was (mean ± SD) 96.88 ± 27.46 and 111.23 ± 23.03 mL/min, respectively (n.s.). One case of nephrotoxicity was observed after amphotericin B, and it was included in the treatment.

Table 2 Vancomycin Serum Concentrations (mean ± SD)

	Peak (mg/L)	Trough (mg/L)
Without anphotericin B	23.63 ± 5.78 (n = 30)	9.46 ± 4.63 (n = 34)
With anphotericin B	33.03 ± 13.41 (n = 10)	16.86 ± 7.74 (n = 10)
Total	25.98 ± 9.01 (n = 40)	11.14 ± 6.20 (n = 44)

Table 3 Vancomycin Pharmacokinetics (mean ± SD) in Neutropenic Cancer Patients Treated with Vancomycin, Amikacin, and Ceftazidime

Patients (n)	17
CL_{cr} (mL/min)	111.2 ± 23.03
CL (mL/h/kg)	86.6 ± 22.29
$T_{1/2}$ (h)	6.1 ± 2.37
V_c (L/kg)[a]	0.188 ± 0.020

[a]V_c = volume of distribution to the central compartment.

Table 4 Changes in Vancomycin Pharmacokinetics in the Group of Patients to Whose Therapy Amphotericin B Was Added

	Without Amphotericin	With Amphotericin
Patients (n)	7	7
CL_{cr} (mL/min)	108.5 ± 22.84	81.2 ± 35.92
CL (mL/h/kg)	87.0 ± 23.78	56.7 ± 24.52
$T_{1/2}$ (h)	5.5 ± 2.26	9.3 ± 5.87
V_c (L.kg)[a]	0.175 ± 0.018	0.179 ± 0.018

[a]V_c = volume of distribution to the central compartment.

The vancomycin clearance showed no correlation with age ($r = -0.1255$), while a weak but significant association with creatinine clearance was observed ($r = 0.4894$; $p < 0.05$) (Figure 1).

In the seven patients to whose therapy amphotericin B was added after a mean of 6.9 days of treatment, a significant change in vancomycin clearance (87.05 versus 56.69 mL/kg/h; $p < 0.05$) was observed. The mean ± SD decreases found in creatinine clearance (108.51 ± 22.83 versus 81.25 ± 35.92 mL/min) and the mean ± SD increases found in vancomycin elimination half-life ($T_{1/2}$) (5.54 ± 2.26 versus 9.26 ± 5.87 h), were not statistically significant (Figure 2).

V. DISCUSSION

The mean vancomycin serum peak concentrations observed were within the therapeutic range. Although the mean trough levels were slightly above recommended concentrations, the only case of nephrotoxicity detected after amphotericin B was added was attributed to prerenal failure.

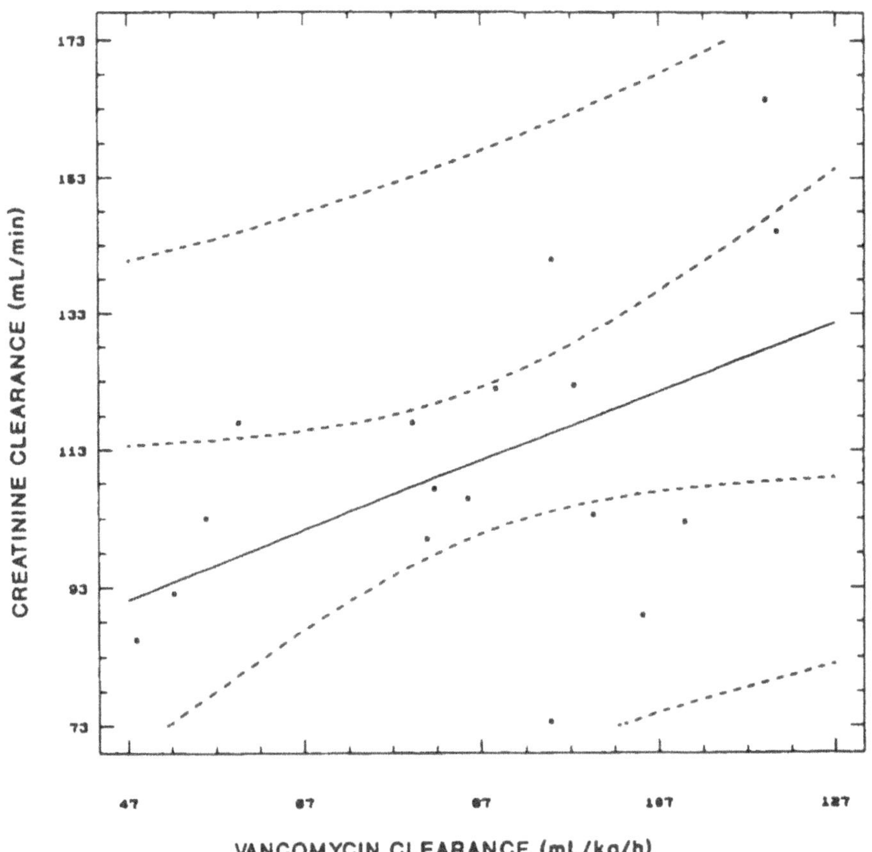

Figure 1 Correlation between creatinine clearance and vancomycin clearance.

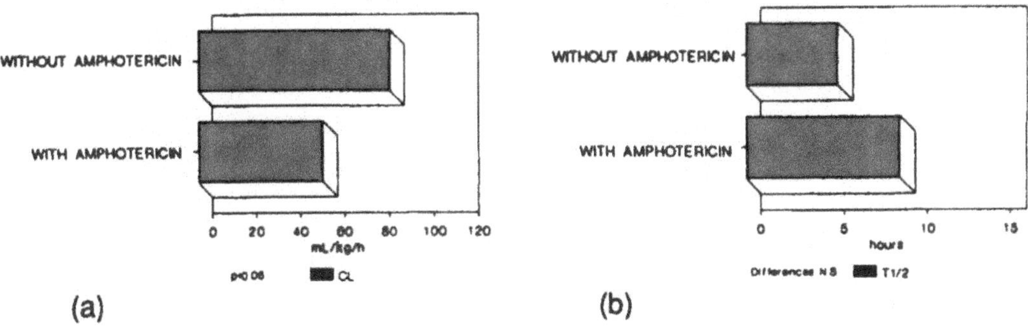

Figure 2 (a) Vancomycin clearance and (b) terminal elimination half-life in neutropenic cancer patients before and after amphotericin B was added to the treatment.

These results indicate that the incidence of nephrotoxicity in vancomycin treatments containing aminoglycosides is very low or absent when vancomycin and aminoglycoside levels, and renal function, are regularly controlled, and there are no other causes of renal disfunction.

Vancomycin pharmacokinetic results were similar to others published (5). Therefore it seems that the neutropenic cancer population does not undergo a different pharmacokinetic profile than other populations with normal renal function.

The correlation observed between vancomycin clearance and creatinine clearance was weakly significant. This could be explained by a certain nonrenal type of elimination previously described (5,6).

REFERENCES

1. R. S. Griffith, Introduction to vancomycin. *Rev. Inf. Dis.*, 3(Suppl. 1):S200 (1981).
2. F. V. Cook and W. E. Farrar, Jr. Vancomycin revisited. *Ann. Intern. Med.*, *88*:813 (1978).
3. J. C. Wade, S. C. Schimpff, K. A. Newman, and P. H. Wiernik. Staphylococcus epidermidis: An increasing cause of infection in patients with granulocytopenia. *Ann. Intern. Med.*, 97:503 (1982).
4. D. W. Cockcroft and M. H. Gault. Prediction of creatinine clearance from serum creatinine. *Nephron.*, *16*:31 (1976).
5. J. C. Rotschafer, K. Crossley, D. E. Zaske, K. Mead, R. J. Swachuk et al. Pharmacokinetics of vancomycin: observations in 28 patients and dosage recommendations. *Antimicrob. Agents Chemother.*, 22:391 (1982).
6. N. Brown, D. H. W. Ho, K. L. Fong, L. Bogerd, A. Maksymiuk, et al. Effects of hepatic function on vancomycin clinical pharmacology. *Antimicrob. Agents Chemother.*, 23:603 (1983).

35

Rifampicin Binding to Serum Proteins in Normal and Brucellosic Individuals

Gallardo L. Pérez, Jimenez M. L. Blanco,
Calvo H. Soria, Rovira F. Iesern *Faculty of Medicine, Valladolid, Spain*
Marcén J. F. Escanero *Faculty of Medicine, Zaragoza, Spain*

I. INTRODUCTION

Rifampicin is a semisynthetic antibiotic, whose biological properties were first described in 1960 (1). It is a first-choice antituberculosis agent, and it has also been used for infections with brucellas (2).

The binding of rifampicin to serum proteins has been investigated by various authors, and the reported percentage bound has varied from 75% to 90% (3,6) over the range of therapeutic concentrations of rifampicin. In previous clinical studies, significant reduced serum protein binding of rifampicin was reported. Serum protein binding of the drug was significantly reduced in undernourished compared to well-nourished patients (5); the binding rates of rifampicin were lower in tuberculous patients and the reproducibility of pooled serum was relatively good (4); the binding of rifampicin to serum proteins amounts to 65% in patients with bronchopulmonary affections (8).

The intention of this communication is to report some preliminary results on experiments that we have made to detect rifampicin binding to serum proteins from patients with brucellosis.

II. MATERIALS AND METHODS

Rifampicin (Homo-Chemicals), dissolved in dimetylformamide, diluted with acetonitrile/2-propanol (1/1) and stabilized with ascorbic acid, was added to serum samples obtained from normal volunteers to obtain eight concentrations ranging from 10 to 150 μmol/L.

After allowing 2 h at 37°C for equilibration, a 2-mL aliquot of each concentration was subjected to ultrafiltration, using membrane cones (Centriflo CF25). The samples were centrifuged at 2000 rpm for 10 min. Immediately after centrifugation, the serum was removed and the unbound fraction of rifampicin was determined by ultraviolet spectrophotometry at 350 nm (Spectronic 2000, Baush & Lomb).

Samples obtained from patients with brucellosis, to whom no drug had been administered for at least 24 h before study, were subjected to the same processing. Before study the pooled sera from healthy volunteers as well as sera from the brucellosic patients were subjected to electrophoretic treatment.

Results are expressed in a graph of bound rifampicin to serum proteins versus total rifampicin. Rifampicin bound was calculated from the equation

$$B = T - U$$

where B is rifampicin bound to serum proteins, T is total rifampicin in each sample, and U is ultrafiltrated rifampicin. The relation between concentration and bound rifampicin was calculated by least-squares regression analysis from no less than eight points. Slopes were analyzed statistically by Student's t-test.

III. RESULTS

Table 1 shows the results of serum electrophoretic patterns in the pooled sera from healthy volunteers as well as the sera from the brucellosic patients. A slight decrease in the albumin fraction and a slight increase in the gamma globulin fraction are observed in the pooled sera from the brucellosic patients, in comparison to pooled sera from healthy volunteers.

Figures 1 and 2 show rifampicin bound to serum proteins in drug concentrations of 10 to 150 μmol/L. Serum proteins binding rates in pooled sera from healthy volunteers ranged between 94% and 70% with a slope calculated as 0.9409 ± 0.028, and from brucellosic patients ranged between 94% and 71% with a slope calculated as 0.9763 ± 0.052 in the concentration range studied. Slopes are similar ($t = 1.940$, $p < 0.5$).

IV. DISCUSSION

The results of previous studies of protein binding of rifampicin have been summarized in Table 2.

The results of the present study show a bound fraction of 91% to 70% (Fig. 1) and are within the upper part of the published results over the range of therapeutic concentrations of rifampicin.

Binding in patients was similar to that in healthy volunteers, 90 to 70% (Fig. 2). One explanation for this binding in the pool of brucellosic patients might be that,

Table 1 Serum Protein Concentrations

	Pooled Sera from	
	Brucellosic	**Healthy**
Albumin (g/dL)	4.40	3.44
α_1-Globulins (g/dL)	0.25	0.24
α_2-Globulins (g/dL)	0.80	0.86
β-Globulins (g/dL)	0.80	0.90
γ-Globulins (g/dL)	0.70	1.50

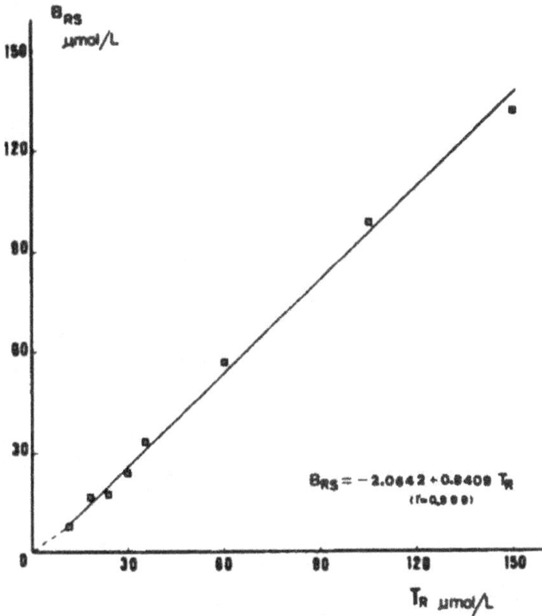

Figure 1 Correlation between total rifampicin (T_R) and bound rifampicin (B_{RS}) to serum proteins from healthy volunteers. Each point is the mean of four determinations.

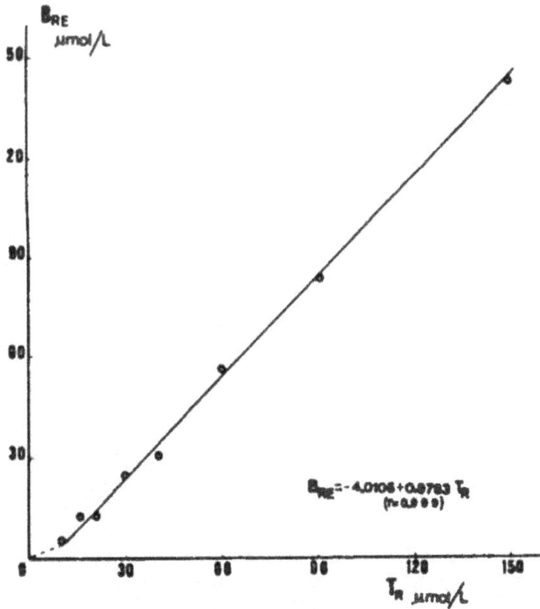

Figure 2 Correlation between total rifampicin (T_R) and bound rifampicin (B_{RE}) to serum proteins from brucellosic patients. Each point is the mean of four determinations.

Table 2 Previous Reports of Protein Binding of Rifampicin[a]

Method	Protein	Temp. (°C)	Rifampicin (μg/mL)	Percent Rifampicin Bound	Ref.
ED	HS	4	10	8–41	3
ED	HP	37	10	84–91	4
ED	HP	37	10	75–79	5
ED	MP	37	10	90	6
UF	HP	—	0.045–1.360	69–71	7
UF	HS	37	8.23–123.5	94–70	PE

[a]Abbreviations: ED, equilibrium dialysis; UF, ultrafiltration; HS, human serum; HP, human plasma; MP, murine plasma; PE, present study.

although albumin is considered the sole binding protein for most drugs, this is incorrect for rifampicin, which is bound to gamma-globulins (9), and the lower serum albumin levels in pooled sera from the brucellosic patients also implied greater levels of gamma-globulins (Table 1), other serum proteins which bind rifampicin.

Although the binding rates of rifampicin do not vary in pooled sera from brucellosic patients in comparison to sera from healthy volunteers, the highest binding rate of rifampicin was observed with an individual serum from a patient with macroglobulinemia (3). Consequently, we are now preparing to undertake experiments designed to shed light on the influence exerted by macroglobulinemia in individual brucellosic patients with high levels of gamma-globulins.

V. SUMMARY

The binding rate of rifampicin to serum proteins was examined in sera from brucellosic patients and from human volunteers (control group). The measurement of serum protein binding rates was carried out by ultrafiltration. Different solutions of rifampicin were mixed with serum proteins to obtain concentrations ranging from 10 to 150 μmol/L, and were filtrated by centrifugation for 10 min. The concentrations of rifampicin in the ultrafiltrates show that serum proteins binding rates in pooled sera from healthy volunteers ranged between 94% and 70% with a slope estimated of 0.9409 ± 0.028, and from brucellosic patients ranged between 94% and 71% with a slope estimated of 0.9763 ± 0.052 in the concentration range studied.

REFERENCES

1. P. Sensi, A. M. Greco, and R. Ballotta. Rifomycin I. Isolation and properties of rifomycin B and rifomycin complex. *Antibiot. Ann.*, 7:262–270 (1960).
2. A. M. Philippon, M. G. Plommet, A. Kazmierczak, J. L. Marly, and P. A. Nevot. Rifampin in the treatment of experimental brucellosis in mice and guinea pigs. *J. Infect. Dis.*, *136*(4): 482–487 (1977).
3. T. Aoyagi. Protein binding of rifampicin to different individual sera. *Scand. J. Resp. Dis., Suppl. 84*:44–49 (1973).
4. G. Boman. Serum concentration and half-life of rifampicin after simultaneous oral administration on aminosalicylic acid or isoniazide. *Eur. J. Clin. Pharmacol.*, 7:217–225 (1974).

5. K. Polasa, K. J. Murthy, and K. Krishnauswamy. Rifampicin kinetics in undernutrition. *Br. J. Clin. Pharmacol.*, *17*(4):481–484 (1984).
6. J. J. Hoogeterp, H. Mattie, A. M. Krul, and R. Furth. The efficacy of rifampicin against staphylococcus aureus in vitro and in an experimental infection in normal and granulocytopenic mice. *Scand. J. Infect. Dis.*, *20*(6):649–656 (1988).
7. M. H. Skinner, M. Hsieh, J. Torseth, D. Pauloin, G. Bhatia, S. Harkonen, T. C. Merigan, and T. F. Blaschke. Pharmacokinetics of rifabutin. *Antimicrob. Agents Chemother.*, *33*(8):1237–1242 (1989).
8. H. Becuwkes, H. J. Buytendijk, and F. P. Maesen. Der wert des rifampicin bei der behandlug von patient en mit bronchopulmonarem affektionen. *Arzneimittelforsch.*, *19*:1283–1285 (1969).
9. B. S. Nilsson, G. Boman, and W. W. Bullock. Detection of rifampicin binding immunoglobulins. *Scand. J. Dis., Suppl.* *84*:50–52 (1973).

36

Comparison by Computer Simulation of Five Different Infusion Regimens of Thiopental as Primary Agent to Reduce Intracranial Hypertension

Daan J. Touw and Sander A. Vinks *The Hague Hospitals Central Pharmacy, The Hague, The Netherlands*

I. INTRODUCTION

Increased intracranial pressure (ICP) is a common and serious problem in patients with severe head injury and other diseases of the brain. ICP to some extent was found in 80% of all patients with major head injury. In over half of the patients, ICP was significantly elevated (1). Increased ICP accounts for 50% of all head injury deaths (2). While the majority of cases of increased ICP are due to head injury, elevated ICP is also seen as a secondary effect of a variety of other conditions such as brain tumors, Reye's syndrome, and intracerebral hemorrhage.

Space occupying lesions will first be compensated by displacement of cerebrospinal fluid (CSF) to the spinal canal and by compression of the venous system (Figure 1) (3). When these compensatory mechanisms are exhausted, the ICP will rise and the cerebral perfusion pressure will fall, resulting in cerebral ischemia. Increased ICP may also result in herniation of the brainstem with irreversible damage. Thus aggressive early treatment of increased ICP is essential to improve outcome (4). The methods used for treatment of increased ICP depend on its etiology. The modalities available include controlled hyperventilation, drainage of CSF, removal of mass lesions, and drug therapy with osmotic diuretics (mannitol), corticosteroids (dexamethasone), and the barbiturates, pentobarbital and thiopental (5). Using high doses of intravenous barbiturates, a considerable reduction of the cerebral metabolism, oxygen requirement, and cerebral blood flow is achieved. Interest in high-dose barbiturate therapy was stimulated by the report of Shapiro et al. (6).

Thiopental pharmacokinetic kinetics can be described with an open three-compartment pharmacokinetic model with first-order elimination from the central

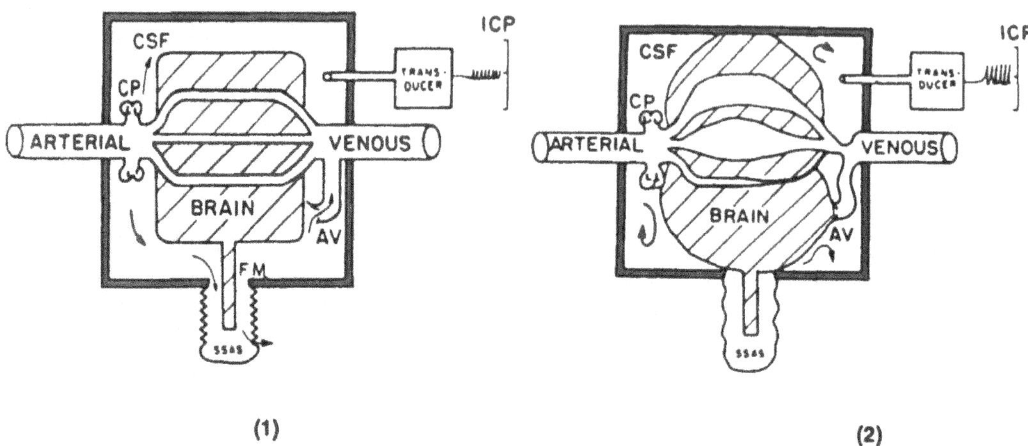

Figure 1 Schematic representation of the intracranial space during normal (1) and increased (2) pressure.

compartment (Figure 2) (7–9). Equilibrating with the central compartment is an infinitesimally small effect compartment from which EEG and ICP are derived. After a single bolus injection, thiopental kinetics consist of three phases (7,9,10):

1. Fast initial distribution over the central compartment including the central nervous system and distribution over peripheral compartments
2. Distribution over deeper compartments and metabolic elimination
3. Metabolic elimination and redistribution from the deeper compartments

Thiopental is metabolized mainly to a carboxylic acid analog for which ICP reducing properties have not yet been established. The second metabolite is pentobarbital (11), which reduces ICP as well (12). Therefore, in kinetic dosing the influence of the serum concentration of pentobarbital has to be taken into account.

Thiopental metabolism is saturable. The Michaelis-Menten constant is reported to be 26.7 ± 22.9 mg/L and maximum enzymatic capacity 405 ± 180 mg/h (13), so at high doses, limited enzymatic capacity will reduce clearance.

Serum concentrations of thiopental and pentobarbital and their effect are well correlated. Reduction of the increased intracranial pressure and an isoelectric EEG are usually seen at serum concentrations ranging from 20 to 40 mg/L. Above serum concentrations of 30 mg/L, arterial hypertension frequently occurs.

There is a need for an efficient mode of administration of this intravenous agent in order to rapidly achieve and maintain stable serum concentrations without undesired serum concentration fluctuations. When thiopental is administered by repetitive bolus doses, serum concentrations undergo rapid fluctuations due to the drug's distributive properties and may result in cardiovascular depression at high drug concentrations and unresponsiveness at low drug concentrations. Uncontrolled infusion may result in unnecessary delay in therapeutic effects.

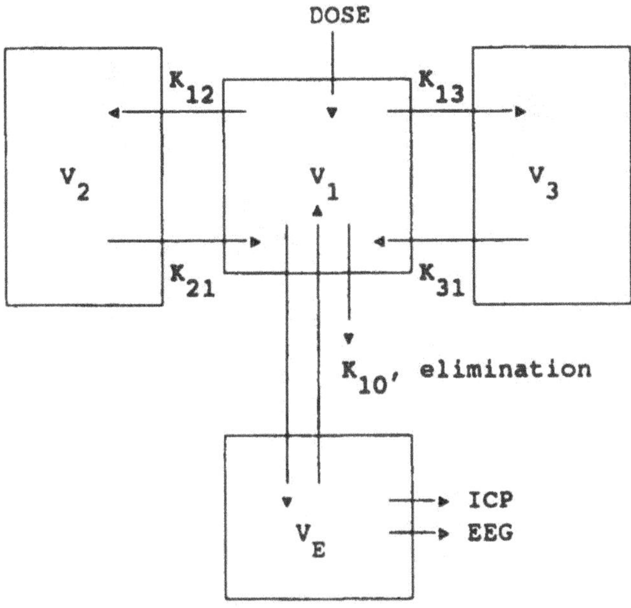

Figure 2 Pharmacokinetic-pharmacodynamic model of thiopental. V_1, V_2, and V_3 represent the central and the rapidly and slowly equilibrating compartments respectively. K_{10} is the rate constant for total drug clearance from the body. K_{12}, K_{21}, K_{13}, and K_{31} are the rate constants that characterize the intercompartment transfer of the drug. Linked to the central compartment is an infinitesimally small effect compartment (V_E) from which ICP and EEG are derived.

Several infusion regimens for chronic intravenous thiopental administration have been described (Table 1) (13–20). In our hospital we use a thiopental dosing scheme (Table 2) based on published population pharmacokinetic parameters derived from surgical patients (21). Due to both the rapid and slow distributive properties of thiopental, dosing regimens comprising at least two different rate loading doses followed by a maintenance infusion appear to be most suitable. A multistage infusion regimen, or a computer-controlled, exponentially decreasing infusion regimen, is superior to a two-stage infusion regimen in maintaining stable serum thiopental concentrations (22). This study, however, included only surgical patients, and the duration of thiopental anaesthesia lasted from 40 min to 6 h (22). For our purpose, thiopental administration up to 72 h may be required. This is when the later phases of thiopental kinetics become important.

To evaluate the different dosing regimens, a computer simulation was carried out to compare the published dosing schemes with our scheme using pharmacokinetic kinetic parameters of our population of traumatic patients.

II. MATERIALS

For simulation, an IBM AT-compatible microcomputer with a 20-Mbyte hard disk equipped with the pharmacokinetic computer program MW-Pharm (Mediware,

Table 1 Published Thiopental Infusion Regimens in Chronological Order

Number	Author	Induction Dose (mg)	Loading Dose 1 (mg/kg/h)	Loading Dose 2 (mg/kg/h)	Maintenance Dose
1.	Carlon, 1978	150			150 mg/h
2.	Frost/Tabaddor, 1981				150 mg/h
3.	Cunitz, 1982	400			150 mg/h
4.	Quandt, 1982		20/1	60/6	3 mg/kg/h
5.	Lundar, 1983				2–4 mg/kg/h
6.	Turcant, 1985	2 mg/kg			3–5 mg/kg/h
7.	Eyre, 1986		10/0.5	10/1	2 mg/kg/h
8.	Goldberg, 1986		15/0.5	30/1.5	5–0.75 mg/kg/h

Groningen, The Netherlands) was used. Thiopental pharmacokinetic parameters were obtained from our own patient population, comprising patients suffering from elevated intracranial pressure due to closed head trauma (Table 3) (23). Simulation was carried out for a patient weighing 70 kg. The program calculates the theoretical serum concentration using a linear three-compartment open model with elimination from the central compartment and gives a graphic representation of the serum concentration-versus-time profile.

III. METHODS

The aim of the infusion regimen is to quickly reach and maintain a stable serum concentration of thiopental ranging from 20 to 30 mg/L. The published infusion regimens listed in Table 1 that are marked with an asterisk were evaluated. In those regimens where thiopental doses are given in milligrams per kilogram body weight, we used the total body weight of 70 kg.

IV. RESULTS

The thiopental serum concentration-versus-time profiles for the simulations are shown in Figure 3. Time to steady state differed strongly among the different dosing schemes, and remarkable fluctuations in serum concentrations occurred.

Table 2 Infusion Regimen for Thiopental

Bolus dose	10 mg/kg in 6 min
High-rate infusion	20 mg/kg in 30 min
Maintenance dose	2.5 mg/kg in 60 min

Table 3 Pharmacokinetic Parameters of Thiopental Derived from Traumatic Patients Suffering Intracranial Hypertension

V1 (L/kg)	0.037 ± 0.015
Kelm (1/uur)	2.71 ± 1.53
Kelr (1/uur)	0
K12 (1/uur)	21.3 ± 14.6
K21 (1/uur)	4.8 ± 3.0
K13 (1/uur)	4.5 ± 3.4
K31 (1/uur)	0.26 ± 0.11

V. DISCUSSION

Regimens 2 and 5 use only a constant rate infusion of 150 mg/h and 2 to 4 mg/kg/h, respectively. Simulation of regimen 2 (Figure 3b) shows a delay in reaching a therapeutic serum concentration of at least 1 day because steady state is reached after at least four times the half-life of the drug.

Regimens 1, 3, and 6 consist of a thiopental dosing scheme which involves the administration of two consecutive doses to compensate for loss due to distribution. The first dose is a bolus dose to achieve therapeutic serum concentrations, the second dose is a maintenance dose designed to replace exactly what is eliminated. Simulation of these regimens (Figures 3a, 3c, and 3e) shows adequate serum concentrations initially but, due to distribution, the serum concentration rapidly declines to subtherapeutic levels. The subsequent maintenance dose only replaces the amount cleared by metabolic processes, and the steady-state situation is attained only after 24 h.

Thiopental infusion regimen 4 is a three-stage regimen. To rapidly achieve therapeutic concentrations, a primary bolus dose is administered, followed by a high-rate infusion to compensate the loss of thiopental due to distribution, and finally a maintenance dose to replace what is eliminated. Simulation (Figure 3d) shows serum concentrations which are initially high with great fluctuations. Serum concentration declines to a steady-state level of 40 mg/L within 10 to 12 h after initiation of the therapy.

Thiopental administered according to regimen 7 was used to protect the brain of severely asphyxiated newborn babies. It also consists of a three-stage thiopental infusion regimen. Simulation of this regimen (Figure 3f) shows initial serum concentrations of 35 to 40 mg/L, which declines to a level just below 20 mg/L by 4 to 6 h after initiation. Steady state is reached 12 h after initiation.

Regimen 8 was designed for the same purpose and comprises a multistage infusion regimen. Therapy is started with two loading infusions followed by a gradually decreasing maintenance infusion. Simulation of this regimen (Figure 3g) shows serum concentrations which are initially high with great fluctuations and which slowly decline to a steady-state level of 25 to 30 mg/L 6 h after start of the therapy.

Simulation of our own regimen (Figure 3h) shows initial serum concentrations which are also initially high but decline within 3 h to a steady-state concentration of 25 mg/L.

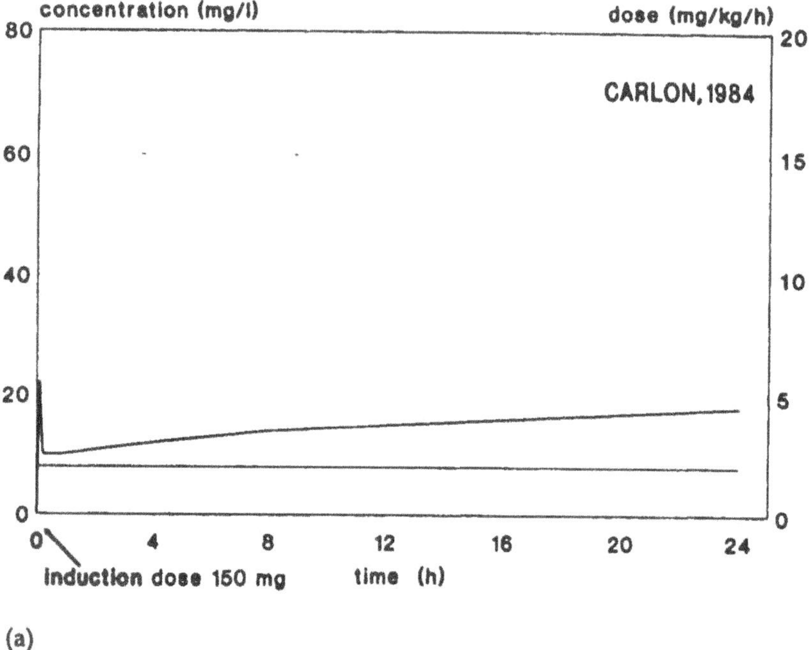

(a)

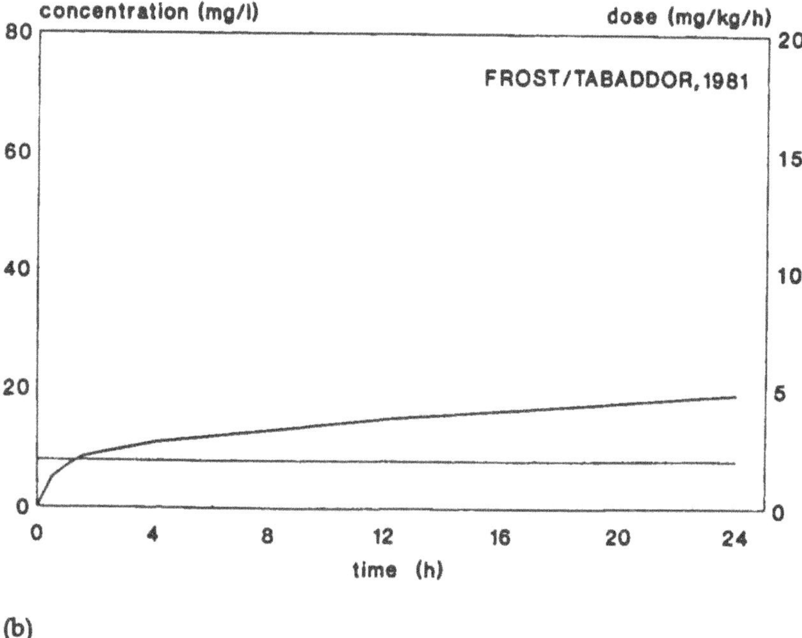

(b)

Figure 3 Thiopental serum concentration-time profiles (solid line, left-hand scale) and infusion profiles of thiopental (right-hand scale) as obtained by computer simulation.

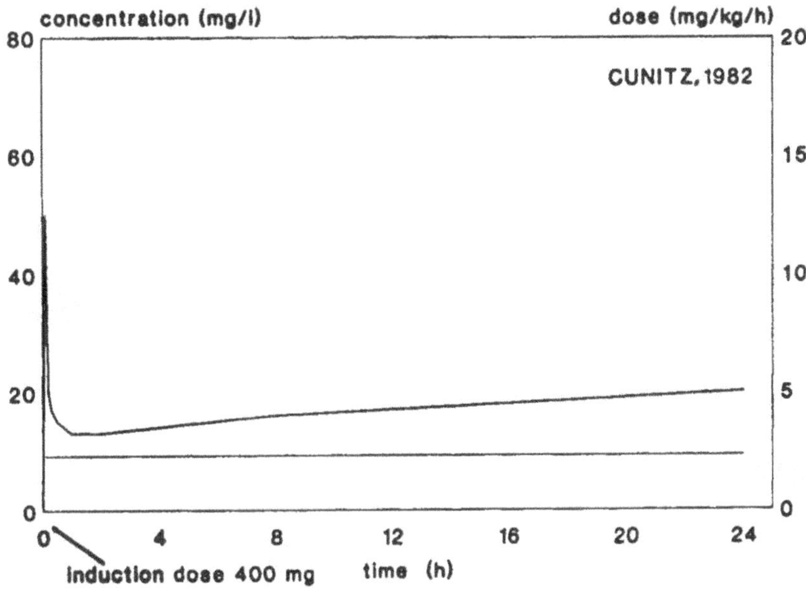
concentration (mg/l)
dose (mg/kg/h)
CUNITZ, 1982
80
60
40
20
0
20
15
10
5
0
0
4
8
12
16
20
24
induction dose 400 mg
time (h)

(c)

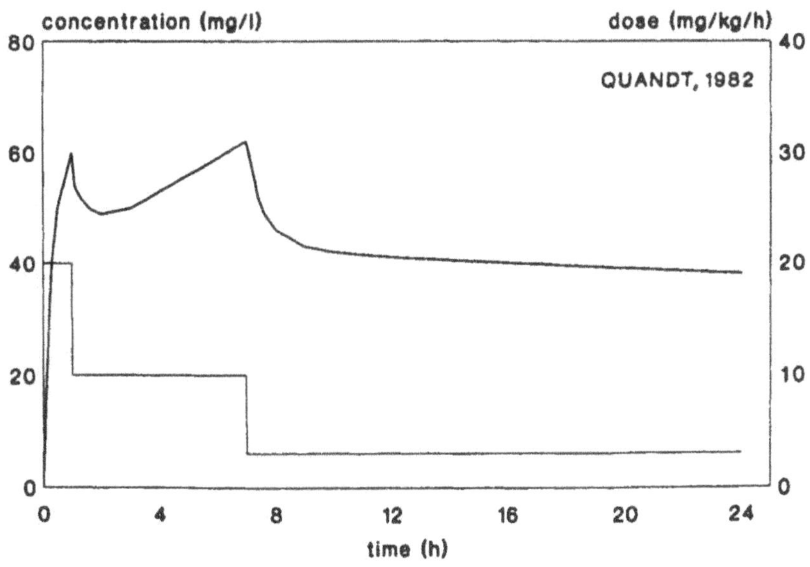
concentration (mg/l)
dose (mg/kg/h)
QUANDT, 1982
80
60
40
20
0
40
30
20
10
0
0
4
8
12
16
20
24
time (h)

(d)

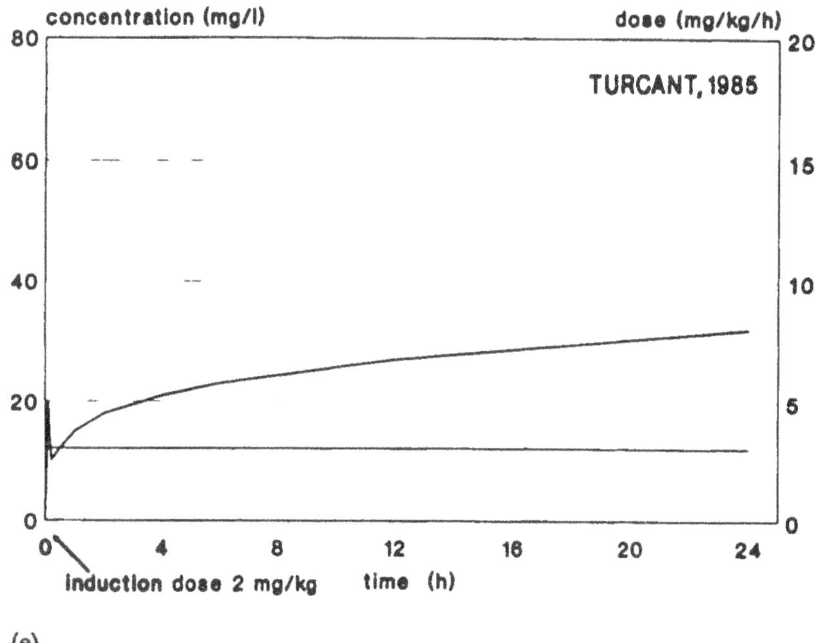
concentration (mg/l)
dose (mg/kg/h)
TURCANT, 1985
80
60
40
20
0
20
15
10
5
0
0
4
8
12
16
20
24
induction dose 2 mg/kg
time (h)

(e)

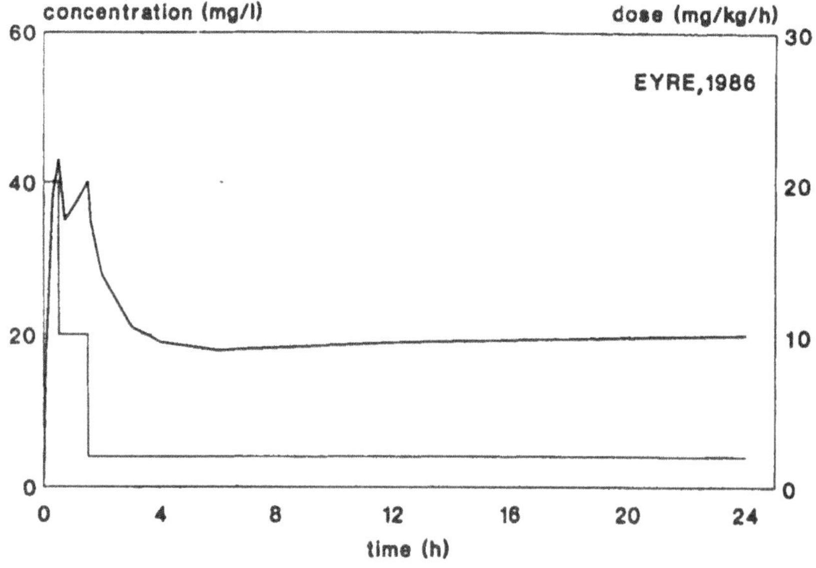
concentration (mg/l)
dose (mg/kg/h)
EYRE, 1986
60
40
20
0
30
20
10
0
0
4
8
12
16
20
24
time (h)

(f)

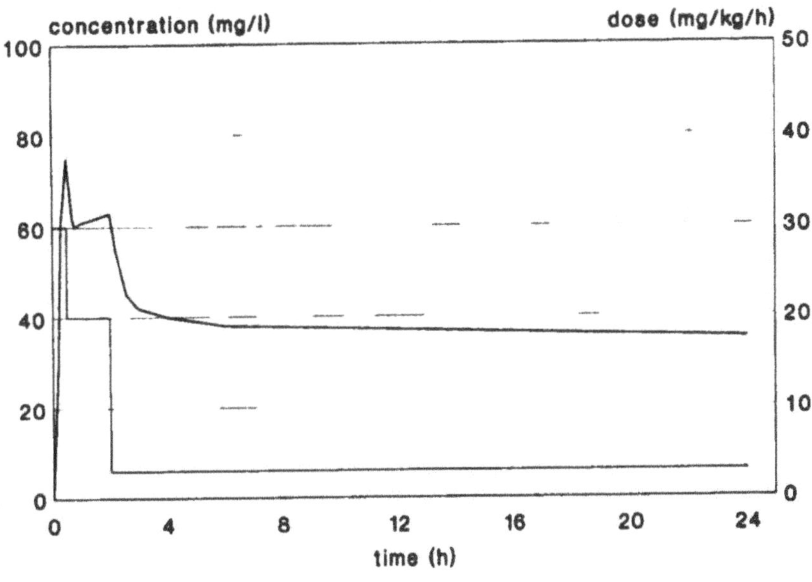
concentration (mg/l)
dose (mg/kg/h)
100
80
60
40
20
0
50
40
30
20
10
0
0
4
8
12
16
20
24
time (h)

(g)

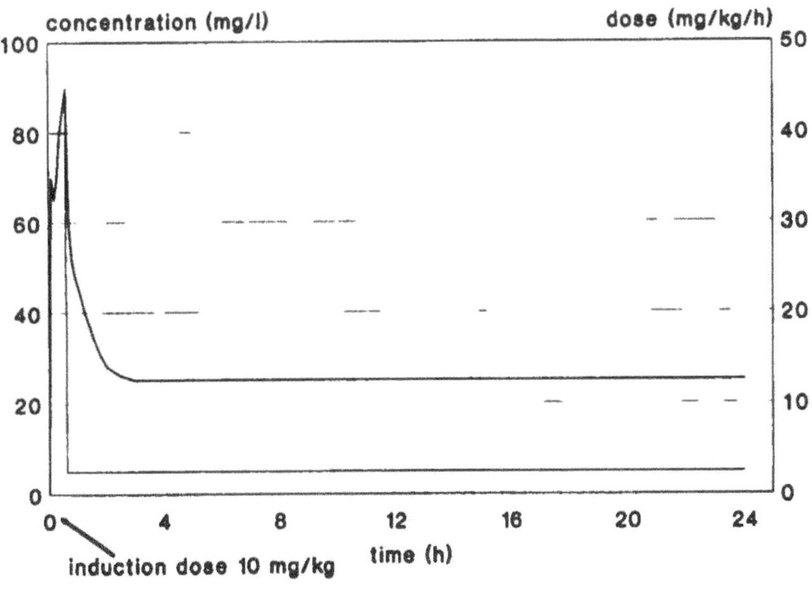
concentration (mg/l)
dose (mg/kg/h)
100
80
60
40
20
0
50
40
30
20
10
0
0
4
8
12
16
20
24
time (h)
induction dose 10 mg/kg

(h)

VI. CONCLUSION

The primary aim of this study was to compare our developed infusion regimen with those published. Computer simulation of our regimen shows initially high serum thiopental concentrations but rapidly achieves stable therapeutic serum concentration with a thiopental steady-state concentration of 25 mg/L within 3 h. Barbiturates have great influence on arterial blood pressure. For hemodynamically instable patients such as the group described, it is important to have minimal fluctuations in the serum concentration during administration. Also, a steady-state situation must be achieved as soon as possible. During the loading phase it is hardly possible to avoid fluctuations of the serum concentration. Our scheme shows very rapid achievement of a steady-state situation.

In practice, this method of comparing dosing regimens has given considerable insight into the serum-concentration time course of thiopental when administered as a continuous infusion. The computer simulations have shown that a steady-state situation can be rapidly achieved when large loading doses are given. In the individual patient a determination of the serum thiopental concentration has to be carried out to assure whether the concentration is in the therapeutic range. Three hours after initiation of the therapy a serum sample can be drawn and analyzed for thiopental and its metabolite, pentobarbital. Based on the analytical results, correction of the maintenance dose can follow.

REFERENCES

1. D. P. Becher, J. D. Miller, J. D. Ward, R. P. Greenberg, H. F. Young, and R. Sakalas. Outcome from severe head injury with early diagnosis and intensive management. *J. Neurosurg.*, 47:491–502 (1972).
2. J. D. Miller, D. P. Becher, J. D. Ward, H. G. Sullivan, W. E. Adams, and M. J. Rosner. Significance of intracranial hypertension in severe head injury. *J. Neurosurg.*, 47:503–516 (1977).
3. H. M. Shapiro. Intracranial hypertension: Therapeutic and anesthetic considerations. *Anesthesiology*, 43:445–471 (1975).
4. T. G. Saul and T. B. Ducker. Effect of intracranial pressure monitoring and aggressive treatment on mortality in severe head injury. *J. Neurosurg.*, 56:498–503 (1979).
5. J. E. McGillicuddy. Cerebral protection: Pathophysiology and treatment of increased intracranial pressure. *Chest*, 87:85–93 (1985).
6. H. M. Shapiro, A. Galindo, S. R. Wyte, and A. B. Harris. Rapid intraoperative reduction of intracranial pressure with thiopentone. *Br. J. Anaesth.*, 45:1057–1061 (1973).
7. P. G. Burch and D. R. Stanski. The role of metabolism and protein binding in thiopental anaestesia. *Anesthesiology*, 58:146–152 (1983).
8. J. H. Christensen, F. Andreasen, and J. A. Jansen. Pharmacokinetics of thiopentone in a group of young women and a group of young men. *Br. J. Anaesth.*, 52:913–918 (1980).
9. M. M. Ghoneim and M. J. van Hamme. Pharmacokinetics of thiopentone: Effects of enflurane and nitrous oxide anaesthesia and surgery. *Br. J. Anaesth.*, 50:1237–1242 (1978).
10. P. Duvaldestin. Pharmacokinetics in intravenous anaesthetic practice. *Clin. Pharmacokin.*, 6:61–82 (1981).
11. L. C. Mark. Metabolism of barbiturates in man. *Clin. Pharmacol. Ther.*, 4:504–530 (1963).
12. C. M. Quandt, R. A. de los Reyes, and F. G. Diaz. Barbiturate-induced coma for the treatment of cerebral ischemia: Review of outcome. *Clin. Pharm.*, 50:549–550 (1982).
13. A. Turcant, A. Delhumeau, A. Premel-Cabic, et al. Thiopental pharmacokinetics under conditions of long-term infusion. *Anesthesiology*, 63:50–54 (1985).
14. E. A. M. Frost, K. Tabaddor, and B. Y. Kim. Induced barbiturate coma: Results in 20 cases. *Anesth. Analg.*, 60:247 (1981).

15. K. Tabaddor, E. A. M. Frost, and B. Kim. Iatrogenic barbiturate coma—Results in 28 cases. *Br. J. Anaesth.*, *53*:303P–304P (1981).
16. T. Lundar, T. Ganes, and K.-F. Lindegaard. Induced barbiturate coma: Methods for evaluation of patients. *Crit. Care Med.*, *11*:559–562 (1983).
17. G. Cunitz. Control of intracranial pressure (ICP) during intensive care. *Br. J. Anaesth.*, *54*:248P–249P (1982).
18. C. M. Quandt and R. A. de los Reyes. Pharmacologic management of acute intracranial hypertension. *Drug Intell. Clin. Pharm.*, *18*:105–112 (1984).
19. J. A. Eyre and A. R. Wilkinson. Thiopentone induced coma after severe birth asphyxia. *Arch. Dis. Child.*, *61*:1084–1089 (1986).
20. R. N. Goldberg, P. Moscoso, C. R. Bauer, et al. Use of barbiturate therapy in severe perinatal asphyxia: A randomized controlled trial. *J. Pediatr.*, *109*:851–856 (1986).
21. G. Heinemeyer. Clinical pharmacokinetic considerations in the treatment of increased intracranial pressure. *Clin. Pharmacokin.*, *13*:1–25 (1987).
22. D. P. Crankshaw, N. E. Edwards, G. L. Blackman, M. D. Boyd, H. N. J. Chan, and D. J. Morgan. Evaluation of infusion regimens for thiopentone as a primary anaesthetic agent. *Eur. J. Clin. Pharmacol.*, *28*:543–552 (1985).
23. D. J. Touw, J. T. J. Tans, and A. A. T. M. M. Vinks. clinical pharmacokinetics of high-dose intravenous thiopental as primary agent to reduce increased intracranial pressure (abstr.). *Quim. Clin.*, *9*:361 (1990).

37

Influence of Cisplatin Infusion Rate on Local Availability of the Drug

J. Erviti, J. J. Rodríguez, R. Mangues, A. Idoate, and J. Giráldez *University Clinic of Navarra, Pamplona, Spain*

I. INTRODUCTION

Chemotherapy for cancer is in a state of evolution. It has assumed an important role as an adjuvant to other modalities, including both surgery and radiation therapy. An essential aspect of bolus chemotherapy is the intermittency of the schedule. A potential disadvantage of this is the possibility of regrowth of tumor during the absence of drug. This clinical phenomenon is most likely to occur in the setting of high-growth-fraction tumors. Prolonging the exposure time of the tumor cell to drugs is a concept that was addressed in the early phases of the development of chemotherapy.

Three bases for applying the infusion schedule for delivery of individual agents are (1):

1. *Cytokinetic profile of tumors.* Cisplatin blocks LoVo cells in S and G_2 phase, suggesting that the drug may be phase-specific. Cisplatin availability at a time when the maximal number of in-cycle cells is present could increase drug effectiveness.
2. *Pharmacokinetics of antineoplastic agents.* For most tumors, the in-cycle cell population is small and the plasma half-life for most of the chemotherapeutic agents employed is short.
3. *Drug transport through the cell membrane.* Drug transport through the cell membrane occurs by diffusion and active transport. The latter may depend on the drug concentration and time the cell is exposed to the drug. Infusion schedule may lead to higher exposition (concentration × time integrals) values.

The purpose of this study is to show the relationship between the cisplatin infusion rate on local availability of the drug after intraarterial administration, comparing a 3-h infusion with a bolus infusion schedule.

II. MATERIALS AND METHODS

Ten female beagle dogs (weight range 19 to 24 kg; mean 21.3 kg) were included in the randomized study. Animals were treated under anesthesia and maintained with O_2/air (60:40) during the experiments. All dogs were administered cisplatin (Placis, Wasserman) through the femoral artery at a 2-mg/kg dose and mannitol at a 2.5-mg/kg dose. Two randomized groups (five dogs each) were established. The first group (BG) received a "bolus infusion" of a 1 mg/mL cisplatin solution (40 mg/min). The infusion group (IG) was given a 40-μg/mL cisplatin solution in a 3-h infusion (6.5 mL/min). Blood samples were obtained from the femoral (local) vein draining the infusion region and from a peripheral vein at multiple sequential times (0, 3, 7, 20, 60, and 120 min) after drug administration. Three additional blood samples were taken at 30, 60, and 120 min after the beginning of drug administration in the IG. Blood samples were centrifuged to obtain plasma. Non-protein-bound Pt species were measured in ultrafiltrate (cutoff MW < 50,000 dalton/Amicon MPS-1 system).

Three hours after the end of the infusion, the animals were sacrificed and muscle tissue samples were obtained in all cases. In each dog, muscle samples were taken from the infusion area and contralateral extremity. A solution of HNO_3 (70%)/$HClO_4$ (70%) (1:1) was utilized for digestion of muscle tissue samples. Four milliliters of that solution was drawn into every test tube (containing 1 g of dry muscle tissue each). All test tubes were heated to 280°C for 10 days, and then 8 mL of HCl (1 N) was added to every tube. As a result, a colorless solution was obtained whose Pt concentrations were measured.

All samples were analyzed for Pt concentration using a flameless atomic absorption spectrophotometric assay (Video 11 Instrumentation Laboratory model). Total Pt tissue concentrations were compared with those in the BG and IG. All data were evaluated with t-test, and a value of $p < 0.05$ was considered to be of statistical significance. On the other hand, correlations between exposition (concentration × time integrals) or C_{max} (peak concentrations) and Pt muscle tissue levels were established. In this case data were evaluated with Pearson's test and a value of $p < 0.05$ was considered to be of statistical significance.

III. RESULTS AND DISCUSSION

A. Infusion Region

A rapid plasma Pt level decrease (39.96 μg/mL to 8.56 μg/mL) is observed within the first 3 min after drug administration in the BG, and then a slow elimination rate is observed.

In the IG, steady state is reached 60 min after the beginning of the infusion. When the infusion is over, Pt elimination becomes slower than in the BG (3.04 μg/mL to 1.18 μg/mL within the first 3 min) (Table 1).

Peak concentrations in the IG are about 10-fold lower than in the BG (3.57 ± 1.33 μg/mL versus 39.96 ± 21.80 μg/mL) (Figures 1 and 2; Table 2), and exposition data in the BG are about twofold lower than in the IG (255.08 ± 20.78 μg/mL versus 524.38 ± 175.26 μg/mL min) (Table 3). Higher Pt tissue levels values are determined in the IG (2.56 ± 0.98 μg/g versus 1.47 ± 0.67 μg/g) (Figure 3 and Table 4). Statistically significant differences are observed between expositions (IG)-expositions (BG) ($p < 0.05$) and peak concentrations (IG)-peak concentrations (BG) ($p < 0.01$). Correlation between expositions and Pt tissue levels ($p < 0.05$) is observed (Figure 4). No correlation

Table 1 Pt Plasma Concentration (μg/ml), Infusion Region

	Infusion Time (min)			Postinfusion Time (min)					
	–150	–120	–60	0	3	7	20	60	120
(BG) 1.				32.30	7.59	4.50	1.48	0.58	0.13
2.				29.86	7.02	3.48	1.75	0.65	0.12
3.				76.65	17.41	11.32	3.86	0.95	0.24
4.				20.12	4.35	3.08	2.05	0.41	0.15
5.				40.87	6.44	3.55	2.74	0.53	0.16
X ± SD				39.96	8.56	5.18	2.37	0.62	0.16
				±21.80	±5.09	±3.47	±0.95	±0.20	±0.05
(IG) 1.	2.40	3.01	3.54	3.28	1.71	1.53	0.48	0.38	0.12
2.	2.48	3.07	3.56	5.00	1.01	0.74	0.82	0.41	0.14
3.	4.67	4.81	3.29	3.66	1.50	1.94	0.79	0.25	0.11
4.	1.32	1.90	2.11	2.27	0.82	0.49	0.24	0.18	0.12
5.	1.87	2.23	1.83	0.99	0.87	0.56	0.58	0.17	0.12
X ± SD	2.54	3.00	2.86	3.04	1.18	1.05	0.58	0.28	0.12
	±1.27	±1.13	±0.83	±1.51	±0.40	±0.65	±0.24	±0.11	±0.01

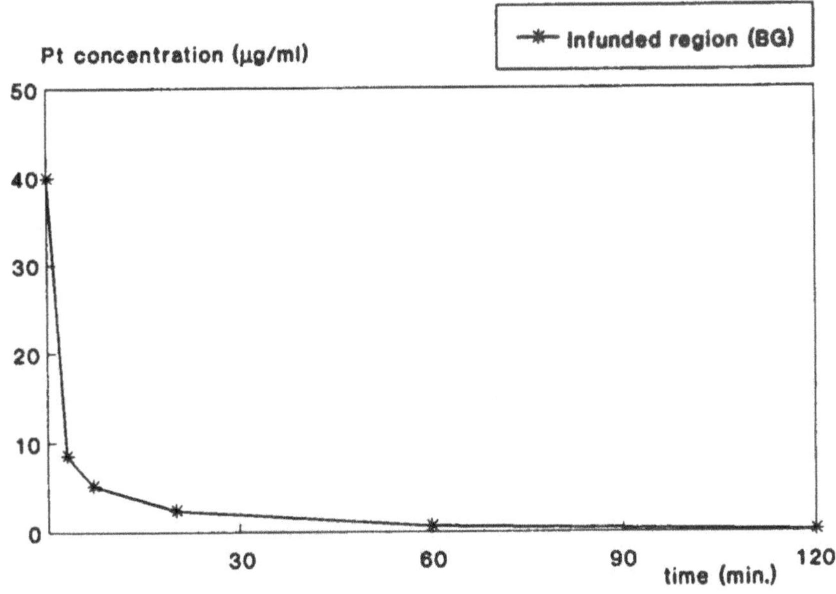

Figure 1 Pt concentration after bolus administration.

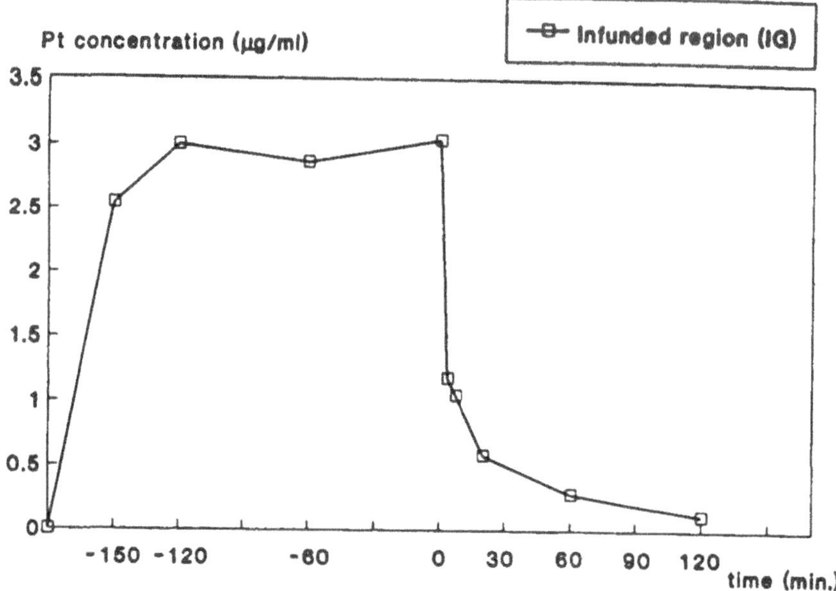

Figure 2 Pt concentration on a 3-h infusion schedule.

Table 2 Peak Concentrations (BG and IG)

	Dog	Pt Peak Concentration (μg/ml)	
		Infused Region	Contralateral Extremity
(BG)	1.	32.30	4.55
	2.	29.36	5.18
	3.	76.65	6.03
	4.	20.12	3.13
	5.	40.87	3.92
	X ± SD	39.96 ± 21.80	4.56 ± 1.12
(IG)	1.	3.54	2.85
	2.	5.00	0.93
	3.	4.81	0.70
	4.	2.27	1.30
	5.	2.23	0.81
	X ± SD	3.57 ± 1.33	1.32 ± 0.89

Table 3 Pt Plasma Concentration (μg/ml), Contralateral Extremity

		Infusion Time (min)			Postinfusion Time (min)					
		−150	−120	−60	0	3	7	20	60	120
(BG)	1.				2.05	7.59	1.49	1.71	0.45	0.19
	2.				2.72	7.02	1.87	1.38	0.49	0.14
	3.				5.48	17.41	3.91	2.11	0.61	—
	4.				3.63	4.35	2.55	1.80	0.43	0.14
	5.				3.67	6.44	2.38	1.36	0.49	0.11
	X ± SD				3.51 ±1.29	8.56 ±5.09	2.44 ±0.92	1.67 ±0.31	0.49 ±0.07	0.14 ±0.03
(IG)	1.	1.38	1.08	1.40	1.47	1.71	0.94	0.48	0.32	0.15
	2.	0.71	0.63	0.56	0.77	1.01	0.61	0.37	0.26	0.12
	3.	0.31	0.35	0.60	0.70	1.50	0.55	0.36	0.15	0.10
	4.	0.58	0.76	0.94	0.71	0.82	0.50	0.43	0.20	0.10
	5.	0.40	0.50	0.52	0.37	0.87	0.34	0.38	0.15	0.09
	X ± SD	0.68 ±0.42	0.66 ±0.27	0.80 ±0.37	0.80 ±0.40	1.18 ±0.40	0.59 ±0.22	0.40 ±0.05	0.22 ±0.07	0.11 ±0.02

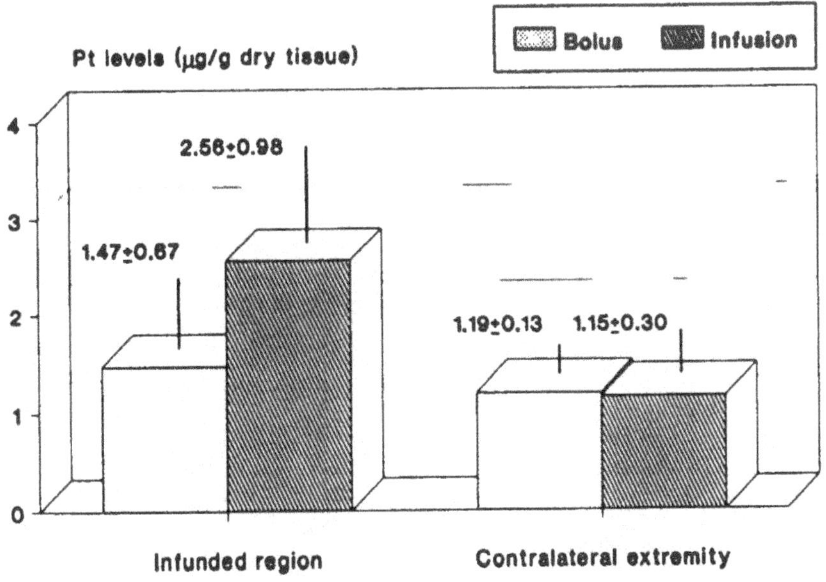

Figure 3 Pt tissue levels (BG) and (IG).

Table 4 Pt Tissue Levels (IG and BG)

	Dog	Concentration (μg/g dry tissue)	
		Infused Region	Contralateral Extremity
(BG)	1.	1.88	1.05
	2.	2.48	1.30
	3.	1.67	1.27
	4.	0.67	1.03
	5.	1.25	1.28
	X ± SD	1.47 ± 668	1.19 ± 0.134
(IG)	1.	2.25	1.59
	2.	4.27	1.10
	3.	2.34	0.79
	4.	1.79	1.28
	5.	2.14	1.23
	X ± SD	2.56 ± 0.980	1.15 ± 0.301

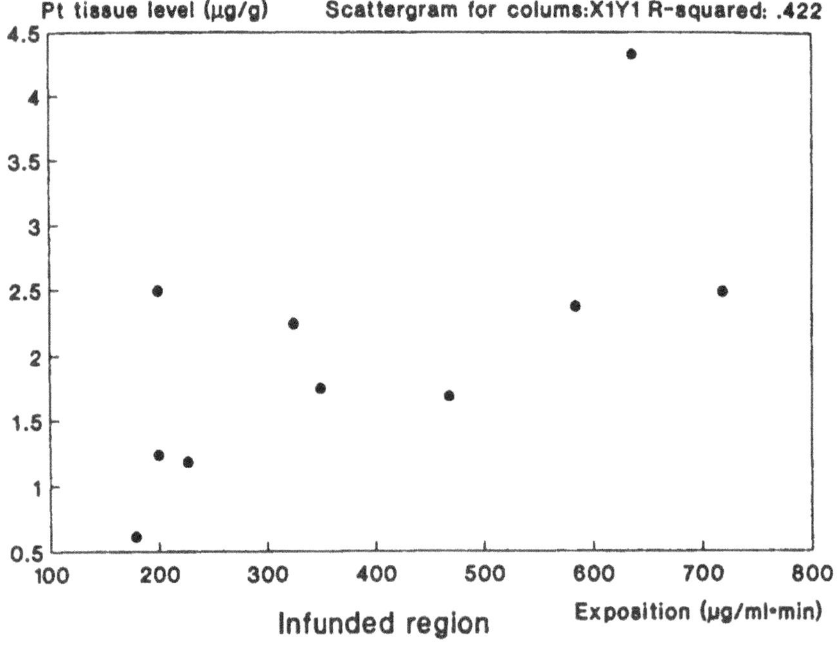

Figure 4 Correlation between exposition and Pt tissue levels.

Table 5 Pt Exposition Data (BG and IG)

	Dog	Exposition (μg/ml/min) Infused Region	Contralateral Extremity
(BG)	1.	195.65	99.49
	2.	196.07	101.01
	3.	467.40	154.05
	4.	177.35	114.03
	5.	238.35	104.74
	X ± SD	255.08 ± 120.78	114.67 ± 22.73
(IG)	1.	577.49	309.66
	2.	640.04	145.65
	3.	722.46	180.06
	4.	349.17	106.67
	5.	332.71	115.67
	X ± SD	524 ± 175.26	176.66 ± 82.39

between peak concentrations-Pt tissue levels or expositions-peak concentrations is observed.

B. Contralateral Extremity

It appears that Pt plasma concentrations and expositions are lower in the contralateral extremity than in the infusion region at any time (Tables 3 and 5). Statistically significant peak concentration (BG versus IG) differences are observed ($p < 0.01$), while Pt tissue levels are similar in the BG and IG and no statistically significant differences are observed. No correlation between expositions and Pt tissue levels was observed.

IV. CONCLUSION

We conclude there is no influence of infusion rate on systemic availability of cisplatin; but considering local availability of the drug, an advantage for the 3-h infusion over the bolus infusion is observed. Moreover, the exposition (concentration ± time integrals) may be the most representative parameter of the cisplatin local availability.

REFERENCE

1. N. J. Vogelzang. Continuous infusion chemotherapy: A critical review. *J. Clin. Oncol.*, 2(11): (1984).

38

Determinants of Urine Caffeine Concentration: Implications for Monitoring Caffeine Intake During Sports Events

John O. Miners and Donald J. Birkett *Flinders Medical Centre, Bedford Park, Adelaide, Australia*

I. INTRODUCTION

Caffeine (1,3,7-trimethylxanthine) is one of the most widely consumed pharmacological agents throughout the world. Apart from its well-documented effects on the central nervous system, cardiovascular system, smooth muscle, gastrointestinal tract, and kidneys (1), there is a perception that caffeine enhances athletic performance. However, the literature regarding this issue is conflicting (2), and there is still no clear evidence to support the view that caffeine is an ergogenic aid. Despite this, caffeine is considered by the International Olympic Committee to be a doping agent, and intake at international sporting events is monitored by the measurement of urine caffeine concentration. The current urine caffeine concentration limit is 12 mg/L.

The rationale for the urine caffeine limit is not clear. There are little or no data correlating plasma caffeine concentration with athletic performance or urine caffeine concentration with blood concentration and/or caffeine intake. Thus, it is currently not possible to assess what level of caffeine intake is "safe" in relation to the urine concentration limit. Caffeine is extensively metabolized by hepatic cytochromes P-450, with less than 3% of a dose being excreted unchanged in urine. Hence, for a given caffeine intake, steady-state plasma caffeine concentration will depend on metabolic clearance. This parameter exhibits up to 15-fold variability in the population (3–5). It would be anticipated that urine caffeine concentration could show greater variability as the parameter might depend on caffeine renal clearance and urine flow rate as well as caffeine plasma concentration.

The studies outlined here and elsewhere (6,7) were designed to assess urine caffeine concentrations achieved during controlled and uncontrolled steady-state caffeine intake in healthy subjects. In addition, the intra- and interindividual variability in plasma and urine caffeine concentrations were determined, as were effects of urine flow rate on urine caffeine concentration.

II. METHODS

A. Study 1

Full details of this study have been published elsewhere (6). Briefly, six healthy subjects were administered 150 mg of caffeine BP in a hard gelatin capsule every 8 h for 6 days. Following the attainment of steady state on day 3, urine was collected over each 8-h dosage interval, giving a total of 11 urine samples for each subject. Additionally, on days 4, 5, and 6, a blood sample was collected at the midpoint of the morning and afternoon dosage intervals. During the study period, subjects abstained from all caffeine-containing foods and beverages.

B. Study 2

Twenty-two healthy subjects were studied while maintaining their usual intake of caffeine. On the study day, subjects collected three successive 4-h urines (0700 to 1100, 1100 to 1500, 1500 to 1900), and a blood sample was collected at the midpoint of each of these urine collection periods. Apart from providing a sample of each caffeine-containing beverage consumed on the study day to allow calculation of the actual caffeine intake, each subject maintained a diary detailing the type and amount of each drink consumed for 3 days to confirm steady-state caffeine intake. Again, full details of the study protocol have been published elsewhere (7).

Approval to perform both studies was granted by the Clinical Investigation and Drugs and Therapeutic Advisory Committees of Flinders Medical Centre.

C. Sample Analysis

Plasma, urine, and beverage caffeine concentrations were determined by a specific high-performance liquid chromatographic procedure (8).

D. Data Analysis

Caffeine renal clearance was calculated as the caffeine excretion rate divided by the midpoint plasma caffeine concentration. Correlations between parameters were assessed using linear regression analysis. Results are expressed as mean ± SD.

III. RESULTS

A. Study 1

The range of caffeine concentrations measured was 0.7 to 11.1 mg/L and 1.5 to 12.1 mg/L in urine and plasma, respectively. Plasma caffeine concentrations were similar to those in urine; the mean plasma-urine ratio was 1.33 ± 0.21, with individual subject mean plasma-urine ratios ranging from 1.10 to 1.74. Caffeine concentrations in plasma and urine were highly correlated ($r = 0.93$, $p < 0.01$) (Figure 1A). There was also a good correlation between caffeine renal clearance and urine flow rate ($r = 0.89$, $p < 0.01$), but urine caffeine concentration and urine flow rates were not significantly correlated ($r = 0.15$, $p > 0.05$).

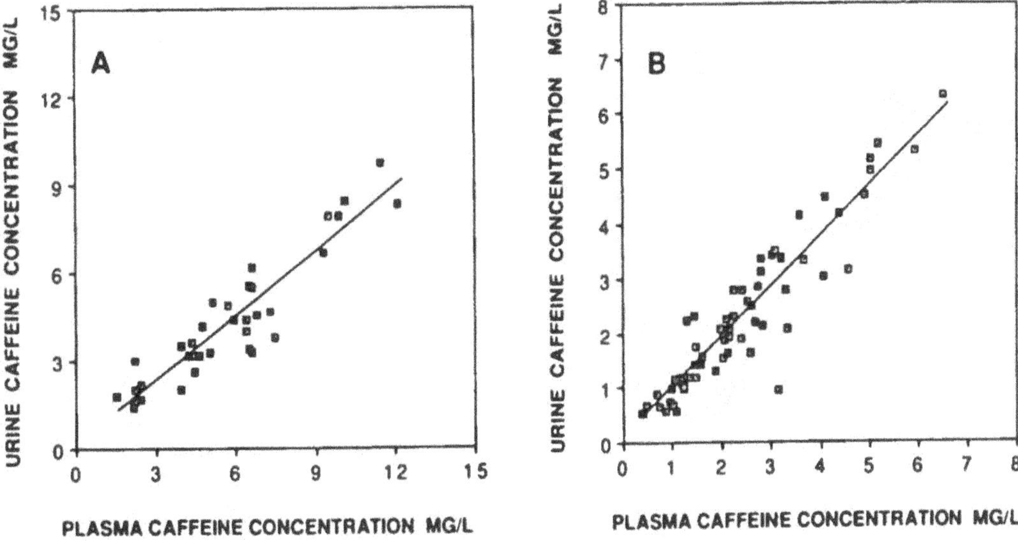

Figure 1 Relationship between plasma and urine caffeine concentrations: (A) data from study 1; (B) data from study 2.

B. Study 2

The mean caffeine intake for the 22 subjects over the study day was 186.2 ± 137.7 mg, ranging from 41.8 to 671.3 mg. Plasma and urine caffeine concentrations ranged from 0.3 to 6.3 mg/L and 0.4 to 6.5 mg/L, respectively. The mean plasma-urine caffeine concentration ratio was 0.95 ± 0.25 and, as in the first study, plasma and urine caffeine concentrations were highly correlated ($r = 0.93$, $p < 0.01$) (Figure 1B). Similarly, there was a good correlation between caffeine renal clearance and urine flow rate ($r = 0.82$, $p < 0.001$), but urine flow rate and urine caffeine concentration were not significantly correlated ($r = 0.26$, $p > 0.05$).

IV. DISCUSSION

Plasma and urine concentrations of caffeine have been determined at steady state in two groups of healthy subjects. One group of subjects was administered a fixed dose of caffeine in capsule form (study 1), while the second group followed their normal intake of caffeine-containing beverages (study 2). The caffeine dose administered in study 1 (450 mg/day) represents a moderate caffeine intake, equivalent to five to six cups of average-strength brewed coffee per day (9). Intersubject intake during study 2 ranged from one to six caffeine-containing beverages per day. The extrapolated mean caffeine intakes required to produce a urinary caffeine concentration over the current international sporting limit of 12 mg/L were 1550 mg/day and 1021 mg/day in studies 1 and 2, respectively. However, substantial interindividual variability in urine caffeine concentration was observed in both studies, and it was apparent that some subjects could exceed the 12-mg/L limit with a caffeine intake under 400 mg/day (i.e. about five cups of brewed coffee). In a more diverse population, where the variability is likely to be even greater, some individuals could well exceed the caffeine urine limit with an even lower intake.

The mean plasma-to-urine caffeine concentrations ranged from 1.33 in study 1 to 0.95 in study 2. Taken together with the high correlation ($r > 0.90$) between urine flow rate and caffeine renal clearance, this suggests that caffeine is reabsorbed from the renal tubule almost to equilibrium with unbound drug in plasma. Such a conclusion is consistent with the known lipophilicity of caffeine. Reabsorption of caffeine to equilibrium with unbound fraction in plasma accounts for the high degree of correlation between plasma and urine caffeine concentrations found in both studies ($r > 0.80$) and the lack of effect of urine flow rate (and volume) on urine caffeine concentration.

Although the present studies involved steady-state caffeine intake, it can be estimated from the known volume of distribution of caffeine (3) and the plasma-to-urine ratios found here that a single caffeine dose of approximately 500 mg would give a urine concentration around 12 mg/L.

In summary, the studies described here have demonstrated that the caffeine concentration measured in urine is a reasonable index of plasma caffeine concentration. There is marked interindividual variability in urine caffeine concentrations for subjects administered either the same fixed dose of caffeine or for those following their usual intake of caffeine-containing beverages. Some individuals may consequently be at risk of exceeding the current international urine caffeine concentration limit of 12 mg/L following modest intake of caffeine-containing foods and drinks.

ACKNOWLEDGMENTS

This work was supported by grants-in-aid from the International Life Sciences Institute (Australia) and the Flinders Medical Centre Research Foundation.

REFERENCES

1. P. W. Curalto and D. Robertson. The health consequences of caffeine. *Ann. Intern. Med., 98*:641 (1983).
2. G. R. Wenger. CNS stimulants and athletic performance. In *Drugs, Athletes and Physical Performance* (J. A. Thomas, ed.), Plenum, New York, p. 217 (1988).
3. A. Lelo, J. O. Miners, and D. J. Birkett. Comparative pharmacokinetics of caffeine and its primary demethylated metabolites paraxanthine, theobromine and theophylline in man. *Br. J. Clin. Pharmacol., 22*:177 (1986).
4. D. C. May, D. H. Jarboe, A. B. Van Bakel, and W. M. Williams. Effects of cimetidine on caffeine disposition in smokers and non-smokers. *Clin. Pharmacol. Ther., 31*:656 (1982).
5. M. D. Parsons and A. H. Neimis. Effect of smoking on caffeine clearance. *Clin. Pharmacol. Ther., 24*:40 (1978).
6. D. J. Birkett and J. O. Miners. Caffeine renal clearance and urine caffeine concentrations during steady state dosing. *Br. J. Clin. Pharmacol. 31*:405–408 (1991).
7. Z. Mohamed, J. O. Miners, and D. J. Birkett. Determinants of urine caffeine concentration. *Br. J. Sports Med.* (submitted).
8. T. Foenander, D. J. Birkett, J. O. Miners, and L. M. H. Wing. The simultaneous determination of theophylline, theobromine and caffeine in plasma by high performance liquid chromatography. *Clin. Biochem., 13*:132 (1980).
9. A. Lelo, J. O. Miners, R. A. Robson, and D. J. Birkett. Assessment of caffeine exposure: Caffeine content of beverages, caffeine intake, and plasma concentrations of methylxanthines. *Clin. Pharmacol. Ther., 39*:54 (1986).

39

Monitoring of Methotrexate Given in Continuous i.v. Infusion to Pediatric Cancer Patients

Lucja Skibińska, Urszula Radwańska, Danuta Michalewska, and Dariusz Boruczkowski *Medical Academy, Poznan, Poland*

I. INTRODUCTION

Methotrexate (MTX) is widely used in the treatment of cancer. Doses of MTX range from conventional oral low doses (~ 20 mg/m^2) to extremely high i.v. doses (~ 12 g/m^2). At standard low doses, monitoring of serum concentrations is not usual. However, serum concentrations are necessary in patients receiving high doses of MTX. In these patients the risk of toxicity increases (1). Moderate or high doses of MTX are given in continuous i.v. infusion with leucovorin rescue to prevent toxic effects. The monitoring of serum MTX concentrations is important to determine the leucovorin dosage regimen. Any delay in the leucovorin dosage and/or too small dose may result in irreversible MTX toxicity. However, the antitumor effect of MTX may be decreased when a dose of leucovorin is given too early (2,3).

The object of this investigation was to evaluate the pharmacokinetics of MTX in children with cancer who were receiving moderate or high i.v. doses of MTX, and to calculate pharmacokinetic parameters for optimizing MTX therapy and leucovorin rescue.

II. PATIENTS AND METHODS

A. Patients

Twenty-eight pediatric cancer patients (aged 4 to 17 yrs) who were treated with moderate or high i.v. MTX doses participated in the study (Table 1). The children were treated according to the BFM program (4). The dose of MTX ranged between 0.5 and 5 g/m^2. Time of i.v. infusion ranged from 24 to 36 h. Complete information about the protocol, especially with regard to hydration, during alkalinization (pH $\geq$ 7.5), and leucovorin rescue, have been previously reported. Additionally, all the children received MTX (12 mg) intrathecally at 1 h after the onset of the infusion.

Table 1 Childrens' Characteristics

No.	Child	Diagnosis[a]	Age (years)	Body Weight (kg)	Dose MTX (g/m^2)
1	M.S.	NHL	8	22	1.0
2	D.Z.	NHL	8	29.5	1.0
3	S.A.	NHL	9	35	1.0
4	Ag.M.	ALL	5	19	0.5
5	G.T.	ALL	9	39	0.5
6	R.T.	ALL	9	37	1.0
7	S.K.	ALL	9	37	1.0
8	P.K.	ALL	8	28	1.0
9	D.H.	ALL	4	18	1.0
10	W.S.	ALL	9	31	1.0
11	R.G.	NHL	6	20	5.0
12	A.M.	NHL	6	21	1.0
13	L.K.	ALL	5	22	1.0
14	M.B.	ALL	5	30	1.0
15	E.K.	ALL	4	20	1.0
16	J.M.	ALL	9	25	1.0
17	J.Mo.	ALL	15	75	1.0
18	A.D.	NHL	14	52	5.0
19	M.Ba.	NHL	14	42	2.0
20	M.R.	NHL	16	56	1.0
21	M.Rz.	ALL	12	60	3.0
22	A.J.	ALL	14	53	1.0
23	S.P.	NHL	15	62	0.5
24	T.F.	NHL	12	33	0.5
25	K.K.	ALL	13	48	1.0
26	P.W.	ALL	24	70	1.0
27	A.P.	ALL	17	68	3.0
28	S.S.	ALL	14	37	1.0

[a]NHL, lymphoma; ALL, leukemia.

B. Sample Assay

Venous samples were taken at 8, 4, 2, 1, and 0.5 h before the end of the infusion, at the end of the infusion, and at 6, 12, 24, 48, 72 h after the end of the infusion. All blood samples were centrifuged and subsequently frozen at –20°C until analysis. Serum MTX concentration was measured by an enzymatic method (5) and fluorescence-polarization immunoassay (FPIA) by means of an automatic TD_x analyzer. The lower limits of sensitivity of these methods were 0.02 μmol/L. Coefficients of variation for duplicate determinations were less than 5%.

RESULTS

The rate constants were calculated according to the open two-compartment body model. In some cases the terminal phase was not found and then a one-compartment body model was applied. The following parameters were calculated: steady-state concentration of MTX in the serum (C_{ss}); systemic clearance (CL_s); elimination rate constants (K and k_{10} for the two- and one-compartment models, respectively); half-life for the first 24 postinfusion hours ($t_{0.5\alpha}$) and the terminal elimination half-life ($t_{0.5\beta}$); distribution rate constants from blood to tissues and from tissues to blood (k_{12} and k_{21}, respectively). The systemic clearance was calculated according to the following equation:

$$Cl_s = \frac{k_0}{C_{ss}}$$

where k_0 is the zero-order infusion rate constant.

Individual and mean parameter estimates are presented in Tables 2 and 3. Pharmacokinetic and statistical calculations were carried out on a Schneider CPC-6128 microcomputer. Comparison of parameters between younger and older children was performed according to a nonpaired t-test. A comparison of pharmacokinetic parameters in patients with relapse of the disease and with the initial phase of the disease are presented in Table 4.

IV. DISCUSSION

MTX disappears from serum after intravenous infusion according to a two-compartment body model. However, an early distribution of half-life after intravenous bolus is not seen in the case of longer intravenous infusion. The distribution throughout the central compartment is virtually complete during administration. The serum MTX

Table 2 Pharmacokinetic Parameters for MTX (Children Aged 4 to 9 Years)

Child	C_{ss} (μmol/L)	K (h^{-1})	k_{12} (h^{-1})	k_{21} (h^{-1})	k_{10} (h^{-1})	$t_{0.5\alpha}$ (h)	$t_{0.5\beta}$ (h)	Cl_s (mL/min/m^2)
W.S.	9.11	—	0.054	0.026	0.145	3.3	38.1	111.8
D.H.	12.39	0.064	—	—	—	10.8	—	82.2
P.K.	48.57	0.272	—	—	—	2.5	—	21.0
S.K.	10.98	—	0.286	0.013	0.125	1.6	69.3	92.6
R.T.	11.16	—	0.055	0.036	0.252	2.2	24.1	91.0
G.T.	28.77	0.286	—	—	—	2.4	—	26.6
Ag.M.	16.35	0.241	—	—	—	2.9	—	46.7
S.A.	14.86	—	0.018	0.115	0.163	3.2	8.2	68.6
R.G.	50.70	—	0.018	0.015	0.184	3.4	51.7	100.5
A.M.	27.89	—	0.060	0.039	0.235	2.3	23.1	35.5
L.K.	16.60	—	0.173	0.032	0.123	2.2	55.8	61.4
M.B.	13.70	—	0.258	0.034	0.133	1.7	63.5	74.3
E.K.	31.40	0.144	—	—	—	4.8	—	32.4
J.M.	21.40	0.276	—	—	—	2.5	—	47.6
D.Z.	17.70	0.198	—	—	—	3.5	—	57.6
M.S.	23.53	—	0.134	0.029	0.156	2.3	46.8	43.3
Mean ± SEM	17.56 ± 2.0[a]	0.211 ± 0.030	0.117 ± 0.010	0.038 ± 0.016	0.168 ± 0.016	3.2 ± 0.5	42.3 ± 6.8	62.1 ± 7.0

[a]Mean was calculated for 1-g/m^2 dose of MTX.

Table 3 Pharmacokinetic Parameters for MTX (Children Aged 10 to 17 Years)

Child	C_{ss} (μmol/L)	K (h^{-1})	k_{12} (h^{-1})	k_{21} (h^{-1})	k_{10} (h^{-1})	$t_{0.5\alpha}$ (h)	$t_{0.5\beta}$ (h)	Cl_s (mL/min/m^2)
M.R.	22.36	0.110	—	—	—	6.3	—	45.6
M.Ba.	192.95	0.199	—	—	—	3.5	—	10.6
A.D.	65.12	—	0.038	0.048	0.036	2.0	16.3	78.2
J.Mo.	57.50	—	0.001	0.031	0.077	8.8	22.5	17.7
M.Rz.	152.60	—	0.008	0.013	0.053	11.0	65.4	20.0
A.J.	22.30	—	0.052	0.016	0.308	1.9	51.7	43.7
S.P.	27.20	—	0.017	0.036	0.058	8.0	28.9	28.1
T.F.	18.40	0.075	—	—	—	9.2	—	41.5
K.K.	20.64	—	0.080	0.019	0.060	4.6	69.3	49.3
P.W.	24.10	0.086	—	—	—	8.1	—	42.3
A.P.	83.24	0.141	—	—	—	4.9	—	36.7
S.S.	13.80	0.274	—	—	—	2.5	—	73.8
Mean ± SEM	29.58 ± 7.0[a]	0.148 ± 0.030	0.033 ± 0.012	0.027 ± 0.006	0.144 ± 0.052	5.9 ± 0.9	42.4 ± 9.3	40.6 ± 6.9

[a]Mean was calculated for 1-g/m^2 dose of MTX.

concentration time profile for some children was evidently monophasic according to the one-compartment body model. The terminal phase was not observed in these children. When MTX concentrations were below the sensitivity limits of the method, we tried to explain this by a delay of the terminal phase. The one-compartment model was assumed for calculation in these children (Tables 2 and 3).

Steady-state concentration of MTX in serum has been established from venous blood samples taken before the end of the infusion and at the end of the infusion (6). According to Ewans et al. (7,8), at high-dose MTX therapy there are clinical risk factors (such as impairment of renal function, pleural effusion, gastrointestinal tract obstruction, ascites) and pharmacokinetic risk factors (half-life within the first 24 postinfusion hours greater than 3.5 h and 12- or 24-h serum MTX concentration greater than 10 μmol/L and 5 μmol/L, respectively). These factors permit earlier identification of high-risk-toxicity patients (within the first 24 postinfusion hours). If

Table 4 Methotrexate Pharmacokinetics Parameters in Children with an Initial Phase of the Disease and with a Relapse of the Disease[a]

Parameter	Initial Phase	Relapse
$\overline{Cl}_s$ ± SEM	38.3 ± 5 (n = 11)	57.6 ± 6.7 (n = 15)
$t_{0.5}$ ± SEM	5.0 ± 0.8 (n = 11)	2.9 ± 0.5 (n = 14)
$\overline{C}_{ss}$ ± SEM (dose of MTX 1 g/m^2)	30.5 ± 7.1 (n = 5)	18.4 ± 1.9 (n = 10)

[a]n = number of children.

any of the factors are observed, the monitoring serum concentration of MTX should be continued and dosage of leucovorin should be corrected to maintain equimolar leucovorin and MTX serum concentrations. Leucovorin administration should be continued until the serum MTX concentrations have declined to 0.01 μmol/L. The results presented in Tables 2 and 3 indicate that risk of the toxicity occurred in children, M.R., J.Mo., M.Rz., S.P., T.F., K.K., P.W., A.P., E.K., and D.M., whose half-lives ranged from 4.6 to 11 h. Essentially, these are older children. MTX serum concentration greater than 5 μmol/L within the first 24 postinfusion hours was observed only in one child (M.Rz.). An investigation (9) has demonstrated that the cytotoxic effect of MTX is a function of both the concentration of the drug and the duration of the exposure to it. This study indicated that exposure to concentrations of 0.5 μmol/L for 48 h or 0.05 μmol/L for 72 h produced the same cytotoxic effect as exposure to 10 μmol/L for 12 h. Children D.N., J.Mo., K.K., A.P., W.S., D.H., M.P., and M.S. were in this category. In the rest of the children, MTX concentration at 72 postinfusion hours was lower or below the limits of the sensitivity of the method. The terminal phase half-life appears to correlate best with toxicity. These values were very variable and ranged from 8.2 to 69.3 h. A similar high intersubject variability in the terminal elimination half-life was reported in adults (10).

The normal renal clearance of MTX has been reported to be about 64 $mL/min/m^2$ and may be compared with calculated systemic clearance, because renal excretion is the major route of MTX elimination. The clearance values were very variable in the children (Tables 2 and 3). The delayed MTX clearance observed in some children may be a result of impairment of renal function. A decrease of the MTX clearance is possible at high doses of MTX as well (11,12). When delayed clearance is observed, the dosage of leucovorin should be escalated and/or the duration of leucovorin administration should be extended. The leucovorin dosage in these children was calculated according to the BFM protocols. In these programs the dosage of leucovorin was dose- and MTX concentration-dependent and was continued not longer than 72 h from the end of the infusion. In those children whose cytotoxic concentration of MTX is higher than 0.05 μmol/L after 72 postinfusion hours, the duration of leucovorin administration should be extended.

The pharmacokinetics of high-dose MTX may be age-related (13). However, since the number of subjects in that study was very small (three children and six adults), the clinical significance of these results cannot be appreciated. In this study we observed that pharmacokinetics of MTX was dependent on the age of the children. Lower steady-state concentration (after 1 g/m^2 dose of MTX), shorter initial half-life, and faster systemic clearance were observed in younger children (Tables 2 and 3). The above differences are statistically significant. The elimination rate constants and terminal half-lives are not age-related. The children treated in our hospital were in various stages of the disease. Some of them were in the initial phase of the disease, whereas the remaining children had suffered a relapse of the disease. Differences in the pharmacokinetic parameters were observed in these children (Table 4). Higher steady-state MTX concentrations, longer initial half-lives, and delayed systemic clearances have been observed in children with an initial phase of the disease. The above differences are also statistically significant. This is a very important observation, because in the same patient the pharmacokinetics of MTX may also be dependent on the stage of the disease. From a clinical point of view, the children who tolerated MTX therapy especially poorly were identified. In these children (G.T., M.Ba., M.Rz.), a decrease of the systemic clearance has been observed. Additionally, MBa had a very high steady-state MTX concentration and in MRz the serum cytotoxic concentration of MTX was retained for a long time.

Administration of leucovorin at higher doses than 15 mg/m^2 for a longer period of time was required in these children.

V. SUMMARY

Pharmacokinetics of methotrexate (MTX) have been studied in children with cancer receiving moderate or high i.v. doses. Pharmacokinetic parameters were calculated. The pharmacokinetics of high doses of MTX is different in younger and older children. The difference of the pharmacokinetics parameters was ascertained in patients at various stages of the disease. The pharmacokinetic risk factors were identified in some children. Either a dose of leucovorin should be increased or the duration of its administration should be extended in these children.

REFERENCES

1. W. A. Bleyer. The clinical pharmacology of methotrexate. New applications of an old drug. *Cancer, 48*:36–51 (1978).
2. R. A. Bender, L. A. Zwelling, and J. H . Doroshow. Antineoplastic drugs: Clinical pharmacology and therapeutic use. *Drugs, 16*:46–87 (1978).
3. F. M. Sirotnak, D. M. Moccio, and D. M. Dorick. Optimization of high-dose methotrexate with leucovorin rescue therapy in L 1210 Leukemia and sarcoma 180 murine tumor models. *Cancer Res., 38*:345–353 (1978).
4. U. Radwanska, D. Michalewska, and J. Armata. Acute lymphoblastic leukemia therapy in Poland: A report from the Polish childrens' leukemia/lymphoma study group. *Folia Haematol. Leipzig,* 2:199–210 (1989).
5. L. Skibinska, P. Daszkiewicz, D. Michalewska, and U. Radwanska. Pharmacokinetics of methotrexate given intrathecally to children with acute lymphoblastic leukemia. *Pol. J. Pharmacol. Pharm., 40*:135–143 (1988).
6. J. D. Borsi and P. J. Moe. Systemic clearance of methotrexate in the prognosis of acute lymphoblastic leukemia in children. *Cancer, 60*:3020–3024 (1987).
7. W. E. Ewans, W. K. Crom, and M. Abramowith. Clinical pharmacodynamics of high-dose MTX in acute lymphocytic leukemia. *N. Engl. J. Med., 314*:471–477 (1986).
8. W. E. Ewans, C. B. Pratt, R. B. Taylor, L. E. Barker, and W. R. Crom. Pharmacokinetic monitoring of high-dose methotrexate: Early recognitions of high-risk patients. *Cancer Chemother. Pharmacol.,* 3:161–166 (1979).
9. H. M. Pinedo and B. A. Chabner. Role of drug concentration, duration of exposure and endogenous metabolite in determining methotrexate cytotoxicity. *Cancer Treat. Rep., 61*:709–715 (1977).
10. A. Awidi, W. Al-Turk, F. Madanat, S. Othman, and O. Shaheen. Pharmacokinetics of high and moderate intravenous doses of methotrexate. *Res. Commun. Chem. Pathol. Pharmacol., 59*:411–414 (1988).
11. W. E. Ewans, C. F. Steward, and C. H. Chen. Methotrexate systemic clearance influences probability of relapse in children with standard risk acute lymphocytic leukemia. *Lancet, 1*:359–362 (1984).
12. P. Bore, R. Bruno, N. Lena, R. Favre, and J. P. Cano. Methotrexate and 7-hydroxymethotrexate pharmacokinetics following intravenous bolus administration and high-dose infusion of methotrexate. *Eur. J. Cancer Clin. Oncol.,* 23:1385–1390 (1987).
13. Y. M. Wang, W. W. Sutow, and M. M. Romsdahl. Age related pharmacokinetics of high-dose methotrexate in patients with osteosarcoma. *Cancer Treat. Rep., 63*:405–410 (1979).

40

Chirality in Pharmacokinetics and Therapeutic Drug Monitoring

E. J. Ariëns *Nijmegen, The Netherlands*

I. INTRODUCTION

An important factor in the failure of pharmacotherapy is uncertainty in the relationship between doses and effect. This factor can be divided into (a) the uncertainty in the relationship between doses and plasma concentration, and (b) the uncertainty in the relationship between plasma concentration and the effect.

The development of highly sensitive and selective analytical methods has opened up possibilities for exact estimation of the levels of drugs and drug metabolites in body fluids, particularly the concentration of the active agent, which may be the drug applied or bioactive metabolites thereof. The latter are essential in the case of prodrugs but may also be important as active components in addition to the parent compound. In both the study of pharmacokinetics in general and for therapeutic drug monitoring, selective methods are essential. Problems arise with the many mixtures of stereoisomers—composite chiral drugs (CCDs) such as racemates which constitute ±25% of the therapeutics on the market. What is said for racemates here mostly holds true also for CCDs in general.

II. ISOMERIC BALLAST IN RACEMIC THERAPEUTICS

Few scientists will deny that stereoselectivity in action is a common and important phenomenon. It concerns the relationship between dose and plasma level as well that between plasma level and effect, pharmacokinetics as well as pharmacodynamics, the latter for the desired as well as undesired actions (1–20). For the former particularly, the rate of enzymatic conversion, transport by carriers, and protein binding and thus distribution and elimination often, if not mostly, are stereoselective. With regard to pharmacodynamics, the misleading distinction between "active" and "inactive" enantiomer is used. The so-called inactive enantiomer does not contribute or hardly contributes to the therapeutic action but often contributes to and sometimes even is primarily responsible for the side effects; further, it may interact with the therapeutically effective stereoisomer and interfere with the action of other drugs. The enantiomer that is most potent for a particular component in the action, e.g., the desired action, is called the "eutomer." The other one—often poorly active but only exceptionally fully inactive—is called the "distomer." As has been noted, it often clearly contributes to the undesired actions and can never be regarded as being fully harmless.

Even in the case where two stereoisomers are nearly pharmacodynamically equiactive, one has to take into account that they will only rarely be equivalent in all components in terms of therapeutic action and side effects and even less so in the various aspects of pharmacokinetics. A stereoisomer that does not or only minimally contributes to the desired action—in general, the distomer—may be indicated as being "isomeric ballast" in the therapeutic product, but it can never be regarded as fully harmless. There is no reason why it should not be involved, for instance, in allergic reactions and formation of active radicals and thus be potentially mutagenic and carcinogenic, this besides the possibility of various undesired pharmacological actions.

It is time that biopharmaceutical experts understand that racemates in fact are fixed-ratio combinations of chemicals which differ in their biological characteristics. They are comparable to the classical fixed-ratio combinations in which two different therapeutic agents—for instance, an antihypertensive β-blocker and a saluretic—are combined. The main difference is that in the case of a classical fixed-ratio combination, the proportion of the two components in the mixture is optimalized, whereas in a racemic compound it is 1:1, which implies that even if there is cooperation between the two stereoisomers in a racemate, this proportion will only very rarely be optimal. In the case of indacrinon, the racemic compound is composed of a eutomer with a diuretic action but that causes uric acid retention as a side effect, and a distomeric stereoisomer with a uricosuric action. Thus the diastomer counteracts the side effect of the diuretic eutomer. It turns out that for proper control of uric acid excretion, the eudismic proportion (EP) has to be 1:8 (21).

III. PHARMACODYNAMICS AND KINETICS OF RACEMIC THERAPEUTICS

Two important parameters have to be taken into consideration.

1. *The eudismic ratio (ER)*: the ratio of the activities, (dynamics) of the eutomer and distomer (15). This usually differs for those components in the desired and undesired actions generated on different sites of action (receptors). It is specific for a particular racemate in relation to a particular action and is only rarely one.
2. *The eudismic proportion (EP)*: the proportion of the concentrations of eutomer and distomer ([eut]/[dist]) in, for instance, plasma. This is of particular significance in kinetics. For a racemate the eudismic proportion (EP) of the enantiomers is 1 by definition. After absorption, due to the usually different rates of metabolic conversion for the enantiomers, the EP in the body fluids gradually changes. Exceptionally, e.g., in case of rapid in-vivo racemization, the racemate is found in plasma. If the elimination dominates for the eutomer, the EP, 1 for the racemate in the tablet, in plasma gradually decreases, e.g., in the case of verapamil (22), to about 1/4 (±80% distomer). If the elimination of the distomer dominates, the EP in plasma increases, e.g., in case of nivaldipine (23), to about 3 (±75% eutomer). In fact the composition, EP, of the "racemate" in the plasma continuously changes until steady state is reached for each of the stereoisomers. Therefore, the time needed differs for the isomers. Without chiral assays the eudismic proportion (EP) in the plasma remains unknown. If there is an extensive presystemic first-pass loss by metabolic conversion in the liver, the plasma concentrations of the racemate obtained after oral and after parenteral application will differ in EP and thus be bioinequivalent (22). The plasma

concentration after application of a racemate, measured by nonchiral assays, gives information only on the sum of the eutomer and distomer and since the eudismic proportion changes with time, is not related in a clear way to the response. The various pharmacokinetic constants, such as half-life, bioavailability, persistence, etc., derived from "racemate" concentrations based on nonchiral assays are fiction and comparable to parameters such as the age or body weight of a married couple. A broad spectrum of publications on the problems of drug chirality have appeared (1–20), such that there is no excuse for the generation of nonsense, "nonscience," polluting the scientific literature.

IV. ETHICAL PROBLEMS IN THE APPLICATION OF RACEMIC THERAPEUTICS

Two major ethical problems arise:

1. Avoidance of "pollution" of the literature on pharmacology and experimental therapeutics by preventing neglect of the racemic nature of the drugs studied; in other words, stereoselective assays are required.
2. Avoidance of medicinal "pollution" of patients by reduction to the scientifically feasible, technologically possible, and economically acceptable limitation of isomeric ballast in chiral therapeutics.

In the "World Medical Association Declaration of Helsinki" (24), under recommendations guiding medical doctors in biomedical research involving human subjects, it is stated under basic principles:

> 1.1 Biomedical research involving human subjects must confirm to generally accepted scientific principles and should be based on adequately performed laboratory and animal experimentation and on a thorough knowledge of the scientific literature.
>
> 1.8 Reports of experimentation not in accordance with the principles laid down in this declaration should not be accepted for publication.

There is something to be learned here by the ethical committees (institutional review boards or IRBs) that supervise evaluation of drugs on humans. If racemic therapeutics or therapeutic isomeric mixtures in general are involved—unless stereospecific assays are used—invalid data are generated. Taking into account the involvement of patients, this is unethical and thus not acceptable. Are the patients informed about this situation? Is there honest "informed consent"? In particular, hospital pharmacists and clinical pharmacologists are invited to review their contributions to science in the clinical evaluation of the many racemic preparations with which they deal. This will increase the awareness of the pitfalls of nonchiral assays in the study of CCDs.

In relation to the problems under discussion one should be aware of the principle that "exposure of nature (including man) and its environment to xenobiotics (including drugs) is only justified if the desirable actions adequately compensate for the undesirable actions and the never fully excluded risks" (12). This implies, among other things, that one should "avoid isomeric ballast in therapeutics and the inherent medicinal pollution of patients" (18). A simple and effective way to stop the steady flow of deficient data generated in the study of CCDs is the extension of the instructions to authors with an indication that:

The composite character of drugs which are mixtures of stereoisomers must be brought to the attention of the reader. The prefix (RS)- or rac-, e.g., rac-propranolol, in the case of racemates and (Z/E)- or cis/trans- in the case of that type of isomers is obligatory. The implications of the composite and chiral nature of the drugs for the interpretation of the data measured and the conclusions drawn must be made explicit

V. SOURCES OF ASTEREOGNOSIS

The persistent neglect of chirality and its implications for drug action as a matter of fact has its roots in the fact that hardly any of the (highly recommended) clinically orientated textbooks on pharmacology, pharmacokinetics, or toxicology principles such as stereoselectivity, stereospecificity, stereochemistry, etc, emphasize or even mention chirality. Pharmacokinetic parameters, particularly half-life times and bioavailabilities for racemic therapeutics calculated from plasma "racemate" concentrations measured with nonchiral assays, are amply presented (9,12). What do they mean? The reader may consult the library and see for himself.

Goodman and Gilman's classical textbook (30) presents several pages of pharmacokinetic constants for therapeutics, roughly a quarter of which are racemates. With few exceptions, the constants are based on plasma levels determined by nonchiral assays. Students should be spared the false idea that there really exists something like "plasma concentrations and pharmacokinetic parameters of racemates." What is real are plasma concentrations, etc., of the individual stereoisomers and their continuously changing eudismic proportions. For the hundreds of therapeutics registered in the past 20 years, regulatory authorities have required pharmacokinetic data—and also for the hundreds of racemates among them. These data, with few exceptions, are based on nonchiral assays. What is the point? Informative and illustrative are some recent (1989) references (25–27) on racemic calcium channel blockers, which present a broad spectrum of kinetic constants. They clearly demonstrate where astereognosis leads. Characteristics of the situation is the fact that the Ref. 26 was awarded the Pharmatec prize for its outstanding scientific qualities! This under the auspices of the Féderation Internationale Pharmaceutique (FIP). Data on changes in clinical parameters such as blood pressure, cardiac output, liver functions, etc., as a function of dose and time have their own intrinsic value as do CCSs and fixed-ratio combinations of drugs in general. In this regard, Refs. 26 and 27 are correct. Pharmacokinetic data on CCDs based on nonchiral assays are unacceptable because they conceal the composite character of the drugs studied.

Table 1 (19) summarized the inadequacy in the generation of nonscience in the study of pharmacokinetics of racemic therapeutics. It is rather closely matched by Ref. 25.

VI. REGULATORY AUTHORITIES IN ACTION

The climate is changing. Japan, in a leading position, in 1987 adapted requirements for racemic products. Full pharmacology and toxicology are required for each of the enantiomers and for the racemate (28). As a matter of fact, only pharmacokinetic data based on chiral assays should and probably will be accepted.

The European Economic Community (EEC) in *Rules governing medicinal products in the European Community*, Volume II (1989) extended the section on stereoisomerism. It states:

Table 1 The Cascade of Inadequacy in the Generation of Nonscience in the Study of Racemic Drugs (19)

1. Pharmaceutical industry
   Concealment of the racemic fixed-ratio character of products presented for research and evaluation
2. Investigators, clinicians, and/or hospital-pharmacists
   Proposals for research projects scientifically deficient due to the neglect of the racemic nature of the drug(s) to be studied
3. Funding institution
   Poor judgment in the evaluation of and thus granting of funds for scientifically deficient research projects by advisory experts and peers
4. Ethics committees
   Overlooking the fact that generation of invalid data—due to disregard of stereoselectivity in action—in the evaluation of racemic therapeutics in studies on patients is unethical, as is the implicit "misinformed consent" obtained from the patient.
5. Investigators, clinicians, and/or hospital-pharmacists
   Neglect of stereoselectivity in pharmacodynamics and kinetics in the study of racemic fixed-ratio mixtures
   Calculation of pharmacokinetic constants and derivation of multicompartment systems for such mixtures on the basis of mostly nonexistent plasma concentrations of the racemate with the use of equations valid only for single agents
   Presentation of dubious and strongly biased research reports and manuscripts for publication due to disregard of the racemic fixed-ratio nature of the therapeutic(s) studied
6. Editorial boards and referees of scientific journals
   Lack of expertise required for proper evaluation of manuscripts dealing with studies on racemic drugs
7. Prize-awarding agencies
   Promotion of nonscience by erroneous awarding of astereognosis in the study of racemic agents
8. IUPHAR (International Union of Pharmaclogy, FIP (Fédération Internationale Pharmaceutique), and APhA (American Pharmaceutical Association)
   Negligence in not taking adequate steps to stop or at least discourage pollution of scientific literature in the field of pharmacology and therapeutics due to neglect of stereoselectivity in the action of drugs
9. Teachers and textbook writers
   Neglect of stereochemistry in educational programs, thus closing the vicious circle in the maintenance of a steady flow of astereognostic nonscience in the field of pharmacology and therapeutics
10. Regulatory authorities
    Acceptance of invalid data based on nonchiral assays in the admission of racemic therapeutics, thus causing a big waste of money and research capacity and facilitating continuation of medicinal pollution of patients via isomeric ballast in their medicines

> Possible problems relating to stereoisomerism, which should be discussed in the appropriate Expert Report and cross referenced, should include: the batch to batch consistency of the ratio of stereoisomers in the various batches used—the toxicological issues—the pharmacological aspects (including evidence on which stereoisomers have the desired pharmacological properties)—*pharmacokinetics including information on the relative metabolism of the stereoisomers*—extrapolation of the pre-clinical data (paying particular attention to possible problems relating to species differences in handling of the stereoisomers)—the significant clinical issues.

This at least opens the possibility of requiring essential information on each of the enantiomers or stereoisomers in general as well as on the racemate or composite chiral products. The EEC Report states that: "where a mixture of stereoisomers has previously been marketed and it is now proposed to market a product containing only one isomer, full data on this isomer should be provided." The requirements for a single isomer (the eutomer) among new drug applications should be analogous to those for a single compound in general. U.S. authorities (the Food and Drug Administration, FDA) came to an approach comparable to that in Japan and the EEC to be effectuated in 1992.

This change in policy will reduce the flow of heavily biased data on CCDs in the scientific literature as far as new drugs are concerned. Unless the editorial boards of the scientific journals join the regulatory authorities in their wisdom, however, for the hundreds of CCDs already on the market, the contestable nonchiral kinetics will stay with us for many years to come.

Therapeutics containing "isomeric ballast" should be withdrawn from the market as soon as ballast-free preparations are available at an acceptable price. Often only a small fraction of the price paid in the pharmacy is related to that of the drug substance in the product. The approach indicated should hold true for generics as well as brand-name products. The resulting incentive for the innovative industry and the rapid development of chiral techniques in chemistry will help to reduce chemical ballast in drug therapy. The patient, if aware of it, would definitely appreciate this.

VII. CONCLUSIONS

1. Acceptance by regulatory authorities of composite chiral drugs such as racemates should, wherever possible, be based on proper information on the individual stereoisomers as well as the mixture. It is preferrable to avoid chemical ballast. A question of weighing advantages against disadvantages include the never fully absent risks of exposure to chemical ballast.
2. The study of pharmacology, dynamics, and in particular the kinetics of CCDs such as racemates requires chiral assays. The investigator has the moral duty to inform the reader of the composite nature of the drugs studied and the implications thereof for the interpretation of the data.
3. The ethical committees, among other things, have the duty to implement in clinical research the recommendations of the "World Medical Association Declaration of Helsinki."
4. Instructors are invited to discuss with their pupils or at least bring to their attention the arguments brought forward here. As an exercise in open-minded reasoning, students might then be asked to write—in the light of the foregoing—a short comment on the books on pharmacology, therapeutics, pharmacokinetics, and

biopharmacy recommended to them. The subject concerns ±25% of the therapeutics available, ±50% of the ones most used.

The recognition of these problems had to wait for hundred years after Pasteur generated the basic insights (29). It should not take another hundred years to eliminate astereognosis from general and clinical pharmacology.

VIII. SUMMARY

Racemic therapeutics are in fact fixed-ratio mixtures of stereoisomers that should be regarded as biologically different compounds. Usually only one of the isomers fully contributes to the therapeutic action, whereas the other is often classified as "isomeric ballast." Due to differences in turnover and pharmacokinetics, the proportion of the enantiomers—1:1 in the racemate—continuously changes in plasma. The implications of neglecting stereoselectivity on the various levels in the investigation of racemic drugs are elucidated, discussed, and summarized in Table 1 (19).

The fact that clinical investigations, ethical committees, and regulatory authorities have accepted for decennia invalid pharmacokinetic data on ±25% of the therapeutics, the racemates in use, makes the benefit and necessity of kinetics in general questionable.

Exposure of patients to the "isomeric ballast" present in about 50% of the most frequently used drugs probably will go on for many decennia. As a result of a change in attitude of the regulatory authorities, for new drugs in the choice between the racemic therapeutic or the single isomeric ballast-free drug will be based largely in the future on a critical evaluation of the chiral characteristics with regard to therapeutic, toxicologic, as well pharmacokinetic aspects.

REFERENCES

1. P. Jenner and B. Testa. The influence of stereochemical factors on drug disposition. *Drug Metab. Rev.*, 2:117–184 (1973).
2. L. K. Low and N. Castagnoli, Jr. Enantioselectivity in drug metabolism. *Ann. Rep. Med. Chem.*, 13:304 (1978).
3. E. J. Ariëns, W. Soudijn, and P. B. M. W. M. Timmermans, *Stereochemistry and Biological Activity of Drugs*, Blackwell, Oxford, pp. 1–190 (1983).
4. E. J. Ariëns. Stereochemistry, a basis for sophisticated nonsense in pharmacokinetics and clinical pharmacology. *Eur. J. Clin. Pharmacol.*, 26:663–668 (1984).
5. K. Williams and E. Lee. Importance of drug enantiomers in clinical pharmacology. *Drugs*, 30:333–354 (1985).
6. D. E. Drayer. Pharmacodynamic and pharmacokinetic differences between drug enantiomers in humans: An overview. *Clin. Pharmacol. Ther.*, 40:125–133 (1986).
7. E. J. Ariëns. Stereochemistry: A source of problems in medicinal chemistry. *Med. Res. Rev.*, 6:451 (1986).
8. A. Abbott. Series on chirality. *Trends in Pharmacol. Sci.*, 7:20–24, 60–65, 112–115, 155–158, 200–205, 227–230, 281–301 (1986).
9. E. J. Ariëns and E. W. Wuis. Bias in pharmacokinetics and clinical pharmacology. *Clin. Pharmacol. Ther.*, 42:361–363 (1987).
10. I. W. Wainer and D. E. Drayer. *Drug Stereochemistry. Analytical Methods and Pharmacology*. Marcel Dekker, New York (1988).
11. A. M. Evans. Stereoselective drug disposition: Potential for misinterpretation of drug disposition data. *Br. J. Clin. Pharmacol.*, 26:771–780 (1988).

12. E. J. Ariëns, E. W. Wuis, and E. J. Veringa. Stereoselectivity of bioactive xenobiotics. A pre-Pasteur attitude in medicinal chemistry, pharmacokinetics and clinical pharmacology. *Biochem. Pharmacol.*, *37*:9–18 (1988).
13. E. J. Ariëns, J. J. S. van Rensen, and W. Welling, *Stereoselectivity of Pesticides—Biological and Chemical Problems*. Elsevier, Amsterdam (1988).
14. E. J. Ariëns. Enzymes in stereospecific analysis. In *Chiral Separations by HPLC: Application to Compounds of Pharmacological Interest* (A. M. Krstulovic, ed.), John Wiley/Ellis Horwood, Chichester, pp. 69–75 (1988).
15. F. Jamali, R. Mehvar, and F. M. Pasutto. Enantioselective aspects of drug action and disposition: Therapeutic pitfalls. *J. Pharm. Sci.*, *78*:695–715 (1989).
16. D. F. Smith. The stereoselectivity of drug action. *Pharmacol. Toxicol.*, *65*:321–331 (1989).
17. Campbell. Stereoselectivity in clinical pharmacokinetics and drug development. *Eur. J. Drug Metab. Pharmacokinet.*, *15*:109–125 (1990).
18. E. J. Ariëns. Stereoselectivity in pharmacodynamics and pharmacokinetics. *Schweiz. Med. Wschr.*, *120*:131–134 (1990).
19. E. J. Ariëns. Racemic therapeutics—problems all along the line. In *Chirality in Drug Design and Synthesis*, Academic Press, New York, p. 29 (1990).
20. M. Eichelbaum and S. Gross. Stereoselectivity in drug action and disposition. *N. Engl. J. Med.* (in press).
21. J. A. Tobert and V. J. Cirillo. Enhancement of the enantiomer ratio. *Clin. Pharmacol. Ther.*, *29*:344–350 (1981).
22. B. Vogelgesang, H. Echizen, E. Schmidt, and M. Eichelbaum. Stereoselective first-pass metabolism of highly cleared drugs: Studies of the bioavailability of L- and D-verapamil examined with a stable isotope technique. *Br. J. Clin. Pharmacol.*, *18*:733–740 (1984).
23. Y. Tokuma, T. Fujiwara, and T. Niwa. Stereoselective disposition of nivaldipine, a new dihydropyridine calcium antagonist, in the rat and dog. *Res. Commun. Chem. Pathol. Pharmacol.*, *63*:249–262 (1989).
24. R. J. Levine. *Ethics and Regulation of Clinical Research*, 2nd ed. Urban & Schwarzenberg, Baltimore, pp. 393–429 (1986).
25. P. A. Soons, A. G. de Boer, P. van Brummelen, and D. D. Breimer. Oral absorption profile of nitrendipine in healthy subjects: A kinetic and dynamic study. *Br. J. Clin. Pharmacol.*, *27*:179–189 (1989).
26. J. van Harten, J. Burggraaf, G. J. Lichthart, P. van Brummellen, and D. D. Breimer. Single- and multiple-dose nisoldipine kinetics and effects in the young, the middle-aged, and the elderly. *Clin. Pharmacol. Ther.*, *45*:600–607 (1989).
27. P. H . J. M. Dunselman, B. Edgar, A. H. J. Scaf, C. E. E. Kuntze, and H. Wesseling. Pharmacokinetics of felodipine after intravenous and chronic oral administration in patients with congestive heart failure. *Br. J. Clin. Pharmacol.*, *28*:45–52 (1989).
28. R. L. Smith and J. Caldwell. Racemates towards a New Year solution? *TIPS*, *9*:75–77 (1988).
29. L. Pasteur. On the asymmetry of naturally occurring organic compounds, the foundations of stereochemistry. In *Memoirs by Pasteur, Van't Hoff, Le Bel and Wislicenus* (G. M. Richardson, ed.), American Book Co., New York, pp. 1–33 (1901).
30. L. S. Goodman and A. Gelman, eds. *The Pharmacological Basis of Therapeutics*, Macmillan, New York, 1985.

41

A Stereoselective HPLC Method for the Determination of Total and Free Disopyramide in Serum Using an α_1-Acid Glycoprotein Column

H. R. Angelo and J. P. Kampmann *Bispebjerg Hospital, Copenhagen, Denmark* **J. Bonde** *KAS Herlev, Copenhagen, Denmark* **L. E. Pedersen and B. M. Tholstrup** *The Royal Danish School of Pharmacy, Copenhagen, Denmark*

I. INTRODUCTION

Disopyramide (DP) is a type IA antiarrhythmic drug used in the suppression of both supraventricular and ventricular arrhythmias. DP is a racemate, consisting of R(–) and S(+) enantiomer, when administered in commercial available preparations. Each of the enantiomers possesses distinct pharmacokinetic and pharmacodynamic characteristics (1).

Cimetidine is a well-known inhibitor of drug metabolism and renal secretion of a variety of drugs (2). Being a basic compound, metabolized in the liver, and actively secreted by the tubules (3), a potential interaction between DP and cimetidine is anticipated.

With the purpose of examining for a possible stereoselective disposition of DP and interaction with cimetidine, we needed a simple analytical method for the measurement of the total and free concentrations of each of the enantiomers of DP and its main dealkylated metabolite (MND) in serum and urine.

Stereoselective analysis of enantioners of DP and MND can be achieved using a column packed with human α_1-AGP (4,5). The present method allows the simultaneous determination of DP and MND enantiomers within a relatively short time of analysis.

Equipment:	Waters HPLC automatic system (WISP 710 sample processor, Model 6000 A pumps, Model 720 System Controller and Model 730 Data Module). Waters Model 450 variable wavelength UV spectrophotometer detector, operating at 262 nm.
Columns:	ChromSep 10 cm Spherisorb CN analytical column (Chrompack) coupled in series with a chiral column, CHIRAL-AGP (Chrom Tech).
Mobile phase:	8 mM sodium phosphate buffer, containing 5% (v/v) of 2-propanol and 0.7% (v/v) dimethyloctylamine; Flow rate 0.9 ml/min.

Figure 1 Analytical procedure.

II. METHODS

Dispyramide, 150 mg, was given intravenously as a bolus injection to seven healthy male volunteers in a open, randomized crossover study before and after 2 weeks of treatment with oral cimetidine, 800 mg daily. Blood and urine samples were collected at appropriate times, 5 to 1440 min following the bolus administration. The analytical procedure is outlined in Figure 1, and the high-performance liquid chromatography (HPLC) system used is described in Figure 2.

Extraction
1.0 ml SERUM
1.0 ml water
500 μl I M NaOH
8.0 ml tert-BUTYL METHYL ETHER
1. Extraction for 5 min by HETO-shaker
2. Centrifugation for 5 min at 1300 g
3. Cooling in a dry ice-acetone bath for 1 min
4. Decanting of organic phase
500 μl 100 mM FORMIC ACID
1. Extraction for 5 min by HETO-shaker
2. Centrifugation for 5 min at 1300 g
3. Cooling in a dry ice-acetone bath for 1 min
4. Aspirating of organic phase
5. Evaporating to dryness at 50° C under nitrogen
200 μl mobile phase
HPLC (40 μl injected)
Ultrafiltration (6)
2 ml serum, adjusted to pH 7.3 - 7.5
15 cm Dialysis tubing (Union Carbide, 8/32)
Centrifugation at 800 g (20° C) for one hour
HPLC (125 μl filtrate injected)

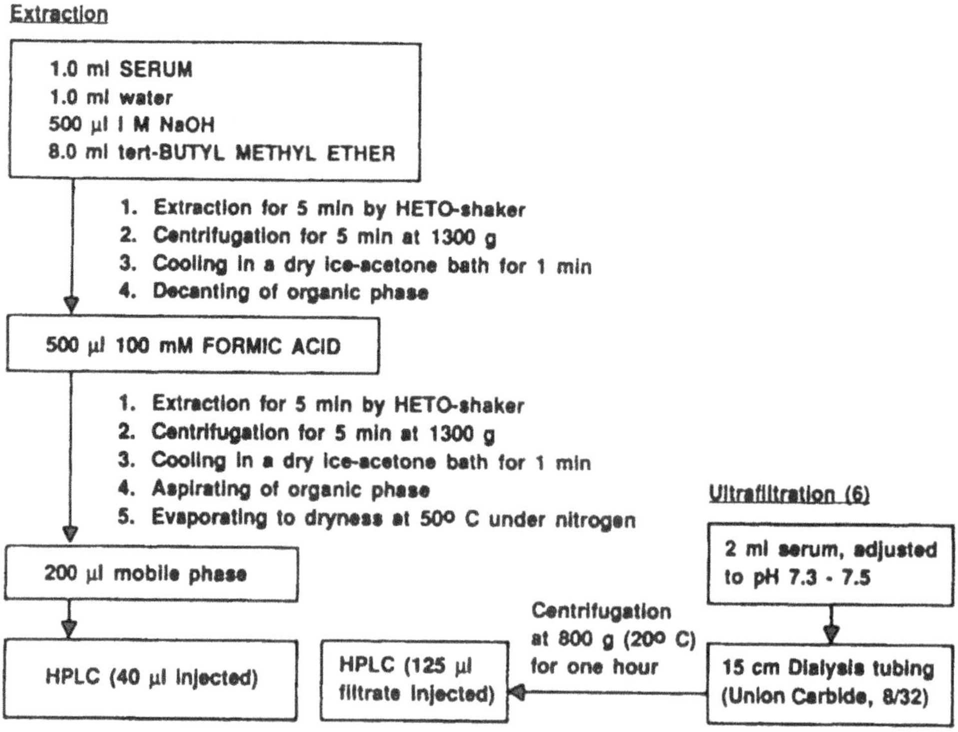

Figure 2 HPLC system employed.

III. ANALYTICAL RESULTS

Analytical results are given in Figures 3, 4, and 5, and in Table 1.

IV. CLINICAL RESULTS

1. Elimination clearance and volume of distribution (Vd_{β}) in terms of total drug were significantly ($p < 0.001$) larger for the R(–) enantiomer than for the S(+) enantiomer (7.9 versus 4.6 L/h and 89 versus 50 L, respectively), whereas no significant differences in half-lives could be demonstrated (7).
2. Elimination clearance in terms of free drug of the S(+) enantiomer was significantly higher ($p < 0.05$) than that of the R(–) enantiomer, whereas no significant differences in terms of volume of distribution could be demonstrated (7).
3. The protein binding of the S(+) enantiomer was significantly ($p < 0.001$) higher than that of the R(–) enantiomer at all plasma levels (7).
4. Co-administration of cimetidine did not alter any of the kinetic parameters (7).

V. DISCUSSION

The cyano-bonded phase provides good selectivity for separating many basic compounds and their metabolites (8). Used in combination with the second-generation α_1-AGP column, CHIRAL-AGP (ChromTech, Sweden), satisfactory resolution factors were obtained for DP and MND enantiomers (2.6 and 1.6, respectively), within a relatively short time (22 min). Using ultraviolet detection at 262 nm, no interference from endogenous compounds or other drugs was observed.

In our earlier described HPLC method for determination of DP and MND, we used 33-nM phosphoric acid in the extraction procedure. In this analysis the acid caused

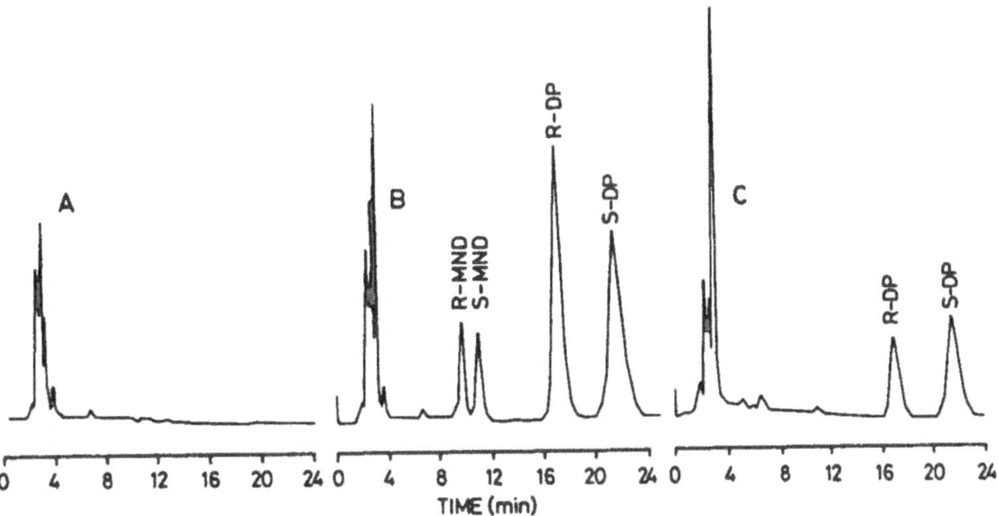

Figure 3 Chromatograms of human serum extracts, analyzed as described above. A, blank serum; B, serum to which R- and S-disopyramide and R- and S-monodesisopropyldisppyramide were added (10.0, 10.0, 2.0, and 2.0, respectively); C, serum from a patient treated therapeutically with disopyramide (R- and S-disopyramide were determined to be 3.0 and 5.1 μmol/L, respectively). MND = monodesisopropyldisopyramide; DP = disopyramide.

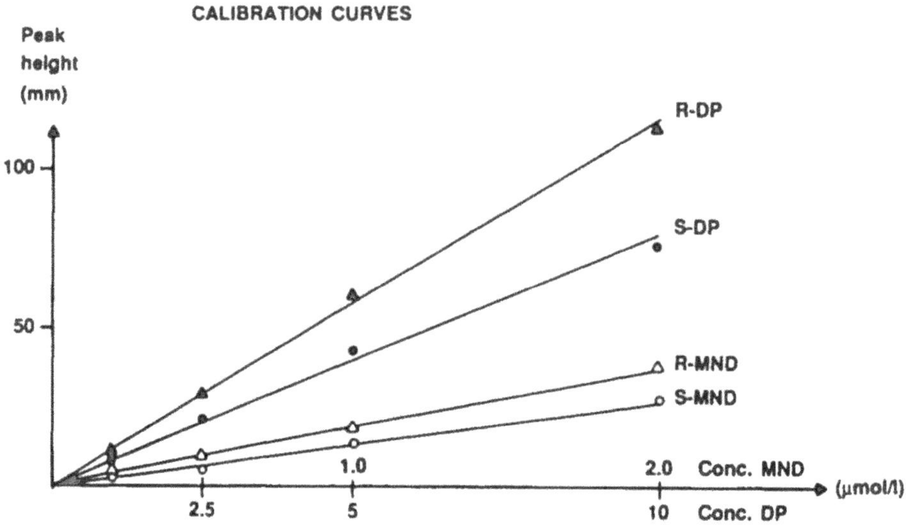

Figure 4 Calibration curves. MND = monodesisopropyldisopyramide; DP = disopyramide.

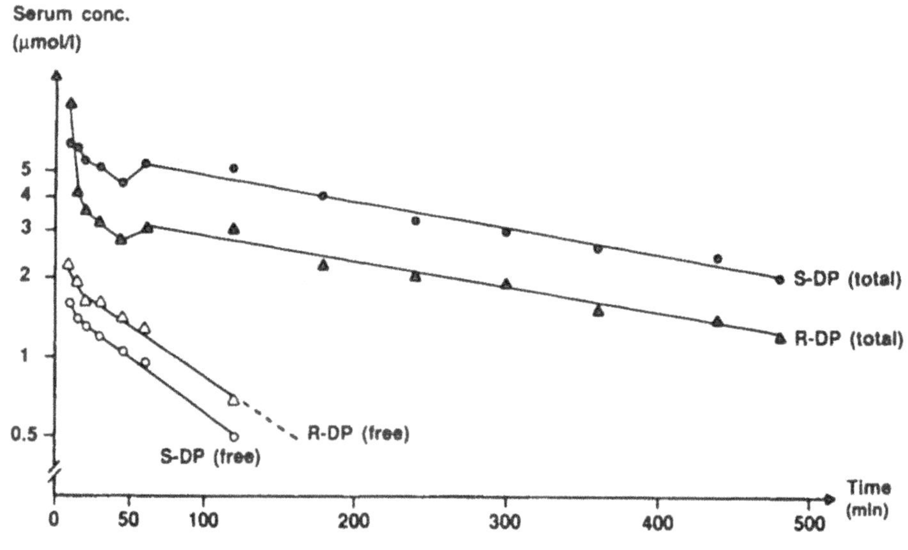

Figure 5 Concentrations of total and free R- and S-DP in serum from a male volunteer, 5 to 480 min following a bolus administration of 150 mg of disopyramide (DP).

Table 1 Reproducibility of Replicate Analysis of Disopyramide and Monodesisopropyldisopyramide Enantiomers Added to Human Serum

Compound	Serum Conc.[a] (μmol/L)	Coefficient of Variation (%)
R-disopyramide	2.6	3.6
	7.6	1.6
S-disopyramide	2.6	3.2
	7.6	1.6
R-monodesisopropyldisopyramide	0.5	5.2
	2.6	3.6
S-monodesisopropyldisopyramide	0.5	5.5
	2.6	3.2

[a]Mean values from eight duplicate samples of each concentration.

splitting of the peaks. Using formic acid instead, as described by Corre et al. (4), the splitting disappeared.

For the determination of the free concentration of (R)- and (S)-DP, it was not necessary to extract the filtrates before the HPLC analysis. In some patients endogenous compounds appeared in the same region as R- and S-MND. In our study, however, the concentration of these metabolites was too low to be detected.

REFERENCES

1. J. J. Lima and H. Boudoulas. Stereoselective effects of disopyramide enantiomers in humans. *J. Cardiovasc. Pharmacol.*, 9:594–600 (1987).
2. A. Samogyi and M. Muirhead. Pharmacokinetic interactions of cimetidine. *Clin. Pharmacokinet.*, 12:321–366 (1987).
3. J. Bonde, N. M. Jensen, L. E. Pedersen, N. A. Graudal, H. R. Angelo, and J. R. Kampmann. Elimination kinetics and urinary excretion of disopyramide in human healthy volunteers. *Pharmacol. Toxicol.*, 62:298–301 (1988).
4. P. L. Corre, D. Gibassier, P. Sado, and R. L. Verge. Direct enantiomeric resolution of disopyramide and its metabolite using chiral high-performance liquid chromatography. Application to stereoselective metabolism and pharmacokinetics of racemic disopyramide in man. *J. Chromatogr.*, 450:211–216 (1988).
5. M. Enquist and J. Hermansson. Comparison between two methods for the determination of the total and free (R)- and (S)-disopyramide in plasma using an α_1-acid glycoprotein column. *J. Chromatogr.*, 494:143 (1989).
6. L. P. Pedersen, K. Hermansen, H. P. Olesen, and S. N. Rasmussen. The pharmacokinetics and protein binding of disopyramide in pigs. *Acta Pharmacol. Toxicol.*, 58:282–288 (1986).
7. J. Bonde, L. E. Pedersen, E. Nygaard, T. Ramsing, H. R. Angelo, and J. P. Kampmann. Stereoselective pharmacokinetics of disopyramide; Interaction with cimetidine. *Br. J. Clin. Pharmacol.*, 31:708–710 (1991).
8. H. R. Angelo, J. Bonde, J. P. Kampmann, and J. Kastrup. A HPLC method for the simultaneous determination of disopyramide, lidocaine and their monodealkylated metabolites. *Scand. J. Clin. Lab. Invest.*, 46:623–627 (1986).

42

Direct Determination of Selenium by Graphite Furnace Atomic Absorption Spectrometry with Deuterium Background Correction

M. V. Seijas, A. Gil, and M. T. Muñoz *Hospital Ramón y Cajal de Madrid, Madrid, Spain* **R. Lozano** *Universidad Complutense de Madrid, Madrid, Spain*

I. INTRODUCTION

Selenium is considered to be an essential trace element for animals and for humans. Indeed, selenium is a constituent of Se-dependent glutathione peroxidase (Se-GPX).

Low blood levels of selenium have been observed in numerous pathological conditions, but only severe deficits appear to provoke observable symptoms, particularly myopathies and/or cardiomyopathies in geographic zones which are poor in selenium (Keshan disease) and in some isolated cases.

Absolute or relative deficit of selenium has been implicated in the development of cancers, coronary atheroma, and degenerative neuropathies.

The determinations of selenium in human blood may become a regular test required by medical practitioners. This practice will require careful monitoring of human blood selenium levels to maintain an effective yet nontoxic level of selenium.

The analytical problems associated with the determination of Se in the human blood by atomic absorption spectrometry with a graphite furnace are spectral interferences, chemical interferences, and thermal preatomization losses.

II. MATERIALS AND METHODS

A Varian SpectrAA 30 graphite furnace, with a deuterium continuum background corrector, was used for this study. We achieved better reproducibility by using a Juniper lamp (S&J Juniper Co., Harlow, Essex, U.K.) at a working current of 6 mA, 196-nm wavelength, and 1.0-nm slit width. We used pyrolytically coated partition graphite tubes throughout.

The samples were diluted by mixing one volume of serum with four volumes of diluent containing 10 mM of KI, 0.5% of Triton X100, and 0.2% of Antifoam B per liter. $PdCl_2$ (50 mg/100 mL of palladium) was used as matrix modifier. We used pooled serum as control; the sera were used directly, without previous treatment.

We used standard additions for calibration. The standard selenium solution was 200 µg/L.

The graphite furnace parameters used for the determination are shown in Table 1.

Table 1 Graphite Furnace and Sampler Parameters

Furnace Parameters					
Step No.	Temperature (°C)	Time (s)	Gas Flow (L/min)	Gas Type	Read Command
1	75	5.0	3.0	Normal	No
2	90	70.0	3.0	Normal	No
3	120	5.0	3.0	Normal	No
4	300	20.0	3.0	Normal	No
5	1200	10.0	3.0	Normal	No
6	1200	34.0	3.0	Normal	No
7	1200	1.0	0.0	Normal	No
8	2700	0.8	0.0	Normal	Yes
9	2700	0.5	0.0	Normal	Yes
10	2700	2.0	3.0	Normal	No

Sampler Parameters Volumes (µL)				
	Standard	Sample	Blank	Modifier
Blank	—	—	45	16
Addition 1	4	40	6	16
Addition 2	6	40	4	16
Addition 3	8	40	2	16
Addition 4	10	40	0	16
Sample	—	40	10	16

Recalibration rate: 1

Multiple inject: No   Hot inject: No   Pre-inject: yes
Last dry step: 1

Instrument mode: absorbance
Calibration mode: standard additions
Measurement mode: peak height
Lamp position: 1
Lamp current: 7 mA
Slit width: 1.0 nm
Slit height: normal

Wavelength: 196.0 nm
Sample introduction: sampler automixing
Time constant: 0.05
Measurement time: 1.0 s
Replicates: 2
Background correction: on

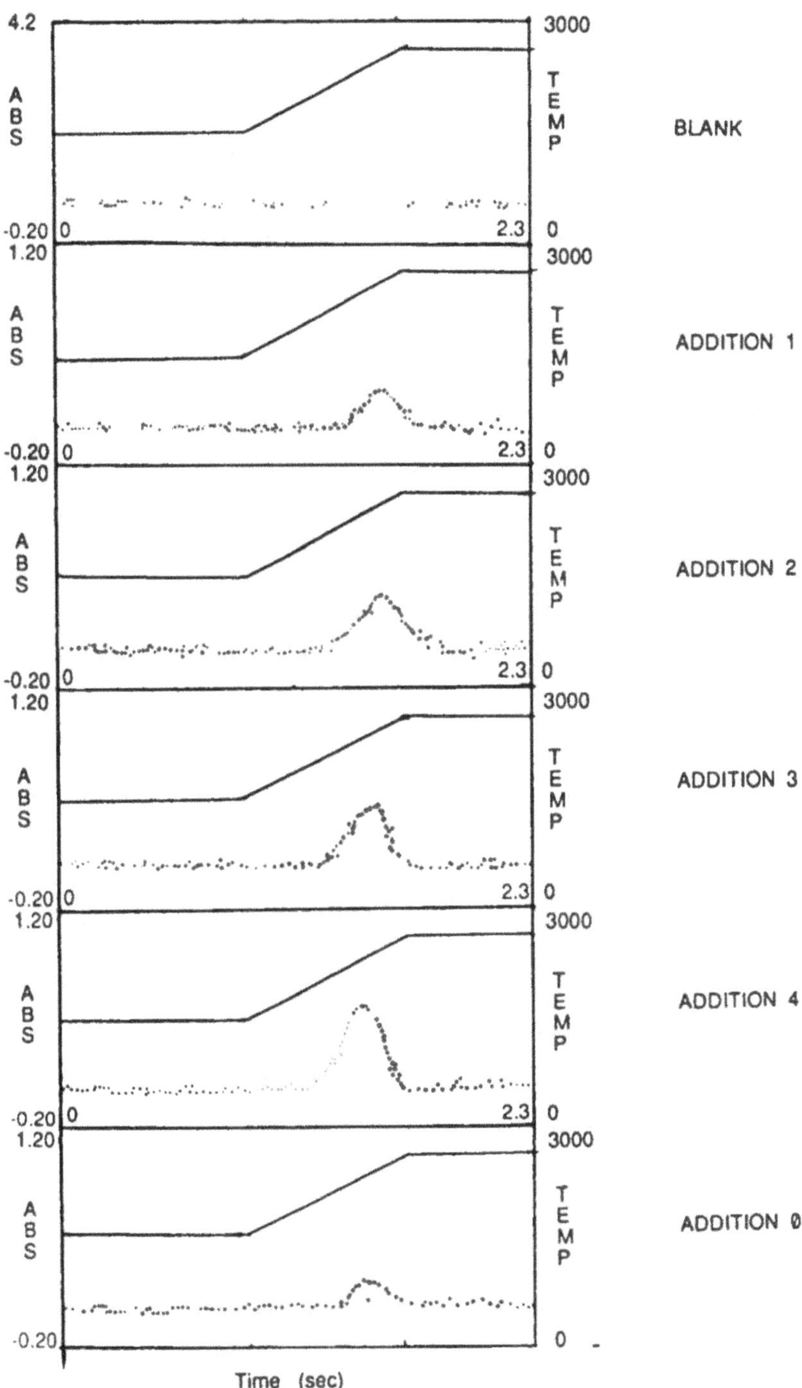

Figure 1 Peaks for all calibration points.

III. RESULTS

Figure 1 shows the peaks for all calibration points. The palladium modifier produces a selenium peak height higher than other modifiers. Blank measurements showed that there was no selenium present in the modifier. The furnace program involved a ramp to ash for 10 s and a hold temperature (1200°C) of 34 s. The levels of background and atomic absorbance are shown.

Calibration curves with standard additions ($r > 0.929$) were linear in the range 0 to 90 μg/L. Intra- and interassay CVs were 11.0% and 11.3% for the pooled serum control.

IV. DISCUSSION

The signals were obtained using an ashing temperature of 1200°C and an atomization temperature of 2700°C. The palladium modifier with KI reduces selenium and palladium to the metallic state, and reduced palladium enhances thermal stability (ashing temperature of 1200°C), allowing breakdown of the complex matrix. So, spectral interferences from iron are not evident with this method. The pre-injection of the palladium modifier tends to reduce carbon buildup in the graphite tube. Confirmation of selenium serum methodology has been difficult because of the absence of certified reference material for human blood. The described method is rapid and permits direct determinations of selenium on as little as 45 μL of serum or plasma; also, available deuterium background correction instrumentation is used.

BIBLIOGRAPHY

1. B. Welz, M. Melcher, and J. Nève. *Anal. Chim. Acta, 165*:131–140 (1984).
2. G. Alfthan. *Anal. Chim. Acta, 165*:187–194 (1984).
3. B. Welz. *Atomic Absorption Spectroscopy*, Verlag Chemie, New York (1976).
4. O. Oster and W. Prellwitz. *Clin. Chim. Acta, 124*:277–286 (1982).
5. J. Nève, F. Vertongen, and L. Molle. *Clin. Endocrinol. Metab., 14*:269–276 (1985).

Part III

Selected High-Performance Liquid Chromatographic Methods

43

HPLC Reference Method for Theophylline: Calibration by Use of Standard Addition

D. Petersen, G. Schumann, M. Oellerich, R. Klauke, and N. Wittner *Medizinische Hochschule Hannover, Hannover, Germany*

I. INTRODUCTION

Control materials with reference values (RMVs) are gaining increasing importance in clinical chemistry and have become mandatory in Germany. According to the legal regulations in the FRG, theophylline (THP) is the first drug for which quality assessment based on reference method values is required. So far, however, a reference method for theophylline has not been established.

Different analytical principles are desirable for the determination of highly accurate reference method values, especially if a definitive method is lacking. In this study we report on two different methods for the determination of THP reference method values: high-performance liquid chromatography (HPLC) and isotope-dilution mass spectrometry (ID-MS). In the HPLC procedure, calibration is performed by use of standard addition. RMVs were assigned to control materials, and method-dependent values were compared with RMVs.

II. MATERIALS AND METHODS

A. HPLC Method

1. *Reagents*

Chloroform, dichloromethane, KH_2PO_4, tetrahydrofurane, NaOH (2 mol/L) were supplied by Merck (Darmstadt, FRG). Triethylamine (purity: >99.5%) and tetraethylammonium bromide (purity: >99%) were purchased from Fluka Chemika (Neu Ulm, FRG). Theophylline reference standard and β-hydroxyethyl-theophylline were supplied by Sigma (Deisenhofen, FRG). All reagents were HPLC grade; water was bidistilled.

2. *HPLC Equipment*

The HPLC analysis is performed on a liquid chromatography system HP 1090 (Hewlett-Packard) equipped with a dual solvent-delivery system, an automated sample injector, a column-switching valve, and a photodiode-array UV detector set at 270 nm.

Chromatograms were recorded and the peak areas integrated by use of a HP 3392 integrator (Hewlett-Packard, Bad Homburg, FRG). A guard column, 20 mm × 3 mm (I.D.), contained Zorbax ODS, the analytical column, 250 mm × 3 mm (I.D.) contained ODS Hypersil (5 µm). Both columns were delivered by Bischoff (Leonberg, FRG).

3. *Mobile Phases*

Two mobile phases (A and B) contained KH_2PO_4 (0.01 mmol/L), tetraethylammonium bromide (12.42 mmol/L), tetrahydrofurane (9 ml/L), and triethylamine (250 µl/L). The pH of mobile phases A and B was adjusted to pH 7.0 and pH 7.5, respectively, by use of NaOH (2 mol/L). Both mobile phases were freshly prepared prior to each HPLC run.

4. *Standard Solutions*

Aqueous calibrators used for standard addition contained THP and internal standard (IS, β-hydroxyethyltheophylline). The standard concentrations used are listed in Table 1. The concentrations of THP and the IS were matched with the expected THP concentration of the sample. The standards were prepared using officially calibrated volumetric flasks and pipettes.

5. *Standard Addition and Sample Pretreatment*

The lyophilized control material was reconstituted according to the manufacturer's recommendations. Five analytical portions were drawn and added volume to volume to standards 1 to 5. Of each sample, 0.5 mL was added to 20 mL dichloromethane/isopropanol and extracted by shaking for 30 s on a Vibromix. The upper phase was discarded. The lower organic phase was transferred into conical glass vials and evaporated to dryness at 60°C under a stream of nitrogen. The remainder was dissolved in 0.25 mL of mobile phase, centrifuged for 5 min at 3000 rpm, and transferred into capped HPLC vials.

6. *Liquid Chromatography*

The column temperature was set at 50°C. The flow was adjusted to 1.2 mL/min. The volume was injected 10 µL. The elution was isocratic. In a prerun, the relation of mobile phases A and B was adjusted to achieve baseline separations of THP and

Table 1 Solutions Used for Standard Addition in HPLC

Range of Expected THP Concentration of Sample (mg/L)	Concentration of THP and Internal Standard [ ] in Solution Used for Standard Addition (mg/L)				
1–10	0 [15]	2.5 [15]	5.0 [15]	7.5 [15]	10.0 [15]
11–20	0 [30]	5.0 [30]	10.0 [30]	15.0 [30]	20.0 [30]
21–30	0 [45]	5.0 [45]	10.0 [45]	20.0 [45]	25.0 [45]

internal standard. Each chromatogram required approximately 10 min to run. Figure 1 shows the complete separation of THP and IS from various other purines in aqueous solutions. Figure 2 shows an extracted serum sample. Figure 3 shows an extracted serum sample with THP standard addition.

7. *Calibration and Calculation*

In the calibration curve, the concentration of the added standards versus peak area ratios (theophylline/internal standard) of corresponding samples are plotted (Figure 4). The linear regression line was calculated. The extrapolated negative intercept on the x axis represents the original theophylline concentration of the sample [x]. In all 20 measurements (four replicates at five concentration levels) contributed to each reference method value.

B. ID-MS Method

1. *Reagents*

1-Iodopentane, N,N-dimethylacetamide, and tetramethylammonium hydroxide were purchased from Aldrich (Steinheim, FRG). Methanol, chloroform, isopropanol, sodium acetate, and hydrochloric acid (2 mol/L) were supplied by Merck (Darmstadt, FRG). Methanol, chloroform, isopropanol, sodium acetate, and hydrochloric acid (2 mol/L) were supplied by Merck (Darmstadt, FRG). Theophylline reference standard was purchased from Sigma (Deisenhofen, FRG). [1,3-^{15}N,2-^{13}c]theophylline was purchased from Cambridge Isotope Laboratories (Cambridge, Massachusetts).

2. *Apparatus*

The instrument used was a Finnigan 1020 quadrupole mass spectrometer coupled with a Sigma 3 gas chromatograph (Perkin Elmer, Überlingen, FRG). A fused-silica

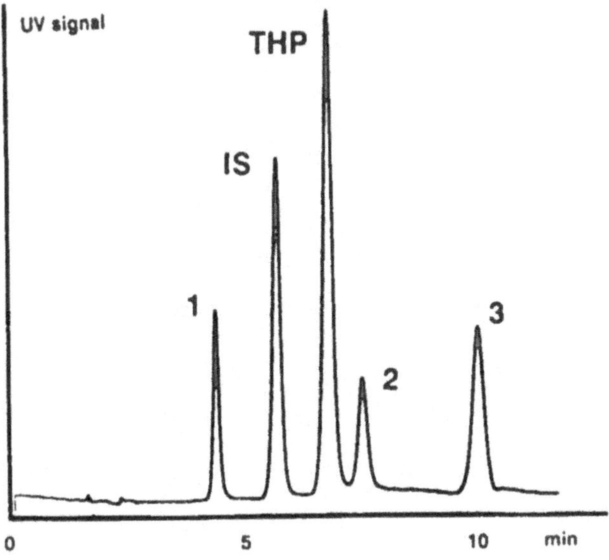

Figure 1 Chromatogram of an aqueous mixture of purines. 1, paraxanthine; IS, β-hydroxyethyl-THP; THP, theophylline; 2, paraxanthine; 3, proxyphylline.

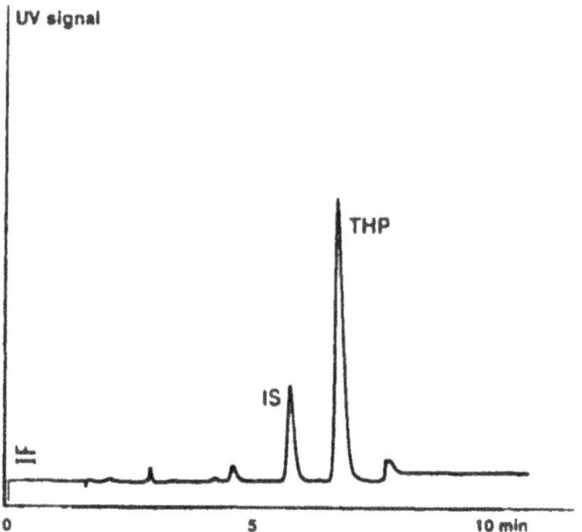

Figure 2 Chromatogram of an extracted serum sample.

capillary column (SE 52S) of 25 m length was employed. A moving needle (WGA, Düsseldorf, FRG) was used for sample injection.

3. *Standard Solutions*

Two sets of three calibrators were prepared. According to the bracketing scheme, THP concentrations were 10% below, 10% above, and close to the expected sample concentration. The concentration of the internal standard ([1,3-^{15}N,2-^{13}C]theophylline) was also in the range of the expected sample concentration.

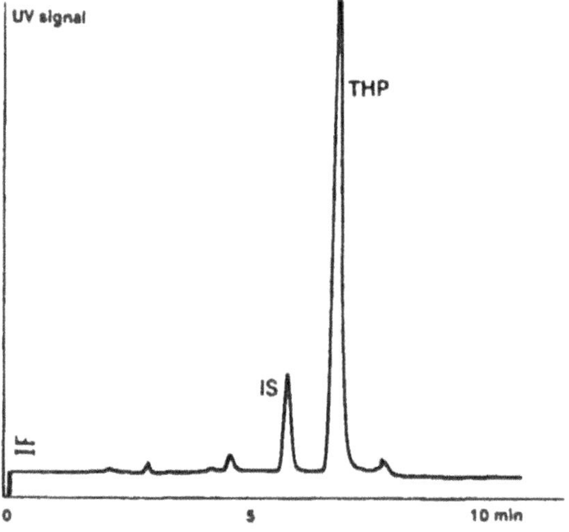

Figure 3 Chromatogram of an extracted serum sample with standard addition.

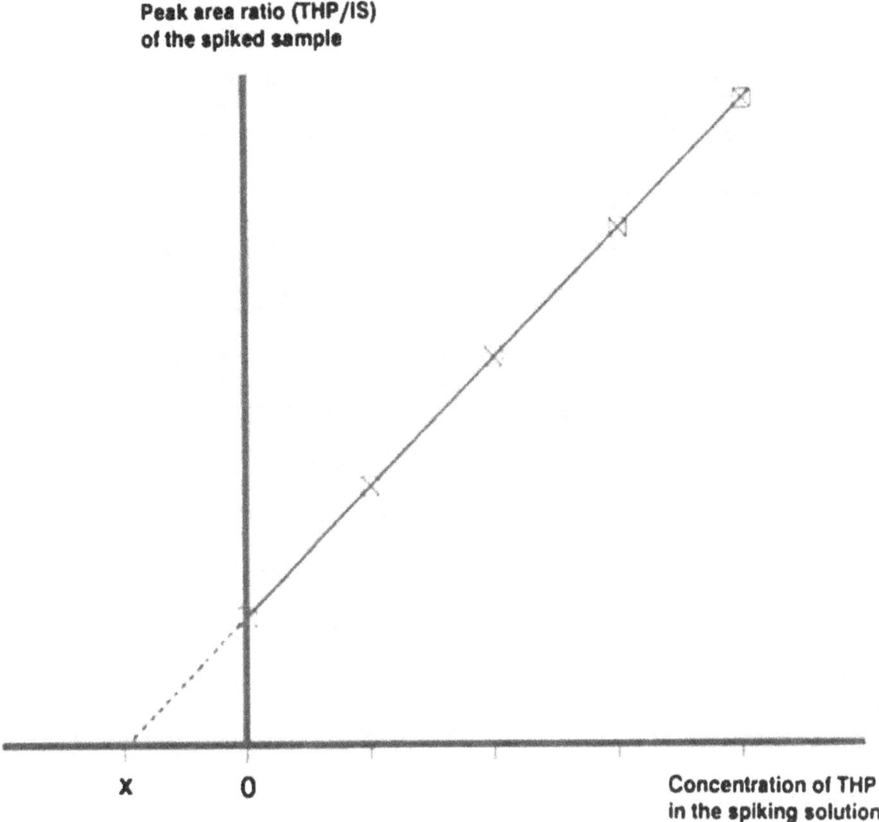

Figure 4 Calculation of THP concentration. Extrapolation of the linear regression line: (X), THP concentration in the sample.

4. Sample Pretreatment

Internal standard, 0.5 mL, was added to 0.5 mL of reconstituted control material resp. THP standard. Aliquots of 0.5 mL were diluted volume to volume with sodium acetate (0.1 mol/L), which was adjusted with hydrochloric acid (2 mol/L) to pH 5.0. These samples were extracted by adding 10 ml of a solution containing 95% chloroform and 5% isopropanol. After mixing (1 min) and centrifugation (5 min, 3000 rpm), the upper phase was discarded and the lower organic phase was evaporated to dryness under a stream of nitrogen at 60°C. The remainder was dissolved in 0.5 mL of N,N-dimethylacetamide and 0.1 mL of tetramethylammonium hydroxide/methanol (25/75).

5. Derivatization

Derivatization of theophylline and the internal standard into 7-pentyl-theophylline was accomplished with 1-iodopentane (0.2 mL) (1–3). After mixing, incubation (15 min) at ambient temperature, and centrifugation (5 min, 3000 rpm, 4°C), the upper organic phase was drained and evaporated to dryness under a stream of nitrogen at 40°C. The residue was dissolved in 0.2 mL cyclohexane.

6. Chromatography and Detection

The injector temperature was adjusted to 250°C; the column temperature and separator temperature were set at 200°C (isotherm) and 240°C, respectively. The manifold temperature was adjusted to 60°C. Helium was used as carrier gas. A 1-μL sample containing THP derivatives in cyclohexane solution was injected manually using the moving needle injector after evaporating the cyclohexane in the stream of carrier gas.

Under these conditions theophylline and internal standard eluated from the column approximately 4.5 min after injection. The monitored ions were m/z 250 for theophylline and m/z 253 for the internal standard.

7. Calibration and Measuring Design

Two analytical portions which had been spiked separately with the internal standard and had undergone separate derivatization and extraction were each measured five times. Every measurement was calibrated individually. The whole procedure was repeated in a second run using the second set of calibrators. In all, 20 individually calibrated measurements contributed to each reference method value.

8. Calculation

The theophylline concentration was calculated according to Siekmann and Breuer (4). This calculation procedure takes into account that the stable isotope used as internal standard may contain interfering trace amounts of natural theophylline and vice versa.

III. RESULTS

Tables 2 and 3 summarize THP reference method values determined in different control materials by HPLC and ID-MS. Two of the control materials are commercially available (Lyphocheck 1 and 2). Each reference method value, whether determined by HPLC or ID-MS, is based on 20 individually calibrated measurements. The corresponding method-dependent values determined by immunoassays are listed for comparison. Except for one control material (IV), the deviations between ID-MS and HPLC results are smaller than 2%. Regarding the HPLC method, the coefficient of correlation between peak area ratio and concentration of added standard is greater than 0.992. With the ID-MS procedure the coefficient of variation is less than 2% for all the control materials examined. Studies on the reproducibility of the ID-MS results are shown in Table 4. In five different control materials (I to V), the RMVs were determined twice by repeating the whole ID-MS procedure. The deviation between the first and second determinations of RMV was less than 1.6% for the control materials investigated.

The deviations of method-dependent values from RMVs determined by HPLC are in the range from +1.27% to –5.04% with a median of –1.31%. The deviations of method-dependent values from RMVs determined by ID-MS are in the range from 0% to –5.04% with a median of –2.25%.

IV. DISCUSSION

The accuracy and high specificity of the ID-MS method is due to the highly efficient gas chromatographic separation and single-ion monitoring. Assuming identical behavior of THP and [1,3-^{15}N, 2-^{13}C]THP in all analytical steps of the described

Table 2 Reference Method Values for Theophylline Determined by HPLC

Control Material	RMV (mg/L)	n[a]	r[b]	Method-Dependent Value (mg/L)
I	4.74	20	0.9996	4.8 (immunoassay)
II	7.39	20	0.9928	7.2 (immunoassay)
III	11.45	20	0.9985	11.3 (immunoassay)
IV	13.48	20	0.999	12.8 (immunoassay)
V	27.09	20	1.0000	27.3 (immunoassay)
Lyphocheck 1 (Bio Rad)	5.37	20	0.9995	5.5 (immunoassay) 5.0 (immunoassay)
Lyphocheck 2 (Bio Rad)	30.48	20	0.9998	28.5 (HPLC) 29.3 (FPIA)[c]

[a]Number of HPLC measurements.
[b]Coefficient of correlation of linear regression line used for calibration.
[c]Fluorescence polarization immunoassay.

Table 3 Reference Method Values for Theophylline Determined by ID-MS

Control Material	RMV (mgL)	n	cv (%)	Method-Dependent Value (mg/L)
I	4.80	20	0.756	4.8 (immunoassay)
II	7.26	20	1.330	7.2 (immunoassay)
III	11.56	20	1.915	11.3 (immunoassay)
IV	13.48	20	1.290	12.8 (immunoassay)
V	28.31	20	0.916	27.3 (immunoassay)

Table 4 Reproducibility of ID-MS Theophylline RMVs

Control Material	First RMV for THP (mg/L)	Second RMV for THP (mg/L)
I	4.80	4.73
II	7.26	7.20
III	11.56	11.38
IV	13.48	13.27
V	28.31	28.29

procedure, THP containing stable isotopes is the superior internal standard. The intraassay precision of the ID-MS procedure is adequate. The interassay reproducibility, as far as it has been tested, is excellent. The ID-MS procedure is the superior technique to determine RMVs for THP. Control samples with reference method values can be applied to quality assurance with a satisfactory agreement of RMVs and method-dependent values.

Reference method values for theophylline determined by the described HPLC procedure are in good agreement with those obtained by ID-MS. In order to achieve these results with a HPLC method, several conditions have to be fulfilled. The standard addition technique, well known in spectroscopy (5), has to be used for calibration. It allows aqueous primary standards to be used for calibration. This is necessary to achieve maximum accuracy, which is demanded in reference methodology. The major advantage of this technique is that different extraction rates of theophylline and internal standard from aqueous solutions and serum matrices can be overcome. However, as indicated by preliminary results, the reproducibility of the HPLC procedure is not as good as that of the ID-MS method. This may be attributed to incomplete extraction or different extraction rates of THP from serum due to strong protein binding of the drug. Further studies are necessary on this subject. Other requirements that have to be fulfilled in order to achieve good agreement between ID-MS and HPLC are a column with high separation efficiency and fresh preparation of the mobile phases prior to analysis. Furthermore, pH adjustment of the mobile phases is necessary to achieve baseline separation of various purines.

By following these guidelines, the relatively time-consuming HPLC procedure can be used for the determination of reference method values of theophylline. This is important because more laboratories have access to HPLC equipment than to a mass spectrometer.

V. CONCLUSION

The data of the present study suggest that both described methods fulfill the requirements of reference methods for theophylline. Reliable reference method values can be assigned to control materials. Furthermore, these methods can be used to assess the THP concentration in secondary calibrators and in patient samples with suspected interference.

REFERENCES

1. R. H. Greeley. Rapid esterification for gas chromatography. *J. Chromatogr.*, *88*:229–233 (1974).
2. G. J. Johnson and W. A. Dechtiaruk. Gas chromatography determination of theophylline in human serum and saliva. *Clin. Chem.*, *21*:144–147 (1975).
3. M. Desage and J. Soubeyrand. Automated theophylline assay using gas chromatography and a mass-selective detector. *J. Chromatogr.*, *336*:285–291 (1984).
4. L. Siekmann and H. Breuer. Determination of cortisol in human plasma by isotope dilution mass spectrometry. *J. Clin. Chem. Clin. Biochem.*, *20*:883–892 (1982).
5. M. Sharaf, D. Illman, and B. Kowalski (series eds.). Chemometrics. In *Chemical Analysis* (P. J. Elving, ed.), Monogr. Ser. Vol. 82, John Wiley, New York (1986).

44

Comparison of Vancomycin Concentration Measured in Serum Versus Plasma

C. Martínez, A. Aldaz, P. Fossa, I. Castro, and J. Giráldez
University Clinic of Navarra, Pamplona, Spain

I. INTRODUCTION

The terms "serum concentration" and "plasma concentration" are often used interchangeably in therapeutic drug monitoring (TDM). In recent years some research works (1–4) have studied the correlation between "serum" and "plasma" concentration when monitoring some particular drugs and the influence of the collection tube on concentration. We decided to study vancomycin, due to its wide use in our hospital and the increasing interest of some physicians in the clinical importance of monitoring this drug (5–7).

Monitoring the relationship between serum and plasma concentration of vancomycin poses some problems to our routine work (Table 1). In our hospital, vancomycin, associated with other antibiotics, is widely prescribed by oncologists, in a standard protocol for immunocompromized patients. Furthermore, there is an increasing emergence of meticillin-resistant *Staphylococcus aureus* strains which are usually treated with vancomycin.

The purposes of this study are the following:

1. To determine whether there are differences between serum and plasma concentration of vancomycin
2. To propose a regression formula in order to predict serum concentrations when measuring plasma levels of vancomycin
3. To establish the influence of freezing plasma and serum samples on the results

II. MATERIALS AND METHODS

Forty-two plasma and serum samples corresponding to 24 patients from four medical services (oncology, pediatric oncology, pediatrics, and internal medicine) were processed. Every sample was divided as follows: 5 mL of blood was drawn into a standard red-stoppered vacutainer tube containing no anticoagulant (serum assay); an additional 5 mL of blood was collected into a green-stoppered tube containing sodium heparin as the anticoagulant. All serum samples were left for 30 min at room

Table 1 Problems in Routine Work

1. Interest in reducing the number of samples because of technical and ethical considerations
2. Convenience of a shorter processing time for a scheduled assay
3. Easier handling of plasma samples in neonates and, in general, when sample volume is small

temperature for clotting. Plasma samples were left at room temperature for 1 h before vancomycin assay and then processed as above. Every serum and plasma sample was centrifuged for 1 min. Any clot that formed during the process was removed. The time until the next scheduled assay never exceeded 4 h.

All samples were analyzed once using TDx fluorescence-polarization immunoassay (Abbott Cientifica). A control was included in every assay to ensure the reliability of this method. The coefficient of variation for the assay is less than 5% (8). A volume of every sample was frozen (–20°C) and analyzed 7 and 14 days later. Because of technical considerations, only 33 samples were analyzed 7 days after freezing, and 28 were analyzed 14 days after freezing. A normality test was performed (Sigma Program by Horus Hardware, 1986). Statistical analysis was performed for differences in measured concentrations for each plasma/serum pair.

The correlation between serum and plasma concentration of vancomycin was determined, and a regression formula was proposed. In all cases, a value of $p < 0.05$ was considered to be of statistical significance.

III. RESULTS

Serum concentrations ranged from 1.81 mg/L to 60.32 mg/L and plasma concentrations ranged from 3.31 mg/L to 69.98 mg/L.

The influence of freezing on serum samples is shown in Figure 1. No statistical significant differences were observed after freezing for 14 days. Plasma samples can only be stored for 7 days. Nevertheless, differences were observed in plasma samples after freezing for 14 days but no difference could be reported after freezing for 7 days (Figure 2). Statistically significant differences were observed between serum and plasma samples throughout the study (Figure 3).

Serum concentration appeared to underestimate plasma concentrations. The relationship between plasma concentration and serum concentration is shown in Figure 3, demonstrating a high correlation between media (serum = 0.59 + 0.9 plasma; $r^2 = 0.993$) (Figure 4).

IV. DISCUSSION AND CONCLUSIONS

Statistically significant differences between serum and plasma concentration were observed throughout the study. However, these differences are unlikely to be of clinical importance, since the limits of vancomycin therapeutic range are not well established (9–12).

There is a high correlation between serum and plasma concentration. Therefore, using the regression formula above, we can predict serum concentration when measuring plasma samples. This concentration difference may be due, in part, to

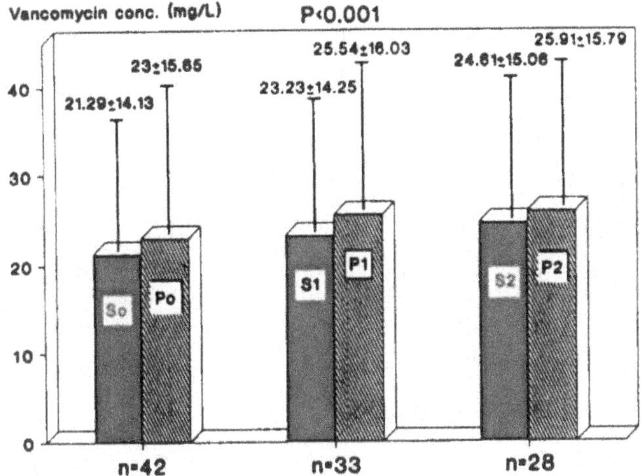

Figure 1 Differences between media (serum/plasma). SO, serum concentration after sample removal; S1, serum concentration after 7-day freezing; S2, serum concentration after 14-day freezing; PO, plasma concentration after sample removal; P1, plasma concentration after 7-day freezing; P2, plasma concentration after 14-day freezing.

differences in protein content between serum and plasma, but it is unlikely to be of importance as vancomycin demonstrates minimal protein binding. There was also an interference between heparin and vancomycin in the analytical assay, but it appeared to be less than 5% when using the TDx (8). However, the reasons for this difference must await further studies.

From a statistical standpoint, serum and plasma vancomycin samples cannot be used interchangeably. Therefore it is necessary to establish whether serum or plasma

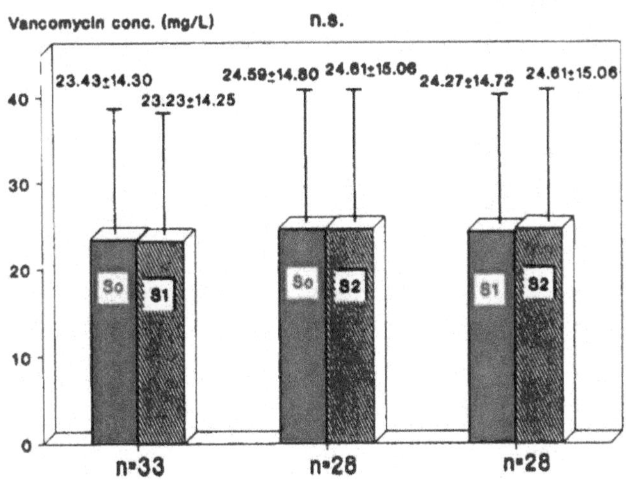

Figure 2 Influence of freezing on serum concentration.

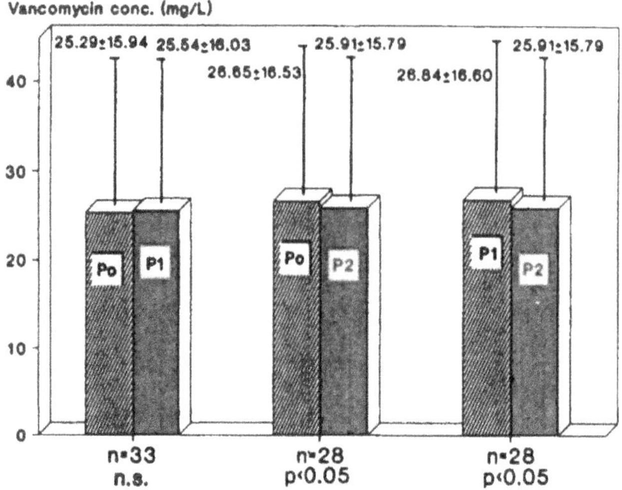

Figure 3 Influence of freezing on plasma concentration.

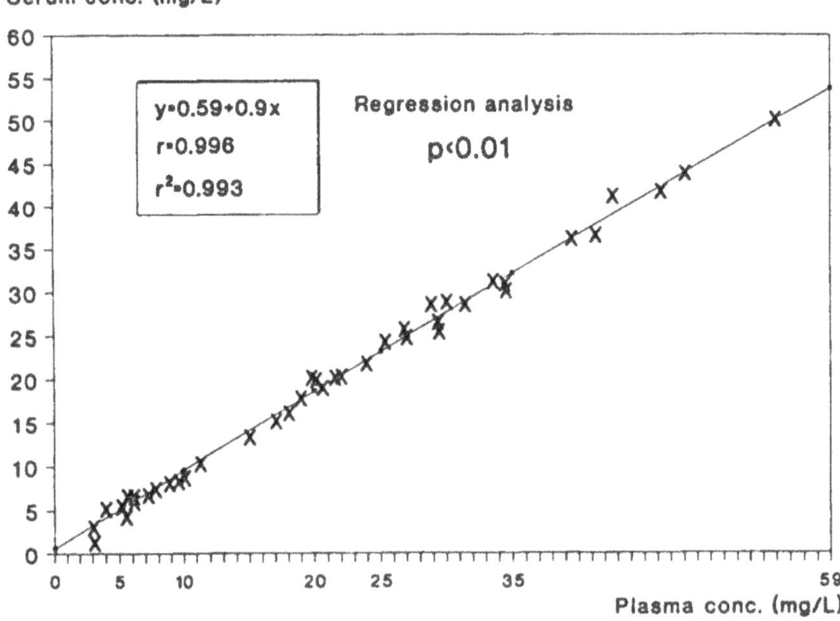

Figure 4 Relationship between serum and plasma concentration.

samples are to be used in both routine and research work. In order to predict serum concentration when measuring plasma samples, a regression formula is proposed.

REFERENCES

1. S. C. Ebert, M. Leroy, and B. Darcey. Comparison of aminoglycoside concentrations measured in plasma versus serum. *Ther. Drug Monitor.*, *11*:44–46 (1989).
2. C. G. Tarasidis, W. R. Garnett, B. J. Kline, and J. M. Pellock. Influence of tube type, storage time and temperature on the total and free concentration of valproic acid. *Ther. Drug Monitor.*, *8*:373–376 (1986).
3. G. Nyberg and E. Martensson. Preparation of serum and plasma samples for determination of tricyclic antidepressants: Effects of blood collection tubes and storage. *Ther. Drug Monitor.*, *8*:478–482 (1986).
4. D. N. Bariley, J. J. Coffec, and J. R. Briggs. Stability of drug concentrations in plasma stored in serum separator blood collection tubes. *Ther. Drug Monitor.*, *10*:352–354 (1988).
5. D. J. Edwards and S. Pancorbo. Routine monitoring of serum vancomycin concentrations: Waiting for proof of its value. *Clin. Pharm.*, *6*:652–654 (1987).
6. K. A. Rodvold, H. Zokufa, and J. C. Rotschafer. Routine monitoring of serum vancomycin concentrations: Can waiting be justified? *Clin. Pharm.*, *6*:655–658 (1987).
7. J. F. B. Sayers and R. Shimasaki. Routine monitoring of serum vancomycin concentrations: The answer lies in the middle. *Clin. Pharm.*, *7*:18 (1988).
8. *Vancomycin. Antibiotic Drug Assays*. Abbott Laboratories Diagnostics Division, 1986.
9. M. Wenk, S. Vozeh, and F. Folloth. Serum level monitoring of antibacterial drugs. *Clin. Pharmacokinet.*, *9*:475–492 (1984).
10. B. H. Ackerman. Evaluation of three methods for determining initial vancomycin doses. *DICP*, *23*:123–128 (1989).
11. K. D.Lake and C. D. Peterson. Comment: Vancomycin dosing method. *DICP*, *23*:618–619 (1989).
12. W. E. Fitzsimmons, M. J. Postelnick, and P. V. Tortorice. Survey of vancomycin monitoring guidelines in Illinois Hospitals. *DICP*, *22*:598–600 (1988).

45

Pharmacokinetics of Oral Gemfibrozil in Healthy Volunteers

Franjo Plavšić and Alka Wolf Čoporda *Zagreb University, Zagreb, Croatia, Yugoslavia*

I. INTRODUCTION

Gemfibrozil is a new hypolipemic agent. Like other related hypolipemic drugs, gemfibrozil is a carboxylic acid, but its absorption is relatively independent of external factors which are otherwise characteristic of acidic drugs (1). It is very rapidly absorbed from the digestive system and intensely metabolized into several products (2), which results in its fast elimination from the body. Until recently, gemfibrozil was determined by gas chromatography requiring derivatization, which made its analysis quite complicated. At present, high-performance liquid chromatography (HPLC) (3) appears to be the method for choice for the analysis of this drug. It is a simple method, using ultraviolet detection.

II. MATERIALS AND METHODS

A. Trial

Bioequivalence of two oral gemfibrozil preparations was tested in an open crossover randomized trial. The trial protocol was discussed and approved by the Clinical Hospital Center Drug Commission on December 10, 1989.

The trial was planned and performed in keeping with the Helsinki Declaration of Human Rights and its Tokyo supplement. On the day of the trial, fasting subjects were administered either of the two drugs, according to the list of randomization, in a dose of 900 mg taken with 100 mL of water at 8:00 a.m. Blood samples were withdrawn using a cannula, immediately before the drug dose and then at 0.5, 1, 1.5, 2, 2.5, 3, 4, 6, 8, 12, and 24 h after the dose.

B. Subjects

Ten male volunteers aged 20 to 35 years were included in the trial. Biochemical parameters indicating liver and kidney functions, i.e., the organs important for gemfibrozil pharmacokinetics, and blood counts, were determined and found to be within normal limits. The subjects underwent a check-up by an internist, and all were found to meet the trial conditions. Study subjects were given proper explanation of the trial protocol and objective by the principal investigator, who was available to them for any additional information throughout the study. The subjects were asked to drink no alcohol for a week, and to use no drugs for two weeks preceding the trial. They also gave their consent to follow the trial instructions.

C. Analytical Method

1. *Reagents*

Acetonitrile for liquid chromatography was purchased from Merck (Darmstadt), and cyclohexane and inorganic chemicals for mobile phase preparation from Kemika (Zagreb). Gemfibrozil standard and internal bezafibrate standard were obtained from Lek (Ljubljana), as analytical specimens.

A Lichrosorb RP-18 column for liquid chromatography was obtained from Merck (Darmstadt).

2. *Instrumentation*

A series 10 pump and an LCI-100 Laboratory Computing Integrator manufactured by Perkin Elmer were used for the high-performance liquid chromatography. Mobile-phase changes were measured on an LKB 2151 variable UV detector.

3. *Extraction*

To a 12-mL centrifuge test tube, 0.5 mL of plasma and 10 μL of internal standard solution (concentration, 1 g/L) were added. The test tube was stirred, and then 1 mol/L of acid chloride (100 μL) and cyclohexane (5 mL) were added. After stirring for several minutes and centrifugation at 2000 ×g for 10 min, the cyclohexane phase (4.5 mL) was transferred into a clean test tube and the solvent was evaporated. The dry residue was dissolved in the mobile phase (0.3 mL).

4. *Chromatographic Procedure*

Chromatography was performed with a mobile phase consisting of equal volumes of acetonitrile and 0.2% orthophosphoric acid in redistilled water. Before mixing, the aqueous solution was filtered through a micropore filter. The mobile phase flow was 1.3 mL/min. Changes in the effluent were observed at 225 nm. Retention times for the internal standard and gemfibrozil were 5.3 and 6.3 min, respectively. The analytical method sensitivity for gemfibrozil was 0.05 μg/L, and precision throughout the concentration range was less than 10%, as expressed by coefficients of variation, which is similar to the original method (3) used.

D. Calculations

An IBM-compatible Olivetti M24 computer with two disk units was used for all calculations.

Statistical processing included the analysis of variance of all gemfibrozil concentrations measured at test times following the dose administered, the time of absorption delay, times to peak concentrations, peak concentrations, areas under the concentration curves, and both individual and mean values of beta-phase curve slopes.

III. RESULTS

In Figure 1, chromatogram of the determination of gemfibrozil in a trial sample is presented. Sensitivity of the method was 0.05 µg/L, and precision, over all the concentrations measured, was below 10%, as expressed by coefficients of variation. No interference of lipemic, icteric, and hemolyzed samples was observed. Table 1 shows mean concentration values for both preparations. Figure 2 shows the mean measured concentrations of gemfibrozil for two trial samples, and simulated concentration curves. Table 2 presents pharmacokinetic parameters of the two preparations studied.

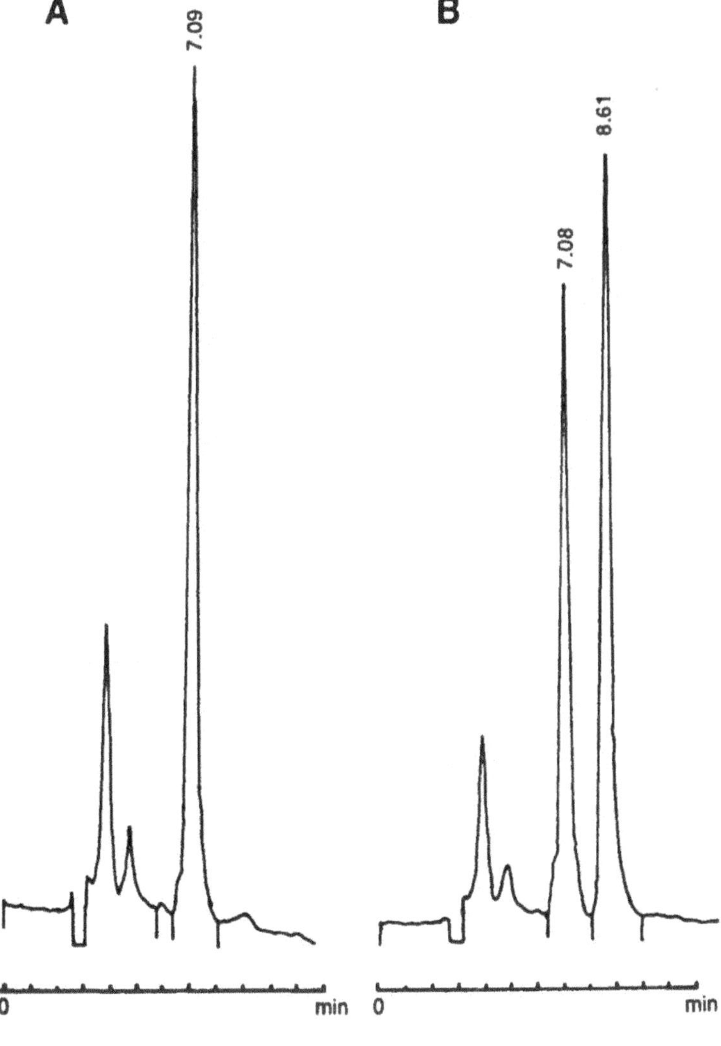

Figure 1 Representative chromatogram of Gemfibrozil determination in biologic material. A, blank serum with internal standard added; B, a trial sample.

Table 1 Mean Values of Gemfibrozil Concentrations After Administration of Two Preparations in Doses of 900 mg

	Gemfibrozil Preparation			
	Lek		Parke Davis	
Time	Mean	SD	Mean	SD
0	0	0	0	0
0.5	13.309	10.13671	11.507	8.374382
1	19.778	8.788087	17.748	9.155979
1.5	24.56	12.4528	23.893	10.86951
2	25.925	10.70603	25.905	10.45587
2.5	23.31	7.625303	24.694	9.945839
3	21.102	6.762503	16.739	10.74687
4	16.263	5.789966	15.205	8.138498
6	8.341001	3.154669	8.07	6.333735
8	3.947	2.620157	4.282001	1.884684
12	2.3999	1.645808	1.832	2.875107
24	1.061	1.32903	1.842	2.648777

IV. SUMMARY

In an open crossover randomized trial in 10 healthy volunteers, the bioequivalence of two oral gemfibrozil preparations (Gevilon tablets, Parke Davis, Berlin) and gemfibrozil tablets (Lek, Ljubljana), both given in a daily dose of 2 × 450 mg, was assessed. Blood samples were taken before and 0.5, 1, 1.5, 2, 2.5, 3, 4, 6, 8, 12, and 24 h after the administration of the dose. Gemfibrozil concentrations were determined by high-performance liquid chromatography.

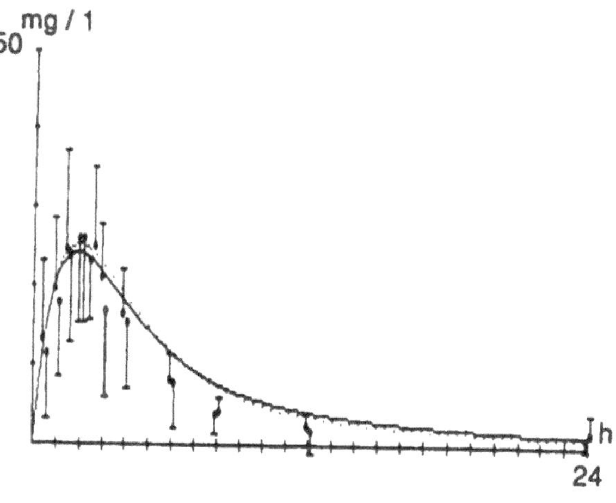

Figure 2 Mean levels of Gemfibrozil with standard deviations for two trial samples, and simulated concentration curves. Full line, Lek preparation; dotted line, Parke Davis preparation.

Table 2 Pharmacokinetic Parameters as Calculated from Mean Gemfibrozil Concentrations for Two Trial Samples

	Gemfibrozil Preparation	
Parameter	Lek	Parke Davis
T_{lag} (h)	0.08	0.18
T_{max} (h)	2.0	2.1
$T_{A1/2}$ (h)	0.85	0.80
$T_{\alpha 1/2}$ (h)	2.18	1.91
$T_{\beta 1/2}$ (h)	11.45	10.89
MRT (h)	7.9	8.51
C_{max} (mg/L)	25.4	25.42
V (L)	15.21	11.39
V_1 (L)	1.44	1.39
V_2 (L)	13.8	9.99
C_1 (L/h)	3.04	3.73
$AUC_{0\text{-}inf}$ (mg × h/L)	221.1	227.9
Simulation r	0.96	0.97

During the observation period, no statistically significant difference was recorded in mean gemfibrozil concentrations between the two preparations studied. Likewise, no statistically significant differences were observed for areas under the concentration curve (Gevilon, Parke Davis, 158.7, 61.9 mg × h/L; Lek preparation, 154.4, 4.97 mg × h/L; $p > 0.05$), peak concentrations (Gevilon, Parke Davis, 28.1, 9.2 mg/L; Lek preparation, 28.8, 9.2 mg/L; $p > 0.05$), and time to peak concentrations (Gevilon, Parke Davis, 1.9, 0.5 h; Lek preparation, 2.25, 0.8 h; $p > 0.05$).

The time of absorption delay was 0.29, 0.22 h and 0.51, 0.37 h for Gevilon (Parke Davis) and Lek preparation, respectively, which was not statistically significant ($p > 0.01$).

REFERENCES

1. A. K. Jain, J. R. Rain, W. S. S. Lacorte, and I. G. McMahan. *Clin. Pharm. Ther.*, *29*:254 (1981).
2. R. A. Okerholm, F. J. Keely, F. E. Peterson, and A. J. Glazko. *Fed. Proc.*, 35:675 (1976).
3. H. Hengy and E. U. Kolle. *Arzneim.-Forsch./Drug Res.*, 35:1637 (1985).

46

Evaluation of a Fluorescence Polarization Immunoassay for Use in Monitoring Plasma Dothiepin Levels in Overdose Cases

L. P. Hackett, L. J. Dusci, and K. F. Ilett *Queen Elizabeth II Medical Centre, Nedlands, Western Australia*

I. INTRODUCTION

Dothiepin is a tricyclic antidepressant which is widely prescribed in Australia. The drug is used for the treatment of both endogenous and reactive depressions and, in our hospital, some 27% of patients requiring antidepressant therapy are receiving dothiepin. It is therefore not surprising that dothiepin is now often encountered in overdose situations.

Monitoring of plasma tricyclic antidepressant levels in both therapeutic and overdose cases is essential, as it has been well documented that high plasma drug levels can cause cardiac arrhythmias (1–3). Hence, reliable and rapid analytical methods for measuring these drugs must be available. In our laboratory this has been accomplished by using the Abbott TD_x fluorescence polarization immunoassay technique. The cross-reactivity for different tricyclic antidepressants on this system varies widely. At present, there is little published data on the cross-reactivity of dothiepin (4) or its metabolites, and the aim of the present study was therefore to determine whether the TD_x assay for tricyclic antidepressants could be used to quantify the concentration of dothiepin in plasma samples from both overdose patients and from patients taking therapeutic doses of dothiepin. In our study we have measured the cross-reactivity of dothiepin and its major metabolites, nordothiepin, dothiepin-S-oxide, and nordothiepin-S-oxide, in the TDM_x assay. In addition, plasma tricyclic antidepressant concentrations were measured by the TD_x method and by a specific high-performance liquid chromatographic method in 140 patients receiving therapeutic doses of dothiepin and in five patients who had overdosed on dothiepin.

II. STUDY DESIGN AND METHODS

A. Cross-Reactivity of Dothiepin and Its Major Metabolites in the Abbott TD_x Assay

Various concentrations of dothiepin (102 to 971 μg/L), nordothiepin (85 to 814 μg/L), dothiepin-S-oxide (91 to 870 μg/L), and nordothiepin-S-oxide (90 to 865 μg/L) were added to blank plasma. Assays were performed in quadruplicate using the standard Abbott TD_x procedure (5) on tricyclic antidepressant reagent kit (lot number 356868V). A blank plasma sample was inserted between each test sample to eliminate any possibility of carryover. The cross-reactivity was expressed as a percentage of the response to the calibrator drug, imipramine.

B. Comparison of the Abbott TD_x Tricyclic Antidepressant Assay with a Specific High-Performance Liquid Chromatography (HPLC) Assay for Dothiepin and Its Metabolites

Plasma samples from 140 patients taking therapeutic doses of dothiepin and multiple plasma samples from five patients who had overdosed on dothiepin were assayed by both Abbott TD_x and HPLC. The TD_x assay was carried out as specified in the company literature (5) and the results expressed relative to imipramine.

The HPLC analytical techniques used were as follows:

1. Dothiepin and Nordothiepin

One milliliter of plasma was mixed with 0.2 mL of 1 M NaOH and 85 ng N-desmethyl-doxepin (internal standard) and extracted with 10 mL of hexane containing 1% isoamyl alcohol. After centrifugation, 9 mL of the organic phase were removed to a clean tube and extracted with 0.2 mL of 0.05 M HCl. Aliquots (0.08 mL) of the HCl extract were injected onto the HPLC. The assay used a Waters Associates, μ-Bondapak phenyl column (30 cm × 4 mm I.D.) with a solvent of 40% CH_3CN in 0.01% H_3PO_4 and 0.01% NaCl. The flow rate was 1.5 mL/min, with detection by ultraviolet absorption at 230 nm. Results were interpolated from a standard curve extracted as above with each batch of unknown samples. The coefficients of variation for dothiepin at 25 and 250 μgL were 4.9 and 1.3%, respectively (n = 5) and for nordothiepin at 25 and 250 μg/L were 1.2% and 1.7%, respectively (n = 5).

2. Dothiepin-S-Oxide and Nordothiepin-S-Oxide

One milliliter of plasma was mixed with 0.2 mL of 1 M NaOH and 2000 ng of disopyramide (internal standard) and extracted with 10 mL of ethyl acetate containing 2% isoamyl alcohol. After centrifugation, 9 mL of the organic phase were evaporated to dryness, redissolved in 9 mL of hexane containing 1% isopropyl alcohol, and extracted with 0.2 mL of 0.05 M HCl. Aliquots (0.025 mL) of the HCl extract were injected onto the HPLC. The assay used a Waters Associates, μ-Bondapak Phenyl column (30 cm × 4 mm I.D.) with a solvent of 18% CH_3CN in 0.01% H_3PO_4 and 0.01% NaCl. The flow rate was 1.5 mL/min, with detection by UV absorption at 210 nm. Results were interpolated from a standard curve extracted as above with each batch of unknown samples. The coefficients of variation for dothiepin-s-oxide at 50 and 500 μg/L were 1.3% and 2.5% (n = 5), respectively, and for nordothiepin-S-oxide at 50 and 500 μg/L were 1.0% and 2.5%, respectively (n = 5).

III. RESULTS

The Abbott TDx tricyclic assay had significant cross-reactivity with both dothiepin and nordothiepin (Figure 1). The mean cross-reactivity for four spiked samples of dothiepin varied from 101% at 102 µg/L to 83% at 971 µg/L, while that for nordothiepin varied from 110% at 85 µg/L to 78% at 814 µg/L. The cross-reactivity for dothiepin-S-oxide and nordothiepin-S-oxide was less than 1.7% at concentrations as high as 870 µg/L, indicating that the contribution of the S-oxides to the total TDx tricyclic assay result would be very small even at plasma concentrations occurring after overdoses of dothiepin.

The Abbott TDx assay data for plasma samples from patients receiving therapeutic doses of dothiepin correlated significantly only with dothiepin and nordothiepin concentrations measured by HPLC (Figures 2A-2D). The correlations for dothiepin and nordothiepin were high (r^2 = 0.759 and 0.796, respectively), while those for dothiepin-S-oxide and nordothiepin-S-oxide were poor (r^2 = 0.12 and 0.262, respectively). For practical purposes, the TDx levels were best represented by a combination of dothiepin and nordothiepin levels measured by HPLC [Figure 3A; TDx = 31.23 + 1.128 (dothiepin + nordothiepin); $r^2 = 0.876$, $p < 0.001$].

The characteristics of the overdose patients are shown in Table 1. Individual plasma profiles for the TDx tricyclic level and the HPLC assay of dothiepin and its metabolites for each patient are shown in Figure 4. When TDx and HPLC measurements were compared for the overdose cases (17 plasma samples taken from five patients), good correlations were obtained between TDx and dothiepin ($r^2 = 0.895$, $p < 0.001$) or nordothiepin ($r^2 = 0.47$, $p = 0.002$) alone as well as between TDx and the sum of dothiepin and nordothiepin [Figure 3B; TDx = 235.3 + 0.603 (dothiepin + nordothiepin); $r^2 = 0.929$, $p < 0.001$].

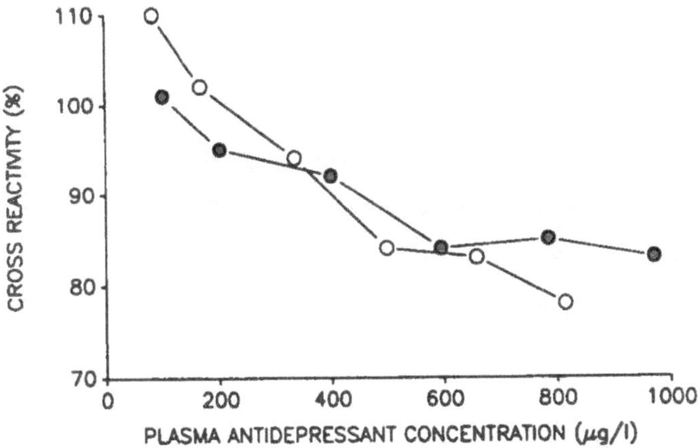

Figure 1 Cross-reactivity (percent of response to calibrator drug imipramine) of dothiepin (●—●) and nordothiepin (○—○) in the TDx tricyclic antidepressant assay. Each point is the mean of quadruplicate determinations. Standard deviations were less than 3.4% for all points.

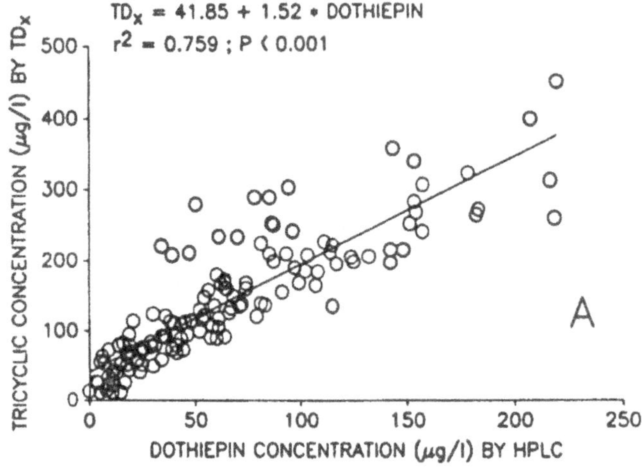

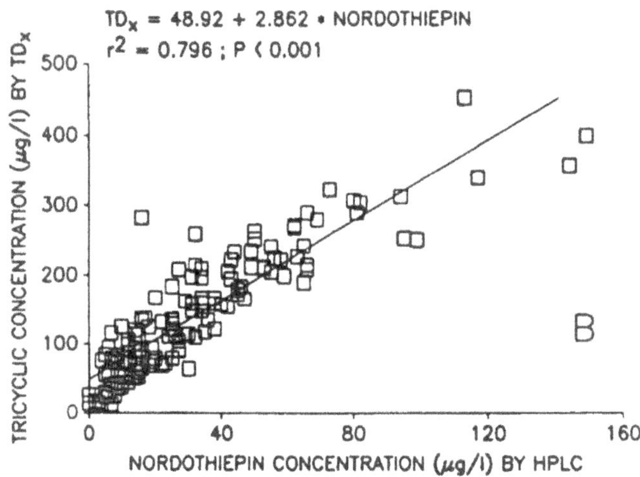

Figure 2 Correlation between TD_x assay result and HPLC assay result for (A) dothiepin, (B) nordothiepin, (C) dothiepin-S-oxide, and (D) nordothiepin-S-oxide in plasma from 140 patients receiving therapeutic doses of dothiepin.

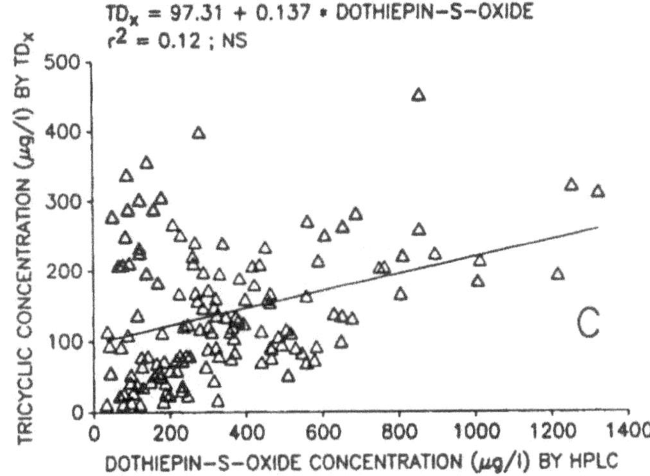
TD_x = 97.31 + 0.137 • DOTHIEPIN-S-OXIDE
r^2 = 0.12 ; NS
TRICYCLIC CONCENTRATION (μg/l) BY TD_x
0
100
200
300
400
500
0
200
400
600
800
1000
1200
1400
DOTHIEPIN-S-OXIDE CONCENTRATION (μg/l) BY HPLC
C

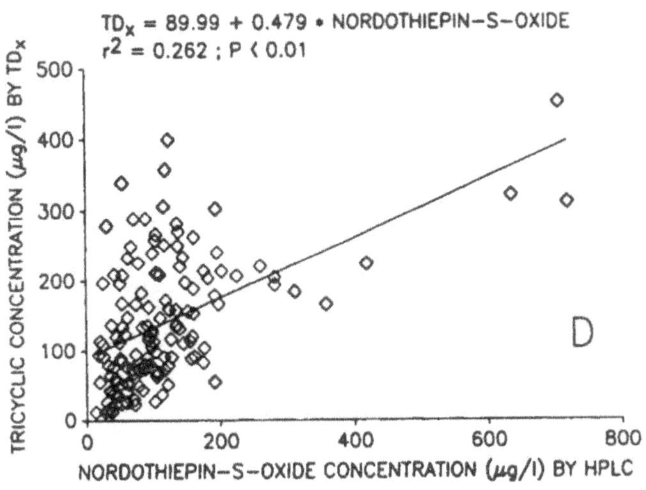
TD_x = 89.99 + 0.479 • NORDOTHIEPIN-S-OXIDE
r^2 = 0.262 ; P < 0.01
TRICYCLIC CONCENTRATION (μg/l) BY TD_x
0
100
200
300
400
500
0
200
400
600
800
NORDOTHIEPIN-S-OXIDE CONCENTRATION (μg/l) BY HPLC
D

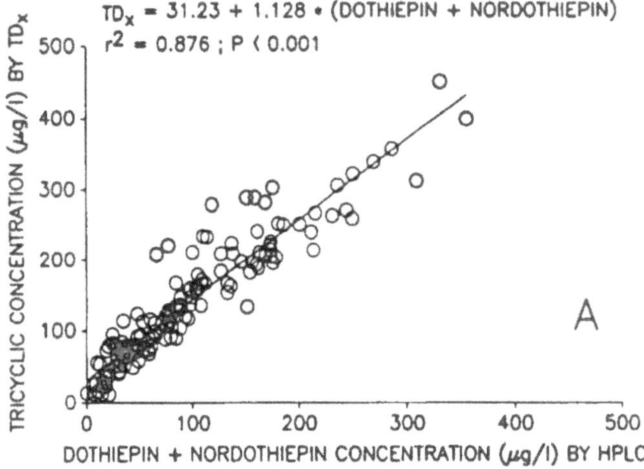

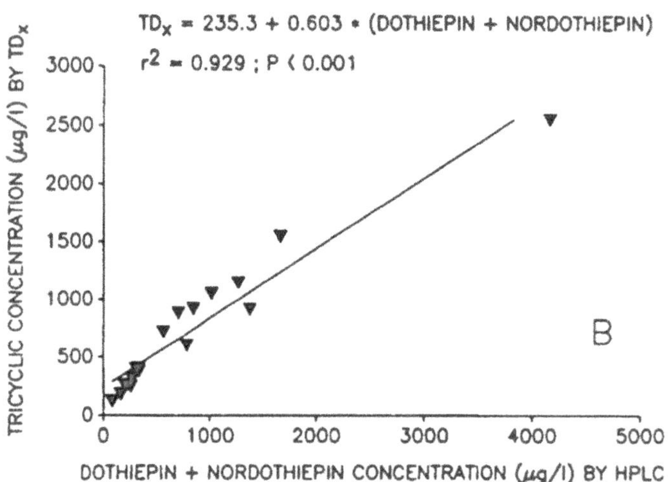

Figure 3 Correlation between TD_x assay result and HPLC assay result for (A) the sum of dothiepin and nordothiepin in single plasma sample from 140 patients receiving therapeutic doses of dothiepin and (B) multiple plasma samples from five dothiepin overdose patients.

Table 1 Characteristics of Overdose Patients

Patient No.	Age (yr)	Sex	Amount of Dothiepin Taken (g)	Other Drugs Taken Concomitantly
1	25	F	4.5	Alcohol
2	20	F	1.0	None
3	26	F	1.25	Alcohol
4	30	M	4.5	Heroin
5	1.75	M	Unknown	None

IV. DISCUSSION

The TD_x tricyclic assay had high cross-reactivity with both dothiepin and nordothiepin but was virtually insensitive to their S-oxide metabolites. Cross-reactivity for dothiepin and nordothiepin was concentration-dependent and ranged from 101% to 110% at low concentrations (< 100 μg/L) to 78% to 83% at high concentrations (800 to 1000 μg/L). Both TD_x and HPLC methods were used to test plasma samples from patients taking therapeutic doses of dothiepin and from patients who had overdosed on the drug. The results for both groups showed good correlations between the TD_x level and the level of dothiepin or nordothiepin measured by HPLC. In practical terms, the most useful correlation was that between TD_x and the sum of the HPLC concentrations for dothiepin and nordothiepin. This correlation explained 87.6% and 92.9% of the variance in the data for therapeutic and overdose patients, respectively (Figures 3A and 3B). In both of the latter plots there was a positive intercept on the y axis. This was greater in the overdose patients (235.3 μg/l) than in the patients receiving therapeutic doses of dothiepin (31.2 μg/L). The higher intercept (and smaller slope) in the overdose group undoubtedly results from their much higher plasma drug concentrations (Figures 4A–4D) compared to those seen after therapeutic doses (Figures 2A–2D) and the fact that the TD_x assay has a lower cross-reactivity (approximately 80%) at high dothiepin and nordothiepin plasma concentrations.

The mean concentration of the sum of dothiepin and nordothiepin for the overdose cases was 840 μg/L for the HPLC assay compared with 742 μg/L for the TD_x assay. Adjusting for cross-reactivity (approximately 85%) resulted in a TD_x concentration of 873 μg/L. The mean concentration for the sum of dothiepin and nordothiepin in therapeutic cases was 96 μg/dL compared with the 140 μg/L by the TD_x assay. Adjusting for cross-reactivity (approximately 105%) resulted in a TD_x concentration of 133 μg/L.

We conclude that the Abbott TD_x assay can be used to give a reliable indication of the combined plasma concentration of the active antidepressants dothiepin and nordothiepin following either therapeutic doses or overdoses of dothiepin. Our particular interest has been in evaluating the TD_x assay for assessment of overdose cases, and in this regard the ease and speed with which results can be obtained is particularly helpful. The lower cross-reactivity of the antibodies used in the TD_x assay should always be borne in mind when interpreting high tricyclic results such as those occurring after an overdose.

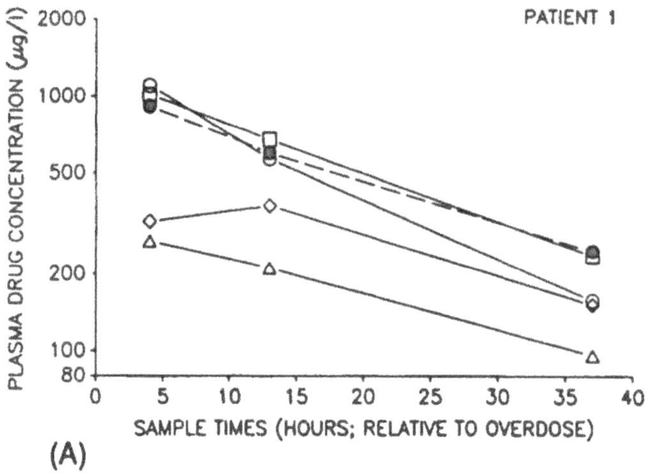

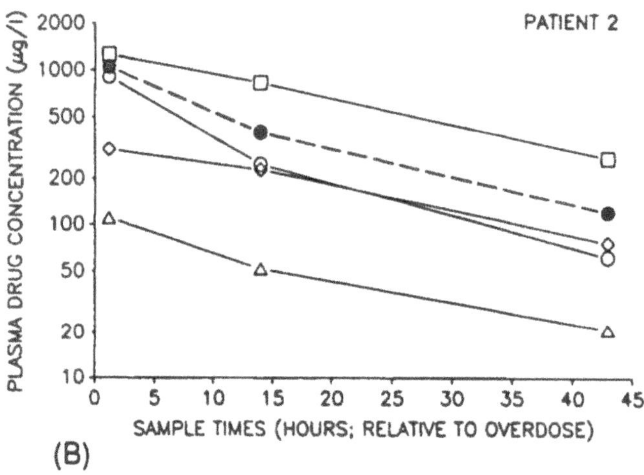

Figure 4 Plasma tricyclic antidepressant concentration-versus-time profiles in five patients who had overdosed on dothiepin. Samples were assayed by the TD_x method for total tricyclic antidepressants (●—●) and by HPLC for dothiepin (○—○), nordothiepin (Δ—Δ), dothiepin-S-oxide (□—□), and nordothiepin-S-oxide (◇—◇).

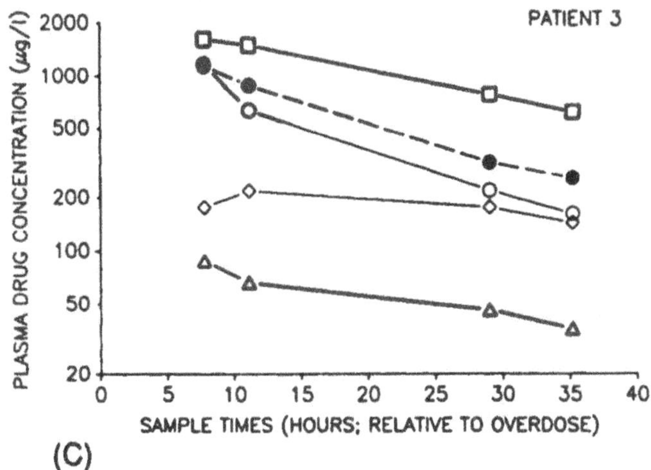
PATIENT 3
PLASMA DRUG CONCENTRATION (μg/l)
2000
1000
500
200
100
50
20
0
5
10
15
20
25
30
35
40
SAMPLE TIMES (HOURS; RELATIVE TO OVERDOSE)

(C)

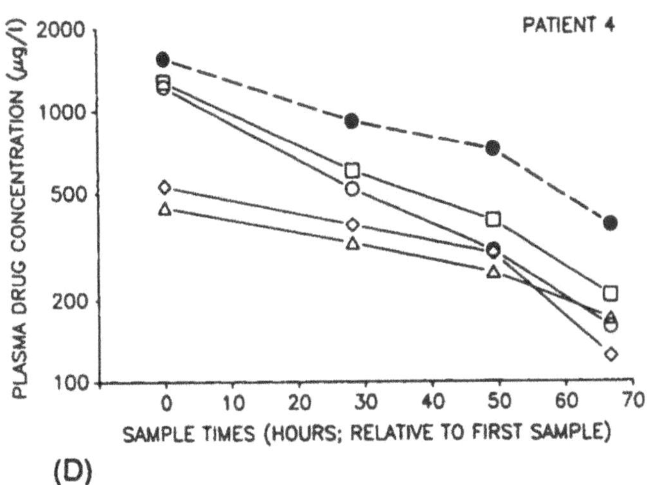
PATIENT 4
PLASMA DRUG CONCENTRATION (μg/l)
2000
1000
500
200
100
0
10
20
30
40
50
60
70
SAMPLE TIMES (HOURS; RELATIVE TO FIRST SAMPLE)

(D)

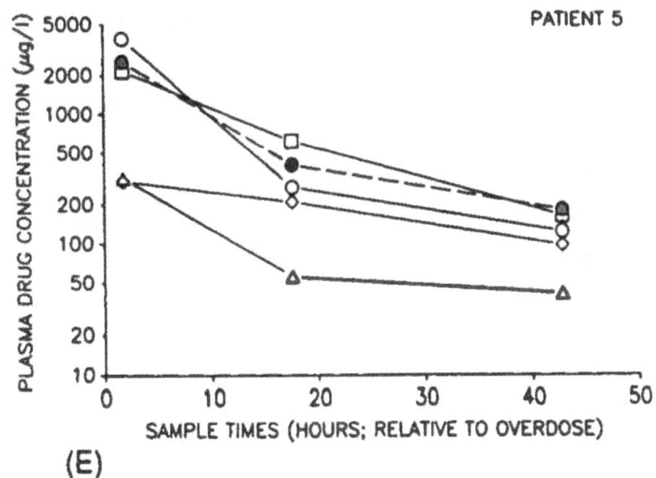
PATIENT 5
PLASMA DRUG CONCENTRATION (μg/l)
5000
2000
1000
500
200
100
50
20
10
0
10
20
30
40
50
SAMPLE TIMES (HOURS; RELATIVE TO OVERDOSE)

(E)

REFERENCES

1. D. G. Spiker, A. N. Weiss, S. S. Chang, J. F. Ruwitch, Jr., and J. T. Biggs. Tricyclic antidepressant overdose. Clinical presentation and plasma levels. *Clin. Pharmacol. Ther., 18*:539 (1975).
2. J. M. Petit, D. G. Spiker, J. F. Ruwitch, V. E. Ziegler, A. N. Weiss, and J. T. Biggs. Tricyclic antidepressant plasma levels and adverse effects after overdose. *Clin. Pharmacol. Ther., 21*:47 (1977).
3. G. D. Burrows, S. Vohra, D. Hunt, J. G. Sloman, B. A. Scoggins, and B. Davies. Cardiac effects of different tricyclic antidepressant drugs. *Br. J. Psychiatr., 129*:335 (1976).
4. P. Nebinger and M. Koel. Specific data of tricyclic antidepressants assay by fluorescent polarization immunoassay. *J. Anal. Toxicol., 14*:219 (1990).
5. Abbott Laboratories, U.S.A. *TD_x System Assays Manual, Toxicology Assays–Tricyclic Antidepressants*, Abbott Laboratories, North Chicago, pp. 22.45–22.59 (1985).

47

Determination of Fluoxetine and Norfluoxetine in Serum by Liquid Chromatography with Fluorescence Detection

Ram N. Gupta *St. Joseph's Hospital, Hamilton, Ontario, Canada*

I. INTRODUCTION

Fluoxetine (Flu) is a new bicyclic antidepressant drug which enhances serotoninergic neurotransmission through potent and selective inhibition of neuronal reuptake of serotonin (1). Norfluoxetine (N-Flu), formed in humans by N-demethylation, also inhibits serotonin reuptake. Preliminary clinical trials have shown that Flu is similar in therapeutic efficacy to traditional tricyclic antidepressants but has significantly fewer side effects. No significant correlation has been observed between serum fluoxetine concentrations and therapeutic response with the increased use of the drug (2). However, the demand for serum Flu assay to monitor compliance or unexpected toxic concentrations after chronic use of Flu is increasing.

Fluoxetines, at therapeutic concentrations, have been determined in plasma, by gas chromatography with electron capture detection (3,4), by flame ionization, and by nitrogen-selective detection at toxic concentrations (5,6). There is an increasing trend toward using liquid chromatography (LC) for monitoring therapeutic concentrations of antidepressants (7) and LC procedures for the determination of fluoxetines have been described (2,8,9). The purpose of this study was to select conditions which may further improve the specificity and sensitivity for the LC determination of fluoxetines.

II. EXPERIMENTAL

Stock fluoxetine solution (1 mg/mL) was prepared by dissolving 11.2 mg of fluoxetine hydrochloride and 14 mg of norfluoxetine maleate in 10 mL of methanol. A serum standard of 2 mg/L was prepared by diluting 50 μL of the stock solution with 25 mL of drug-free pooled serum. This standard was serially diluted to a low standard of 31.25 μg/L.

Bond Elut 1-mL C-18 columns were washed with one full column volume of 1 N HCl and two volumes of methanol. A 0.25-mL aliquot of the working IS solution, 0.25 mL of 0.1 N sodium carbonate solution, and 0.5 mL of the sample (standard or unknown), were applied to the washed columns. The liquid was sucked through the column at a slow rate (<1 mL/min). The column was then washed with two volumes of water, one volume of 50% aqueous methanol, and one volume of acetonitrile. The columns were placed on correspondingly labeled 12 × 75 glass tubes washed with methanol and eluted with 0.25 mL of a mixture of 0.1 N perchloric acid and acetonitrile (1:3) by centrifugation. The eluate was vortex-mixed, and an aliquot of 10 μL of the eluate was injected into the liquid chromatograph.

The HPLC procedure used a 15 cm × 4.6 mm (I.D.) reversed-phase Ultrasphere ODS column packed with 5-μm particles (Beckman). This analytical column was protected by a 1.5-cm RP-18 guard cartridge (Brownlee). The column was maintained at ambient temperature. The mobile phase, consisting of acetonitrile (400 mL) + water (600 mL) + tetramethylammonium perchlorate (1.5 g) + 70% perchloric acid (0.1 mL), was pumped at a flow rate of 1.6 mL/min. The sample was injected with a Rheodyne model 7125 syringe loading valve injector equipped with a 20-μL loop. The peaks were detected with a Shimadzu Model RF 535 fluorescence detector (e_x = 235 nm, e_m = 310 nm). The peaks were recorded and peak areas determined with a Shimadzu model C-R6A integrator plotter.

III. RESULTS AND DISCUSSION

Fluoxetines have been extracted from serum by a four-step liquid-liquid extraction in a number of procedures (3,8). In the most recent report (9), the extraction procedure has been simplified by avoiding the second extraction into organic phase to obviate the need for evaporation. However, liquid-liquid extraction is still tedious, as it requires two sets of extraction tubes, and it is slow due to the mixing and centrifugation steps. The recovery of fluoxetines by liquid-liquid extraction is relatively low and variable due to their variable adsorption on glass surface. There is an increasing trend toward using solid-phase extraction columns to isolate drugs from biological matrices. A number of procedures for the solid-phase extraction of antidepressant drugs have been described (10,11). Solid-phase extraction of fluoxetines described in this report proved highly convenient and efficient. Fluoxetines and the internal standard are recovered in more than 90% yield. Aliquots of 25 μL of the eluate can be injected directly without further dilution. There is no change in the ratio of the areas of Flu/IS or N-Flu/IS when aliquots of an aqueous mixture of the three compounds are injected before and after extraction with the described procedure.

The described extraction procedure is highly specific, as a large number of acidic, neutral, and weakly basic foreign and endogenous compounds are removed during washes with 50% methanol and acetonitrile. However, strongly basic compounds, including commonly prescribed tricyclic antidepressants, are co-extracted with fluoxetines with equal efficiency. Drugs such as amitriptylines and imipramines interfere with the LC determination of fluoxetines using UV absorbance detection. Separation of these drugs from fluoxetines by changing chromatographic parameters makes the chromatographic run time too long.

In an attempt to find a selective mode of detection for fluoxetines, native fluorescence of fluoxetines was examined. It was observed that both Flu and N-Flu showed intense fluorescence at e_x = 235 nm and e_m = 310 nm. The areas of fluorescence peaks of fluoxetines are approximately 30 times the areas of absorbance peaks at 227 nm. The fluorescence detector used in this study has a weak excitation energy at 235 nm. As a result, the baseline of the fluorescence detector was quite noisy and required plotter attenuation of 5 for a smooth baseline, whereas the absorbance detector (Shimadzu Model SPD-6AV) gave a smooth baseline at the plotter attenuation of 2. The heights of the fluorescence peaks of fluoxetines are still approximately four times than those of the absorbance peaks when the two detectors are connected in series. Protriptyline, maprotiline, and desmethylmaprotiline are the only compounds among the traditional antidepressants which show fluorescence response under the conditions selected for the detection of fluoxetines. Protriptyline was selected as the internal standard, as it is a rarely prescribed drug and shows extraction and detection properties similar to those of fluoxetines.

A variety of reversed-phase silica columns, e.g., cyano (2), phenyl (8), and C-18 (9), have been used for the determination of fluoxetines. In the initial experiments an Ultrasphere C-8 silica column was used. However, C-18 silica columns are being used increasingly for drug analysis. C-8 and C-18 columns provide similar separation and peak shape using the same mobile phase but with a slight adjustment of flow rate and the percentage of organic modifier. There was an increase in the peak height and a 50% saving in the consumption of the mobile phase when a 10 cm × 2 mm (I.D.) C-8 silica column packed with 3-μm particles (Bio-Rad) or a 15 cm × 2 mm (I.D.) C-8 column packed with 5-μm particles (Beckman) was used for the determination of fluoxetines. However, reproducibility of peak areas with the use of narrow-bore columns was poor as compared to that shown by 4.6-nm (I.D.) C-8 or C-18 silica columns packed with 5-μm particles.

Chromatography is performed at ambient temperature because of the use of the fluorescence detector. However, with the use of the described mobile phase and absorbance detection, there was no improvement in peak shape or separation when the temperature of the C-8 or C-18 column was raised to 55°C in gradual increments.

The described procedure is linear from 30 μg/L to 2000 μg/L of each fluoxetine. Figure 1A shows a chromatogram of an extract of a serum standard spiked with 500 μg/L of each fluoxetine. The peaks are sharp and well separated from the solvent peaks and from each other. Figure 1B shows a chromatogram of a patient receiving 40 mg of fluoxetine per day for 8 weeks. Figure 1C is the chromatogram of the extract of TheraChem High therapeutic drug control (Fisher). This bovine serum-based material has been spiked with acetaminophen (160 mg/L), amikacin (32 mg/L), amitriptyline (0.5 mg/L), carbamazepine (15 mg/L), chloramphenicol (25 mg/L), desipramine (0.5 mg/L), digoxin (4 μg/L), disopyramide (5 mg/L), ethosuximide (120 mg/L), gentamicin (10 mg/L), imipramine (0.5 mg/L), lidocaine (8 mg/L), lithium (2 meq/L), methotrexate (0.7 μmol/L), N-acetylprocainamide (14 mg/L), netilmicin (16 mg/L), nortriptyline (0.5 mg/L), phenobarbital (60 mg/L), phenytoin (30 mg/L), primidone (16 mg/L), procainamide (12 mg/L), quinidine (6 mg/L), salicylate (360 mg/L), theophylline (30 mg/L), tobramycin (8 mg/L), valproic acid (110 mg/L), and vancomycin (40 mg/L). This specimen gave only one additional peak which is well separated from the desired peaks of the fluoxetines. Selective extraction

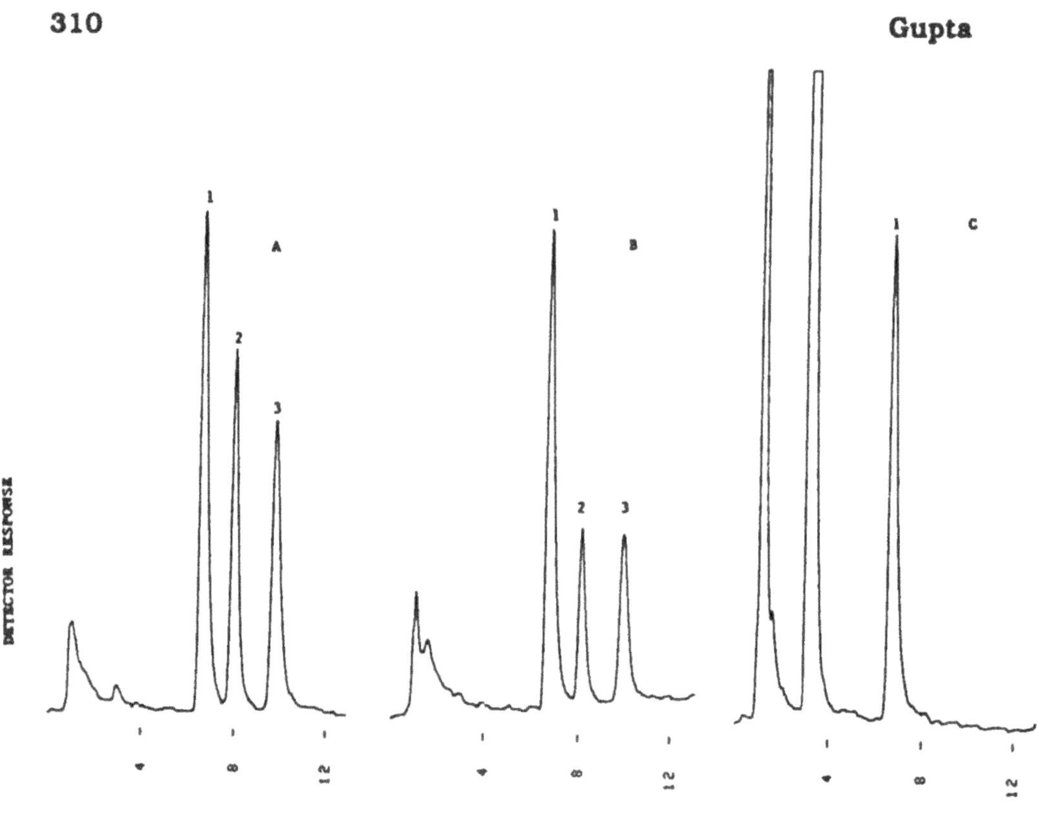

Figure 1 Chromatograms of 10-μL injections of extracts of (A) serum standard of 500 μg/L; (B) serum of a patient receiving fluoxetine (norfluoxetine = 245 μg/L, fluoxetine = 310 μg/L); (C) TheraChem TDM control. Fluorescence detection: e_x = 235 nm, e_m = 310 nm; sensitivity, high; response, slow; detector output, 1 V. Plotter attenuation, 5; chart speed, 5 mm/min. Peak 1, protriptyline; peak 2, norfluoxetine; peak 3, fluoxetine.

and fluorescence detection ensures high specificity. Between batch precision for both fluoxetines is less than 7% for both high (1000 μg/L) and low (125 μg/L) concentrations.

In conclusion, the described procedure is quite simple and suitable for use in clinical laboratories. The procedure allows the detection of 25 μg/L of either of the fluoxetines using only 0.5 mL of the sample. Aliquots of 20 μL of the extract are injected for the determination of lower concentrations.

REFERENCES

1. P. Benfield, R. C. Heel, and S. P. Lewis. Fluoxetine: A review of its pharmacodynamic and pharmacokinetic properties, and therapeutic efficacy in depressive illness. *Drugs*, 32:481 (1986).
2. M. W. Kelly, P. J. Perry, S. G. Holstad, and M. J. Garvey. Serum fluoxetine and norfluoxetine concentrations and antidepressant response. *Ther. Drug Monitor.*, *11*:165 (1989).

3. J. F. Nash, R. J. Bopp, R. H. Carmichael, K. Z. Farid, and L. Lemberger. Determination of fluoxetine and norfluoxetine in plasma by gas chromatography with electron-capture detection. *Clin. Chem.*, *28*:2100 (1982).
4. S. Caccia, M. Cappi, C. Fracasso, and Garatlini. Influence of dose and route of administration on the kinetics of fluoxetine and its metabolite norfluoxetine in the rat. *Psychopharmacology*, *100*:509 (1990).
5. T. P. Rohrig and R. W. Prouty. Fluoxetine overdose: A case report. *J. Anal. Toxicol.*, *13*:305 (1989).
6. J. R. Roettger. The importance of blood collection site for the determination of basic drugs: A case with fluoxetine and diphenhydramine overdose. *J. Anal. Toxicol.*, *14*:191 (1990).
7. S. H. Y. Wong. Measurement of antidepressants by liquid chromatography: A review of current methodology. *Clin. Chem.*, *34*:848 (1988).
8. P. J. Orsulak, J. T. Kenny, J. R. Debus, G. Crowley, and P. D. Wittman. Determination of the antidepressant fluoxetine and its metabolite norfluoxetine in serum by reversed-phase HPLC, with ultraviolet detection. *Clin. Chem.*, *34*:1875 (1988).
9. S. H. Y. Wong, S. S. Dellafera, R. Fernandes, and H. Kranzler. Determination of fluoxetine and norfluoxetine by high performance liquid chromatography. *J. Chromatogr.*, *499*:601 (1990).
10. G. Carfagnini, A. D. Corcia, M. Marchetti, and R. Samperi. Antidepressants in serum determined by isolation with two on-line sorbent cartridges and liquid chromatography. *J. Chromatogr.*, *530*:359 (1990).
11. M. Mazhar and S. R. Binder. Analysis of benzodiazepines and tricyclic antidepressants in serum using a common solid-phase clean-up and a common mobile phase. *J. Chromatogr.*, *497*:201 (1989).

48

HPLC Analysis of Gabapentin: A New Anticonvulsant Drug

Albert D. Fraser and Wallace MacNeil *Dalhousie University, Halifax, Nova Scotia, Canada*

I. INTRODUCTION

Gabapentin [1-(aminomethyl) cyclohexane acetic acid], GP, is a novel amino acid (Figure 1) designed to mimic the steric conformation of the inhibitory neurotransmitter γ-aminobutyric acid (GABA). It has been administered to patients with spasticity, migraine, and epilepsy in clinical research studies. The drug is currently undergoing active investigation for the treatment of symptomatic generalized epilepsy in children.

GP is structurally unlike any currently available anticonvulsant agent. It is eliminated primarily in the urine as unchanged drug. GP does not stimulate or inhibit the mixed function oxidase system or bind to any plasma protein. A recent review on GP summarizes its pharmacodynamics and pharmacokinetics in humans and animals (1).

A HPLC method for quantitation of GP in plasma has been published (2). The objective of this study was to optimize the extraction, purification, and chromatographic analysis of GP prior to analysis of clinical specimens. The method developed is based on precolumn derivatization of GP after extraction, purification by acetic acid precipitation of derivatized GP, and quantitation by reversed phase HPLC with detection at 340 nm.

The method was also validated by analysis of blind plasma specimens obtained from Parke-Davis Pharmaceutical Research Division.

II. METHODS AND MATERIALS

A. Reagents and Standards

GP and the internal standard 1-(aminomethyl) cycloheptane acetic acid were obtained from Parke Davis Pharmaceutical Research Division, Ann Arbor, Michigan.

Sodium tetraborate and acetone were obtained from Mallinckrodt Chemicals. Acetonitrile (HPLC grade) was purchased from Calendon Laboratories. The

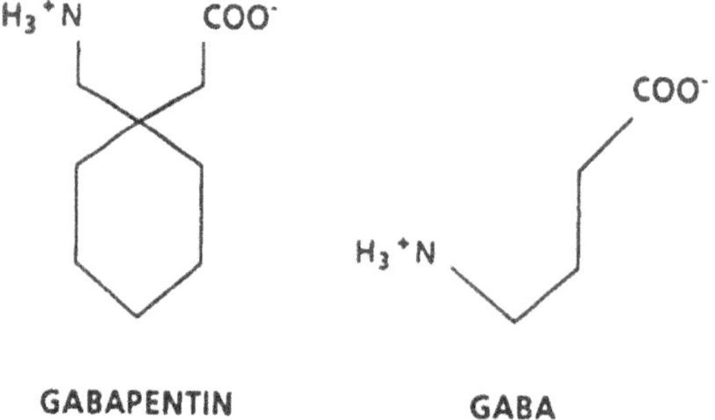

Figure 1 Chemical structure of gabapentin and γ-aminobutyric acid (GABA).

derivatizing agent 2,4,6-trinitrobenzene sulfonic acid (TNBSA) was obtained from Pierce Chemical Co. Two hundred and fifty milligrams of TNBSA was dissolved in 25 mL of distilled water.

A stock standard of GP was prepared by weighing 100 mg of GP and dissolving in 100 mL of distilled water. Working standards were prepared in drug-free bovine serum to give final concentrations ranging from 200 to 1000 ng/mL. A wider range of standards was prepared to determine the linearity of the method.

The internal standard was prepared by dissolving 10 mg of 1-(aminomethyl) cycloheptane acetic acid in 10 mL of acetone. This solution was further diluted with acetone to give a final concentration of 2000 ng/mL.

Sodium tetraborate (0.05 M) was prepared by weighing out 0.48 g of the powder and dissolving in 25 mL of distilled water.

B. Analytical Method

A 0.5-mL aliquot of serum or standard is placed in a 1.5-mL microcentrifuge tube and serum proteins are precipitated by addition of 0.5 mL of acetone containing the internal standard. After mixing and centrifugation for 5 min, a 0.5-mL aliquot of the supernatant is taken off and evaporated under N_2 in another tube at 50°C in a heating block. The residue is then dissolved in 50 μL of 1% TNBSA and 50 μL of 0.05 M sodium tetraborate. After mixing, the solution is allowed to stand at room temperature for 1 h for derivatization to occur.

GP and the internal standard are purified by addition of 20 μL of 5 M acetic acid prior to mixing and centrifugation. The supernatant is decanted to waste and the precipitate (containing GP and the internal standard) is dissolved in the mobile phase. The mixture was centrifuged again prior to injection of the supernatant onto the HPLC.

C. HPLC Analysis

Separation and quantitation of GP was performed on a Spectra Physics SP 8770 isocratic pump/controller, SP 8780 autosampler, and eluant monitoring at 340 nm on a Hewlett Packard Series 1050 variable-wavelength UV detector.

Analysis was performed on a 250 × 4.6-mm-I.D. RP-8 column obtained from Brownlee Labs (5 μm particle size).

All separations were performed using an isocratic mobile phase of acetonitrile/acetate buffer (5 M). The mobile phase was prepared by measuring 500 mL of acetonitrile, 492 mL of distilled water, and 8 mL of acetate buffer (pH 4.6). Thirty to fifty microliters of each standard or unknown were injected onto the column. Approximate retention times for GP and the internal standard were 8 and 10 min, respectively.

III. RESULTS

The method was linear from 50 to 5000 ng/mL, although the routine standard curve ranged from 200 to 2000 ng/mL. The limit of detection was 50 mg/mL based on a signal/noise ratio of 3:1 at that concentration.

The reproducibility of the method was evaluated by analyzing serum-based controls at 250 and 750 ng/mL on 10 separate occasions. The coefficients of variation were 4.2% and 5.5%, respectively.

Recovery was assessed by comparing peak heights of aqueous GP standards following derivatization relative to serum standards taken through the entire procedure. Over a concentration range of 200 to 2000 ng/mL, recovery ranged from 90% to 100%.

The method was also validated by analysis of 18 blind serum standards provided by Parke Davis Pharmaceutical Company. For the 15 specimens where quantitative results were compared, the correlation coefficient was 0.991, the slope was 1.027, the mean value for this laboratory was 2646 ng/mL, and the mean reference value was 2657 ng/mL. The individual results obtained for these specimens ar given in Table 1.

IV. DISCUSSION

The quantitation of low amounts of gabapentin in serum requires derivatization with a UV-absorbing chromophore. TNBSA (Figure 2) has been used previously for derivatization of GP (3) and other amino acids (2). Certain batches of TNBSA used in this study contained impurities which eluted near the GP peak. These impurities (unidentified) were removed by extracting the aqueous TNBSA solutions with chloroform.

Table 1 Results of Blind Specimen Analysis for Gabapentin

Specimen Number	VGH (ng/mL)	Reference (ng/mL)
1–3	<200	95
4–6	500	499
7–9	700 × 1	736
	800 × 2	736
10–12	2000	2010
13–15	2100	2180
16–18	7700 × 2	7860
	8200	7860

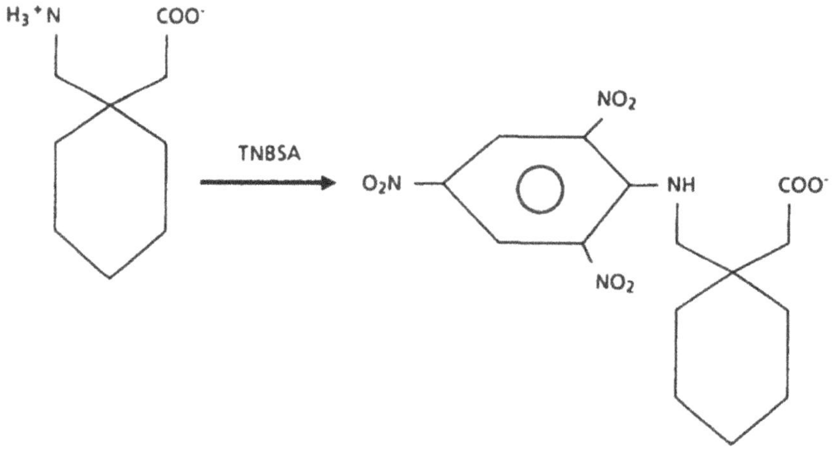

Figure 2 Derivatization of gabapentin with TNBSA.

The derivatization of GP by TNBSA is not specific. All primary serum amino acids and other endogenous amino compounds are also derivatized. The acetic acid precipitation-purification step removed a lot of these derivatized compounds. Chromatograms of blank serum and a GP serum-based standard without acetic acid treatment are found in Figures 3 and 5. By comparison, in Figures 4 and 6 one notes a much cleaner chromatogram in the first 4 min after injection and a smooth baseline by 8 to 10 min, where GP and the internal standard elute.

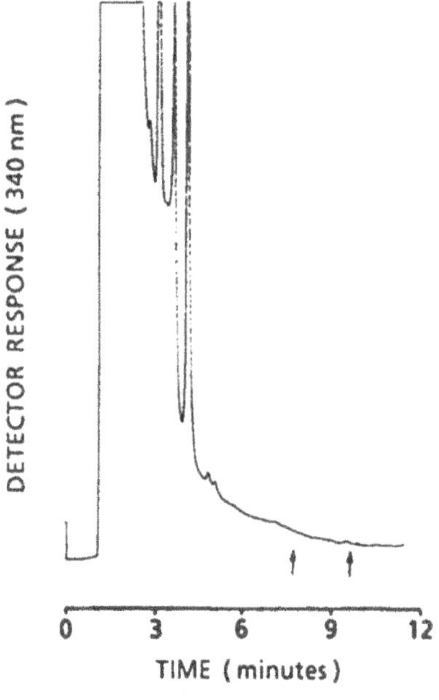

Figure 3 Chromatogram of a blank serum extract without acetic acid addition.

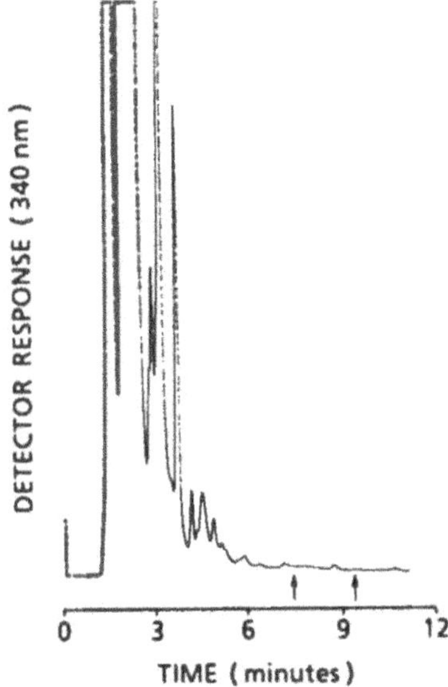

Figure 4 Chromatogram of a blank serum extract after acetic acid addition.

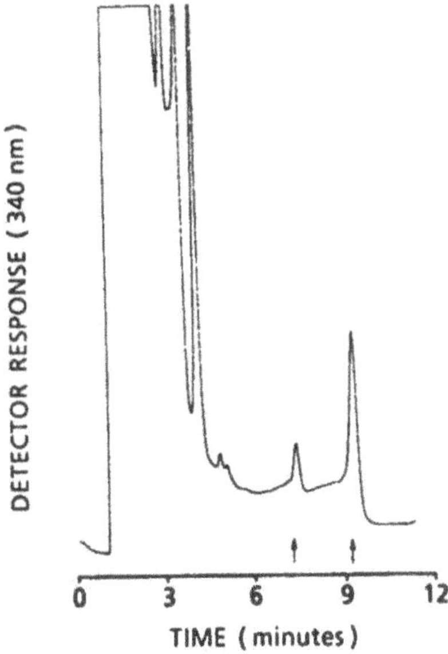

Figure 5 Chromatogram of a gabapentin serum standard (500 ng/mL) without acetic acid addition.

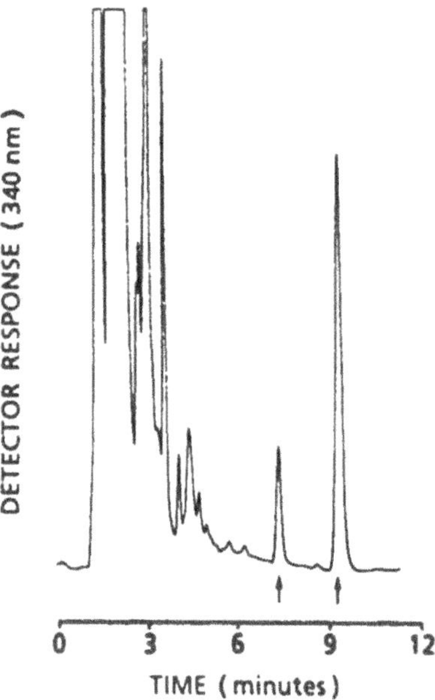

Figure 6 Chromatogram of a gabapentin serum standard (500 ng/mL) after acetic acid addition.

The pH and composition of the HPLC mobile phase were also investigated, since the previously published method (3) used an acidic mobile-phase pH. In Figures 7 and 8, one observes that at the very acidic pH of 3.4, more substances are retained on the column and a large interfering peak elutes just before GP. By increasing the mobile-phase pH to 5.7, one obtains a much cleaner chromatogram and a flat baseline when GP elutes at 8 min.

The wavelength used for eluant monitoring was not given in the previous GP method (3). The optimal wavelength for monitoring GP was determined using a diode array detector. The optimum wavelength was 340 nm, not 250 nm as previously determined by a UV scanning spectrophotometer where the blank solution contained TNBSA.

Precision was assessed by analyzing serum-based standards at 250 and 750 ng/mL. The coefficients of variation (both <6%) are acceptable for this type of analysis.

The results for the 18 blind plasma specimens provided by the pharmaceutical company are given in Table 1. The three specimens with a weighed-in concentration < 200 ng/mL were not included in the statistical evaluation, since our lowest calibrator was 200 ng/mL. The specimens reading greater than the highest calibrator were diluted prior to quantitation. There was excellent correlation between the weighed-in plasma standard values and the measured results obtained.

In conclusion, this modified HPLC method for GP following precolumn derivatization and purification by acetic acid precipitation provides improved chromatographic separation (relative to a previous HPLC method) and is precised and accurate over a wide concentration range.

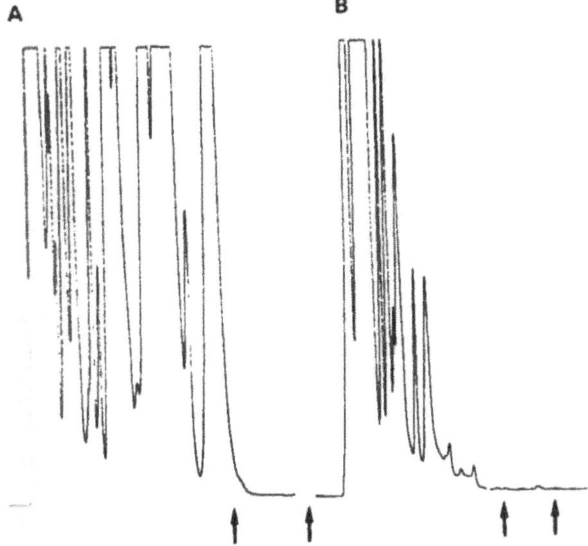

Figure 7 Effect of mobile phase pH on serum blank for gabapentin: A, pH = 3.4; B, pH = 5.7.

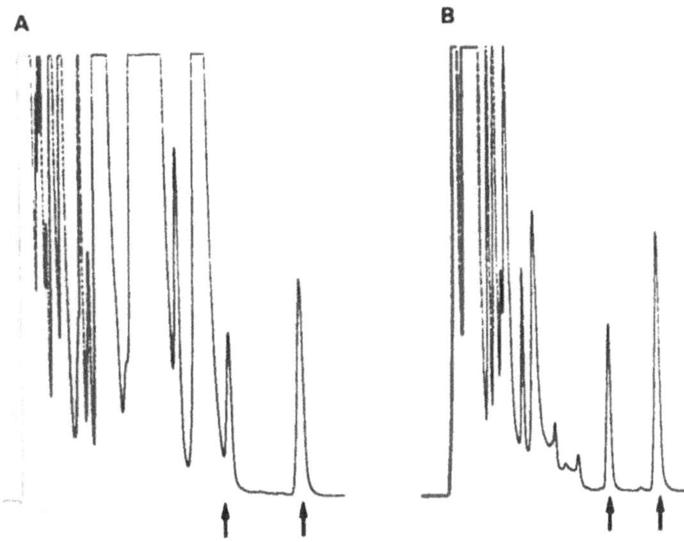

Figure 8 Effect of mobile-phase pH on separation of a 1000 ng/mL gabapentin serum standard: A, pH = 3.4; B, pH = 5.7.

This method will allow quantitation of GP in clinical and pharmacokinetic studies involving this drug.

To date, there is no definitive information available indicating whether therapeutic drug monitoring is beneficial for patients receiving this drug for seizure control.

V. SUMMARY

Gabapentin (GP) is a structural analog of the neurotransmitter γ-aminobutyric acid and is currently under investigation as a drug for the treatment of symptomatic generalized epilepsy. The objective of this study was to develop an HPLC method for the analysis of GP in serum. After addition of the internal standard [1-(aminomethyl) cycloheptane acetic acid], GP and the internal standard are extracted from 0.5 mL of serum into acetone. Following mixing, centrifugation, and evaporation of acetone, GP is derivatized with trinitrobenzene sulfonic acid. The derivative is purified by precipitation with acetic acid. The precipitate is dissolved in the mobile phase prior to injection onto a RP-8 (5-μm) HPLC column. The mobile phase (CH_3CN/H_2O/5 M) acetate buffer (500:492:8) is pumped at a flow rate of 1.6 mL/min with UV detection at 340 nm. Retention times are 8 and 10 min for GP and the internal standard, respectively. The standard curve is linear from 50 to 10,000 ng/mL. The method was also validated by analysis of 18 specimens provided by a pharmaceutical company.

In conclusion, this HPLC method provides reliable quantitation of GP in serum over a wide concentration range. The method is suitable for use in pharmacokinetic studies involving GP.

ACKNOWLEDGMENTS

Parke Davis Pharmaceutical Research Division, Ann Arbor, Michigan, provided pure standards of GP and the internal standard. The assistance of Lee Hayes, Research Coordinator, Pharmacokinetics/Drug Metabolism at Parke Davis, is acknowledged for providing the blind specimens for assay validation.

REFERENCES

1. B. Schmidt. Potential antiepileptic drugs. In *Antiepileptic Drugs*, Raven Press, New York, pp. 925–935 (1989).
2. W. L. Caudill, G. P. Houck, and R. M. Wightman. Determination of γ-aminobutyric acid by liquid chromatography with electrochemical detection. *J. Chromatogr.*, *227*:331–339 (1982).
3. H. Hengy and E-U. Kolle. Determination of gabapentin in plasma and urine by high performance liquid chromatography and precolumn labelling for ultraviolet detection. *J. Chromatogr.*, *341*:473–478 (1985).

49

Clonazepam and 7-Aminoclonazepam in Human Plasma

James H. Edge and Philip D. Walson *Ohio State University Children's Hospital, Columbus, Ohio*

I. INTRODUCTION

A. Background

The benzodiazepine clonazepam was first introduced in Europe in 1966 as an oral and intravenous antiepileptic medication under the trade name Rivotril and was approved for oral use in the United States as Klonopin in 1976. It has been used either alone or in combination with other antiepileptic drugs to treat resistant childhood seizures including atypical absence seizures and myoclonic epilepsy. Initially it is very effective, especially in seizures which are refractory to other medications. Unfortunately, after chronic use it may become necessary to increase doses to produce continuing effects, and frequently side effects or toxicity occurs with prolonged use or when the dosage is increased or decreased too rapidly.

It has been claimed that blood levels do not to correlate well with seizure control, side effects, or toxicity. Good clinical responses have been seen when blood concentrations exceed published "therapeutic ranges," but side effects have also occurred with blood levels in the subtherapeutic range.

We hypothesize that the lack of correlation between plasma levels and clinical effects is due, at least partly, to collection of blood samples at random times after dosing.

B. Structure

Clonazepam is composed of a benzene ring with a nitro group which is attached to a 7-member benzodiazepine ring connected to a second benzene ring with a chlorine attached.

Clonazepam is extensively metabolized, mainly by nitro-reduction to 7-aminoclonazepam and then, to a lesser extent, acetylated in the liver to 7-acetamidoclonazepam. A small portion of these compounds are also hydroxylated.

Clonazepam and the major metabolite 7-aminoclonazepam are the only compounds found in significant quantities in human plasma.

The elimination half-life of clonazepam is reported to be 24 to 36 h. This is based on non-steady-state plasma radioassay clonazepam measurements following administration of a single radiolabeled dosed (4). However, using standard assays, blood levels are usually unmeasurable at 12 to 24 h after a single dose.

C. Objective

Our objective was to measure plasma clonazepam and 7-aminoclonazepam concentrations in children as a function of time after dosing.

II. METHODS

Heparinized plasma was obtained from samples obtained from pediatric epileptic patients during regular visits. Accurate demographic, dosing, and collection-time information was obtained according to the protocol that has been used effectively in our therapeutic drug monitoring (TDM) laboratory for the past 10 years (2).

Plasma clonazepam and 7-aminoclonazepam concentrations were measured by high-pressure liquid chromatographic (HPLC) methods:

1. For clonazepam measurements, one of two HPLC methods was used: using a reverse-phase C6 column (9), or using reverse-phase C18 column (8). Extraction, mobile phase, and detection were essentially identical.
2. All 7-aminoclonazepam (7-ACZP) concentrations were determined by the method of Petters et al. (8).

III. RESULTS

Preliminary studies showed:

1. Acceptable correlation (r=0.983) between the two clonazepam methods (8,9).
2. Good recovery of both clonazepam and 7-ACZP at both normal and elevated levels (97% to 105%).
3. Acceptable stability of both CZP (1 year) and 7-ACZP (8 weeks) when stored at 4°C and below.
4. Acceptable assay reproducibility (4% to 20%) with a low coefficient of variation (4%) at the high ranges around 100 ng/mL and a higher variation (20%) equivalent ± 1 ng at the 5-ng/mL range. The limits of detection of all assays was 5 ng/mL.

Clonazepam and 7-aminoclonazepam concentrations were measured in 36 samples from 26 different pediatric epilepsy patients. Dosing and demographic information was obtained simultaneously.

There were 13 males and 13 females, who varied in age from 3 months to 20 years (mean 7.0) and whose weight ranged from 8 to 85 kg (mean 25). Clonazepam dosages ranged from 0.01 to 0.25 mg/kg/day (mean 0.09) administered at either 8- (n=18), 12- (n=13), or 24- (n=5) h intervals as prescribed by their physicians. Blood collection times ranged from 1.5 to 15 h after the last administered clonazepam dose.

IV. DISCUSSION

All of our samples were from children receiving chronic clonazepam dosing who were at steady state. The recovery studies, inter- and intraassay precision studies, and stability studies confirmed that the methods used produced valid, reproducible results on stored samples assayed within an acceptable time after sample processing.

Few pharmacokinetic studies of clonazepam have been reported in the literature (1,3,4,5,6,10), and only one evaluated children (3). The early pharmacokinetic studies were performed after single doses (non-steady-state) given to healthy adult volunteers (4). Although a few later reports indicated dosages and dosage frequency, they did not provide blood collection times or include metabolite measurements. Only two studies (7,10) reported simultaneously measured clonazepam and 7-aminoclonazepam concentrations and were accompanied by any dosing information. Even though one of these contained pediatric patients (7), neither of them indicated any time after dosing information and therefore were not comparable to our study.

Plasma clonazepam concentrations in our patients ranged from 0 to 174 ng/ml with a mean of 34. As can be seen in Figure 1, our samples showed a typical concentration/time after dosing distribution with the highest values occurring during the first 4 h after dosing. These high concentrations in the early postdosing phase are not associated with either the highest dosages administered nor with the frequency of dosing.

Plasma 7-ACZP concentrations (Figure 2) in our patient ranged from 0 to 93 ng/mL. As with clonazepam (Figure 1), there is a tendency toward higher values soon after dosing, but this is less pronounced than for clonazepam.

V. CONCLUSION

CZP concentrations appear to reflect time after dosing more than interindividual pharmacokinetic differences. Although it should be very obvious, it is unfortunately a frequently overlooked fact that sample collection should be avoided during the rapid distribution/absorption phase (up to 4 h after the last dose). Since our results came

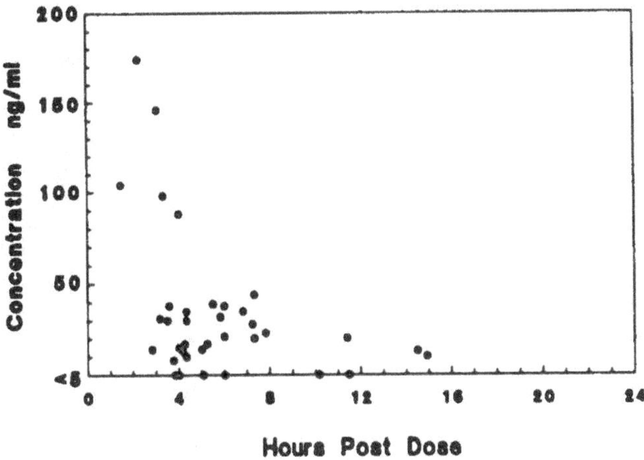

Figure 1 Plasma clonazepam concentration versus time after dosing.

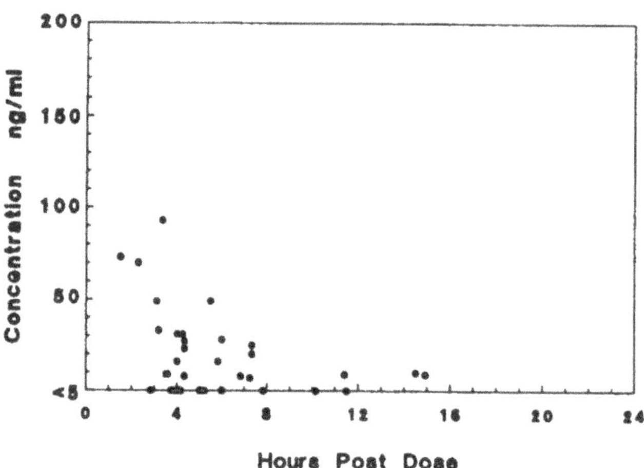

Figure 2 Plasma 7-aminoclonazepam concentrations versus time after dosing.

from samples submitted for analysis by clinicians treating patients, our study confirms that this fact is often ignored.

If collection times can be standardized, preferably just prior to dosing (i.e., predose or trough levels) but at least 4 h after dosing, more useful information may be obtained for study of the dose/concentration relationship.

REFERENCES

1. A. Berlin and H. Dahlstrom. Pharmacokinetics of the anticonvulsant drug clonazepam evaluated from single oral and intravenous doses and by repeated oral administration. *Eur. J. Clin. Pharmacol.*, 9:155–159 (1975).
2. S. Cox and P. D. Walson. Providing effective therapeutic drug monitoring services. *Ther. Drug Monitor.*, 11:310–322 (1989).
3. F. E. Dreifuss, J. K. Penry, S. W. Rose, H. J. Kupferberg, P. Dyken, and S. Sato. Serum clonazepam concentrations in children with absence seizures. *Neurology*, 25:255–258 (1975).
4. E. Eschenhof. Untersuchungen uber das Schicksal des Antikonvulsivums Clonazepam im Organismus der Ratte, des Hundes und des Menschen. *Arzneimittel-Forschung*, 23:390–399 (1973).
5. S. A. Kaplan, K. Alexander, M. J. Jack, C. V. Puglisi, J. A. F. de Silva, T. L. Lee, and R. E. Weinfeld. Pharmacokinetic profiles of clonazepam in dog and humans and of flunitrazepam in dog. *J. Pharm. Sci.*, 63:527–532 (1974).
6. H. J. Knop, E. van der Kleijn, and L. C. Edmunds. In *Clinical Pharmacology of Anti-Epileptic Drugs* (H. Schneider, D. Janz, C. Gardner-Thorpe, H. Meinardi, and A. L. Sherwin, eds.), Springer-Verlag, New York, pp. 247–259 (1975).
7. B. H. Min and W. A. Garland. Determination of clonazepam and its 7-amino metabolite in plasma and blood by gas chromatography-chemical ionization mass spectrometry. *J. Chromatogr.*, 139:121–133 (1977).
8. I. Petters, D.-R. Peng, and A. Rane. Quantitation of clonazepam and its 7-amino and 7-acetamido metabolites in plasma by high-performance liquid chromatography. *J. Chromatogr.*, 306:241–248 (1984).
9. V. Rovei and M. Sanjuan. Simple and specific high performance liquid chromatographic method for the routine monitoring of clonazepam in plasma. *Ther. Drug Monitor.*, 2:283–287 (1980).
10. O. Sjo, E. F. Hvidberg, J. Naestoft, and M. Lund. Pharmacokinetics and side-effects of clonazepam and its 7-amino-metabolite in man. *Eur. J. Clin. Pharmacol.*, 8:249–254 (1975).

50

Analysis of Benzodiazepines and Tricyclic Antidepressants in Whole Blood Using a Common Solid-Phase Clean-Up and a Common Mobile Phase

Mohammed Mazhar and Steven R. Binder *Bio-Rad Laboratories, Hercules, California*

I. INTRODUCTION

Benzodiazepines are the most widely prescribed class of drugs, due to their broad spectrum of pharmacological actions (1). The widespread use of these drugs has led to misuse and overuse, and has provoked considerable concern relevant to their liability for abuse (2). Epidemiological data of their use in emergency and medical examiners' cases (3), and their potential to affect behavior adversely, particularly in combination with alcohol and other CNS depressants (4–6), warrant a systematic approach to their analysis. Tricyclic antidepressants (TCAs) are frequently encountered in forensic samples in nonsuicidal deaths resulting from drug overdose (7,8). A 10-year survey report (1975 to 1984) from the American Association of Poison Control Centers ranked TCAs from 1 to 4 as a main cause of death due to drug ingestion (9,10).

Amitriptyline was often found as a common drug in polydrug deaths (3). Combinations of benzodiazepines and TCAs are frequently reported in intentional and nonintentional suicide episodes in several medical examiners' cases (3). In order to evaluate forensic samples in the above-mentioned circumstances, a versatile method with good specificity and sensitivity is required to analyze benzodiazepines, tricyclic antidepressants, and their active metabolites in whole blood.

Reported methods for analysis of whole blood involving immunoassay techniques (11–15) may be utilized to screen these drugs as a group. They lack the sensitivity and specificity needed to distinguish and quantitate individual drugs and metabolites. Gas chromatographic (GC) and GC-Mass Spectroscopic methods generally require derivatization. Furthermore, these methods have limitations due to the thermal instability of some of these drugs (16–19). High-pressure liquid chromatography (HPLC) has been applied successfully to analyze whole-blood samples containing benzodiazepines and TCAs (20–24). Sample preparation was performed by solvent extraction, precipitation of proteins, or by solid-phase extractions. Such reported methods have the potential of co-extraction and co-elution of TCAs in the benzodiazepine procedure and benzodiazepines in the TCA procedure. These methods do not have the advantage of adopting a single common sample-preparation procedure for specific separation and analysis of both benzodiazepines and TCAs in the same sample.

We report here a modification of our published method (25) to achieve specific sample clean-up and analysis of benzodiazepines, TCAs, and their major metabolites in forensic whole-blood samples.

II. EXPERIMENTAL

A. Materials

Benzodiazepines (clonazepam, chlordiazepoxide, oxazepam, desalkylflurazepam, nordiazepam, and diazepam) and TCAs (nordoxepin, doxepin, desipramine, imipramine, nortriptyline, amitriptyline, and trimipramine) were purchased from Alltech-Applied Science (State College, Pennsylvania). Alprazolam and triazolam were supplied by Upjohn (Kalamazoo, Michigan). Fludiazepam was synthesized in-house by methylation of desalkylflurazepam with methyl iodide under alkaline conditions and purified by repeated crystallization from methanol. The melting point corresponded to the value reported in the literature, and the compound showed a single peak upon analysis by HPLC. HPLC-grade solvents (methanol and acetonitrile) and ACS-certified chemicals (potassium dihydrogen phosphate, ammonium acetate, potassium bicarbonate, and potassium hydroxide) were purchased from Fisher (Pittsburgh, Pennsylvania). Drug-free whole blood was obtained from Southern Biologics, Inc. (Tallahassee, Florida). N,N-Dimethyloctylamine was purchased from Aldrich (Milwaukee, Wisconsin) and water used was purified in-house using an ion-exchange system. Bonded-phase C_{18} extraction columns with 100 mg capacity and 40-μm particle size were obtained from Bio-Rad Laboratories (Hercules, California).

The first reagent for conditioning the solid-phase columns was prepared by dissolving 77 mg of ammonium acetate in 100 mL of HPLC-grade methanol and stored at room temperature. The same reagent was also used for the elution of TCAs. The second conditioning agent was prepared by dissolving 1 g of potassium bicarbonate in 100 mL of 10% acetonitrile in water. The wash reagent was prepared by the addition of 18 mL of acetonitrile to 82 mL of distilled water. The benzodiazepines' elution reagent was made by mixing 20 mL of distilled water with 30 mL of methanol and 30 mL of acetonitrile.

B. Standards and Controls

A 1-mg/mL solution of individual drugs in methanol was prepared by dissolving 10 mg in HPLC-grade methanol. The working composite standards of different concentrations of the benzodiazepines studied, containing oxazepam, chlordiazepoxide, desalkylflurazepam, nordiazepam, and diazepam, was prepared by combining an aliquot of each stock solution and diluting with drug-free whole blood. A similar procedure was adopted for the standards containing nordoxepin, doxepin, desipramine, imipramine, nortriptyline, and amitriptyline. A mixture containing different concentrations of clonazepam, alprazolam, and triazolam was also prepared by the same procedure. These composite standards of various concentrations in drug-free whole blood were used for linearity precision and recovery studies. Two different standards were prepared to be used regularly as calibrators for the assay. The first standard contained the major benzodiazepines [oxazepam (500 ng/mL), chlordiazepoxide (500 ng/mL), desalkylflurazepam (200 ng/mL), nordiazepam (500 ng/mL), diazepam (500 ng/mL), and 200 ng/mL of all the TCAs listed above in drug-free whole blood. The second standard contained 100 ng/mL each of clonazepam, alprazolam, and triazolam and 200 ng/mL of all the above-listed TCAs in drug-free whole blood. The standards were stored at 4°C (stable for 15 days at 4°C).

From a 1-mg/mL methanolic solution of fludiazepam (internal standard for benzodiazepines) and trimipramine (internal standard for TCAs) was prepared a working internal standard solution in 18% acetonitrile solution containing a mixture of fludiazepam (5 μg/mL) and trimipramine (3.75 μg/mL).

C. Solid-Phase Extraction

Solid-phase columns fitted with a reservoir were positioned in luer-lock fittings on the cover of a vacuum box. The columns were conditioned by washing with 3 mL of the first conditioning reagent under a vacuum of 254 mmHg. A 3-mL solution of the second conditioning reagent was applied and allowed to drain under vacuum. To 1.0 mL of whole blood sample (standard, control, or patient sample) was added 200 μL of working internal standard and this mixture was diluted with 1 mL of distilled water, 2 mL of 30% acetonitrile in water was added to the diluted sample, and allowed to stand for 2 min at room temperature. It was vortex-mixed for 20 s and left at room temperature for 3 min. The samples were centrifuged at 2000 xg for 5 min, poured on the columns, and allowed to drain under a vacuum of 254 mmHg. The columns were washed two times with 2 mL of wash reagent. After the last wash, the columns were left under vacuum for 2 min. The reservoirs were disconnected and the undersurface of the vacuum box cleaned with water. The benzodiazepines were eluted with two 400 μL portions of benzodiazepines elution reagent. The combined eluates were mixed with 1 mL of distilled water and vortexed, and 50 μL were injected on the column. TCAs, which were retained on the column, were subsequently eluted with two 300 μL portions of the TCA elution reagent. The combined eluates were diluted with 300 μL of distilled water and mixed, and 50 μL were injected on the column.

D. Mobile-Phase Preparation

Potassium dihydrogen phosphate (1.36g, 0.01 M) was dissolved in 1000 mL of distilled water and 100 μL of N,N-dimethyloctylamine were added while stirring. The pH of the solution was adjusted to 6.4 by the careful addition of 2 M potassium hydroxide. The mobile phase was prepared by mixing 700 mL of this buffer with 300 mL of acetonitrile.

E. HPLC Instrumentation

The chromatographic system (Bio-Rad) was composed of a model 1330 dual-piston pump, a model 1306 variable-wavelength UV detector, a column heater, and a model AS-48 autosampler with a 50-μL fixed loop. A model HP 3392 integrator (Hewlett Packard, Avondale, Pennsylvania) was used for the analysis of the data collected by monitoring the eluent at 242 nm. A 100 × 2.1-mm I.D., 3-μm-particle-size C_8 analytical column (Bio-Rad) with a 30 × 2.1-mm-I.D., 10-μm-particle-size guard cartridge was used for the chromatographic separation.

F. HPLC Conditions

For the analysis of the major benzodiazepines, the analytical and guard columns were maintained at 55°C with a flow rate of 0.3 mL/min and the UV detector was set at 242 nm with 0.02 a.u.f.s. Alprazolam, clonazepam, and triazolam were chromatographed while maintaining the column temperature at 35°C with a flow rate of 0.3 mL/min and the detector set at 242 nm with 0.01 a.u.f.s. The TCAs were run with the column temperature at 35°C with a flow rate of 0.6 mL/min and monitored at 242 nm with 0.01 a.u.f.s. The chromatograms of benzodiazepines and TCAs after extraction from spiked whole blood samples, are presented in Figures 1 and 2. The chromatogram of extracted whole blood from a forensic case showing toxic concentration of doxepin is illustrated in Figure 2B.

III. RESULTS

The assay for major benzodiazepines was linear in the range 50 to 2000 ng/mL and for alprazolam in the range 10 to 200 ng/mL. Regression analysis was performed between the concentration and the peak-height ratio and showed a correlation coefficient greater than 0.999. TCAs were assessed for linearity over the range 20 to 800 ng/mL. A strong linear relationship was observed by comparison of concentration to the peak-height ratio. The correlation coefficient for all the tricyclic antidepressants was above 0.99. Recovery studies were conducted at two different concentrations for major benzodiazepines (50 and 500 ng/mL) and for alprazolam (10 and 50 ng/mL); the recovery ranged from 84% to 106% based on the comparison of peak heights of

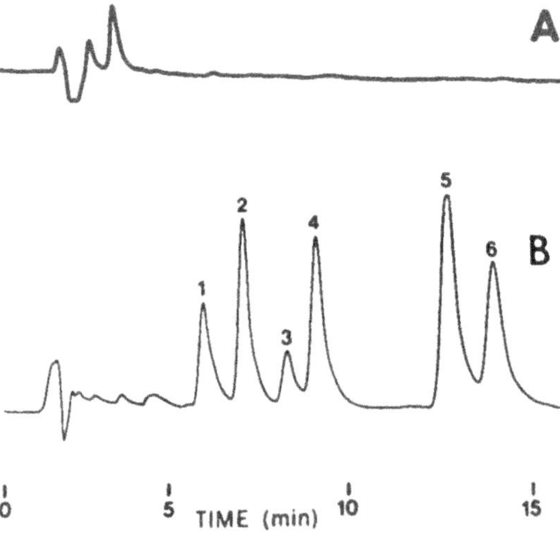

Figure 1 Chromatograms of 50 μl of (A) extracted drug-free whole blood and (B) extracted benzodiazepine-spiked whole blood. Peaks: 1, oxazepam (500 ng/mL); 2, chlordiazepoxide (500 ng/mL); 3, desalkylflurazepam (200 ng/mL); 4, nordiazepam (500 ng/mL); 5, fludiazepam (internal standard, 1000 ng/mL); 6, diazepam (500 ng/mL). Flow rate 0.3 mL/min at 55°D. Detection, 242 nm (0.02 a.u.f.s.).

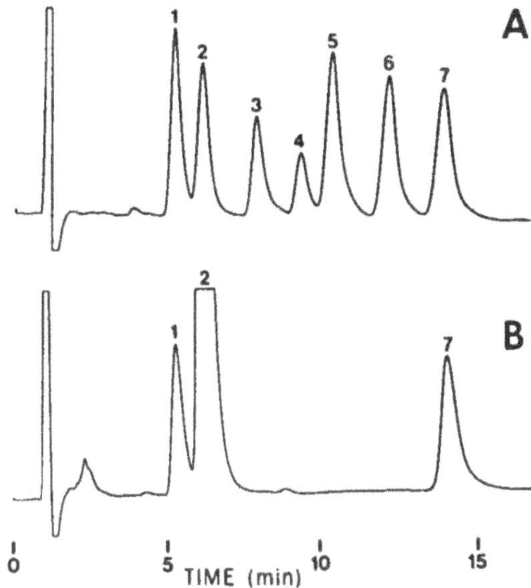

Figure 2 Chromatograms of (A) extracted TCA-spiked whole blood. Peaks: 1, nordoxepin (200 ng/mL); 2, doxepin (200 ng/mL); 3, desipramine (200 ng/mL); 4, imipramine (200 ng/nL); 5, nortriptyline (200 ng/mL); 6, amitriptyline (200 ng/mL); 7, trimipramine (750 ng/mL). Flow rate, 0.6 mL/min at 35°C. Detection, 242 nm (0.01 a.u.f.s.). (B) A forensic whole-blood sample containing nordoxepin (peak 1=165 ng/mL), doxepin (peak 2 = 1680 ng/mL), and internal standard (peak 7).

extracted spiked whole-blood samples against unextracted aqueous standards. The recoveries for TCAs were determined similarly at two levels (50 and 500 ng/mL) and were lower (62% to 69%) than for the benzodiazepines. Tables 1, 2, and 3 show the recovery data.

Within-run precision studies of benzodiazepines were performed on the two levels of spiked whole-blood samples. The data showed coefficients of variation (CV) of less than 7% (Tables 4 and 5). TCAs were measured in replicate at two levels using spiked whole-blood samples for the six components, and showed a variation of less than 7% as presented in Table 6.

Several benzodiazepines, antidepressants, phenothiazines, and other related drugs were evaluated for potential interference. Retention time for these drugs are listed in Tables 7 and 8. Phenothiazines and other basic drugs elute from the solid-phase columns along with TCAs and thus did not interfere with benzodiazepines. Several of the commonly prescribed phenothiazines had longer retention times than the tricyclic antidepressants analyzed by this method. Extracted drug-free blood did not show any endogenous peaks interfering with either the benzodiazepines or the TCA analyses.

IV. DISCUSSION

The method described here for the analysis of benzodiazepines and tricyclic antidepressants involves three crucial steps. The first step is the sample dilution step.

Table 1 Recovery of Major Benzodiazepines (N = 4)

Drug	Spiked Amount (ng/ml)	Recovery (%)	Spiked Amount (ng/ml)	Recovery (%)
Oxazepam	500	91	50	89
Chlordiazepoxide	500	100	50	92
Desalkylflurazepam	500	100	50	100
Nordiazepam	500	100	50	94
Diazepam	500	99	50	93

Table 2 Recovery of Minor Benzodiazepines (N = 4)

Drug	Spiked Amount (ng/ml)	Recovery (%)	Spiked Amount (ng/ml)	Recovery (%)
Clonazepam	50	94	10	84
Alprazolam	50	104	10	106
Triazolam	50	100	10	100
				85

Table 3 Recovery of Tricyclic Antidepressants (N = 4)

Drug	Spiked Amount (ng/ml)	Recovery (%)	Spiked Amount (ng/ml)	Recovery (%)
Nordoxepin	500	62	50	60
Doxepin	500	64	50	65
Desipramine	500	68	50	69
Imipramine	500	62	50	64
Nortriptyline	500	65	50	66
Amitriptyline	500	63	50	65

Table 4 Within-Run Precision Data for Major Benzodiazepines

Drug	Spiked Conc. (ng/nl)	N	CV (%)	Spiked Conc. (ng/ml)	N	CV (%)
Oxazepam	500	9	0.8	50	9	4.0
Chlordiazepoxide	500	9	0.5	50	9	6.5
Desalkylflurazepam	500	9	0.7	50	9	7.0
Nordiazepam	500	9	1.1	50	9	2.6
Diazepam	500	9	0.9	50	9	2.3

Table 5 Within-Run Precision Data for Minor Benzodiazepines

Drug	Spiked Conc. (ng/nl)	N	CV (%)	Spiked Conc. (ng/ml)	N	CV (%)
Clonazepam	50	10	2.6	10	7	5.7
Alprazolam	50	11	3.5	10	8	4.7
Triazolam	50	11	2.9	10	8	5.8

Table 6 Within-Run Precision Data for Tricyclic Antidepressants

Drug	Spiked Conc. (ng/nl)	N	CV (%)	Spiked Conc. (ng/ml)	N	CV (%)
Nordoxepin	400	7	4.5	100	7	5.3
Doxepin	400	7	5.7	100	7	7.2
Desipramine	400	7	5.5	100	7	6.0
Imipramine	400	7	5.2	100	7	3.6
Nortriptyline	400	7	4.1	100	7	4.7
Amitriptyline	400	7	3.6	100	7	3.4

Table 7 Retention Times of Benzodiazepines and Related Drugs

Compound	Retention Time (min)	Compound	Retention Time (min)
Bromzaepam	3.28	Triazolam	7.20
Demoxepam	3.67	Desalkylflurazepam	7.28
Carbamazepine	3.92	Temazepam	7.97
Norchlordiazepoxide	4.19	Nordiazepam	8.07
Methaqualone	4.46	Clobazam	8.47
Nitrazepam	4.69	Trazodone	9.10
Oxazepam	5.14	Buspirone	10.20
Clonazepam	5.34	Fludiazepam (IS)	11.35
Desmethylclobazam	5.47	Midazolam	12.29
Lorazepam	5.81	Diazepam	12.55
Chlordiazepoxide	6.10	Flurazepam	15.49
Alprazolam	6.63	Medazepam	31.75
Flunitrazepam	6.50	Halazepam	33.27

HPLC conditions: flow rate, 0.3 ml/min; column temperature, 55°C.

Table 8 Retention Times of Tricyclic Antidepressants and Related Drugs

Compound	Retention Time (min)	Compound	Retention Time (min)
Quinidine	2.02	Cyclobenzaprine	7.88
10-Hydroxynortriptyline	2.06	Maprotiline	9.02
2-Hydroxydesipramine	2.51	Methadone	9.0
Propranolol	2.73	Imipramine	9.2
Disopyramide	2.84	Nortriptyline	10.1
2-Hydroxyimipramine	2.90	Propiomazine	11.2
10-Hydroxyamitriptyline	3.10	Amitriptyline	12.0
Benztropine	4.62	Propoxyphene	12.0
Trazodone	4.69	Thiothixene	12.6
Nordoxepin	4.94	Chlorpromazine	12.9
Doxepin	5.90	Trimipramine (IS)	13.9
Methadone Metabolite[a]	5.93	Perphenazine	16.12
Desipramine	7.71	Fluphenazine	27.15
Amoxapine	7.71	Thioridazine	28.18

[a]2-Ethylidene-1, 5-dimethyl-3, 3-diphenylpyrrolidine.
HPLC conditions: flow rate, 0.6 ml/min; column temperature, 35°C.

Several procedures were tested which involved the precipitation of proteins from whole blood by adding methanol or zinc sulfate or acetone, before solid-phase extraction. These procedures, in our experience, gave poor recovery, impure sample extracts, slowed flow, and produced clogging of solid-phase columns. Dilution of the blood sample with optimum concentration and volume of acetonitrile gave a solution with very little precipitation of proteins (26). Centrifugation of the sample and subsequent extraction on the solid-phase column gave no problem with the sample flow. The second step involves the selective adsorption and sequential elution from column. This step and the final step of chromatographic separation of benzodiazepines and tricyclic antidepressants were performed as described earlier (25), except for a minor modification of TCA elution reagent. Ammonium acetate in methanol was used to elute TCAs from the solid-phase column instead of methanol containing diethylamine (25), due to the instability of the latter reagent at room temperature. The new elution reagent was stable for at least 15 days at room temperature.

V. CONCLUSION

The method described here permits the quantitative analysis of several benzodiazepines and tricyclic antidepressants along with their major active metabolites in a whole-blood sample. We feel that the simplicity, specificity, and sensitivity of this method, along with the utilization of common reagents for the specific sample preparation and HPLC analysis, offers a wide range of applications in the forensic setting, where the analysis of both these classes of drugs are required in antemortem or postmortem blood samples.

VI. SUMMARY

A rapid and specific clean-up procedure was developed for the isolation of benzodiazepines and tricyclic antidepressants from 1 mL of whole blood. The sample was initially diluted with water and 30% acetonitrile and purified by solid-phase extraction. Benzodiazepines were eluted from the column using a mixture of water-methanol-acetonitrile (2:3:3), then tricyclic antidepressants were eluted with methanol containing 0.01 M ammonium acetate. Chromatographic separation of both classes of drugs were performed by our previously published protocol (25). Detection was carried out at 242 nm. The sensitivity limit of the assay was 25 ng/mL with a recovery of 84 to 106% for benzodiazepines and 62 to 69% for tricyclic antidepressants. Analysis of both classes of drugs have good within-run precision with a coefficient of variation of less than 7.2%. The method described here is suitable for the analysis of benzodiazepines and tricyclic antidepressants in postmortem forensic blood specimens.

ACKNOWLEDGMENT

The authors thank Upjohn Company, Kalamazoo, Michigan, for the generous gift of alprazolam and triazolam. The authors also wish to thank Yanru Zhang for excellent technical assistance.

REFERENCES

1. M. Linnoila. In *The Benzodiazepines—From Molecular Biology to Clinical Practice* (E. Costa, ed.). Raven Press, New York, p. 267 (1983).

2. J. H. Woods, J. L. Katz, and G. Winger. Abuse liability of benzodiazepines. *Pharmacol. Rev.*, *39*:251 (1987).
3. A decade of DAWN: Benzodiazepine related cases. *National Institute of Drug Abuse Statistical Series, Series H, No. 4* (1976–1985).
4. R. Battegay, R. Muhlemann, and R. Zehnder. Comparative investigations of the abuse of alcohol, drugs, and nicotine for a representative group of 4082 men of age 20. *Comp. Psychiatr.*, *16*:247 (1975).
5. R. L. Bauer. Traffic accidents and minor tranquilizers. A review. *Public Health Report*, *99*:573 (1984).
6. E. M. Sellers and U. Busto. Benzodiazepines and ethanol: Assessment of the effects and consequences of psychotropic drug interactions. *J. Clin. Psychopharm.*, *2*:249 (1982).
7. D. A. Frommer, K. W. Kulig, J. A. Marx, and B. Rumack. Tricyclic antidepressants overdose: A review. *J. Am. Med. Assoc.*, *257*:521 (1987).
8. R. G. Kathol and F. A. Henn. Tricyclics: The most common agent used in potentially lethal overdoses. *J. Nerv. Ment. Dis.*, *171*:250 (1983).
9. T. L. Litovitz and J. C. Veltri. 1984 annual report of the American Association of Poison Control Centers national data collection system. *Am. J. Emergency Med.*, *3*:423 (1985).
10. T. L. Litovitz and J. C. Veltri. 1983 annual report of the American Association of Poison Control Centers national data collection system. *Am. J. Emergency Med.*, *2*:420 (1984).
11. W. M. Asselin, J. M. Leslie, and B. McKinley. Direct detection of drugs of abuse in whole hemolyzed blood using the EMIT d.a.u. urine assays. *J. Anal. Toxicol.*, *12*:207 (1988).
12. L. T. Lewellen and H. H. McCurdy. A novel procedure for the analysis of drugs in whole blood by homogeneous enzyme immunoassay (EMIT). *J. Anal. Toxicol.*, *12*:260 (1988).
13. C. P. Goddard, A. H. Stead, P. A. Mason, B. Law, A. C. Moffat, M. McBrien, and S. Cosby. A 1-125 radioimmunoassay for benzodiazepines in blood and urine. *Analyst.*, *111*:525 (1986).
14. E. L. Slighton, J. C. Cagle, H. H. McCurdy, and F. Castagna. Direct and indirect homogeneous enzyme immunoassay of benzodiazepines in biological fluids and tissues. *J. Anal. Toxicol.*, *6*:22 (1982).
15. S. Spector, W. L. Spector, and M. P. Almeida, Jr. Radioimmunoassay for desmethylimipramine. *Psychopharm. Commun.*, *1*:421 (1975).
16. R. C. Baselt, C. B. Stewart, and S. J. Franch. Toxicological determination of benzodiazepines in biological fluids and tissues by flame-ionization gas chromatography. *J Anal. Toxicol.*, *1*:10 (1977).
17. J. A. F. Desilva, I. Bekersky, and C. V. Puglisi, et al. Determination of 1,4 benzodiazepines and 1-diazepine-2-ones by electron capture gas liquid chromatography. *Anal. Chem.*, *48*:10 (1976).
18. J. R. Joyce, T. S. Bal, R. E. Ardrey, H. M. Stevens, and A. C. Moffet. The decomposition of benzodiazepines during analysis by gas-chromatography mass-spectrometry. *Biomed. Mass Spectrom.*, *11*:284 (1984).
19. P. Hartvig and B. Naslund. Simultaneous determination of amitriptyline and nortriptyline as trichloroethyl carbamates by electron capture gas chromatography. *J. Chromatog.*, *133*:367 (1977).
20. S. N. Rao, A. K. Dhar, H. Kutt, and M. Okamato. Determination of diazepam and its pharmacologically active metabolites in blood by Bond Elute column extraction and reversed-phase high-performance liquid chromatography. *J. Chromatogr.*, *231*:341 (1982).
21. J. B. F. Lloyd and D. A. Parry. Forensic applications of the determination of benzodiazepine in blood samples by microcolumn cleanup and high performance liquid chromatography with reductive mode electrochemical detection. *J. Anal. Toxicol.*, *13*:163 (1989).
22. M. A. Evenson and D. A. Engstrand, A Sep-Pak HPLC method for tricyclic antidepressant drugs in human vitreous humor. *J. Anal. Toxicol.*, *13*:322 (1989).
23. S. H. Y. Wong. Measurement of antidepressants by chromatography: A review of current methodology. *Clin. Chem.*, *34*:848 (1988).
24. R. J. Hughes and M. D. Osselton. Comparison of methods for the analysis of tricyclic antidepressants in small whole blood samples. *J. Anal. Toxicol.*, *13*:77 (1989).

25. M. Mazhar and S. R. Binder. Analysis of benzodiazepines and tricyclic antidepressants in serum using a common solid-phase cleanup and a common mobile phase. *J. Chromatogr.*, *497*:201 (1989).
26. G. L. Lensmeyer and B. L. Fields. Improved liquid chromatographic determination of cyclosporine with concomitant detection of a cell-bound metabolite. *Clin. Chem.*, *31*:196 (1985).

51

Detection of Urinary Benzodiazepines by Radioreceptor Assay

Takashi Nishikawa, Satomi Suzuki, Hideki Ohtani, Noriko Watanabe, Sadanori Miura, and Rumiko Kondo
Kitasato University, Kitasato, Sagamihara, Kanagawa, Japan

I. INTRODUCTION

Benzodiazepines (BZs) constitute a class of versatile and widely prescribed CNS depressants. Their widespread use creates widespread misuse and abuse. BZ overdose patients comprise 25% to 50% of the drug-poisonings seen in the emergency room. For the diagnosis of an overdose, measurement of BZ level in blood or urine is required as an emergency test. BZ level is also measured for the confirmation of BZ ingestion, as a drug screening test, and for the management of psychotic disorders.

However, there are some problems in measuring BZ levels by chromatographic or immunoassay methods. First, more than 30 BZ derivatives and even non-BZ agonists of the BZ receptor (e.g., zopiclone) are now in clinical use. Second, active metabolites are present for some BZs (e.g., diazepam and flurazepam). Third, therapeutic levels of potent BZs (e.g., triazolam) are very low (less than 10 ng/mL of plasma). We have been studying a BZ receptor assay to ascertain if the receptor assay might overcome the analytical problems described above. The levels of any active BZs, including non-BZ agonists and active metabolites, would be measured by the receptor assay in terms of the receptor binding activity. Very low levels of potent BZ could be measured because the binding activity of potent BZ is very high. We had already established the radioreceptor assay method for BZ in blood (1). Here we describe the establishment of the radioreceptor assay method for BZ in urine and its clinical application.

II. MATERIALS, METHODS, AND SUBJECTS

Materials and methods were similar to those in our previous report (1). The receptor suspension was prepared from rat brain homogenate by centrifugation. The labeled ligand, ^{3}H-flunitrazepam, was obtained from New England Nuclear Research Products (Boston, Massachusetts). The enzymes β-glucuronidase from *E. coli*, bovine liver, and helix pomatia) were obtained from Sigma Chemical Company (St. Louis, Missouri).

Urine and serum were analyzed immediately after collection or stored frozen until analysis. Serum (0.5 mL) was mixed with pH 9 borate buffer (0.5 mL) and extracted with ethyl acetate (5 mL), and a 0.5 mL portion of the solvent extract was evaporated to dryness. The residue was subjected to the receptor assay. The assay procedure was: (a) mix the sample with the receptor suspension and then with ^{3}H-flunitrazepam, (b) incubate, (c) filter the mixture through a glass fiber filter (Whatman International, Maidstone, England), and (d) count radioactivity on the filter. Urine could be analyzed without pretreatment by extraction as will be described later. Unknown sample concentrations were usually estimated from a standard curve of diazepam-spiked specimen and expressed as diazepam equivalents concentration.

The healthy volunteers were given a single clinical dose of BZ, and aliquots of their blood, urine, or both were collected. Specimens were also collected from patients with psychotic disorders such as depression, manic-depression, neurosis, and schizophrenia when they came to the hospital. These patients had been taking one or more of the benzodiazepines orally every day for more than 1 month (long-term users). Routine clinical laboratory tests for liver function, kidney function, and urine analyses were done. Specimens were also collected from subjects suspected of BZ overdose in the emergency room.

III. RESULTS

A. Standard Curve of Urine Benzodiazepine

The direct addition of less than 50 μL of urine to the assay solution caused little interference. We usually added 50 μL of more-than-twofold-diluted urine. All the blank values of the volunteers before the administration were below the detection limit (<40 ng/mL).

B. Effect of β-Glucuronidase

The urine specimens of the long-term users of diazepam were analyzed. Their BZ activities increased 1.5 to 2 times over the original activity after treatment with any of three enzymes. We usually diluted urine with phosphate buffer, added the solution of the *E. coli* enzyme, incubated at 3°C for 3 h, and then performed the receptor assay.

C. BZ Concentrations in Volunteers

After a single clinical dose of clotiazepam, etizolam, flurazepam, flutazolam, or triazolam, high BZ activity was detected in urine of the volunteers. All of the "active BZ excretion type" were the BZs with ultrashort- or short-half life in blood. Even after serum BZ decreased to negative level (<20 ng/mL), the urine BZ tests were positive. However, the urine test usually turned negative within 24 h.

Alprazolam, bromazepam, clonazepam, cloxazolam, estazolam, flunitrazepam and nitrazepam were the "non-excretion-type" BZs, which gave little or no BZ activity in urine after a single clinical dose. All of these BZs had intermediate or long half-lives.

Some urine specimens showed a significant increase of BZ activity after treatment with the enzyme. These "conjugated-(active BZ)-excretion-type" BZs include diazepam.

D. BZ Concentrations in Blood and Urine of Long-Term Users

The 99 psychotic patients were taking one of the following BZs: alprazolam, bromazepam, clonazepam, cloxazolam, diazepam, estazolam, etizolam, flunitrazepam, flurazepam, lorazepam, medazepam, nitrazepam, and triazolam. The distribution of the serum levels was: 0 to 250 ng/mL, 83 patients (83.8%); 250 to 500 ng/mL, 9 (9.1%); over 500 ng/mL, 7 (7.1%). In 53 patients receiving two or three BZs, 36 patients (67.9%) had 0 to 250 ng/mL, 11 (20.8%) showed 250 to 500 ng/mL, and 6 (11.3%) showed over 500 ng/mL. We concluded that the serum level is below 500 ng/mL in most (about 90%) of these patients on BZ therapy.

The results of the users' BZ analyses were in accordance with those obtained with the volunteers. However, high activity was occasionally found in the urines of "non-excretion-type" BZ user. In those patients, the serum level was also relatively high.

E. BZ Concentrations in Blood and Urine of Overdose Patients

We had patients who overdosed with brotizolam, chlordiazepoxide, diazepam, estazolam, nitrazepam, and other BZs. Their serum BZ concentrations were greater than 700 ng/mL. At serum levels of 700 to 2000 ng/mL, they showed no responses when their names were called. High BZ activity was found in the urine of two patients who overdosed with nitrazepam, a "non-excretion type." HPLC analysis showed that the parent drug, nitrazepam, was present in their urine specimens. It is suggested that the urine test possibly gives a positive result when overdose occurs with "non-excretion-type" BZs.

IV. DISCUSSION

We consider that therapeutic levels (diazepam equivalents in serum in most subjects on BZ therapy) is below 500 ng/mL, toxic levels are over 1000 ng/mL, and 500 to 1000 ng/mL is the overlapping range. However, we consider that any dose adjustment should be based on the patients' clinical symptoms, not on the serum level.

Some immunoassay kits for urine BZs are on the market in the United States. According to the manufacturers, the cutoff value is 200 to 300 ng/mL of desmethyldiazepam or oxazepam. Though we have not had a chance to evaluate any of the kits and do not know exactly what is meant by the word "cutoff value," this level could be in the assay range of the receptor assay.

We should take two precautions in the interpretation of the receptor assay results. First, if antagonists (e.g., flumazenil) or inverse agonists are present, the assay value should not be used as an index of BZ pharmacological activity. Second, an endogenous BZ-like substance, which binds to the receptor as well as to the anti-BZ antibody is present in patients with hepatic encephalopathy (2) and would interfere with the assay.

Further clinical studies may be needed, but the present study indicates that the receptor assay for urinary BZs is useful as a preliminary, screening test and as an emergency test for the assessment of ingestion of certain kinds of BZs.

REFERENCES

1. T. Nishikawa, A. Nishida, H. Ohtani, W. Sunaoshi, H. Miura, Y. Sudo, and H. Kubo. Radioreceptor assay of clonazepam and diazepam in blood for therapeutic drug monitoring. *Ther. Drug Monitor.*, *11*:483–486 (1989).
2. K. D. Mullen, K. M. Szauter, and K. Kaminsky-Russ. "Endogenous" benzodiazepine activity in body fluids of patients with hepatic encephalopathy. *Lancet*, *336*:81–83 (1990).

52

Comparison of Specific Radioimmunoassays for Cyclosporine Measurement

I. Sabaté, S. Gracia, B. Arranz, J. Cortés, J. Valero, J. Bover, J. M. Griñó, A. M. Castelao, A. Gonzalez, and A. Rivera *Hospital de Bellvitge, University of Barcelona, Barcelona, Spain*

I. INTRODUCTION

Cyclosporine (CsA) therapeutic drug monitoring has become an essential component of successful CsA therapy. Measurement of CsA concentrations has been complicated by several issues, reviewed by the Task Force on Cyclosporine Monitoring (1), and recently by the Consensus Document: Hawk's Cay Meeting on Therapeutic Drug Monitoring of Cyclosporine (2).

Many clinical laboratories utilize radioimmunoassay (RIA) for CsA measurements because of its adaptability on a large number of samples. Two commercial RIA kits use a monoclonal antibody that is selective for parent CsA molecule. One kit (Sandimmun) involves ^{3}H-labeled tracer, while the other (CYCLO-Trac-SP) involves a ^{125}I-labeled-CsA ligand.

In this study, we compare whole-blood CsA concentrations from transplant patients measured by both kits. In order to assess the difference in concentrations attributed to standards (3), we used the CYCLO-Trac-SP with two different standard calibration curves.

II. MATERIALS AND METHODS

A. Clinical Samples

One hundred thirty whole-blood samples, collected in EDTA, were obtained from renal and liver transplant patients receiving CsA. Samples for CsA measurement were taken 11 to 12 h after the previous dose.

B. Methods

The two RIA methods (Sandimmun, Sandoz, and CYCLO-Trac-SP, INCSTAR Corp.) were performed according to the manufacturers' instructions. The main difference

Table 1 Interassay Precision of cyclosporine

	n	x (ng/mL)	CV (%)
Sandimmun			
Blood-1	20	106	6.5
Blood-2	20	395	3.6
Blood-3	20	72	8.4
Blood-4	20	220	13.2
Cyclo-Trac-SP(A)			
Blood-1	20	112	9.4
Blood-2	20	445	7.7
Blood-3	20	71	11.2
Blood-4	20	246	7.6
Cyclo-Trac-SP(B)			
Blood-1	20	104	5.7
Blood-2	20	408	4.7
Blood-3	20	69	10.6
Blood-4	20	228	3.8

between the methods lies in the separation step: Sandimmun requires the use of charcoal, while the CYCLO-Trac is based on double-antibody competitive binding.

CYCLO-Trac-SP was evaluated using two standard calibration curves. The first was performed using the calibrators supplied in the kit (CYCLO-Trac-SP-A), while the second was prepared in-house using the Sandimmun CsA standard in the same matrix as the samples.

Data were analyzed by logit-log reduction. For the Sandimmun standard curve we used whole-blood CsA standards of 25, 50, 100, 200, 400, 800, and 1600 µg/L. In one typical standard curve, the following values were generated: nonspecific binding 0.5%; binding at 0 concentration, 45%; $r = 0.9998$. For the CYCLO-Trac-SP(A) standard curve we used standards provided in the kit of 25, 65, 165, 388, and 1264 µg/L; the values were as follows: nonspecific binding, 2%; binding at 0 concentration, 52%; $r = 0.9994$. For the CYCLO-Trac-SP(B) the values were as follows: nonspecific binding, 3%; binding at 0 concentration, 51%; $r = 0.9995$.

C. Control Samples

We used four separate control samples: two blood controls prepared and provided in the Sandimmun kit (Blood-1 and Blood-2), and two sample pools from renal transplant patients receiving CsA (Blood-3 and Blood-4).

D. Statistical Analysis

Data obtained were analyzed by regression analysis.

Table 2 Dilution Parallelism Study

Sample no. (dilution)	Sandimmun	Cyclo-Trac SP(A)	Cyclo-Trac (SP(B)
1 (1/1)	800	841	825
1 (1/2)	410	431	391
1 (1/3)	270	271	264
1 (1/4)	190	208	204

III. RESULTS

The detection limit of the Sandimmun assay was 16 μg/L and 18 and 19 μg/L for the CYCLO-Trac-SP(A) and (B), respectively.

Recoveries were studied by adding different standard concentrations to a patient sample and ranged from 94 to 106%.

Interassay precisions are shown in Table 1, and the dilution parallelism study is shown in Table 2.

A correlation study was performed between Sandimmun and CYCLO-Trac-SP(A) and (B) on 130 whole-blood samples from transplant patients. The scattergrams and the equations obtained by regression analysis are shown in Figures 1, 2, and 3.

IV. DISCUSSION

Our results show a similar precision, linearity, and parallelism for both methods evaluated. On the other hand, an excellent correlation was observed between the Sandimmun and CYCLO-Trac-SP assays.

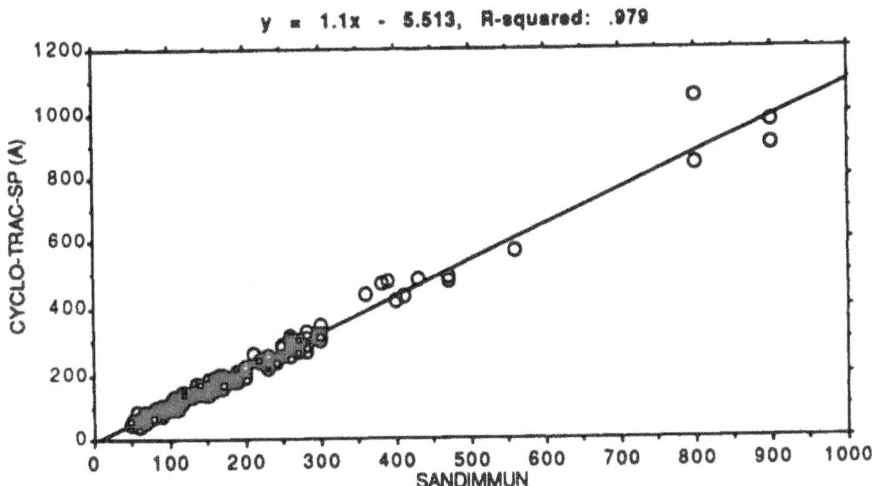

Figure 1 Relationship between CsA concentrations (μg/L) measured by Sandimmun and CYCLO-Trac-SP(A).

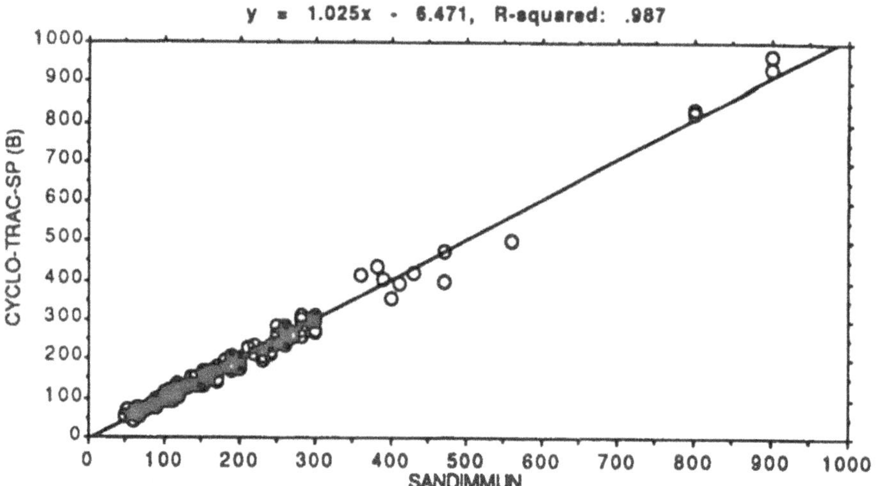

Figure 2 Relationship between CsA concentrations (μg/L) measured by Sandimmun and CYCLO-Trac-SP(B).

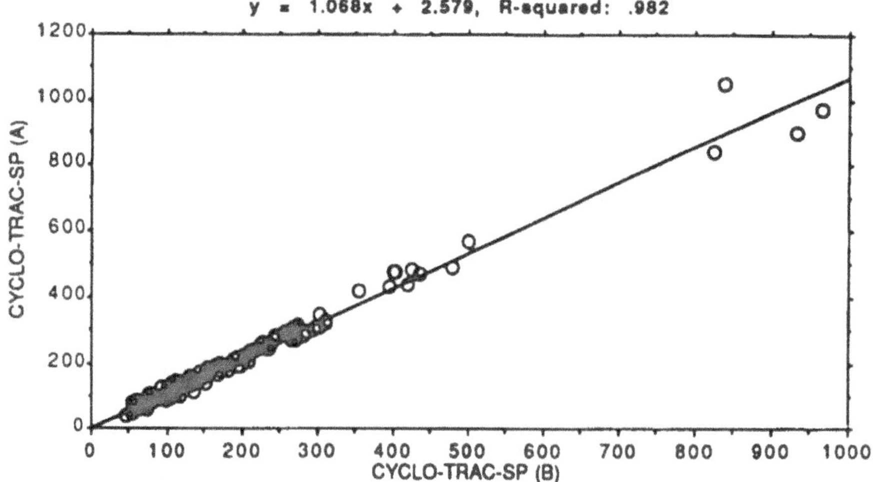

Figure 3 Relationship between CsA concentrations (μg/L) measured by CYCLO-Trac-SP(A) and CYCLO-Trac-SP(B).

Although Keown et al. (3) describe a disparity of 8 to 28% between the concentrations of CsA measured by ^{125}I-RIA(CYCLO-Trac-SP) and HPLC between September 1988 and April 1989 (as a result of different reagent batches), our discrepancy is less than 10% in concentrations lower than 500 µg/L, when using different calibration curves, probably due to the introduction of new standards in the period studied by our group (April to May 1990).

We conclude that the CYCLO-Trac-SP is a reasonable substitute for the Sandimmun kit with a substantial reduction in assay time.

REFERENCES

1. L. M. Shaw, L. Bowers, L. Demers, et al. Critical issues in cyclosporine monitoring: Report of the Task Force on Cyclosporine Monitoring. *Clin. Chem.*, 33:1269–1288 (1987).
2. B. D. Kahan, L. M. Shaw, D. Holt, et al. Consensus document: Hawk's Cay meeting on therapeutic drug monitoring of cyclosporine. *Clin. Chem.*, 36:1510–1516 (1990).
3. P. A. Keown, J. Glenn, J. Denegri, et al. Therapeutic monitoring of cyclosporine: Impact of a change in standards on ^{125}I-monoclonal RIA performance in comparison with liquid chromatography. *Clin. Chem.*, 36:804–807 (1990).

53

Whole-Blood Cyclosporine Concentration Measured by Radioimmunoassay Using Polyclonal and Monoclonal Antibodies and Risk of Acute Graft-Versus-Host Disease After Allogeneic Bone Marrow Transplantation

R. Deulofeu, N. Villamor, E. Carreras, S. Roman, A. Grañena, C. Rozman, and A. Ballesta
Hospital Clínic i Provincial de Barcelona, Barcelona, Spain

I. INTRODUCTION

Since 1983, we have used cyclosporine (CsA) following allogeneic bone marrow transplantation for graft-versus-host disease (GVHD) prophylaxis with a significant decrease on the number and severity of this rejection process (1). Several authors have studied the correlation between the trough CsA concentration and the risk that patients would develop an acute GVHD (2–4). All these analyses were performed in plasma, serum, or whole blood using radioimmunoassay (RIA) techniques with a polyclonal, nonspecific antibody (PAb), yielding controversial results.

In our center, morning, trough, whole-blood CsA concentration was monitored using RIA with a polyclonal nonspecific antibody until 1988, when a new monoclonal-specific assay was introduced. In the present work we have evaluated retrospectively the relation between whole-blood CsA concentration and acute GVHD, in 91 consecutive allogeneic BMT patients, using both immunoassays.

II. PATIENTS AND METHODS

All patients received marrow grafts from genotypically HLA-identical sibling donors, after treatment with cyclophosphamide plus fractionated total-body irradiation as the conditioning regimen.

Intravenous CsA and methotrexate on a short schedule were used, as a prophylaxis for GVHD.

Cyclosporine was given parenterally in a 12-h infusion during the first 2 or 3 weeks until oral administration was possible. The CsA doses were adjusted to maintain levels higher than 400 ng/mL when we used a polyclonal antibody and higher than 200 ng/mL with the monoclonal one. The daily dose was reduced if side effects such as renal dysfunction or severe hepatotoxicity appeared.

GVHD diagnosis was always demonstrated histologically and its severity graded according to the Seattle group's criteria.

Fifty-five patients were monitored by the polyclonal-nonspecific (PAb), and 36 by the monoclonal-specific (MoE) antibody.

Blood samples were obtained in the morning, three times a week, before the next dose was given. CsA was measured by RIA in whole blood after hemolysis by freeze-thawing, using kits provided by Sandoz (Basel, Switzerland), with the corresponding antibody.

To analyze the results, the mean trough CsA concentration for each patient was calculated during 7 and 10 days before the onset of GVHD. In those with no signs of illness, mean CsA concentration was calculated on the same days prior to the mean time necessary to develop acute GVHD, 19 days in our patients.

Three groups were established according to the severity of GVHD. Group O-I, for those with no illness or mild symptoms, group II for those with moderate GVHD, and group III-IV for those with severe GVHD. Patients were also grouped according to the presence or absence of illness, group O-I versus group II-IV.

Statistical analysis were performed by SPSS, adapted for a PC.

III. RESULTS

Of the 55 patients monitored by the polyclonal nonspecific antibody, 60% presented a GVHD degree O-I, 18% degree II, and 22% degree III-IV, proportions that were not statistically different from those obtained from patients monitored by specific antibody, which were 50%, 31%, and 19%, respectively.

According to the presence or absence of acute GVHD, group O-I versus II-III, of the patients monitored by polyclonal antibody, 78% had no illness; for the monoclonal one 81% had no illness, although this difference was not statistically significant. This results agree with those published by other authors (5,6).

Figure 1 shows the mean trough CsA concentration for the 7 days prior to the appearance of GVHD for both antibodies. Among patients monitored with the polyclonal Ab, no differences were found relating to severity of disease: 406 ± 198 ng/mL for group O-I, 398 ± 223 ng/mL for group II, and the highest levels, 567 ± 319 ng/mL, for those with more severe GVHD. In patients controlled by the specific assay, the highest levels were observed in groups O-I, 242 ± 114 ng/mL, and II, 276 ± 108 ng/mL, while group III-IV showed the lowest levels, 118 ± 116 ng/mL.

Similar results were obtained studying the mean CsA concentrations 10 days prior to GVHD (Figure 2).

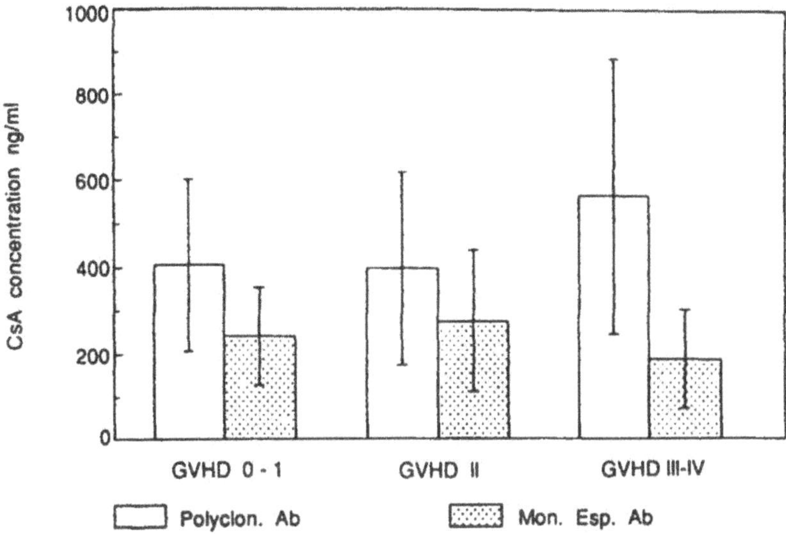

Figure 1 Mean cyclosporine concentration on the 7 days prior to acute GVHD, as a function of severity. Differences in severity were not statistically significant for any group.

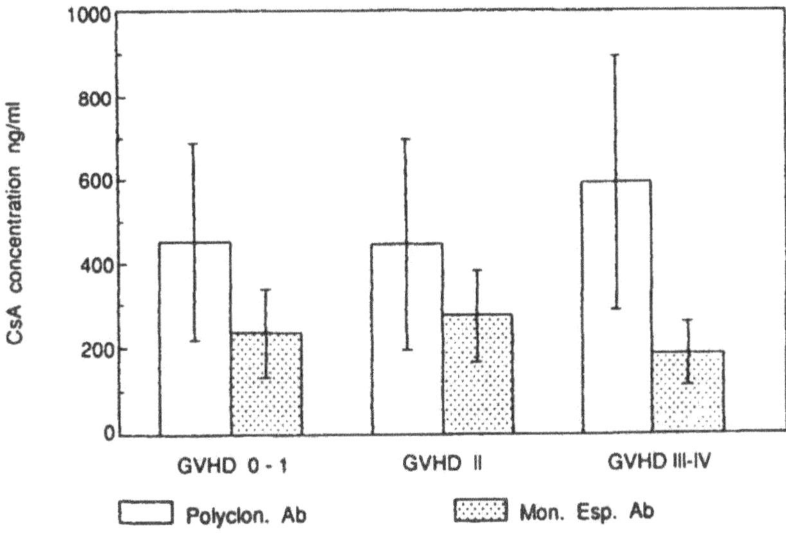

Figure 2 Mean cyclosporine concentration on the 10 days prior to acute GVHD, as a function of severity. Differences in severity were not statistically significant for any group.

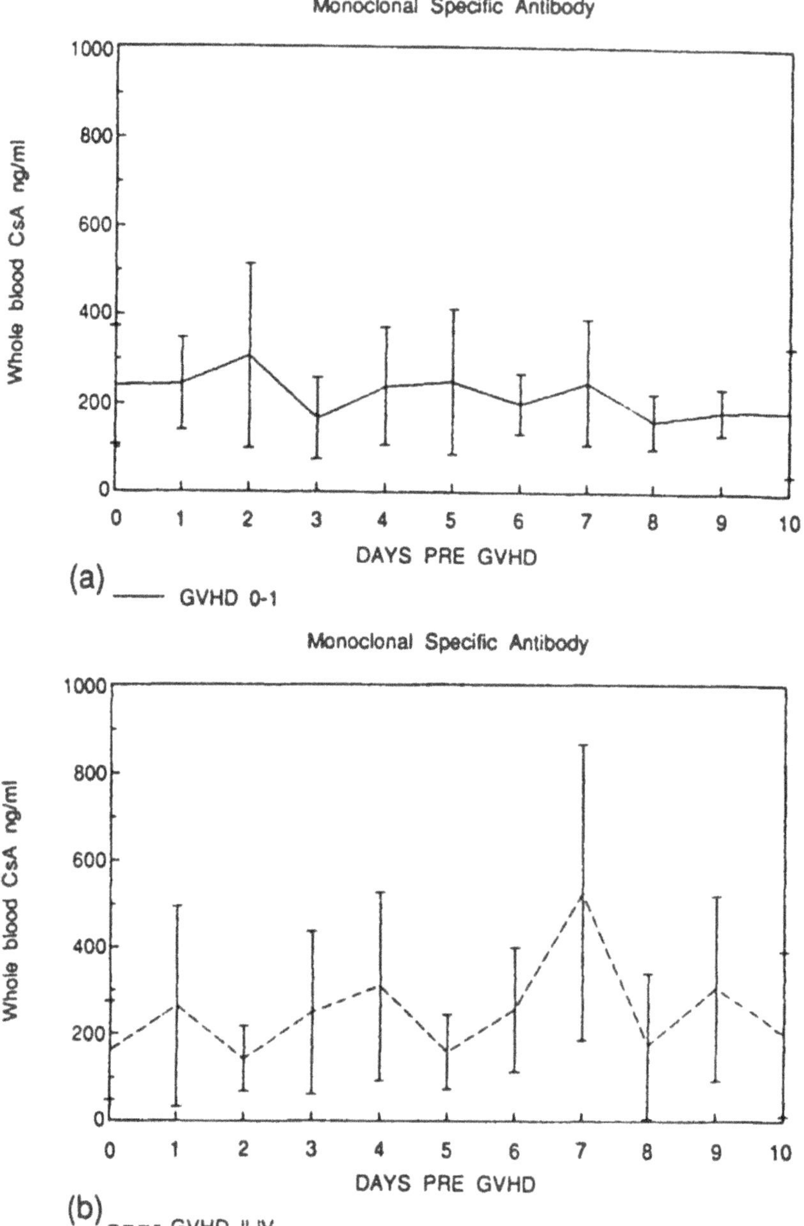

Figure 3 Evolution profile of mean cyclosporine concentrations on patients monitored by the monoclonal-specific antibody according to the presence (a) or absence (b) of acute GVHD.

Mean day-to-day CsA levels on the 10 days before the onset of GVHD in patients monitored by the polyclonal antibody were irregular, with a high dispersion. This was similar in all the groups studied, contrasting with those of patients monitored by the specific antibody. In this case a regular profile was seen in patients with GVHD O-I (Figure 3), while for those with a GVHD II-IV, a fluctuating pattern was observed. Furthermore, when considered individually, patients with GVHD III-IV presented CsA concentrations that were more frequently lower than 200 ng/mL than patients with no GVHD. Despite this, the differences were not statistically significant.

IV. DISCUSSION

In our experience with the polyclonal antibody, we often found fluctuations of patients' CsA levels, which could be produced by different causes: alterations of liver function, drug interferences, and sometimes wrong sampling time or contamination from indwelling lines (6). These could not be distinguished until all the information on the patient was available. Having a specific antibody gave us hope of improving CsA monitoring.

The results obtained for the mean CsA concentration during the 7 days before the onset of GVHD show that there was no correlation between CsA concentration and GVHD severity, and when using the PAb, the highest levels were for those with more severe disease, GVHD degree III-IV, while the group monitored by the MoE showed the lowest mean levels, and were lower than the desired value of 200 ng/mL. For the 10 days prior to GVHD, there was again no correlation between the mean CsA concentration and severity of GVHD, because the coefficient of variation of the mean CsA levels was very high in all groups. When using the polyclonal antibody, the highest concentrations and dispersion were found in the group with more severe disease, 591 ± 304 ng/mL, while for groups O-I and II, mean levels were 453 ± 235 ng/mL and 446 ± 250 ng/mL, respectively. These results are different from previous reports (2), which found lower levels for the more severe GVHD, probably because the CsA dosage was not adjusted according to levels.

The mean cyclosporine concentration of patients controlled by the monoclonal-specific antibody showed the lowest levels for group GVHD III-IV, 186 ± 74 ng/mL, and the widest spread was found in group II, with a mean plus standard deviation of 275 ± 108 ng/mL. Levels in the group with no illness or very light symptoms were 236 ± 104 ng/mL, always higher than 200 ng/mL, which correlates better with immunosuppression, but differences were not statistically significant.

V. CONCLUSIONS

In our experience, the use of a monoclonal-specific antibody improves CsA monitoring in terms of technical aspects, especially reducing interferences, but is of little predictive value in the individual patient because of the considerable overlap among patients.

REFERENCES

1. L. M. Shaw et al. Critical issues in cyclosporine monitoring: Report of the Task Force on Cyclosporine Monitoring. *Clin. Chem.* 33:1269–1288 (1987).
2. G. Bondini, E. Strocchi, P. Ricci, et al. Cyclosporine A: Correlation of blood levels with acute graft versus host disease after bone marrow transplantation. *Acta Haematol.*, *78*:6–12 (1987).

3. H. Schmidt, G. Ehninger, R. Dopfer, et al. Correlation between low CsA plasma concentration and severity of acute GvHD, in bone marrow transplantation. *Blut, 57*:139–142 (1988).
4. A. Gratwohl, B. Speck, M. Wenk, I. Forster, M. Müller, B. Osterwalder, C. Nissen, and F. Follat. Cyclosporine in human bone marrow transplantation. *Transplantation 36*:40–44 (1988).
5. L. Bowers, D. Canafax, J. Singh, R. Seifedlin, R. Simmons, and J. Najarian. Studies of cyclosporine blood levels: Analysis, clinical utility, pharmacokinetics, metabolites and chronopharmacology. *Transplant. Proc. XVIII*(Suppl. 6):137–143 (1986).
6. E. Carreras, M. Lozano, R. Deulofeu, S. Roman, A. Grañena, and C. Rozman. Influence of different indwelling lines on the measurement of blood cyclosporin A levels. *Bone Marrow Transplant.* 3:637–639 (1988).

54

Fluorescence Polarization Immunoassay for Measurement of Cyclosporine A

J. M. Vergara, M. Benjumeda, M. L. Meca, A. Plata, J. Márquez, and L. Campos *Hospital del Servicio Andaluz de Salud de Cadiz, Cadiz, Spain*

I. INTRODUCTION

Cyclosporine A (CsA) is a cyclic polypeptide with 11 amino acids. It is a potent immunosuppressant that is not myelotoxic. CsA selectively inhibits interleukin-2, resulting in decreased activity of cytotoxic T-lymphocytes while sparing T-suppressor cells. In addition, it inhibits other lymphokines including γ-interferon and macrophage migration inhibition factor (1).

CsA is frequently required in patients who are receiving organ transplants and treatment for autoimmune diseases (2–4).

The dose of CsA administered does not correlate well with the clinical response for several reasons. Oral absorption is very variable. Blood concentrations are influenced by factors such as transplant type, the patient's age, disease state, and concurrent drug therapy. In addition, the optimal range of concentrations in blood for CsA is narrow, and drug toxicity, such as nephrotoxicity and hepatotoxicity, must be minimized (5,6). For all these reasons, knowing the CsA concentration in blood is essential for optimal therapy.

Several methods are available for measuring CsA, some of them specific for this drug and others that measure both CsA and its metabolites (7). The fluorescence polarization immunoassay (FPIA) uses a polyclonal antibody that reacts with both CsA and some its metabolites (8).

In our study we evaluate a FPIA method for CsA and metabolites measurement in whole blood and compare the results obtained in 30 patients who had renal transplants with the results of a specific RIA for CsA.

II. MATERIALS AND METHODS

Reagents consisted of two commercial immunoassay kits:

Table 1 Patients

	Without hepatic disease (N = 26)	With Hepatic disease (N = 4)
Age (years)	30 ± 10 (15–56)	43 ± 3.5 (38–46)
Treatment oral with CsA	1 month to 1 year	3 months to 5 years
Dose (mg/kg/day)	5.0 ± 2.0	3.2 ± 1.9
Mother treatments:		
Steroids	96%	3
Antihypertensives	42%	3
Azathioprine	30%	1

1. The TDx FPIA (Abbott Laboratories) for cyclosporine and metabolites in whole blood
2. The Cyclo-Trac SP (Incstar), RIA method, which uses a monoclonal antibody and an ^{125}I-labeled CsA tracer to measure CsA in blood. Manufacturers' instructions were followed for all kits.

We used whole-blood samples from 30 renal transplant patients who were receiving CsA (Table 1) Specimens were collected in EDTA-containing collection tubes.

In four patients we observed hepatic infection (GGT > 50 U/L and ALT > 50 U/L).

III. RESULTS

The between-run precision was determined by assaying three levels samples (low, medium, and high) on 20 consecutive days.

Recoveries of CsA with the FPIA were determined by adding cyclosporine to different whole-blood samples to obtain concentrations of 250, 500, 1000, and 1500 ng/mL.

The between-run variation and recoveries of CsA by the FPIA are presented in Table 2. The results of the FPIA and RIA analyses are shown in Table 3.

Twenty-six samples from renal transplant patients receiving CsA orally were measured by FPIA and RIA. The least-squares linear regression of the data gives the following equation (Figure 1):

$$\frac{\text{CsA}}{\text{Mets}} = 48.1 + 2.2 \times \text{CsA} \quad (r = 0.822;\ n = 26)$$

IV. DISCUSSION

FPIA for CsA and metabolites in whole blood has advantages over other polyclonal methods, because between-run CVs > 5.5% with the TD_x controls and the analytical recovery from whole blood is acceptable (100.8% to 113.0%).

There was considerable variation of the ratio of the RIA results to the FPIA (0.27 to 0.61) in the patients without hepatic disease. Cyclosporine is extensively metabolized by cytochrome P_{450} in the liver (9). Abbott's antibody appears to be more cross-reactive for metabolites M_{17}, M_{18}, $M_{203\text{-}218}$, M_{21}, and M_{25}, and much less cross-reactive for M_1, M_8, and M_{26} (8). The analytical results of FPIA are highly influenced by these metabolites. In patients with transplanted hearts or livers, it can influence the

Table 2 Accuracy Results

	Between-run precision of FPIA			
	Number	Means	SD	CV
Low	20	345	18	5.2%
Medium	20	1218	46	3.7%
High	20	1740	66	3.8%

	Recovery of CsA by FPIA		
Theoretical CsA (ng/mL)	Found (ng/mL)	Recovery (%)	Hto/Hb[a] (%-g/L)
250	257	102.8	35.7/12.1
500	503	100.8	37.6/12.8
1000	1010	101.0	41.0/14.2
1500	1695	113.0	41.0/14.2

[a]Hto, hematocrit; Hb, hemoglobin.

preponderance of metabolites relative to the drug. In our patients with hepatic disease, the ration of RIA/FPIA was 0.21 (range 0.12 to 0.32), lower than that of other patients (0.39). The toxic and immunosuppressive activities of metabolites are not definitively known.

In conclusion, FPIA is an accurate method for measurement of CsA and metabolites in whole blood, and it has a good recovery of CsA in the levels of concentration assayed.

In patients with hepatic disease it is necessary to use both specific and nonspecific methods for CsA assays, because in these patients their exist larger proportions of metabolites.

V. SUMMARY

The fluorescence polarization immunoassay (FPIA) is a method for monitoring CsA in whole blood. It uses a polyclonal antibody which reacts with both CsA and some of its metabolites. In our study we evaluated the FPIA method for CsA and metabolites in

Table 3 Patient Results

	CsA (RIA) (ng/mL)	CsA (FRIA) (ng/mL)	CsA (CsA + Mets)
Without hepatic disease			
Number	26	26	26
Means	125 (46–209)	325 (117–658)	0.39 (0.27–0.61)
SD	47	126	0.08
With hepatic disease	4	4	4
Number	166 (36–406)	706 (212–1254)	0.21 (0.12–0.32)
Mean	164	455	0.08
SD			

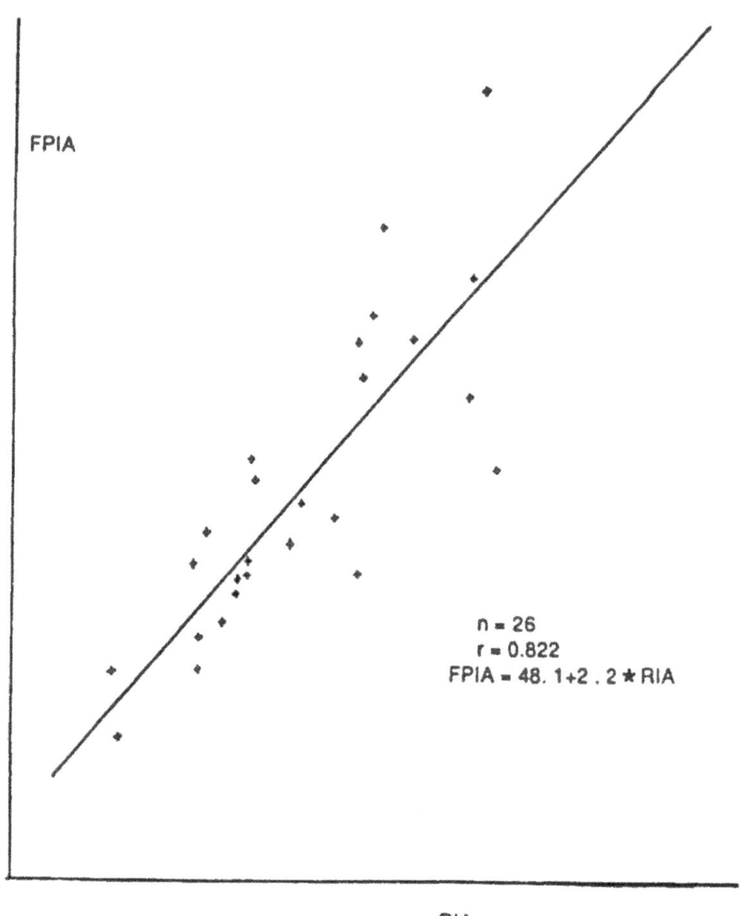

Figure 1 Correlation of RIA and FPIA for CsA.

whole blood and compared it with the results of a specific RIA for CsA (Cyclo-trac SP, Incstar) using 30 patients who had had renal transplants.

Accuracy assays were made at three levels: low (345 ng/mL, CV 5.2%), medium (1218 ng/mL, CV 3.7%), and high (1740 ng/mL, CV 3.8%). Recovery assays were made by adding CsA to several blood samples in order to obtain concentrations of 250, 500, 1000, and 1500 ng/mL. Recoveries varied between 100% (500 ng/mL) and 113% (1500 ng/mL).

Four of 30 patients studied had hepatic abnormalities (GGT > 50 U/L and ALT > 50 U/L). They had a lower proportion of CsA (CsA + Mets) (0.21) than the rest of the patients (0.39). In patients without hepatic abnormalities, the correlation of FPIA versus RIA gave r = 0.822 and a slope of 2.2 in the regression.

REFERENCES

1. C. Piet and M. D. de Groen. Cyclosporine: A review and its specific use in liver transplantation. *Mayo Clin. Proc.*, 64:680–689 (1989).

2. L. M. Shaw, L. Bowers, L. Demers, et al. Critical issues in cyclosporine monitoring: Report of the Task Force on Cyclosporine Monitoring. *Clin. Chem.*, *33*:1269–1288 (1987).
3. E. Carreras, R. Deulofeu, A. Grañena, and C. Rozman. Ciclosporina A en el Trasplante de Médula Osea. *Rev. Diag. Biol.*, *38*:239–243 (1989).
4. N. Talal. Cyclosporine as an immunosuppressive agent for autoimmune disease: Theoretical concepts and therapeutic strategies. *Transplant. Proc.*, *XX*:11–15 (1988).
5. P. A. Keown and R. S. Calvin. Ciclosporina: una espada de doble filo. *Hosp. Prac.* (Ed. Esp.), 3:51–65 (1988).
6. J. P. Bantle, M. S. Paller, R. J. Boudreau, M. T. Olivari, and T. F. Ferris. Long-term effects of cyclosporine on renal function in organ transplant recipients. *J. Lab. Clin. Med.*, *115*:233–240 (1990).
7. E. B. Kenneth, H. M. Sameh, D. F. Henry, and L. G. Ronald. A validation study of selected methods routinely used for measurement of cyclosporine. *Clin. Chem.*, *36*:670–674 (1990).
8. G. L. Lensmeyer, D. A. Wlebe, I. H. Carlson, and D. J. de Vos. Three commercial polyclonal immunoassays for cyclosporine in whole blood compared: 2. Cross-reactivity of the antisera with cyclosporine metabolites. *Clin. Chem.* *36*:119–123 (1990).
9. Y. Hayashi, N. Shibata, T. Minouchi, H. Shibata, T. Ono, and H. Shimakawa. Evaluation of fluorescence polarization immunoassay for determination of cyclosporine in plasma. *Ther. Drug. Monitor.*, *11*:205–209 (1989).

55

A Homogeneous Enzyme Immunoassay for Phenytoin (CEDIA) and Application on the Hitachi 704

T. Morton, S. Vannucci, J. Murakami, V. Hertle, F. Davoudzadeh *Microgenics Corporation, Concord, California* **D. Engel, P. Kaspar** *Boehringer Mannheim GmbH, Tutzing, Germany* **W. Coty, R. Loor, P. Khanna** *Microgenics Corporation, Concord, California*

I. INTRODUCTION

Phenytoin is a highly effective drug used in the treatment of most types of epilepsy. Unlike some drugs, it exerts antiepileptic activity without causing general depression of the central nervous system. Phenytoin exhibits a narrow range of therapeutic effect, and individuals show wide variability in rates of drug absorption, metabolism, and clearance (1–4). In addition, similar symptoms are seen with both under- and overdosage. The toxic effects of phenytoin include central nervous system symptoms such as nystagmus (involuntary, cyclic eye movements), ataxia, diplopia, and vertigo, as well as behavioral alterations, disorders of folate absorption and/or metabolism, and altered vitamin D metabolism and/or action with resulting osteomalacia. As a result, monitoring the serum phenytoin concentrations is essential for achieving optimal therapeutic effects and avoiding drug toxicity (1–3).

We have developed a homogeneous enzyme immunoassay for phenytoin using the CEDIA technology (5,6). In this method, genetically engineered, inactive fragments of *E. coli* β-galactosidase have been designed which can spontaneously recombine to form active enzyme. The sequence of the small fragment, termed enzyme donor (ED), has been modified to allow covalent attachment of phenytoin in a manner that does not affect recombination with the large fragment (enzyme receptor, or EA). However, binding of a phenytoin-specific antibody to the ED-phenytoin conjugate inhibits the reassociation of ED and EA. Phenytoin in a sample can bind to the antibody, preventing binding of the ED-phenytoin conjugate. The resulting unbound conjugate is available for reaction with EA, resulting in enzyme formation. The β-galactosidase formed in the reaction is measured photometrically using a chromogenic substrate, producing an increasing rate of change of absorbance with increasing concentrations of phenytoin in the sample. In this report, we describe the characteristics of the CEDIA phenytoin assay as applied to the Hitachi 704 automated analyzer.

II. MATERIALS AND METHODS

A. Reagents

The CEDIA Phenytoin Assay Kit contains two lyophilized reagents. Reagent 1 contains EA and phenytoin-specific monoclonal antibody; reagent 2 contains ED-phenytoin conjugate and chlorophenolred-β-D-galactopyranoside (CPRG) as substrate. Both reagents contains buffer salts, stabilizers, and antimicrobial agents. Two reagent-specific reconstitution buffers are provided; the kit is sufficient for approximately 50 determinations on the Hitachi 704, and can be used for at least 30 days after reagent constitution.

The CEDIA Phenytoin Assay Kit also includes 0.0 and 40.0 mg/L phenytoin calibrators in a protein matrix containing stabilizers. The kit calibrators were formulated to match calibrators containing U.S.P. reference phenytoin in drug-free human serum; the reference drug was dried under vacuum according to the instructions supplied by U.S.P.

Phenytoin and drugs used for cross-reactivity studies were purchased from Sigma Chemical Co. (St. Louis, Missouri); Aldrich Chemical Co. (Milwaukee, Wisconsin); Applied Sciences (Deerfield, Illinois); and Abbott Laboratories (Abbott Park, Illinois). Patient serum samples were obtained from hospitals and clinical laboratories. Human serum albumin, human gamma-globulin, and bilirubin were from Sigma; triglyceride stock solutions (IntraLipid) were purchased from KabiVitrum, Inc. (Alameda, California). Phenytoin Fluorescence Polarization Immunoassay Kits (Abbott TDx) were obtained from Abbott Laboratories, and were used in conjunction with an Abbott TDx analyzer according to the instructions supplied by the manufacturer.

The CEDIA phenytoin EA and ED reagents were dissolved in 14 and 12 mL of their respective reconstitution buffers, mixed, and chilled to the temperature of the Hitachi reagent compartment. Reagents and calibrators were stored at 2 to 8°C when not in use.

B. Assay Procedure

The CEDIA phenytoin assay was carried out in the Hitachi 704 automated analyzer as follows: Four microliters of calibrator, commercial control, or unknown sample was pipetted into a reaction curvette, followed by addition of 200 µl of reagent 1, containing EA and antiphenytoin antibody. This solution was mixed and incubated at 37°C for 5 min, and then 150 µL of reagent 2, containing ED-phenytoin conjugate and substrate, were added and the incubation was continued for 4.5 min. Photometric measurements at 570 nm were taken at 20-s intervals between 4.5 and 5.5 min after reagent 2 addition, and the rate of substrate hydrolysis in mAU/min was determined by instrument software, after correction for absorbance at the secondary wavelength (660 nm). Phenytoin concentrations in unknown samples were calculated by linear interpolation from a two-point standard curve using the rates obtained with 0.0- and 40.0-mg/L phenytoin calibrators.

III. RESULTS AND DISCUSSION

A. Calibration Curve

In the CEDIA phenytoin assay, sample is first incubated with antibody and EA. During this time, phenytoin present in the sample binds to the antibody; EA does not participate in the reaction. ED-phenytoin conjugate is then added to the reaction mixture, and can either bind to unoccupied antibody binding sites or react with EA to form active enzyme. The amount of β-galactosidase formed is then measured as the rate of CPRG hydrolysis at 570 nm. As shown in Figure 1, the CEDIA phenytoin assay exhibits a linear dose-response relationship between 0 and 40 mg phenytoin/L ($r > 0.999$), with a slope of 5.1 mAU/min/mg phenytoin/L. The concentration of phenytoin is unknown samples can be determined automatically by linear interpolation from the rates and phenytoin concentrations of the 0.0- and 40-mg/L phenytoin calibrators.

B. Sensitivity and Precision

Sensitivity of CEDIA phenytoin assay was estimated as twice the intraassay standard deviation of the 0.0-mg/L phenytoin calibrator. The mean of three determinations on separate instruments was 0.4 mg/L.

Intraassay precision was determined by testing of multiple replicates ($n = 20$) of commercial control sera within the same run. Three concentrations of phenytoin were tested; the results obtained with three different instruments are summarized in Table 1. Average CVs obtained were 2.6%, 1.7%, and 0.9% at 7.2, 12.2, and 26.8 mg phenytoin/L, respectively. Interassay precision was determined by testing single replicates of these three control sera on each of 20 runs performed on three separate instruments during a 3-week period. The results were 7.3 ± 0.2 mg/L (2.7% CV), 12.6 ± 0.4 mg/L (3.2% CV), and 26.8 ± 0.5 mg/L (1.9% CV) for the low, mid, and high controls, respectively.

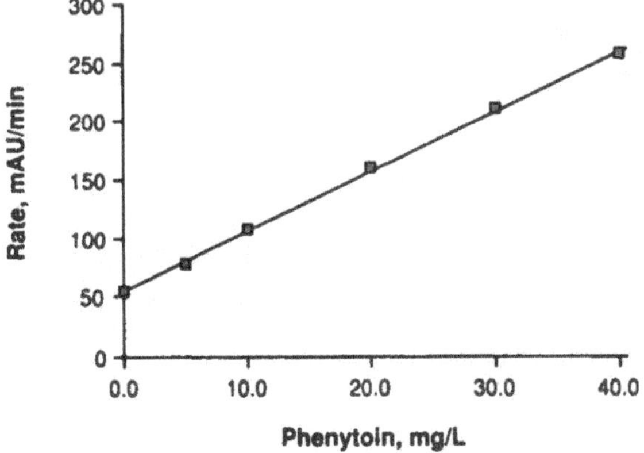

Figure 1 CEDIA phenytoin assay dose-response curve. Calibrators containing varying concentrations of phenytoin were prepared by serial dilution and tested in the CEDIA phenytoin assay. The rates were calculated from the measured phenytoin concentrations of the calibrators and the calibrator rates printed by the analyzer.

Table 1 CEDIA Phenytoin Intraassay Precision

	Instrument A	Instrument B	Instrument C	Average
Control level 1				
Mean value (mg/L)	7.2	7.1	7.4	7.2
Standard deviation (mg/L)	0.16	0.26	0.17	0.19
CV (%)	2.2	3.6	2.3	2.6
Control level 2				
Mean value (mg/L)	12.4	12.1	12.2	12.2
Standard deviation (mg/L)	0.16	0.24	0.22	0.21
CV (%)	1.3	1.9	1.8	1.7
Control level 3				
Mean value (mg/L)	26.8	26.7	26.9	26.8
Standard deviation (mg/L)	0.27	0.23	0.26	0.25
CV (%)	1.0	0.9	1.0	0.9

C. Accuracy

Accuracy of the CEDIA phenytoin assay was assessed by method comparison. One hundred serum samples from patients under treatment with phenytoin were tested by the CEDIA method and by FPIA. The results, shown in Figure 2, indicate an excellent correlation between these two methods ($r = 0.993$), with close agreement between the values obtained: CEDIA = 0.99 • FPIA-1.95 mg/L; $s_{y,x} = 1.1$ mg/L.

Accuracy was also evaluated by dilution of patient samples containing high concentrations of phenytoin with either low-concentration samples or with the 0.0-mg/L phenytoin calibrator. The results, summarized in Table 2, show recoveries ranging from 100.0% to 106.4% of the expected values.

D. Specificity

Specificity was evaluated by testing of 28 phenytoin metabolites, analogs, and other antiepileptic agents for cross-reactivity in the CEDIA phenytoin assay. As shown in Table 3, no compound tested exhibited more than 6% cross-reactivity. The major metabolite of phenytoin is 5-(4-hydroxyphenyl)-5-phenylhydantoin (*p*-hydroxyphenyl phenylhydantoin); about 60 to 70% of the drug is converted to this pharmacologically inactive compound (4). The CEDIA phenytoin assay is highly specific for the parent drug, since the cross-reactivity for this metabolite is less than 2%.

The CEDIA phenytoin assay was also evaluated for sensitivity to patient sample interferences (7). No change in phenytoin quantitation of more than 10% was observed in the presence of up to 100 g/L total protein, 10 mg/L triglyceride, 4 g/L hemoglobin, or 300 mg/L bilirubin.

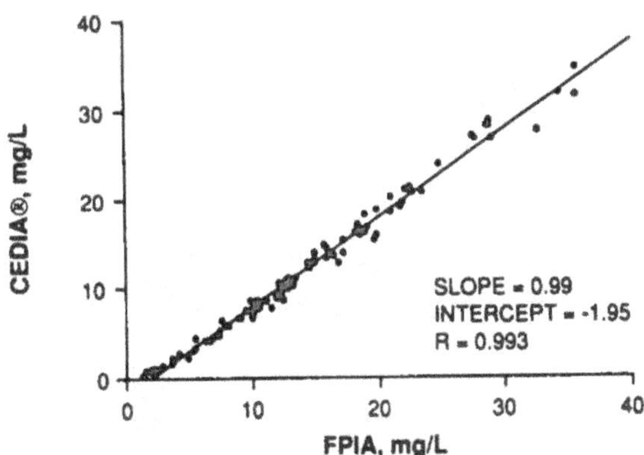

Figure 2 Method comparison study. Serum samples from patients being monitored for phenytoin levels were tested by the CEDIA phenytoin assay and by FPIA (Abbott TDx). The line shown in the figure represents a best fit to the data using least-squares regression analysis.

Table 2 Recovery and Linearity of Dilution

% of high sample	Measured phenytoin (mg/L)	Expected phenytoin (mg/L)	Recovery (%)
Recovery:			
100	38.2		
75	29.2	28.6	102.4
50	20.1	18.9	106.4
25	9.7	9.2	105.3
0	0		
100	33.8		
75	25.8	25.3	102.4
50	17.2	16.7	103.0
25	8.4	8.1	103.1
0	0		
100	30.9		
75	23.2	23.1	100.7
50	15.7	15.2	102.8
25	7.4	7.4	100.7
0	0		
Linearity:			
100	30.9		
75	23.4	23.3	100.6
50	16.0	15.6	102.5
25	8.3	8.0	102.9
0	0.4		

Mixtures were prepared containing varying proportions of high-phenytoin concentration samples with either low-phenytoin concentration samples (recovery) or with the zero phenytoin calibrator (linearity). These mixtures were tested by the CEDIA phenytoin assay, and the measured values were compared with the results expected from the concentrations of the undiluted samples.

Table 3 CEDIA Phenytoin Assay Specificity

Cross-reactant	Cross-reactivity
p-Hydroxyphenylphenylhydantoin (P-HPPH)	1.8
p-HPPH glucoside	<0.1
5-(*p*-Methylphenyl)-5-PH	5.5
5-Ethyl-5-phenyhydantoin	<1.0
Hydantoin	<0.1
Amobarbital	<1.0
Amitryptyline	<0.1
Carbamazepine	<1.0
Chlorazepate	<0.4
Chlordiazepoxide	<1.0
Chlorpromazine	<0.1
Diazepam	1.5
Ethosuximide	<0.5
Ethotoin	<0.5
Glutethimide	7.8
Imipramine	0.25
Mephenytoin	<0.5
Mephobarbital	1.3
Methsuximide	<0.5
2-Phenyl-2-ethylmalonamide (PEMA)	<1.0
Pentobarbital	<0.5
Phenobarbital	<0.4
p-Hydroxyphenobarbital	<1.0
Phensuximide	<0.5
Promethazine	1.2
Secobarbital	<0.5
Valproic acid	<0.1
Primidone	<1.0

Cross-reactivity was calculated as (apparent phenytoin concentration ÷ concentration of cross-reactant) × 100%. Where possible, cross-reactivity was measured at a response equal to that obtained with a sample containing 20 mg/L phenytoin (the ED_{50}). If the highest concentration of cross-reactant gave a response of less than 0.4 mg/L phenytoin, the cross-reactivity was estimated as less than (0.4 mg/L ÷ the highest concentration of cross-reactant tested) × 100%.

IV. CONCLUSION

The data described in this report demonstrate that the CEDIA phenytoin assay is a sensitive, precise, accurate, and specific method for the determination of phenytoin in human serum. When used on conjunction with the Hitachi 704 automated analyzer, the CEDIA method provides a rapid, convenient, and fully automated system for monitoring phenytoin therapy. The linear standard curve of the CEDIA phenytoin assay is a particularly advantageous feature, allowing two-point calibration, and eliminating inaccuracies due to complex mathematical models required for fitting of nonlinear dose-response curves. The linear standard curve also provides improved precision at the extremes of the assay range when compared to methods with sigmoid dose-response curves.

The CEDIA technology has been successfully demonstrated for a variety of analytes, including digoxin, T4, and vitamin B12 (5,6), as well as more recently for theophylline and phenobarbital (in these proceedings). CEDIA methods allow the clinical laboratory to perform a variety of moderate- to high-sensitivity ligand assays without the requirement for dedicated immunoassay equipment—a significant advantage for laboratories with low testing volumes. Laboratories with large numbers of tests can also make use of CEDIA products by application to high-throughput instruments such as the Hitachi series of clinical chemistry analyzers. Thus, methods such as the CEDIA phenytoin assay are uniquely suited to the present needs of the clinical chemistry laboratory.

REFERENCES

1. F. Buchthal and M. A. Lennox-Buchthal. Diphenylhydantoin: Relation of anticonvulsant effect to concentration in serum. In *Antiepileptic Drugs* (D. M. Woodbury, J. K. Penry, and K. P. Schmidt, eds.), Raven Press, New York, pp. 193–209 (1972).
2. H. Kutt. *Diphenylhydantoin: Relation of plasma levels to clinical control. In Antiepileptic Drugs* (D.M. Woodbury, J. K. Penry, and K. P. Schmidt, eds.), Raven Press, New York, pp. 211-218 (1981).
3. F. Buchthal and O. Svensmark. Serum concentrations of diphenylhydantoin (Phenytoin) and phenobarbital and their relation to therapeutic and toxic effects. *Psychiat. Neurol. Neurochir.*, 74:117–136 (1971).
4. W. J. Jusko. Bioavailability and disposition kinetics of phenytoin in man. In *Quantiative Analytic Studies in Epilepsy* (P. Kellaway and I. Petersen, eds.), Raven Press, New York, pp. 115–136 (1976).
5. D. R. Henderson, S. B. Friedman, J. D. Harris, W. B. Manning, and M. A. Zoccoli, CEDIA, a new homogeneous immunoassay system. *Clin. Chem.*, 32:1637–1641 (1986).
6. P. K. Khanna, R. T. Dworschack, W. B. Manning, and J. D. Harris. A new homogeneous enzyme immunoassay using recombinant enzyme fragments. *Clin. Chem. Acta, 1985:* 231–240 (1989).
7. M. R. Glick and K. W. Ryder. *Interferographs: A User's Guide to Interferences in Clinical Chemistry Instruments*. Science Enterprises, Indianapolis, 1987.

56

A Homogeneous Enzyme Immunoassay for Theophylline (CEDIA) and Application on the Hitachi 704

D. Jenkins, S. Horgan, S. Whang, S. Rinne, F. Davoudzadeh *Microgenics Corporation, Concord, California* **D. Engel, P. Kaspar** *Boehringer Mannheim GmbH, Tutzing, Germany* **W. Coty, P. Khanna** *Microgenics Corporation, Concord, California*

I. INTRODUCTION

Theophylline, or 1,3-dimethylxanthine, which acts to stimulate central nervous system function, diuresis, and cardiac muscle, and to relax smooth muscle (1), is used in the treatment of asthma and neonatal apnea. However, the toxic effects of excessive theophylline are significant, necessitating careful control of dosage. Although the absorption and metabolism of theophylline varies widely among individuals (2–4), numerous studies have shown a close relationship between drug concentrations in the blood and therapeutic effectiveness (5,6). High concentrations of theophylline in the circulation are also closely correlated to toxic symptoms, such as headache, palpitation, dizziness, nausea, hypotension, tachycardia, agitation, seizures, and convulsions (4,7–9). As a result, monitoring of serum drug concentrations is considered essential in the safe treatment of patients with theophylline.

To address the need for simple, rapid, and convenient methods for measurement of serum theophylline, we have developed a homogeneous enzyme immunoassay for theophylline using the CEDIA technology (10). In this method, theophylline is conjugated to a small, inactive peptide fragment of *E. coli* β-galactosidase (enzyme donor, or ED). Alone, the ED-theophylline conjugate can recombine with a larger inactive enzyme fragment (enzyme acceptor, or EA) to form fully active β-galactosidase. In the assay, antibody specific for theophylline binds to the ED-theophylline conjugate, preventing its association with EA. Theophylline present in a sample competes for binding to the antibody, making ED-theophylline available for reaction with EA. The resulting enzyme formed is detected by hydrolysis of the chromogenic substrate chlorophenolred-β-D-galactopyranoside (CPRG). The theophylline concentration in an unknown sample can be determined by comparison of the rate of substrate hydrolysis in the presence of sample to a standard curve of rate versus theophylline concentration. The present report describes the performance of the CEDIA theophylline assay as applied to the Hitachi 704 automated analyzer.

II. MATERIALS AND METHODS

A. Materials

The CEDIA theophylline assay kit contains two lyophilized reagents: EA reagent, containing enzyme acceptor and monoclonal antitheophylline antibody; and ED reagent, containing ED-theophylline conjugate, secondary antibody (goat antimouse IgG), and substrate; both reagents contain buffer salts, carrier protein, and stabilizers. Specific buffers are provided for reconstitution of the EA and ED Reagents. Each kit contains sufficient reagents for approximately 75 determinations on the Hitachi 704, and can be used for at least 30 days after reconstitution.

The kit also contains zero and high theophylline calibrators in a protein solution containing stabilizers. The theophylline concentrations of the calibrators were determined by comparison with reference calibrators prepared from U.S.P. standard theophylline dissolved in human serum. The U.S.P. drug used in these calibrators was subjected to vacuum drying according to the instructions provided by U.S.P.

Commercial control materials used in precision studies (Fisher TheraChem) were obtained from Fisher Scientific Co. Human serum specimens were from clinical laboratories and hospitals. Reference theophylline drug was from U.S.P.; other drugs and theophylline analogs used for cross-reactivity studies were from Sigma Chemical Co. (St. Louis, Missouri), except for Terbutaline, which came from Ciba-Geigy Corp. (Suffern, New York). Human gammaglobulin and bilirubin were purchased from Sigma; triglyceride solution (Intralipid) was from KabiVitrum, Inc. (Alameda, California). Human hemoglobin was prepared by freeze-thaw lysis of freshly isolated erythrocytes, and the hemolysate was stored at –70°C until use.

Reagents for the measurement of hemoglobin, total protein, bilirubin, and triglyceride were from Sigma Chemical Co., and these assays were performed on a COBAS BIO analyzer according to the instructions provided by the manufacturer. Theophylline fluorescence polarization immunoassay (FPIA) reagents (TDx theophylline II) were obtained from Abbott Laboratories (Abbott Park, Illinois), and were used in conjunction with an Abbott TDx analyzer according to the directions supplied with the instrument.

B. Preparation of Reagents

The CEDIA theophylline EA and ED reagents were dissolved in 20 and 16 mL of their respective reconstitution buffers, mixed, and chilled to the temperature of the Hitachi reagent compartment. Reagents and calibrators were stored at 2–8°C when not in use.

C. Assay Procedure

Each CEDIA theophylline assay was performed on the Hitachi 704 as follows: Four microliters of sample (calibrator, control, or unknown specimen) were pipetted into a reaction cuvette, followed by 200 µL of EA reagent. This solution was incubated for approximately 5 min at 37±C. Subsequently, 150 µL of ED reagent was added, and the incubation was continued for an additional 4.5 min. Absorbance measurements were taken at 570 and 660 nm, at approximately 20-s intervals between 4.5 and 5.5 min after

ED reagent addition. The rate of substrate hydrolysis was measured at 570 nm of mAU/min, after correction for absorbance at 660 nm. Theophylline concentrations in unknown samples were automatically calculated by instrument software from a two-point standard curve using the rates obtained with zero and high theophylline calibrators. A "stat" result is available with approximately 10 min, and subsequent results are obtained every 20 s.

III. RESULTS

A. Dose-Response Curve

In the CEDIA theophylline assay, sample is first incubated with EA reagent containing antitheophylline antibody. During this step, theophylline present in the sample binds to antibody; EA does not participate in the reaction. Subsequently, ED-theophylline conjugate added in the ED reagent can bind to unoccupied antibody binding sites, or can react with EA to form active β-galactosidase. With increasing concentrations of theophylline in the sample, an increasing amount of ED-theophylline conjugate becomes available for enzyme formation. Thus the rate of substrate hydrolysis is directly proportional to the theophylline concentration in the sample.

As shown in Figure 1, the CEDIA theophylline assay exhibits a linear dose-response relationship between 0 and 40 mg theophylline/L, with a slope of 5.1 mAU/min/mg theophylline/L. As a result, a standard curve can be obtained using only two calibrators, and the concentrations of theophylline in unknown samples can be determined by linear interpolation from the rates and theophylline concentrations of these two calibrators.

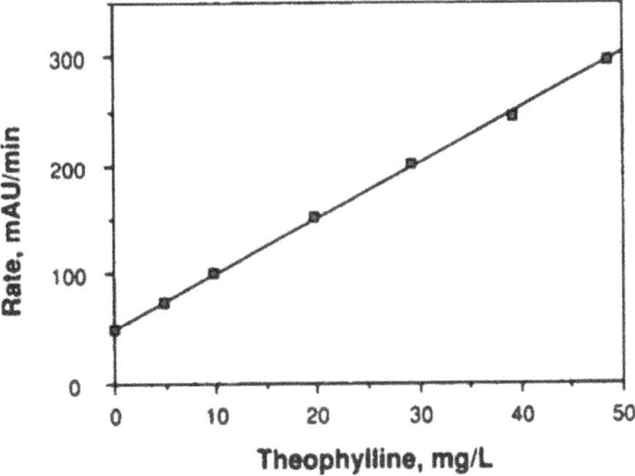

Figure 1 CEDIA theophylline assay dose-response curve. Calibrators containing varying concentrations of theophylline were prepared by serial dilution and tested in the CEDIA theophylline assay. The rates were calculated from the measured theophylline concentrations of the calibrators and the calibrator rates printed by the analyzer.

B. Sensitivity and Precision

Sensitivity of the CEDIA theophylline assay (least detectable analyte concentration at 95% confidence limit) was estimated as twice the intraassay standard deviation of the zero calibrator. Three separate determinations on different instruments gave values of 0.25, 0.27, and 0.54 mg/L, for an average sensitivity of 0.35 mg/L.

Intraassay precision of the CEDIA theophylline assay was determined by 20 replicate measurements of commercial control sera at three concentrations of theophylline; three different instruments were used in each study. The results, summarized in Table 1, show coefficients of variation (CVs) of 2% to 4% at the low control, 2% to 6% at the mid-level, and 1.8% to 2.5% at the high control level.

Interassay precision was determined by 38 replicate measurements of commercial control sera at three concentrations of theophylline. One replicate of each level was tested during each run of the current study over a period of approximately 2.5 weeks using five different instruments. The control values were measured as 5.5 ± 0.25, 15.2 ± 0.8, and 32.4 ± 0.9 mg/L for the low, mid, and high levels, respectively, yielding intraassay CVs of 4.5%, 5.3%, and 2.8%.

C. Accuracy

Accuracy of the CEDIA theophylline assay was evaluated by a method comparison study. Ninety-six serum samples from patients treated with theophylline were tested by the CEDIA and FPIA methods. Analysis by least-squares regression shows a very close relationship between CEDIA and FPIA: CEDIA = 1.06 × FPIA − 0.82 mg/L. The correlation coefficient was 0.997, and the standard error of the Y estimate was 0.72 mg/L.

Accuracy was also evaluated by recovery of analyte when human serum samples containing high concentrations of theophylline were diluted with samples containing low drug concentrations. The results, shown in Table 2, show close agreement between the expected and measured concentrations of theophylline in the various

Table 1 CEDIA Theophylline Intraassay Precision

	Instrument A	Instrument B	Instrument C	Average
Control level 1				
Mean value (mg/L)	5.4	5.4	5.2	5.3
Standard deviation (mg/L)	0.10	0.19	0.15	0.15
CV (%)	1.9	3.5	2.9	2.8
Control level 2				
Mean value (mg/L)	15.0	15.5	15.3	15.3
Standard deviation (mg/L)	0.39	0.39	0.30	0.36
CV (%)	2.6	2.5	2.0	2.4
Control level 3				
Mean value (mg/L)	30.9	33.9	32.4	32.4
Standard deviation (mg/L)	0.77	0.60	0.64	0.67
CV (%)	2.5	1.8	2.0	2.1

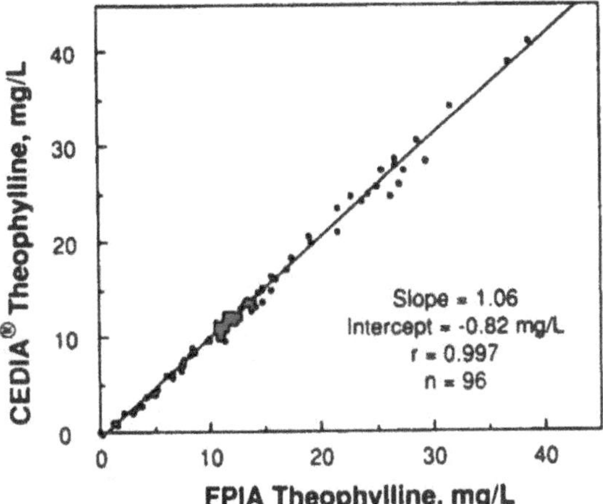

Figure 2 Method comparison study. Serum samples from patients being monitored for theophylline levels were tested by the CEDIA theophylline assay and by FPIA (Abbott TDx). The line shown in the figure represents a best fit to the data using least-squares regression analysis.

dilutions. Similar studies were also performed with the zero calibrator instead of human serum samples; these results show excellent recovery, supporting the accuracy of two-point calibration of the CEDIA method.

Plasma samples prepared using sodium heparin or EDTA as anticoagulants were evaluated in the CEDIA theophylline assay. The mean values of 10 theophylline-free plasma samples were within the average sensitivity determined for the theophylline assay for each anticoagulant. Recovery of theophylline added to heparin or EDTA plasma pools was 103% and 104%, respectively, compared to 99% with a serum pool. Dilutions of the theophylline-containing serum and plasma pools with the corresponding unspiked pools gave recoveries of 100 ± 3%.

D. Specificity

Cross-reactivity studies were performed to assess the specificity of the CEDIA theophylline assay to structural analogs and metabolites of theophylline. As shown in Table 3, the CEDIA theophylline assay shows low cross-reactivity to common theophylline analogs such as caffeine (1,3,7-trimethylxanthene; 2.7%) and theobromine (3,7-dimethylxanthine; 1.5%), as well as a major serum metabolite of theophylline (11), 3-methylxanthine (1.3%).

Specificity was also evaluated by interference studies using the method of Glick and Ryder (12). Human hemoglobin, human gammaglobulin, bilirubin, or triglyceride were added to pools of human serum at three different levels of theophylline (approximately 8, 15, and 23 mg/L), and the effect on recovery by the CEDIA method determined. No interference (defined as a change in theophylline recovery of more than ±10% from the control) was observed with up to 10 g/L hemoglobin, 100

Table 2 Recovery and Linearity of Dilution

% of high sample	Measured theophylline (mg/L)	Expected theophylline (mg/L)	Recovery (%)
Recovery:			
100	29.9		
75	22.9	22.7	100.9
50	15.9	15.5	102.5
25	8.7	8.3	104.8
0	1.1		
100	30.4		
75	24.3	23.9	101.7
50	17.9	17.4	102.9
25	11.1	10.9	101.8
0	4.4		
100	29.3		
75	23.7	23.8	99.6
50	18.7	18.3	102.2
25	13.0	12.8	101.6
0	7.3		
Linearity:			
100	29.0		
75	21.7	21.8	99.5
50	14.6	14.6	100.0
25	7.3	7.3	100.0
0	0.1		

Mixtures were prepared containing varying proportions of high-theophylline-concentration samples with either low-theophylline-concentration samples (recovery) or with the zero theophylline calibrator (linearity). These mixtures were tested by the CEDIA theophylline assay, and the measured values were compared with the results expected from the concentrations of the undiluted samples.

g/L total protein (obtained by addition of human gammaglobulin to samples containing approximately 6.5 g total protein/dL), 20 g/L triglycerides, or 200 mg/L bilirubin.

IV. CONCLUSIONS

The studies described above show that the CEDIA theophylline assay is a reproducible, accurate, and specific method for the determination of theophylline in human serum. Application of this method to the Hitachi 704 clinical chemistry analyzer provides a convenient system for fully automated monitoring of serum theophylline levels. The capabilities of Hitachi analyzers allow flexibility in performing high-volume batch measurement of theophylline, incorporation of individual theophylline assays during the regular workload, and "stat" theophylline determinations when required.

Table 3 CEDIA Theophylline Assay Specificity

Cross-reactant	Cross-reactivity (%)
1,3-Dimethyluric acid	10.8
1-Methylxanthine	1.2
1,7-Dimethyluric acid	<0.2
3,7-Dimethyluric acid	0.35
Proxyphylline (7-[β-hydroxypropyl]theophylline)	0.9
Caffeine	2.7
Xanthine	<0.15
7-Methylxanthine	<0.4
7-(2-Hydroxyethyl)theophylline	1.1
Uric acid	<0.2
3-Methylxanthine	1.3
Terbutaline	<0.1
Xanthosine	<0.2
1,7-Dimethylxanthine (Paraxanthine)	1.4
Urea	<0.2
3-Methyluric acid	<0.4
Theobromine	1.5
8-Chlorotheophyllinle	6.5
Hypoxanthine	<0.4
1,3,7-Trimethyluric acid	0.4
Allopurinol	<0.2
7-Methyluric acid	<0.2
Diprophylline (Diphylline)	0.8
1-Methyluric acid	<0.2

Cross-reactivity was calculated as (apparent theophylline concentration ÷ concentration of cross-reactant) × 100%. Where possible, cross-reactivity was measured at a response equal to that obtained with a sample containing 20 mg/L theophylline (the ED_{50}). If the highest concentration of cross-reactant gave a response of less than 0.4 mg/L theophylline, the cross-reactivity was estimated as less than (0.4 mg/L ÷ the highest concentration of cross-reactant tested) × 100%.

REFERENCES

1. T. W. Wall, The xanthines. In *The Pharmacological Basis of Therapeutics* (A. G. Gilman, L. S. Goodman, and A. Gilman, eds.), Macmillan, New York, pp. 592–607 (1980).
2. R. G. Van Dellen. Clinical pharmacology. Series on pharmacology in practice. *Mayo Clin. Proc.*, 54:733–745 (1979).
3. C. A. Miller, L. B. Slusher, and E. S. Vesell. Polymorphism of theophylline metabolism in man. *J. Clin. Invest.*, 75:1415–1425 (1985).
4. R. I. Ogilvie. Clinical pharmacokinetics of theophylline. *Clin. Pharmacokinet.*, 3:267–293 (1978).
5. L. Hendeles and M. M. Weinberger. Theophylline therapeutic use and serum concentration monitoring. In *Individualizing Drug Therapy: Practical Applications of Drug Monitoring* (W. J. Taylor and A. L. Finn, eds.), Gross Townsend Franc, New York, pp. 31–66 (1981).
6. J. V. Aranda, D. S. Sitar, W. D. Parsons, P. M. Loughnan, and A. H. Neims. Pharmacokinetic aspects of theophylline in premature newborns. *N. Engl. J. Med.*, 295:413–416 (1976).
7. M. H. Jacobs, R. M. Senior, and G. Kessler. Clinical experience with theophylline. Relationships between dosage, serum concentration, and toxicity. *J. Am. Med. Assoc.*, 235:1983–1986 (1976).

8. M. W. Weinberger, R. A. Matthay, E. J. Ginchansky, Ch. A. Chidsey, and T. L. Petty. Intravenous aminophylline dosage. Use of serum theophylline measurement for guidance. *J. Am. Med. Assoc.*, *235*:2110–2113 (1976).
9. C. W. Zwillich, F. D. Sutton, T. A. Neff, W. M. Cohn, R. A. Matthay, and M. M. Weinberger. Theophylline-induced seizures in adults. Correlation with serum concentrations. *Ann. Intern. Med.*, *82*:784–787 (1975).
10. D. R. Henderson, S. B. Friedman, J. D. Harris, W. B. Manning, and M. A. Zoccoli. CEDIA™, a new homogeneous immunoassay system. *Clin. Chem.*, *32*:1637–1641 (1986).
11. R. D. Thompson, H. T. Nagasawa, and J. W. Jenne. Determination of theophylline and its metabolites in human urine and serum by high-pressure liquid chromatography. *J. Lab. Clin. Med.*, *84*:584–593 (1974).
12. M. R. Glick and K. W. Ryder. *Interferographs: A User's Guide to Interferences in Clinical Chemistry Instruments*. Science Enterprises, Indianapolis, IN (1987).

57

A Homogeneous Enzyme Immunoassay for Phenobarbital (CEDIA) and Application on the Hitachi 704

H. Scholz, P. Pedriana, S. Vannucci, N. Mehta, F. Davoudzadeh, J. Fong, M. Krevolin *Microgenics Corporation, Concord, California* **D. Engel, P. Kaspar** *Boehringer Mannheim GmbH, Tutzing, Germany* **W. Coty, R. Loor, P. Khanna** *Microgenics Corporation, Concord, California*

I. INTRODUCTION

Phenobarbital was the first drug found to be effective in the treatment of epilepsy, and it is still commonly used for this purpose. It is used primarily for generalized tonic-clonic (grand mal) and cortical focal seizures. As with many antiepileptic drugs, monitoring phenobarbital concentrations in serum or plasma during therapy is essential to assure patient compliance and to avoid the toxic effects of drug overdosage. Phenobarbital exhibits a narrow therapeutic index and wide variability in individual rates of drug absorption, metabolism, and clearance (1,2). Phenobarbital concentrations of 15 to 40 mg/L in serum are correlated with maximally effective seizure control (3). Toxic effects of phenobarbital occur at concentrations above 40 mg/L (4–7); central nervous system effects include sedation, nystagmus, and ataxia. Behavioral effects include irritability and hyperactivity in children and confusion in the elderly. A high level of phenobarbital can cause increased frequency of seizures. Other toxic effects include scarlatiniform or morbilliform rash, megaloblastic anemia, and osteomalacia.

We have developed a homogeneous enzyme immunoassay for phenobarbital using the CEDIA technology (8,9). In this method, *E. coli* β-galactosidase is split into two inactive fragments which can recombine spontaneously to form active enzyme. The small fragment, enzyme donor (ED), has been genetically engineered to allow covalent attachment of phenobarbital to a specific site that does not interfere with recombination with the large fragment (enzyme acceptor or EA). However, binding of a phenobarbital-specific antibody to the ED-phenobarbital conjugate inhibits enzyme formation. Phenobarbital in the sample displaces ED-phenobarbital conjugate bound to the antibody; the resulting unbound conjugate is free to react with EA to produce active enzyme. The β-galactosidase formed in the reaction is measured photometrically using a chromogenic substrate; the rate of substrate hydrolysis is directly proportional to the amount of phenobarbital in the sample. The concentration of phenobarbital in an unknown sample is determined from a calibration curve of rates versus phenobarbital concentration using calibrators with known concentrations of drug. This report describes the characteristics and performance of the CEDIA phenobarbital assay as applied to the Hitachi 704 automated analyzer.

II. MATERIALS AND METHODS

A. Reagents

The CEDIA phenobarbital assay kit contains two lyophilized reagents. Reagent 1 contains ED-phenobarbital conjugate, phenobarbital-specific monoclonal antibody, and chlorophenolred-β-D-galactopyranoside (CPRG) as substrate; Reagent 2 contains EA and secondary antibody (goat antimouse IgG); both reagents contain buffer salts, stabilizers, and antimicrobial agents. Two reagent-specific reconstitution buffers are provided; the kit is sufficient for approximately 50 determinations on the Hitachi 704, and can be used for at least 30 days after reagent reconstitution.

The CEDIA phenobarbital assay kit also includes 0.0- and 80.0-mg/L phenobarbital calibrators in a protein matrix containing stabilizers. The kit calibrators were formulated to match calibrators containing U.S.P. reference phenobarbital in drug-free human serum; the reference drug was dried under vacuum and prepared according to the instructions supplied by U.S.P.

Phenobarbital and drugs used for cross-reactivity studies were purchased from Sigma Chemical Co. (St. Louis, Missouri); Aldrich Chemical Co. (Milwaukee, Wisconsin); Applied Sciences (Deerfield, Illinois); and Abbott Laboratories (Abbott Park, Illinois). Patient serum samples were obtained from hospitals and clinical laboratories. Human serum albumin, human gammaglobulin, and bilirubin were from Sigma; triglyceride stock solutions (IntraLipid) were purchased from Kabi-Vitrum, Inc. (Alameda, California). Phenobarbital fluorescence polarization immunoassay kits (FPIA; Abbott TDx) were from Abbott Laboratories, and were used in conjunction with and the Abbott TDx analyzer according to the instructions supplied by the manufacturer.

B. Preparation of Reagents

The CEDIA phenobarbital ED and EA reagents were dissolved in 14 and 12 mL of their respective reconstitution buffers, mixed, and chilled to the temperature of the Hitachi reagent compartment. Reagents and calibrators were stored at 2 to 8°C when not in use.

C. Assay Procedure

The CEDIA phenobarbital assay was carried out in the Hitachi 704 automated analyzer as follows: Four microliters of calibrator, commercial control, or unknown sample were pipetted into a reaction cuvette, followed by addition of 200 μL of Reagent 1 containing substrate and a preformed complex of antibody and ED-phenobarbital conjugate. This solution was mixed and incubated at 37°C for 5 min and then 150 μL of reagent 2 containing EA were added and the incubation was continued for 4.5 min. Photometric measurements were taken at 20-s intervals between 4.5 and 5.5 min after reagent 2 addition, and the rate of substrate hydrolysis in mAU/min was determined by instrument software, after correction for absorbance at the secondary wavelength (660 nm). Phenobarbital concentrations in unknown samples were calculated by linear interpolation from a two-point standard curve using the rates obtained with 0.0- and 80.0-mg/L phenobarbital calibrators.

III. RESULTS AND DISCUSSION

A. Calibration Curve

In the CEDIA phenobarbital assay, sample is first incubated with a preformed complex of antibody and ED-phenobarbital conjugate; during this step, phenobarbital present in the sample equilibrates with the antibody ED-conjugate complex. Subsequently, unbound conjugate is available to react with EA in the second reagent to form active β-galactosidase. The enzyme formed was them measured as the rate of CPRG hydrolysis at 570 nm. As shown in Figure 1, the CEDIA phenobarbital assay exhibits a linear dose-response relationship between 0 and 80 mg/L phenobarbital ($r > 0.999$), with a slope of 3.1 mAU/min/mg phenobarbital/L. Thus the concentration of an unknown sample can be determined by linear interpolation from the rates of the 0.0- and 80-mg/L phenobarbital calibrators.

B. Sensitivity and Precision

Sensitivity of the CEDIA phenobarbital assay was estimated as two times the intra-assay standard deviation ($n = 20$) of the 0.0-mg/L phenobarbital calibrator. The mean of three separate determinations on different instruments was 0.6 mg/L.

Intraassay precision was determined by within-run measurement of commercial control sera ($n = 20$) at three concentrations of phenobarbital; the results obtained with

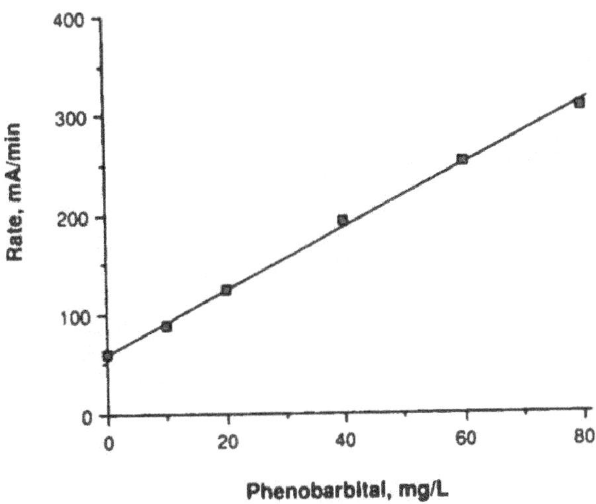

Figure 1 Phenobarbital calibration curve. Calibrators containing varying concentrations of phenobarbital in human serum were prepared by serial dilution and tested in the CEDIA phenobarbital assay.

Table 1 CEDIA Phenobarbital Intraassay Precision

	Instrument A	Instrument B	Instrument C	Average
Control level 1				
Mean value (mg/L)	9.1	9.1	8.9	9.0
Standard deviation (mg/L)	0.23	0.36	0.36	0.32
CV (%)	2.5	4.0	4.0	3.6
Control level 2				
Mean value (mg/L)	17.2	16.9	16.9	17.0
Standard deviation (mg/L)	0.45	0.62	0.37	0.48
CV (%)	2.6	3.7	2.2	2.8
Control level 3				
Mean value (mg/L)	62.9	62.0	58.9	61.2
Standard deviation (mg/L)	2.79	1.97	0.98	1.91
CV (%)	4.4	3.2	1.7	3.1

three different instruments are summarized in Table 1. Average CVs obtained were 2.8%, 3.1%, and 3.6% at approximately 9, 17, and 61 mg/L phenobarbital, respectively. Interassay precision was determined by testing single replicates of these three control sera on each of 23 runs performed on three separate instruments during a 3-week period. The results were 9.1 ± 0.5 mg/L (5.5% CV), 17.2 ± 0.9 mg/L (5.2% CV), and 62.3 ± 1.9 mg/L (3.0% CV) for the low, mid, and high controls, respectively.

C. Accuracy

Accuracy of the CEDIA phenobarbital assay was evaluated by method comparison. One hundred serum samples from patients under treatment with phenobarbital were assayed by the CEDIA and FPIA methods. An excellent correlation was obtained (r = 0.994), with close agreement of CEDIA and FPIA values: CEDIA = 1.02 • FPIA - 2.44 mg/L; Syx = 1.14 mg/L.

Accuracy was also evaluated by dilution of patient samples containing high concentrations of phenobarbital with low-concentration samples and with the 0.0-mg/L phenobarbital calibrator (Table 2). These studies show recovery of 94.2% to 104.7% in all cases, and thus confirm the linearity of the dose-response curve.

D. Specificity

Cross-reactivity studies were performed to assess the specificity of the CEDIA phenobarbital assay. Results obtained with 27 compounds, including phenobarbital metabolites, other barbiturates, antiepileptic agents, and related drugs, are summarized in Table 3. With the exceptions of structurally related compounds such as mephobarbital (> 100%), amobarbital (28%), and aprobarbital (11%), no compound gave more than 10% cross-reactivity. Of particular importance is the low cross-reactivity observed for *p*-hydroxyphenobarbital, the major metabolite of phenobarbital.

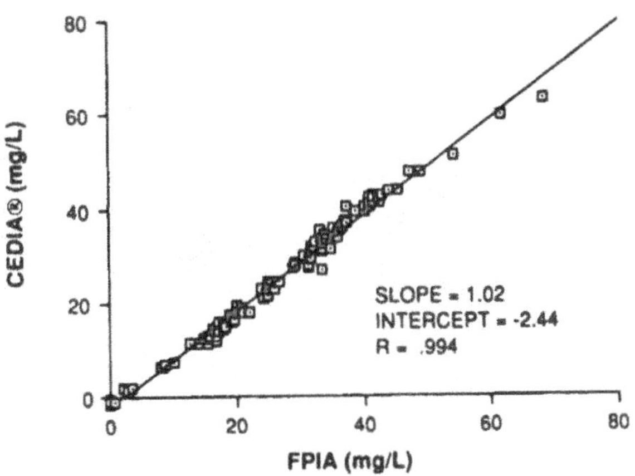

Figure 2 Method comparison study. Serum samples from patients being treated with phenobarbital were tested by the CEDIA and FPIA (Abbott TDx) methods. The best-fit line was estimated by least-squares regression analysis.

Table 2 Recovery and Linearity of Dilution

% of high sample	Measured phenobarbital (mg/L)	Expected phenobarbital (mg/L)	Recovery (%)
Recovery:			
100	74.1		
75	57.4	57.0	100.5
50	39.8	40.0	99.6
25	21.5	22.9	94.2
0	5.8		
100	56.7		
75	42.7	42.5	100.5
50	27.0	28.3	95.3
25	13.9	14.1	99.1
0	0		
100	60.9		
75	47.0	46.6	100.8
50	31.4	32.4	97.0
25	17.8	18.1	98.3
0	3.9		
Linearity:			
100	81.4		
75	63.9	61.0	104.7
50	41.0	40.6	100.8
25	20.9	20.2	103.5
0	0		

Mixtures were prepared containing varying proportions of high-phenobarbital-concentration samples with either low-phenobarbital-concentration samples (recovery) or with the zero phenobarbital calibrator (linearity). These mixtures were tested by the CEDIA phenobarbital assay, and the measured values were compared with the results expected from the concentrations of the undiluted samples.

Table 3 CEDIA Phenobarbital Assay Specificity

Cross-reactant	Cross-reactivity
Amitryptiline	<1%
Carbamazepine	<1%
Chlorazepate	<1%
Chlorpromazine	<1%
Diazepam	<1%
Ethosuximide	<1%
Ethotoin	<1%
5-Ethyl-5-phenylhydantoin	<1%
Glutethimide	<1%
5-(*p*-Hydroxyphenyl)-5-phenylhydantoin	<1%
Imipramine	<1%
Mephenytoin	<1%
Mephobarbital	>100%
Methosuximide	<1%
2-Phenyl-2-ethylmalonamide	<1%
Phenytoin	<1%
Primidone	<1%
Promethazine	<1%
Valproic Acid	<1.5%
p-hydroxyphenobarbital	<1%
Amobarbital	<2.8%
Aprobarbital	<11%
Butabarbital	<3.6%
1.3-Dimethylbarbitaric Acid	<1%
Secobarbital	<4.7%
Pentobarbital	<1%
Barbital	<1.5%

Cross-reactivity was calculated as (apparent phenobarbital concentration + concentration of cross-reactant) × 100%. Where possible, cross-reactivity was measured at a response equal to that obtained with a sample containing 40 mg/L phenobarbital (the ED_{50}). If the highest concentration of cross-reactant gave a response of less than 0.8 mg/L phenobarbital, the cross-reactivity was estimated as less than (0.8 mg/L + the highest concentration of cross-reactant tested) × 100%.

Specificity was also evaluated by interference studies with hemoglobin, gamma-globulin, bilirubin, and triglycerides (10). Interfering substances were added to each of three human serum pools containing varying concentrations of phenobarbital (10, 40, and 60 mg/L), and the concentrations of phenobarbital were determined by the CEDIA phenobarbital assay. Total protein (up to 130 g/L), triglyceride (up to 15 mg/L), hemoglobin (up to 4 g/L), and bilirubin (up to 300 mg/L) gave no interference, defined as a change in phenobarbital quantitation of more than ±10% from that of the control.

IV. CONCLUSION

The results presented in this report demonstrate that the CEDIA phenobarbital assay is an accurate, precise, and specific method for the determination of phenobarbital in human serum. Application of this assay to the Hitachi 704 automated analyzer provides rapid, convenient, and fully automated monitoring of phenobarbital concentrations during the treatment of epilepsy.

The CEDIA phenobarbital assay provides unique advantages when compared to existing methods. For example, the linear calibration curve eliminates inaccuracy due to mathematical fitting of nonlinear standard curves, such as are obtained with FPIA (11) and EMIT (12) methods. The linear CEDIA dose-response curve also minimizes imprecision due to "flat" regions in a nonlinear curve, which are also observed with the FPIA and EMIT methods. Moreover, use of a linear standard curve allows calibration with only two standards, as compared to the six used in other methods.

Tests are currently available for measurement of total T4 and thyroxine binding capacity; assays for other hormones and therapeutic drugs are under development. These assays, when combined with the high throughput, random-access capability, and full walk-away capability of Hitachi clinical chemistry analyzers, will create an efficient system for special chemistry testing and therapeutic drug monitoring without the requirement for dedicated immunoassay equipment.

REFERENCES

1. H. Kutt and K. Penry. Usefulness of blood levels of antiepileptic drugs. *Arch. Neurol.*, 3:283–288 (1974).
2. W. J. Waddell and T. C. Butler. The distribution and excretion of phenobarbital. *J. Clin. Invest.*, 36:1217–1226 (1957).
3. F. Buchthal and O. Svensmark. Serum concentrations of diphenylhydantoin (phenytoin) and phenobarbital and their relation to therapeutic and toxic effects. *Psychiatr. Neurol. Neurochir.*, 74:117–136 (1971).
4. J. K. Penry and M. E. Hewmark. The use of anti-epileptic drugs. *Ann. Intern. Med.*, 90:207–218 (1979).
5. F. Buchthal and O. Svensmark. Aspects of the pharmacology of phenytoin (dilantin) and phenobarbital relevant to their dosage in the treatment of epilepsy. *Epilepsia*, 1:373–384 (1960).
6. M. J. Eadie and J. H. Tyrer. *Anticonvulsant Therapy: Pharmacological Basis and Practice.* Livingstone, London (1974).
7. C. E. Dent, A. Richens, D. J. E. Rowe, and T. C. B. Stamp. Osteomalacia with long-term anticonvulsant therapy in epilepsy. *Br. Med. J.*, 4:69–72 (1970).
8. D. R. Henderson, S. B. Friedman, J. D. Harris, W. B. Manning, and M. A. Zoccoli, CEDIA®, a new homogeneous immunoassay system. *Clin. Chem.*, 32:1637–1641 (1986).
9. P. L. Khanna, R. T. Dworschack, W. B. Manning, and J. D. Harris. A new homogeneous enzyme immunoassay using recombinant enzyme fragments. *Clin. Chem. Acta*, 185:231–240 (1989).
10. M. R. Glick and K. W. Ryder. *Interferographs: A User's Guide to Interferences in Clinical Chemistry Instruments*. Science Enterprises, Indianapolis, IN (1987).
11. M. Lu-Steffes, G. W. Pittluck, M. E. Jolley, H. N. Panas, D. L. Olive, C. H. J. Wang, D. D. Nystrom, C. L. Keegan, T. P. Davis, and S. D. Stroupe. Fluorescence polarization immunoassay IV. Determination of phenytoin and phenobarbital in human serum and plasma. *Clin. Chem.*, 28:2278–2282 (1982).
12. K. E. Rubinstein, R. S. Schneider, and E. F. Ullman. Hemogeneous enzyme immunoassay. *Biochem. Biophys. Res. Commun.*, 47:846–851 (1972).

58

An HPLC Method for the Quantitation of Propafenone and Its Metabolites in Pediatric Samples

Zul Verjee and Esther Giesbrecht *The Hospital for Sick Children, Toronto, Ontario, Canada*

INTRODUCTION

Propafenone HCl (Rythmol) is a new antiarrhythmic agent, with fast sodium channel-blocking activity, associated with weak β-adrenergic receptor and calcium channel-blocking actions. It has recently been approved for oral use for the management of supraventricular and ventricular arrhythmias. Propafenone suppresses nonsustained ventricular arrhythmias, whereas its effect is less consistent in patients with sustained ventricular arrhythmias (1). The drug is metabolized in the liver, into two active metabolites, 5-hydroxypropafenone and N-depropylprepafenone, with pharmacologic activity comparable to that of propafenone (2).

Interpreting the relationship between its plasma concentration and its action is complicated by the variable, genetically determined formation of the two metabolites. With increased use of the drug in therapeutics, a need is being realized for a reliable assay for these compounds in therapeutic drug monitoring (TDM) service. A high-performance liquid chromatography (HPLC) method for the simultaneous detection of propafenone and its metabolites in pediatric samples, requiring short run time and convenient extraction procedure, has been investigated in our laboratory; the results are reported in this paper.

II. MATERIALS

Propafenone hydrochloride, 5-OH propafenone and N-despropylpropafenone were supplied by Knoll Pharmaceuticals (Markham, Ontario). C_{18} Bond Elut Cartridges, 1-ml size, were from Analytichem International (Harbor City, California); the Whatman Partisil 5 ODS RAC column (100 mm × 9.4 mm) was from Whatman, Inc. (Clifton, New Jersey); all other reagents and solvents were of analytical grade, including KH_2PO_4 from Fischer (New Jersey), Trizma base from Sigma Chemical Co. (St. Louis); and ethyl acetate, methanol, and acetonitrile from Baxter (Illinois).

HPLC analysis was performed using Waters HPLC instrumentation, which included pump model 6000A, Lambda-Max model 481 spectrophotometer, WISP 710 automatic sampler, and QA1 Data System Integrator.

Precolumn filters were purchased from Upchurch Scientific SPE (Rexdale, Ontario).

III. METHOD

In a 1.5-mL eppendorf vial, pipette: 100 μL serum, 10 μL internal standard, 50 μL Tris buffer, pH 12, and 1 mL ethyl acetate. Vortex the mixture for 1 min; centrifuge for 2 min (TDx centrifuge). Prewash C_{18} Bond Elut cartridges (1-mL size) three times with methanol. Apply supernatant to the washed cartridge. Wash with 0.5 mL of methanol. Elute with 500 μL of 95% methanol/5% buffer, 10 mM phosphate, pH 2.5. Dry off under N_2 and reconstitute with 100 μL of acidified water. Allow to sit at least 5 min to ensure dissolution. Inject 75 μL.

HPLC conditions are as follows:

Whatman column with in-line filter
Mobile phase: 58% buffer (10 mM phosphate, pH 2.5), 42% acetonitrile
Flow rate: 2 to 3 mL/min, back pressure < 1000 psi
Detection: 214 mm, 0.005 AUFS

Technical notes:

Methanol wash should not exceed 0.5 to 1.0 mL.
Recovery during drying off is optimal when the temperature is kept below 50°C and eluate volume is not more than 0.5 mL.
Solid-phase extraction with C_{18} cartridges is required to remove interfering polar substances that co-elute with 5-OH propafenone.

IV. RESULTS AND DISCUSSION

This paper describes an isocratic HPLC method for the separation of propafenone and its two metabolites, 5-hydroxypropafenone and N-depropylpropafenone. Quantitation of propafenone and 5-hydroxypropafenone is also described. The assay requires 100 μL of serum, compared to much larger volumes in previous assays (3). It has relatively short retention times, with a total run of 11.0 min.

Resolution of propafenone and its two metabolites from each other and interfering peaks was optimal with a mobile phase containing 58% buffer/42% acetonitrile. In order to achieve a run time of 10 to 11 min, the flow rate required was 2 to 3 mL/min (Figures 1 and 2). At ambient temperature, the Whatman C_{18} column could handle this flow rate adequately and with a reasonable back pressure (<1000 psi). Other column types, such as the Supelco C_{18} column, 3-μm packing, and the Beckman C_{18}ODS column, with the same mobile phase and flow rate, gave unacceptably high back pressures. These columns might be suitable with high temperatures but low flow rate, resulting in increased run time.

The method gave good resolution between propafenone and its two metabolites. The limit of sensitivity for propafenone was 30 μg/L. Linearity up to 2500 μg/L was satisfactory, with recoveries of 80% for propafenone and the 5-hydroxy metabolite.

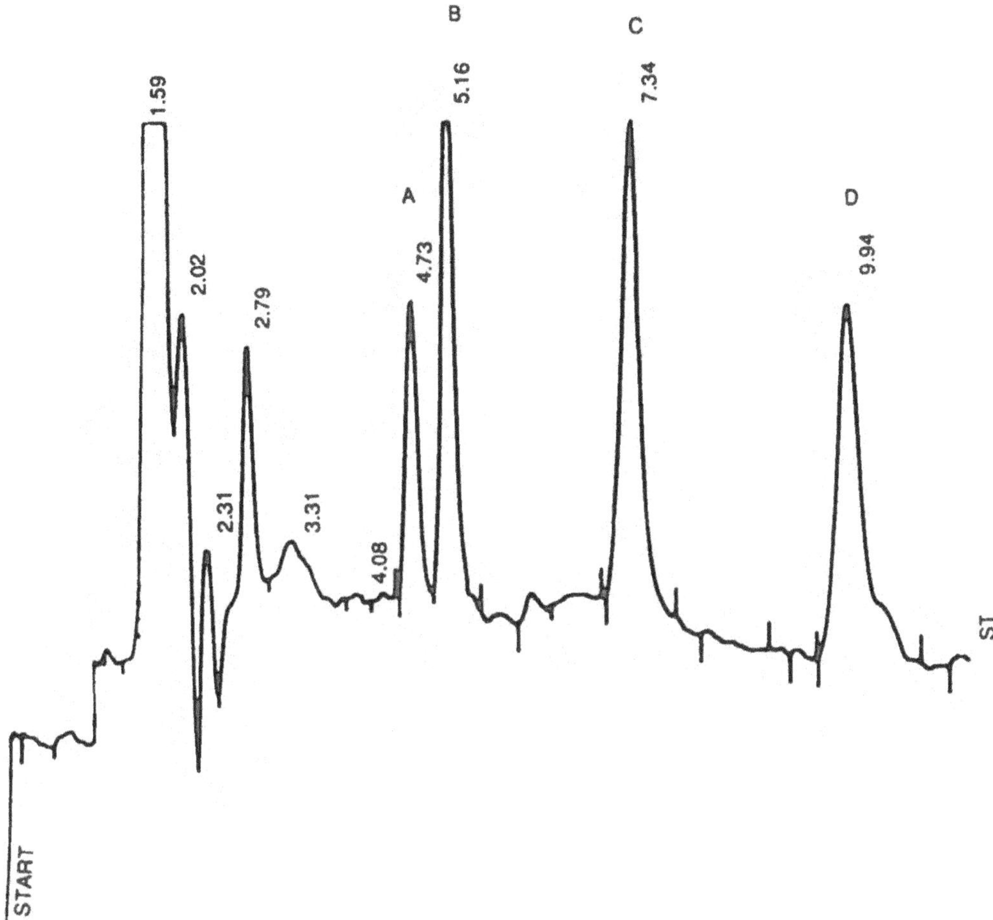

Figure 1 Aqueous standards. (A) N-despropylpropafenone (B) 5-OH propafenone (C) Internal Standard (D) Propafenone.

Within-batch precision of 2.7% for 5-OH propafenone and 3.3% for propafenone at concentration of 450 µg/L was obtained.

Serum extraction with ethyl acetate was more convenient than extraction with methylene chloride. No interference was noted from other drugs or their metabolites in the serum of patients on multiple drugs.

Table 1 shows some preliminary data on patient samples, illustrating that the method described can measure fairly low levels of both propafenone and 5-OH propafenone. These few cases suggest that neonates have limited capability to metabolize the parent drug because they have delayed maturation of the liver.

Recently it has been shown that the formation of 5-OH propafenone in humans, is dependent on whether the subjects are poor or extensive metabolizers (4). Hydroxylation of propafenone parallels that of debrisoquine, suggesting that the same mechanism is involved in the process. This suggests, therefore, that approximately 10% of the population would be poor metabolizers of propafenone. They would

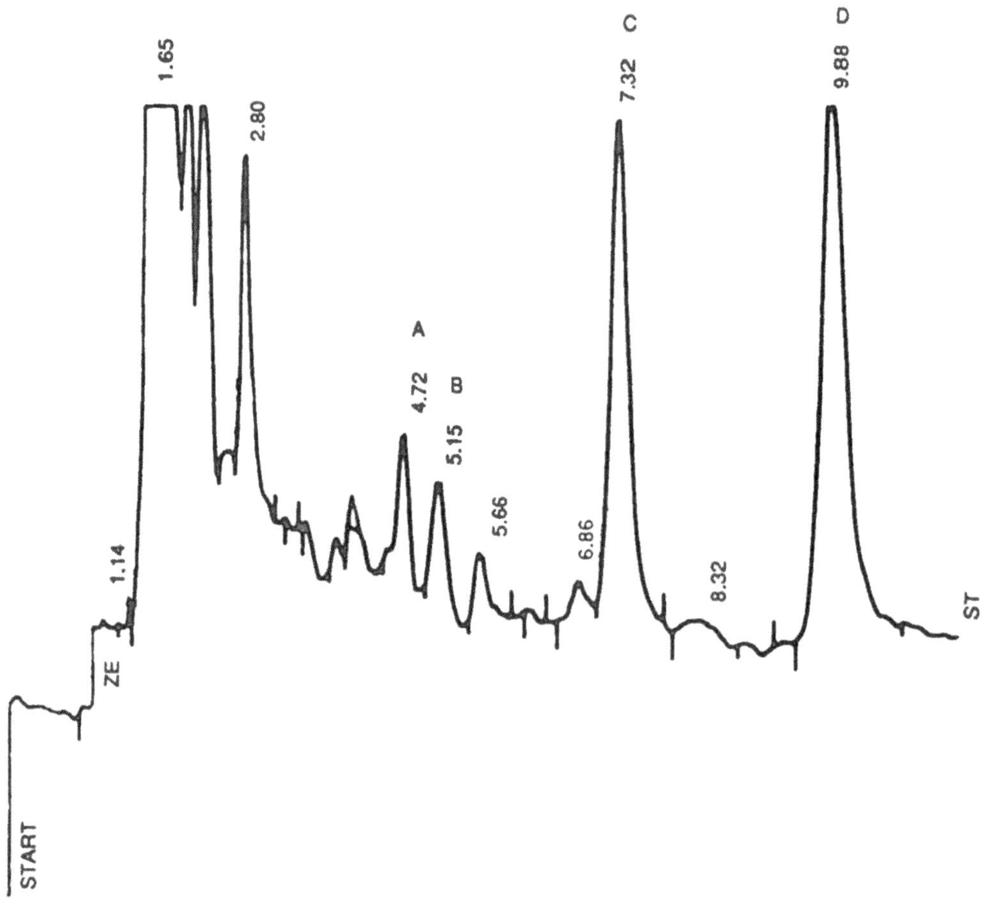

Figure 2 Patient during treatment. (A) N-despropylpropafenone (B) 5-OH propafenone (C) Internal Standard (D) Propafenone.

achieve higher steady-state blood levels compared to extensive metabolizers. Therapeutically, this would lead to higher blood levels in poor metabolizers, with increased risk of side effects. Therefore, it would be advantageous to monitor patient serum levels of propafenone and 5-OH propafenone, since the later has been shown to have greater antiarrhythmic activity but less β-blocking activity than propafenone (3).

The method described here offers the ability to simultaneously and rapidly detect and quantitate the levels of propafenone and its metabolites in serum samples from our pediatric population.

V. SUMMARY

Propafenone is a new antiarrhythmic agent with fast sodium channel-blocking activity, associated with weak β-adrenergic receptor and calcium-blocking actions. It has recently been shown to be effective in the control of chronic recurrent

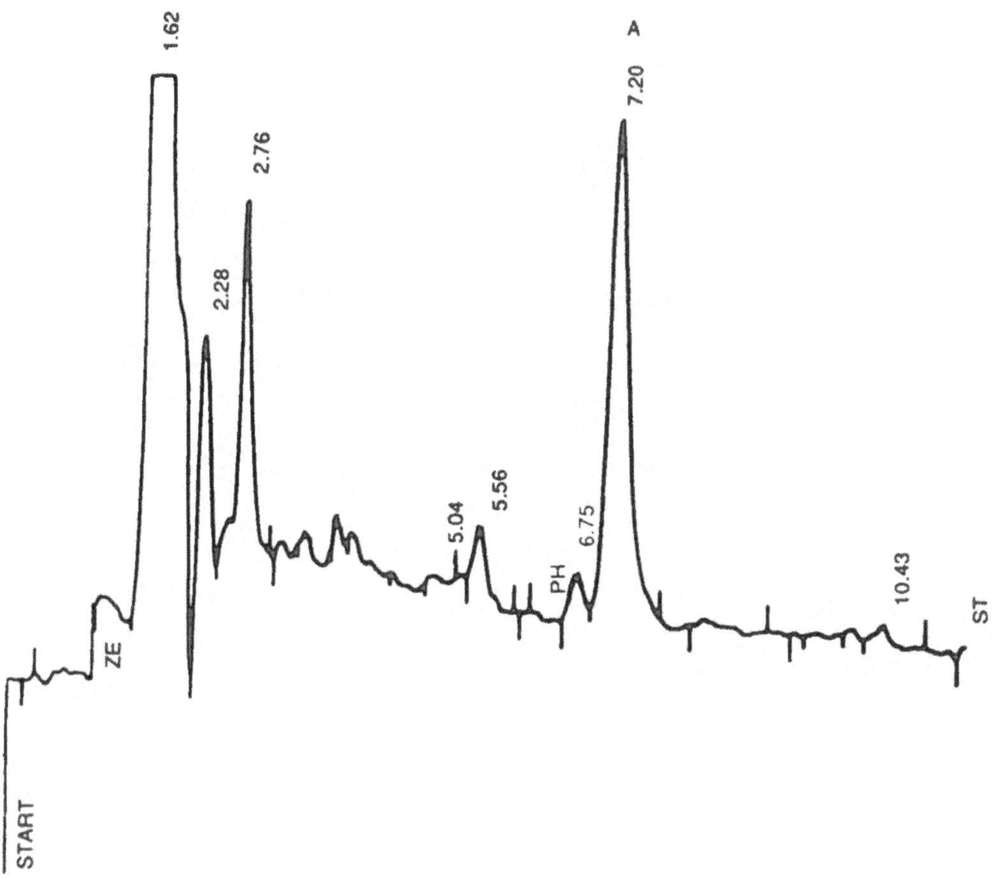

Figure 3 Patient before treatment. (A) Internal Standard.

supraventricular and ventricular arrhythmias. The drug is metabolized in the liver, into two active metabolites, 5-hydroxypropafenone and N-depropylpropafenone, with pharmacologic activity comparable to that of propafenone. With increased use of the drug in therapeutics, a need for a reliable assay for these compounds in TDM service is being realized. An HPLC method for pediatric samples requiring short run time and convenient extraction procedure is under investigation in our laboratory.

Of the several column types tested (Supelco C_{18} 3μ packing, Beckman C_{18} ODS 5-μm packing, and Whatman C_{18} cartridge), with different mobile phases, the most suitable combination achieved was the Whatman column run at room temperature with a mobile phase of 42% acetonitrile in 10 nM phosphate buffer, pH 2.5. At a flow rate of 2.3 mL/min, chromatography was complete in 10 to 11 min. The back pressure was <1000 psi. Serum samples (100 μL) were extracted at alkaline pH with ethyl acetate and the extract applied to a C_{18} extraction column; interfering polar compounds were washed off with acetonitrile. The drugs and metabolites were eluted with 95% methanol/5% buffer. Detection was made at 214 nm, with a UV spectrophotometer.

Table 1 Preliminary Patient Data

Patient	Drug concentration (μg/L)		Comments
	5-OH propafenone	Propafenone	
CD	136	665	5 years old; trough
CJ	40	1832	Neonate; trough
MM	<30	254	7 years old; continuous infusion
HBG	Nil	209	Neonate
	Nil	226	Neonate
	Nil	106	Neonate
WC	103	747	Infant; 3 h postdose
	71	446	Trough?
MJ	Nil	3014	Neonate, continuous infusion
	Nil	3625	Neonate, continuous infusion
	Nil	1642	Neonate; trough

The method is sensitive to 30 μL and detects both propafenone and its 5-hydroxypropafenone derivative simultaneously with good resolution. Linearity up to 2500 μg/L was satisfactory, with recoveries of 80% for both compounds. Within-batch precision of 2.7% for 5-OH propafenone and 3.3% for propafenone at a concentration of 450 μg/L was obtained. No interference was noted from other drugs or their metabolites in the serum of patients on multiple drugs. Application of this assay to TDM is in progress.

ACKNOWLEDGMENT

Financial support from Knoll Laboratories, Ontario, Canada, is gratefully acknowledged.

REFERENCES

1. C. Funck-Brentano, H. K. Kroemer, J. T. Lee, and D. M. Roden. *N. Engl. J. Med.*, 322:518–525 (1990).
2. R. E. Kates, Y. G. Yee, and R. A. Winkle. *Clin. Pharmacol. Ther.*, 37:610–614 (1985).
3. G. L. Hoyer. *Chromatographia*, 25:1034–1038 (1988).
4. J. T. Lee, H. K. Kroemer, D. J. Silberstein, C. Funck-Brentano, M. D. Lineberry, A. J. J. Wood, D. M. Roden, and R. L. Woosley. *N. Engl. J. Med.*, 322:1764–1768 (1990).

Part IV

Antiepileptics

59

Distribution of Phenytoin in Different Regions of the Brain and in the Serum: Analysis of Autopsy Specimens from 25 Epileptic Patients

B. Rambeck, *R. Schnabel, T. May, U. Jürgens, and *R. Villagrán *Gesellschaft für Epilepsieforschung e.V., and *Institute of Neuropathology, Bethel, Bielefeld, Germany*

I. INTRODUCTION

Postmortem brain and serum specimens from 25 epileptic patients, who had been treated for a long time with phenytoin, were analyzed in order to answer the following questions:

To what extent does the free and the total concentration in the serum correlate with the concentration in the cortex?

How uniformly is phenytoin distributed over the various regions of the brain?

Are there, in all or individual cases, lowered phenytoin concentrations in specific regions of the cortex, which could possibly explain therapy resistance in spite of therapeutic serum concentrations?

Are there, in all or individual cases, elevated phenytoin concentrations in specific regions of the cerebellum which could explain occasionally observed damage of the cerebellum?

Are there any peculiarities in the case of patients, who died suddenly and unexpectedly, the cause of death not being sufficiently explainable upon postmortem examination (mors subita)?

II. PATIENTS, MATERIALS, AND METHODS

The autopsy samples were from 17 male and 8 female epileptic patients. The age at death ranged from 19 to 86 years, with a median of 64 years. The patients were mainly long-term patients of the epilepsy center, Bethel, Bielefeld. The epileptic syndromes included simple focal, complex focal, and secondarily generalized seizures and grand mal seizures. Most patients had been stabilized on phenytoin medication for many years.

Twenty-one patients died from chronic diseases, which included severe infection of the liver, chronic nephritis, metastatic carcinoma of the mamma, gallbladder, and lung, as well as encephalitis and tuberculosis of the lymphatic nodes. The immediate causes of death of the 21 patients were most cardiorespiratory insufficiency, shock, coronary diseases, and pulmonary disorders. Mors subita—that is, unexpected and sudden death, which cannot be satisfactorily explained either pathologically or toxicologically, occurred in four cases.

Samples of the brain and venous femoral blood were taken at the autopsy. The samples included cortex and white matter from the frontal, occipital, and temporal cortex and the cerebellum. In five cases further special regions of the temporal cortex and cerebellum were examined. The determination of the phenytoin concentration in brain samples by high-performance liquid chromatography (HPLC) is described elsewhere (1). The determinations of the total serum concentration, and of the free serum concentration after ultrafiltration at 25°C, were also performed by HPLC (1).

The in-day coefficient of variation of the phenytoin concentration of a homogenate from patient samples of brain tissue was less than 2%. Storage of the specimens over 3 days at 4°C did not change the phenytoin concentration of the tissues and the protein binding in the serum by more than 4%. A detailed statistical analysis of our data showed that protein binding in the serum as well as the cortex-to-serum ratio were not significantly influenced by the time interval between death and autopsy.

III. RESULTS AND DISCUSSION

The correlation of the concentration in the frontal cortex and the total concentration of phenytoin in the serum is shown in Figure 1. The slope b of the regression line is 2.09. This means that the concentrations in the cortex are about twice the concentrations in the serum. The correlation is improved if the concentration in the frontal cortex is correlated with the free concentration instead of the total concentration, Figure 2. The concentration in the cortex is about nine times that of the free serum concentration. This agrees with the results of Friel and co-workers in 1989 (2). They found, in 18 patients who underwent cortical resection for intractable seizures, that the phenytoin concentration in the frontal cortex was about 11 times higher than the free serum concentration.

In our study the protein binding of phenytoin in the serum of the 21 patients who died after a chronic illness was, at 23.8%, relatively high. In the four patients with mors subita, an average protein binding of 11.4% was found. This value agrees with the protein binding of 10.4% found in a control group of 87 otherwise healthy epileptic patients. Sudden death apparently did not lead to slow inactivation of the kidneys and other organs with the ensuing enrichment of endogenous substances, which can displace phenytoin from the protein binding sites. This problem has already been investigated in another study (3). Our results confirm studies of Reidenberg et al. (4), Hooper et al. (5), Boucher et al. (6,7), and other authors, who showed that patients with severe diseases of the kidneys and liver, as well as critically ill trauma patients, had a clearly increased free fraction of phenytoin in the serum.

The essentially better correlation of the free serum concentration with the frontal cortex underlines the need to measure the free concentration in severely ill patients,

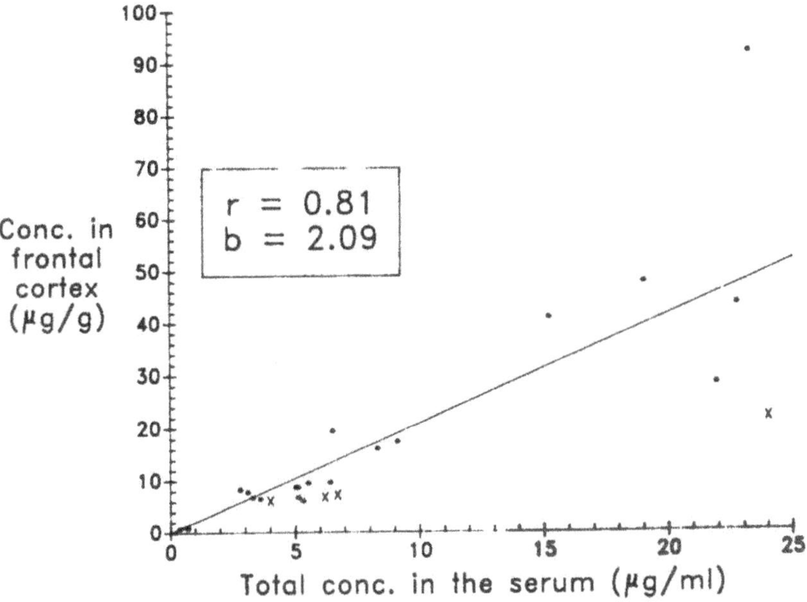

Figure 1 Relation between the concentrations in the frontal cortex and total concentrations in the serum. The slope b of the regression line and the Pearson correlation coefficient r are given. Cases with mors subita are marked with a cross.

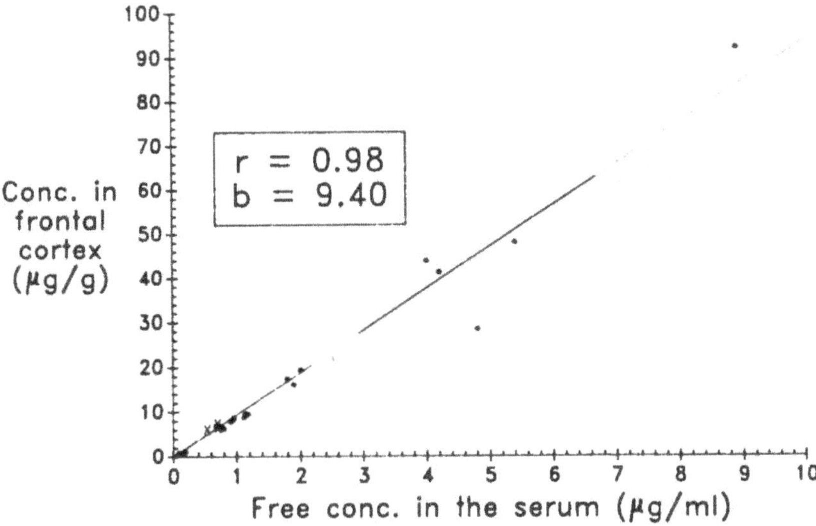

Figure 2 Relation between the concentrations in the frontal cortex and free concentrations in the serum.

because in these cases the total concentration does not correlate sufficiently with the decisive concentration in the brain.

The correlation between the temporal cortex and the frontal cortex for the 25 patients on phenytoin is shown in Figure 3. There is very good agreement between the values in all instances. The correlation between the occipital cortex and the frontal cortex is shown in Figure 4. Again there is very good agreement. Even the ratios of the mors-subita patients agree with those of the other patients.

The correlation between the concentration of phenytoin in the frontal white matter and the frontal cortex is shown in Figure 5. The concentration in the white matter is, apparently because of its higher lipid content, 1.5 times higher than in the cortex. This agrees with studies of Sherwin et al. (8,9) and other authors, who found comparable ratios. We found only one patient who had a lower concentration in the white matter than expected from the concentration in the cortex, perhaps because of losses in lipids in the white matter. The corresponding concentrations of the temporal white matter compared with the frontal white matter are shown in Figure 6. The slope of 0.82 shows that concentrations in the temporal white matter are somewhat lower than in the frontal white matter, possibly because of differences in their lipid contents. The difference between temporal white matter and frontal white matter is statistically significant. The concentrations of the occipital white matter compared with the frontal white matter are shown in Figure 7. In summary, the data demonstrate that the concentrations of phenytoin in the white matter are higher than in the cortex. Otherwise a rather uniform distribution of phenytoin is found in the cerebrum. Our results agree with the data of Sironi et al. (10) and other authors, but in these studies only a very few patients were included. No striking features were found in mors-subita patients.

The correlation of the cerebellum with the frontal cortex is shown in Figure 8. The correlation coefficient and the slope indicate that concentrations in the cerebellum and

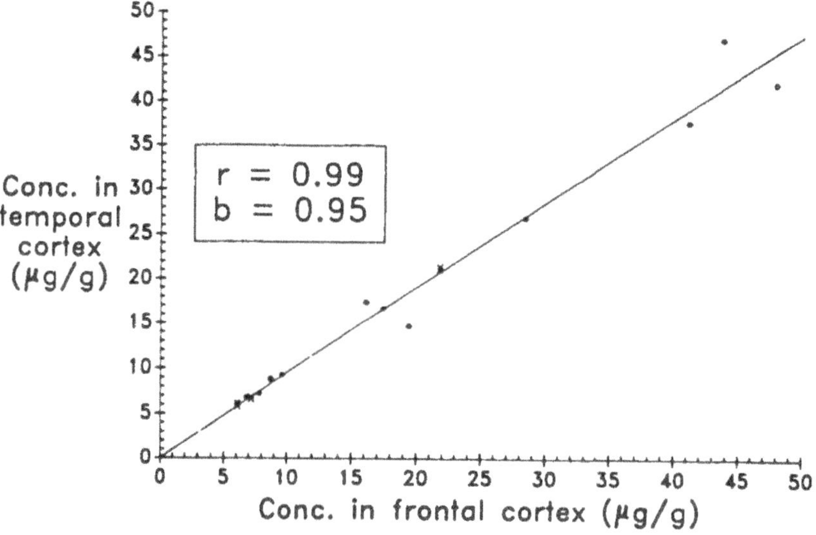

Figure 3 Relation between the concentrations in the temporal cortex and in the frontal cortex.

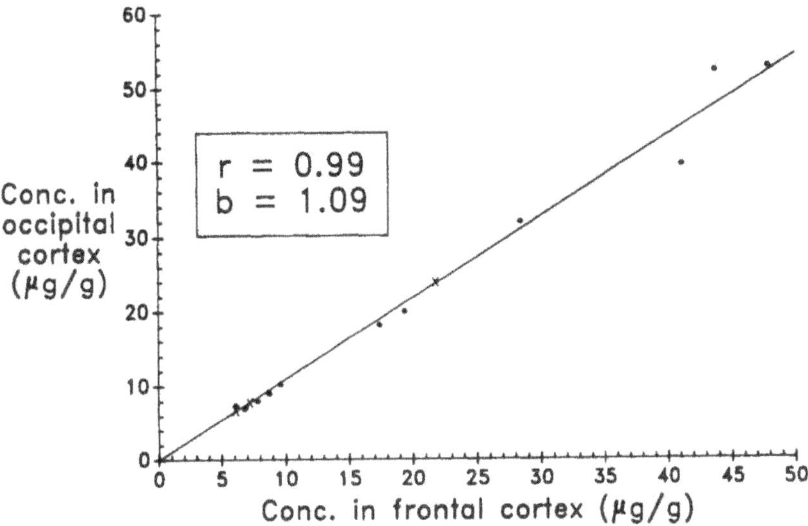

Figure 4 Relation between the concentrations in the occipital cortex and in the frontal cortex.

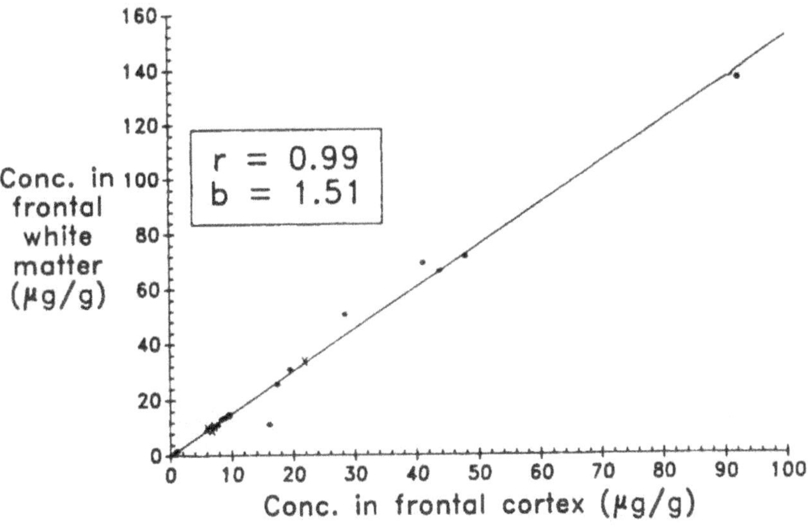

Figure 5 Relation between the concentrations in the frontal white matter and in the frontal cortex.

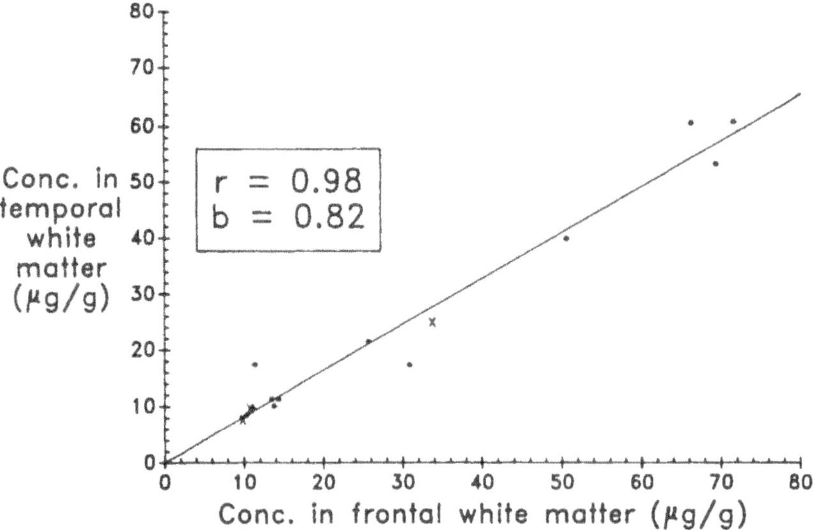

Figure 6 Relation between the concentrations in the temporal white matter and in the frontal white matter.

the cortex are comparable. There are no cases of elevated tissue concentrations in the cerebellum relative to the cortex.

It is supposed in certain epilepsy syndromes that specific regions of the temporal lobe are of particular significance in the development of a seizure. Furthermore, it is known that damage of the cerebellum can be induced by phenytoin. For this reason regions of the temporal lobe and of the cerebellum were in five cases further differentiated, Figure 9. Here the mean concentration ratios of selected regions to the frontal cortex, as a reference, are given for five patients. In addition to the frontal and occipital

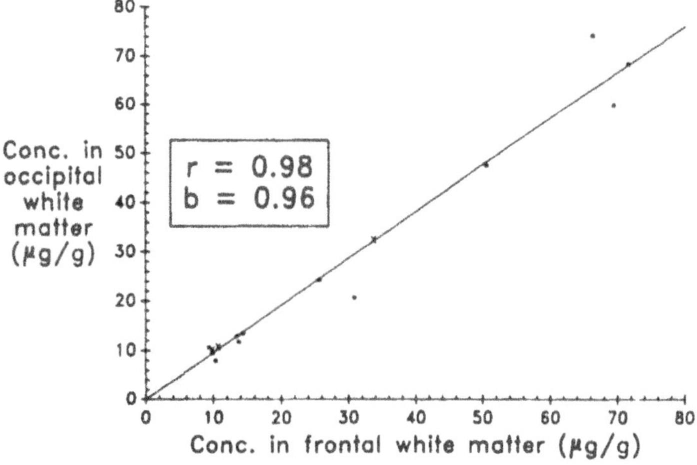

Figure 7 Relation between the concentrations in the occipital white matter and in the frontal white matter.

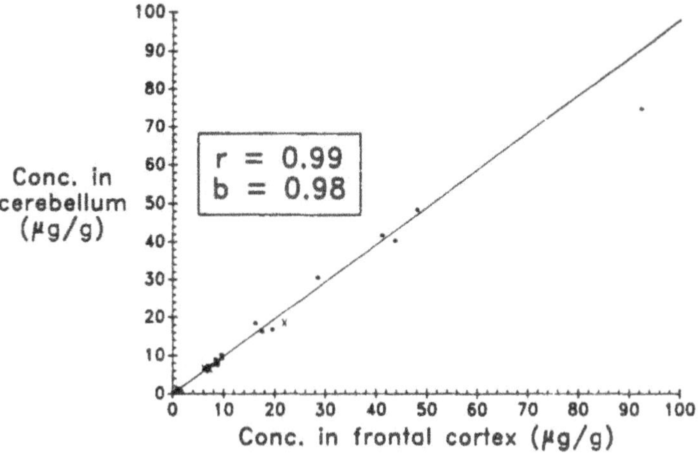

Figure 8 Relation between the concentrations in the cerebellum and in the frontal cortex.

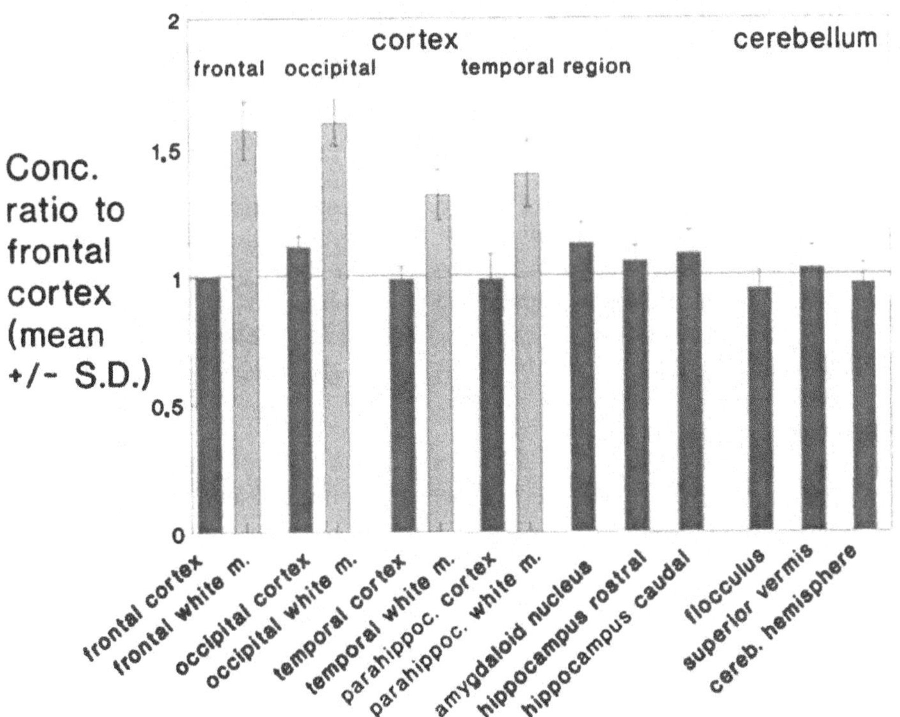

Figure 9 Concentration ratios of selected regions to the frontal cortex as a reference. Mean ± standard deviation of five patients.

region, the temporal cortex and white matter, para-hippocampal cortex and white matter, the amygdaloid nucleus, and the rostral and caudal hippocampus from the temporal lobe are considered. From the cerebellum the ratios for the flocculus, superior vermis, and cerebellar hemisphere are given.

As already seen from Figure 5, the concentrations in the white matter are generally higher than in the cortex. It is interesting that, as already demonstrated in Figure 6, in the white matter of the temporal region the concentration is lower than in the white matter of the frontal or occipital regions. No individual or interindividual peculiarities were found either in the temporal lobe or in the cerebellum.

Further studies are planned in which specific tissue samples from deceased epileptic subjects with cerebellar atrophy as well as samples which are surgically removed from a suspected epileptogenic focus will be examined.

IV. CONCLUSIONS

In postmortem specimens of 25 epileptic patients, the phenytoin concentration in the frontal cortex correlates better with the free serum concentration than with the total serum concentration.

Phenytoin is uniformally distributed throughout all the investigated regions of the cortex and the cerebellum. The concentration in the white matter is about 50% higher than in the cortex.

The uniform distribution of phenytoin throughout all the examined regions of the cortex, especially the temporal lobe, indicates that therapy resistance, even with therapeutic serum concentrations, can supposedly not be explained by locally lowered concentrations in the cortex.

The very high correlation between the concentrations in the cortex and in the cerebellum, as well as the uniform distribution throughout the cerebellum, shows that damage of the cerebellum by phenytoin can hardly be attributed to locally elevated concentrations in the cerebellum.

The serum protein binding of four patients who died suddenly and unexpectedly is comparable with the protein binding of otherwise healthy epileptic patients on phenytoin therapy. On the other hand, patients who died after a long illness had greatly elevated free fractions of phenytoin in the serum.

There were no peculiarities in cases of mors subita with respect to phenytoin concentrations in the brain.

ACKNOWLEDGMENT

The author are greatly indebted to G. S. Macpherson for translating the paper.

REFERENCES

1. U. Jürgens and B. Rambeck. Sensitive analysis of antiepileptic drugs in very small portions of human brain by microbore HPLC. *J. Liquid Chromatogr.*, *10*:1847–1863 (1987).
2. P. N. Friel, G. A. Ojemann, R. L. Rapport, R. H. Levy, and G. Van Belle. Human brain phenytoin: Correlation with unbound and total serum concentrations. *Epilepsy Res.*, 3:82–85 (1989).
3. B. Rambeck, R. Schnabel, T. May, U. Jürgens, and R. Villagran. Postmortem serum protein binding and brain concentrations of antiepileptic drugs in autopsy specimens from 45 epileptic patients. *Ther. Drug. Monitor.*, in press.

4. M. M. Reidenberg, I. Odar-Cederlöf, C. von Bahr, O. Borga, and F. Sjöqvist. Protein binding of diphenylhydantoin and desmethylimipramine in plasma from patients with poor renal function. *N. Engl. J. Med.*, *285*:264–267 (1971).
5. W. D. Hooper, F. Bochner, M. J. Eadie, and J. H. Tyrer. Plasma protein binding of diphenylhydantoin. Effects of sex hormones, renal and hepatic disease. *Clin. Pharmacol. Ther.*, *15*:276–282 (1973).
6. B. A. Boucher, J. H. Rodman, T. C. Fabian, et al. Disposition of phenytoin in critically ill trauma patients. *Clin. Pharmacol.*, *6*:881–887 (1987).
7. B. A. Boucher, J. H. Rodman, G. S. Jaresko, N. S. Rasmussen, C. B. Watridge, and T. C. Fabian. Phenytoin pharmacokinetics in critically ill trauma patients. *Clin. Pharmacol. Ther.*, *44*:675–683 (1988).
8. A. L. Sherwin, A. A. Eisen, and C. D. Sokolowski. Anticonvulsant drugs in human epileptogenic brain. *Arch. Neurol.*, *29*:73–77 (1973).
9. A. L. Sherwin, C. D. Harvey, and I. E. Leppik. Quantitation of antiepileptic drugs in human brain. In *Quantitative Analytic Studies in Epilepsy* (P. Kellaway and I. Petersen, eds.), Raven Press, New York, pp. 171–182 (1976).
10. V. A. Sironi, G. Cabrini, M. G. Porro, L. Ravagnati, and F. M. Marossero. Antiepileptic drug distribution in cerebral cortex, Ammon's horn and amygdala in man. *J. Neurosurg.*, *52*: 686–692 (1980).

60

Prediction of the Free Serum Concentrations of Phenytoin, Phenobarbital, Carbamazepine, and Carbamazepine-10,11-epoxide in Patients on Valproic Acid Co-medication

T. May and B. Rambeck *Gesellschaft für Epilepsieforschung, Bielefeld, Germany*

I. INTRODUCTION

Serum protein binding of antiepileptic drugs can be influenced by various factors, such as pregnancy, certain illnesses, or co-medication (1–7). In our study we investigated the influence of co-medication with valproic acid on the free fraction of other antiepileptic drugs. The following questions were of particular interest:

To what extent does co-medication with valproic acid (VPA) influence the free fraction of phenytoin (PT), phenobarbital (PB), carbamazepine (CBZ), and carbamazepine-10,11-epoxide (CE)?
Is the free fraction of PT, PB, CBZ, and CE dependent on the VPA serum concentration?
Can the free concentration of PT, PB, CBZ, and CE be estimated from the total serum concentration and the VPA concentration?

In particular, is there a simple and reliable method for predicting the free concentration of PT?

Phenytoin is of special interest because PT, like VPA, binds strongly to plasma proteins—namely, about 90%. It has been shown in various studies (1,4,5) that, in combination treatment, PT is displaced by VPA. The free fraction of PT is thus considerably increased. In these instances the monitoring of free PT concentration gives a better assessment than the total PT concentration, because only the free PT is able to cross the blood-brain barrier and hence correlates better with the clinical effects or possible side effects.

But it is also of interest, in the case of PB, CBZ, and CE, to determine to what extent the free fraction is influenced by VPA.

II. METHODS AND PATIENTS

The data of patients on a combination therapy with VPA were analyzed and compared with corresponding control groups not on VPA (Table 1). The patients were inpatients of the epilepsy center, Bethel, Bielefeld, Germany. The ages of the patients ranged from 5 to 86 years. Patients who suffered from diseases of the liver or the kidneys were excluded.

The serum concentrations of PT, PB, CBZ, and CE were determined by high-performance liquid chromatography, and the VPA concentrations by gas chromatography (8). The assay of the unbound concentration was carried out at 25°C following ultrafiltration using the Amicon-Centrifree micropartition system. Day-to-day coefficients of variation were less than 4% for the total and free serum concentrations. It is to be noted that the free fraction is temperature dependent.

In order to predict the free concentrations of PT, PB, CBZ, and CE from the corresponding total serum concentration and the VPA concentration, the following regression equation was used:

$$\text{AEDf} = a \times \text{AEDt} + b \times \text{AEDt} \times \text{VPA} \tag{1}$$

where

AEDf = free serum concentration (of PT, PB, CBZ, CE)
AEDt = total serum concentration (of PT, PB, CBZ, CE)
VPA = (total) VPA serum concentration

The parameters "a" and "b" were estimated according to the method of least squares. The estimated parameter "a" corresponds to the protein binding in the absence of VPA.

A similar regression model for PT (and CBZ) was also used in the studies of Haidukewych et al. (5,7) and Scheyer et al. (6). In order to determine the accuracy of the estimated regression equation for PT, it is necessary to check the predictions for an additional control group. For this reason we investigated a further control group (II) of 33 patients, determining the total and free PT concentrations at 8.00 h, and in

Table 1 Influence of VPA on the Unbound Fraction of PT, PB, CBZ, and CE (Mean ± SD)[a]

	Without VPA			With VPA			
	N	F_u (%)		F_u (%)	N	r	VPA (μg/ml)
PT	36	7.78 ± 0.64	$<^b$	11.72 ± 2.34	84	0.82[b]	57.1 ± 29.2
PB	39	53.99 ± 3.61	$<^b$	58.22 ± 4.37	28	0.59[b]	65.8 ± 32.9
CBZ	38	20.11 ± 2.01	$<^b$	22.01 ± 1.85	36	0.56[b]	58.6 ± 26.7
CE	36	40.34 ± 5.33	$<^b$	44.51 ± 5.17	36	0.49[b]	58.6 ± 26.7

[a] r = Pearson correlation coefficient of the VPA concentration with the unbound fraction of PT, PB, CBZ, CE; F_u = unbound (free) fraction; N = number of patients.
[b] $p < 0.01$.

addition at 11.00, 14.00, and 16.00 h. The measured and predicted free serum concentrations, the absolute mean prediction error (MAE), and the root-mean-square prediction error (RMSE) were calculated (9).

III. RESULTS AND DISCUSSION

The influence of valproic acid on the free fraction of PT, PB, CBZ, and CE is summarized in Table 1. Table 1 shows that the free fractions of the four substances were significantly (Mann-Whitney test, $p < 0.01$) elevated by VPA. As expected, the influence of VPA is most marked in the case of PT. With VPA the free fraction of PT is, on average, 50% higher. Also to be noted is the considerably higher standard deviation for the free fraction of PT in co-medication with VPA. The free fraction of other substances was, on average, only slightly increased by VPA (about 10% higher).

The free fractions of these antiepileptic drugs correlate significantly with the VPA concentration. The correlation for PT is clearly higher than for the other substances. The relation between the total and free PT concentration is shown in Figure 1. In some cases the free PT concentrations of patients on VPA co-medication were more than twice as high as those of patients on PT monotherapy, although the total PT concentrations were comparable. Total PT concentration can in this instance be confusing, because free PT concentration can be markedly higher than estimated from the total concentration.

The influence of VPA on the free fraction of PB (Figure 2) is not so obvious, but some patients with VPA have clearly higher free PB concentrations than expected from the regression line for patients without VPA. This also applies to CBZ and CE.

Next investigated was whether the free concentrations of PT, PB, CBZ, and CE can be predicted from the corresponding total serum concentration and the VPA concentration. The estimated regression parameters are summarized in Table 2. The high correlations r, the low standard errors of estimate, and the mean absolute prediction

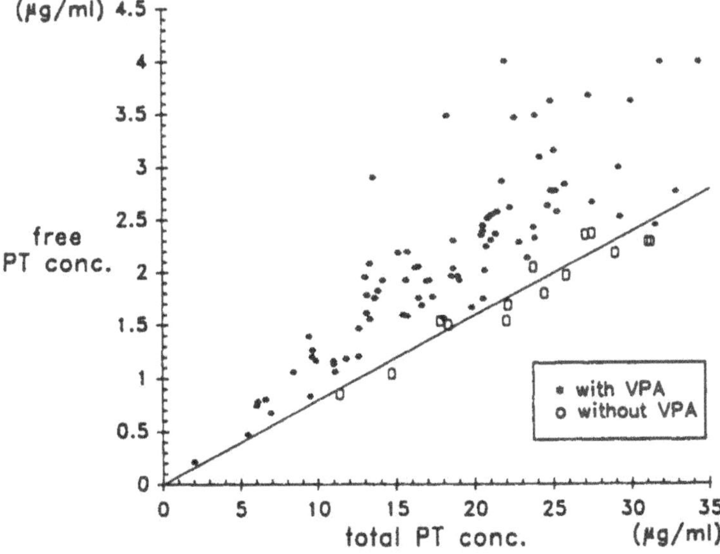

Figure 1 Relation between total and free phenytoin serum concentration. The line is the regression line for patients without VPA.

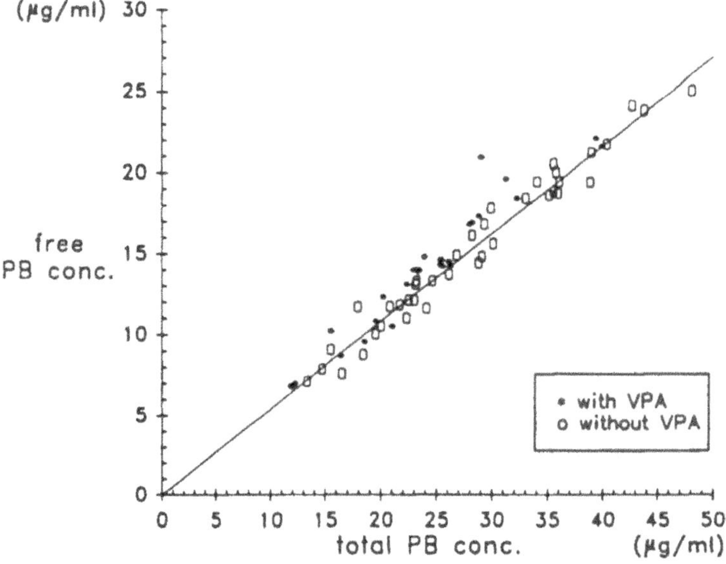

Figure 2 Relation between total and free phenobarbital serum concentration. The line is the regression line for patients without VPA.

errors show that the measured and estimated free concentrations are in good agreement. A consistent over- or underestimation was not found (compare low mean prediction errors). The estimated parameters a agree well with calculated mean values for protein binding in the absence of VPA (compare Table 1).

Thus the results of Haidukewych et al. (5,7) can in principal be confirmed for PT and CBZ. But, probably due to a different method of determination of free serum concentration (e.g., different temperature), certain differences with respect to their regression equations occur. The regression coefficients of Haidukewych et al. (5,7),

$$PTf = 0.095 \times PTt + 0.00100 \times PTt \times VPA$$

$$CBZf = 0.247 \times CBZt + 0.00062 \times CBZt \times VPA$$

are clearly higher than our parameters (compare Table 2). Using their regression equations, the free concentration of PT and CBZ would be considerably overestimated. Better agreement for PT is shown with the estimated regression parameters ($a = 0.080$, $b = 0.00073$) of Scheyer et al. (6), but they evaluated data on only a few patients.

In order to determine the accuracy of our regression equation for PT, the predictions for the additional control group (II) were checked. The measured and predicted free serum concentrations, the absolute mean prediction error, and the root-mean-square prediction error are given in Table 3. The results show that the predicted free PT concentrations agree well with the measured concentrations at different times of day.

Table 2 Estimated Regression Parameters[a]

AED	a	b	r	SE (μg/ml)	MAE (μg/ml)	ME (μg/ml)	X (μg/ml)
PT	0.079	0.000636	0.96	0.23	0.18	0.02	2.11
PB	0.526	0.000841	0.98	0.85	0.68	0.02	14.02
CBZ	0.200	0.000360	0.95	0.14	0.10	−0.01	1.76
CE	0.409	0.000588	0.97	0.11	0.08	0.01	1.12

[a]r = correlation coefficient of predicted and measured concentration; SE = standard error of estimate; MAE = mean absolute prediction error; ME = mean prediction error; X = mean measured free concentration.

In order to simplify the prediction of the free PT concentration a nomogram based on the above regression equation was constructed. The nomogram and an example are given in Figure 3. Without the use of a calculator, the estimated free PT concentration can be read off in dependence on total PT concentration and VPA concentration. For orientation, the total phenytoin concentration, which corresponds to the estimated free phenytoin without VPA co-medication, can also be read off. This orientation value can give the practitioner a better indication of possible therapeutic effects or side effects than the actually measured total PT concentration. The nomogram is a simple and useful aid for drug monitoring in patients on a combined therapy of PT and VPA.

IV. SUMMARY AND CONCLUSIONS

The free fractions of PT, PB, CBZ, and CE are significantly increased by VPA and correlate significantly with the VPA concentration. The increase of the free fraction of PT is especially important (7.8% without VPA versus 11.72% with VPA). The mean fractions of PB, CBZ, and CE are only slightly increased by VPA. Thus, in general the determination of the free concentration of these substances is not necessary in co-medication with VPA. However, since the increase of the free fraction depends on the VPA concentration, determining the free serum concentration of patients with high or very high VPA concentrations may be useful.

Table 3 Predicted and Measured Free PT Concentrations for the Control Group II (mean ± SD)[a]

Time	PT free measured (μg/ml)	PT free predicted (μg/ml)	MAE (μg/ml)	RMSE (μg/ml)	r	N
8.00	2.33 ± 0.66	2.34 ± 0.64	0.19 ± 0.15	0.24	0.93	33
11.00	2.61 ± 0.71	2.59 ± 0.71	0.20 ± 0.15	0.25	0.94	31
14.00	2.59 ± 0.69	2.58 ± 0.75	0.22 ± 0.18	0.27	0.93	33
16.00	2.47 ± 0.70	2.51 ± 0.72	0.26 ± 0.17	0.30	0.30	32

[a]MAE = mean absolute prediction error; RMSE = root-mean-square prediction error; r = Pearson correlation coefficient of predicted and measured free PT concentration; N = number of blood samples.

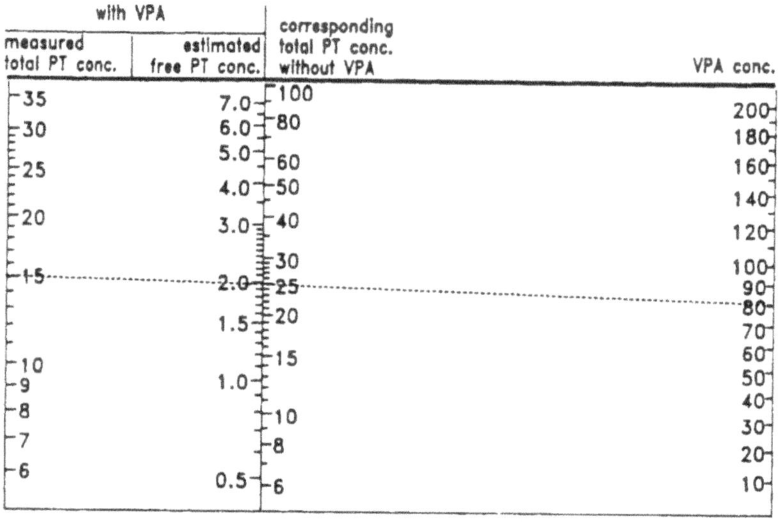

with VPA
measured total PT conc. | estimated free PT conc. | corresponding total PT conc. without VPA | VPA conc.

35 30 25 20 15 10 9 8 7 6
7.0 6.0 5.0 4.0 3.0 2.0 1.5 1.0 0.5
100 80 60 50 40 30 25 20 15 10 8 6
200 180 160 140 120 100 90 80 70 60 50 40 30 20 10

Figure 3 Nomogram for the prediction of free phenytoin concentration. The measured phenytoin total concentration (left-hand scale) is connected by a line with the corresponding VPA concentration (right-hand scale). The point where this line cuts the center scale gives the predicted free phenytoin concentration. For orientation, the total phenytoin concentration which corresponds to the estimated free phenytoin without VPA comedication can also be read off.

Example: If a total phenytoin concentration of 15 μg/ml and VPA concentration of 80 μg/ml was measured, then a free phenytoin concentration of ca. 2.0 μg/ml is predicted. Without VPA the free phenytoin concentration of 2.0 μg/ml would correspond to a total phenytoin concentration of ca. 25 μg/ml.

Note: The prediction of free PT concentration is less reliable for patients with certain illnesses, such as of the liver or kidneys, and in pregnant patients, because these factors can further change the protein binding.

The free serum concentrations of PT, PB, CBZ, and CE can be estimated from the total serum concentration and the VPA concentration.

A nomogram was constructed in order to predict the free PT concentration. This nomogram is a simple and useful aid for the drug monitoring of patients on a combined therapy of PT and VPA.

ACKNOWLEDGMENT

The authors are greatly indebted to G. S. Macpherson for translating the paper, and to N. Nothbaum and N. Ehnert for computational assistance.

REFERENCES

1. J. Barre, F. Didey, F. Delion, and J.-P. Tillement. Problems in therapeutic drug monitoring: Free drug level monitoring. *Ther. Drug Monitor.*, *10*:133–143 (1988).
2. E. Perucca. Plasma protein binding of phenytoin in health and disease: Relevance to therapeutic drug monitoring. *Ther. Drug Monitor.*, 2:331–344 (1980).

3. M. S. Yerby, P. N. Friel, K. McCormick, M. Koerner, M. van Allen, A. M. Leavitt, C. J. Sells, and J. A. Yerby. Pharmacokinetics of anticonvulsants in pregnancy protein binding. *Epilepsy Res.*, *5*:223–228 (1990).
4. R. H. Mattson, J. A. Cramer, and P. D. Williamson. Valproic acid in epilepsy: Clinical and pharmacological effects. *Ann. Neurol.*, 3:20–25 (1978).
5. D. Haidukewych, E. A. Rodin, and J. J. Zielinski. Derivation and evaluation of an equation for prediction of free phenytoin concentration in patients co-medicated with valproic acid. *Ther. Drug Monitor.*, *11*:134–139 (1989).
6. R. D. Scheyer, J. A. Cramer, and R. H. Mattson. Valproate-induced variable phenytoin binding (abstract). *Epilepsia*, *30*:64 (1989).
7. D. Haidukewych, J. J. Zielinski, and E. A. Rodin. Derivation and evaluation of an equation for prediction of free carbamazepine concentration in patients co-medicated with valproic acid. *Ther. Drug Monitor.*, *11*:528–532 (1989).
8. U. Juergens, T. May, K. Hillenkötter, and B. Rambeck. Systematic comparison of three basic methods of sample pretreatment for high-performance liquid chromatographic analysis of antiepileptic drugs using gas chromatography as a reference method. *Ther. Drug Monitor.*, *6*:334–343 (1984).
9. L. B. Sheiner and S. L. Beal. Some suggestions for measuring predictive performance. *J. Pharmacokinet. Biopharm.*, *8*:246–256 (1981).

61

Steady-State Pharmacokinetics and Dosing of Phenytoin and Phenobarbital in Patients

Jovan Popović *University of Novi Sad, Novi Sad, Yugoslavia*

I. INTRODUCTION

The well-known nonlinear, individually variable pharmacokinetics and narrow therapeutic index of phenytoin make its dosage regulation both difficult and important (1). The therapeutic range of serum levels (10 to 20 mg/L) must be maintained for a long period with an optimal, individually variable, multiple dose (2). A number of investigators have developed methods, based on Michaelis-Menten principles, to predict the phenytoin dose required to achieve a given target steady-state serum concentration (3–6). But predictions are much more complicated if the phenytoin is used in combination with other antiepileptics. A combination of phenytoin and phenobarbital is marketed in Yugoslavia (Hydanphen, Pliva Pharmaceuticals, capsules = 50 mg phenytoin sodium + 100 mg phenobarbital acid). Since a large number of epileptic patients are treated with this combination, the opportunity was taken to assess the impact of phenytoin-phenobarbital co-medication on the calculation of Michaelis-Menten pharmacokinetic parameters and the validity of phenytoin-level prediction based on these data.

II. METHODS

The study was performed on 15 adult epileptic outpatients, with normal hepatic and renal function, receiving Hydanphen. Additional increments of phenytoin, because of incomplete seizure control, were obtained using Difetoin (Pliva) tablets (100 mg of phenytoin sodium). Three different steady states were achieved by three different multiple doses of phenytoin (at least 6 weeks treatment per dose), but one phenobarbital multiple dose. Capillary blood samples were taken at each steady-state period, by finger stick, just before the morning dose. Phenytoin and phenobarbital steady-state concentrations in serum were measured by high-performance liquid chromatography (HPLC) according to our modification of the Soldin-Gilbert method (7).

A Varian 5000 HPL chromatograph with a Valco universal injection system and a temperature-control block was used. The detector was the Vari-Chrom UV-visible variable spectrophotometer. A Varian model 9176 strip-charge recorder and CDS 111 L chromatography data system were used for peak determination.

Our modification involves precipitation of serum proteins with an equivolume amount of acetonitrile, centrifugation, and reverse-phase chromatography of 10 μL of supernatant on a 4 mm × 30 cm column containing Micro Pak MCH-10. For determination of phenytoin, the mobile phase consisted of potassium phosphate buffer (10 mmol/L, pH 8.0) and acetonitrile (60:40, by volume), the mobile phase flow was 0.8 mL/min, column temperature was 30°C, and column eluate was monitored at 215 nm. For phenobarbital determination, the mobile phase consisted of potassium phosphate buffer (10 mmol/L, pH 8.0) and acetonitrile (85:15, by volume), the mobile phase flow was 1.6 mL/min, column temperature was 20°C, and column eluate was monitored at 254 nm. Concentrations were determined by measuring peak heights.

The modification provides accuracy, with an analytical recovery of phenytoin at 94.2% to 104.4%, and good precision with the coefficient of variation 1.5 to 6.5% for measured subtherapeutic and therapeutic concentrations. For phenobarbital the modification also provides accuracy, with an analytical recovery at 106.6% to 118.6% and good precision with the coefficient of variation 1.4% to 7.9% for measured subtherapeutic and therapeutic (15 to 40 mg/L) concentrations.

The pharmacokinetic technique used here for phenytoin concentrations has been described previously (8). Briefly, the Michealis-Menten equation can be expressed as

$$\frac{D}{R} = \frac{V_{max}C_{ss}}{K_m + C_{ss}} \tag{1}$$

where D/R is dose rate (mass/time), V_{max} is the maximal velocity of metabolism (mass/time), K_m is the serum concentrations at $1/2V_{max}$ (mass/volume), and C_{ss} is the steady-state serum concentration (mass/volume). The estimation of K_m and V_{max} requires a minimum of two measurements of D and C_{ss}, and solution of two simultaneous equations obtained from Eq. (1):

$$V_m - \frac{K_m D_1}{C_{ss1}R} = \frac{D_1}{R} \tag{2}$$

$$V_m - \frac{K_m D_2}{C_{ss2}R} = \frac{D_2}{R} \tag{3}$$

$$K_m = \frac{D_1 - D_2}{D_2/C_{ss2} - D_1/C_{ss1}} \tag{4}$$

$$V_m = \frac{D_2 + K_m D_2/C_{ss2}}{R} \tag{5}$$

These K_m and V_{max} values can be used again in Eq. (1) to estimate the D for desired C_{ss} and vice versa:

$$C_{ss} = \frac{DK_m}{V_{max} - D} \tag{6}$$

Instead of average C_{ss} values, minimum phenytoin steady-state concentrations (just before the next dose) may be used (9). This approach has been applied successfully in numerous studies dealing with individualization of phenytoin therapy (1,5,6,10–13). The percentage of saturation of the enzyme system metabolizing phenytoin is calculated, for C_{ss} = 10 mg/L, according to the formula

$$\% = \frac{100D}{V_{max}R} = \frac{100C_{ss}}{K_m + C_{ss}} \tag{7}$$

Phenobarbital clearance is obtained as

$$CL = \frac{D}{C_{ss}R} \tag{8}$$

since this drug possesses linear kinetics for the therapeutic range of serum concentrations.

III. RESULTS

The phenytoin concentrations observed in this study were nonlinearly related to the dose (Figure 1). The phenobarbital concentrations are also presented in Figure 1. Using Student's t-test for paired observations (t_d), the following phenobarbital

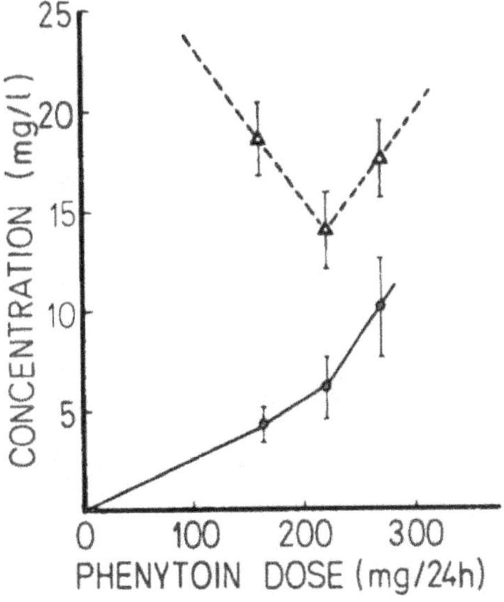

Figure 1 Steady-state serum phenytoin (●) and phenobarbital (Δ) concentrations measured at three different multiple doses of phenytoin but the same multiple phenobarbital dose (150 ± 66 mg/24 h). Blood was obtained just before the next dose. Values shown are means ± SD from 15 patients. Differences between the first and the second and between the second and third mean that phenobarbital concentrations are significant.

Table 1 Individual Patient Data Including Estimates of K_m and V_{max} Values for Phenytoin and CL Values for Phenobarbital

Patient	Age (yr)	Weight (kg)	Sex	K_m (mg/L)	V_{max} (mg/kg/24 h)	C (mg/kg/h)
1	53	56	M	4.55	10.09	8.34
2	48	61	F	5.76	4.88	5.66
3	44	60	M	5.37	6.69	3.73
4	39	55	F	5.33	6.63	8.21
5	33	67	M	6.86	8.14	5.76
6	44	80	M	2.06	4.77	7.96
7	34	90	F	3.12	7.34	9.72
8	37	48	F	9.79	7.92	2.76
9	55	65	F	17.32	10.55	6.35
10	31	55	F	7.09	10.91	11.36
11	23	45	F	9.93	10.00	6.94
12	33	68	M	3.36	4.89	4.68
13	57	81	F	7.95	6.02	2.57
14	45	63	F	3.42	6.92	3.40
15	27	58	F	13.21	15.54	3.51
Mean ± SD	40 ± 10	63 ± 12		7.01 ±4.15	8.09 ± 2.92	6.06 ± 2.67

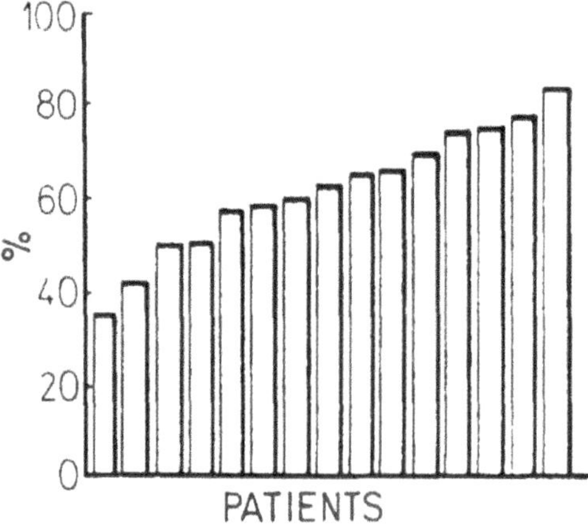

Figure 2 Percent saturation of the enzyme system for phenytoin, obtained from 15 patients, according to Eq. (7) with C_{ss} = 10 mg/L.

concentrations differ significantly: first and second ($t_d = 3.99$, $p < 0.01$) and second and third ($t_d = 4.06$, $p < 0.01$). A pronounced interindividual variations in K_m, V_{max}, and the percent saturation parameters for phenytoin and the CL values for phenobarbital were found (Table 1, Figure 2). The extreme phenytoin K_m values differ by 8 times, V_{max} values by 3 times, and the percent saturation by 2 times. The extreme phenobarbital CL values differ by 4 times.

By employing the individual K_m and V_{max} values obtained with the first two phenytoin doses and Eq. (6), a steady-state serum concentrations was predicated for the third dose of that drug. The predicated serum concentration was compared with the observed level. In the 15 patients for whom the third increment of phenytoin dose was obtained, the predicated steady-state concentration agreed closely with the observed levels ($r = 0.961$, $p < 0.01$). The absolute average error of the calculated versus observed concentrations was 7.11% (range 0 to 20.32%). Also, by employing the individual K_m and V_{max} values and Eq. (1), an optimal phenytoin dose that would result in a desired minimum serum concentration of 10 mg/L was calculated for the 15 patients. The calculated optimal phenytoin doses ranged from 190.4 to 504.0 mg/24 h (mean ± SD = 299.4 ± 83.0 mg/24 h).

No side effects were noticed during the study.

IV. DISCUSSION

Different pharmacokinetic techniques based on Michaelis-Menten principles have been reported as a means of facilitating the attainment of optimal phenytoin dosage (14,15). In the usual clinical situation, however, where only two reliable steady-state serum concentrations at different dosing rates are available, these methods are mathematically equivalent: The problem is reduced to the solving of two simultaneous algebraic equations with two unknowns, K_m and V_{max} (8), such as Eqs. (2) and (3) in the present study.

The K_m and V_{max} parameters calculated in this study are approximate but not identical to the true K_m and V_{max} values obtained from average steady-state serum concentrations only (9). The latter require several blood samples, but the former require only one sample (e.g., just before the next dose) per dosage interval.

Previous studies (1,5,6,10–13) reported similar approximate K_m and V_{max} values and correlation coefficients for observed versus predicated serum phenytoin concentrations, such as obtained here for phenytoin-phenobarbital co-medication. Our data indicate that Eqs. (1) to (6) can be employed successfully for individualization of phenytoin therapy when phenobarbital is co-medicated.

Some steady-state serum phenobarbital concentrations measured at different multiple doses of phenytoin but at the same multiple phenobarbital doses differed statistically significantly. This probably occurs because, in addition to inducing, phenytoin also competes with phenobarbital for the enzyme, and the net effect then depends on the balance between induction and inhibition.

The conclusions may be drawn that apparent K_m and V_{max} values for phenytoin, as determined in the present study, by using minimum steady-state serum concentrations, are operationally useful parameters for phenytoin-level prediction, even in the case of phenobarbital co-administration. Finally, the situation of phenobarbital co-medication with phenytoin, which exhibits nonlinear kinetics, requires a careful approach based on serum concentration measurements and determinations of individual kinetic parameter for both drugs.

V. SUMMARY

Three different steady-state serum phenytoin and simultaneous phenobarbital concentrations were measured by high-performance liquid chromatography (HPLC) in each of 15 epileptic adults, treated with a combination of these two drugs. Doses of phenobarbital were not changed when phenytoin doses were adjusted. Apparent K_m and V_{max} for phenytoin were estimated. High correlation between phenytoin serum levels calculate from individual K_m and V_{max} values and the observed levels at third adjustment of phenytoin daily dose ($r = 0.961$, $p < 0.01$) were found. Statistically significant differences were obtained in some steady-state serum phenobarbital concentrations measured at different daily doses of phenytoin but the same phenobarbital daily dose. It has been suggested that phenytoin-phenobarbital co-medication requires a careful approach based on serum concentration measurements and determination of individual kinetic parameters for both drugs.

REFERENCES

1. S. Vozeh, T. K. Muir, B. L. Sheiner, and F. Follath. Predicting individual phenytoin dosage. *J. Pharmacokinet. Biopharm.*, 9:131–146 (1981).
2. E. Martin, N. T. Tozer, B. L. Sheiner, and S. Riegelman. The clinical pharmacokinetics of phenytoin. *J. Pharmacokinet. Biopharm.*, 5:579–596 (1977).
3. A. Richens and A. Dunlop. Serum phenytoin levels in management of epilepsy. *Lancet*, 2:247–248 (1975).
4. M. T. Ludden, W. D. Hawkins, P. J. Allen, and F. S. Hoffman. Optimum phenytoin dosage regimens. *Lancet*, 7:307–308 (1976).
5. M. T. Ludden, P. J. Allen, A. W. Valutsky, et al. Individualization of phenytoin dosage regimens. *Clin. Pharmacol. Ther.*, 21:287–293 (1977).
6. W. P. Mullen. Optimal phenytoin therapy: A new technique for individualizing dosage. *Clin. Pharmacol. Ther.*, 23:228–232 (1978).
7. J. S. Soldin and J. Gilbert-Hill. Rapid micromethod for measuring anticonvulsant drugs in serum by high-performance liquid chromatography. *Clin. Chem.*, 22:856–859 (1976).
8. M. T. Ludden. Individualization of phenytoin therapy. *J. Pharm. Pharmacol.*, 32:152 (1980).
9. G. J. Wagner. Time to reach steady-state and prediction of steady-state concentrations for drugs obeying Michaelis-Menten elimination kinetics. *J. Pharmacokinet. Biopharm.*, 6: 209–225 (1978).
10. K. Chiba, T. Ishizaki, H. Miura, and K. Minagawa. Michaelis-Menten pharmacokinetics of diphenylhydantoin and application in the pediatric age patient. *J. Pediatr.*, 96:479–494 (1980).
11. G. P. Blain, C. J. Mucklow, J. C. Bacon, and D. M. Rawlins. Pharmacokinetics of phenytoin in children. *Br. J. Clin. Pharmacol.*, 12:659–661 (1981).
12. E. W. Dodson. Nonlinear kinetics of phenytoin in children. *Neurology*, 32:42–48 (1982).
13. G. Koren, N. Brand, and M. S. MacLeod. Influence of bioavailability on the calculated Michaelis-Menten parameters of phenytoin in children. *Ther. Drug Monitor.*, 6:11–14 (1984).
14. W. P. Mullen and W. P. Foster. Comparative evaluation of six techniques for determining the Michaelis-Menten parameters relating phenytoin dose and steady-state serum concentrations. *J. Pharm. Pharmacol.*, 31:100–104 (1979).
15. E. G. Schumacher. Using pharmacokinetics in drug therapy VI: Comparing methods for dealing with nonlinear drugs like phenytoin. *Am. J. Hosp. Pharm.*, 37:128–132 (1980).

62

Valproic Acid Enhances Lipid Peroxidation in Rat Liver, in Vivo

D. Cotariu, S. Evans, J. L. Zaidman, and A. Yulzari
Tel Aviv University, Zerifin, Israel

I. INTRODUCTION

Valproic acid (VPA) is a very effective anticonvulsant agent that is widely used in the management of various forms of epilepsy. However, treatment with the drug has occasionally been shown to induce severe or even fatal hepatotoxicity. The mechanisms underlying the liver injury are still ill-defined (1–3).

One of the mechanisms by which drugs may evoke toxicologic responses is depletion of hepatic glutathione. The latter, in its reduced (GSH) and oxidized (GSSG) forms, is the major thiol redox system of the cell, providing defence for cellular structures against peroxidative damage. Along with selenium-dependent glutathione peroxidase, it reduces both hydrogen peroxide and organic hydroperoxides. It is also involved in the scavenging of electrophilic intermediates generated during biotransformation of xenobiotics via the cytosolic enzymes glutathione transferases.

The major cellular targets at risk following drug-induced depletion of liver glutathione are membrane lipids. The molecular damage is expressed as the autocatalytic process of lipid peroxidation, resulting in impairment of membrane functioning.

In previous work we have provided evidence that exposure of rats to VPA results in changes in liver GSH and pyridine nucleotide phosphate levels as well as in the activity of enzymes involved in the maintenance of these reducing equivalents (4). In this report we show that VPA enhances lipid peroxidation rate when administered to rats in vivo.

II. MATERIALS AND METHODS

Female Charles River rats weighing between 150 and 200 g were randomly assigned to treatment groups (6 to 12 rats per group). Animals were maintained on a normal light-dark cycle and had free access to a standard diet. Experiments were started between 8 and 9 a.m. Rats were given a single i.p. injection of 100, 300, 500, or 750 mg/kg VPA in sterile saline. Control animals received equivalent volumes of sterile saline. At various times following treatment, animals were killed by cervical dislocation, and livers were quickly excised, weighed, blotted dry, and homogenized in five volumes of ice-cold buffer containing 50 mM Tris-HCl, pH 7.4, 100 mM KCl, and 1 mM EDTA.

The rate of lipid peroxidation was determined in 10% (w/v) tissue homogenates according to the method of Ohkawa et al. (5) as modified by Jamall and Smith (6), by estimating malondialdehyde (MDA) formed with 2-thiobarbituric acid. Values are expressed as micromoles of MDA per milligram of protein.

GSH content was measured after extraction with 10% metaphosphoric acid (w/v) containing 10 mM EDTA, by an enzymatic method, as described by Davies et al. (7).

Protein content was determined by the method of Lowry et al. (8), with bovine albumin as standard.

Serum and liver VPA was measured by a commercial kit (Emit, Syva, Palo Alto, California), and results are summarized in Table 1.

Statistics: Results are given as the mean ± SD. Data were analyzed by one-way analysis of variance followed by Bonferroni's test. Values of p less than 0.05 were considered significant.

III. RESULTS

The animals tolerated the drug and survived the time of experiments (6 h) with doses up to 500 mg/kg. The mortality increased up to 50% at 750 mg/kg within 3 h after drug exposure. Table 2 shows the effect of varying concentrations of VPA on liver GSH content. At the end of 30 min, GSH level was reduced by 30%, 49%, and 58% in the 300-, 500-, and 750-mg/kg groups, respectively. After 3 h, GSH values either returned to control limits (300 mg/kg VPA) or tended to revert to normal, but were still significantly reduced (500 mg/kg VPA). No effect on GSH level was seen with nontoxic doses (100 mg/kg/VPA).

The dose- and time-response relation of MDA formation versus VPA is given in Figure 1. An elevation of about twofold in MDA level occurred within 1 h after administration of 100 mg/kg in vivo (nontoxic dose). The MDA levels stayed elevated for at least 3 h and returned to control levels within 6 h. At higher doses, a significant elevation occurred within 30 min and MDA levels remained increased for at least 6 h.

IV. DISCUSSION

The present study provides evidence that VPA is able to stimulate the rate of lipid peroxidation within a short time following a single exposure in vivo, as evidenced by

Table 1 Serum and Liver Valproic Acid Content After i.p. Administration (mean ± SEM; number of animals in parentheses)

	Serum (μg/ml)		Liver (μg/g)	
VPA (mg/kg)	30 min	60 min	30 min	60 min
100	91 ± 16 (5)	—	38 ± 1.5 (4)	—
300	339 ± 22 (4)	157 ± 6 (5)	123 ± 36 (6)	61 ± 9 (5)
500	470 ± 27 (7)	402 ± 15 (7)	220 ± 37 (5)	126 ± 24 (7)
750	531 ± 41 (7)	507 ± 41 (8)	683 ± 58 (5)	294 ± 26 (7)

Table 2 Rat liver content of Reduced Glutathione after i.p. administration of Valproic Acid (mean ± SD; number of animals in parentheses)

VPA treatment (mg/kg body weight)	Reduced glutathione (μmol/g)		
	30 min	60 min	180 min
None	4.9 ± 0.14 (18)	—	—
100	4.74 ± 0.32 (18)	4.34 ± 0.64 (6)	4.56 ± 0.51 (7)
300	3.4 ± 0.78[a] (6)	2.7 ± 0.39[a] (5)	4.6 ± 0.52 (7)
500	2.5 ± 0.25[a] (5)	1.42 ± 0.22[a] (5)	3.39 ± 0.29[a] (6)
750	2.08 ± 0.27[a] (5)	2.58 ± 0.33[a] (5)	—

[a]$p < 0.01$.

MDA formation in liver homogenates. Lipid peroxidation in the liver is a function of dose and time. The hepatotoxicity of many xenobiotics has been previously associated with the depletion of hepatic GSH (9–12). But the levels of GSH had to drop below critical values (20% of the initial value) in order for peroxidation to occur (9,10). However, in other instances, no relationship between the intensity and timing of GSH levels and promotion of lipid peroxidation could be evidenced (13). In this study there appears to be significant lipid peroxidation in the liver at a VPA dose (100 mg/kg) at

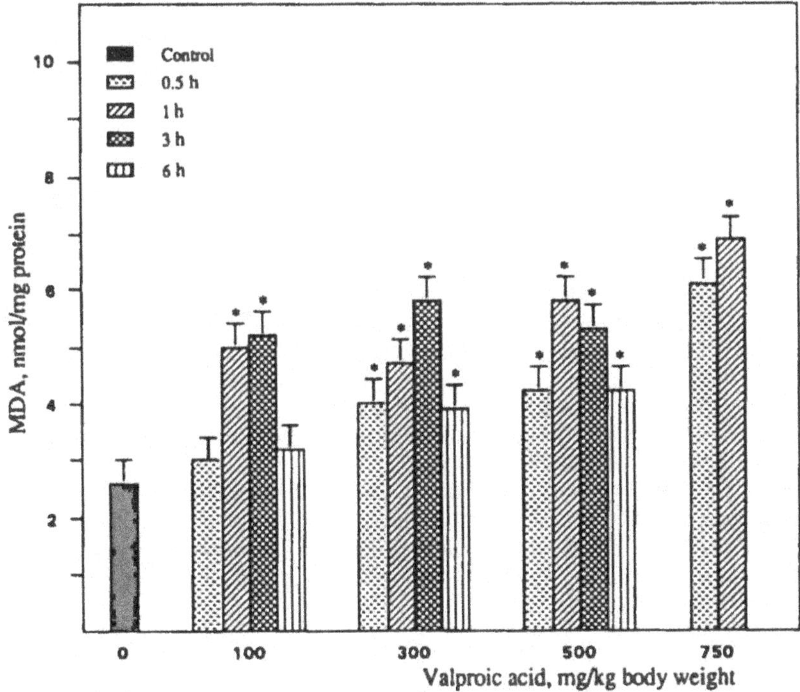

Figure 1 Effect of valproic acid on liver lipid peroxidation. Each bar represents the mean ± SD of determination from 5 to 12 animals.
*$p < 0.05$ with respect to saline-injected control animals.

which hepatic GSH is still within control values, but parallel courses for decreased hepatic GSH and enhanced lipid peroxidation are apparent at other VPA doses. In a previous study (4) we found, along with the GSH reduction, a drop in glutathione peroxidase activity, ranging with the various doses from 30% to 55%, even at low VPA doses. The prominent role of glutathione peroxidase in preventing tissue peroxidation has been well documented (14). Apparently, the suppression of glutathione peroxidase activity that we observed shortly after a single dose of VPA indicates a loss in the liver's capacity for disposing of peroxides. Therefore, promotion in the liver of lipid peroxidation might be explained by a reduced antioxidant activity of the tissue at low doses, which further impairs at high doses with the shift in the redox potential of liver tissue to less reducing conditions (4). However, the relative importance of lipid peroxidation in the subsequent development of clinically significant pathologic changes in liver during VPA therapy remains to be substantiated by further research.

V. SUMMARY

Lipid peroxidation and reduced glutathione (GSH) levels were studied in livers of rats following a single i.p. administration of 0, 100, 300, or 750 mg/kg of valproic acid (VPA). The latter elicits a dose- and time-dependent stimulation of the rate of lipid peroxidation. Parallel courses for decreased hepatic GSH and enhanced lipid peroxidation are apparent at high VPA doses.

REFERENCES

1. F. E. Dreifuss, N. Santilli, D. H. Langen, K. P. Sweeney, K. A. Moline, et al. Valproic acid hepatic fatalities: A retrospective review. *Neurology, 37*:379–385 (1987).
2. D. Cotariu and J. L. Zaidman. Valproic acid and the liver. *Clin. Chem., 34*:890–897 (1988).
3. M. J. Eadie, W. D. Hooper, and R. G. Dickinson. Valproate-associated hepatotoxicity and its biochemical mechanisms. *Med. Toxicol., 3*:85–106 (1988).
4. D. Cotariu, S. Evans, J. L. Zaidman, and O. Marcus. Early changes in hepatic redox homeostasis following treatment with a single dose of valproic acid. *Biochem. Pharmacol., 40*:589–593 (1990).
5. H. Ohkawa, N. Ohishi, and K. Yagi. Assay for lipid peroxides in animal tissues by thiobarbituric reaction. *Anal. Biochem., 95*:351–358 (1979).
6. I. S. Jamall and J. C. Smith. Effects of cadmium on glutathione peroxidase, superoxide dismutase and lipid peroxidation in the rat heart: A possible mechanism of cadmium cardiotoxicity. *Toxicol. Appl. Pharmacol., 80*:33–42 (1985).
7. M. H. Davies, D. F. Birt, and R. C. Schnell. Direct enzymatic assay for reduced and oxidized glutathione. *J. Pharmacol. Meth., 12*:191–194 (1984).
8. O. H. Lowry, N. J. Rosebrough, A. L. Farr, and R. J. Randall. Protein measurement with the Folin phenol reagent. *J. Biol. Chem., 193*:265–275 (1951).
9. S. P. Srivastava, M. Das, and P. K. Seth. Enhancement of lipid peroxidation in rat liver on acute exposure to styrene and acrylamide, a consequence of glutathione depletion. *Chem. Biol. Interact., 45*:373–380 (1983).
10. M. Younes and C. P. Sigers. Formation of ethane in vitro by rat liver homogenates following glutathione depletion in vivo. *Toxicol. Lett., 15*:213–218 (1983).
11. D. J. Kornbrust and J. S. Bus. Glutathione depletion by methyl chloride and association with lipid peroxidation in mice and rats. *Toxicol. Appl. Pharmacol., 72*:388–399 (1984).
12. M. Dubin, S. G. Goijman, and A. O. M. Stoppani. Effect of nitroheterocyclic drugs on lipid peroxidation and glutathione content in rat liver extracts. *Biochem. Pharmacol., 33*:3419–3423 (1984).

13. L. Hernandez and W. Lijinsky. Glutathione and lipid peroxide levels in rat liver following administration of methapyrilene and analogs. *Chem.-Biol. Interact.*, 69:217–224 (1989).
14. T. W. Simmons and I. S. Jamall. Significance of alterations in hepatic antioxidant enzymes. Primacy of glutathione peroxidase. *Biochem. J.*, 251:913–917 (1988).

63

Carbamazepine Drug Interactions: The Influence of Concurrent Drug Therapy on Serum Concentrations of Carbamazepine and Its Epoxide Metabolite

Rong-Bor Wang, Li-Ting Liu, Chun-Hing Yiu, and Tzu-Yao Chang *Veterans General Hospital-Taipei, Republic of China*

I. INTRODUCTION

Carbamazepine (CBZ), an effective anticonvulsant, is metabolized in the liver by the mixed-function oxidase system to the active metabolite carbamazepine-10,11-epoxide (CBZ-E) (1). This metabolite is known to relieve the pain of trigeminal neuralgia and to cause adverse effects at very high serum levels (2–6). Studies have shown that a poor correlation exists between the CBZ dose and the serum concentration of CBZ and CBZ-E (7,8). Monitoring of drug levels therefore seems to be essential. However, in a review of antiepileptic drug monitoring, it was concluded that clinical relevance for free (unbound) CBZ and CBZ-E concentration monitoring in the presence of co-medication has not been established (9). Since the final action of anticonvulsants takes place within the brain and only free drugs can exert their action at the appropriate site, measurement of the free concentrations of CBZ and CBZ-E can provide a better basis for the evaluation of the pharmacological effect of CBZ than measurement of the total concentrations. Although previous studies have indicated that the concomitant use of other antiepileptics has the potential to reduce the binding of CBZ to plasma proteins and to disproportionately increase the free fraction, their results are in part contradictory for lack of the co-medication serum levels (7,8,10).

The aim of the present study was to examine the effect of co-medication such as phenytoin (PHT), valproate (VA), phenobarbital (PB), and primidone (PRM) on the metabolism of CBZ in epileptic patients. In addition to the associated drug therapy, the effect of CBZ dose on the serum CBZ and CBZ-E levels was also examined. The clinical usefulness of routine monitoring of CBZ and CBZ-E concentrations is discussed.

II. METHODS

A. Patients

One hundred and thirty-five otherwise healthy epileptic adults (aged 15 to 45), who had been on a stable CBZ monotherapy or combined therapy for at least 2 months, were included. CBZ was usually dosed three times daily. To avoid pooling heterogeneous data together, patients were divided into five groups according to the co-medications they received. The characteristic data for each group of patients are listed in Table 1. The plasma albumin levels of all included cases were within the normal range (i.e., 3.7 to 6.3 g/dL). Fasting blood specimens were drawn at 8:00 a.m., before medication, to check steady-state trough levels. Each patient had at least two consecutive samples drawn at semimonthly intervals; 5 mL of blood specimens were collected in Venoject red tubes. The serum was stored frozen at –20°C and analyzed within 1 week.

B. Sample Preparation and Assay Technique

Ultrafiltration (Centrifree micropartition system, Amicon) was used to separate the free (unbound) drugs from the protein-bound drugs in serum. This separation process was achieved with the help of a Sorvall RT6000B refrigerated tabletop centrifuge set at 25 ± 1°C (1500 × g, for 30 min).

Fluorescence polarization immunoassay (TDx, Abbott) was used to determine the total serum concentrations of PHT, VA, PB, and PRM (TDx system, no. 9507, 9514, 9500, 9513).

A high-performance liquid chromatographic method (HPLC) was adapted for the simultaneous determination of CBZ and CBZ-E concentrations using a Hewlett Packard HP 1084 liquid chromatograph (11). Standard powders of CBZ, CBZ-E, and 10-methoxycarbamazepine were gifts from Dr. R. Heckendorn of Ciba-Geigy. Acetonitrile (HPLC grade) and methyl-*t*-butyl ether were purchased from E. Merck. Sodium hydroxide as supplied by Fisher Scientific Co. The assay for CBZ and CBZ-E was carried out on 250 μL of plasma standard or test sample in a glass tube, to which was added 1.25 μg of 10-methoxycarbamazepine as internal standard and 250 μL of 1.0 N sodium hydroxide. The mixture was extracted with 3.0 mL of methyl-*t*-butyl

Table 1 Clinical Data on Patients Who Received Carbamazepine With or Without Other Anticonvulsants[a]

Type of therapy	No. of patients	Age (years)	Body weight (kg)	CBZ daily dose (mg/kg)	CBZ clearance (1 h/kg)
CBZ alone	22	21.9 ± 6.0[b]	58.8 ± 16.3	14.28 ± 7.44	0.07 ± 0.05
CBZ + PHT	57	27.7 ± 7.9	54.8 ± 6.5	14.83 ± 6.09	0.11 ± 0.03[c]
CBZ + VA	37	25.5 ± 5.6	58.1 ± 5.3	19.32 ± 3.93	0.10 ± 0.03[d]
CBZ + PB	8	25.4 ± 6.7	52.9 ± 5.7	12.07 ± 4.42	0.10 ± 0.02
CBZ + PRM	11	26.8 ± 4.7	53.4 ± 4.1	17.34 ± 7.83	0.13 ± 0.04[d]

[a]CBZ, carbamazepine; PHT, phenytoin; VA, sodium valproate; PB, phenobarbitol; PRM, primidone.
[b]Mean ± SD.
[c]Statistically different from CBZ monotherapy at [c]$p < 0.001$, [d]$p < 0.01$.

ether, vortex mixed for 1 min, and then shaken continuously for 5 min on a rotator. The mixture was centrifuged at 2100 × g for 5 min. The organic phase was transferred into a new glass tube, and the methyl-*t*-butyl ether phase was evaporated to dryness under a stream of nitrogen. The residue was reconstituted in 250 μL of mobile phase, and 20 μL was injected. The column used was a reverse-phase LiChrospher RP-Select B (4 mm I.D. × 250 mm, 5 μm, E. Merck), and the mobile phase was water/acetonitrile (68:32, v/v). The flow rate was 1.0 mL/min, with the detector wavelength set at 212 nm, the temperature maintained at 37°C, and the whole chromatogram completed within 15 min. The detection limit for either CBZ or CBZ-E was 0.05 μg/mL. The coefficient of variation for intraday and interday determinations of both compounds was within ±10%. The physical recovery of CBZ and CBZ-E was 98% and 95%, respectively. Figure 1 shows the chromatograms obtained in extracted blank serum (A), extracted blank serum with internal standard (B), extracted spiked serum (C), and patient serum (D).

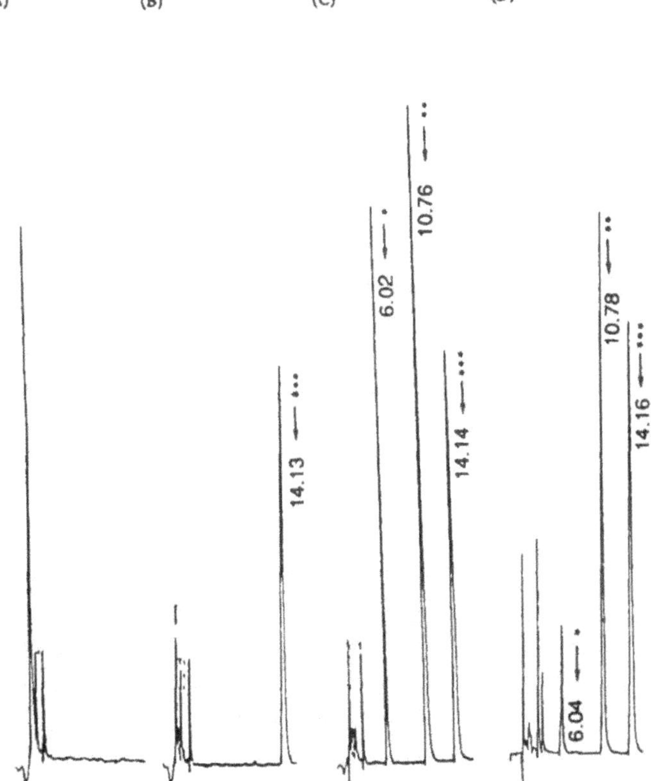

Figure 1 High-performance liquid chromatograms showing extracted blank serum (A), extracted blank serum with internal standard (B), extracted spiked serum (CBZ 6.0 mg/L), CBZ-E 4.0 mg/L) (C), and patient serum (D). The peaks shown are *CBZ' **CBZ; and ***internal standard.

C. Statistical Evaluations

Statistical analyses were carried out using the fixed-effect two-way analysis of variance, the Student's t-test, or the nonparametric test of Mann and Whitney whenever appropriate (12). Gaussian distribution hypothesis was tested according to the chi-square analysis for goodness of fit. Variances were compared by the F-test. Linear regression analysis was performed by the classical least-squares regression analysis. The analysis of covariance was employed to test the equality of slopes. For all these tests, a computer software Statistical Analysis System (SAS) was used (13).

III. RESULTS AND DISCUSSION

A. Serum Concentrations of CBZ and CBZ-E in Patients Receiving CBZ Monotherapy

The mean serum levels of CBZ and CBZ-E in each group of patients are listed in Table 2. Our results confirmed the dose-dependent autoinduction of CBZ metabolism (14,15). A disproportionately small rise in the serum concentration of CBZ and CBZ-E was found in patients receiving CBZ monotherapy. There was a statistically significant, nonlinear relationship between the dose and the total CBZ-E serum levels. Similar to CBZ-E, but to a lesser extent, a relationship existed between the logarithm of CBZ dose and the CBZ serum levels ($y = 5.90 + 1.09\,x, r = 0.23, n = 22, p > 0.1$). The relationship between the logarithm of the oral CBZ dose (in milligrams per kilogram) and the steady-state serum levels of CBZ and CBZ-E is shown in Figure 2. Since a wide scattering was evident, these dose-concentration relationships have no practical value for predicting the serum CBZ and CBZ-E levels in individual patients. The same results have also been reported previously (16,17).

Linear regression analyses of results in all 22 monotherapy subjects demonstrated a high correlation between total and free CBZ concentrations, and between total and free CBZ-E concentrations (Figure 3). The bound fraction of CBZ and CBZ-E averaged 78.30 ± 3.69% and 52.43 ± 9.83%, respectively. The calculated mean CBZ-E/CBZ ratios (total and free) of each group of patients are shown in Table 3. Due to the lower degree of protein binding of the epoxide, CBZ-E/CBZ ratios were significantly higher for the free than for total levels in all groups. The CBZ-E free concentrations averaged 0.48 ± 0.20 of the CBZ free concentrations. This high metabolite/parent drug ratio in plasma

Table 2 Free and Total Serum CBZ and CBZ-E Levels[a]

Type of therapy	Total CBZ (mg/L)	Free CBZ (mg/L)	Total CBZ-E (mg/L)	Free CBZ-E (mg/L)
CBZ alone	8.69 ± 2.14[b]	1.86 ± 0.68	1.83 ± 0.68	0.87 ± 0.38
CBZ + PHT (12.01 ± 7.66)[c]	6.02 ± 2.26	1.21 ± 0.40	1.65 ± 0.66	0.76 ± 0.41
CBZ + VA (48.06 ± 17.59)	8.24 ± 2.26	2.01 ± 0.55	3.20 ± 1.25	1.73 ± 0.76
CBZ + PB (12.77 ± 8.05)	5.20 ± 1.91	1.24 ± 0.44	1.37 ± 1.00	0.82 ± 0.67
CBZ + PRM (6.63 ± 3.24)	5.69 ± 2.09	1.22 ± 0.43	2.34 ± 1.56	1.12 ± 0.90

[a]CBZ, carbamazepine; CBZ-E, carbamazepine-10,11-epoxide; PHT, phenytoin; VA, sodium valproate; PB, phenobarbitol; PRM, primidone.
[b]Mean ± SD.
[c]( ): averaged total serum level of the comedication ± SD (mg/L).

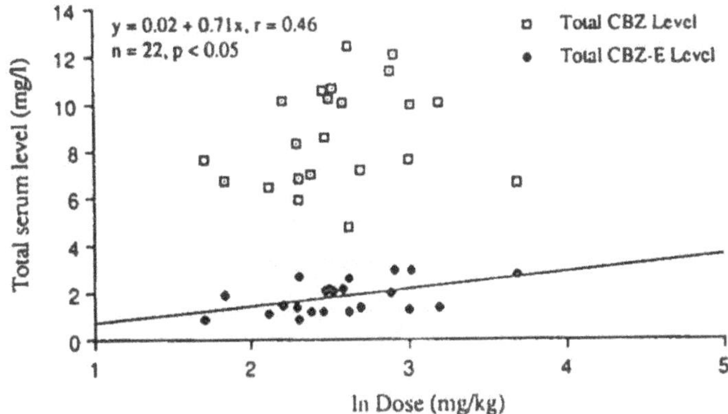

Figure 2 Correlation between the natural logarithm of CBZ dose and the serum levels of CBZ and CBZ-E in the CBZ monotherapy patients.

water is clinically important when we consider that the CBZ-E itself is a potent anticonvulsant in animals models (18,19). Furthermore, regardless of whether or not total or free serum concentrations were used, mean CBZ-E/CBZ ratios were significantly higher in subjects taking either VA, PHT, or PRM in addition to CBZ as compared with those on CBZ monotherapy.

B. Influence of Other Antiepileptic Drugs on the Serum Concentration of CBZ

As previously reported by other investigators, we also found that the serum concentration of CBZ was lower when CBZ was given in combination with PHT, VA,

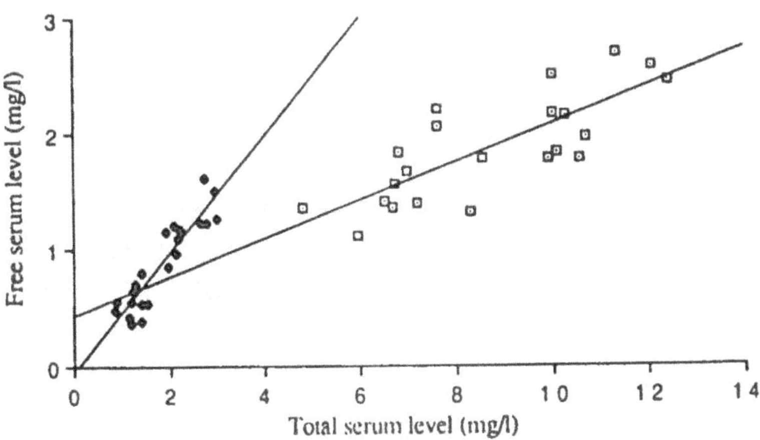

Figure 3 Correlation between total CBZ and CBZ-E levels with their respective free levels. ▫ Free CBZ level; $y = 0.43 + 0.16x$, $r = 0.78$, $n = 22$, $p < 0.01$. • Free CBZ-E level; $y = -0.05 + 0.50x$, $r = 0.91$, $n = 22$, $p < 0.01$.

Table 3 Serum CBZ-E/CBZ Ratios and Free Fractions of CBZ and CBZ-E[a]

	CBZ-E/CBZ ratio		Free fraction	
Type of therapy	Total	Free	CBZ	CBZ-E
CBZ alone	0.22 ± 0.08[b]	0.48 ± 0.20	0.22 ± 0.04	0.48 ± 0.10
CBZ + PHT	0.29 ± 0.13[c]	0.67 ± 0.39[e]	0.21 ± 0.04	0.49 ± 0.27
CBZ + VA	0.40 ± 0.15[c]	0.87 ± 0.35[c]	0.26 ± 0.09[f]	0.54 ± 0.15[f]
CBZ + PB	0.26 ± 0.14	0.64 ± 0.45	0.24 ± 0.03	0.61 ± 0.26[e]
CBZ + PRM	0.44 ± 0.21[c]	0.88 ± 0.50[d]	0.22 ± 0.06	0.44 ± 0.16

[a]CBZ, carbamazepine; CBZ-E, carbamazepine-10,11-epoxide; PHT, phenytoin; VA, sodium valproate; PB, phenobarbitol; PRM, primidone.
[b]Mean ± SD.
[c,d,e,f]Statistically different from CBZ monotherapy at [c]$p < 0.0001$, [d]$p < 0.01$, [e]$p < 0.05$, [f]$p < 0.10$.

PRM, or PB than when it was given alone (7,8,20,21). As an indicator of the CBZ disposition, the apparent clearance of CBZ (L/h/kg) was calculated using the CBZ dose (mg/kg/h) divided by the steady-state serum CBZ level (mg/L). PRM appeared to increase the apparent CBZ clearance most of all (by 86%), whereas the influence of VA was relatively small (by 43%), as shown in Figure 4. The reduction of the CBZ levels due to the co-medications, such as PHT, PB, and PRM, could be explained by the induction of the epoxide-diol pathway. Since VA is not an enzyme inducer, metabolic induction could not explain the small but significant increase in the apparent CBZ clearance by this drug. A protein-binding competition was a reasonable explanation (22). As shown in Table 3, VA competed with CBZ for plasma protein-binding sites, resulting in an increase in the free CBZ concentration ($p < 0.10$) and thus enhanced the possibility of CBZ elimination.

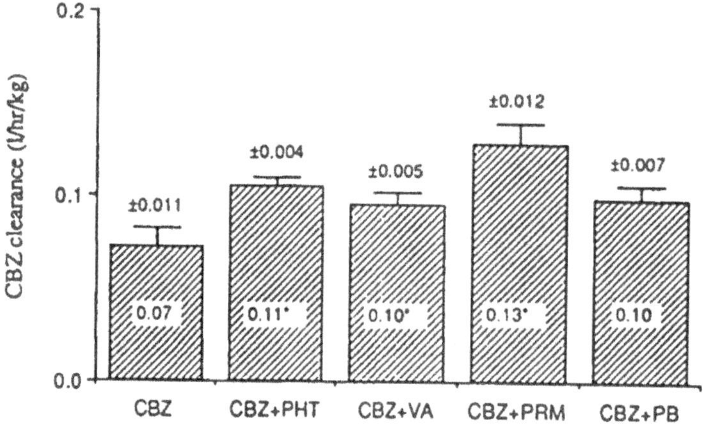

Figure 4 CBZ apparent clearance in patients who received CBZ with or without other anticonvulsants (mean ± SEM). *Significantly different from the CBZ monotherapy group ($p < 0.01$).

C. Influence of Other Antiepileptic Drugs on the Serum Concentration of CBZ-E

The free fraction of CBZ-E was much higher than the free fraction of CBZ and, especially in VA- or PRM-co-medicated patients, the free CBZ-E/CBZ ratio almost approached unity. Figure 5 and 6 show that valproate causes a marked increase in the concentration of CBZ-E, supposedly due to inhibition of the epoxide hydrolase. This was in agreement with the studies that also found an increase in CBZ-E caused by VA (10,23). Patients on PRM had markedly higher CBZ-E concentrations than the other groups (apart from the VA group), and the greatest increase in the CBZ-E/CBZ ratio was observed in this group of patients. PRM is metabolized into phenylethylmalonamide and PB in the liver, and the ratio between PRM and PB in blood is around 1:1.5 to 1:3.0 after the drug reaches its steady state (24). The decrease in serum CBZ level caused by concurrent PRM treatment could be associated with the enzyme-inducing effect of its PB metabolite. However, our patients on PB and CBZ combined therapy had much lower CBZ-E concentrations and CBZ-E/CBZ ratios than those receiving PRM and CBZ together. In this instance it must be realized that with PRM, the CBZ-E metabolism was inhibited and this inhibitory mechanism was independent of its PB metabolite since PB itself could influence the CBZ-E metabolism in a different way.

D. Influence of Serum Concentrations of Other Antiepileptic Drugs on the Simultaneous CBZ and CBZ-E Concentrations

To investigate the influence of serum concentration of PHT, VA, PB, and PRM on the concentrations of CBZ and CBZ-E, the CBZ and CBZ-E concentrations in the groups of

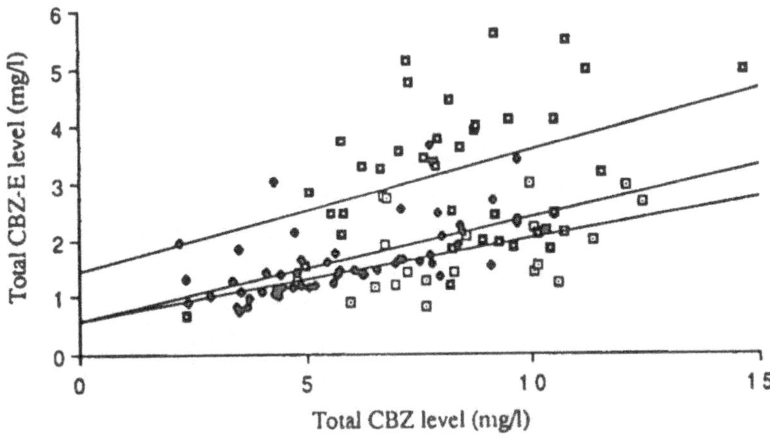

Figure 5 Relationship between the total CBZ level and the total CBZ-E level in patients receiving three different treatments. ▫ CBZ alone; $y = 0.55 + 0.15x$, $r = 0.46$, $n = 22$, $p < 0.05$. • CBZ + PHT; $y = 0.56 + 0.18x$, $r = 0.65$, $n = 57$, $p < 0.01$. ▪ CBZ + VA; $y = 1.42 + 0.22x$, $r = 0.39$, $n = 37$, $p < 0.05$.

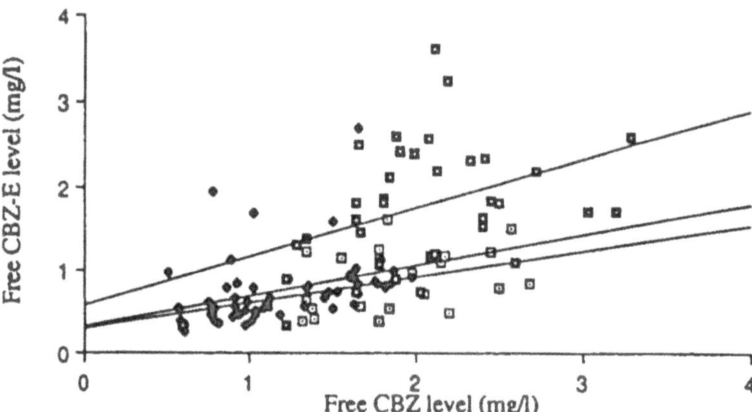

Figure 6 Relationship between the free CBZ level and the free CBZ-E level in patients receiving three different treatments. ▫ CBZ alone; $y = 0.29 + 0.31x$, $r = 0.37$, $n = 22$, $p < 0.10$. • CBZ + PHT; $y = 0.32 + 0.37x$, $r = 0.36$, $n = 57$, $p < 0.01$. ▪ CBZ + VA; $y = 0.55 + 0.59x$, $r = 0.43$, $n = 37$, $p < 0.01$.

patients, who were given either PHT, VA, PB, or PRM as the co-medication were correlated with the corresponding serum concentration of the respective co-medication, as shown in Table 4. The correlations indicated that the decrease in the CBZ concentration depended on the PHT and VA concentrations ($p < 0.05$), whereas the concentration of PB and PRM appeared to have no significant influence on the CBZ concentration. The VA concentration was also highly correlated with the increase of both total and free CBZ-E concentrations ($p < 0.01$). As demonstrated in Figure 7, the CBZ-E/CBZ ratio increased as the VA concentration increased. Our data suggest that VA inhibits the metabolism of CBZ-E in a concentration-dependent way (25). Further clinical study is in progress to examine if it is possible to derive an equation for prediction of free CBZ concentration in patients co-medicated with VA.

IV. CONCLUSION

The dose of CBZ cannot be used as a reliable index for predicting the serum CBZ or CBZ-E level. Monitoring of CBZ and CBZ-E, especially in patients on polytherapy, can therefore assist in optimization and individualization of CBZ therapy. However, for normoalbuminemic patients who have no liver or renal dysfunction, monitoring of free CBZ or free CBZ-E concentration appears to be unnecessary when these patients are not co-medicated with other antiepileptics (10,26). The significant correlation between both total CBZ and CBZ-E concentrations and their respective free concentrations suggests that binding capacities of both CBZ and CBZ-E are essentially constant within the therapeutic range for CBZ monotherapy.

Table 4 Regression Parameters (Least-Squares Analyses) for Co-medication Serum Levels Versus Total and Free Serum CBZ and CBZ-E Levels[a]

Type of therapy	Total CBZ			Free CBZ			Total CBZ-E			Free CBZ-E		
	a	b	r	a	b	r	a	b	r	a	b	r
CBZ + PHT	7.06	−0.087	0.29[c]	1.45	−0.020	0.39[b]	1.56	0.009	0.10	0.79	0.002	0.04
CBZ + VA	10.22	−0.041	0.33[c]	1.93	0.002	0.06	1.25	0.041	0.52[b]	0.42	0.027	0.64[b]
CBZ + PB	5.83	−0.049	0.21	1.22	0.001	0.03	0.70	0.053	0.43	0.13	0.054	0.65[d]
CBZ + PRM	5.96	−0.040	0.06	1.29	−0.011	0.09	1.67	0.115	0.24	0.69	0.066	0.24

[a]Equation for linear regression is y = a + bx, r = correlation coefficient.
[b,c,d]Statistically significant at $^{b}p < 0.01$, $^{c}p < 0.05$, $^{d}p < 0.10$.

Other antiepileptic drugs can influence the metabolism of CBZ in various ways. In general, patients co-medicated with PHT, VA, PRM, or PB require higher CBZ dosages than patients on CBZ monotherapy to achieve serum CBZ levels within the optimal range. Nevertheless, the interaction between VA and CBZ is much more complicated than the enhancement of CBZ clearance (27,28). VA can cause a remarkable increase in serum CBZ-E concentrations, and the elevation of CBZ-E/CBZ ratios is related to the VA concentration. In some of our patients their serum CBZ-E levels doubled after VA was introduced. Since both compounds will compete for the plasma protein-binding sites, free drug monitoring is clinically important for patients taking VA and CBZ concurrently.

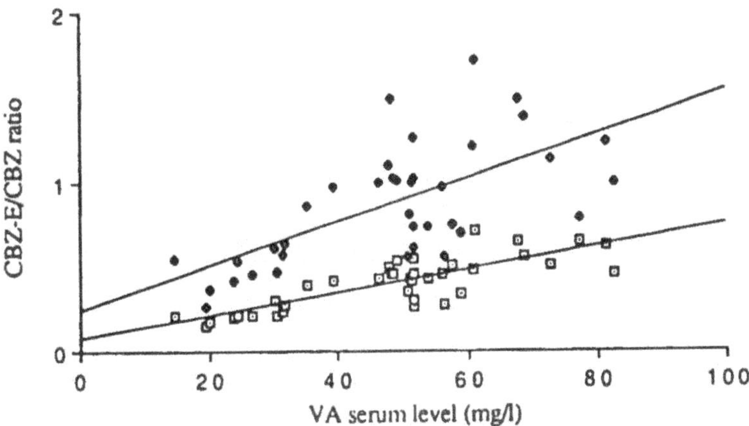

Figure 7 Correlation between the total and free CBZ-E/CBZ ratios with the simultaneous VA level in patients receiving carbamazepine and valproate polytherapy. ▫ Total CBZ-E/CBZ ratio; y = 0.08 + 0.01x, r = 0.81, n = 37, $p < 0.01$. • Free CBZ-E/CVZ ratio; y = 0.25 + 0.01x r = 0.65, n = 37, $p < 0.01$.

REFERENCES

1. J. W. Faigle and K. F. Feldmann. Carbamazepine: Biotransformation. In *Antiepileptic Drugs*, 2nd ed. (D. M. Woodbury, J. K. Penry, and C. E. Pippenger, eds.), Raven Press, New York, pp. 483–495 (1982).
2. T. Tomson and L. Bertisson. Potent therapeutic effect of carbamazepine-10,11-epoxide in trigeminal neuralgia. *Arch. Neurol.*, *41*:598–601 (1984).
3. E. Pisani, A. Fazio, and G. Oteri. Sodium valproate and valpromide: Differential interactions with carbamazepine in epileptic patients. *Epilepsia*, *27*:548–552 (1986).
4. F. Pisano, A. Fazio, G. Oteri, et al. Carbamazepine-viloxazine interaction in patients with epilepsy. *J. Neurol. Neurosurg.*, *49*:1142–1145 (1986).
5. P. N. Patsalos, T. S. Stephenson, S. Krishna, A. A. Elyas, P. T. Lascelles, and C. M. Wiles. Carbamazepine-10,11-epoxide induced side effects. *Lancet*, 2:496 (1985).
6. L. Bertilsson and T. Tomson. Clinical pharmacokinetics and pharmacological effects of carbamazepine and carbamazepine-10,11-epoxide—an update. *Clin. Pharmacokinet.*, *11*:177–198 (1986).
7. D. Battino, L. Bossi, D. Droci, et al. Carbamazepine plasma levels in children and adults: Influence of age, dose, and associated therapy. *Ther. Drug Monitor.*, 2:315–322 (1980).
8. L. McKauge, J. H. Tyrer, and M. J. Eadie. Factors influencing simultaneous concentrations of carbamazepine and its epoxide in plasma. *Ther. Drug Monitor.*, *3*:63–70 (1981).
9. D. Schmidt. Clinical relevance of free antiepileptic drug monitoring. In *Advances in epileptology. The XVth Epilepsy International Symposium* (R. J. Porter, R. H. Mattson, A. A. Ward, and M. Dam, eds.), Raven Press, New York, pp. 155–160 (1984).
10. B. Rambeck, T. May, and U. Juergens. Serum concentrations of carbamazepine and its epoxide and diol metabolites in epileptic patients: The influence of dose and comedication. *Ther. Drug Monitor.*, *9*:298–303 (1987).
11. R. D. Chelberg, S. Gunawan, and D. M. Treiman. Simultaneous high-performance liquid chromatographic determination of carbamazepine and its principal metabolites in human plasma and urine. *Ther. Drug Monitor.*, *10*:188–193 (1988).
12. J. H. Zar. Two-sample hypotheses. In *Biostatistical Analysis*, 2nd ed. (J. H. Zar, ed.), Prentice-Hall, Englewood Cliffs, NJ, pp. 122–149 (1984).
13. *SAS User's Guide: Statistics*. SAS Institute, NC (1982).
14. T. Tomson, J. O. Svensson, and P. Hilton-Brown. Relationship of intraindividual dose to plasma concentration of carbamazepine: Indication of dose-dependent induction of metabolism. *Ther. Drug Monitor.*, *11*:533–539 (1989).
15. L. Bertilsson, B. Hojer, G. Tybring, J. Osterloh, and A. Rane. Autoinduction of carbamazepine metabolism in children examined by a stable isotope technique. *Clin. Pharmacol. Ther.*, *27*:83–88 (1980).
16. A. Kumps and Y. Mardens. A retrospective study on epileptic patients treated with carbamazepine: Interaction between age and co-medication on the drug disposition. *Pharm. Acta Helv.*, *58*:160–164 (1983).
17. A. A. Elyas, P. N. Patsalos, O. A. Agbato, E. M. Brett, and P. T. Lascelles. Factors influencing simultaneous concentrations of total and free carbamazepine and carbamazepine-10,11-epoxide in serum of children with epilepsy. *Ther. Drug Monitor.*, *8*:288–292 (1986).
18. P. S. Albright and J. Bruni. Effects of carbamazepine and its epoxide metabolite on amygdala-kindled seizures in rats. *Neurology*, *34*:1383–1386 (1984).
19. B. F. D. Bourgeois and N. Wad. Individual and combined antiepileptic and neurotoxic activity of carbamazepine and carbamazepine-10,11-epoxide in mice. *J. Pharmacol. Exp. Ther.*, *231*:411–415 (1984).
20. M. Furlanut, G. Montanari, P. Bonin, and G. L. Casara. Carbamazepine and carbamazepine-10,11-epoxide serum concentrations in epileptic children. *J. Pediatr.*, *106*:491–495 (1985).
21. D. Lindhout, R. J. E. A. Hoppener, and H. Meinardi. Teratogenicity of antiepileptic drug combinations with special emphasis on epoxidation (of carbamazepine). *Epilepsia*, *25*:77–83 (1984).
22. R. H. Levy and W. H. Pitlick. Carbamazepine: Interactions with other drugs. In *Antiepileptic Drugs*, 2nd ed. (D. M. Woodbury, J. K. Penry, and C. Pippenger, eds.), Raven Press, New York, pp. 497–505 (1982).

23. J. F. Schoeman, A. A. Elyas, E. M. Brett, and P. T. Lascelles. Altered ratio of carbamazepine-10,11-epoxide/carbamazepine in plasma of children: Evidence of anticonvulsant drug interaction. *Dev. Med. Child. Neurol.*, 26:749–755 (1984).
24. E. H. Reynolds, G. Fenton, P. Fenwick, et al. Interaction of phenytoin and primidone. *Br. Med. J.*, 2:594–595 (1975).
25. G. J. A. Macphee, J. R. Mitchell, L. Wiseman, et al. Effect of sodium valproate on carbamazepine disposition and psychomotor profile in man. *Br. J. Clin. Pharmac.*, 25:59–66 (1988).
26. R. Riva, M. Contin, F. Albani, E. Perucca, G. Procaccianti, and A. Baruzzi. Free concentration of carbamazepine and carbamazepine-10,11-epoxide in children and adults: Influence of age and phenobarbital co-medication. *Clin. Pharmacokinet.*, 10:524–531 (1985).
27. E. Perucca. Free level monitoring of antiepileptic drugs—clinical usefulness and case studies. *Clin. Pharmacokinet.*, 9(Suppl. 1):71–78 (1984).
28. R. H. Levy and D. Schmidt. Utility of free level monitoring of antiepileptic drugs. *Epilepsia*, 26:199–205 (1985).

64

Comparison of Carbamazepine and Phenytoin Effects on Multimodal Evoked Potentials

A. Rysz and K. Gajkowski *IMK MSW, Warsaw, Poland*

I. INTRODUCTION

It has been known for a long time that some antiepileptic drugs can slow peripheral nerve conduction (1). Their possible effects on sensory pathways within the central nervous system (CNS) can be detected by evoked potentials (EPs). The aim of this study was to investigate the effects of monotherapy with carbamazepine (CBZ) or phenytoin (PHT) on the visual evoked potentials (VEPs), brainstem evoked potentials (BAEPs) and somatosensory evoked potentials (SEPs).

II. PATIENTS AND METHODS

We selected 21 subjects (7 males, 14 females, age range 12 to 66) with newly diagnosed epilepsy—complex partial seizures or partial seizures evolving to secondarily generalized seizures. Patients with brain pathology upon CT scan or progressive neurologic disease and patients likely to be noncompliant were excluded. Serum drug levels were measured for CBZ or PHT (at least once) and were within therapeutic range in all subjects.

Electrophysiologic studies included VEPs after monooculs full-field pattern-reversal stimulation, click-stimulus BAEPs, and median nerve short-latency somatosensory evoked potentials (MN-SSEPs), which were recorded before the treatment, according to American Electroencephalographic Society guidelines for clinical EPs studies (2).

The patients were randomly divided into two groups according to the drug administered: CBZ (12 patients, 3 males and 9 females, mean age 28), PHT (9 patients, 4 males and 5 females, mean age 39).

Electrophysiologic evaluation was repeated after at least 2 to 3 months of monotherapy. The results on the same group of patients before and after treatment were compared. The patients' values at baseline were compared to those of the control groups. The controls consisted of 26 healthy subjects (ages 19 to 69, mean 39) in VEPs, 22 healthy subjects (ages 20 to 62, mean 38) in BAEPs, and 25 healthy subjects (ages 19

Table 1 VEP Peaks in Controls and Patients, Before and After CBZ Monotherapy

	Controls	Patients (n = 11)			
		Before treatment		After treatment	
	(n = 26)	(mean, ms)	SD	(mean, ms)	SD
N2	71.6 (6.64)	72.3[a]	4.18	76.1[b]	7.2
P2	93.0 (6.24)	95.4[a]	6.9	105.1[c]	10.7
N3	124.1 (7.69)	124.9[a]	4.8	133.3	11.8

VEP, visual evoked potentials; CBZ, carbamazepine.
[a]$p>0.10$, Student's two-tailed t-test. Control group versus patients before CBZ monotherapy.
[b]$p<0.025$; [c]$p<0.01$, Student's t-test for paired samples. Patient group, before versus after CBZ monotherapy.

to 60, mean 38) in MN-SSEPs. All EPs data were analyzed statistically using Student's t-test.

III. RESULTS

The patients' mean values of VEPs peak N2, P2, and N3, BAEP interpeak latencies (IPL)-I-III, III-V, I-V, MN-SSEPs peak N9 and IPLs N9-N13, N9-N20, and central conduction time (CCT) did not differ from those of the control group.

Patients after several months of CBZ monotherapy (Table 1) showed significantly prolonged VEPs latencies as compared to patient means before treatment. MN-SSEPs N9 peak and all IPIs were prolonged except N9-N13 IPL (Table 2). BAEP IPLs (Table 3) were also significantly prolonged after CBZ monotherapy. Similar effects were observed after PHT monotherapy on VEPs and BAEPs (Tables 4 and 6). MN-SSEPs showed only prolonged CCT means after PHT monotherapy (Table 5).

Table 2 Median SEP Latencies in Controls and Patients, Before and After CBZ Monotherapy

	Controls	Patients (n = 42)			
		Before treatment		After treatment	
	(n = 25)	(mean, ms)	SD	(mean, ms)	SD
N9	10.0 (0.70)	9.9[a]	0.64	10.4[b]	0.78
N9-N13	3.6 (0.46)	3.4[a]	0.52	3.6	0.50
N9-N20	9.4 (0.56)	9.1[a]	0.57	9.5[b]	0.58
CCT					
N13-N20	5.85 (0.51)	5.6[a]	0.35	6.1[b]	0.41

SEP, somatosensory evoked potentials; CBZ, carbamazepine; CCT, central conduction time.
[a]$p>0.05$, Student's two-tailed t-test. Control group versus patients before CBZ monotherapy.
[b]$p<0.005$, Student's t-test for paired samples. Patient group, before versus after CBZ monotherapy.

Table 3 BAEP Interpeak Latencies in Controls and Patients, Before and After CBZ Monotherapy

		Patients (n = 12)			
	Controls	Before treatment		After treatment	
	(n = 22)	(mean, ms)	SD	(mean, ms)	SD
I-III	2.15 (0.13)	2.08[a]	0.10	2.43[b]	0.11
III-V	1.87 (0.12)	1.87[a]	0.15	1.93	0.12
I-V	4.0 (0.16)	3.94[a]	0.16	4.04[c]	0.16

BAEP, brainstem auditory evoked potentials; CBZ, carbamazepine.
[a]p>0.10, Student's two-tailed t-test. Control group versus patients before CBZ monotherapy.
[b]p<0.025; [c]p<0.05, Student's t-test for paired samples. Patient group, before versus after CBZ monotherapy.

Table 4 VEP Peaks in Controls and Patients, Before and After PHT Monotherapy

		Patients (n = 9)			
	Controls	Before treatment		After treatment	
	(n = 26)	(mean, ms)	SD	(mean, ms)	SD
N2	71.6 (6.64)	72.5[a]	3.78	75.0[b]	3.6
P2	93.0 (6.24)	96.9[a]	4.86	99.9[c]	4.83
N3	124.1 (7.69)	126.9[a]	8.43	131.0[c]	7.14

VEP, visual evoked potentials; PHT, Phenytoin.
[a]p>0.10, Student's two-tailed t-test. Control group versus patients.
[b]p<0.005; [c]p<0.02, Student's t-test for paired samples. Patient group before versus after PHT monotherapy.

Table 5 SEP Component Latencies in Controls and Patients, Before and After PHT Monotherapy

		Patients (n = 9)			
	Controls	Before treatment		After treatment	
	(n = 25)	(mean, ms)	SD	(mean, ms)	SD
N9	10.0 (0.7)	9.84[a]	0.96	10.1	1.02
N9-N13	3.63 (0.47)	3.62[a]	0.47	3.75	0.49
N9-N20	9.42 (0.58)	9.31[a]	0.51	9.58	0.54
CCT					
N13-N20	5.85 (0.51)	5.78[a]	0.29	6.11	0.41

SEP, somatosensory evoked potentials; CCT, central conduction time; PHT, phenytoin.
[a]p>0.20, Student's two-tailed t-test. Control group versus patients.
[b]p<0.01, Student's t-test for paired samples. Patient group, before versus after PHT monotherapy.

Table 6 BAEP Interpeak Latencies in Controls and Patients, Before and After PHT Monotherapy

	Controls (n = 22)	Patients (n = 9)			
		Before treatment		After treatment	
		(mean, ms)	SD	(mean, ms)	SD
I-III	2.15 (0.13)	2.2[a]	0.17	2.25[b]	0.16
III-V	1.87 (0.12)	1.97[a]	0.22	2.0[b]	0.21
I-V	4.01 (0.16)	4.16[a]	0.26	4.25[c]	0.27

BAEP, brainstem auditory evoked potentials; PHT, phenytoin.
[a]$p>0.05$, Student's two-tailed t-test. Control group versus patients.
[b]$p<0.025$; [c]$p<0.01$, Student's t-test for paired samples. Patient group, before versus after PHT monotherapy.

IV. DISCUSSION

Little information is available on the effects of PHT or CBZ on EPs. It concerns mainly patients receiving chronic anticonvulsant therapy (3–7). Green et al. (3) demonstrated that PHT produces an increase in the interpeak latencies of SEPs and BAEPs, but found no alteration of SEPs and BAEPs in patients with CBZ monotherapy. Mervaala et al. (4) showed significantly prolonged SEP peak latencies and major VEP peak latencies after CBZ monotherapy. Similar influence on BAEPs was obtained by Medgalini et al. (5). A prospective study of patients with newly diagnosed epilepsy (6,7) treated with CBZ or PHT monotherapy showed no significant modification of SEPs.

Our data, obtained in a small group of patients, showed significantly prolonged multimodal EPs after CBZ or PHT monotherapy as compared to results before treatment. None of the patients evidenced any prolongation of EPs peak or interpeak latencies 3 SD above the control mean.

REFERENCES

1. C. Geraldini, M. T. Faedda, and S. Sideri. Anticonvulsant therapy and its possible consequences on peripheral nervous system: A neurographic study. *Epilepsia*, 25:502–505 (1984).
2. American Electroencephalographic Society. Guidelines for clinical evoked potential studies. *J. Clin. Neurophysiol.*, 1:3–53 (1984).
3. I. B. Green, M. R. Walcoff, and I. F. Lucke. Comparison of phenytoin and phenobarbital effects on full-field auditory and somatosensory evoked potential interpeak latencies. *Epilepsia*, 23:417–421 (1982).
4. E. Mervaala, I. Partonen, U. Nousiainen, et al. Electrophysiologic effects of vinyl-GABA and carbamazepine. *Epilepsia*, 30:189–193 (1989).
5. S. Medaglini, M. Filippi, L. Smirne, et al. Effects of long-lasting antiepileptic therapy on brainstem auditory evoked potentials. *Neuropsychobiology*, 19:104–107 (1988).
6. N. C. Borah and M. C. Matheshwari. Effect of antiepileptic drugs on short-latency somatosensory evoked potentials. *Acta Neurol. Scand.*, 71:331–333 (1985).
7. L. Carenini, M. Bottacchi, S. D'Alessandro, et al. Carbamazepine does not affect short latency somatosensory evoked potentials: A longitudinal study in newly diagnosed epilepsy. *Epilepsia*, 29:145–148 (1988).

65

Antiepileptic Effect and Serum Levels of Zonisamide in Epileptic Patients with Refractory Seizures

Takashi Mimaki and Makoto Mino *Osaka Medical College, Osaka, Japan*
Tateo Sugimoto *Kansai Medical University, Osaka, Japan* **Ryosuke Murata** *Osaka City Medical School, Osaka, Japan*

I. INTRODUCTION

Despite the development of new anticonvulsants and other experimental treatments for epilepsy, many patients with intractable epilepsy continue to have seizures. In epileptic children, secondary generalized epilepsy and partial epilepsy (simple partial, complex partial, and partial seizures evolving to generalized tonic clonic seizures) are among the most difficult to control with currently available antiepileptic drugs.

Zonisamide (1,2-benzisoxazole-3-methanesulfonamide) is a new antiepileptic drug developed in Japan (1,2) (Figure 1). In experimental animals, this compound was found to have a strong inhibitory effect on convulsions of cortical origin by suppressing both focal spiking and spread of secondarily generalized seizures, whether induced by electrical or chemical stimuli (2–4). These animal studies indicate that the anticonvulsant profile of zonisamide is similar to that of phenytoin or carbamazepine (2,5). In fact, a double-blind placebo-controlled study revealed the efficacy of zonisamide on medically refractory complex partial seizures (6). Others (7–9) have also demonstrated the clinical efficacy of zonisamide in epileptic patients with refractory partial seizures.

The present study was conducted to evaluate further the use of zonisamide in epileptic patients with a variety of refractory seizures. Serum zonisamide levels were also determined, and the correlation between serum zonisamide levels and clinical response is described.

II. SUBJECTS AND METHODS

The study was conducted at three pediatric neurology clinics, the departments of pediatrics at Osaka University Hospital, Kansai Medical University Hospital, and

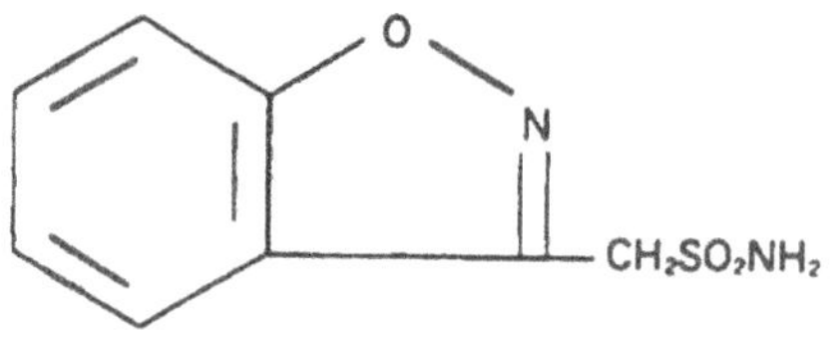

Figure 1 Chemical structure of zonisamide (1,2-benzisoxazole-3-methane sulfonamide).

Osaka City University Hospital. Seventy-eight patients (48 males and 30 females) were selected because of persistent, frequent seizures that failed to be controlled with standard antiepileptic drugs (i.e., phenobarbital, primidone, phenytoin, carbamazepine, sodium valproate, clonazepam, diazepam, or nitrazepam). Informed consent was obtained following a full explanation of the procedures to be undertaken. Patients were admitted to the study independently of the seizure type or etiology.

Of these 78 patients, 4 had primary generalized epilepsy, 26 had secondary generalized epilepsy (West syndrome 1, Lennox-Gastaut syndrome 17, others 8), and 48 had partial epilepsy (temporal epilepsy 15, nontemporal epilepsy 22, others 11). Seventeen patients were aged 4 years or less, 21 were from 5 to 9 years, 26 from 10 to 14 years, 8 from 15 to 19 years, and 6 from 20 to 26 years. Thus, patients under the age of 10 years accounted for almost one-half of the population (Table 1). Prior medication history was recorded, and medications were adjusted if necessary to achieve therapeutic blood levels before zonisamide therapy. The basic design was an add-on study, in which doses of standard antiepileptic drugs were maintained. The number of clinical seizures were charted daily. Most patients had attended each of these three pediatric neurology clinics for periods of several years, and thus their histories and seizure calendars were continued throughout the study.

Zonisamide was usually administered twice daily at an initial dose of 2 mg/kg/day. Dosage was adjusted at intervals of 2 weeks until maximal clinical response was obtained or until side effects appeared. Response to treatment was considered "markedly improved" if seizure frequency decreased by more than 75%, "improved" if there was a 50% to 75% decrease, "slightly improved" if less than 50% decrease, and "unchanged or aggravated" if there was no decrease or an increase in seizure frequency.

Serum samples were obtained before the morning dose after at least 4 weeks of maintenance zonisamide treatment. Serum zonisamide levels were determined by a high-performance liquid chromatographic technique. Serum levels of other antiepileptic drugs were measured by the usual laboratory routine procedures, and the results were recorded.

III. RESULTS

Overall, 11 patients (15.5%) were markedly improved, 13 patients (18.3%) were moderately improved, 16 patients (22.5%) were slightly improved, and 31 patients were unchanged or aggravated (Figure 2).

Improvements by type of epilepsy is shown in Figure 2. Of 45 patients with partial epilepsy, marked improvement was seen in 9 patients (20.0%), moderate

Table 1 Patient Characteristics

	Number of cases (%)
Sex:	
Male	48 (61.5)
Female	30 (38.5)
Age:	
0–4 (years)	17 (21.8)
5–9	21 (26.9)
10–14	26 (33.3)
15–19	8 (10.3)
20–26	6 (7.7)
Epilepsy type (1):	
Partial epilepsy	48 (61.5)
Primary generalized epilepsy	4 (5.1)
Secondary generalized epilepsy	26 (33.3)
Epilepsy type (2):	
Temporal lobe epilepsy	15 (19.2)
Nontemporal lobe epilepsy	22 (28.2)
Other partial epilepsy	11 (14.1)
Primary generalized epilepsy	4 (5.1)
West syndrome	1 (1.3)
Lennox syndrome	17 (21.8)
Other secondary generalized epilepsy	8 (10.3)
Duration of epilepsy:	
0–1 (years)	5 (6.5)
1–5	34 (44.2)
5–10	16 (20.8)
10–15	14 (18.2)
15–20	4 (5.2)
20–25	4 (5.2)

improvement in 11 patients (24.4%), slight improvement in 7 patients (15.5%), and no benefit was seen in 18 patients (40.0%). Of 4 patients with primary generalized epilepsy, 2 showed moderate improvement. Of 22 patients with secondary generalized epilepsy, marked improvement was seen in 2 patients (9.1%), moderate improvement in 2 patients (9.1%), and partial improvement in 7 patients (31.8%). No benefit was seen in 11 patients (50.0%).

Table 2 shows maximal doses and serum zonisamide levels in each patient response group. Serum zonisamide levels in markedly improved patients were 7.3 to 22.7 μg/mL (mean 13.0 μg/mL), those in moderately improved patients were 6.7 to 40.0 μg/mL (mean 14.5 μg/mL), and those in partially improved patients were 4.6 to 39.4 μg/mL (mean 20.6 μg/mL). Serum zonisamide levels in patients with no benefit were 1.1 to 62.1 μg/L (mean 26.8 μg/mL). These results suggest that in responders the

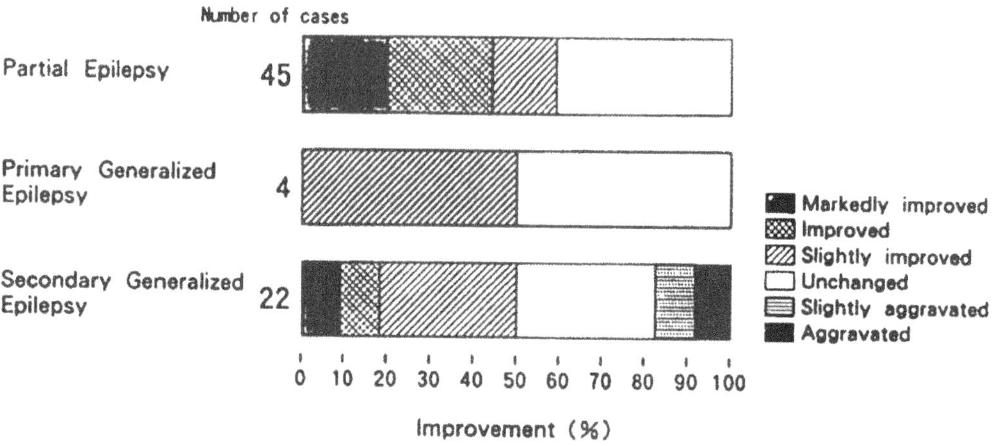

Figure 2 Overall improvement by type of epilepsy.

therapeutic range of serum zonisamide appears to be between 10 and 40 μg/mL. However, the relationship between serum zonisamide levels and clinical response was weak. After zonisamide co-medication, serum phenytoin levels increased in three patients to approximately twice as high as before zonisamide add-on therapy.

Adverse effects of zonisamide included drowsiness in 19 patients (24.3%). Other mild untoward effects seen in a few patients included ataxia (8 patients; 10.3%), loss of body weight (4 patients; 5.1%), decreased activity (2 patients; 2.6%), mental slowing (2 patients; 2.6%), loss of appetite (2 patients; 2.6%), nausea and vomiting (2 patients; 2.6%), stomatitis (one patient; 1.3%). No blood, liver, or renal toxicity occurred. The relationship between serum zonisamide levels and side effects was weak (Table 3).

IV. DISCUSSION

A pilot study of zonisamide by Sackellares et al. (7) indicated that zonisamide is effective in patients with refractory partial seizures. Wilensky et al. (8) also reported the clinical efficacy of zonisamide in five of eight patients with uncontrolled partial seizures. The present study confirms the efficacy of zonisamide in epileptic children, particularly those with refractory partial seizures.

Table 2 Overall Improvement Dosage, and Serum Levels of Zonisamide[a]

Clinical response	Maximum dose (mg/kg/day)		Serum levels (μg/mL)	
	Mean ± SD	Range	Mean ± SD	Range
Markedly improved	4.7 ± 1.3 (11)	2.9-7.5	13.0 ± 5.6 (10)	7.3-22.7
Improved	7.1 ± 4.9 (13)	1.6-18.2	14.5 ± 10.9 (8)	6.7-40.0
Slightly improved	6.4 ± 3.0 (16)	1.8-11.5	20.6 ± 11.9 (13)	4.6-39.4
Unchanged—aggravated	7.6 ± 4.7 (31)	0.7-18.6	26.8 ± 19.9 (14)	1.1-62.1

[a]Number of cases in each group is given in parentheses.

Table 3 Side Effects and Serum Zonisamide Levels[a]

	Dose (mg/kg/day)		Serum levels (μg/mL)	
Side effects	Mean ± SD	Range	Mean ± SD	Range
(−)	6.1 ± 3.9 (50)	0.7-18.6	17.4 ± 15.0 (31)	1.1-62.1
(+)	4.6 ± 3.8 (27)	1.0-18.5	12.2 ± 10.7 (17)	1.2-35.7

[a]Number of cases in each group is given in parentheses.

In the present study, maximal doses of zonisamide in patients with some improvement ranged from 1.6 mg/kg/day to 18.6 mg/kg/day (mean 6.4 mg/kg/day for responders). Therapeutic effects of zonisamide occurred between 10 and 40 μg/L, but there was no direct relationship between serum zonisamide levels and clinical response. These results are consistent with those of Wilensky et al. (8), who reported maximal response to zonisamide at doses approximating 6 mg/kg/day, with plasma levels of 20 to 30 μg/mL.

Furthermore, some patients showed a decrease in seizure frequency soon after initiating zonisamide add-on therapy at relatively low serum zonisamide levels, and three patients showed apparent nonlinear zonisamide pharmacokinetics with doses around 10 mg/kg/day. Wagner et al. (10) also observed nonlinear pharmacokinetics of this drug in adult epileptic patients after zonisamide co-medication. These aspects of both clinical response and pharmacokinetics of zonisamide resemble those of phenytoin.

The doubling of serum phenytoin levels seen in our three patients after zonisamide add-on therapy seems clinically very important. Of these, one patient showed marked improvement, another showed moderate improvement, and the other showed drowsiness after zonisamide co-medication. Thus the effect of both zonisamide itself and increased blood phenytoin concentrations should be taken into account when assessing the clinical response to zonisamide co-medication.

In conclusion, zonisamide has a broad antiepileptic spectrum. It is particularly effective in children with refractory partial seizures. Therapeutic drug monitoring of zonisamide as well as other antiepileptic drugs seems necessary to assess clinical response and to adjust dosage.

REFERENCES

1. Y. Masuda, Y. Utsui, T. Shiraishi, et al. Relationships between plasma concentrations of diphenylhydantoin, phenobarbital, carbamazepine, and 3-sulfamoylmethyl-1,2-benzisoxazole (AD-810), a new anticonvulsant agent, and their anticonvulsant or neurotoxic effects in experimental animals. *Epilepsia, 20*:623–633 (1979).
2. Y. Masuda, T. Karasawa, Y. Shiraishi, et al. 3-Sulfamoylmethyl-1,2-benzisoxazole, a new type of anticonvulsant drug. *Arzneim.-Forsch. Drug Res., 30,* 3:477–483 (1980).
3. T. Ito, M. Hori, Y. Masuda, et al. 3-Sulfamoylmethyl-1,2-benzisoxazole, a new type of anticonvulsant drug. Electroencephalographic profile. *Arzneim.-Forsch. Drug Res., 30, A*:603–609 (1980).
4. T. Ito, M. Hori, and T. Kadokawa. Effects of zonisamide (AD-810) on tungstic acid gel-induced thalamic generalized seizures and conjugated estrogen-induced cortical spike-wave discharge in cats. *Epilepsia,* 27:367–374 (1986).

5. C. Kamei, M. Oka, Y. Masuda, et al. Effects of 3-sulfamoylmethyl-1,2-benzisoxazole (AD-810) and some antiepileptics on the kindled seizures in the neocortex, hippocampus and amygdala in rats. *Arch. Int. Pharmacodyn. Ther.*, 249:164–176 (1981).
6. R. E. Ramsay, B. J. Wilder, J. C. Sackellares, et al. Multicenter study on the efficacy of zonisamide in the treatment of medically refractory complex partial seizures. *Epilepsia,* 25:673 (1984).
7. J. C. Sackellares, P. D. Donofrio, J. G. Wagner, et al. Pilot study of zonisamide (1,2-benzisoxazole-3-methanesulfonamide) in patients with refractory partial seizures. *Epilepsia,* 26:206–211 (1985).
8. A. J. Wilensky, P. N. Friel, L. M. Ojemann, et al. Zonisamide in epilepsy: A pilot study. *Epilepsia,* 26:212–220 (1985).
9. H. Shuto, T. Sugimoto, A. Yasuhara, et al. Efficacy of zonisamide in children with refractory partial seizures. *Curr. Ther. Res.,* 45:1031–1036 (1989).
10. J. G. Wagner, J. C. Sackellares, P. D. Donofrio, et al. Nonlinear pharmacokinetics of CI-912 in adult epileptic patients. *Ther. Drug Monitor.,* 6:277–283 (1984).

66

Metabolism and Pharmacokinetic Study of the Antiepileptic Drug Halonal

V. M. Okujava, B. G. Chankvetadze, M. M. Rogava, and Z. D. Chitiashvili *Tbilisi State University, Tbilisi, USSR*

I. INTRODUCTION

Several very effective drugs have been introduced in antiepileptic therapy during the last two decades (1). Nevertheless, in many cases, seizures still remain therapeutically uncontrollable. Thus, elaboration and pharmacological investigation in clinical practice of new antiepileptic drugs (AEDs) appears to be a topical problem.

In the USSR (2–4) compounds exhibiting AED properties were described for barbituric acid, acyclic urea, indol, imidazole, succinimide and glutharimide, and derivatives of other heterocyclics. Some of these compounds—for example, benzonal, benzobamyl, and puphemid—were included in the National Pharmacopoeia of the USSR for treatment of various forms of epilepsy (2), and some others (halonal, halodiph) have recently been recommended for this purpose (3,5,6). Many investigations have been carried out to evaluate the antiepileptic effectiveness of these drugs (3–6). On the basis of these studies, halonal (HL) was chosen as one of the most promising AEDs.

Halonal is 1-(*o*-fluorinbenzoyl)-5-ethyl-6-phenyl barbituric acid ($C_{19}H_{15}N_2O_4F$), where structure is given in Figure 1. The molecular weight of HL is 354. It is a white crystalline substance that is insoluble in water and soluble in chloroform, acetone, and ether. It has been established that HL is a potent anticonvulsant, and in specific activity it exceeds phenobarbital (PB). In experimental models of epilepsy, this substance prevents seizures induced by maximal electroshock, corazol, camphor, strychnine, nicotine, arecolin, and thiosemicarbazide. Electroencephalographic study of this compound showed its inhibitory effect on CNS, in particular on the motor cortex, intralaminar thalamic nuclei, caudate nucleus, and mesencephalic reticular formation. This compound does not act on structures of the limbic system. HL is characterized by low toxicity and does not show any adverse effect on the functioning of the liver, kidneys, thyroid gland, cardiovascular, and central nervous systems (6). In all of above-mentioned studies, less attention was paid to the

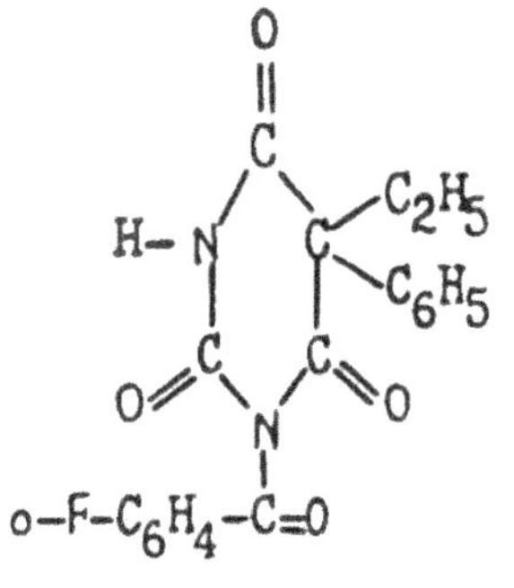

Figure 1 Structure of halonal (HL).

metabolism profiling and pharmacokinetics of HL. The latter problems were the aim of the present investigation.

II. MATERIALS AND METHODS

A. Equipment

Chromatographic analyses were carried out with a Milikhrom series (Nauchribor, USSR) microcolumn liquid chromatograph equipped with a diode matrix-type UV detector set in the wavelength range of 190 to 360 nm. The stainless steel columns (62 × 2.1 mm I.D.) were packed with commercially available 5-μm Silasorb-600 (Lachema, Czechoslovakia). Mass spectra were made by means of the Chrom-Mass system Ribermag R10-10B (France), IR spectra by an IR-75 spectrophotometer, and UV spectra with a UV-VIS spectrophotometer (both Carl Zeiss, GDR). Elemental analysis of the HL metabolite was also carried out. Tissue samples were ground and then homogenized using the homogenizer MPW-324 (Poland). HL, PB, and phenazepam (PZ) were kindly supplied by the Drug Design Laboratory of the Tomsk Polytechnic Institute (Tomsk, USSR). Hexane, isopropyl alcohol, and chloroform were of analytical reagent grade and were used without further purification. Pharmacokinetic data were processed by means of the PHAKIN program. This Fortran-program is written for IBM PC XT/AT compatible computers and requires 256K RAM. Pharmacologic experiments were carried out in guinea pigs with body weights of 550 ± 50 g.

B. Sample Preparation

To prepare samples, 150 μL of 0.25 M hydrochloric acid, 100 μL of internal standard solution, and 1 mL of chloroform were added to 100 μL of serum sample in a 10-mL screw-capped centrifuge tube and mixed for 3 min. The extraction tubes were centrifuged for 10 min at 3000 ×g at room temperature. Then 1 mL of the organic layer was transferred to a conical glass tube and evaporated under nitrogen at 35°C. The residue was reconstituted with 30 μL of eluent, and 3 to 5 μL were injected into the chromatograph. The tissue samples had been previously ground, then mixed with isotonic saline (1 mL saline to 0.5 g of tissue material) and homogenized using the homogenizer for 10 min. Then 200 mg of homogenate were taken and analyzed the same way, as the blood serum samples.

C. Chromatographic Conditions

The mobile phase was a hexane-isopropanol-chloroform mixture (70:8:22 v/v) with a flow rate of 100 μL/min. Eluates were detected at 230 nm. Preparative collection of HL metabolite for further analysis was carried out after its analytical separation under the above-mentioned conditions.

D. Quantification

Quantification was carried out using the peak height method. Phenazepam (PZ) was used as an internal standard as a 4 mg/L solution in chloroform. Calibration graphs were obtained first using various amounts of stock solutions and the same amount of internal standard. The intra- and interassay variations were determined by analyzing 10 serum samples spiked with drugs under study.

E. Interferences

Fifteen drug-free serum samples were extracted and analyzed for possible interferences by endogenous constituents. The retention of other AEDs was also determined in order to avoid interferences.

F. Study of Gastrointestinal Metabolism of HL

Gastrointestinal metabolism of HL was experimentally investigated in guinea pigs (n = 30). A 100-mg/kg dose of HL was administered orally, and various subgroups of animals were decapitated at 1-, 3-, 7-, and 12-h intervals. The parent drug and metabolites were extracted from: (a) various parts of gastrointestinal tract; (b) tissue homogenates of liver, kidney, and brain; and (c) blood serum. The separation of parent drug and its metabolites from each other and from other endogenous substances was accomplished using the above-described HPLC method. The structure of metabolites of HL was established using elemental analysis, UV, IR, and mass spectra, as described in (7).

G. Pharmacokinetic Study of the HL Metabolite

The pharmacokinetic study of PB as parent drug and PB as metabolite of HL was carried out in guinea pigs in the following way: One group of animals (n = 10) was treated orally with 100-mg/kg doses of PB, and the other group (n = 10) was treated with an equivalent dose (152.6 mg/kg) of HL. Blood samples were collected at various times from the auricle and stored at 37°C for 1 h. Drug concentration in serum was determined in the above-described way. Pharmacokinetic parameters of PB were estimated using a one-compartment linear pharmacokinetic model following the equation

$$C = \frac{k_{ab}D}{V(k_{el} - k_{ab})}[\exp(-k_{ab}t) - \exp(-k_{el}t)]$$

where C is the concentration of PB in serum at time t, D is a single dose, and V is the blood volume. Calculation of the main pharmacokinetic parameters [constants of absorption (k_{ab}) and elimination (k_{el}), area under C-t curve (AUC), specific values for volume of distribution (V_s), and body clearance (CL_s)] were performed as described in (7) using the pharmacokinetic program PHAKIN. A 2-week chronic pharmacokinetic study of PB and HL was also carried out using 31.6-mg/kg HL and 20.7 mg/kg PB dosage twice daily. CL_s and V_s in this case were estimated on the basis of the following equations:

$$CL_s = \frac{D}{C_{ss}\tau} \quad V_s = \frac{CL}{K_{el}}$$

using the program PHAKIN, C_{ss} is the steady-state concentration of drugs, τ is the dosage interval.

III. RESULTS AND DISCUSSION

Our previous studies on the HL analog benzonal (7) led us to the idea of examining the possibility of hydrolytic degradation of HL in the intestinal tract. Therefore we worked out a microcolumn HPLC method for the simultaneous quantitative determination of PB and HL. The chromatogram in Figure 2 shows the separation of these compounds and the internal standard, PZ. The linearity of the method was studied for each drug in spiked plasma samples. Regression lines were obtained by plotting the peak height ratio of each drug to that of the internal standard against their concentration in the spiked plasma samples. The linear range of these relationships was 0.5 to 70 μg/mL. The analytical recoveries of PB and HL were measured at seven different concentrations of the linearity range and ranged from 97% to 99%. The precision within each assay and from day to day was also established within the linearity limits.

The chromatographic analyses of ventricular and intestinal contents and of the hepatic, renal, and brain tissue homogenates and blood serum are presented in Figure 3. The parent drug, HL, cannot be detected in tissue homogenates (except in

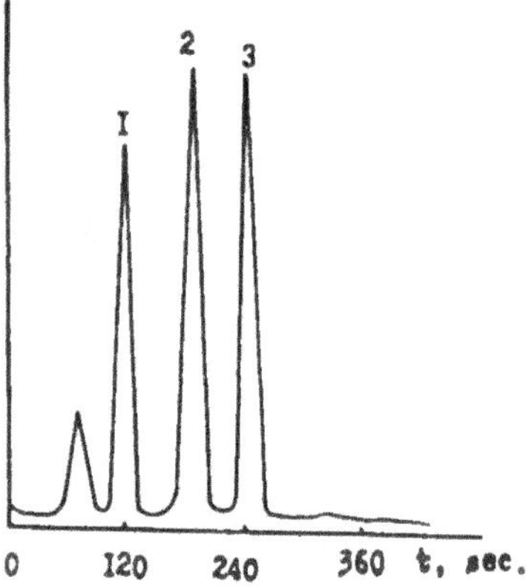

Figure 2 Chromatogram of donor serum sample spiked with HL (1), PB (2), and PZ (3).

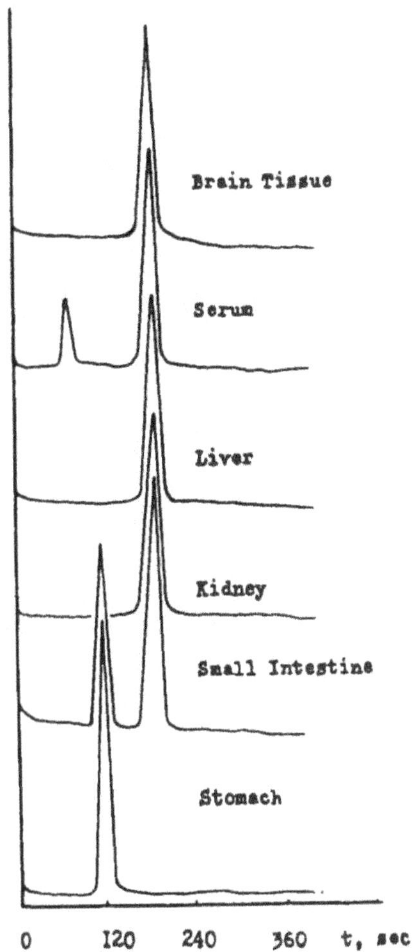

Figure 3 Chromatogram of extracts of the stomach and intestinal contents, as well as of the renal, hepatic, and brain tissue homogenates and of the serum of guinea pigs that received a 100-mg/kg single dose of HL.

the intestinal tract) or in serum. Metabolism of HL does not take place in the stomach; it starts only in the small intestine. Spectral analysis (Figure 4) of the main metabolic product showed that the metabolite must be PB, which has been widely used in antiepileptic therapy since 1912. PB in this case might be produced as a result of hydrolytic degradation of HL, in conformity with Figure 5. Thus, on the basis of these studies it may be concluded that the AED HL acts as a prodrug substance, but not as an original parent drug.

In order to establish the probability of any pharmacokinetic differences between the parent drug PB and PB, formed in the intestinal tract as a metabolite of HL, we performed a thorough pharmacokinetic study of these drugs.

There were no remarkable differences between the pharmacokinetic properties of these two compounds. Comparative pharmacokinetic study of HL and PB during

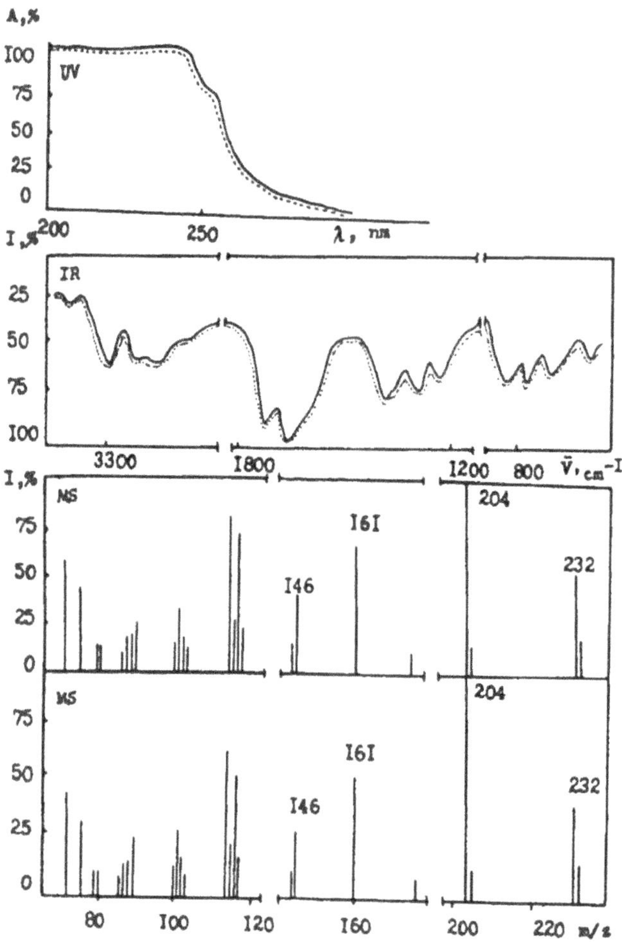

Figure 4 Ultraviolet, infrared, and mass spectra of HL metabolite and PB.

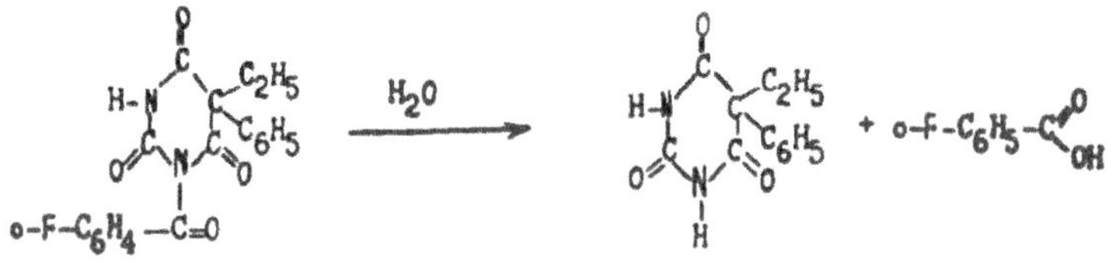

H-N, C, O, C_2H_5, C_6H_5, o-F-C_6H_4—C=O $\xrightarrow{H_2O}$ H-N, C, O, C_2H_5, C_6H_5, N-H + o-F-C_6H_5-C(=O)OH

Figure 5 Schema of intestinal metabolism of HL.

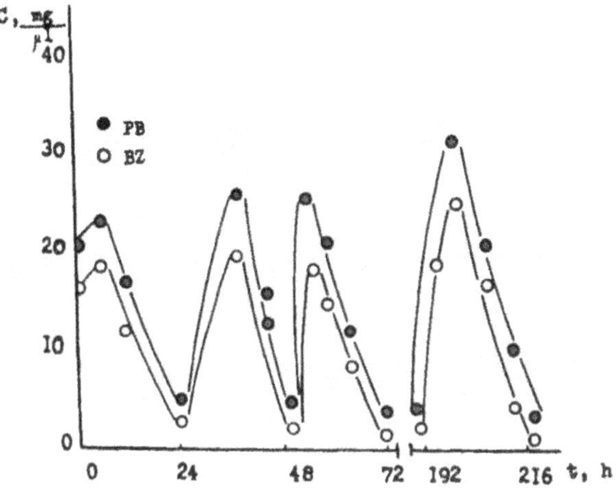

Figure 6 Long-term C-t curves of PB and HL.

long-term dosing in guinea pigs was also performed. The results (Figure 6) showed identical C-t dependence for these drugs. Good correlation of the values of C_{ss} (predicted on the basis of single-dose study of these drugs) with the real C_{ss} (measured during chronic dosing), 15.3 and 14.5 μL, respectively, showed the adequacy of the chosen pharmacokinetic model and the reliability of the estimated parameters.

IV. CONCLUSIONS

1. The antiepileptic drug halonal undergoes hydrolytic degradation in the gastrointestinal tract of guinea pigs, producing phenobarbital and *o*-fluorin benzoyc acid. Such degradation of halonal starts only in the small intestine.
2. The parent drug halonal does not appear in blood serum, kidneys, liver, or at receptor sites. The pharmacologically active metabolite of halonal, phenobarbital, appears to be fully responsible for the therapeutic effect of halonal.
3. No remarkable differences exist between the pharmacokinetic parameters of phenobarbital as a parent drug and phenobarbital as a metabolite of halonal.

REFERENCES

1. B. S. Meldrum and R. J. Porter (eds.). *New Antiepileptic Drugs*. John Libbey (1986).
2. M. D. Mashkovskii. *Drug Substances* (in Russian). Medizina, Moscow, p. 145 (1985).
3. V. K. Gorshkova and M. I. Naydenova. New antiepileptic drugs—Cyclic and noncyclic urea derivatives. In *Recent Problems of Pharmacology and Development of New Drugs Pharmacology and Development of New Drugs* (in Russian) (E. D. Goldberg, ed.), Tomsk, USSR, vol. 2, pp. 15–21 (1986).
4. L. O. Mnjoyan, S. A. Avetisyan, D. A. Akopyan, D. A. Gerasimyan, I. A. Jagazpanyan, and S. A. Pashinyan. New antiepileptic drug puphemid (in Russian). *Khim. Pharm. Zurn.*, 17:757 (1983).
5. M. I. Naydenova. New antiepileptic drug halonal. In *Problems of Theoretical and Clinical Medicine* (E. D. Goldberg, ed.), Tomsk, USSR, vol. 8. p. 23 (1980).

6. M. I. Naydenova, M. I. Smagina, and V. K. Gorshkova. New antiepileptic drug halonal (in Russian). *Farmacol. Toxicol.*, *6*, p. 23 (1988).
7. V. M. Okujava, B. G. Chankvetadze, M. D. Rukhadze, M. M. Rogava, and N. B. Tkesheliadze. Use of normal phase HPLC in study of hydrolytic stability, Metabolic profiling and pharmacokinetic of antiepileptic drug benzonal. *J. Pharmaceut. Biomed. Anal.*, in press.

67

Metabolic and Pharmacokinetic Study of the Antiepileptic Drug Phuphemid

B. G. Chankvetadze, V. M. Okujava, Z. D. Chitiashvili, and A. S. Mikautidze *Tbilisi State University, Tbilisi, USSR*

I. INTRODUCTION

The antiepileptic drug (AED) puphemid (PPH) was synthesized in the early 1970s (1) and was admitted to clinical use in the USSR in 1978.

PPH is a *p*-isopropoxyphenyl derivative of succinimide (Figure 1). The molecular weight of PPH is 233. PPH is a white crystalline substance that is insoluble in water and soluble in ethanol, acetone, and ether. Its melting point is 101 to 103°C. PPH possesses on asymmetric carbon atom in the succinimide ring and exists as a racemic mixture of two enantiomers, which may differ dramatically in their pharmacokinetic and pharmacodinamic properties. Although enantiomers of the widely used AED ethosuximide were resolved (2), no data have been published concerning the enantioselectivity of the metabolism and the pharmacokinetics of this drug, as well as of other succinimide derivatives.

PPH is characterized by a wide spectrum of antiepileptic activity. It prevents convulsive seizures elicited by corasol, strychnine, camphor, and maximal electroshock. PPH is less active against nicotine and picrotoxin-type seizures (3).

Clinical investigation of PPH has shown its effectiveness in treatment of petit mal and temporal seizures. This drug is characterized by low toxicity. In therapeutic doses it does not exhibit sedative, hypnotic, or other undesirable side effects (4). Although PPH has been introduced into clinical practice, less attention has been paid to the metabolic and pharmacokinetic study of this compound (5,6). One of the reasons for this is the absence of a reliable method for the quantitative determination of PPH and its metabolites in biological specimens.

A proposed spectrophotometric method for the determination of PPH in drug formulations (7) is unacceptable for the determination of PPH in biological samples. Recently, however, a GLC method for quantitative determination of PPH in plasma and urine was described (5), but the use of this method in long-term pharmacokinetic study seems doubtful due to the considerable waste of biological specimens.

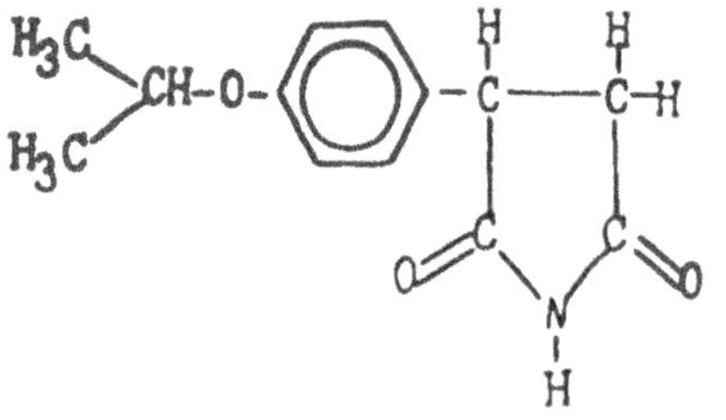

Figure 1 Structure of puphemid ($C_{13}H_{15}NO_3$).

In the present report a normal-phase microcolumn HPLC method for the quantitative determination of PPH and its major metabolite, *p*-hydroxyphenylsuccinimide (*p*-HPS), is proposed and its potential in pharmacokinetic and metabolic studies in guinea pigs is demonstrated. A method is also proposed for the chiral separation of PPH enantiomers using reverse-phase HPLC with a Pirkle-type chiral stationary phase.

II. MATERIALS AND METHODS

A. Chemicals

Hexane (Synthetic Products, Novocherkask, USSR), isopropyl alcohol (Chemical Reactives, Shostka, USSR), and chloroform (Chempharm, Kiev, USSR) were of analytical reagent grade and were used without further purification. PPH and *p*-HPS were supplied by the Institute of Fine Organic Synthesis of the Academy of Sciences of the Armenian SSR (Yerevan, USSR). Phenazepam (PZ) was extracted from the drug formulation and purified by multiple recrystallization from various solvents.

B. Sample Preparation

Samples for both nonchiral and chiral separation were prepared the same way. In each case, 150 μL of 0.25 M hydrochloric acid, 100 μL of internal standard solution, and 1 mL of chloroform were added to 50 μL of serum sample in a 10-mL screw-capped centrifuge tube and mixed for 3 min. The extraction tubes were centrifuged for 10 min at 3000 ×g at room temperature. Then 1 mL of the organic layer was transferred to a conical glass tube and evaporated under nitrogen, at 35°C. The residue was reconstituted with 30 μL of eluent, and 3 to 5 μL were injected into the chromatograph.

C. Chromatographic Conditions

Nonchiral chromatographic analyses were carried out using a Milikhrom microcolumn liquid chromatograph (Nauchpribor, USSR) equipped with a diode matrix-type UV detector. Separation of the compounds was performed on a stainless steel column (2 × 62 mm) filled with a commercially available sorbent, Silasorb-600 (Lachema, Czechoslovakia). The mobile phase was a hexane-isopropyl alcohol-chloroform mixture (70:8:22 v/v) with a flow rate of 100 μL/min. Detection of eluates was carried out at 230 nm. Preparative collection of PPH and *p*-HPS for further analysis was done by repeated analyses under the above conditions.

Chiral separation of PPH enantiomers was carried out using Tswet-306 HPLC equipment with a fixed-wavelength UV detector at 254 nm. Separation of enantiomers was accomplished on a Bakerbond DNBPG column, ionic type (250 × 4.6 mm), particle size 5 μm(Baker, Germany). A hexane-isopropyl alcohol mixture (80:20 v/v) was used as eluent with a flow rate of 1.0 mL/min.

D. Quantification

Quantification was carried out using the peak height method. A 20-mg/L solution of PZ in chloroform was used as an internal standard. Calibration graphs were obtained first using various amounts of stock solutions and the same amount of internal standard. The intra- and interassay variations were determined by analyzing 10 serum samples spiked with PPH and *p*-HPS.

E. Interferences

Fifteen drug-free serum samples were extracted and analyzed for possible interferences by endogenous constituents. The retention times of other AEDs (phenobarbitone, carbamazepine, phenytoin, ethosuximide, primidone) were also determined in order to ascertain if they would interfere.

F. Identification of Chromatographic Peaks

Identification and evaluation of the purity of chromatographic peaks were carried out be means of preparative-scale collection of substances corresponding to these peaks and their further analysis using ultraviolet, infrared, and mass spectra.

Ultraviolet spectra were recorded using a Milikhrom chromatograph and Specord UV-VIS spectrophotometer (Carl Zeiss, GDR). Infrared spectra were taken in vaseline oil using an IR-75 spectrophotometer (Carl Zeiss, GDR).

Mass spectra were detected using the Nermag 10.10B GLC-MS system (Ribermag, France) by means of direct introduction of the sample into the ion source of the spectrometer at 70 eV.

Identification of *p*-HPS was carried out using only retention time and mass spectra.

G. Pharmacokinetic Study

Pharmacokinetic study of PPH was carried out in guinea pigs with body weights of 450 to 500 g. In each case, 100 mg/kg of PPH was administered orally to the animals. Blood samples were collected from the auricle at various times and stored at 37°C for 1 h. Drug concentration in serum was determined as described above. Pharmacokinetic parameters of PPH were estimated using a one-compartment linear pharmacokinetic model following the equation

$$C = \frac{k_{ab}D}{V(k_{el} - k_{ab})}[\exp(-k_{ab}t) - \exp(-k_{el}t)]$$

where C is the concentration of PPH in serum at time t, D is single dose, and V is blood volume. Calculation of main pharmacokinetic parameters [constants of absorption (k_{ab}) and elimination (k_{el}), area under the C-t curve (AUC), specific volume of distribution (V_s), and specific body clearance (CL_s)] were performed as described in

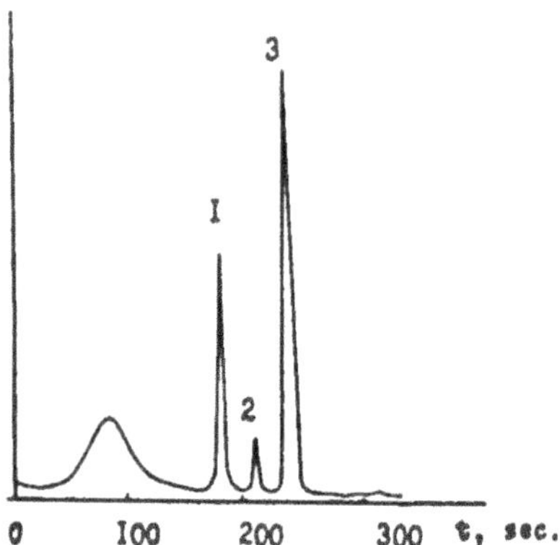

Figure 2 Chromatogram of serum sample of guinea pig treated with PPH. Sample was taken 2 h after dosing. 1, PPH; 2, *p*-HPS; 3, PZ.

(8) using a pharmacokinetic program written in BASIC and realized using a CM-1300 computer system (USSR).

III. RESULTS AND DISCUSSION

Two chromatographic peaks were detected in serum samples of guinea pigs receiving a 100-mg/kg dose of PPH (Figure 2). Their retention times correspond to that of PPH and *p*-HPS.

Retention time, absorbency in UV spectra at 222 nm in ethanol solution, and appearance in IR spectra of absorption bands at 3190, 3060, 1750, 1700, 1685, and 1665 cm^{-1}, as well as molecular ion with $m/z = 233$ (relative intensity = 10), and other ions with $m/z = 191(25)$, 149(12), 129(12), and 120(100) (Figure 3) in mass spectra, confirmed conclusively that peak 1 in Figure 2 belongs to PPH.

The retention time and mass spectrum (Figure 4) of the substance corresponding to peak 2 (Figure 2) suggest that the latter is *p*-HPS the major metabolite of PPH.

The usefulness of the above-described method of quantification of PPH and *p*-HPS for pharmacokinetic investigation was examined in guinea pigs. Figure 5 shows that PPH is rapidly absorbed from the intestinal tract after oral administration of this drug. Main pharmacokinetic parameters are AUC = 248 ± 40 μg/mL/h; $k_{ab} = 0.56 \pm 0.10\ h^{-1}$, peak time of concentration 2.5 ± 0.5 h; peak value of concentration is 50.0 ± 1.5 μg/mL. Elimination of PPH is sufficiently well described using a one-compartment linear model with the following parameters of elimination: $k_{el} = 0.42 \pm 0.10\ h^{-1}$; $V_s = 0.95 \pm 0.22$ mL/g; $T_{1/2} = 1.66 \pm 0.22$ h; $CL_s = 0.41 \pm 0.05$ mL/(g·h).

As shown in Figure 6, under the described conditions enantiomers of PPh can be resolved with $\alpha = 1.29$ and $K_1' = 4.05$. It was impossible at present to collect enantiomers separately and determine their configuration. Identification of enantiomer peaks was made using mass spectra. A preliminary enantioselective pharmacokinetic study of PPH showed that enantiomeric ratio does not change through the whole interval of time studied (1 to 9 h, Figure 7). More detailed study of this problem is in progress now.

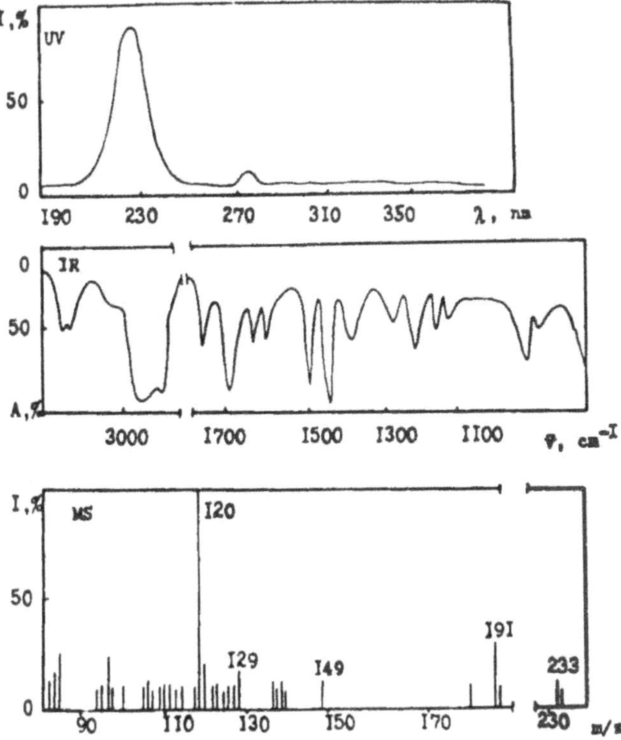

Figure 3 UV, IR, and mass spectra of the substance corresponding to the chromatographic peak of PPH.

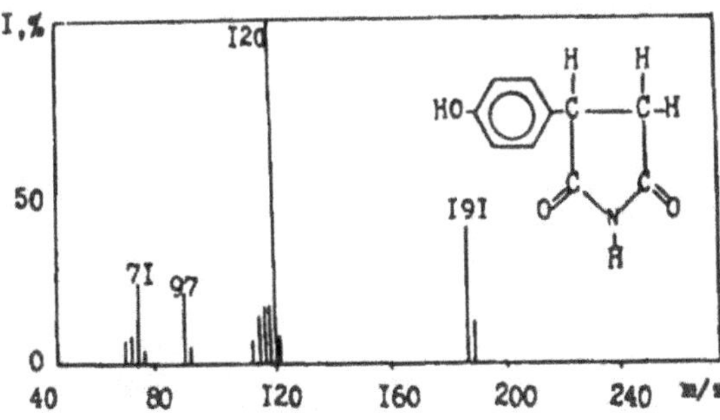

Figure 4 Mass spectrum of the substance corresponding to the chromatographic peak of *p*-HPS.

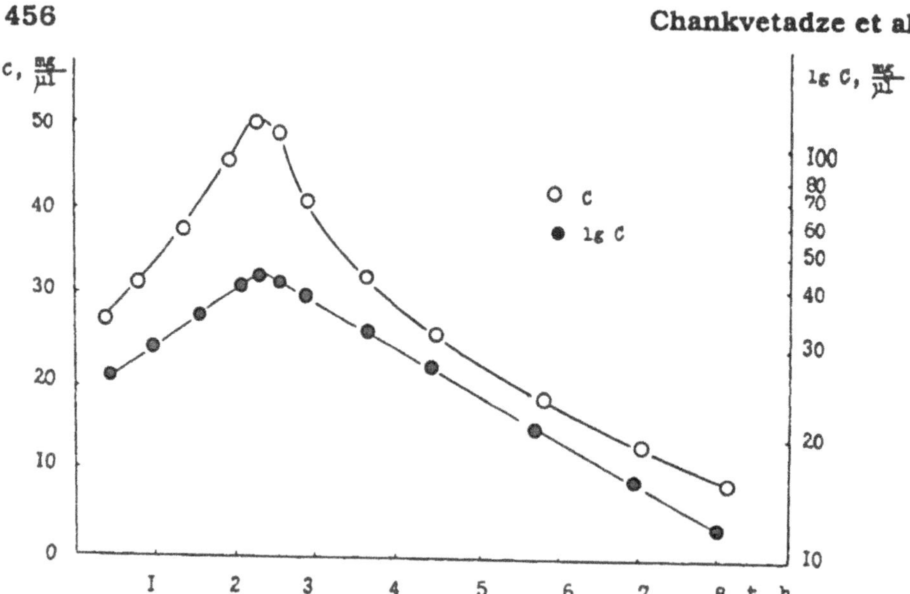

Figure 5 C-t and log c-t curves of PPH in serum of guinea pigs that received 100 mg/kg of this drug.

IV. CONCLUSIONS

The described method for the quantitative determination of the antiepileptic drug puphemid, and its major metabolite *p*-hydroxyphenylsuccinimide, is characterized by high sensitivity and good reproducibility as well as by low expenditure of eluents and sorbent materials.

The value of this method for metabolic and pharmacokinetic studies of puphemid has been demonstrated, and the main pharmacokinetic parameters of this drug in guinea pigs have been determined.

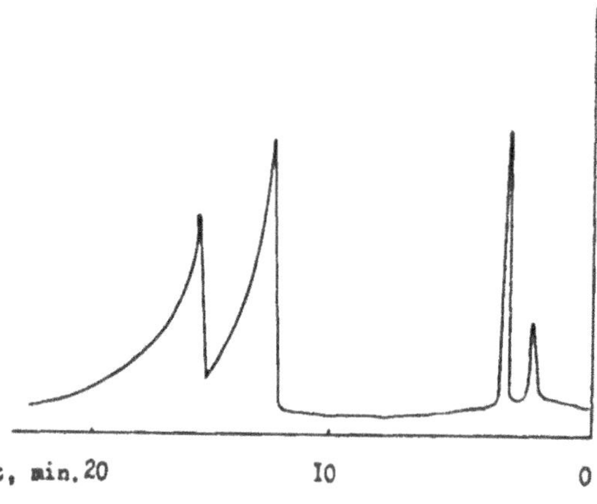

Figure 6 Chromatogram of PPH enantiomers.

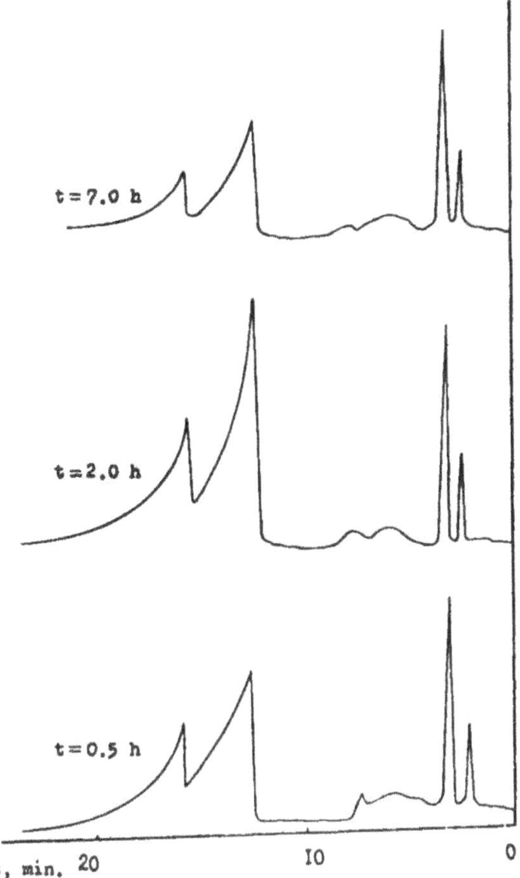

Figure 7 Chromatogram of PPH enantiomers extracted from serum of guinea pigs after various treatment times.

A method is proposed for the chiral separation of PPH enantiomers. Preliminary study shows that there are no remarkable differences in elimination of PPH enantiomers.

REFERENCES

1. Soviet Patent 689675 (1979).
2. Zh. Yang, S. Barkan, Ch. Brunner, J. D. Weber, Th. D. Doyle, and I. W. Wainer. Application of high performance liquid chromatographic chiral stationary phases to pharmaceutical analysis. Resolution of enantiomeric barbiturates, succinimides and related molecules on four commercially available chiral stationary phases. *J. Chromatogr.*, 324:444 (1985).
3. L. O. Onjoyan, S. A. Avetisyan, D. A. Akopyan, D. A. Gerasimyan, I. A. Jagazpanyan, and S. A. Pashinyan. New antiepileptic drug—Puphemid (in Russian). *Khim. Pharm. Zurn.*, 6:757 (1983).
4. P. V. Melnichuk and N. P. Nechkina. Clinical use of puphemid for treatment of petit mal epilepsy in children. In *Puphemid* (L. O. Mnjoian, ed.), Publishing Agency of the Armenian Academy of Sciences, Yerevan, USSR, p. 43 (1980).

5. I. A. Jagazpanyan, N. E. Suleimanyan, B. A. Odabashyan, A. P. Rodionov, and V. P. Zherdev. Peculiarity of puphemid excretion and metabolism in rat (in Russian). *Khim. Pharm. Zurn.*, *9*:1037 (1988).
6. N. E. Suleimanyan, I. Nazaryan, I. A. Jagazpanyan, and B. A. Odabashyan. Study of puphemid pharmacokinetics in rat (in Russian). *Khim. Pharm. Zhurn.*, *8*:910 (1982).
7. S. O. Sarkisyan, A. A. Sarkisyan, I. V. Persianova, and M. S. Goizman, Spectrophotometric determination of puphemid in drug formulations (in Russian). *Chim. Pharm. Zurn.*, *4*:486 (1986).
8. O. M. Gibaldi and D. Perier. *Pharmacokinetics*, 2nd ed. Marcel Dekker, New York (1982).

68

Performance of Techniques for Measurement of Therapeutic Drugs in Serum: A Comparison Based on External Quality Assessment Data

J. F. Wilson, L. M. Tsanaclis,* J. E. Perrett,* J. Williams, J. F. C. Wicks,* and A. Richens *University of Wales College of Medicine and *Cardiff Bioanalytical Services, Ltd., Cardiff, Wales*

I. INTRODUCTION

A range of analytical techniques are in widespread use for the routine measurement of therapeutic drug concentrations in serum. The techniques fall into two broad classes. There are chromatographic techniques using either gas-liquid (GLC) or high-pressure liquid (HPLC) methods, and immunological techniques which are available in a variety of commercial kits. New techniques are frequently validated by intra-laboratory comparison with existing techniques. A more realistic assessment can, however, be made by consideration of data reported to external quality assessment schemes (EQAS), where the evaluation is based on the performance of techniques when implemented by a number of laboratories as part of their routine clinical service. We report here such a comparison based on data reported to the Heathcontrol U.K.-EQAS in the 18-month period from November 1988 to April 1990.

II. MATERIALS AND METHODS

The samples were prepared by adding pure drug compounds into drug-free serum (1–4). The serum was accurately dispensed in small aliquots and freeze-dried for distribution by post to participating laboratories. All samples except those containing digoxin were prepared using newborn calf serum (Gibco, Ltd., Paisley, U.K.). Samples containing digoxin were prepared in human serum. The concentrations of drugs were chosen to span the therapeutic range, with two-thirds of sample concentrations being within the therapeutic range, one-sixth at subtherapeutic, and the remaining one-sixth at toxic concentrations. The 15 analytes included in the samples are listed in Table 1. For each analyte, the table gives the number of samples included in the survey and the

Table 1 Analytes Included in the Heathcontrol TDM Survey, November 1988 to April 1990

Analyte	Number of samples	Mean number of reports/sample
Antiepileptic scheme		
1. Phenytoin	36	260
2. Phenobarbitone	36	251
3. Primidone	18	127
4. Carbamazepine (CBZ)	36	254
5. CBZ-10,11-epoxide	18	20
6. Ethosuximide	18	103
7. Valproic acid	18	228
8. Clonazepam	18	19
Respiratory scheme		
9. Theophylline	36	159
10. Caffeine	6	26
Cardiac scheme		
11. Digoxin	18	74
Antidepressants scheme		
12. Amitriptyline	18	16
13. Nortriptyline	18	16
14. Imipramine	18	16
15. Desipramine	18	16

mean number of reports received for each sample. Two samples were circulated each month containing the antiepileptic drugs. One sample contained the first eight analytes and the other phenytoin, phenobarbitone, and carbamazepine. There were two theophylline-containing samples each month, and latterly caffeine was included in these samples. Digoxin was prepared as a single analyte sample. The 18 samples included in the analysis were from the most recent 6 months. This followed a change in serum matrix from newborn calf serum to human serum in order to avoid the unacceptable interference by nonhuman serum in digoxin immunoassays (5). The antidepressant drugs were presented as two monthly samples. One sample was a mixture of amitriptyline with nortriptyline, and the other was imipramine with desipramine.

The Heathcontrol schemes have an international membership of over 350 laboratories from 24 countries (Table 2). United Kingdom laboratories dominate in the scheme, being 60% of the total membership. For each drug concentration measurement reported to the scheme, laboratories were asked to report the analytical technique used following the classifications shown in Table 3. There are six immunoassay groups. Radioimmunoassay is used almost exclusively for measurement of digoxin. The enzyme immunoassay group is composed of those techniques based on the Syva EMIT assays. From our surveys, this is a varied group with many modifications in

Table 2 International Membership of Heathcontrol External Quality Assessment Schemes for Therapeutic Drug Monitoring

Country	Antiepileptic scheme	Respiratory scheme	Cardiac scheme	Antidepressants scheme
Australia	3	3	3	—
Austria	2	—	—	—
Belgium	6	3	2	—
Brazil	2	—	—	—
Denmark	3	3	2	1
Eire	6	8	3	1
Finland	10	5	2	3
France	9	5	1	1
Germany	17	7	1	2
Italy	36	8	5	2
India	1	—	—	—
Kuwait	1	—	—	—
Netherlands	4	—	—	2
Norway	2	1	—	2
Poland	1	—	—	—
Portugal	1	—	—	—
Saudi Arabia	13	13	13	—
South Africa	3	—	1	—
Spain	1	1	1	—
Sweden	3	—	1	1
Switzerland	5	3	2	1
United Kingdom	195	134	62	13
United States	1	—	—	—
Yugoslavia	1	—	—	—
Total	326	194	99	29

Table 3 Heathcontrol Technique Classifications

Technique	Abbreviation
1. Radioimmunoassay	RIA
2. Enzyme immunoassay	EMIT
3. Fluorescence immunoassay	FIA
4. Polarisation fluoroimmunoassay	PFIA
5. Nephelometry	Neph
6. Turbidimetric immunoassay	Turbid
Gas-liquid chromatography	
7. With derivatization	GLC-D
8. Without derivatization	GLC-ND
High-pressure liquid chromatography	
9. Straight-phase column	HPLC-SP
10. Reverse-phase column	HPLC-RP

use. For example, laboratories may dilute the commercial reagents further than the manufacturer's recommendations, and there are a variety of automated analysers in use. FIA is the Ames TDA assay, and PFIA is the Abbott TDx technique. Nephelometry refers to the Beckman ICS technique, and the turbidimetric assay is the most recent addition to the range of commercial kits from Beckman. We subdivide the two chromatographic technique groups as shown. Gas-liquid chromatography is split into methods using a derivatization step and those without derivatization. HPLC is divided into systems using straight-phase columns and those with reverse-phase columns.

The statistical analysis of the drug measurement data was in two stages. The data for each sample were first analyzed in total without regard to assay technique to reject outliers that were greater than 3 standard deviations from the sample mean (6). The number or frequency of these outliers for each technique was used as one measure of assay precision. Analysis of the frequencies was by chi-square.

The second stage of the analysis was to consider the remaining nonrejected measurements. The data were grouped by assay technique and individual technique means and standard deviations calculated. The standard deviation of measurements expressed as the coefficient of variation was used as a second measure of assay precision. The percentage difference of the technique mean from the spike value was used as a measure of technique accuracy. The latter two parameters were analyzed by one-way analysis of variance followed by the least-significant-difference test.

III. RESULTS

Of the potential total of 150 drug and technique group combinations, 70 were in use. A number of examples have been chosen to illustrate either general trends in the survey data or specific points of interest.

A. Rejected Observations

Figure 1 presents data on the number of measurements of phenytoin rejected as outliers. Each column is the number of rejects expressed as a percentage of the total number of measurements by the particular technique, which is identified by the abbreviation given in Table 3. The number above each column is the mean number of measurements on a single sample for each technique. This varied from 87 for EMIT measurements to an average of just two laboratories using nephelometry. The techniques are displayed in ascending order, and the solid bars under the columns bracket techniques that did not differ significantly with a probability value of >0.05.

It is interesting to note that significant between-technique differences are present in this data. We know from past experience that a proportion of the rejected measurements are due to transcription and other errors not related to the analytical technique. Nevertheless, significant between-technique differences were present for 8 of the 15 analytes. Users of the Abbott TDx kit showed particularly well on this parameter, being in the statistically better group of techniques for all drugs to which the technique was applied. HPLC and other immunoassays usually formed a middle band of techniques, with from 4% to 6% of rejects. The GLC techniques performed less well. The derivatized approach always gave numerically more outliers than the nonderivatized technique.

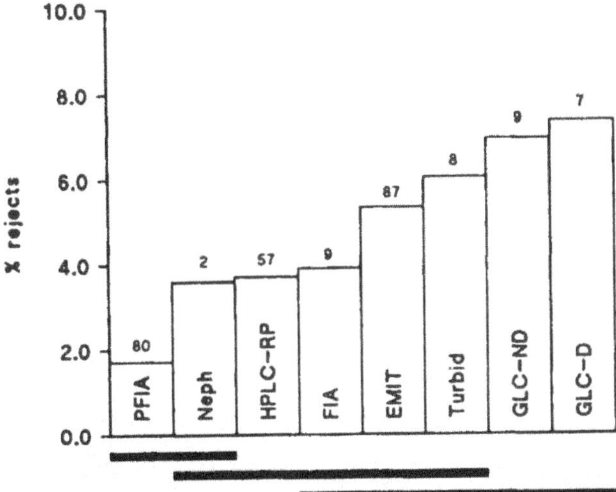

Figure 1 Mean percentage of rejected measurements >3 SD from the sample mean for measurements of phenytoin by members of the Heathcontrol External Quality Assessment Scheme. The number above each bar gives the mean number of measurements per sample by each technique. Techniques not significantly different ($p > 0.05$) are underlined. See Table 3 for key to technique abbreviations.

B. Coefficient of Variation

Figure 2 displays data for the second measure of assay precision. Each column is the mean coefficient of variation of measurements, with the I-bar showing the standard error of the mean. The number above each bar is the number of samples providing data for the analysis. The number for turbidimetric assay of phenytoin is low, as this was a new technique classification introduced partway through the study. The techniques are presented in ascending order, with each identified by its abbreviation (Table 3). Groups of techniques not significantly different with a probability of >0.05 are underlined by the solid bars.

The pattern of between-technique differences was similar to that for analysis of rejected observations. The Abbott PFIA technique was in the statistically best group for all drugs to which the technique was applied. HPLC and other immunoassays also had generally very satisfactory levels of performance. The exception among the immunoassays was nephelometry, which, as here for phenytoin and also for phenobarbitone and theophylline, performed statistically less well than other techniques. GLC and, in particular, GLC with a derivatization step, performed less well. GLC-D was significantly more variable than other techniques for six analytes: phenytoin, phenobarbitone, primidone, carbamazepine, clonazepam, and theophylline.

Two exceptions to the general trends in technique precision are shown in the lower two panels of Figure 2. The middle panel displays the coefficient-of-variation data for valproic acid. This analyte is not very suitable for analysis by HPLC, and reverse-phase HPLC performed poorly with a coefficient of variation greater than

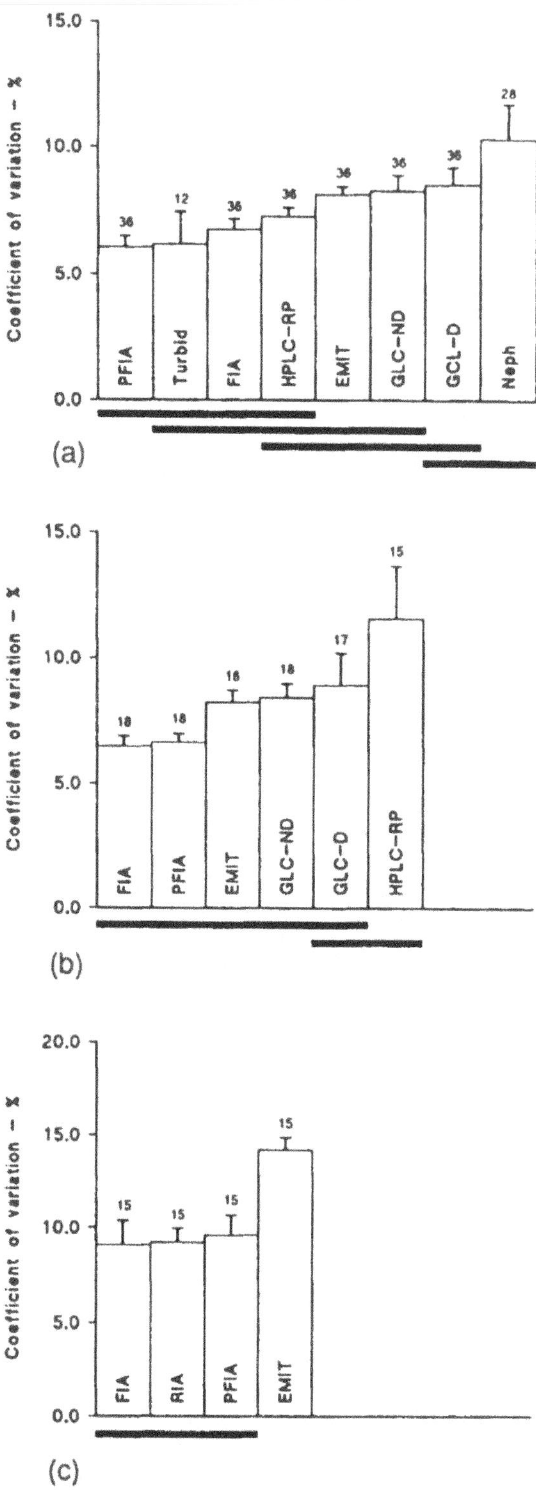

Figure 2 Mean coefficient of variation of measurements of phenytoin (a), valproic acid (b) and digoxin (c) by members of the Heathcontrol External Quality Assessment Scheme. Each bar is the mean ± SEM with the number of samples. Techniques not significantly different ($p > 0.05$) are underlined. See Table 3 for key to technique abbreviations.

10%. The final panel of Figure 2 shows the data from the four immunoassay techniques applied in the measurement of digoxin. The EMIT group in this case gave significantly more variability and a higher coefficient of variation than the other techniques. (*Editor's note: Syva has replaced the reagents used in this study with a new product, which needs evaluation.*)

C. Percent Difference from Spike Value

Analysis of the difference of technique means from the spike value expressed as a percentage of the spike value demonstrated significant differences in accuracy between techniques. Figure 3 shows the data for primidone and carbamazepine. Each column is the mean, with the I-bar showing the standard error of the mean. Techniques are displayed in ascending order. Histogram columns projecting downward indicate negative bias, and those projecting upward indicate positive bias. Techniques not significantly different with a probability of >0.05 are underlined by the solid bars.

The accuracy data was more variable, with fewer significant between-technique differences than for the precision data. The data for primidone was typical, with bias values of less than 5%. The data for carbamazepine shows two interesting exceptions. First, the GLC techniques both produce results with significant negative bias. This has been a consistent finding over many years (2) and is attributed to the loss of carbamazepine owing to its instability at high temperatures. An even greater negative bias, however, was seen for Abbott TDx measurements of carbamazepine, and a comparable problem existed for PFIA measurements of valproic acid. This results from the use of nonhuman serum in the preparation of the samples. Our own comparisons of human to bovine or newborn calf serum have demonstrated that the bias is the result of some as-yet-unidentified interaction with the serum matrix (5).

IV. DISCUSSION

The present study has demonstrated a high level of both precision and accuracy for the majority of assay techniques when employed by routine clinical laboratories to measure concentrations of therapeutic drugs in external quality assessment samples. Certain specific exceptions have been described. Care must be exercised, however, when interpreting these findings. Uniform and clean freeze-dried material containing spiked compounds as distributed in this study is clearly not identical to heterogeneous patient samples of variable quality. The negative bias shown by the Abbott TDx assay in measurements of carbamazepine is, as noted above, a product of the use of a nonhuman matrix in sample preparation. Data from external quality assessment schemes, although valuable, are thus only part of the information to be considered in comparisons of different analytical techniques. Commercial considerations such as technical difficulty and cost in relation to sample throughput are of equal importance.

V. SUMMARY

Ten assay techniques were compared for the measurement in serum of eight antiepileptic drugs, four antidepressants, theophylline, caffeine, and digoxin using data reported to the Heathcontrol U.K. External Quality Assessment Scheme between November 1988 and April 1990. Samples were freeze-dried aliquots of newborn calf serum (digoxin in human serum) spiked with weighed-in concentration of drugs that

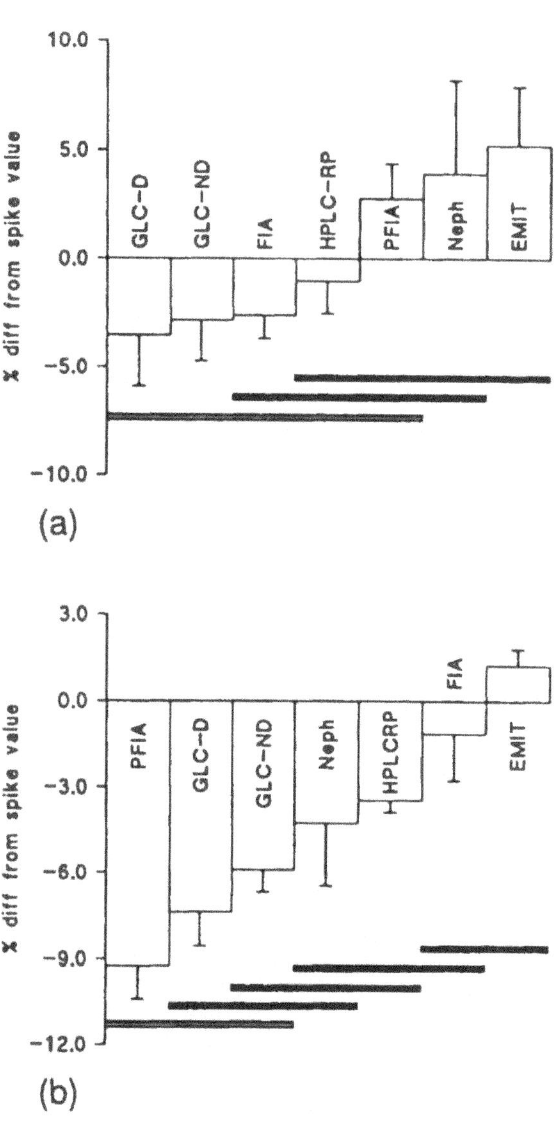

Figure 3 Mean percentage difference of sample mean from the spike value for measurements of primidone (a) and carbamazepine (b) by members of the Heath-control external Quality Assessment Scheme. Each bar is the mean ± SEM. Techniques not significant different ($p > 0.05$) are underlined. See Table 3 for key to technique abbreviations.

ranged from subtherapeutic to toxic. Sample measurements were screened to reject outliers >3 SD from the consensus mean. Three measures of performance were studied: frequency of outliers, the coefficient of variation (CV) of sample measurements, and difference of sample mean from the spike value. Frequencies were compared using chi-square tests ($p = 0.05$), and CV and difference measures by one-way analysis of variance and the least-significant-difference test ($p = 0.05$).

Measurements by polarization fluoroimmunoassay (PFIA) were of consistently high precision, being in the group of methods that were significantly less variable and produced significantly fewer outliers than other techniques. PFIA, however, gave significant negative bias in measurements of carbamazepine and valproic acid. Other immunoassay techniques, including a new turbidimetric assay, also performed well. Nephelometry was significantly more variable than other techniques for measurement of phenytoin, phenobarbitone, and theophylline. Reverse-phase high-pressure liquid chromatography (HPLC) produced precise and accurate results for a wide range of analytes with mean CV and bias values in the significantly better technique groups. The exception was valproic acid, where HPLC measurements were significantly less precise than with other techniques. Measurements by gas-liquid chromatography (GLC) were in many instances significantly more variable than by other techniques. In particular, GLC with derivatization gave significantly higher CV values and significantly more rejects for several analytes.

REFERENCES

1. J. F. Wilson, J. Williams, L. M. Tsanaclis, J. E. Tedstone, and A. Richens. Evaluation of chromatographic and kit immunoassay techniques for the measurement of theophylline in serum: A study based on external quality assurance measurements. *Ther. Drug Monitor., 10*:438–445 (1988).
2. J. F. Wilson, L. M. Tsanaclis, J. Williams, J. E. Tedstone, and A. Richens. Evaluation of assay techniques for the measurement of antiepileptic drugs in serum: A study based on external quality assurance measurements. *Ther. Drug Monitor., 11*:185–195 (1989).
3. J. F. Wilson, L. M. Tsanaclis, J. Williams, J. E. Tedstone, and A. Richens. External quality assurance of tricyclic antidepressant measurements in serum: Eight years of progress? *Ther. Drug Monitor., 11*:196–199 (1989).
4. J. J. Wilson, L. M. Tsanaclis, J. Williams, J. E. Tedstone, and A. Richens. Comparison of nonisotopic and radioimmunoassay techniques for digoxin: A study based on external quality assurance measurements. *Ther. Drug Monitor., 11*:477–479 (1989).
5. L. M. Tsanaclis, J. F. Wilson, J. Williams, J. E. Perrett, and A. Richens. Comparison of human, bovine, and newborn calf serum in the preparation of external quality assurance samples for therapeutic drugs. *Ther. Drug Monitor., 12*:373–377 (1990).
6. M. J. R. Healy. Outliers in clinical chemistry quality control schemes. *Clin. Chem., 25*: 675–677 (1979).

69

Simultaneous Determination of Clobazam and Its Metabolite Desmethylclobazam in Serum by Gas Chromatography with Electron-Capture Detection

H. Vandenberghe,*† and J. C. MacDonald† **St. Joseph's Health Center and †University of Western Ontario, London, Ontario, Canada*

I. INTRODUCTION

Clobazam (Figure 1) is a 1,5-benzodiazepine with marked antiepileptic action (1). It binds to the benzodiazepine receptor, although the exact mechanism is still unclear. Clobazam (CLB) is metabolized and its N-desmethylclobazam (DMCLB) (Figure 1) metabolite contributes significantly to the anticonvulsant effect of the drug (1). They are active against partial and generalized seizures, and in epilepsy of different etiologies. The drug seems to be safe, with minimal side effects, mostly sedation, dizziness, and fatigue. It is used in children and adults primarily as adjuvant therapy (1). Long-term use may cause tolerance. Discontinuation of the drug can lead to rebound seizures (1). Desmethylclobazam has a longer half-life, two to three times longer than clobazam, and serum concentrations of the metabolite can be 10-fold higher than the parent drug. Despite being an emergency release drug in Canada, clobazam is being used increasingly. Analysis of clobazam and its metabolites has been performed by gas chromatography (2–4) and by high-performance liquid chromatography (HPLC) (5,6). We describe a sensitive and specific gas chromatographic assay with electron-capture detection (GC-ECD) for the analysis of clobazam and desmethylclobazam using only 100 μL of serum. Using 300 to 500 μL of serum, the same extraction procedure can be used for analysis by HPLC.

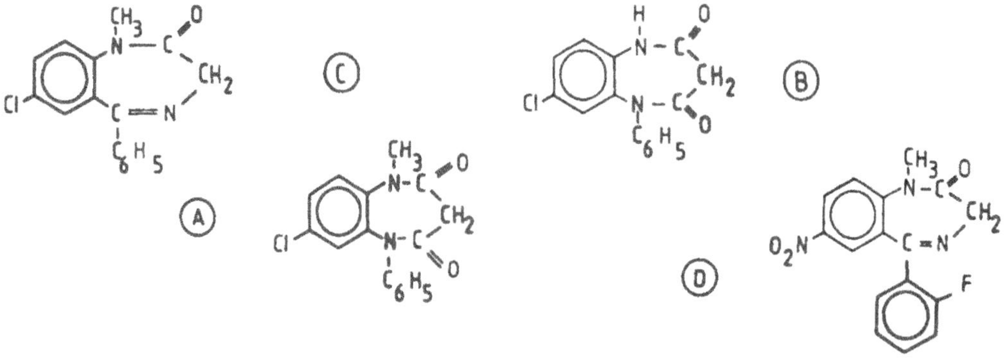

Figure 1 Chemical structure of (A) clobazam, (B) desmethylclobazam, (C) diazepam, and (D) flunitrazepam.

III. MATERIALS AND METHODS

A. Apparatus

Throughout the development of the method, we used a HP 5830 gas chromatograph with electron-capture detector (Hewlett Packard) equipped with a 6 ft × 2 mm I.D. glass column packed with 5% OV-25 on 100/120 Supelcoport (Supelco).

B. Reagents

Boric acid, potassium chloride, and sodium carbonate, analytical-grade chemicals, and methanol, cyclohexane, methylene chloride, toluene, isoamylalcohol, and acetonitrile, HPLC-grade solvents, were purchased from Fisher Scientific Canada.

Sodium borate buffer was prepared by mixing 630 mL of solution A and 320 mL of solution B. Solution A is prepared by dissolving 61.8 g of boric acid and 74.6 g of potassium chloride in 1 L of deionized water, while solution B contains 106 g of sodium carbonate in 1 L of deionized water. If necessary, we adjust the pH to 9 with solution B. The sodium borate buffer is kept in a waterbath at 37°C to prevent boric acid from crystallizing out. The extraction solvent consists of 60% cyclohexane and 40% methylene chloride.

C. Standards

Pure standard samples of clobazam and desmethylclobazam were provided by Hoechst Company, Canada, and the internal standard, diazepam, by Hoffman-La Roche, Inc., Canada. Stock solutions of 10 mmol/L were prepared in methanol for CLB and the internal standard diazepam. DMCLB (batch dependent) was dissolved in a mixture of ethyl acetate/methylene chloride/cyclohexane (80/10/10) to obtain a concentration of 10 mmol/L. A substock solution of 100 µmol/L of CLB and 500 µmol/L of DMCLB was prepared in methanol. Working standards ranging from 0.25 to 10 µmol/L for CLB and 1.25 to 40 for DMCLB were prepared in stripped serum by appropriate dilutions of the substock. Samples were aliquoted and stored at –20°C.

D. Extraction Procedure and Conditions of Analysis

To 100 μL of standard, control, or patient serum, add 50 μL of internal standard, diazepam 4 μmol/L, in a 16 × 125 mm screw-capped tube. Alkalinize the serum with 1 mL of borate buffer, pH 9.0, and extract the three compounds into 5 mL of extraction solvent. Shake on a multivortexer for 5 min and centrifuge for 5 min at 1500 × g. Transfer the organic layer and dry in a 60°C water bath under a stream of air. Reconstitute the dry residue in 100 μL of toluene:isoamylalcohol (85:15). Inject 5 μL into the gas chromatograph.

The conditions of analysis are shown in Table 1, and a typical chromatogram from a serum sample of a patient containing 0.1 μmol/L clobazam and 60.0 μmol/L desmethylclobazam is illustrated in Figure 2.

III. RESULTS AND DISCUSSION

The liquid-liquid extraction described circumvents usage of toxic chemicals such as benzene (3,4) and is also applicable to extraction of other benzodiazepines. The extraction solvent mixture, cyclohexane and methylene chloride, is effective only in the proportions used. The same extraction procedure was also used for HPLC analysis, but 400 to 500 μL of sample was required and flunitrazepam was used as the internal standard. The analytical recoveries obtained with this extraction procedure at concentrations ranging from 0.25 to 10 μmol/L for CLB and 1.25 to 40 μmol/L for DMCLB averaged 95% to 100%. The detection limit of the method using the extraction as described and, taking losses of recovery into accounts, is 0.1 μmol/L for CLB and 0.4 μmol/L for DMCLB. Figure 2 illustrates a representative chromatogram of a human serum blank supplemented with the internal standard and one of an extracted patient sample. No interference from endogenous substances was encountered in over 20 patient samples assayed. Table 2 lists the compounds evaluated for possible interference with the assay. Desmethylchlordiazepoxide elutes at the same time as CLB, but thus far we have not encountered patients taking both drugs. Climazolam does not cause a problem, as it is not available in Canada. Other antiepileptic drugs are either not extracted with the presented method, or have extremely long retention times. The extraction procedure is linear up to 10 μmol/L for CLB and up to 20 μmol/L for DMCLB. To have reliable results for DMCLB and extend the concentration range, a curvilinear curve is used up to 40 μmol/L. The HP 5830 gas chromatograph used for this analysis allows only a one-point calibration; with an

Table 1 Conditions for Chromatographic Analysis

Parameter	Conditions
Column	5% OV-25 on 100/120 Supelcoport packed in a 6 ft × 2 mm I.D. glass column
Carrier gas	Argon methane, 90/10
Flow rate	18 mL/min
Temperature	280°C, isothermal
Detector	Electron-capture detector
Internal standard	Diazepam
Chromatography time	6 min
Sample volume	100 μL

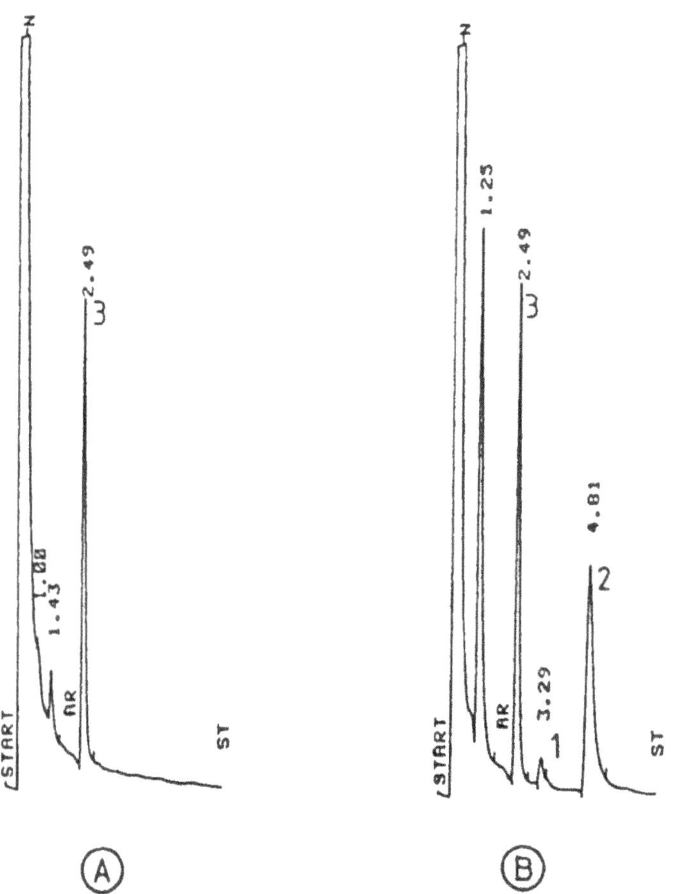

Figure 2 Chromatograms for (A) a blank serum and (B) a patient sample containing 0.1 μmol/L CLB (peak 1), 6.0 μmol/L DMCLB (peak 2), and the internal standard diazepam (peak 3).

updated model a multipoint calibration curve could be used, extending the linearity to 40 μmol/L. The within-run precision was determined by analyzing 10 aliquots of the serum pools with four different concentrations in the same run. Between-run precision was obtained by analyzing aliquots of serum pools with four different concentrations in eight analytical runs over a 2-month period. The results are presented in Table 3.

Figure 3 illustrates the concentrations of clobazam and its metabolite observed in 30 different patients. There is a wide variability in the concentration of the metabolite, while the concentrations of the parent drug falls in a lower concentration range. These results coincide with published studies (2).

IV. SUMMARY

We have described a reliable micro method for simultaneous quantitation in serum of clobazam (CLB) and N-desmethylclobazam (DMCLB), its major metabolite, by gas

Table 2 Interference Study

Compound	Relative Retention Time to Internal Standard
Oxazepam	0.75
Lorazepam	0.79
Diazepam (internal standard)	1.00
Desalkylflurazepam	1.15
Demoxepam	1.32
Desmethylchlordiazepoxide	1.37
Clobazam	1.37
Midazolam	1.39
Flunitrazepam	1.46
Ro-79957	1.51
Flurazepam	1.52
Temazepam	1.76
Desmethylclobazam	1.81
Climazolam	1.83
Ro-79749	1.89
Bromazepam	2.01
Nitrazepam	2.20
Clonazepam	2.44
Alprazolam	2.73
Triazolam	3.09

chromatography with electron-capture detection. Clobazam, its metabolite, and the internal standard are isolated by a liquid-liquid extraction using 100 μL serum, and the three compounds are separated on a packed column with 5% OV 25 on 100/20 Supelcoport (6 ft × 2 mm). The analytical recovery of clobazam and desmethylclobazam at concentrations ranging from 0.25 to 10 μmol/L and 1.25 to 40 μmol/L, respectively, averaged 95% to 100%. The analysis is linear up to 10 μmol/L for CLB and up to 20 μmol/L of DMCLB. The determination of DMCLB could be

Table 3 Precision Data

Concentration μmol/L	Clobazam CV%	Desmethylclobazam CV%
Within-run		
0.5	7	10
1.0	7	6
5	2	5
10	4	3
Between-run		
0.5	8	15
1.0	7	6
5	5	4
10	5	4

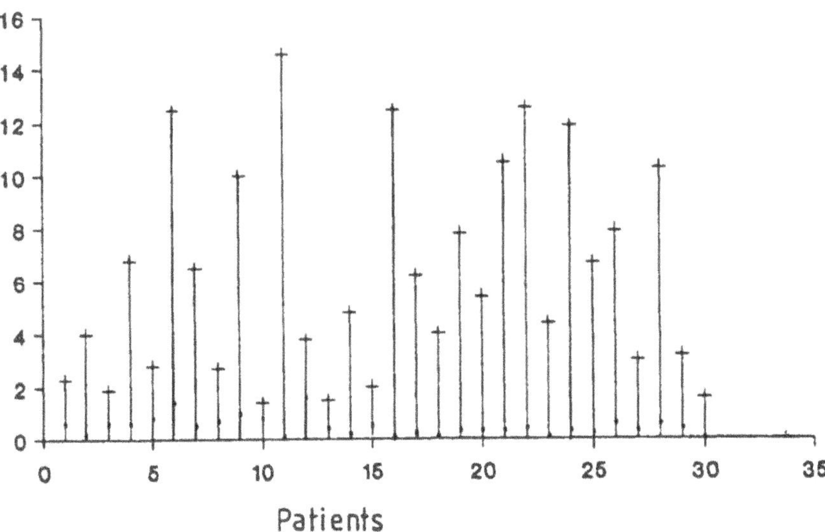

Figure 3 Concentrations of (●) clobazam and (+) desmethylclobazam in 30 patient samples.

extended to 30 or 40 μmol/L when a curvilinear graph was used. The between-day precision at concentrations of 1.0, 5.0, and 10.0 μmol/L for clobazam and desmethylclobazam was less than 10% (CV). The procedure is routinely used for therapeutic drug monitoring of clobazam and its metabolite in adults and children.

REFERENCES

1. R. H. Levy, F. E. Dreifuss, R. H. Mattson, B. S. Meldrum, and J. K. Penry, eds. *Antiepileptic Drugs*, 3rd ed. Raven Press, New York.
2. N. R. Badcock and G. D. Zoanetti. Micro-determination of clobazam and N-desmethylclobazam in plasma or serum by electron capture gas chromatography. *J. Chrom.*, *421*:147–154 (1987).
3. S. Caccia, M. Ballabio, G. Guiso, and M. G. Zanini, Gas-liquid chromatographic determination of clobazam and N-desmethylclobazam in plasma. *J. Chrom.*, *164*:100–105 (1979).
4. R. Riva, T. Gioacchino, A. Florenzo, and B. Agostino. Quantitative determination of clobazam in the plasma of epileptic patients by gas-liquid chromatography with electron-capture detection. *J. Chrom.*, *225*:219–224 (1981).
5. M. Zilli and N. Giuseppe. Simple and sensitive method for the determination of clobazam, clonazepam and nitrazepam in human serum by high-performance liquid chromatography. *J. Chrom.*, *378*:492–497 (1986).
6. W. Gazdzik, J. Podlesny, and M. Filipek. HPLC method for simultaneous determination of clobazam and N-desmethylclobazam in human serum, rat serum and rat brain homogenates. *Biomed. Chromatogr.*, *3*:79–81 (1989).

70

Analysis of Fluoxetine and Its Metabolite Norfluoxetine in Serum by Reversed-Phase HPLC, with Ultraviolet Detection

H. Vandenberghe*† and E. Bassoo* *St. Joseph's Health Centre and †University of Western Ontario, London, Ontario, Canada*

I. INTRODUCTION

Fluoxetine hydrochloride, or N-methyl-8-[4-(trifluoromethyl)phenoxy]benzenepropanamine, is a second-generation, nontricyclic propylamine antidepressant that has recently been released in Canada (Prozac, Eli Lilly Company). Fluoxetine and its N-demethylated metabolite, norfluoxetine (Figure 1), selectively inhibit presynaptic serotonin reuptake and enhance serotonergic neurotransmission (1,2). The drug has been used in treatment of major depression, in patients with obsessive-compulsive disorders, and for intention myoclonus (2). Fluoxetine has milder anticholinergic effects and no adverse cardiac effects (1). Fluoxetine and its metabolite have been determined by electron-capture gas chromatography (3) and by high-performance liquid chromatography (HPLC) with ultraviolet spectrophotometric detection (4,5).

We have developed a simple and reliable procedure for quantitative analysis of fluoxetine and norfluoxetine in serum by reversed-phase HPLC with ultraviolet detection.

II. MATERIALS AND METHODS

A. Apparatus

Throughout the development of this method, we used an HPLC system consisting of a solvent-delivery module 114M (Beckman), model 210 injector (Beckman), a Gilson Holochrome UV detector (Gilson), and a Kipp-Zonen strip-chart recorder.

B. Reagents

Sodium phosphate, hydrochloric acid, and acetic acid were analytical-grade chemicals. All solvents used were HPLC grade, purchased from Fisher Scientific, Canada. Sodium phosphate buffer 0.5 mol/L, pH 10, was prepared by dissolving 77.99 g of

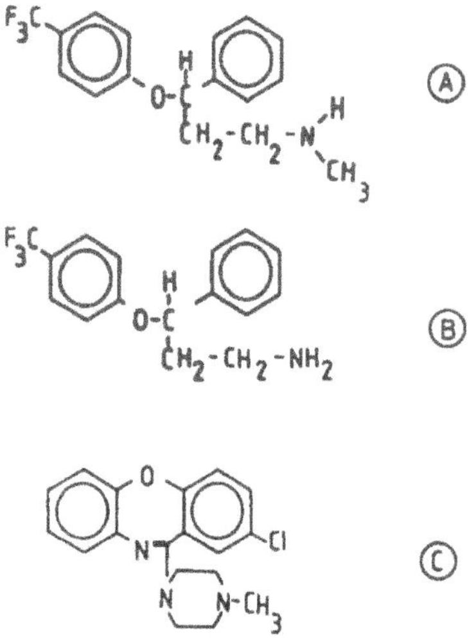

Figure 1 Chemical structure of (A) fluoxetine, (B) norfluoxetine, and (C) the internal standard, loxapine.

sodium phosphate, monobasic dihydrate, into 1 L of deionized water and adjusting to pH 10 with 10 N sodium hydroxide solution. Acidic methanol was prepared by mixing equal volumes of HPLC-grade methanol and 0.1 mol/L hydrochloric acid. The mobile phase was obtained by mixing 350 mL of acetonitrile with 650 mL of reagent A. Reagent A was prepared by addition of 1 mL of triethylamine to 1 L of deionized water after adjusting to pH 5.6 with glacial acetic acid.

C. Standards

Pure standard samples of fluoxetine and norfluoxetine were provided by Eli Lilly Company, Canada, and the internal standard, loxapine, by Lederle Cyanamid Canada, Inc. Stock solutions of 1 g/L were prepared in methanol. Substock solutions (10 mg/L) were prepared of loxapine in methanol and the parent drug and metabolite in deionized water. Working standards ranging from 50 to 1000 μg/L were made up in stripped serum by appropriate dilutions of the substock. These were aliquoted and stored at −20°C.

D. Extraction Procedure and Conditions of Analysis

To 1 mL of standard, control, or patient serum, add 25 μL of internal standard, loxapine (10 mg/L), in a 16 × 125 mm screw-capped tube. Alkalinize the serum with 1 mL of 0.5 mol/L sodium phosphate buffer, pH 10, and vortex-mix for 5 s. Extract the compounds into 6 mL of hexane:isoamyl alcohol (98:2). Shake on the multivortexer for 10 min, and centrifuge for 10 min at 1000 × g. Transfer the organic layer to a screw-capped conical tube and add 200 μL of acidic methanol. Cap with a Teflon-lined screw cap. Shake on the multivortexer for 5 min, and centrifuge for 5 min at 1000 × g. Aspirate the top organic layer, and rinse the sides of the tube with 200 μL of methanol and vortex-mix for 20 s. Evaporate the solvent under a stream of air in a water bath at

Table 1 Conditions of Analysis

Parameter	Assay conditions required
Column	Supelco C-18 DB 5 μm (15 cm × 4.6 mm I.D.)
Mobile phase	35/65 mixture, acetonitrile/reagent A
Flow rate	2 mL/min
Temperature	Ambient temperature
Internal standard	Loxapine
Chromatography time	25 min
Absorbance full scale	0.01
Sample volume	1 mL
Wavelength	226 mm

56°C. Reconstitute the dry residue in 200 μL of mobile phase and inject 100 μL onto the column.

The conditions of analysis are shown in Table 1, and a typical chromatogram for a serum sample of a patient containing 200 μg/L fluoxetine and 1250 μg/L of norfluoxetine is illustrated in Figure 2.

III. RESULTS AND DISCUSSION

The single-step liquid-liquid extraction procedure with back-extraction into acidic methanol is fast and reliable. This procedure yielded consistent recoveries of 80 ± 5% for both fluoxetine and norfluoxetine at serum concentrations of 200 μg/L.

The back-extraction into acidic methanol is essential for removing endogenous interfering peaks, but a second extraction into organic solvent (4) did not appear necessary with the procedure described. In our hands, the reconstitution of the dried residue after a single solvent extraction (5) resulted in additional serum peaks interfering with the compounds of interest, hence the back-extraction into acidic methanol. No peaks with retention times similar to fluoxetine, its metabolite, or the internal standard were identified in the analysis of 20 samples from patients known not to be taking any drugs. Table 2 lists the antidepressants evaluated for interference in the assay. The between-day precision of the procedure (n = 10) was assessed by repeated analysis of serum samples containing various concentrations of fluoxetine and norfluoxetine over a period of 10 consecutive days. The method yielded coefficients of variation of less than 10% for concentrations of 200, 500, and 1000 μg/L.

The detection limit of the method as described corresponds to 30 μg/L (a signal-to-baseline noise ratio of about 2).

Figure 2 illustrates representative chromatograms of a human serum blank supplemented with internal standard, standards containing 200 and 1000 μg/L of fluoxetine and norfluoxetine, extracted from serum and a patient's sample.

The extraction procedure is linear for both fluoxetine and norfluoxetine over a concentration range of 50 to 1000 μg/L and is presented in Figure 3.

IV. SUMMARY

We have described a simple and reliable method for simultaneous quantitation of fluoxetine and its desmethyl metabolite norfluoxetine by reversed-phase HPLC with

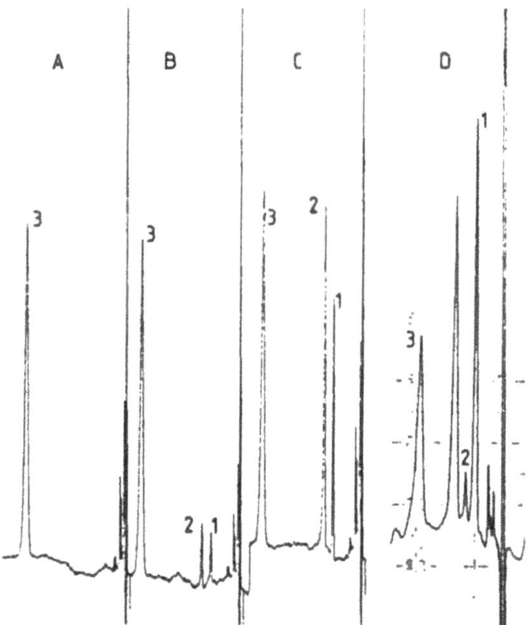

Figure 2 Chromatogram for a blank serum (A), for extracted standards containing 200 µg/L (B) and 1000 µg/L (C) of both fluoxetine and norfluoxetine, and for a patient sample containing 200 µg/L fluoxetine and 1250 µg/L norfluoxetine. Peaks 1, 2, and 3 correspond to norfluoxetine, fluoxetine, and the internal standard, respectively.

ultraviolet spectrophotometric detection at 226 nm. Fluoxetin, norfluoxetine, and the internal standard, loxapine, were isolated by a two-step liquid-liquid extraction from serum. The percentage analytical recovery of fluoxetine and norfluoxetine was 80 ± 5%. The within- and between-day precision of the procedure at 100, 200, and 500 µg/L was less than 10%. The extraction procedure is linear for both fluoxetine and its metabolite over a concentration range of 50 to 1000 µg/L. The lowest concentration that can be detected accurately with this method is 30 µg/L for both fluoxetine and norfluoxetine.

Table 2 Interference Study (at 226 nm)

Compound	Retention time (min)
Desipramine	6.0
Norfluoxetine	6.5
Fluoxetine	8.0
Nortriptyline	8.5
Maprotiline	11.5
Loxapine	18
Imipramine	20
Amitriptyline	25

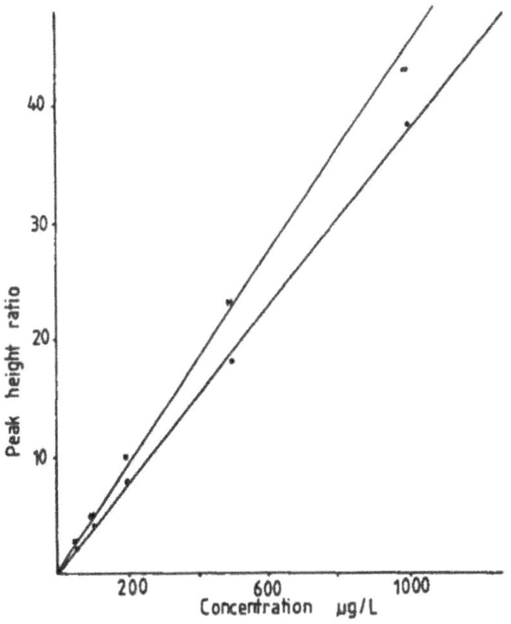

Figure 3 Linearity:peak height ratio versus concentration of fluoxetine (✱) and norfluoxetine (●) from extracted serum samples.

REFERENCES

1. P. Benfield, R. C. Heel, and S. P. Lewis. Fluoxetine: A review of its pharmacodynamic and pharmacokinetic properties and therapeutic efficacy in depressive illness. *Drugs*, *32*: 481–508 (1986).
2. L. Lemberger, R. F. Brigstrom, R. I. Wolen, et al. Fluoxetine: Clinical pharmacology and physiologic disposition. *J. Clin. Psychiatr.*, *46*:14–19 (1985).
3. J. F. Nash, R. J. Bopp, and R. H. Carmichael. Determination of fluoxetine and norfluoxetine in plasma by gas chromatography with electron-capture detection. *Clin. Chem.*, *28*: 2100–2102 (1982).
4. P. J. Orsulak, J. T. Kerney, J. R. Debus, C. Crowley, and P. D. Wittman. Determination of the antidepressant fluoxetine and its metabolite norfluoxetine in serum by reversed-phase HPLC, with ultraviolet detection. *Clin. Chem.*, *34*:1857–1878 (1988).
5. M. W. Kelly, P. J. Perry, S. G. Holstad, and M. J. Garvey. Serum fluoxetine and norfluoxetine concentrations and antidepressant response. *Ther. Drug Monitor.*, *11*:165–170 (1989).

Part V

Substances (Drugs) Subject to Abuse

71

A Rapid Alcohol Test for On-Site Testing

Steven S. Bachand, Stephen K. Schultheis, and C. Michael O'Donnell *TOXI-LAB, Inc., Irvine, California*

I. INTRODUCTION

A simple, portable alcohol test based on microdiffusion was evaluated in a field-site testing situation where volunteer subjects donated specimens before, during, and after drinking ethanol for the purpose of testing with On-Site ALCOHOL. The intent of the evaluation was to determine the ability of the On-Site ALCOHOL method to qualitatively screen a variety of specimen types for alcohol in field-site or on-site testing.

The On-Site ALCOHOL procedure incorporates the use of a specimen-absorbing matrix which quickly distills or releases the alcohol from the specimen. The released alcohol vapor is focused or concentrated onto an acid dichromate reagent suspended in an inert matrix. The concentrated alcohol vapor is oxidized at this site, in turn reducing dichromate detection reagent. The reduction of the dichromate detection reagent is evidenced by the appearance of a blue-green spot in the middle of the unreduced yellow-orange detection reagent pad.

The acid dichromate detection reagent is provided in a sealed glass ampule for easy dispensing and long shelf life, making this test ideal for clinical laboratory settings or in field-site testing situations where portability, long shelf life, simplicity, and minimal equipment requirements are critical needs.

II. SAMPLE COLLECTION AND HANDLING

Saliva and breath samples were collected within 5 min of each other, and blind tests were performed immediately with On-Site ALCOHOL and Intoxilyzer, respectively. Urine and serum samples were also collected within 5 min of each other and blind-tested with On-Site ALCOHOL. Following testing, urine and serum specimens were sealed tightly, chilled, and transported to the Sheriff-Coroner Department of Orange County for gas chromatography confirmation. The samples were stored under refrigeration prior to analysis.

II. MATERIALS AND METHODS

A. On-Site ALCOHOL Method (Microdiffusion)

Reducing volatiles such as ethanol will reduce chromium trioxide (CrO_3) to chromic oxide (Cr_2O_3) (1). In the presence of vapor from a reducing alcohol, the yellow chromium trioxide will change to blue chromic oxide. The intensity of the blue is proportional to ethanol concentration.

On-Site ALCOHOL relies on this color change for detection of alcohol in biological specimens. The procedure calls for the application of 1 drop of the alcohol detection reagent (acid dichromate) to an inert pad located in a well on the On-Site ALCOHOL card designated "Detection Reagent." This is followed by the specimen application step. After specimen collection, 3 drops of urine or serum are dispensed into the well on the On-Site ALCOHOL card marked "Specimen," using the disposable plastic pipet provided with each test. Alternatively and noninvasively, saliva can be collected with each test with the disposable swab also provided with each test.

Following collection, the saliva-saturated swab is inserted into the specimen well and locked into position for the duration of the test. The On-Site ALCOHOL card is allowed to react for 3 min before the results are read (Figure 1). Positive specimens produce a blue-green spot in the center of the yellow-orange detection reagent pad in less than 3 min (Figure 2).

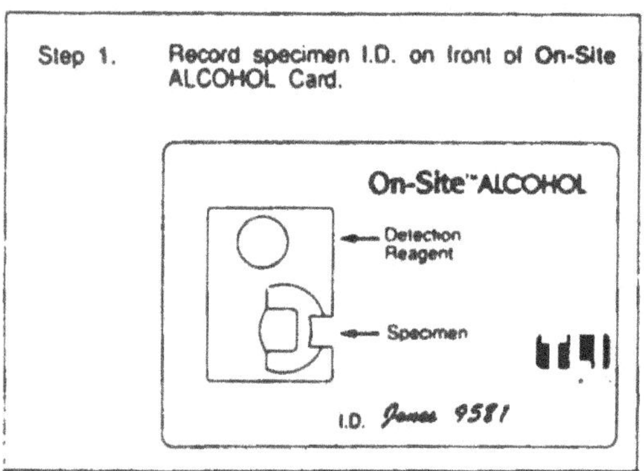

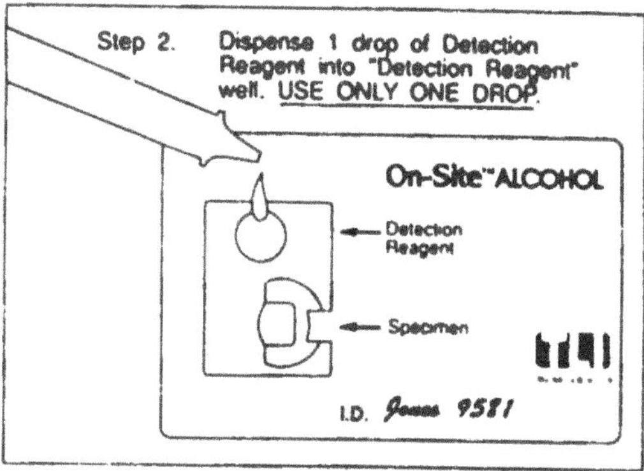

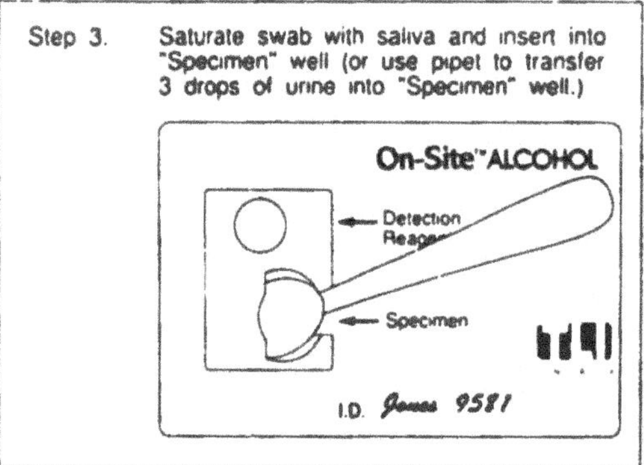

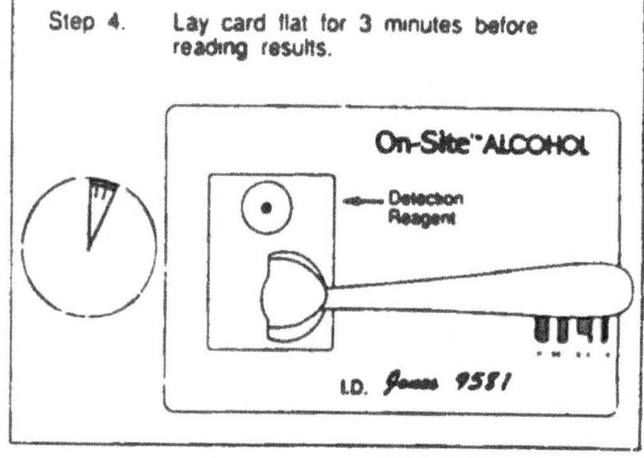

Figure 1

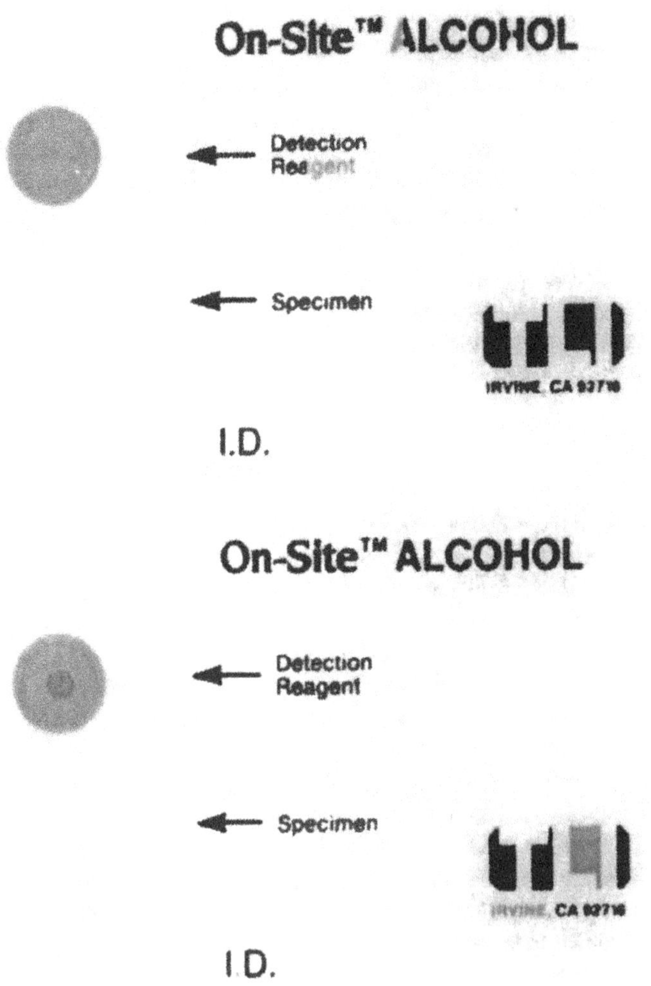

Figure 2

In negative specimens the detection reagent pad remains yellow-orange, with no development of a blue-green spot. Any blue-green spot in the center of the detection reagent pad is recorded as a presumptive positive. Further analysis with a different analytical method would be required for differentiation and quantitation.

B. Breath Analysis and Gas Chromatography Analysis

Four volunteer subjects were tested for breath alcohol at seven intervals ranging from a baseline breath alcohol 5 min before the start of alcohol intake to 2 h after the last alcohol intake. The breath alcohol instrument used was an Intoxilyzer 5000, manufactured by CMI, Inc. (Minturn, Colorado). The Intoxilyzer 5000 analyzes alcohol by infrared absorption. The instrument and a qualified operator were provided by the Sheriff-Coroner Department, County of Orange, California.

Urine and serum specimens were collected from each volunteer at four different intervals ranging from baseline serum and urine samples to samples taken 1.5 h after the last alcohol intake. The specimens were stored continuously at 4°C in tightly closed containers for 1 week prior to gas chromatography confirmation. The specimens were tested at the Sheriff-Coroner Department, County of Orange, California, with a Sigma 2000 GC with HS 100 Headspace Sampler manufactured by Perkin-Elmer Corporation (Irvine, California). The principle analysis was headspace, using a 1/30 dilution of the questioned sample with *n*-propanol internal standard. Each questioned sample was analyzed in duplicate, usually by two different analysts on two different instruments; a minimum of seven different standards and controls were used in each analysis.

IV. RESULTS

Specimens with ethanol concentrations as low as 0.015% w/v produced small, faint blue-green spots surrounded by a yellow-orange background on the On-Site ALCOHOL test card. Concentrations over 0.05% produced distinct blue-green spots that appeared quickly and were easily observable against the yellow-orange background.

Test results for the On-Site ALCOHOL test, Intoxilyzer breath test, and gas chromatography analysis are listed in Table 1.

A comparison of the On-Site ALCOHOL method with the Intoxilyzer breath test and gas chromatography confirmation showed excellent agreement. Of the 28 saliva samples tested, the On-Site ALCOHOL method agreed with Intoxilyzer breath results (qualitative) in all 28 samples tested. Of the 16 urine samples tested, On-Site ALCOHOL results were confirmed in 15 of the 16 samples. The one discrepancy could be traced to a urine sample reported to be positive at 0.003% w/v ethanol by GC analysis. This level is below the detection limit of On-Site ALCOHOL. For serum samples, the 16 results generated from On-Site ALCOHOL were confirmed by the GC analysis, which included 12 positive results and 4 negative results.

Table 1 On-Site ALCOHOL Test Correlation with Intoxilyzer Breath Analysis and Gas Chromatography Analysis Results

	Positive	Negative
On-Site ALCOHOL saliva test results	24	4
Intoxilyzer breath analysis results	24	4
	Range: 0.024–0.098%	
On-Site ALCOHOL urine test results	10	6
GC urine analysis results	11	5
	Range: 0.003–0.092%	
On-Site ALCOHOL serum test results	12	4
GC serum analysis results	12	4
	Range: 0.014–0.092%	

Table 2 Interfering Substances

The following substances have been tested for reaction with On-Site ALCOHOL. Ethanol is assigned a value of 100% reactivity.

Alcohol at 0.1% w/v	Percent reactivity
Ethanol	100
Methanol	100
Acetaldehyde	100
Isopropanol	40
Acetone	0
Glycerol	0
Ethylene glycol	0

Several samples spiked with reducing alcohols were tested and found to react with the On-Site ALCOHOL detection reagent (Table 2). These reducing substances included methyl alcohol, acetaldehyde, and isopropyl alcohol (2).

V. CONCLUSION

A self-contained, simple-to-use device has been developed to detect ethanol in urine, saliva, and serum. The procedure requires only minimal technical training and provides accurate, objective results at any location in approximately 3 min.

The variety of specimens that can be tested using On-Site ALCOHOL allows for more options in determining patient alcohol status information. In the postabsorptive stage, urine alcohol concentration will be approximately 25% higher than blood and serum alcohol levels (3). Urine specimens will test positive long after breath and blood specimens are negative. Saliva (salivary ethanol) has been shown to be a good index of blood ethanol concentration provided the saliva is obtained at least 20 min after the ingestion of alcohol (4).

The On-Site ALCOHOL test in a simulated field-site testing situation proved to be an easy-to-use, convenient, and reliable method for the determination of reducing alcohol in biological fluids.

On-Site ALCOHOL urine and serum test results were confirmed by gas chromatography analysis, and On-Site ALCOHOL saliva correlated 100% with the Intoxilyzer breath test.

On-Site ALCOHOL may be used in chemical dependency units, probation and parole offices, public and private transportation systems, and schools, whereas blood specimens are not easily obtainable as urine and saliva.

REFERENCES

1. *Manual on Alcohol*, Council on Forensic Pathology, American Society of Clinical Pathology, p. 2 (1967).
2. Data on file, TOXI-LAB, Inc., Irvine, California.

3. P. H. Meyers, E. Jawetz, and A. Goldfien, eds. *Review of Medical Pharmacology*, 4th ed. Lange Medical, Los Altos, CA, pp. 236–238 (1974).
4. K. E. L. McColl, B. Whiting, M. R. Moore, and A. Goldberg. Correlation of ethanol concentration in blood and saliva. *Clin. Sci.*, *56*:283–286 (1979).

72

Therapeutic Monitoring of Alcohol Withdrawal During Diazepam Loading-Dose Treatment

Halina Matsumoto,* Mirosław Rewekant,† Dariusz Wasilewski,* and Witold Gumułka†
**Warsaw Medical Academy and †Institute of Physiological Sciences, Warsaw, Poland*

I. INTRODUCTION

In the treatment of acute alcohol withdrawal syndrome (AWS), benzodiazepines are the drugs of choice (1,2,9,15). In Poland, diazepam is most frequently used. Its high therapeutic efficacy is due, among other reasons, to its anticonvulsive action, since seizures are the most severe symptoms in AWS patients and are life threatening (4,7). Diazepam is typically given p.o. or i.m., b.i.d. or t.i.d. over 5 to 7 days, the total dose ranging from 10 to 30 mg daily. However, this routine diazepam treatment has not been found more effective than nonpharmacologic "supportive care" (12,14). Besides, both diazepam and its major metabolite, N-desmethyldiazepam, are long-acting. Repeated diazepam administration at a constant dosage would result in cumulation of the drug and its metabolite in the patient's body (3,11). Then unwanted symptoms such as excessive drowsiness, lethargy, ataxia, diplopia, and confusion may occur (5,7,12). For this reason attempts have been made to find alternative techniques of diazepam administration (2,7,9,12,15). Recently, Sellers and his colleagues demonstrated an oral loading-dose technique of diazepam administration which seems to be the best one from the therapeutic efficacy point of view (11).

The aim of this study was to estimate the clinical efficacy of diazepam swift loading-dose treatment in acute alcohol withdrawal syndrome. Both the therapeutic effects and serum diazepam and N-desmethyldiazepam concentration have been monitored.

II. SUBJECTS AND METHODS

The study was carried out in the Department of Psychiatry, Warsaw Medical Academy, on 11 male patients admitted with symptoms of acute alcohol syndrome (predelirium or delirium tremens). Their mean age was 37.8 ± 6.9, their mean body weight was 77.5 ± 5.8 kg, and their mean length of alcohol abuse was 13.7 ± 5.4 years. The average time that had passed since the last alcohol ingestion was less than 12 h in four patients, more than 12 h and less than 24 h in five patients, and more than 48 h but less than 72 h in two patients. Every patient was clinically interviewed. The intensity of AWS was quantitatively assessed by Shaw's CIWA-A scale (14). The clinical state was estimated before pharmacological treatment, 1 to 2 h after successive doses of diazepam, and on the third day of hospitalization. Diazepam was given orally in 10- to 20-mg doses, every 1 to 2 h. The treatment was discontinued when the clinical condition of the patient improved—that is, when his clinical state was judged to be 10 points or less by CIWA-A scale (11). In all the patients, the blood ethanol level (BAL) was measured (6,8). The first dose of diazepam was administrated when BAL ≤ 10 mg/dL. Serum concentrations of diazepam (D) and N-desmethyldiazepam (DD) were measured before the treatment (as a control study), 1 h after diazepam loading was completed, and on the third day of hospitalization. Both compounds were determined by high-performance liquid chromatography (HPLC) with ultraviolet detection at 226 nm (8,16). The therapeutic effect (E_t) was defined as the difference in the CIWA-A point scores the patient obtained before diazepam administration and 1 h after the last diazepam dose on the third day of hospitalization. As the maximum therapeutic effect (E_{max}), the hypothetical value of the maximum difference in point score (that is, if the score on the CIWA-A scale fell to zero) was accepted.

III. RESULTS

Table 1 shows the clinical and pharmacometric parameters in the group of 11 AWS patients treated by the diazepam swift loading method. The average time during which the objective clinical improvement occurred was 3 h.

The mean rate of diazepam loading was calculated to be 0.128 mg/kg of body weight/h (Figure 1). This means that a patient weighing 70 kg, with the symptoms of acute alcohol withdrawal syndrome, should be given ca. 10 mg of diazepam every hour in order to achieve clinical improvement as measured according to the CIWA-A scale.

Table 2 shows the serum diazepam and N-desmethyldiazepam concentrations in patients' blood 1 h after diazepam loading had been completed and on the third day of hospitalization, when no drug had been administered. No significant linear correlations were found between serum diazepam, N-desmethyldiazepam, or the sum of the parent drug and its metabolite and the therapeutic (E_t) or maximum therapeutic (E_{max}) effect. In the case of serum diazepam concentration and therapeutic effect measured 1 h after the last dose of the drug, the tendency toward correlation at α = 0.05 ($r = 0.61$, $p = 0.94$) was observed.

Table 3 presents the results of comparing two methods used for treating the symptoms of acute alcohol withdrawal syndrome: the traditional method and the swift loading-dose method. The degree of clinical improvement, expressed in percentage change in TSA or CIWA-A score, was twice as high for the swift loading-dose method in comparison with the traditional method. On the third day of hospitalization the average serum diazepam concentration was 1.5 times higher for the traditional method. On the other hand, the average serum N-desmethyldiazepam concentration was ca. 4 times higher for the swift diazepam loading-dose treatment.

Table 1 Characteristics of 11 AWS Patients During Diazepam Loading-Dose Treatment (E_{max} = Maximum Therapeutic Efficacy, E_t = Therapeutic Efficacy)

Patient no.	Patient's initials	Body weight	CIWA-A score		Total diazepam dose (mg)	Duration of diazepam loading (h)	Dynamics of diazepam administration (mg/kg/h)	E_{max}	E_t
			Before treatment	1 h after loading					
1	B.W.	82	23	0	15	1	0.18	23	23
2	F.L.	80	25	5	40	2	0.10	25	20
3	M.M.	70	19	4	30	2	0.21	19	15
4	Sz.M.	70	31	2	60	6.5	0.13	31	29
5	M.M.	75	22	4	10	1	0.13	22	18
6	P.S.	80	18	10	40	2	0.1	18	8
7	T.T.	73	36	8	40	3	0.18	36	26
8	S.M.	79	14	6	15	1.5	0.12	14	8
9	P.J.	75	24	6	30	3	0.13	24	18
10	R.A.	90	22	6	30	4.5	0.07	22	16
11	S.E.	79	27	8	30	6.5	0.06	27	19
$\bar{x}$	77.5	23.7	23.7	5.36	30.9	3.0	0.128	23.7	18.2
± SD	± 5.8	± 6.1	± 6.1	± 2.8	± 14.3	± 2.0	± 4.6E-0.2	± 6.1	± 6.5

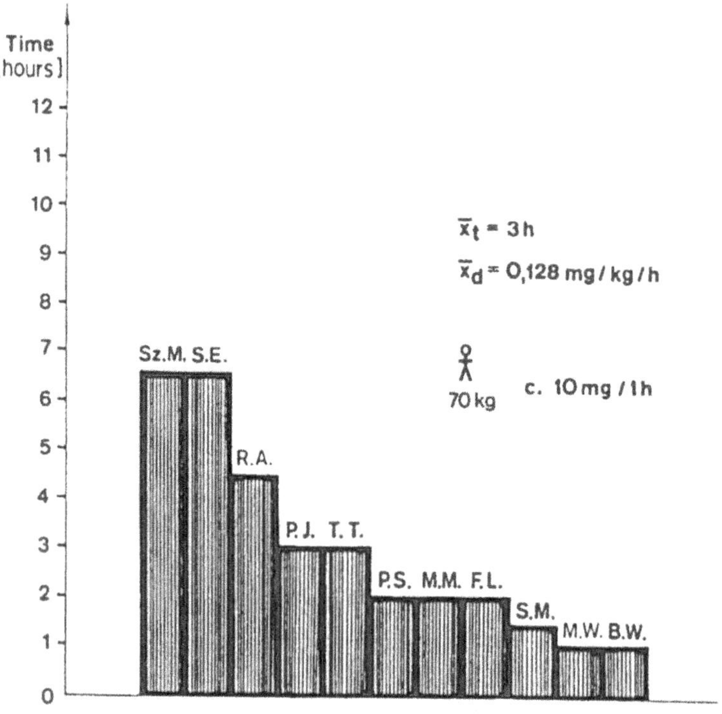

Figure 1 Time from first diazepam dose to objective clinical improvement in 11 patients with AWS treated with diazepam loading technique; $\bar{x}_t$, mean time value; $\bar{x}_d$, mean diazepam dose.

Table 2 Serum Diazepam (D) and N-Desmethyldiazepam (DD) Concentrations in AWS Patients 1 h After the Last Diazepam Dose Administration and on the Third Day of Hospitalization (No Drug Administrated)

Time of blood sampling	Serum concentration (ng/ml) D	Serum concentration (ng/ml) DD	D/DD ratio
1 h after the last diazepam dose (n = 10)			
$\bar{x} \pm$ SD	367.59 ± 329.68	672.18 ± 540.51	0.70 ± 0.68
(Range)	(21.16–982.70)	(258.83–1690.15)	(0.05–1.80)
Third day of hospitalization			
$\bar{x} \pm$ SD	253.88 ± 52.60	698.32 ± 477.92	0.60 ± 0.58
(Range)	(198.53–351.21)	(190.29–1431.38)	(0.18–1.84)

Table 3 Comparison of Clinical Efficacy, Serum Diazepam (D), and N-Desmethyldiazepam (DD) Concentrations in AWS Patients Treated with Diazepam According to the Traditional and Loading-Dose Methods

Method of treatment	Assessment of AWS[a]	Clinical estimation: Before treatment	Clinical estimation: On the third day of hospitalization	Clinical improvement (percentage change score)	Serum D and DD concentrations (ng/ml) on the third day of hospitalzation: D	Serum D and DD concentrations (ng/ml) on the third day of hospitalzation: DD	D/DD ratio
Traditional	TSA score	18 ± 12.03	11.1 ± 5.9	40.9	394.3 ± 211.4	167.66 ± 77.33	2.35 ± 0.92
Loading-dose treatment	CIWA-A score	23.16 ± 4.95	1.83 ± 0.81	92.1	253.88 ± 52.6	698.32 ± 477.92	0.6 ± 0.58

[a]There is a high correlation between the TSA and CIWA-A scales (12).

IV. DISCUSSION

Results of previous studies (8) demonstrated an evident irrationality of the traditional or routine method of giving diazepam to AWS patients. We have shown, among others, an inappropriate rate of serum diazepam level rise during this treatment, with the highest level reached on the last days of the treatment, and that the concentrations of diazepam in alcohol-dependent subjects are lower than in the population of healthy subjects, which is in agreement with the observations of other authors (10,13). In addition, we observed that at high concentrations of diazepam a correlation between diazepam concentrations and therapeutic effect appeared ($r = 0.85$, $p < 0.01$) which had the character of an exponential function. This suggested that significantly better therapeutic effects could be expected over a certain threshold of drug concentration. Since in AWS a high therapeutic effect should be achieved within the first 24 to 48 h after ethanol withdrawal, this is the time when a high diazepam concentration in the blood is particularly important. Simulating diazepam concentrations in blood, we have shown that the optimum dosing interval should be 1 to 2 h, with the diazepam dose being at least 10 mg.

The results of present investigations have shown that the swift diazepam loading-dose method is really much more effective in treating acute alcohol withdrawal syndrome than the traditional method. It has been confirmed that the swift loading diazepam dose treatment results in the fast relief of alcohol withdrawal symptoms—as early as during the first 24 h of treatment (11,12).

The patients have not shown any intoxication symptoms which were likely to appear considering the narrow interval of diazepam dosing in a relatively large dose. With the traditional method the patient receives 10 to 30 mg of diazepam per 24 h, while with the swift loading-dose method he receives a dose of 10 to 20 mg every 1 to 2 h. The results obtained suggest that diazepam metabolism and diazepam and N-desmethyldiazepam pharmacokinetics are different in these two treatment options. This can be seen in the diazepam/N-desmethyldiazepam ratio, whose average value calculated on the third day of treatment was 2.35 with the traditional method and 0.58, 4 times lower, with the swift loading-dose method. Perhaps this is due to the tapering of the swift dose (the longest duration of diazepam loading was only 6.5 h) to the accelerated process of N-demethylation in the conditions of suitably high serum diazepam level, or perhaps the result of the nonlinear kinetics of N-desmethyldiazepam (3). With the swift loading-dose method, no diazepam therapeutic effect tolerance has been developed, although tolerance could be seen with the traditional method (3,8). The definitely changed progression of serum diazepam and N-desmethyldiazepam concentrations with the swift loading-dose treatment guarantees their high levels just in the period when the patient's life is exposed to the greatest danger—that is, during the first 24 to 48 h after alcohol withdrawal. During the hospitalization it was possible to see the faster elimination of diazepam than N-desmethyldiazepam, which agrees with the longer half-life of the major active diazepam metabolite (3,10,11). To achieve more precise dosing of diazepam we suggest that the swift loading-dose method be used with a dynamically expressed diazepam dose, that is, mg/kg body weight/h. The swift loading-dose method has been approved at our clinic. However, more clinical material is needed to get better acquainted with the mechanism of high therapeutic efficacy of this method and show the significance of the tendencies observed.

V. CONCLUSIONS

1. The method of swift diazepam loading-dose treatment is more effective in acute alcohol withdrawal syndrome management then the traditional method.
2. The swift diazepam loading-dose treatment is simple, economical, and has a pharmacokinetic and a pharmacodynamic rationale.

REFERENCES

1. H. Busch and A. Frings. Pharmacotherapy of alcohol withdrawal syndrome in hospitalized patients. *Pharmacopsychiatry, 21*:232–237 (1988).
2. R. Castaneda and P. Cushman. Alcohol withdrawal: A review of clinical management. *J. Clin. Psychiatr., 50*:278–284 (1989).
3. W. A. Colburn and M. L. Jack. Relationship between CSF drug concentration, receptor binding characteristics, and pharmacokinetics and pharmacodynamic properties of selected 1,4-substituted benzodiazepines. *Clin. Pharmacokinet., 13*:179–160 (1987).
4. D. J. Greenblatt and R. J. Shader. Benzodiazepines. *N. Engl. J. Med., 291*:1011–1015 (1974).
5. D. Greenblatt, R. Shader, and D. Abernethy. Current status of benzodiazepines. *N. Engl. J. Med., 309*:354–358 (1983).
6. H. Matsumoto, M. Rewekant, W. Gumulka, and H. Wardaszko-Lyskowska. Clinical efficacy and serum levels of diazepam in alcohol withdrawal. In *Advances in Therapeutic Drug Monitoring* (K. Tanaka, C. E. Pippenger, T. Mimaki, P. D. Walson, and S. Ohgitani, eds.), Enterprise, Tokyo, pp. 154–157 (1990).
7. C. Naranjo and E. Sellers. Clinical assessment and pharmacotherapy of alcohol withdrawal syndrome. In *Recent Developments in Alcoholism* Vol. IV, pp. 265–281 (1986).
8. M. Rewekant, H. Matsumoto, R. Wilczak-Szadkowska, and J. Krzyzowski. Serum diazepam and N-desmethyldiazepam monitoring during detoxification of alcohol dependent patients. *Proc. XXXVI Congr. Polish Psychiatric Association*, Vol. 2, pp. 375–382 (1990) (in Polish).
9. A. Rosenbloom. Emerging treatment options in the alcohol withdrawal syndrome. *J. Clin. Psychiatr., 49*(Suppl.):28–31 (1988).
10. H. Schutz. *Benzodiazepines*. Springer-Verlag, Berlin, pp. 209–218 (1982).
11. E. M. Sellers, C. A. Naranjo, M. Harrison, P. Devenyi, C. Roach, and K. Sykora. Diazepam loading: Simplified treatment of alcohol withdrawal. *Clin. Pharm. Ther., 34*:822–826 (1983).
12. E. M. Sellers and C. A. Naranjo. New strategies for the treatment of alcohol withdrawal. *Psychopharmacol. Bull., 22*:88–92 (1986).
13. R. Sellman, A. Pekkarinem, L. Kangas, and E. Raijola. Reduced concentrations of plasma diazepam in chronic alcoholic patients following an oral administration of diazepam. *Acta Pharmacol. Toxicol., 36*:25–32 (1975).
14. J. M. Shaw, G. S. Kolesar, E. M. Sellers, H. L. Kaplan, and P. Sandor. Development of optimal treatment tactics for alcohol withdrawal. I. Assessment and effectiveness of supportive care. *J. Clin. Psychopharm., 1*:382–389 (1981).
15. M. A. Suchckit. Alcoholism: Acute treatment. In *Drug and Alcohol Abuse*, Plenum Medical Book Company, New York, London, pp. 77–95 (1989).
16. J. Wallace, H. Schwertner, and E. Shimke. Analysis for diazepam and nordiazepam by EC-GC and LC. *Clin. Chem., 25*:1296–1300 (1979).

73

Clinical Monitoring of Ethanol Metabolism in Humans

Donald T. Forman *University of North Carolina, Chapel Hill, North Carolina*

I. INTRODUCTION

The classical way of monitoring elimination of alcohol from the human body is to measure blood alcohol concentrations following alcohol consumption to give blood alcohol clearance curves (1). These curves reflect the sequence of events as alcohol is absorbed into the body, equilibrated through the water compartments of all tissues, and eliminated via enzymatic oxidation, a process which takes several hours for moderate or large doses of ethanol. Figure 1 gives an example of blood, urine, and breath alcohol curves for a subject given a dose of ethanol orally—sufficient to raise the blood alcohol levels to about 60 mg/dl. The figure also demonstrates that the urine alcohol curve parallels and follows the blood alcohol curve.

Since blood contains about 80% water and urine is nearly 100% water, the urine/blood ratio should be theoretically about 1.25. Urine is not a recommended specimen for the analysis of alcohol for any purpose other than simply demonstrating the presence of alcohol in a subject. Its insurmountable shortcomings include a highly variable urine/blood ratio, inability of the analyst to determine the effect of, or to compensate for, bladder pooling, and inadequate correlation of urine-alcohol concentrations with an associated impairment. Breath alcohol analyses are also shown in Figure 1. In estimation of blood alcohol concentrations from breath measurements, it is necessary to use a partition coefficient for alcohol between water and air, usually at 34°C. The partition coefficient used in the original calculation was 2100:1 (2). This factor was used to convert the measured breath and alcohol concentration into the presumed blood level. Current practice is to calibrate all quantitative breath alcohol analyzers in grams of alcohol per 210 L of breath and, in most instances, to designate the number obtained by analysis as the corresponding blood alcohol concentration in g/dL (= %w/v). The units "g/210 L" in breath and g/dL or %w/v in blood are numerically identical when a conversion factor of 2100:1 for the blood/breath alcohol relation is used. The result actually measured is g/210 L of breath alcohol. An improved correlation with blood ethanol measurements has been obtained using the ratio 2300:1 (3). Breath alcohol analyses are now of legal significance in many countries (4).

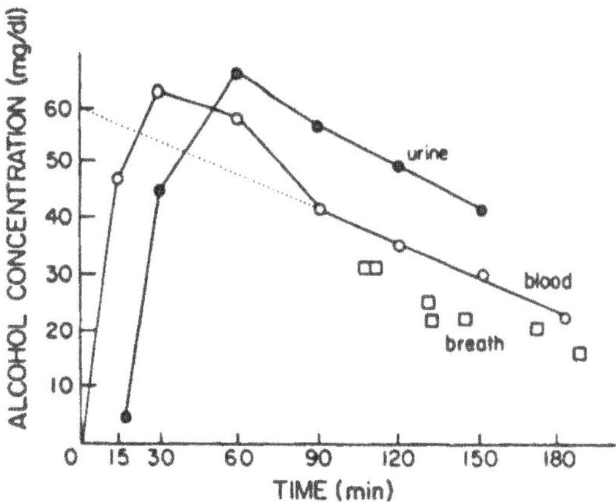

Figure 1 Blood, breath, and urine alcohol curves for a human subject given a dose of ethanol orally-sufficient to raise the blood alcohol levels to approximately 60 mg/dL.

II. MEASUREMENT OF ETHANOL ELIMINATION RATES

Accurate measurement of the ethanol elimination rate is important for both forensic and experimental purposes. In the latter case, it is important for studying the metabolic rates of different ethnic groups and between alcoholics and nonalcoholic subjects, and in elucidating the factors which regulate rates of ethanol metabolism in humans. The rate of ethanol elimination from the body is usually calculated from the decreasing phase of a blood (or breath) ethanol-versus-time curve (Figure 2). There is still debate as to the best way of performing such calculations. Controversy exists concerning the validity of considering the decreasing phase of the blood alcohol curve as being linear (zero order), as was originally assumed by Widmark (5), or whether a Michaelis-Menten kinetic (first-order) model should apply (6). For medicolegal purposes, the zero-order model is often assumed valid, while many basic researchers use the multicompartment, multienzyme, Michaelis-Menten kinetic model.

The maximal elimination rate of ethanol in human subjects has been determined in numerous studies (7,8). As an average, rates of 100 mg/kg/h (range 70–130) (elimination rates: 10–20 mg/dL/h), have been reported. Daily alcohol intake in human subjects is also known to result in metabolic tolerance and decreased sensitivity to increasing dosages of alcohol (e.g., 300 g/day and more). That these amounts can be metabolized over longer periods of time has also been demonstrated in controlled experimental studies in a subpopulation of chronic alcoholics (8).

III. MEASUREMENT OF ALCOHOL IN BLOOD AND BREATH FOR LEGAL PURPOSES

Breath and blood are acceptable body fluids for alcohol determination. Whenever the patient can deliver a suitable breath sample, breath alcohol analysis is the preferred procedure for clinical purposes. It is inherently simple, rapid, and noninvasive, and it reflects the alcohol content of the arterial circulation, which is physiologically and

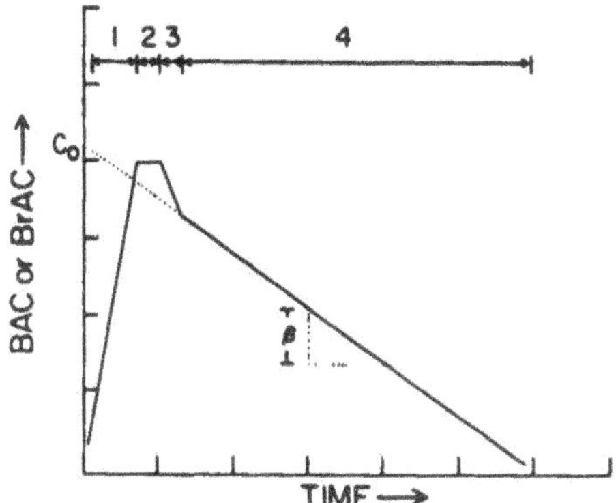

Figure 2 Theoretical blood alcohol (BAC) or breath alcohol curve (BrAC), after Widmark (5). 1 = absorption phase, 2 = plateau, 3 = diffusion-equilibration, 4 = elimination phase, Co = intercept of the ordinate which represents the theoretical instantaneously absorbed and distributed dose, β = slope representing the rate of ethanol metabolism.

clinically more significant than the venous blood alcohol concentration (BAC), and it mirrors the alcohol concentration reaching the brain.

The desired breath specimen consists substantially of expired alveolar air in which the alcohol concentration has reached a typical plateau or physiological equilibration (i.e., usually 1 to 1.5 h after ingestion). In practice, such a specimen can most easily be obtained as an end-expiratory breath sample, by collecting the terminal portion of a full, steady, and uninterrupted expiration after a normal inspiration.

Blood is an acceptable specimen when breath is not available. Plasma or serum are physiologically more appropriate specimens than whole blood, since alcohol is stored in the body water and its diffusion from and into the vascular system occurs through the plasma. However, statutory interpretation of the results of analyses of blood for law enforcement and some other forensic applications are universally based on the alcohol content in whole blood.

The collection of plasma and whole blood requires the use of anticoagulants. Disodium EDTA, sodium or potassium citrate or oxalate, and heparin are suitable for samples to be analyzed immediately. For specimens to be retained, or whose analysis is expected to be delayed for more than an hour, potassium oxalate (5 mg/mL blood) and sodium fluoride (1.5 mg/mL blood) constitute an appropriate anticoagulant preservative.

Currently available instrumentation for breath alcohol analysis includes dedicated devices employing gas chromatography, infrared absorptiomety, fuel-cell catalysis, solid-state gas sensing, and chemical oxidation photometry (Table 1).

The two prevailing methods for the analysis of alcohol in blood are gas chromatography and enzymatic oxidation. Both can be effectively employed with fully or partially automated instruments. Recently, specialized devices and instruments have been developed for rapid alcohol determination on small volumes of blood or saliva (e.g., Q.E.D. Saliva Alcohol Test, Enzymatics, Inc., Horsham, Pennsylvania) using enzymatic oxidation with immobilized alcohol dehydrogenase (ADH) or alcohol oxidase in the form of an alcohol electrode.

Table 1 Development of Instruments and Devices for Breath Alcohol Analysis

Instrument	Year	Principle for alcohol detection
Breathanalyzer (Photometric end point)	1954	Oxidation with potassium dichromate
		FID gas chromatography
GC—Intoximeter	1970	GC gas chromatography
Alco-Analyzer	1971	IR absorption spectrophotometry
Intoxilyzer	1971	Electrochemical oxidation (fuel cell)
Alcolmeter	1974	Metal-oxide semiconductor
ALERT	1974	

Nevertheless, headspace gas chromatography (GC) remains the method of choice for quantitative and qualitative analysis of alcohol and related low-molecular volatiles in blood. The application of capillary columns appears feasible for blood alcohol analysis, but they seem to offer no distinct advantage over traditional packed columns. The widely used graphitized carbon-black packing materials, such as Carbopak B or C, have high efficiency in terms of number of theoretical plates and give excellent separation of the ethanol peak on the chromatogram from closely related low-molecular-weight species such as methanol, acetaldehyde, acetone, and isopropanol within about 3 to 4 min. The overall time of analysis is important to consider, because blood alcohol analysis occupies a large part of the workload at forensic laboratories.

Over the past decade, the most significant developments in GC methods of blood alcohol analysis are related to computer-aided analytical systems with options for data processing of results and on-line statistical analysis and quality control. Microprocessor-based control systems are gradually becoming standard features. A wide selection of computer programs are available for GC analysis, and these allow for controlling the entire analytical run including sample identification, updating of the chromatographic parameters, integration of peaks, as well as matching of replicate analytical results.

Quality assurance and proficiency tests of blood alcohol methodology have become important aspects of good laboratory practice. An example of interlaboratory collaborative studies in the Nordic countries is given in Table 2. Although both between-run and within-run components of variance are included in the standard deviations, the coefficients of variation are mostly below 2%, implying high precision.

Table 2 Interlaboratory Proficiency Survey Conducted During 1986 of Blood Alcohol Analysis

	Ethanol concentration in blood, mean ± SD (mg/g)		
Country	Spec 1	Spec 2	Spec 3
Denmark	2.99 ± 0.015	2.23 ± 0.008	1.38 ± 0.015
Finland	3.01 ± 0.079	2.24 ± 0.048	1.37 ± 0.034
Norway	3.08 ± 0.067	2.32 ± 0.072	1.41 ± 0.024
Sweden	3.00 ± 0.025	2.24 ± 0.014	1.39 ± 0.014

Headspace GC was used for the analysis, and the reported results represent the mean ± SD of three to four determinations made on each of two consecutive days.

REFERENCES

1. R. D. Batt. Absorption, distribution, and elimination of alcohol. In *Human Metabolism of Alcohol*, Vol. 1 (K. E. Crow and R. D. Batt, eds.), CRC Press, Boca Raton, FL, pp. 6–8 (1989).
2. M. F. Mason and K. M. Dubowski. Breath-alcohol analysis: Uses, methods and some forensic problems—Review and opinion. *J. Forensic Sci.*, *21*:9–41 (1976).
3. K. M. Dubowski and B. O'Neill. The blood/breath ratio of ethanol. *Clin. Chem.*, *25*:1144 (1979).
4. E. Martin, W. Moll, P. Schmid, and L. Deltli. The pharmacokinetics of alcohol in human breath, venous and arterial blood after oral ingestion. *Eur. J. Clin. Pharmacol.*, *26*:619–626 (1984).
5. E. M. P. Widmark. *Die theoretischen Grundlagen und die praktische Verwendbarkeit der gerichtlich-medizinishen Alkohofestimmung*. Urban & Schwarzenberg, Berlin, 1932.
6. J. G. Wagner, P. K. Wilkinson, A. J. Sedman, D. R. Kay, and D. J. Weidler. Elimination of alcohol from human blood. *J. Pharm. Sci.*, *65*:152–154 (1976).
7. A. W. Jones. Interindividual variations in the disposition and metabolism of ethanol in health men. *Alcohol*, *1*:385–390 (1984).
8. M. Kopun and P. Propping. The kinetics of ethanol absorption and elimination in twins and supplementary repetitive experiments in singleton subjects. *Eur. J. Clin. Pharmacol.*, *11*: 337–341 (1977).
9. C. L. Winek and K. C. Murphy. The rate and kinetic order of ethanol elimination. *Forensic Sci. Inter.*, *25*:159–162 (1984).

74

Development of Sensitive HPLC Assays for the Analysis of Steroids in In-Vitro and In-Vivo Studies

R. S. Gardner, M. Walker, and P. A. D. Edwardson
Bristol-Myers Squibb, Clwyd, Wales

I. INTRODUCTION

Analysis of steroids within pharmaceutical products is mainly carried out using chromatographic techniques, the chief of which is high-performance liquid chromatography (HPLC). Several reviews concerning HPLC analysis of steroids used in the pharmaceutical industry have appeared over the past decade (1–3).

The majority of assays in the pharmaceutical industry concerned with steroid formulations—i.e., stability studies, content uniformity, etc.—usually involve the measurement of microgram-per-milliliter concentrations of steroid. Assaying at these concentrations requires relatively low sensitivity on the part of the detection unit, normally an ultraviolet detector. With the present choice of HPLC systems, columns, and detectors there are very few problems involving steroids in pharmaceutical preparations that cannot be solved using HPLC.

A moderate UV absorbance is a characteristic of many naturally occurring and synthetic steroids. The synthetic corticosteroids, with which this report is concerned, are no exception to this feature. They owe their UV absorbance, in the main, to the presence of a 1,4-diene-3-one system in the A ring (3), which is responsible for an absorption maxima at approximately 240 nm.

The two synthetic corticosteroids examined in the present investigations are triamcinolone acetonide (TACA) and halcinonide (HALC), whose structures are given in Figure 1. TACA was among the first of the corticosteroids to be assayed by HPLC (4–8) and is one of the steroids whose HPLC assay is now included in the U.S. Pharmacopoeia (9). HALC has also been successfully assayed in pharmaceutical preparations by HPLC (8,10). However, far less research has been devoted to their determination in biological materials.

The in-vitro and in-vivo investigations reported here are concerned with the percutaneous absorption of TACA and HALC in order to determine the influence of various factors on the permeation rates of the two corticosteroids. The percutaneous

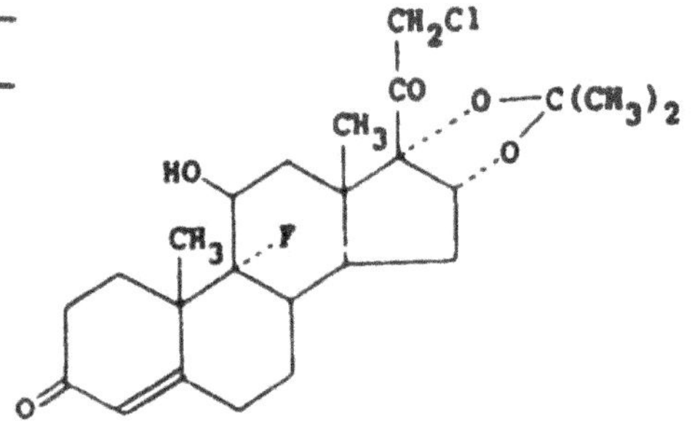

Figure 1 Structures of triamcinolone acetonide (I) and halcinonide (II).

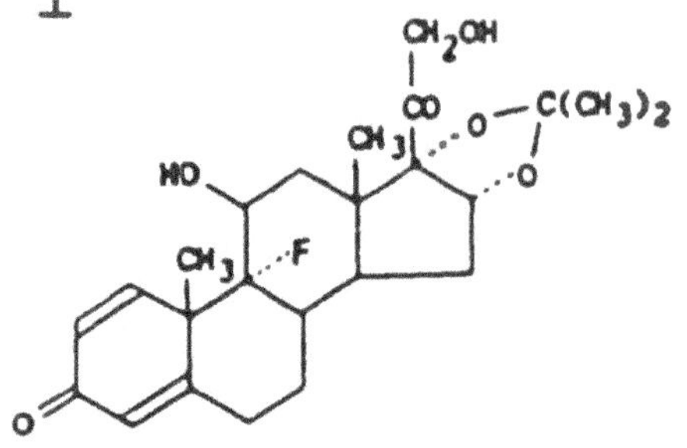

absorption of TACA has been examined using radiolabeled TACA (11,12). However, due to the lack of suitable methodology to determine the permeation rates and skin content of unlabeled TACA and HALC, it was decided to develop appropriate HPLC methodology.

To determine the above values, two techniques have been exploited: first, measurement of the steroid released into the receptor phase of a horizontal glass diffusion cell (13), which enables the in-vitro penetration of corticosteroids through the epidermis to be determined; and second, measurement of the corticosteroid content of the stratum corneum using tape stripping (14). Results from this technique will yield information on the in-vivo distribution of corticosteroid within the layers of the stratum corneum as well as the total corticosteroid content. Since the above techniques are concerned with the movement, and therefore measurement, of nanogram quantities of corticosteroid through and within the epidermis and stratum corneum, respectively, the HPLC assays must be relatively sensitive and highly specific (15).

II. INSTRUMENTATION

The HPLC system consisted of a Kontron model 420 pump, a Kontron model 460 autosampler, and a Hewlett Packard model 3396A integrator. Detection was carried out using a Kratos model Spectroflow 773 UV detector.

III. CHROMATOGRAPHIC CONDITIONS

All separations were carried out using narrow-bore, 25 cm × 2 mm columns, containing 5-μm Spherisorb ODS1 packing material. The mobile phase used for permeation cell samples consisted of methanol:water (3:1, v/v) at a flow rate of 0.2 mL/min, while the mobile phase for tape-strip samples was acetonitrile:water (1:1, v/v) at a flow rate of 0.16 to 0.30 mL/min. Injection volumes were 10 μL for permeation cell samples and between 10 and 60 μL for tape-strip samples. In all experiments, detection was carried out at 240 nm in order to optimize the assay (16,17).

IV. IN-VITRO STUDIES

All in-vitro permeation studies were carried out using the epidermal membrane from Caucasian human skin obtained following surgical amputation in conjunction with horizontal glass diffusion cells (Figure 2). The epidermis was isolated from whole skin (i.e., epidermis plus dermis) by heat separation at 60°C for 3 s. Once separated, then individual pieces of epidermal membrane could be punched out (27-mm diameter, 5.7 cm^2).

Approximately 4 to 5 mg of the cream under investigation could then be applied to the upper surface of the epidermal membrane within an area of 2.55 cm^2. Following application, the circle of epidermis was placed centrally onto the lower half of the diffusion cell, cream applied surface uppermost, and the two halves of the diffusion cell then clamped together. The receptor chamber was filled with a selected receptor medium, i.e., deionized water, in all of the present investigations.

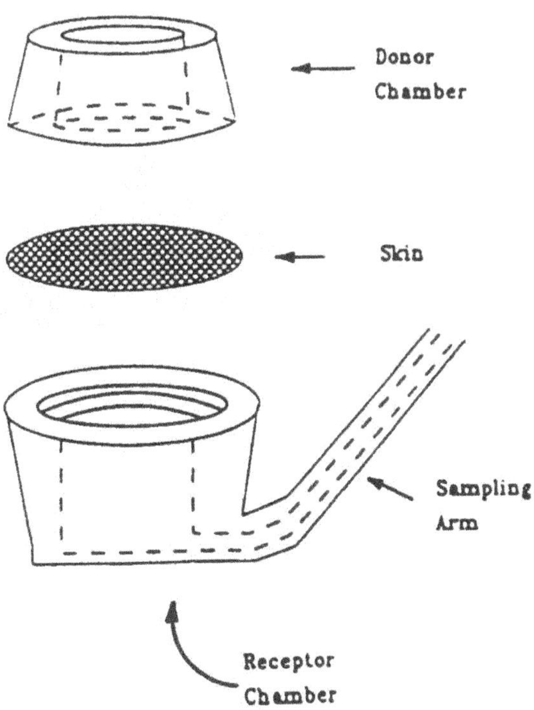

Figure 2 Horizontal glass diffusion cell.

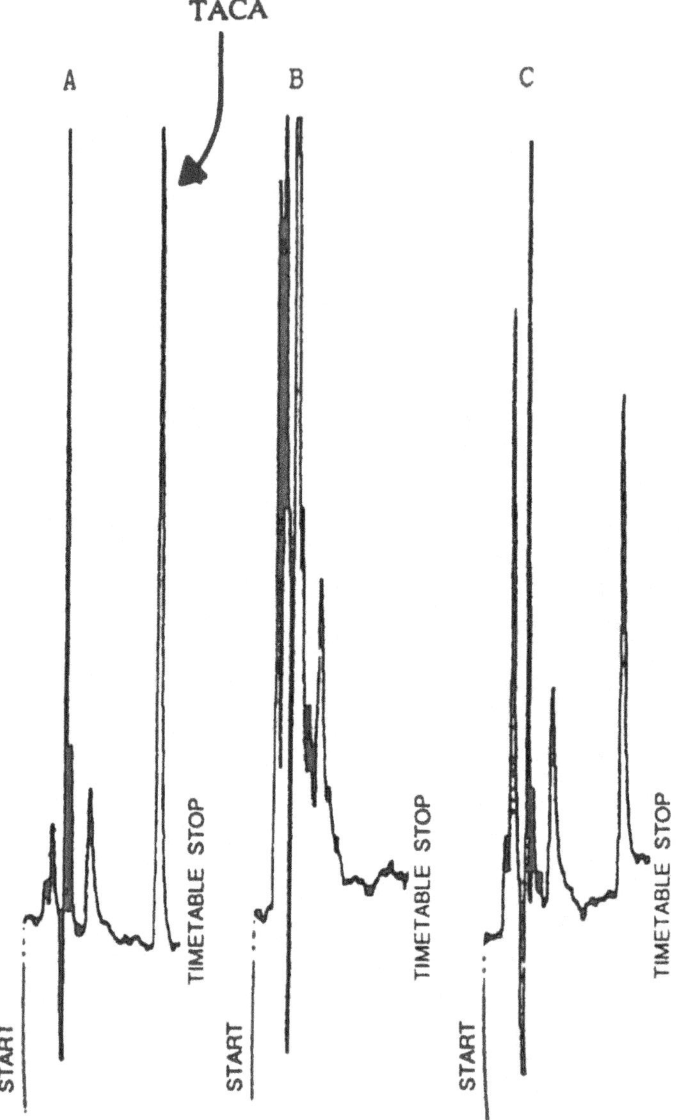

Figure 3 HPLC chromatograms for TACA content of permeation cell samples (in-vitro studies) from (A) TACA standard, 100 ng/mL; (B) control permeation cell with placebo cream; (C) TACA sample containing 52 ng/mL.

Into each receptor chamber was also added a micromagnetic follower as an aid to the maintenance of a homogenous solution. The diffusion cell was placed into a water bath at 32°C; each cell was positioned on a multiplace magnetic stirring plate. Typically, 15 diffusion cells could be placed in one water bath. At predetermined times, 1 mL of the receptor medium was removed and assayed directly by HPLC. The volume removed was replaced by 1 mL of deionized water. Diffusion cell experiments could be carried out for periods of up to 120 h.

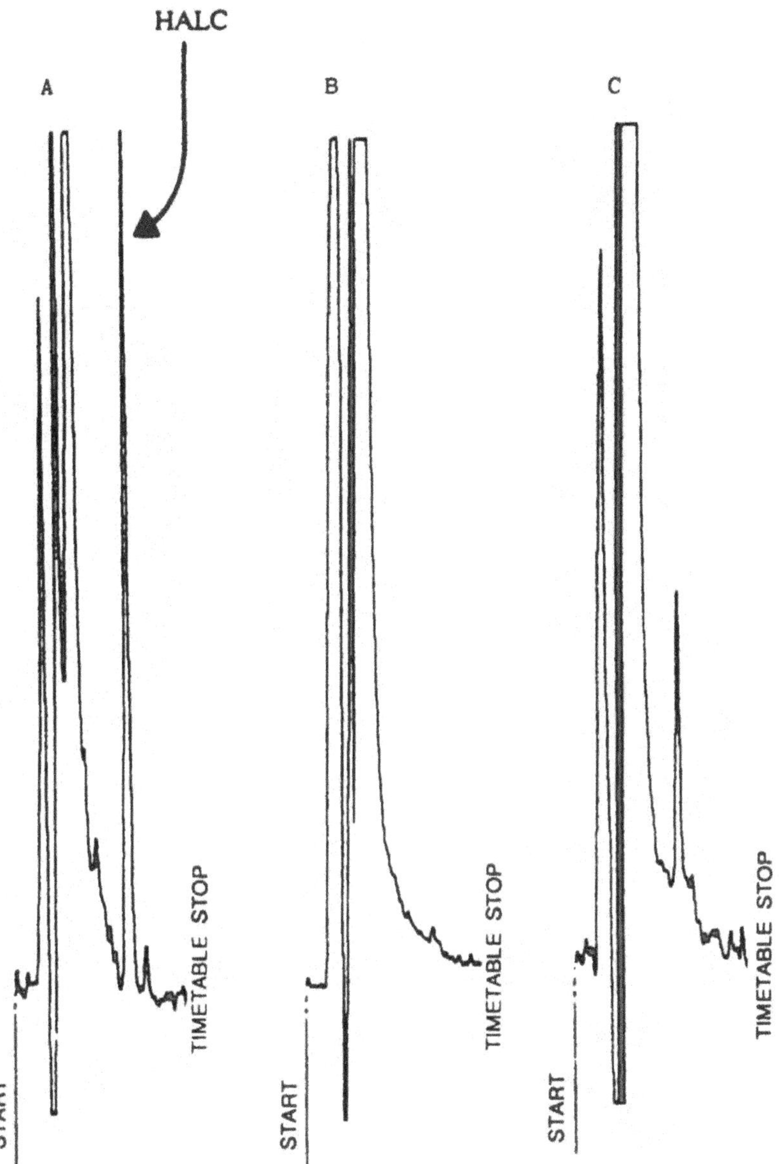

Figure 4 HPLC chromatograms for HALC content of permeation cell samples (in-vitro studies) from (A) HALC standard, 100 ng/mL; (B) control permeation cell with placebo cream; (C) HALC sample containing 30 ng/mL.

V. IN-VIVO STUDIES

All in-vivo permeation studies were conducted on healthy male volunteers. The present study was concerned with the pharmacokinetics of TACA absorption into the skin. Two aspects of this study pertinent to this report are the determination of the total TACA absorbed into the stratum corneum and the distribution of TACA within the layers of the stratum corneum.

All samples were applied to the hair-free volar region of the forearm; the formulation used in this study was Kenalog cream, containing 0.1% TACA (w/w). Approximately 5 mg of cream was placed onto the central area (1.5 cm × 1.5 cm) of an Actiderm dermatological patch (3.5 cm × 3.5 cm). A polythene template was used to define the central area, the cream being spread as evenly as possible using a microspatula. The prepared sample was then positioned onto the forearm and secured to the skin by micropore tape. Up to 10 samples could be applied to each forearm. Samples were removed at predetermined times to assay TACA remaining associated with the Actiderm dermatological patch. Immediately after removal of Actiderm, the sample site area was dry-swabbed using cotton buds. Following this, the sample site was then tape-stripped 10 times; each tape strip being placed into an individual 20-mL glass scintillation vial, which were stored at 4°C until assayed.

Sample preparation of the tape strips was minimal. To each scintillation vial was added 0.1 mL of acetonitrile, which was adequate to cover the tape strip sample. Following mixing of the vial contents on a vortex mixer for 10 s, the samples were left at ambient for 30 min. Afterwards, 0.5 mL of the acetonitrile extract was then transferred to a HPLC autosampler vial containing 0.5 mL of water. The contents were mixed and assayed directly by HPLC.

VI. RESULTS AND DISCUSSION

A. In-Vitro Studies

Both corticosteroids gave a linear response in receptor cell medium. The linear response for TACA was observed over the range 10 to 600 ng/mL (regression coefficient = 0.99990; slope = 49.822; intercept = 8.221), while that for HALC was over the range 10 to 400 ng/mL (regression coefficient = 0.99832; slope = 39.773; intercept = −37.955). In neither case were there found to be any interferences from either the components of the cream formulations or epidermal membrane disks. Typical chromatograms from TACA experiments (Figure 3) and HALC experiments (Figure 4) are presented.

With the ability to quantify very low concentrations of TACA and HALC, further investigations into the various factors affecting corticosteroid permeation rates could be considered. Two preliminary investigations are outlined below.

First, the effect of occlusion on the permeation rate of TACA was explored. Experiments were carried out as already described with this addition: In some samples the applied cream was covered with a disk of Actiderm dermatological patch or Saran Wrap. It was possible to observe an increase in permeation rate in both types of occluded sample, i.e., approximately 1.0 ng/cm^2/h and 1.9 ng/cm^2/h, respectively, compared with an approximate value of 0.2 ng/cm^2/h for the nonoccluded samples (Figure 5). The results are, for Saran Wrap, in good agreement with previous in-vivo studies, where a 10-fold increase in permeation rate of TACA was achieved using occlusion (18). The increased permeation rate observed with Actiderm could indicate

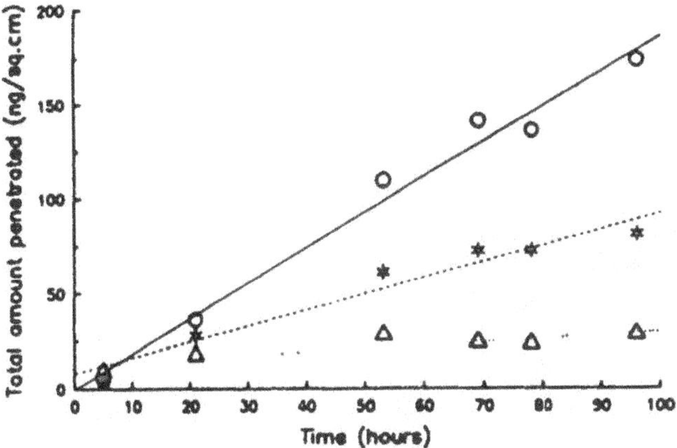

Figure 5 Human epidermal layer permeation studies for TACA from (Δ) 0.1% Kenalog cream; (✱) 0.1% Kenalog cream occluded under Actiderm; (O) 0.1% Kenalog cream occluded under Saran Wrap.

its use in drug delivery under occlusion. Although not to the same extent as Saran Wrap, it does have many characteristics that would be preferable to plastic films (19).

Second, the effect of cream foundation on the permeation rate of HALC was explored. Two different creams, Halog E and Halog USP, both containing 0.1% (w/w) HALC, were compared. They gave permeation rates of approximately 3.7 ng/cm^2/h and 5.7 ng/cm^2/h.

B. In-Vivo Studies

TACA gave a linear response over the range 5 to 100 ng/mL (regression coefficient = 0.99693; slope = 41.16; intercept = −87.20). There were found to be no interferences from either the cellotape or the stratum corneum when acetonitrile was used as the extraction solvent. Recovery experiments determined that 98.1 ± 3.0% of 50-ng TACA spiked into tape strips was extracted. Typical chromatogram depicting tape strip samples are presented (Figure 6).

The TACA content of individual layers of the stratum corneum revealed that a greater portion of the steroid was present in the upper layers (i.e., the first few tape strip samples) than the lower layers (Figure 7). This was especially true at 24 h and noticeably less as time progressed; i.e., there is apparently an equilibration of the corticosteroid within the layers of the stratum corneum. The absolute amount of TACA within the stratum corneum as a whole remained virtually unchanged throughout the study period, i.e., approximately 350 to 500 ng, which is equivalent to approximately 10% of the topically applied steroid. This is in broad agreement with previous in-vivo investigations utilizing radiolabeled TACA (11,12).

VII. CONCLUSIONS

HPLC assays were successfully developed for the determination of very low (ng/mL) concentrations of corticosteroids in skin permeation (in-vitro) and tape-strip (in-vivo) investigations. The methods showed good linearity, specificity, accuracy, and

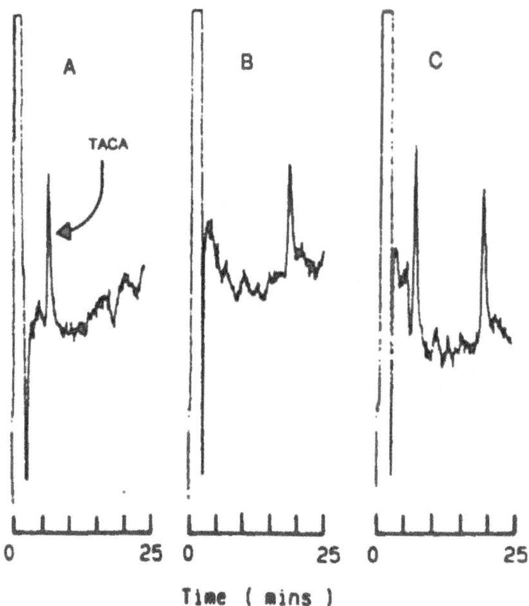

Figure 6 HPLC chromatograms for TACA content of tape strip samples (in-vivo studies) from (A) TACA standard, 25 ng/mL; (B) extracted control tape strip sample; (C) TACA sample containing 31 ng/mL.

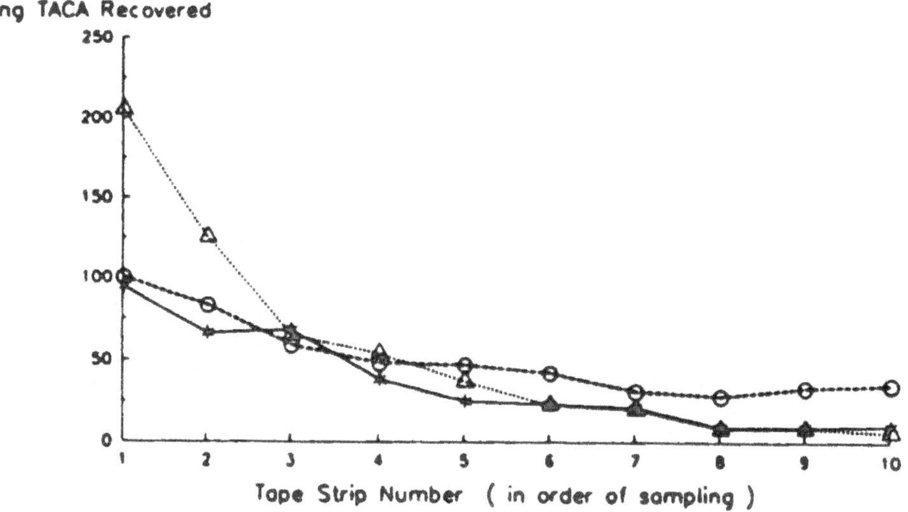

Figure 7 TACA content in individual tape strip samples at (Δ) 24 h; (*) 48 h; (O) 72 h.

precision. Sample preparation was minimal, thereby reducing cumulative errors that might otherwise occur with multistep sample preparation, as well as saving time.

Analysis of very low concentrations of corticosteroids was made possible by the use of a sensitive UV detector in connection with narrow-bore (2-mm) analytical columns. The relative ease of the methods has paved the way for a wide range of studies to be carried out on factors influencing topical drug delivery, which should be transferable to other classes of drugs. It should also be possible for other laboratories to utilize the above methodologies in situ.

REFERENCES

1. F. A. Fitzpatrick. High performance liqüid chromatography of the steroid hormones. *Adv. Chromatogr.*, *16*:16 (1978).
2. S. Görög. Steroid analysis in the pharmaceutical industry. *Trends Anal. Chem.*, 3:157 (1984).
3. S. Görög. In *Steroid Analysis in the Pharmaceutical Industry* (S. Görög, ed.), Ellis Horwood, Chichester (1989).
4. E. Gaetani and C. F. Laureri. Cromatografia liquido-solido ad alta pressione. *Il Farmaco.*, 29:110 (1974).
5. J. W. Higgins. High-pressure liquid chromatographic analysis for triamcinolone acetonide in rice starch. *J. Chromatogr.*, *115*:232 (1975).
6. G. Gordon and P. R. Wood. Determination of triamcinolone acetonide in cream and suspension formulations by high-performance liquid chromatography. *Analyst*, *101*:876 (1976).
7. N. W. Tymes. The determination of corticoids and related steroid analogs by high-performance liquid chromatography. *J. Chromatogr. Sci.*, *15*:151 (1977).
8. G. Gordon and P. R. Wood. Use of reversed-phase high-performance liquid chromatography in the analysis of products containing halcinonide and triamcinolone acetonide. *Anal. Div. Chem. Soc.*, *14*:30 (1977).
9. *The United States Pharmacopeia, Twenty-First Revision*, United States Pharmacopeia Convention, Rockville, MD, p. 1077 (1985).
10. J. Kirschbaum, R. Poet, K. Bush, and G. Petrie. High-performance liquid chromatography of the topical anti-inflammatory steroid halcinonide. *J. Chromatogr.*, *190*:481 (1980).
11. F. D. Malkinson and M. B. Kirschenbaum. Percutaneous absorption of C^{14}-labelled triamcinolone acetonide. *Arch. Dermatol.*, *88*:427 (1963).
12. H. Schaefer, A. Zesch, and G. Stüttgen. Penetration, permeation and absorption of triamcinolone acetonide in normal and psoriatic skin. *Arch. Dermatol. Res.*, *258*:241 (1977).
13. R. L. Bronaugh, R. F. Stewart, E. R. Congdon, and A. L. Giles. Methods for in-vitro percutaneous absorption studies. 1. Comparison with in-vivo results. *Toxicol. Appl. Pharmacol.*, *62*:474 (1982).
14. A. Rougier, D. Dupuis, C. Lotte, R. Roguet, R. C. Wester, and H. I. Maibach. Regional variation in percutaneous absorption in man: Measurement by the stripping method. *Arch. Dermatol. Res.*, *278*:465 (1986).
15. R. S. Gardner, M. Walker, and D. A. Hollingsbee. A sensitive high performance liquid chromatographic method for the assessment of percutaneous absorption of topical corticosteroids. *J. Pharm. Biomed. Anal.*, in press.
16. K. Florey. Triamcinolone acetonide. In *Analytical Profiles of Drug Substances*, Vol. 1 (K. Florey, ed.), Academic Press, New York and London, p. 397 (1972).
17. J. Kirschbaum. Halcinonide. In *Analytical Profiles of Drug Substances*, Vol. 8 (K. Florey, ed.), Academic Press, New York and London, p. 251 (1979).
18. H. I. Maibach. In-vivo percutaneous penetration of corticoids in man and unresolved problems in their efficacy. *Dermatologica*, *152*(Suppl. 1):*11* (1976).
19. J. E. Fairbrother. Beyond occlusion and back again. In *Beyond Occlusion Dermatology Proceedings* (T. J. Ryan, ed.), Royal Society of Medicine Services, London and New York, 1988.

75

Importance of Creatinine for Urine Quality Control in Determination of Abused Drugs

P. Lafolie,* O. Beck,* G. Blennow,† L. Boréus,* S. Borg,† C-E. Elwin,‡ L. Karlsson,§ G. Odelius,* and Paul Hjemdahl*
**Karolinska Institute, Karolinska Hospital, Stockholm, Sweden, †Karolinska Institute, St. Görans Hospital, Stockholm, Sweden, ‡Karolinska Institute, Danderyds Hospital, Danderyd, Sweden, §University Hospital, Linköping, Sweden*

We suggest that the authenticity of incoming urine samples for drug abuse should be checked by measuring the creatinine concentration in the urine prior to the drug assay. This may be done routinely with the same equipment as that used for the drug screening. Additional information is obtained by determination of urine osmolality. We compared the relationship between creatinine and osmolality in urine and found that this was lower in samples with creatinine < 4.3 mmol/L. In healthy volunteers, intake of 0.5 L of water significantly reduced the concentration of creatinine several hours after intake. In 10 randomly selected urines, all "clean" in initial immunological drug testing, showing creatinine < 4.0 mmol/L and osmolality < 200 mosm/kg, five turned out drug-positive after concentration (Figure 1).

In a formerly heavy cannabis smoker, excretion of cannabinoids and creatinine were followed for 93 days. The substances showed very good correlation (Figure 2).

We conclude that urines used in drug abuse testing should be tested for creatinine. If this is < 4.0 mmol/L, osmolality should also be tested. If the osmolality is < 200 mosm/kg, negative results are not valid, and another sample should be requested. Creatinine measurements used together with quantitative testing for cannabinoids may also detect falsely cannabinoid-positive results.

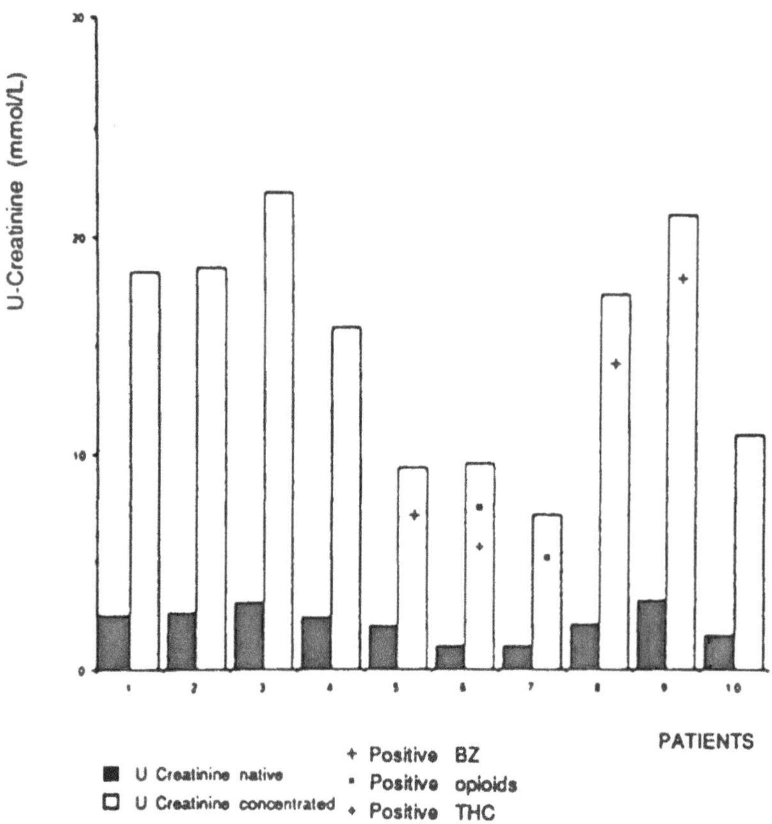

Figure 1

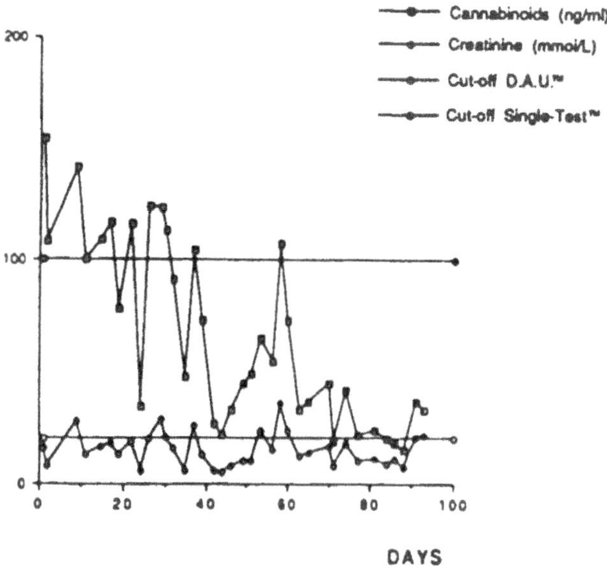

Figure 2

76

Delta-9-Tetrahydrocannabinol: Pharmacokinetics and Pharmacodynamics

Leslie M. Shaw and Judy Edling-Owens *University of Pennsylvania Medical Center, Philadelphia, Pennsylvania*

I. INTRODUCTION

Marijuana is the most commonly used illicit substance in the United States and in many other countries in the world (1). According to the most recent National Household Survey on Drug Abuse, conducted by the National Institute on Drug Abuse in 1988 (2), approximately 66 million Americans have tried marijuana at least once in their lifetime. In 1988, 20 million Americans used either marijuana or hashish on at least one occasion (2).

Marijuana consists of a mixture of more than 60 cannabinoids and several hundred other components (3) and is typically prepared from the flowering tops and adjacent leaves of the female plant of *Cannabis sativa*. Hashish is essentially the resin secreted by the female plant and contains much higher cannabinoid concentrations than marijuana (1). Delta-9-tetrahydrocannabinol (delta-9-THC) is the most prevalent psychoactive compound in marijuana and hashish. The delta-9-THC content of marijuana varies with the source and method of cultivation. Sinsemilla is a marijuana product which results from cultivation of *Cannabis sativa* by removing all male plants from the growth area, with consequent stimulation of bud formation and overall delta-9-THC content (4)—buds are the source of the resin which is rich in delta-9-THC. The potency of domestic "regular" marijuana and sinsemilla in samples seized by Drug Enforcement Administration (DEA) officials is reported annually (4). The average delta-9-THC content in marijuana is currently 3% to 4% by weight, three to four times the average value of 1.2% for "street" marijuana in 1975 (4,5). Since the early 1980s, the average delta-9-THC content of sinsemilla has almost doubled, from 4% to the current average of 6% to 7% (4,5). The delta-9-THC in marijuana is primarily in the form of a delta-9-THC acid derivative, which is converted rapidly and completely to delta-9-THC itself by the heat of the burning cigarette (3). The majority of the other cannabinoids in marijuana are not psychoactive (1).

II. PHARMACOLOGIC EFFECTS OF DELTA-9-THC IN HUMANS

Investigations of the acute pharmacological effects of delta-9-THC in humans, using synthetic delta-9-THC administered orally or by i.v. injection or smoking marijuana cigarettes with known delta-9-THC content, have revealed a number of central nervous system and cardiovascular effects. The major CNS effects are feeling relaxed, sleepiness, impairment of short-term memory, and impairment of the capacity to carry out tasks requiring multiple mental steps (6). The cardiovascular effects are an increased heart rate and a marked reddening of the conjunctivae (6).

Chronic studies have shown several effects that are associated with smoking marijuana cigarettes: apathy; impairments in memory, judgment, and concentration; loss of interest in personal appearance and pursuit of conventional goals; use of more potent illicit drugs—the "gateway" effect (1). A number of other health effects have been attributed to delta-9-THC (7). These will not be discussed here, but the interested reader may wish to consult Ref. 7 for a critical discussion of these. Thus, although commonly used and usually considered the least harmful of the commonly available illicit substances, marijuana has the potential for producing devastating effects on the quality of human life.

A. Effects of Delta-9-THC on Psychomotor Function

Numerous studies of the effects of acute doses of marijuana or delta-9-THC on various psychomotor functions have been performed. In essence, these studies show, providing the dose of drug is high enough and the complexity of the psychomotor tasks are complex enough, that delta-9-THC or marijuana can impair psychomotor function (7). It is uncertain how or whether direct extrapolation from such studies to predicting the effects of delta-9-THC on driving a motor vehicle can be made. Several studies in which the effects of smoking a marijuana cigarette on simulated automobile driving have shown mixed results (7). But studies involving actual driving over a course set up to mimic various traffic problems showed that marijuana smoking (19 mg of delta-9-THC/cigarette) in 80 male marijuana users produced significant impairment of driving performance (8). When both marijuana and alcohol were ingested, the effects were particularly detrimental (8). The effects of marijuana were produced more rapidly than those of alcohol but were somewhat less severe for most of the tasks performed (8).

That marijuana smoking could be a causative factor in automobile accidents is suggested by the significant association of cannabinoid detection in motorists detained for erratic driving behavior in accident victims. In one study, 25% of ethanol-negative blood obtained from erratic motorists tested positive for delta-9-THC (9). In studies of tractor-trailer drivers (10) and in trauma victims, a significant number of whom were involved in a vehicular accident, the most frequently detected substances in urine specimens were alcohol, marijuana, or cocaine, alone or in various combinations (11–13). Fifty-nine subjects smoked marijuana cigarettes until they became "high" and then were tested by highway patrol officers on the roadside sobriety test. Ninety-four percent of these individuals failed the test 90 min after smoking, and 60% failed 2.5 h after the smoking session (14). There is an emerging consensus that marijuana can impair the ability to drive an automobile and, analogous to the situation for alcohol, "smoking marijuana and driving do not mix." According to Hollister (7), the biggest areas of debate are how long the impairment, even though subtle, may last, and how to deal forensically with driving while under the influence of marijuana.

B. Plasma Concentration of Delta-9-THC and Impairment

We are interested in evaluating the effects of delta-9-THC on psychomotor functions necessary for driving a motor vehicle and the relationship between the degree of impairment of these functions and the blood plasma concentrations of delta-9-THC. Very few studies have compared the degree of psychomotor effects of delta-9-THC with plasma concentrations of the parent compound. Several studies have investigated the relationship between degree of impairment and heart rate, or blood carbon monoxide content (7,15). Only a limited number of studies of the relationship between plasma delta-9-THC concentration and degree of psychomotor impairment have been published (14,16). Another way to put this into perspective is to compare the current thinking about impairment by illicit substances such as marijuana with the development of information and policy about ethyl alcohol. In a recent conference on drug concentrations and driving impairment, a consensus panel concluded that: "The development of methods for testing the effects of ethanol on driving skills and relating these to body fluid concentrations took many years. It is obvious that a beginning has just been made to develop similar information about other drugs of similar interest (7)."

III. MEASUREMENT OF DELTA-9-THC

Based on observations that the amount of delta-9-THC absorbed after smoking marijuana cigarettes containing known and consistent amounts of the psychoactive cannabinoid can vary as much as 25-fold (18) it is essential that studies of the effects of delta-9-THC on psychomotor function include careful quantification of the parent drug and primary metabolites in timed biofluid samples. Furthermore, since the concentrations of delta-9-THC are a few nanograms per milliliter and lower in the postabsorption phase of the plasma concentration-versus-time curves, sufficiently sensitive methodology is required. The most commonly used methods for quantification of delta-9-THC and its major metabolites (11-nor-delta-9-THC-9-carboxylic acid and 11-hydroxy-delta-9-THC) have been radioimmunoassays and isotope-dilution gas chromatography/mass spectrometry methods (19–21). To date, radioimmunoassay methods have been used in the majority of studies relating plasma concentrations of delta-9-THC to pharmacological effects. However, these methods suffer from two deficiencies, namely, a lack of specificity for the pharmacologically active parent drug, and limited sensitivity of about 2.5 ng/mL (22). Nevertheless, useful information regarding the time of appearance and disappearance in plasma of the parent compound together with cross-reacting metabolites after marijuana consumption have been revealed by these studies. Furthermore, since these assays cross-react with cannabinoid metabolites, they are useful for analysis of urine specimens in which metabolites but little or no delta-9-THC are excreted (23). Negative-ion chemical ionization isotope-dilution gas chromatography/mass spectrometry is the current method of choice for studies seeking to establish as rigorously as possible the relationship between the concentrations of delta-9-THC and metabolites in plasma, and possibly other biofluids, and impairment of psychomotor functions (20,21). Using the trifluoroacetic anhydride derivatization procedure (20) and negative-ion chemical

Table 1 Sensitivity and Detection Limits for Two Detectors and Negative Chemical Ionization GC/MS[a]

Compound	Ionization mode	Detector	Sensitivity (area cts/pg)	Detection limit (pg)	Plasma conc. (pg/mL)
THC	Neg CI	Standard	4.0	25	500
THC	Neg CI	HED (8 kV)	62.7	5	100

[a]THC is measured as the trifluoroacetylated derivative.

ionization GC/MS (Hewlett Packard 5985B) with a High Energy Dynode detector (Phrasor Scientific, Inc.), we have attained a detection limit of 0.1 ng/mL of plasma (Table 1 and Ref. 21). With further improvements in the surface work function of the High Energy Dynode (23), it should be possible to reach detection limits of <50 pg/mL of plasma. This should make possible the full evaluation of the postabsorption pharmacokinetics of delta-9-THC and primary metabolites in relation to impairment.

REFERENCES

1. M. S. Gold. *Drugs of Abuse. A Comprehensive Series for Clinicians, Vol. 1, Marijuana.* Plenum Medical Book Co., New York (1989).
2. National Institute on Drub Abuse. National Household Survey on Drug Abuse: Main Findings 1988. DHHS Publication No. (ADM) 90-1682, Rockville, MD (1990).
3. National Institute on Drug Abuse. General information on marijuana cigarettes from NIDA and their use in therapeutic programs. NIDA data sheet (1989).
4. R. L. Hawks and R. L. Walsh. Potency of marijuana over the last ten years. Technical Review Brief, Research Technology Branch, Division of Preclinical Research, NIDA (1990).
5. R. H. Schwartz. Marijuana: An overview. *Ped. Clin. N. Am., 34*:305–317 (1987).
6. J. H. Jaffe. Drug addiction and drug abuse. In *The Pharmacological Basis of Therapeutics*, 8th ed. A. G. Gilman, (T. W. Rall, A. S. Nies, and P. Tsylor, eds.), Pergamon Press, New York, pp. 549–563 (1990).
7. L. E. Hollister. Health aspects of cannabis. *Pharm. Rev., 38*:1–20 (1986).
8. R. C. Peck, A. Biasotti, P. N. Boland, C. Mallory, and V. Reeve. The effects of marijuana and alcohol on actual driving performance. *Alcohol, Drugs and Driving,* 2:135–154 (1986).
9. E. G. Zimmerman, E. P. Yeager, J. R. Soares, et al. Measurement of delta-9-THC in whole blood from impaired motorists. *J. Forensic Sci., 28*:957–962 (1983).
10. A. K. Lund, D. F. Preusser, R. D. Blomberg, and A. F. Williams. Drug use by tractor-trailer drivers. *J. Forensic Sci., 33*:648–666 (1988).
11. C. A. Soderman, A. L. Trifillis, B. S. Shankar, et al. Marijuana and alcohol use among 1023 trauma patients. *Arch. Surg., 123*:733–737 (1988).
12. F. P. Rivera, B. A. Mueller, C. L. Fligner, et al. Drug use in trauma victims. *J. Trauma, 29*:462–470 (1989).
13. E. P. Sloan, R. J. Zalenski, N. I. Keys, et al. Toxicology screening in urban trauma patients with mental status alterations: Drug prevalence and relationship to trauma severity. *J. Trauma, 29*:1647–1653 (1989).
14. L. E. Hollister, H. K. Gillespie, A. Ohlsson, et al. Do plasma concentrations of delta-9-THC reflect the degree of intoxication? *J. Clin. Pharmacol., 21*:1715–1775 (1981).
15. S. J. Heishman, M. L. Stitzer, and G. E. Bigelow. Alcohol and marijuana: Comparative dose effect profiles in humans. *Pharmacol. Biochem. and Behavior, 31*:649–655 (1989).
16. R. C. Peck, A. Biasotti, P. N. Boland, C. Mallory, and V. Reeve. The effects of marijuana and alcohol on actual driving performance. *Alcohol, Drugs and Driving: Abstracts and Reviews,* 2:135–154 (1986).

17. R. V. Blanke, Y. H. Caplan, T. Chamberlain, et al. Drug concentrations and driving impairment. Report of the consensus development panel. *J. Am. Med. Assoc.*, 254:2618–2621 (1985).
18. S. Augurell, J-E Lindgrem, A. Ohlsson, et al. Recent studies on the pharmacokinetics of delta-1-THC in man. In *The Cannabinoids: Chemical, Pharmacologic, and Therapeutic Aspects* (S. Agurell, W. L. Dewey, and R. E. Willette, eds.), Academic Press, New York, pp. 165–183 (1984).
19. M. E. Wall, T. M. Harvey, J. T. Bursey, D. R. Brim, and B. Rosenthal. Analytical methods in the determination of cannabinol in biological materials. In *Cannabinoid Assays in Humans* (R. E. Willette, ed.), NIDA Research Monograph, Rockville, MD (1976).
20. R. L. Foltz, K. M. McGinnis, and D. M. Chinn. Quantitative measurement of delta-9-THC and two major metabolites in physiological specimens using capillary column gas chromatography negative ion chemical ionization mass spectrometry. *Biomed. Mass Spectrom.*, *10*:316–323 (1983).
21. L. M. Shaw and J. Edling-Owens. Ultrasensitive measurement of delta-9-THC with a high energy dynode detector and negative ion chemical ionization gas chromatography-mass spectrometry. In preparation.
22. C. E. Cook, H. H. Seltzman, V. H. Schindler, et al. Radioimmunoassays for cannabinoids. In *The Analysis of Cannabinoids in Biological Fluids* (R. L. Hawks, ed.), NIDA Research Monograph, Rockville, MD (1982).
23. J. Perel, J. F. Mahoney, and R. C. Speiser. Mechanisms for improved high mass ion detection (Abstract). *Proc. 38th ASMS Conf. Mass Spectrometry and Allied Topics*, June 3–8, 1990, pp. 510–511.

77

Evaluation of a Thin-Layer Chromatographic Method for the Detection of Cannabinoid (THC) Metabolite Using a Microcolumn/Disk Extraction Technique

David L. King, Perry Fukui,* Stephen K. Schultheis, and C. M. O'Donnell *TOXI-LAB, Inc., Irvine, California, and*
**PharmChem Laboratories, Inc., Menlo Park, California*

I. INTRODUCTION

A new thin-layer chromatographic (TLC) method has been evaluated for the detection of the major urinary marijuana metabolite, 11-nor-Δ^9-tetrahydrocannabinol-9-carboxylic acid (Δ^9-THC-COOH). Using this method, hydrolyzed urine samples are aspirated directly through a small, porous disk composed of bonded silica impregnated within a glass fiber matrix. The THC metabolite is retained by the bonded silica adsorbent through the process of solid-phase extraction. The disk is inserted into a hole in a silica gel/glass fiber TLC plate for subsequent TLC development and detection.

We present the results of three separate studies in which urine samples were analyzed by the TLC method and an immunoassay technique ($EMIT_{dau}$). Gas chromatography/mass spectrometry was employed as the reference method.

II. MATERIALS AND METHODS

A. Thin-Layer Chromatography

The TLC system (TOXI-LAB THC II Screen) was obtained from TOXI-LAB, Inc. (Irvine, California) and was performed according to the manufacturer's instructions (1). A 3-mL aliquot of urine specimen or control (15 ng/mL, Δ^9-THC-COOH) was hydrolyzed with 200 µL of potassium hydroxide (11.8 N) for 5 min at room temperature to liberate the Δ^9-THC-COOH from its glucuronide conjugate. The sample was then acidified with 700 µL of glacial acetic acid.

A SPEC-1 extraction cartridge, containing a filtration disk and a 5/16-in.-diameter bonded-silica extraction disk, was inserted into a vacuum manifold and a sample reservoir was attached to the cartridge (Figure 1). The hydrolyzed sample was aspirated through the cartridge followed by 1 mL of 20% acetic acid wash reagent.

The cartridge was opened and the filter disk was discarded. The SPEC attraction disk was dried by applying vacuum (20 in. Hg) for 10 min.

Sample and control disks from the processed samples were inserted into the holes of a thin-layer chromatogram (TOXI-GRAM THC II-PLUS). A 1/8-in.-diameter standard disk containing approximately 350 ng of Δ^8-TCH-COOH was also included in each chromatogram. Each chromatogram can accommodate nine sample disks, one control disk, and one standard disk.

A total of six chromatograms can be inserted into a developing rack for simultaneous migration. The TLC developing solvent was heptane/acetone/glacial acetic acid (50/50/1), and the development distance was 4.5 cm. Total development time was 2 min.

Chromatograms were dried on an electric warmer for 3 min after migration. Visualization of Δ^9-THC-COOH was achieved by dipping each chromatogram into a solution of Fast Blue BB salt (0.1% in dichloromethane) and then exposing the chromatogram to diethylamine vapor. Δ^9-THC-COOH in the specimens was identified by a narrow red band 3.5 cm from the origin (R 0.78), which became purple upon exposure to HCl vapor and matched the color and position of both the standard and control. For semiquantitative determinations, only those samples that produced red bands whose size and intensity was greater than or equal to the control were classified as positive.

Thin-layer chromatograms were preserved by placing transparent tape (Highland, 3M Co.) across both sides of the TLC media at the *Rf* of Δ^9-THC-COOH. When stored protected from light, chromatograms were stable for at least 1 year.

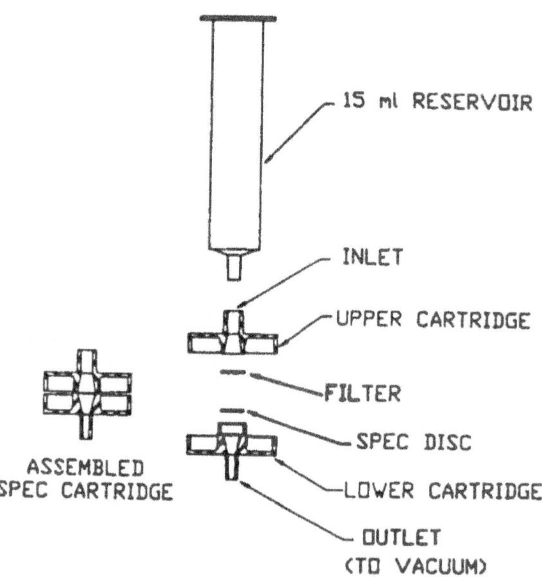

Figure 1 SPEC extraction cartridge.

B. Immunoassay

The EMIT d.a.u. cannabinoid 100-ng assay and the EMIT$_{dau}$ cannabinoid 50-ng assay were obtained from Syva Co. (Palo Alto, California). Both assays were performed according to the manufacturer's instructions (2) at TOXI-LAB using the Syva CP-5000/Guilford Stasar III (manual method). The EMIT d.a.u. cannabinoid 100-ng assay was also performed at PharmChem Laboratories, Inc. (Menlo Park, California) on a Hitachi 717 automatic analyzer (Boehringer Mannheim, Indianapolis, Indiana).

C. Gas Chromatography/Mass Spectrometry

Gas chromatography/mass spectrometry (GC/MS) assays were performed at PharmChem Laboratories using a 3-ion SIM detection of the trimethylsilyl derivative of Δ^9-THC-COOH. Other GC/MS assays were performed at the Center for Human Toxicology (Salt Lake City, Utah) using negative-ion chemical ionization of the hexafluoroisopropyl/pentafluoropropionyl derivative (3).

D. Sample Analysis

Study 1. One hundred urine samples were obtained from PoisonLab, Inc. (San Diego, California) and were analyzed on a blind basis at TOXI-LAB using both EMIT$_{dau}$ 50-ng and 100-ng assays (manual method) and the TLC method. A 15-ng/mL control was employed as the TLC cutoff.

Study 2. A total of 239 samples were analyzed on a blind basis at PharmChem Laboratories using the EMIT$_{dau}$ 100-ng assay (Hitachi analyzer) and the TLC method. For TLC analysis, samples were classified as positive if red bands appeared to be greater than the 15 ng/mL control, (+/−) for red bands less than the 15-ng/mL control, or negative if no red bands were observed.

Study 3. During routine screening at PharmChem Laboratories (April 9, 1990 to June 29, 1990), 5661 urine specimens produced positive EMIT$_{dau}$ 100-ng assay results. These EMIT-positive specimens were analyzed by TOXI-LAB THC II Screen using a 50-ng/mL control as the TLC cutoff.

III. RESULTS

A. Study 1

The results for 100 samples analyzed by the EMIT$_{dau}$ 50-ng and 100-ng assays and the TOXI-LAB THC II Screen are shown in Table 1. Forty samples produced positive TLC results (≥15 ng/mL), and all of these samples were positive by both EMIT assays. The remaining 60 samples were negative by all three assays. No disparate results were observed.

B. Study 2

Of 239 samples analyzed by the EMIT d.a.u. 100-ng assay and TOXI-LAB THC II Screen, 59 samples produced positive results by both methods (Table 2). Multiple GC/MS analyses at Pharmchem Laboratories and the Center for Human Toxicology failed to detect Δ^9-THC-COOH in one EMIT-positive specimen at the minimum detection limit of 1 ng/mL.

Table 1 TOXI-LAB THC-II Veresus $EMIT_{dau}$ (100 Samples)

Method	Results	
	Positives	Negatives
$EMIT_{dau}$ 100-ng assay	40	60
$EMIT_{dau}$ 50-ng assay	40	60
TOXI-LAB	40	60

Of 179 samples that produced EMIT responses below the 100-ng/mL calibrator, 172 were classified as negative by the TLC method (Table 3). Four samples were classified as positive (≥15 ng/mL) by TLC, and three samples that produced red bands whose size and intensity was less than the 15-ng/mL control were classified as (+/−). GC/MS analysis of these seven disparate results detected Δ^9-THC-COOH in all samples with concentrations ranging from 3.9 ng/mL to 19.9 ng/mL.

C. Study 3

In the third study, 5661 samples which produced positive results with the $EMIT_{dau}$ 100-ng assay were analyzed by the TLC method using a 50-ng/mL control as the cutoff. The results are shown in Table 4. A total of 5548 samples produced positive TLC results (≥50 ng/mL). One hundred four samples produced red bands whose size and intensity was less than the 50-ng/mL control and were classified as (+/−). A total of nine samples were found to be unsuitable for analysis after hydrolysis.

IV. DISCUSSION

In study 1, a 100% correlation between the EMIT and TLC methods was observed. In study 2, the TLC method confirmed 98% of the Emit-positive and 95% of the EMIT-negative results. Consistent with other reports (4,5), one EMIT false-positive result was identified; however, no TLC false-positive results were observed. In study 3, the TLC method again confirmed 98% of the EMIT-positive results.

As determined in study 2, the limit of the detection of the TLC method for Δ^9-THC-COOH was less than 10 ng/mL using a 3-mL sample. The throughput of the assay was approximately 50 samples per hour.

Table 2 TOXI-LAB® THC-II vs EMIT® dau™ 100 (60 EMIT Positives)

Method	Results	
	Positives	Negatives
$EMIT_{dau}$	60	0
TOXI-LAB	59	1[a]

[a]Also negative by GC/MS.

Table 3 TOXI-LAB® THC-II vs EMIT® dau ™ 100 (179 EMIT Negatives)

	Results		
Method	Positives (≥15 ng/mL)	Positives (≥15 ng/mL)	Negatives
$EMIT_{dau}$	0	0	179
TOXI-LAB®	4*	3** (±)	172

*GC/MS results (ng/mL): *19.9, 16.1, 14.0, 13.6; **8.3, 5.8, 3.9. (±) red spot less dense than control's.

Table 4 TOXI-LAB® THC-II vs EMIT® dau™
High Volume Population Study (5,661 EMIT Positives)

Results	No. of samples
TOXI-LAB (+); positive ≥50 ng/mL	5548 (98%)
TOXI-LAB (+/–); positive <50 ng/mL	104 (1.8%)
(Sample unsuitable for analysis)	9 (0.2%)

V. CONCLUSION

A new TLC system for the detection of Δ^9-THC-COOH in urine (TOXI-LAB THC II Screen) has demonstrated correlations exceeding 95% when compared with the $EMIT_{dau}$ 100-ng cannabinoid assay. The TLC method produced consistent and reliable results which were preservable and permitted higher sample throughput than conventional TLC techniques.

The method may be used as a sensitive, semiquantitative screening technique or as a confirmation method for immunoassays in selected populations.

REFERENCES

1. *TOXI-LAB THC II Screen Instruction Manual.* TOXI-LAB, Inc., Irvine, CA (1990).
2. $EMIT_{dau}$ cannabinoid 100-ng assay package insert. Syva Co., Palo Alto, CA (1990).
3. K. M. McGinnis, D. S. Barbario, R. Bridges, and R. L.Foltz. A high sensitivity, high throughput assay for the major urinary metabolite of Δ^9-tetrahydrocannabinol. *Proc. 33rd Annual Conf. on Mass Spectrometry and Allied Topics*, San Diego, CA, pp. 464–465 (May 26–31, 1985).
4. R. L. Foltz and I. Sunshine. Comparison of a TLC method with EMIT and GC/MS for detection of cannabinoids in urine. *J. Anal. Toxicol.*, *14*:1–4 (1990).
5. D. L. King, M. J. Gabor, P. A. Martel, and C. M. O'Donnell. A rapid sample-preparation technique for thin-layer chromatographic analysis of 11-nor-Δ^9-tetrahydrocannabinol in human urine. *Clin. Chem.*, *35*:163–165 (1989).

78

Sensitivity and Specificity of Techniques Used to Detect Drugs of Abuse in Urine

J. F. Wilson,* J. Williams,* G. Walker,† P. A. Toseland,‡ B. L. Smith,§ A. Richens,* and D. Burnett¶ **University of Wales College of Medicine, Cardiff, Wales, †University Hospital, Nottingham, England, ‡Guys Hospital, London, England, §Maudsley Hospital, London, England, and ¶St. Albans City Hospital, St. Albans, England*

I. INTRODUCTION

Techniques for the detection and identification of drugs of abuse in urine are applied in a variety of fields, including sports medicine, forensic science, and for employee and preemployment screening. The rational for performing the tests differs among these fields, and hence thresholds for detection also differ. In the clinical field, the question posed is often whether an illicit compound has been taken with the last 24 to 72 h. Limits of detection are therefore higher than those used in preemployment screening. All fields should operate, however, to the same absolute standard of specificity, as the consequences of error can be monitored by performance in an external quality assessment scheme (EQAS). An EQAS for clinical testing for drugs of abuse in urine was established in the United Kingdom in 1987 (Burnett et al., 1990). Data from such schemes provide information on the relative performance of different an analytical techniques as implemented in a routine service environment, which may differ from the optimal conditions used during technique development. We report here a comparison of the sensitivity and specificity of commonly used chromatographic and immunological techniques for the detection of morphine, methadone, amphetamine, cocaine, and benzoylecgonine based on data reported by members of the U.K.-EQAS for drugs of abuse in urine.

II. MATERIALS AND METHODS

Twenty-five samples of freeze-dried 25-mL aliquots of urine were analyzed between April 1987 and December 1989 by members of the U.K.-EQAS for Drugs of Abuse. The urine, which was collected from volunteers and patients, was heat-treated at 60°C for 1-5 h before drying and contained a maximum of 0.1% w/v of sodium azide as a preservative. The samples contained mixtures of analytes (Table 1) and included

Table 1 Composition of Samples[a]

Sample no.	Concentration, mg/L (i = + interferent) Morphine	Methadone	Amphetamine	Cocaine	Benzoylecgonine	Interferents	Other analytes present
01	—	0	0	2	1	—	Morphine, morphine-3-glucuronide, phenobarbitone
02	0i	0	0i	0	0	Pholcodine, ephedrine, phenylpropanolamine	—
03	0	5.4	0	0	0	—	Benzodiazepine, 9-COOH, 11-Nor,δ-8 tetrahydrocannabinol
04	—	0i	0	0	0	Dipipanone	Morphine, morphine-3-glucuronide, codeine, benzodiazepine
05	—	0	0	2	1	—	As sample 01 + additional amitriptyline, nortriptyline
06	0i	0	—	0	0	Dihydrocodeine	Amphetamine, amylobarbitone, lignocaine
07	0	0i	—	0	0	Propoxyphene, norpropoxyphene	Paracetamol, amphetaminelike artifact
08	0i	0	—	0	0	Pholcodine	Ephedrine, phenylpropanolamine, amphetaminelike artifact
09	0	0	—	0	0	—	Ampicillin, lignocaine, amphetaminelike artifact
10	1	—	0	0	0	—	Methadone, benzodiazepine
11	0	0	0i	2	—	Methylamphetamine	Benzoylecgonine, benzocaine, paracetamol
12	—	—	—	0	0	—	Methadone, amphetamine, ephedrine, benzodiazepine
13	—	0i	—	1	4	Norpropoxyphene	Morphine-3-glucuronide, amphetamine, methylamphetamine, paracetamol, chlorpromazine sulfoxide
14	5	—	0	1	0.4	—	Methadone, benzodiazepine
15	—	0	0	0	0	—	Codeine, benzodiazepine
18	0	0	0	0	0	—	Riboflavin, flucloxacillin
19	0	1	5	5	2	—	Amylobarbitone, benzodiazepine
20	5	5	2	2	0.4	—	Morphine-3-glucuronide, amylobarbitone, benzodiazepine, cannabinoid, erythromycin
21	2	2	2	0	0	—	Phenobarbitone, benzodiazepine, cannabinoid, erythromycin
22	0	1	0	—	2	—	Benzodiazepine, cannabinoid, erythromycin
23	—	1	1	0	0	—	Morphine, dihydrocodeine, amylobarbitone, benzodiazepine, erythromycin
24	2	2	—	0	0	—	Amphetamine, methylamphetamine, benzodiazepine, erythromycin
25	0i	0	0i	0	0	Dihydrocodeine, ephedrine, phenylpropanolamine	Phenobarbitone
26	0i	0	—	0	0	Pholcodine	Amphetamine, ephedrine, phenylpropanolamine, phenytoin, carbamazepine, benzodiazepine, cannabinoid
27	1	0	5	5	4	—	—

[a]The five numerical columns show the concentrations of the five study analytes present in samples included in the data analysis. Some zero-concentration samples (0i) contained the interferents specified.

replicate samples containing morphine, methadone, amphetamine, and cocaine at concentrations of 0, 1, 2, and 5 mg/L, and benzoylecgonine at concentrations of 0, 0.4, 1.2, and 4 mg/L. Morphine, amphetamine, cocaine, and benzoylecgonine concentrations were obtained by weighing in pure compounds to a pooled collection of urine samples. The base urine samples were either drug-free material collected from the authors or were first screened by Emit immunoassays (Syva, Palo Alto, California) and shown to be analytically free of the spiked analytes. Methadone, when required, was present in metabolized form in the base urine which was collected from patients undergoing detoxification in a closed clinical unit. The concentration of free methadone in the urine was estimated by $EMIT_{dau}$ and the pool diluted with blank urine to achieve the desired concentration. Assay of material before and after freeze-drying confirmed that no substantial loss of the study analytes occurred during sample preparation. However, an amphetaminelike freeze-drying artifact (1) was detected in three samples, and the amphetamine data for these samples were excluded. Data from samples containing the five study analytes were also excluded where the analyte concentration was either uncertain or where other interfering compounds were present (Table 1). Data from samples numbered 16 and 17 from the scheme were excluded entirely, the samples being prepared from unquantified pools of addict urine.

Laboratories were requested to analyze the samples as if they were clinical drugs of abuse specimens. They reported on the presence or absence of a range of analytes of their own choosing and, for each analyte reported, on the results of the separate analytical tests used to determine the overall report. Results were classified as being correct or incorrect on the basis of the sample formulation.

Statistical comparisons of percentages were made by one- or two-way analysis of variance (anova) after an arcsine transformation. Significant differences between means were located by the Student-Newman-Keuls test. Frequencies were compared by the chi-square test. Where values are said in the text to be significantly different, p was <0.05.

III. RESULTS

A. Reports

Data were received from an average of 95 laboratories (range 88 to 103). The number of correct, false-negative, false-positive, and uncertain reports are given in Table 2 expressed as a percentage of the total number of laboratories responding. There were significantly fewer reports for morphine, methadone, and amphetamine when the analyte was absent from the sample, as a result of some laboratories reporting only on analytes present in the samples or on analyte groups instead of the five individual study analytes. Correct negative report and test results are therefore underestimated in the data presented. Similarly, there were significantly fewer reports for lower versus higher concentrations of analytes for amphetamine and cocaine, demonstrating an underreporting of false-negative results.

B. Techniques Used

The mean number of test results by different techniques are shown in Table 3 for each analyte. Results by thin-layer chromatography (TLC) were divided into those by the commercial TOXI-LAB systems (Mercia Diagnostics Ltd., Guildford, U.K.) and others,

Table 2 Percentage of Correct, False-Negative, False-Positive, and Uncertain Reports for Morphine, Methadone, Amphetamine, Cocaine, and Benzoylecgonine in Samples from the U.K.-EQAS for Drugs of Abuse

Analyte	Concentration (mg/L)	Number of samples	Percent total reports[a]			
			True positive	False negative or uncertain	True negative	False positive or uncertain
Morphine	5	2	75 ± 1	1 ± 1	—	—
	2	2	65 ± 7	8 ± 5	—	—
	1	2	49 ± 2	17 ± 2	—	—
	0	7	—	—	39 ± 1	1 ± 1
	0 + interferent	5	—	—	40 ± 2	9 ± 2
Methadone	>5	2	83 ± 3	3 ± 1	—	—
	2	2	86 ± 0	3 ± 1	—	—
	1	3	83 ± 1	5 ± 1	—	—
	0	12	—	—	59 ± 3	1 ± 0
	0 + interferent	3	—	—	53 ± 10	7 ± 5
Amphetamine	5	2	77 ± 4	1 ± 1	—	—
	2	2	68 ± 4	7 ± 1	—	—
	1	1	59	7	—	—
	0	9	—	—	41 ± 1	1 ± 0
	0 + interferent	3	—	—	36 ± 6	10 ± 3
Cocaine	5	2	54 ± 4	9 ± 5	—	—
	2	4	32 ± 4	17 ± 3	—	—
	1	2	31 ± 3	16 ± 1	—	—
	0	16	—	—	40 ± 2	1 ± 1
Benzoylecgonine	4	2	44 ± 9	3 ± 1	—	—
	2	2	46 ± 2	6 ± 3	—	—
	1	2	21 ± 3	8 ± 0	—	—
	0.4	2	35 ± 9	8 ± 1	—	—
	0	16	—	—	32 ± 3	1 ± 0

[a]Percentages given as mean ± SEM.

Table 3 Mean Number of Test Results by Different Assay Techniques [Values for Samples With and (Without) the Analyte][a]

Analyte	TOXI-LAB TLC	In-house TLC	GC	$EMIT_{dau}$	$EMIT_{st/Qst}$	Abbott TDx	RIA	BCL opiates
Morphine	53 (21)	36 (29)	2 (2)	—	—	—	2 (2)	—
Opiates	—	—	—	25 (17)	11 (12)	8 (5)	—	14 (17)
Methadone	48 (22)	31 (27)	11 (6)	14 (9)	8 (9)	—	—	—
Amphetamine	58 (16)	5 (5)	26 (15)	—	—	—	—	—
Amphetamine group	—	—	—	26 (13)	7 (10)	7 (4)	0 (2)	—
Cocaine	27 (16)	10 (9)	6 (3)	—	—	—	—	—
Benzoylecgonine	20 (11)	3 (3)	—	16 (16)	5 (4)	6 (6)	—	—

[a]TLC = thin-layer chromatography, GC = gas-liquid chromatography, RIA = radioimmunoassay. See Figure 1 for explanation of commercial abbreviations.

which are termed in-house TLC. Results from the Perkin-Elmer polarization fluoroimmunoassay (Perkin-Elmer Ltd., Beaconsfield, U.K.), high-pressure liquid chromatography, and gas-liquid chromatography (GC) with mass spectrometry were too few in number for inclusion in the analysis. Additional data from group-specific immunoassays are included for reports of opiates and the amphetamines group. Means are given separately for samples with and without the presence of the analyte. The difference between the two means indicates, in part, the scale of underreporting of true negative results by the various techniques. The larger differences seen with chromatographic techniques have an additional cause. They result from the reporting of additional confirmatory tests, sometimes by the same technique as the preliminary test, in cases where preliminary tests were positive.

Confirmation of results by use of a second technique was used for morphine (41%), methadone (29%), amphetamine (36%), cocaine (26%), and benzoylecgonine (16%), with frequency given in parentheses. However, use of a confirmatory test significantly reduced the incidence of reporting errors only in the case of samples containing cocaine.

C. Sensitivity

The sensitivity of different analytical techniques was studied by calculating its inverse from the results of tests on those samples containing the analyte in question.

$$\text{Insensitivity} = \frac{\text{number of negative tests} + \text{number of uncertain tests}}{\text{total number of tests}} \times 100\%$$

The percentage insensitivity values are displayed in Figure 1. Data from group-specific immunoassays for opiates and amphetamines are included. It should be noted that the observed underreporting of false-negative tests will result in the values displayed being underestimates. From the differences in reporting frequency at low and high concentrations (Table 2), the maximum underestimate in insensitivity will be 16%. All immunoassays for morphine or the opiate group were significantly more sensitive than chromatographic procedures. Gas-liquid chromatography (GC), the least sensitive method for morphine, was also significantly less sensitive than in-house TLC. In-house TLC was the least sensitive technique for other analyses. It differed significantly from the most sensitive GC and EMIT$_{dau}$ techniques for assay of methadone. For amphetamine, both commercial and in-house TLC techniques were significantly less sensitive than either GC or the nonspecific immunoassays for the amphetamines group. In-house TLC was also significantly inferior compared with commercial TLC. No technique was adequate at the study concentrations for the specific detection of cocaine as opposed to benzoylecgonine, though GC was significantly more sensitive than either TLC group. The TLC techniques were similarly inadequate for the detection of benzoylecgonine. The immunoassay techniques, which did not differ significantly, were all significantly more sensitive than TLC.

D. Specificity

The specificity of different analytical techniques was studied by calculating its inverse from the results of tests on those samples not containing the analyte in question.

$$\text{Nonspecificity} = \frac{\text{number of positive tests + number of uncertain tests}}{\text{total number of tests}} \times 100\%$$

The percentage nonspecificity values are displayed in Figure 2. Data from group-specific immunoassays for opiates and amphetamines are included for those samples without interferents. It should be noted that the observed underreporting of true negative tests will cause the values presented to be overestimates. From the observed reduction in reporting in analyte-free samples (Table 2), the scale of the overestimation is calculated to lie between zero and 32%.

The specificity data were more variable than the sensitivity data, with fewer significant between-technique differences. Among group-specific immunoassays for opiates, the BCL opiates tests [Boehringer Corporation (London) Ltd., Lewes, U.K.] was significantly less specific than other techniques. A statistically significant interaction (anova, $p < 0.05$) in amphetamine assay data was the result of poor specificity of TLC tests in samples containing ephedrine and phenylpropanolamine or methylamphetamine. The commercial TLC assay was also significantly less specific than other techniques for benzoylecgonine. In addition to these between-technique differences, significant general interference by compounds related to the study analytes was demonstrated for pholcodine and dihydrocodeine in assays for morphine; by dipipanone, dextropropoxyphene, and norpropoxyphene in assays for methadone; and by ephedrine, phenylpropanolamine, and methylamphetamine in assays for amphetamine.

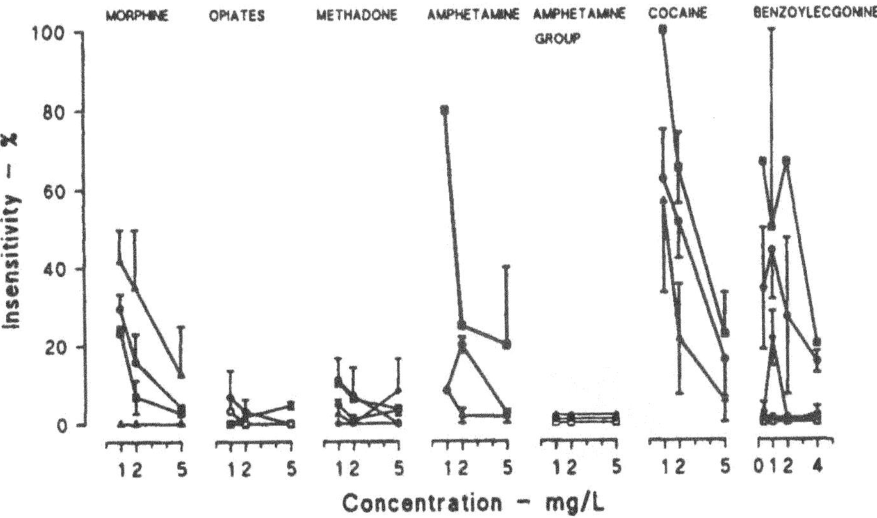

Figure 1 Insensitivity of different techniques for the detection of a range of analytes in samples from the U.K.-EQAS for Drugs of Abuse. Values are plotted as means and standard errors for different analyte concentrations. ●, Commercial thin-layer chromatography (TLC) (TOXI-LAB, Mercia Diagnostics Ltd., Guildford, U.K.); ■, in-house TLC; ▲, gas-liquid chromatography; ◆, EMITdau (Syva, Palo Alto, California); O, Emit st/Qst (Syva); ☐, Abbott TDx (Abbott Laboratories Ltd., Wokingham, U.K.); Δ, radioimmunoassay; and ◊, BCL opiates test [The Boehringer Corporation (London) Ltd., Lewes, U.K.].

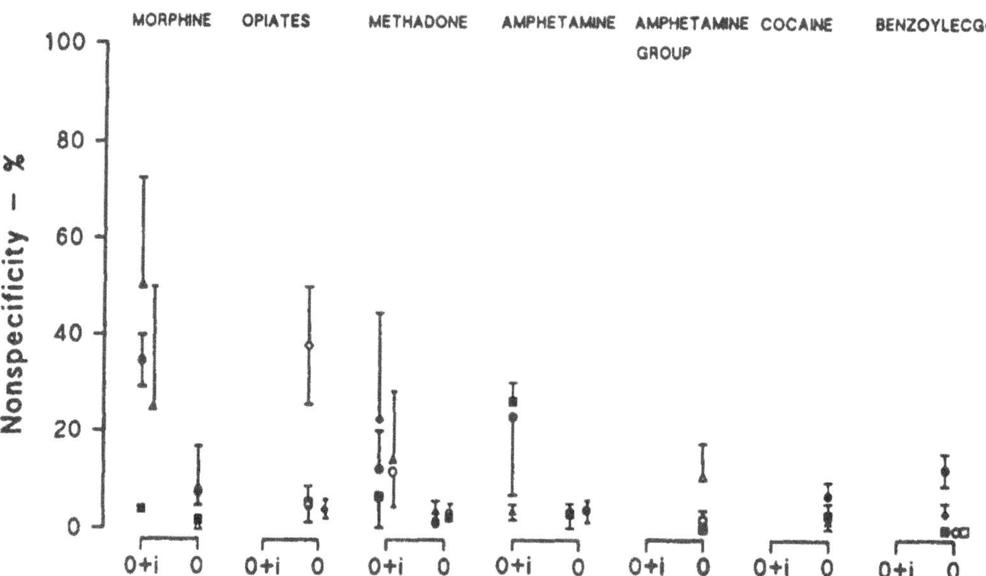

Figure 2 Nonspecificity of different techniques for the detection of a range of analytes in samples from the U.K.-EQAS for Drugs of Abuse. Points show means and standard errors for samples with (0+i) and without (0) known interferents. See Figure 1 for key to symbols.

IV. DISCUSSION

Comparisons of techniques for the analysis of drugs of abuse in urine demonstrated a significant lack of sensitivity for some chromatographic techniques when used to detect morphine, amphetamine, and cocaine at concentrations down to 1 mg/L, and benzoylecgonine down to 0.4 mg/L. These findings are not seriously compromised by a major weakness in the data resulting from underreporting of false-negative results. The observed differences between techniques were greater than the estimated maximum error and, moreover, were consistent over varying analyte concentrations, in contrast to the errors which would be predicted to be concentration-related.

The other major findings concern specificity. Significant interference was demonstrated by structurally related compounds in assays for morphine, methadone, and amphetamine, and a significant lack of specificity was observed in the BCL opiates test and in TLC of amphetamine and benzoylecgonine. The specificity parameter was subject to positive bias resulting from the underreporting of true negative results, which was most prevalent with the chromatographic techniques. The finding concerning the BCL opiates test therefore appears sound, but those relating to TLC require more accurate quantitation.

The present data demonstrate that clinical laboratories are not yet achieving a satisfactory level of performance in screening for drugs of abuse. Since the concentration ranges chosen for the present study were thought appropriate to the clinical field and increasing thresholds of detection to improve performance is unrealistic, we conclude that the use of certain techniques is inappropriate for some analytes. In the case of the identification of cocaine misuse, common chromatographic techniques were clearly inadequate. The immunoassays for benzoylecgonine apparently offer satisfactory sensitivity and specificity.

V. SUMMARY

Twenty-five samples of freeze-dried urine from the U.K. External Quality Assessment Scheme for Drugs of Abuse were analyzed by an average of 95 laboratories between April 1987 and December 1989. Samples contained mixtures of analytes and included replicated concentrations of morphine, methadone, amphetamine, and cocaine of 0, 1, 2, and 5 mg/L, and benzoylecgonine at 0, 0.4, 1.2, and 4 mg/L. Lack of sensitivity was detected in samples containing analytes by two-way analysis of variance of an arcsine transform of the percentage of false-negative and uncertain results. Lack of specificity was detected in analyte-free samples by similar analysis of false-positive and uncertain results. Significant differences were located by the Student-Newman-Keuls test ($p = 0.05$). The sensitivity of some chromatographic techniques was inadequate for detection of morphine, amphetamine, cocaine, and benzoylecgonine at the lower concentrations studied. Gas chromatography (GC), the least sensitive method for morphine, was significantly different from in-house thin-layer chromatography (TLC), while all chromatographic procedures were significantly less sensitive than immunoassays for morphine or opiates. In-house TLC was least sensitive for the other analytes. It differed significantly from the most sensitive GC and $EMIT_{dau}$ (Syva) techniques for methadone and from TOXI-LAB TLC (Mercia) for amphetamine. Both TLC groups were significantly less sensitive than GC and group-specific immunoassays for amphetamine, than GC for cocaine, and than the immunoassays for benzoylecgonine. Few significant differences in specificity were detected between techniques, though significant interference from structurally related compounds was demonstrated in assays for morphine, methadone, and amphetamine. The exceptions were the BCL test for opiates and TLC when applied to amphetamine and benzoylecgonine, where significant nonspecificity was demonstrated.

REFERENCE

1. D. Burnett, S. Lader, A. Richens, B. L. Smith, P. A. Toseland, G. Walker, J. Williams, and J. F. Wilson. A survey of drugs of abuse testing by clinical laboratories in the United Kingdom. *Ann. Clin. Biochem.*, 27:213–222 (1990).

79

A Cost-Effective EMIT Method for Qualitative Drug Abuse Analysis Performed in the Technicon Chem-1 Analyzer

Stephen P. Harrison and Ian M. Barlow *Bradford Royal Infirmary, Bradford, England*

I. INTRODUCTION

About five years ago, a drugs-of-abuse screening service was set up in our laboratory, broadly based on guidelines published by the Department of Health and Social Security (1). Included in this protocol were urine tests for barbiturates, opiates, amphetamines, benzodiazepines, cocaine metabolite, and cannabinoids, all by $EMIT_{dau}$ assays which had been adapted to the RA-1000 (Technicon Instruments Corp., Tarrytown, New York). Gradually, however, as the workload increased, it became obvious that a quicker and more cost-effective solution had to be found. To obviate this problem we adapted the six $EMIT_{dau}$ assays to the Chem-1 analyzer (Technicon Instruments Corp.), a system that was designed primarily for analysis of serum but that has also proved useful for urine analysis (2). This innovative analyzer is a hybrid "continuous flow/random-access" system that produces capsules of sample and reagent divided by bubbles in a continuously flowing single stream within Teflon tubing lined with a thin film of polyfluorinated hydrocarbon liquid that eliminates carryover.

Because the Chem-1 requires only 1 μL of sample and 14 μL of reagent (7 μL each of reagents 1 and 2) for all assays, considerable reagent cost savings are possible. Furthermore, its random-access sampling capability up to maximum of 720 tests per hour would ensure a much faster turnover of samples.

II. MATERIALS

A. Reagents

EMIT kits were purchased from Syva Co. (Palo Alto, California). EMIT buffer and reagents A and B were all reconstituted as recommended by the manufacturer. Working reagents were prepared for all six assays, based on the ratios recommended in the

$EMIT_{dau}$ assay documentation (0.5 mL:3.75 mL). To each batch of reagent A and B we added 5 μL of wetting agent W and left the reagents at least 2 h to equilibrate. As empty Chem-1 cassettes are not available from Technicon, we removed the contents of previously used cassettes and thoroughly cleaned them with 0.1 M HCl followed by deionized water, and left them to dry at room temperature. With Pasteur pipettes, we transferred working reagent B to reagent 1 and working reagent A to reagent 2 positions of the Chem-1 cassette, respectively. Once prepared, the cassettes were left on the system at 7°C.

B. Calibrators

We used Syva Co. calibrators—negative, low, and medium—reconstituted according to the manufacturer's instructions.

III. PROCEDURES

A. Chem-1 Method Development

The Chem-1 has no facility for user-defined chemistries, and so for the EMIT methods we had to reprogram analytical parameters of Technicon-defined assays that were not used on our system. We characterized the absorbance time course of each EMIT assay by analyzing negative-, low-, and medium-level calibrators and obtained absorbance data for each of the nine read stations through the "mark sense" data port. Typical traces for each $EMIT_{dau}$ assay are shown in Figures 1, 2, and 3. As can be seen from these results, the absorbance response for each assay was linear, although sensitivity varied between assays. The barbiturates assay showed the greatest separation between the calibrator and the negative, whereas the cannabinoids assay showed the least.

Because the absorbance response was linear, we chose to operate the assays as zero-order chemistries rather than at equilibrium as with our comparison method below. The analytical parameters we selected are shown in Table 1.

The results we obtained are in arbitrary units but are proportional to the dA/min obtained for each test reaction. Before each batch of samples, we ran reagent baseline corrections for each $EMIT_{dau}$ assay. For routine analysis, we assayed test samples as profiles in random-access mode and analyzed negative-, low-, and medium-level calibrators at the beginning and end of each run of samples to assess any potential baseline drift.

B. RA-1000 Methods

For comparison purposes, we determined $EMIT_{dau}$ assays in the RA-1000 analyzer at equilibrium, by methods we had developed previously.

C. Patients' Samples

Random urine samples were obtained from patients attending the local drug rehabilitation unit and usually stored overnight at 4°C. Before analysis, samples were mixed and centrifuged, and the supernatants were analyzed in variously sized batches in the RA-1000 and then, when practicable, on the same day in the Chem-1. The results from both instruments were compared with those obtained for the calibrators and reported as negative (result < negative calibrator); positive (low

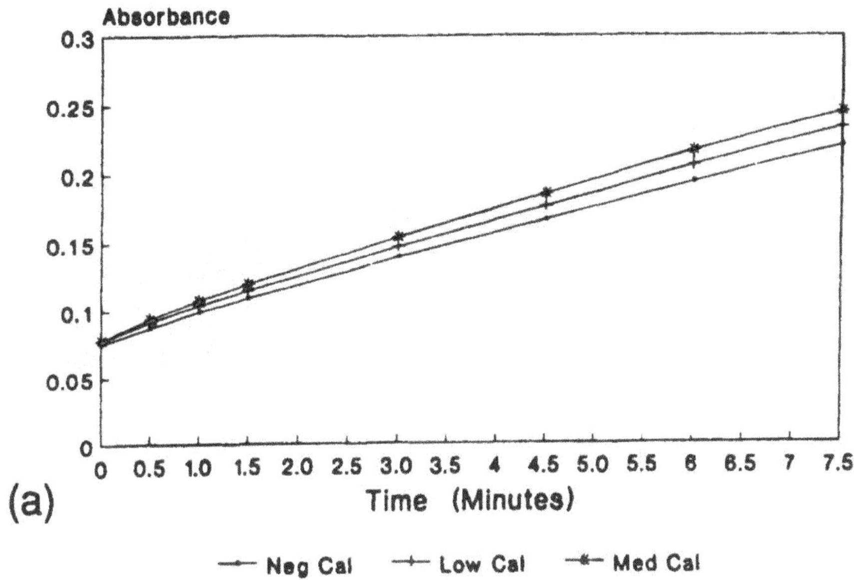

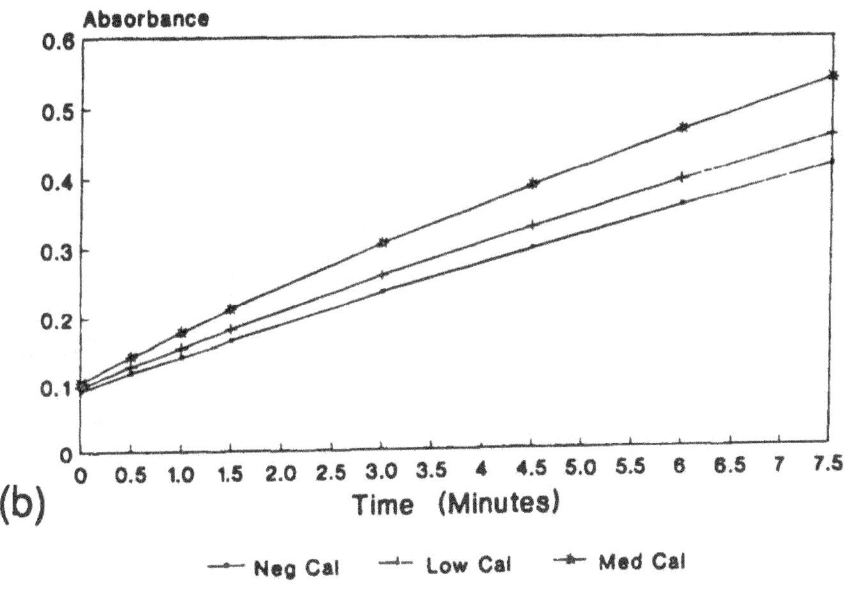

Figure 1 Typical absorbance changes: (a) benzodiazepine; (b) cocaine.

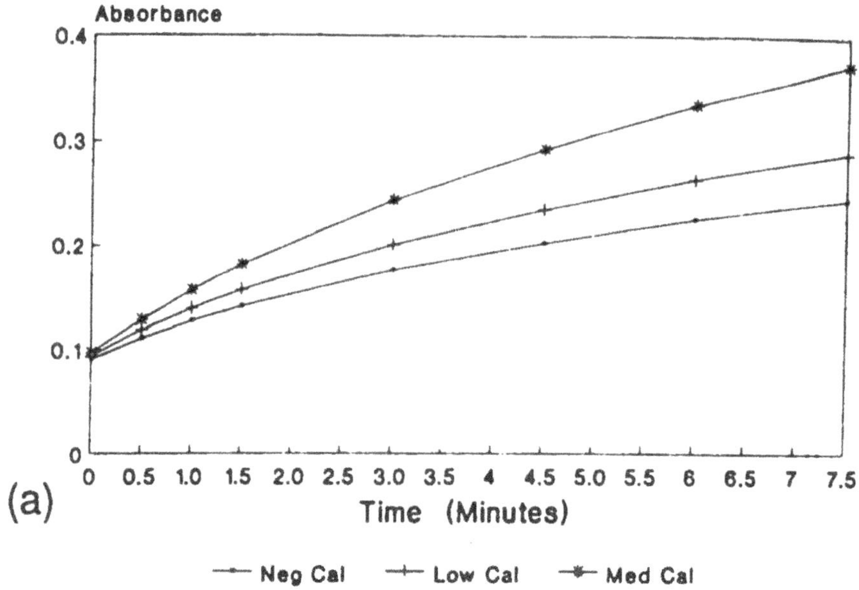

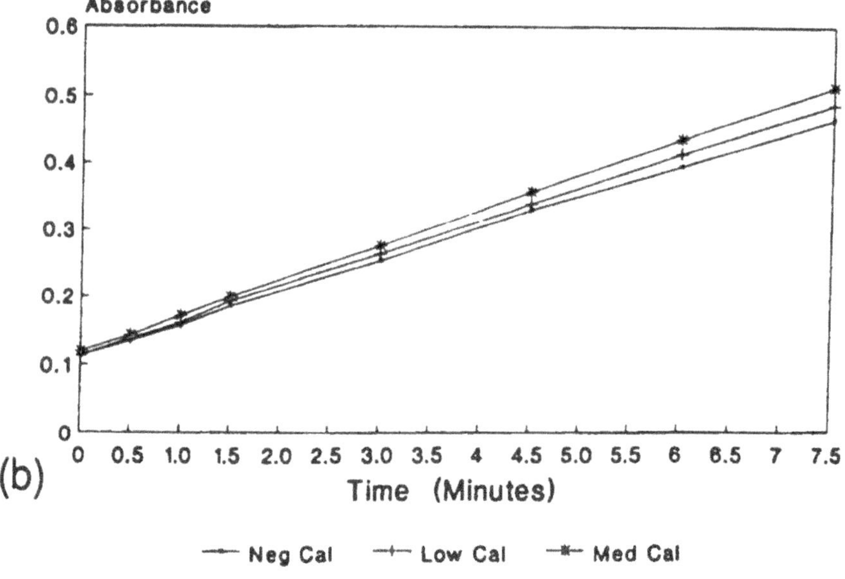

Figure 2 Typical absorbance changes: (a) amphetamine; (b) cannabinoid.

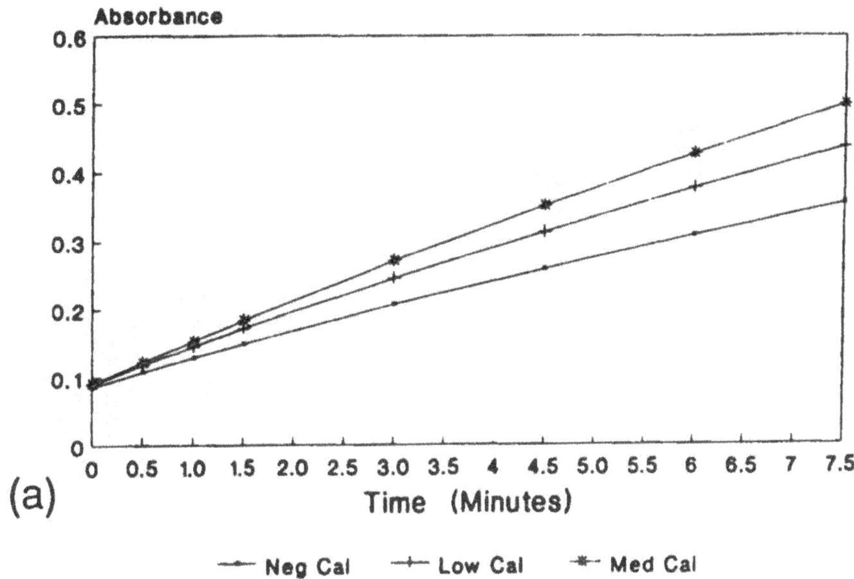

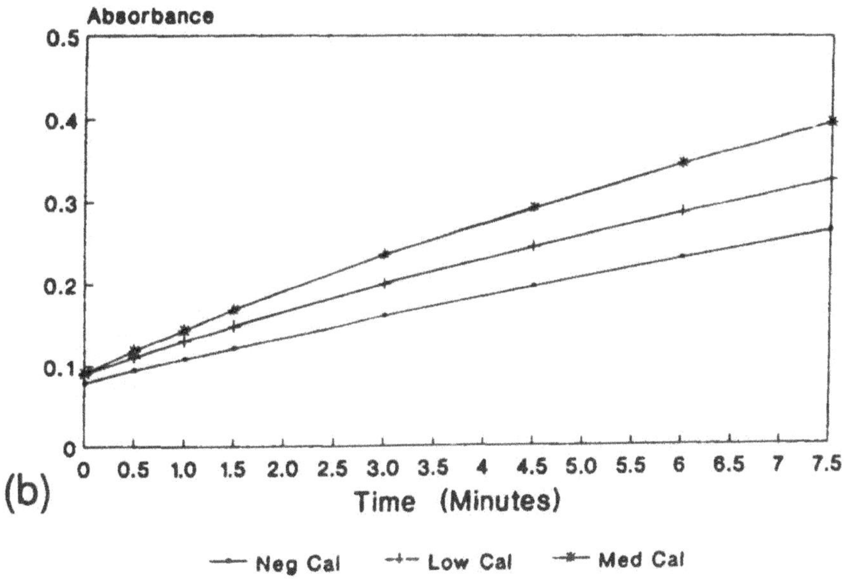

Figure 3 Typical absorbance changes: (a) barbituates; (b) opiate.

Table 1 Instrument Parameters for the Chem-1 $EMIT_{dau}$ Methods

Chemistry	Type Z0	Chem names		Calibrant 0	Impl'ted Y
BREAKPT LIMIT	0.0260	POST WATER	0	RBRATE	0.0266
BREAKPT END FC	6	PRE WATER	0	RBATE RANGE	0.0004
CAL FACT MAX	*				
CAL FACT HI	*	RBL1 AR LIM	0.050	RB SLOPE RANGE	0.0004
CAL FACT CUR	*	RBLN AR LIM	0.050	RB INTER RANGE	0.0300
CAL FACT LO	*			RES/STD ER LIM	4.0
CAL FACT MIN	*	RBL1A HI	0.020	RES DIFF MAX	0.01
CAL SLOPE RANGE	0.1000	RBL1A CUR	*		
CONC MAX	10000	RBL1A LO	–0.005	SAMSTK	0
CONC MIN	0			SB ABS MAX	1.0000
DEPLETION LIM	0	RBL1B HI	0.100	FLOW CELL N	2
DIL	1	RBIB CUR	*	START F CELL	4
DILSTK	0	RBLIB LO	0.020	SD OD ABS MIN	0.0003
FILTER	340			S (10*MXRSS5)	0.0316
INVERSE	N	RBL HI	0.075	S (10*MXRSS4)	0.0316
MULT SET NUM	1	RBLN CUR	*	S (10*MXRSS3)	0.0316
OFFSET SET NUM	3	RBLN LO	0.010	UNITS	U/L

*Varies between assays.

calibrator < result < medium calibrator); and strong positive (result > medium calibrator).

IV. RESULTS

The within-batch precision of each assay was determined at negative, low, and medium concentrations by replicate analyses within the same run with Syva Co. calibrators; the results are shown in Table 2. From this data, 2-SD ranges were calculated for the calibrator values for each assay, and these were used to resolve any sample results that fell in the borderline area.

Reagent carryover between the six EMIT$_{dau}$ assays were assessed by the protocol as outlined by the European Committee for Clinical Laboratory Standards (ECCLS) (3). In all cases, interaction between the EMIT$_{dau}$ assays was negligible (<2%).

Urine samples from patients previously analyzed in the RA-1000 were also analyzed for all six EMIT$_{dau}$ methods in the Chem-1. The results were compared to those obtained for the Syva calibrators and classified as negative, positive, or strong positive as described previously. However, for comparison purposes, the results are shown as positive or negative for the assays on the two analyzers (see Table 3).

The patient results for barbiturates, cocaine, and opiates were exactly the same for both analyzers. But one result for amphetamines and one for benzodiazepines were positive in the Chem-1 but negative in the RA-1000. However, in both cases these

Table 2 Within-Batch Precision of the Six EMIT$_{dau}$ Methods (n = 24)

EMIT$_{dau}$	Mean	SD	CV
Amphet neg	*	*	*
Amphet low	187.3	1.81	0.97
Amphet med	701.9	1.85	0.26
Barbs neg	1207.4	12.59	1.04
Barbs low	3922.8	29.56	0.75
Barbs med	6029.2	18.74	0.31
Benzo neg	19.2	1.01	5.31
Benzo low	38.4	0.82	2.15
Benzo med	62.0	0.62	1.01
Cannab neg	46.4	1.81	3.91
Cannab low	65.4	1.74	2.66
Cannab med	88.3	1.83	2.08
Coca neg	347.0	9.11	2.63
Coca low	1183.8	12.85	1.09
Coca med	3670.4	12.40	0.34
Opiate neg	*	*	*
Opiate low	128.3	1.20	0.94
Opiate med	276.8	1.08	0.39

*Denotes 0 U/L or no reaction.

Table 3 Comparison of Results as Obtained for Six $EMIT_{dau}$ Assays Determined in the RA-1000 and the Chem-1

	Negative	Positive	Total	Negative	Positive	Total
Amphetamine RA-1000				Barbiturate RA-1000		
Chem-1						
Negative	294	0	294	305	0	305
Positive	1	15	16	0	5	5
Total	295	15	310	305	5	310
Benzodiazepine RA-100				Cannabinoids RA-1000		
Chem-1						
Negative	204	0	204	204	0	204
Positive	1	105	106	10	96	106
Total	205	105	310	214	96	310
Cocaine metabolite RA-1000				Opiate RA-1000		
Chem-1						
Negative	289	0	289	248	0	248
Positive	0	21	21	0	62	62
Total	289	21	310	248	62	310

results fell within 2 SD of the low-level calibrator. For the cannabinoids data, 10 results were positive in the Chem-1 and negative in the RA-1000. In 8 cases the results fell in the borderline area of the low-level calibrator. The two remaining results were both repeated in both analyzers with the same results. Unfortunately, there was insufficient sample to perform any confirmatory test, and so these discrepancies remain unexplained.

V. DISCUSSION

We found the six $EMIT_{dau}$ assays adapted to the Chem-1 to be simple to use and precise. Results for urine samples in general compared well with those obtained in the RA-1000, and we recorded no false-negative results. However, false positives were evident for amphetamines, cannabinoids, and benzodiazepines assays, although in most cases these were in the borderline area of result classification. Two cannabinoids results, were definite false positives, but unfortunately we were unable to confirm the results by other means due to lack of sample. The greater number of discrepant results for the cannabinoids assay is in part to be expected, as it is the least sensitive of the six assays in both the RA-1000 and the Chem-1; we are presently trying to improve this by reformulating reagent concentrations. Notwithstanding these differences, we feel that the adaptations in the Chem-1, with their use of zero-order kinetics, are inherently more sound than equilibrium assays in the RA-1000, as more data points are used to calculate the result.

Although the analysis time of the $EMIT_{dau}$ assays in the Chem-1 is slightly longer than that of the RA-1000, the throughput is greater by a factor of three. Furthermore, considerable cost savings are evident, as the reagent consumption in the Chem-1 is reduced by 20-fold over the RA-1000 methods. Indeed, we estimate that a standard 100-assay $EMIT_{dau}$ kit could produce over 6000 tests when used in the Chem-1.

We believe ours is the first published adaptation of EMIT drug assays to the Chem-1, although EMIT thyroid-screening assays have been determined in the instrument by other workers (4). Technicon, originally slow to see the potential of their instrument for the drugs-of-abuse testing market, is now actively investigating the adaptation of 14 $EMIT_{dau}$ assays to the Chem-1. Moreover, with software changes to allow for multipoint calibration and curve-fitting, we see no reason why other EMIT drug assays, such as those for antiepileptics and aminoglycosides, could not be adapted to the system with the resultant attractive cost savings.

VI. SUMMARY

Six Syva $EMIT_{dau}$ assays—amphetamines, barbiturates, benzodiazepines, cannabinoids, cocaine metabolites, and opiates—have been adapted for use in the Technicon Chem-1 analyzer.

In operation, 1 µL of urine is mixed with 7 µL of reagent B, incubated at 37°C for 2 min, after which 7 µL of reagent A is added and the reaction rate (dA/min) is determined between flow cells 5 and 9 over a 6-min period. Samples in which dA/min equals or exceeds that of the Syva Low Level Calibrator are considered positive.

Comparison of results with those obtained in the RA-1000 analyzer were excellent for all assays, although a small number of discrepant results were observed in the borderline area. In most of these cases, reaction rates fell within 2 SD of Low Level Calibrator values.

Because of the small reagent volumes required for these assays, a typical 100-assay kit could produce over 6000 assays, i.e., a 60-fold increase.

In conclusion, the Chem-1, with its random-access capability, 720-test-per-hour throughput, and low reagent consumption, is particularly applicable to routine DAU screening analysis.

REFERENCES

1. Department of Health and Social Security. Guidelines of good clinical practice in the treatment of drug misuse. Report of the Medical Working Group on Drug Dependence. DHSS, London (1984).
2. I. M. Barlow, S. P. Harrison, and G. H. Hogg. Evaluation of the Technicon Chem-1. *Clin. Chem.*, *34*:2340–2344 (1988).
3. European Committee for Clinical Laboratory Standards (ECCLS). Guidelines for the evaluation of analysers in clinical chemistry. *ECCLS Document*, 3(2):1–32 (1986).
4. P. Duffy and C. M. Marshall. Performance of EMIT T4 and T-uptake assays on the Technicon Chem-1 analyser (abstract). *Clin. Chem.*, *6*:1083 (1990).

80

Development, Implementation, and Management of a Drug Testing Program in the Workplace

C. A. Burtis *Oak Ridge National Laboratory,† Oak Ridge, Tennessee*

I. INTRODUCTION

To combat the rising use of drugs in the workplace (1), many American companies have implemented drug testing programs and are testing employees and job applicants for use of illegal drugs (2–5). In addition, on September 15, 1986, Executive Order No. 12564 was issued by President Reagan, which requires all federal agencies to develop programs and policies, one of the goals of which is to achieve a drug-free federal workplace. Included in this Executive Order is the requirement that federal agencies implement drug abuse testing. Thus, in both the private and government sectors in the United States, drug abuse testing has become a prevalent practice as a means to detect and deter drug abuse in the workplace.

Before a drug abuse testing program is implemented, it is imperative that policies and procedures are developed that (a) ensure the accuracy of test results, (b) protect the validity and integrity of the specimen, (c) guarantee due process, and (d) maintain confidentiality. To make certain that these prerequisites were met in the government drug abuse testing programs, the U.S. Department of Health and Human Services (HHS) was directed to develop technical and scientific guidelines for conducting such programs. Consequently, on April 11, 1988, the HHS published a document in the *Federal Register* entitled "Mandatory Guidelines for Federal Workplace Drug Testing

The submitted manuscript has been authored by a contractor of the U.S. Government under contract DE-AC05-84OR21400. Accordingly, the U.S. Government retains a nonexclusive, royalty-free license to publish or reproduce the published form of this contribution, or allow others to do so, for U.S. Government purposes.

†Operated by Martin Marietta Energy Systems, Inc., under contract DE-AC05-84OR21400 with the U.S. Department of Energy.

Programs; Final Guidelines; Notice" (6).* These guidelines have made a considerable impact on the quality and standardization of drug abuse testing programs and have become de facto "standards" to which all government drug abuse testing programs must conform. In addition, the U.S. Nuclear Regulatory Commission (NRC) and the U.S. Department of Transportation (DOT) have modified these guidelines and have issued rules on June 7, 1989 (7) and December 1, 1989 (8), respectively.

It is significant that the NRC and DOT rules require that the essential elements of the NIDA guidelines apply to selected industries in the private sector. To mandate that the NIDA guidelines apply to *all* sectors of the U.S. workplace, current legislation is pending in the U.S. Congress. When, and if, enacted, this legislation will result in a single federal standard to which all drug abuse testing programs must conform.

The NIDA guidelines are organized into three sections. Section 1 is entitled "General," and includes subsections on applicability and definitions. Section 2 discusses a variety of scientific and technical requirements, and Section 3 discusses certification of laboratories engaged in urine drug testing.

This report will discuss several of the key elements found in Sections 2 and 3 of the NIDA guidelines and how these guidelines relate to planning and implementing a drug abuse testing program.

II. SCIENTIFIC AND TECHNICAL ISSUES

A. Drugs

Before implementing a drug testing program, a company or agency should first decide for which drugs they will test. The NIDA guidelines allow a federal agency to test for any drug included in Schedules I and II of the Controlled Substance Act for which no valid prescription has been issued. In practice, however, the guidelines are very specific in that they only allow for testing of the five drugs listed in Table 1. Additional drugs may be added to this group of five by petitioning the Secretary of the Department of Health and Human Services.†

The NIDA guidelines also provide specific concentration levels (Table 1) for the five drugs that are used to designate a specimen as positive or negative. If a specimen is found to have drugs or metabolites at concentration levels equal to or greater than their listed levels, the specimen is deemed positive. Most drug testing programs have adopted these concentration levels, but a recent Consensus Conference (9) has recommended that the screening levels for marijuana and cocaine metabolites be reduced to 50 and 100 ng/ml, respectively.

B. Testing Accuracy

Initially, drug abuse testing in the workplace generated considerable controversy, due primarily to legal and technical considerations, with the latter focusing specifically on the accuracy of the testing process. To ensure accuracy in the testing process, the NIDA guidelines mandate that drug testing be performed in a two-stage process.

---

*NOTE: This document was prepared for HHS by the National Institute on Drug Abuse (NIDA) and is therefore known as the "NIDA Guidelines."

†Many private programs include drugs such as barbiturates, benzodiazepines, methadone, and methaqualone in their testing menu. In addition, it was recommended at a recent Consensus Conference that the NIDA guidelines be revised to include barbiturates and benzodiazepines (9).

Table 1 Drugs of Drug Metabolites and Their Cutoff Levels for Which Testing is Allowed by the NIDA Guidelines

	Cutoff levels (ng/mL)	
Drugs	Screening	Confirmation
Marijuana metabolites	100	15
Cocaine metabolites	300	150
Opiate metabolites	300	300
Phencyclidine	25	25
Amphetamines	1000	500

Initially, each specimen is "screened" for the specified drugs or their metabolites. If the results are positive (i.e., the concentration level is equal to or greater than the concentration levels given in the screening column of Table 1), then the specimen is designated as a presumptive positive. The presence and concentration of the drug or its metabolite(s) must then be confirmed using an analytical technique that employs a different analytical principle having better analytical sensitivity and specificity than the technique used for screening (10). A logical path for this two-tiered approach to testing is shown in Figure 1.

Several techniques are used for screening and include planar chromatography (e.g., thin-layer chromatography) and various types of immunoassays. For confirmation, both liquid and gas chromatography are sometimes used, but gas chromatography/mass spectrometry (GC/MS) is currently the technique of choice.

C. Collection of Specimens

To ensure the validity of the drug testing process, special care is required during specimen collection (6,10). Consequently, a detailed protocol must be developed, and standard operating procedures must be written and adhered to rigorously throughout the collection process. In addition, the collection site must be secured, and facilities that protect the privacy of the individual should be provided, unless observed collection is necessary (6). Precautions must also be taken to guarantee the authenticity of the specimen and to prevent substitution and adulteration. For example, the NIDA guidelines require that the temperature of the specimen has to be measured within 4 min of voiding and has to be within the temperature range 32.5 to 37.7°C. Other physical tests that can be performed to ensure the authenticity of a specimen include specific gravity, pH, and chemical tests (such as creatinine and electrolytes) to confirm that a specimen contains physiological levels of these analytes. To prevent dilution of a specimen, a blue dye is added to the water in the toilet bowl, and other sources of water are turned off during the collection process. A minimum of 60 mL is required for each sample.

D. Documentation

Extensive documentation is required in a drug testing program, beginning with the collection of the specimen and every subsequent step in the collection process. This process of documentation is called the "chain-of-custody," and its purpose is to

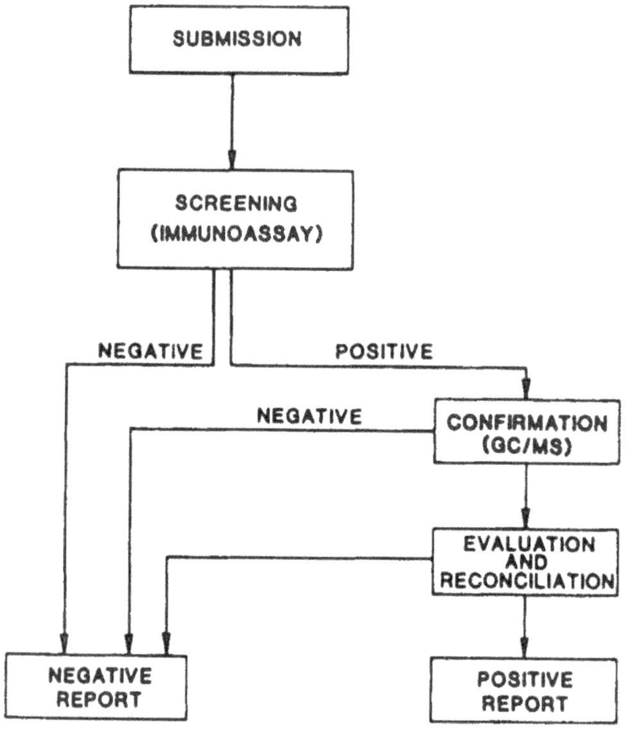

Figure 1 Logic path for processing data from a drug testing program.

demonstrate and document the identity and integrity of a specimen throughout the entire testing procedure. It is a legally defensible record of the individual providing the specimen and every other person who has been responsible for it or aliquots of it.

In addition to chain-of-custody documentation, other documents are needed, such as consent, request, and drug information forms. The consent form provides information about the program and the testing process and gives details about the sequence of events that are followed if a positive result is obtained. The request form specifies the drugs for which each specimen will be tested. The drug information form provides the individual with the opportunity to provide prior information to the medical review officer (MRO) concerning his or her legal drug use.

To consolidate and standardize the documentation required in the collection process and to ensure that the individual providing the specimen (donor) is never identified by name to the testing laboratory, the DOT (8) has developed a comprehensive urine custody and control form (UCCF). This form includes several sections (Table 2) for the identification of the individual providing the specimen, the collection-site person, the MRO, and the testing laboratory. In addition, the UCCF contains a chain-of-custody section, a request section, and a section where the laboratory enters the analytical results. The UCCF is also a multicopy form, with copies 1 and 2 going to the laboratory, copy 3 to the MRO, copy 4 to the donor, copy 5 to the collector, and copy 6 to the employer. After the laboratory has completed the analysis, the results are entered in the appropriate sections of copies 1 and 2, with copy 1 being retained by the

Table 2 Individual Sections Included in the Department of Transportation Urine Custody and Control Form

1. Employer identification
2. Medical review officer (MRO) identification
3. Order request
4. Type of test
5. Temperature of specimen
6. Chain of custody
7. Donor identification (copies 3-6 only)
8. Laboratory results
9. Acknowledged receipt and review by MRO

laboratory and copy 2 sent to the MRO. It should be noted that only on copies 3 through 6 is the donor identified by name—on copies 1 and 2, which go to the laboratory, the donor is identified by number only.

Each UCCF contains a unique number that is printed on each copy of the form. A removable label with this number is also provided on the UCCF. To execute the label, the collection-site person enters the identification number (Social Security or some other I.D. number) of the individual donating the specimen on the label and the donor then initials and dates it. In the presence of the donor, the collection-site person then removes the label and places it on the specimen bottle; thereafter, the specimen is identified by the UCCF number. After receiving copy 2 from the laboratory, the MRO matches its UCCF number with that of copy 3 that he or she had previously received, and both are then placed in the donor's file to await further action.

III. CERTIFICATION OF LABORATORIES

In addition to the type of analytical technologies used to perform accurate drug analyses, the reliability of the laboratory performing the testing is also important in ensuring analytical accuracy. Consequently, the U.S. government requires that testing laboratories be certified before performing drug analyses on specimens obtained from government employees. Currently laboratories may be certified either by the NIDA or by the American Association for Clinical Chemistry (AACC)/College of American Pathologists (CAP), through their joint forensic drug testing program. Both of these certification programs are very rigorous and consist of demonstrated proficiency followed by an extensive on-site inspection. After certification, a continued demonstration of proficiency is required, and unannounced on-site inspections occur.

Criteria that are used to judge a laboratory's capabilities include (a) adequacy of its facilities, (b) expertise and experience of the laboratory personnel, (c) excellence of its quality assurance/quality control program, (d) its performance in proficiency tests, (e) its compliance with standards as reflected in any laboratory inspections, and (f) any other factors affecting the reliability and accuracy of drug tests and reporting done by the laboratory (6). As of October 1, 1990, 55 laboratories had been certified by NIDA and 64 by the AACC/CAP joint program. Several laboratories have been certified by both programs.

A. Quality Assurance

Due to the sensitive nature of a drug testing program and the potentially adverse consequences of a positive drug finding on an individual, it is mandatory that such programs have a comprehensive quality assurance (QA) program accompanying them. The basic component of a QA program is a set of comprehensive documents that are rigorously adhered to. The initial document in this set would be the Drug Testing Policy. Each company and agency should develop such a policy document (11,12) before testing is begun, and it should be available to all individuals who are affected by it.

In addition to the policy document, a series of documents that describe all the details of the collection process (6,13) the preparation and storage of records, the role of the MRO (14), and internal and external quality control are also needed. As with all QA documents, they must show approval by the appropriate supervisory personnel, must indicate the effective date, and must be periodically reviewed. In addition, personnel using these procedures must be trained in their use, and training records should be documented.

To demonstrate that the policy and operation procedures are being followed with rigorous adherence, the program should be audited periodically. In addition, the performance of the laboratory should be routinely tested by a blind proficiency program (8,15) in which either negative or positive specimen-containing drugs and their metabolites at specified concentrations are submitted to the laboratory as genuine specimens. The laboratory must accurately detect and quantitate the presence and absence of drugs in these specimens.

IV. SUMMARY

Drug abuse testing has become a widespread practice in both the private and government sectors of the U.S. workplace. As a result of the availability and wide application of several government documents, a degree of standardization has been bought to drug abuse testing, since these precedents are being used as de facto standards to which all such programs must adhere. In addition, drug abuse testing laboratories must undergo a rigorous certification process that is maintained only by continually demonstrating their proficiency in providing accurate and reliable drug results. As a consequence of these standardization and certification efforts, drug abuse testing in the workplace is increasing, and initial results in many of these programs indicate that such testing is effective in establishing a drug-free work environment.

REFERENCES

1. C. R. Schuster. The United States drug abuse scene: An overview. *Clin. Chem.*, 33:5–12B (1987).
2. J. Castro. Battling the enemy within, *Time*, pp. 52–63 (March 17, 1986).
3. S. W. Gust and J. M. Walsh. Research on the prevalences, impact, and treatment of drug abuse in the workplace. In *Drugs in the Workplace* (S. W. Gust and J. M. Walsh, eds.), National Institute and Drug Abuse, NIDH Research Monograph, Rockville, MD, *91*, pp. 3–16 (1989).
4. J. Sheridan and H. Winkler. An evaluation of drug testing in the workplace. In *Drugs in the Workplace* (S. W. Gust and J. M. Walsh, eds.), National Institute and Drug Abuse, NIDH Research Monograph, Rockville, MD, *91*, pp. 195–216 (1989).

5. J. M. Canella. Drug abuse in the workplace: An industry's point of view. *Clin. Chem.*, 33:61B–65B (1987).
6. Mandatory guidelines for federal workplace drug testing programs; Final guidelines; Notice, *Federal Register*, 53(69):11970–11989 (April 11, 1988).
7. Fitness-for-Duty programs; Final rule and statement of policy. *Federal Register*, 54(108):24468–24508 (June 7, 1989).
8. Procedures for transportation workplace drug testing programs; Final rule and notice of conference. *Federal Register*, 54(230):49854–49884 (December 1, 1989).
9. B. S. Finkle, R. V. Blanke, and J. M. Walsh. Employee drug testing consensus report, *HHS Publication No.* (ADM), National Institute on Drug Abuse, Rockville, MD, pp. 90–1684 (1990).
10. Critical issues in urinalysis of abuse substances: Report of the substance-abuse testing committee. *Clin. Chem.*, 34:605–632 (1988).
11. Model plan for a comprehensive drug free workplace program. *DHHS Publication No. (ADM)* 89-1635, National Institute on Drug Abuse, Rockville, MD (1988).
12. J. D. Carraher, B. S. Nogman, R. E. Willette, and W. C. Collins. Testing for substance abuse: A policy and procedure guide for industry. Syva Corporation, Palo Alto, CA (1986).
13. Urinalysis collection handbook for federal drug testing programs. *DHHS Publication No. (ADM)* 88-1596, National Institute on Drug/Abuse, Rockville, MD (1988).
14. Medical review officer manual—A guide to evaluating urine drug analysis. *DHHS Publication No. (ADM)* 88-1526, National Institute on Drug Abuse, Rockville, MD (1988).
15. C. S. Frings, D. J. Battaglia, and R. M. White. Status of drugs-of-abuse testing in urine under blind conditions: An AACC study. *Clin. Chem.*, 35:891–894 (1989).

81

Hair Analysis: Techniques and Potential Problems

David A. Kidwell *Naval Research Laboratory, Washington, DC*
David L. Blank *Naval Military Personnel Command, Washington, DC*

I. INTRODUCTION

For the past several years the Department of Defense (DoD) has placed increased emphasis on detection and deterrence of drug abuse. The employment of urinalysis as a deterrence for drug use is one of the key components of this emphasis. The Navy, as a component of DoD, has been a strong proponent of urinalysis and as such is probably the world's largest drug screening organization. As emphasis was placed on the urinalysis program, the self-reported drug use rate in the Navy dropped from 33% to under 4% in the past 5 years. A recent survey of substance abuse in the military found that 76% of this population believes that urinalysis testing has reduced drug use in the military (1).

However, despite the success of the Navy's program, there is always room for improvement. For example, the short half-life of many drugs of abuse in urine makes detection of infrequent drug use difficult. About four years ago, due to various reports that were appearing in the literature, we became interested in hair analysis as an adjunct to urinalysis (2–5). Policy dictates that when an individual is identified as positive in the urinalysis program, an administrative hearing or court martial may be held. One application of hair analysis could be to support or refute an individual's testimony at such an administrative hearing or trial. Hair analysis could provide a long-term usage history, so that better decisions could be made regarding the retention of an individual in the Navy. Hair analysis could also be used in posttreatment monitoring. However, before hair analysis could be employed, several important questions needed to be answered, for example: What is the sensitivity of hair analysis? Could a single use of drugs be detected? What if the results of urinalysis and hair analysis disagreed? What confirmation technology is appropriate? What are the mechanisms of incorporation, retention, and loss of drugs of abuse? The last question will be addressed in this report.

II. RESULTS AND DISCUSSION

A. Mechanisms for the Incorporation of Drugs in Hair

How do drugs become incorporated into hair, and how are they retained? Answers to these questions are critical to defining the scope of hair analysis. If drugs can be introduced from an external source, then passive exposure is possible. Passive exposure is also possible for urinalysis, but one major distinction between urine and hair is that the body has active clearing mechanisms for most species. Therefore, the lifetime of a drug in urine is limited to a few days after exposure. Hair does not have such mechanisms. Hair can be exposed to drugs and retain it for a long time, as shown below. Thus, on a random test, the probability of being misidentified via urinalysis is much reduced compared to hair analysis, because the window of detection is reduced.

One mechanism of incorporation of drugs into hair is dynamic equilibrium. Two opposing forces are interacting between incorporation and retention of the drug in the hair and loss of the drug to the environment. Drugs may be incorporated into the hair shaft as one of three possibilities. They may be incorporated from the blood as the hair grows, from sweat, or from an external source. Sweat could be a potential rich source of drugs, since drugs are known to be excreted in sweat (5,7,8) and the concentrations of materials in it would increase as water evaporates.

Does sweat contribute to incorporation of drugs? We had the opportunity to test this hypothesis with a drug-free individual. Hair was obtained from an individual who had taken 135 mg of dihydrocodeine over a 1-week period after an operation. The hair was cut 2 to 4 cm from the scalp (by shaving with a cutter set to a given length) 1 week after the dihydrocodeine was stopped. This individual did not undertake strenuous exercise/work during this 2-week period, and the hair was washed daily through normal hygiene. Analysis showed dihydrocodeine at a level of approximately 0.4 ng/10 mg. Since this hair could not have grown 2 to 4 cm in the week after the dihydrocodeine was stopped, sweat could have been a contributor of drug incorporation. Also, the amount detected was surprisingly high considering that most opiate abusers have only 1 to 10 ng/10 mg.

The retention of drugs in the hair may be by ionic forces, in which case differences in the charge on the drug would affect profoundly the amount retained. For example, negatively charged (anionic) species, such as aspirin, are not well retained in hair (9), whereas cationic species such as hair dyes are well retained. The loss from the hair could be by normal washing, dyeing, perming, bleaching, etc., or by degradation of the drug in the hair shaft.

One model of such a dynamic equilibrium is the dyeing of hair. If one applies red hair dyes to blond hair, the hair will turn red. The incorporation of the dye into the hair is very rapid and the binding of the dye quite tenacious. Yet exposure to the sun for several weeks, or repeated washing of the hair, will decrease the red color intensity. The color would fade either through leaching of the dye from the hair or through gradual degradation of the dye into colorless compounds. If the hair is redyed several weeks later, the red color will be restored. In this example, a rapid uptake of dye is followed by a slow decrease in dye concentration. Interestingly, most hair dyes are cationic compounds, and so are most drugs of abuse (cocaine, amphetamines, opiates, PCP) that are found in hair in high concentrations.

If this dynamic equilibrium mechanism is tenable for drugs of abuse, several concerns arise. Dynamic equilibrium implies both that drugs deposited endogenously into hair can be removed and that drug present exogenously can be incorporated into hair. Therefore, the amount of drug ingested may not be correlated with the amount found in the hair. For example, as hair ages, drugs deposited endogenously will have time to leach out and their concentrations may fall below the detection limit of the assay. On the other hand, drugs deposited in recently grown hair will not have time to leach out and may be detected in hair at high concentrations.

B. Mechanisms for Retention of Drugs in Hair

One mechanism for the retention of drugs in hair is to consider hair as an ion-exchange membrane. Hair contains negatively charged amino acids that may bind the drugs as an ionic pair. This may be one reason why cationic drugs can be found in higher concentrations compared to anionic species such as Δ9-tetrahydrocannabinol carboxylic acid or aspirin.

We investigated the ion-exchange mechanism in the following experiment. We carefully selected two molecules (Figure 1). One was PCP labeled with a fluorescent compound, DANSYL-PCP, and the control was DANSYL-aniline. These materials were chosen because the DANSYL-PCP has only one potentional cationic function, which is absent in the control. Although the DANSYL group does contain a dimethylamino substituent, if this is protonated to form a cationic ammonium salt and retained in the hair, the DANSYL group will not be fluorescent.

We exposed hair to solutions of these compounds, mounted the hair, and sectioned it. Observation of the cross sections under a fluorescence microscope produced some very interesting results. Although hair soaked in a control solution showed some fluorescence, its intensity was much less than that of the labeled PCP. This could indicate that either the control does not bind as well as the labeled PCP or that, when it is bound, it is protonated and therefore not fluorescent. Since the source of the hair

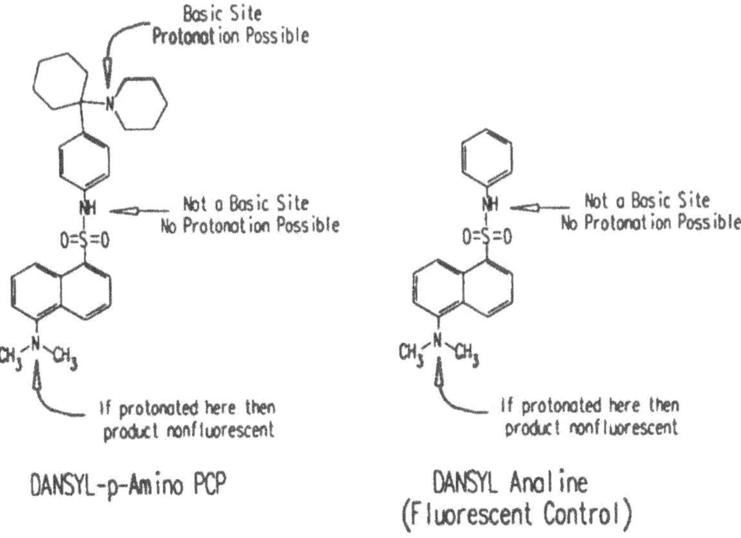

Figure 1 Fluorescently labeled PCP and control.

was known, we knew that the hair had not been damaged by bleaching or perming, which could allow access of the fluorescent compounds to the interior. In many cross sections of the hair shaft the DANSYL-PCP was only on the surface. However, in a few cases the fluorescence was observed throughout the hair, and in some cases in both the center and the outside. These different results were obtained on the same strand of hair. Since a single strand of hair produces such varying results, we might expect even greater variability among hair samples from different individuals.

Theoretically, other mechanisms may exist for retention of drugs in hair, such as hydrophobic/hydrophilic interactions. However, if such interactions played a large role, then compounds such as THC, a highly water-soluble material, would be tightly retained compared to cocaine, a highly water-soluble material. This is contrary to what is observed when both drugs are detected; cocaine is found in larger amounts than is THC.

C. Passive Exposure

Questions arise as to how valid these fluorescent model compounds are to actual drugs of abuse. To obtain a better model system, a sample of hair from a cocaine user was soaked in a solution of *p*-bromococaine, a derivative of cocaine, at 1 μg/mL for 1 h. It was then rinsed and air-dried and extracted according to literature procedures for cocaine (2). The hair was washed once with ethanol, three times with phosphate buffer (pH 7), rinsed with water three times, and then the cocaine remaining in the hair was extracted two times with 0.6 N HCl. Cocaine and *p*-bromococaine were quantitated in all solutions by gas chromatography/mass spectrometry (GC/MS); the results are shown in Figure 2. The wash-out kinetics for the cocaine and the externally introduced *p*-bromococaine are very similar, again implying that external contamination can mimic drug use even at as low a concentration as 1 μg/mL.

Note that even though the hair was not dissolved in the drug extraction step, significant amounts of both cocaine and *p*-bromococaine were found in the final hair extracts. The quantity of *p*-bromococaine was 67 times greater in this extract than in the last wash. This ratio between the quantity extracted from hair and that in the last wash has been suggested by some to distinguish between active and passive exposure (10). Although the criteria are arbitrary, a factor of 10 between the extract and the wash has been suggested. Using this factor, this cocaine user would have been considered a user of *p*-bromococaine. Little added discrimination would be gained by increasing this ratio, since then most users of drugs would be considered negative.

Of course, some external contamination can be washed away. In initial experiments, we exposed hair to vapors of PCP, rather than cocaine, and found that very large amounts of PCP could be removed by washing. However, large amounts of PCP remained when the hair was extracted, more than that observed in PCP users' hair. This cannot be compared directly to the soaking experiments described above, since hair may have a limited capacity to retain drugs. In many of these high-concentration experiments, large amounts of drug are removed by washing, which may give a false impression that washing is effective in removing external contamination instead of the true result that the retention of drugs in hair is limited.

What is the likelihood that an individual could expose his or her hair to a solution of a drug? The answer is unknown. However, people do sweat. Sweat will dissolve drugs readily from the air and concentrate them. When the sweat dries, the hair

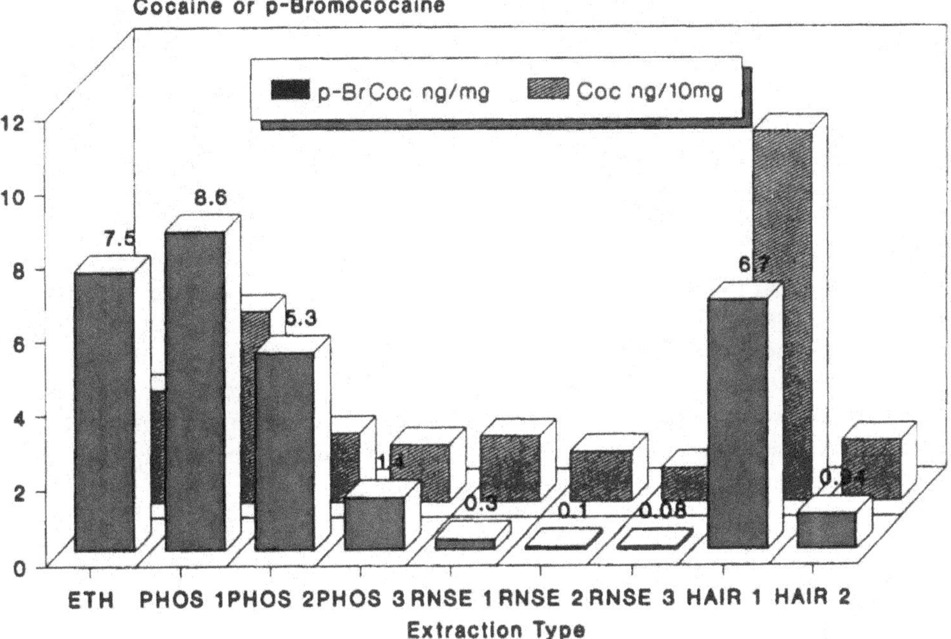

Figure 2 Extraction profile of a cocaine user's hair. Note that the *p*-bromococaine axis is 10 times the cocaine axis. Hair 1 and hair 2 are final extraction steps with acid.

would be "soaked in a solution of drugs." By this mechanism, passive exposure may occur in non-drug users.

D. Metabolite Screening as a Possible Solution to Passive Exposure

One method to distinguish between active ingestion of a drug and passive exposure is to measure the metabolites of the drug rather than the parent compound. Cocaine is one of the few drugs that is excreted primarily as its metabolites. Many of the other drugs of abuse are excreted as the parent drug, in which case passive exposure could be of concern. Presumably, if the metabolites of cocaine were found in the hair, then it would be assumed that the individual had ingested the drug. This is only partially true. The three primary metabolites, benzoylecgonine, methyl ecgonine, and ecgonine, are present in blood and are excreted in urine in varying amounts that differ among individuals. However, these metabolites may also be formed by in-vitro hydrolysis with base. For example, if negative urine is spiked with cocaine and then made basic with sodium carbonate to pH 10, the cocaine will completely degrade to ecgonine in 12 h at room temperature. In-vitro basic conditions will produce *exactly* the same metabolites from the parent drug as will human metabolic processes, depending on time, pH, and temperature.

As is well known, hair is typically washed with detergents in normal personnel hygiene. These detergents are basic, typically pH of 10. In fact, even tap water is pH 8 and is buffered at this pH to prevent lead from leaching from the water pipes. With normal hygiene, hair is constantly being exposed to basic conditions, under which cocaine, if present, may degrade into "metabolites." Whether this actually occurs is unknown. However, some evidence suggests that cocaine is degraded in hair (Figure 3). In this case, sectional analysis was performed on an individual with a self-reported constant use rate of cocaine over a several-year period. We examined this hair with thermal desorption tandem mass spectrometry and quantitated cocaine, benzoylecgonine, and ecgonine (6). In Figure 3, the centimeters from the scalp constitutes a time line of drug use and, perhaps, a time line of degradation for the cocaine. Ecgonine increases with the length (time) of the hair, and ecgonine also would be the final degradation product of cocaine. Unfortunately, a simple explanation of these data is not possible. If the increase in ecgonine is due to degradation, then we should have seen a concomitant decrease in cocaine in distal sections. Such was not the case; therefore, the source of increased ecgonine is not clear. One possibility is that the incorporation/wash-out kinetics of metabolites into/out of hair are different than of cocaine. It is known that if hair is soaked in solutions of cocaine or benzoylecgonine, the uptake rates are different (11).

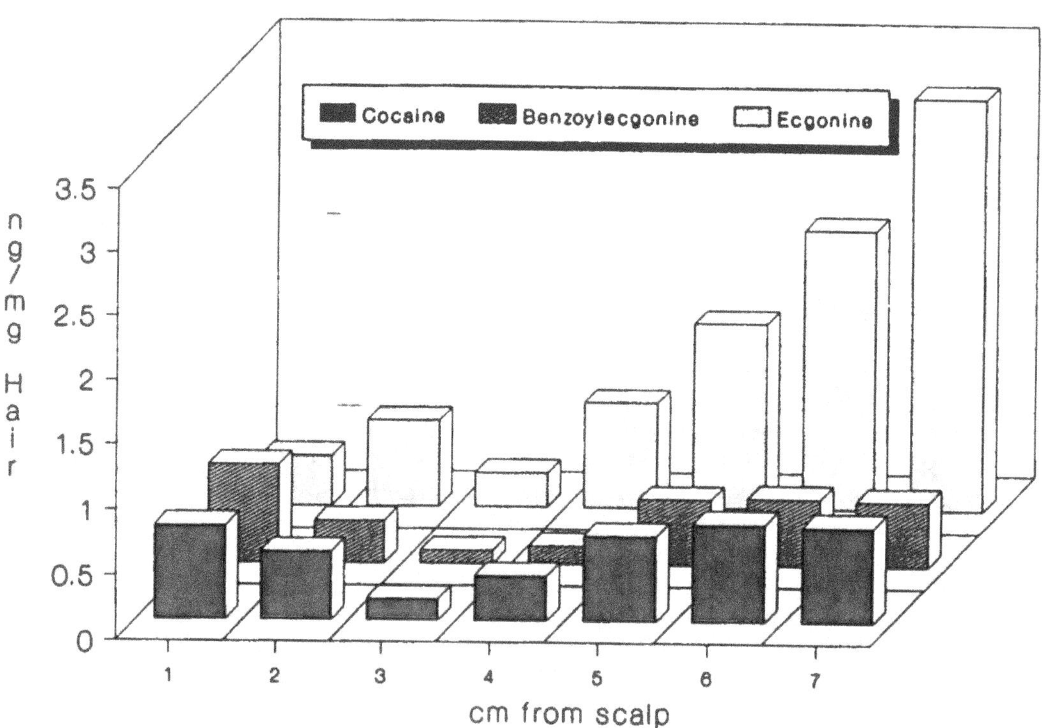

Figure 3 Concentration of cocaine, benzoylecgonine, and ecgonine in a user's hair. The intensity decrease at the third section was probably due to problems in the quantitation. The relative amounts of cocaine and the metabolites at this section would not be affected.

E. Differences Between Hair Types

Hair is an extremely heterogeneous media, which is treated differently by probably every individual on earth. Are there differences among hair types with regard to the uptake and retention of drugs? One answer to this question is to compare the self-reported drug use history between users with two types of hair, black and brown. The use histories were matched as closely as possible and the results are shown in Table 1. Table 1 indicates a dramatic difference between hair types. For example, one individual with black hair indicated use of 250 mg of cocaine on only one occasion and 6.4 ng of cocaine per 10 mg of hair was found. Compare that to an individual with brown hair who also indicated use of 250 mg of cocaine on five separate occasions. In this case the cocaine level in the hair was below the detection limit of about 1 ng/10 mg.

Similarly, data for a heavier user with black hair of 100 mg/week showed a small amount of cocaine in the hair. Compare this to a user with brown hair of 2-1/2 times as much cocaine per week. In this individual the amount was below the detection limit. Incidentally, differences in uptake and retention of drugs with hair coloration also were shown in some controlled animal studies (7,12).

Realizing that a comparison of self-reported use histories and drug amounts would be only preliminary data because of the limited number of samples available and their potential unreliability, the following experiment was conducted. Two types of undamaged and unbleached hair, light brown hair and thick black hair, were soaked in a solution of PCP for 1 h, removed, and allowed to dry. The hair was divided into portions and extracted with a number of different solutions. The residual drugs in the hair was determined by thermal desorption tandem mass spectrometry (6). PCP was used since it would be stable to almost all extraction conditions. The results are shown in Figure 4. The data are plotted relative to water, as some solutions are poorer in removing drugs than is water. As these data indicate, drugs were removed from the brown hair much more readily than from the black hair. The brown hair was treated only with the first three solutions, since these effectively removed the drug. If the concept of dynamic equilibrium described above is tenable, then drugs could be removed faster from an individual with brown hair as compared to one with black hair. The concentrations may then be at too low a level to be detectable with present technology, so such an individual would be negative by hair analysis. Also,

Table 1 Comparison of Cocaine Present in Different Hair Types

Black		Brown	
Use	Amount detected (ng/10 mg)	Use	Amount detected (ng/10 mg)
		0.25 g once	0
0.25 g once	6.4	0.25 g 5x's	0
4–5 g total	3.5	2–5 g total	2.2
		5 g total	0
100 mg/wk	2.2	250 mg/wk	0
500 mg/wk	18	250–500 mg/day	20

Figure 4 shows just how difficult it is to remove externally introduced drugs from some types of hair.

III. CONCLUSIONS

In order to enjoy the confidence of the public served, it is essential that any drug testing program employ drug testing procedures that are accurate, reliable, and relate to the individual's pattern of use of drugs. We believe that accuracy and reliability are not problems for hair analysis, based on our experience in tandem mass spectrometry and other techniques applicable to hair testing. However, based on the observations reported here, validity may be questioned. For example, does hair analysis measure the use of a drug or passive exposure to it? If it only measures exposure, then severe limitations may be placed on the scenarios where hair testing could be employed. Of course, it is possible that metabolites may distinguish between use and exposure, but if metabolites are to be used for such a determination then their stability in various types of hair exposed to numerous types of treatments must be established. If our dynamic equilibrium hypothesis is correct, we need to know the mechanisms that

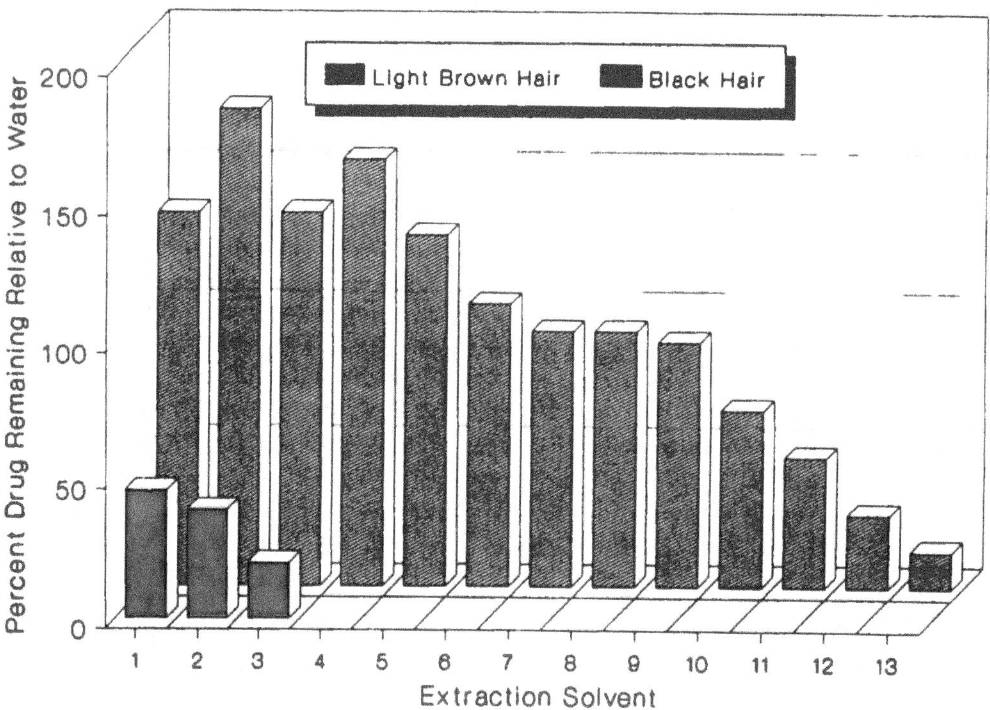

Figure 4 PCP remaining in blank hair soaked in PCP after various wash solutions. The amounts of PCP remaining in the hair are relative to water washing. Only the first three solutions were tested on the brown hair. 1, Triton X100; 2, sodium dodecyl sulfate; 3, dodecyltrimethylammonium chloride; 4, tris-hydrochloride; 5, calcium chloride; 6, triethylammonium chloride; 7, *p*-bromo PCP; 8, methylammonium chloride; 9, magnesium sulfate; 10, sodium chloride (1 M); 11, phosphate buffer pH 7; 12, 0.1 M HCl; 13, sodium carbonate. Except as noted, all concentrations are 0.1 M.

affect it: mechanisms of drug incorporation, retention and loss, and the differences in these mechanisms for different types of hair. If bias in the analysis of drugs is due to interindividual differences, such as hair thickness or color, use of hair analysis will be severely limited.

The results of our studies provide some information on these important issues and suggest that, at our present level of understanding of the process involved, the use of hair analysis should be limited to research applications.

REFERENCES

1. R. M. Bray, M. E. Marsden, L. L. Guess, S. C. Wheeless, V. G. Iannacchione, and S. R. Keesling. *1988 Worldwide Survey of Substance Abuse and Health Behaviors Among Military Personnel*, RT1/4000/06-02FR. Research Triangle Institute, Research Triangle, North Carolina, December 1988.
2. A. M. Baumgartner, P. F. Jones, C. T. Black, and W. H. Blahd. Radioimmunoassay of cocaine in hair. *J. Nuclear Med.*, 23:790 (1982).
3. A. M. Baumgartner, P. F. Johns, and C. T. Black. Detection of phencyclidine in hair. *J. Forensic Sci.*, 26:576 (1981).
4. M. Marigo, F. Tagliaro, C. Polesi, S. Lafisca, and C. Neri. Determination of morphine in the hair of heroin addicts by high performance liquid chromatography with fluorimetric detection. *J. Anal. Toxicol.*, *10*:158 (1986).
5. F. P. Smith and R. H. Liu. Detection of cocaine metabolite in perspiration stain, menstrual bloodstain and hair. *J. Forensic Sci.*, *31*:1269 (1986).
6. D. A. Kidwell. Analysis of drugs of abuse in hair by tandem mass spectrometry. Presented at the 36th American Society of Mass Spectrometry Conference on Mass Spectrometry and Allied Topics, San Francisco, June 6–10, 1988.
7. I. Ishiyama, T. Nagai, and S. Tosida. Detection of basic drugs (methamphetamine, antidepressants, and nicotine) from human hair. *J. Forensic Sci.*, *28*:380 (1983).
8. F. P. Smith and D. A. Pomposini. Detection of phenobarbital in bloodstains, semen, seminal stains, saliva stains, perspiration stains, and hair. *J. Forensic Sci.*, *26*:582 (1981).
9. W. H. Harrison, R. M. Gray, and L. M. Solomon. Incorporation of l-DOPA, 1-α-methylDOPA, and dl-isoproterenol into guinea pig hair. *Acta Dermatovener (Stockholm)*, *54*:249 (1974).
10. W. A. Baumgartner, V. A. Hill, and W. H. Blahd. Hair analysis for drugs of abuse. *J. Forensic Sci.*, *34*:1433 (1989).
11. C. C. Allgood, L. T. Sniegoski, and M. T. Welsh. Determination of drugs of abuse in hair by GC/MS. Presented at the 38th American Society of Mass Spectrometry Conference on Mass Spectrometry and Allied Topics, Tucson, Arizona, June 1990.
12. I. S. Forrest, L. S. Otis, M. T. Serra, and G. C. Skinner. Passage of ^{3}H-chloropromazine and ^{3}H-Δ-tetrahydrocannabinol into the hair (fur) of various mammals. *Proc. West. Pharmacol. Soc.*, *15*:83 (1972).

82

Forensic Hair Analysis: Cocaine

Frederick Paul Smith *University of Alabama at Birmingham, Birmingham, Alabama*

I. INTRODUCTION

Forensic hair analyses for opiates, amphetamines, phencyclidine, methaqualone, and cocaine-related substances have been conducted on drug overdose victims (1) and criminals claiming drug-induced incapacitation (2) in Germany since 1978 using radioimmunoassay (RIA) at first and, later, gas chromatography/mass spectrometry (GC/MS). Similar tests have figured in several U.S. legal proceedings during the 1980s as well (3–5). Among the issues in these disputes were sensitivity of methods, external contamination, interfering substances, controls, interpretation of results, and chain of custody.

This communication summarizes the methods, background, results, opinions, and legal challenges involved in the three U.S. case examples in which the author tested hair for cocaine-related substances.

II. MATERIALS AND METHODS

Hair samples were prepared, extracted, and analyzed using well-documented, previously published materials and methods (4,6). Before any questioned (case sample) hairs were analyzed, negative control hair was extracted, filtered, dried (using the same extraction glassware as the questioned hair), and analyzed as a check for contamination and interference.

III. CASE EXAMPLES AND RESULTS

A. Case 1

In May 1982, Richard Majdic was tried in Kodiak, Alaska, on the charge of sexual assault involving his former paramour. Head hair exemplars were collected from the complainant by police investigators. None of the physical evidence proved useful to the prosecution. Therefore, the complainant's credibility was essential to the state's case. Earlier, the complainant had denied every using cocaine.

Defense attorney Phillip Paul Weidner received court permission for an examination of the evidence by an expert chosen by the defense. Samples sent by the Kodiak police included 41 head hairs.

Radioimmunoassay* revealed 3.2 ng benzoylecgonine (BE) equivalents/10 mg hair. Other samples (perspiration stain and blood stain) revealed amounts of BE equivalents within the linear region of the standard curve. The lowest amount, 570 pg, was found in the bloodstain; this sample was confirmed by GC/MS analyses conducted by prosecution experts. These results were used to support the opinion that they indicated cocaine ingestion.

The state's attorney challenged the hair test results on a number of points. First, the nanogram finding was challenged on the grounds that such a small amount could have resulted from airborne contamination. This argument was countered by citing agreement of duplicates, linearity of the standard curve, and results of controls. Another challenge involved the possibility that the positive results could have been caused by handling evidence with cocaine-contaminated hands. This argument was countered by the finding of BE equivalents extracted from within the pulverized hair, after six successive washings of the intact hair surface. Next, the specificity of the RIA was scrutinized intensively. The manufacturer's reported specificity for a limited number of structurally similar compounds was cited, showing how minor structural variations from cocaine and BE resulted in nearly negligible recognition by the RIA.

After this testimony at trial, the complainant admitted cocaine use. Mr. Majdic was acquitted.

B. Case 2

In August 1986, a bus driven by Perry Jioia of Staten Island, New York, was hit by an at-fault motorist. This was Mr. Jioia's first accident in nearly 20 years as a professional driver, and neither he nor his family had any record of involvement with illicit drug use or any other wrongdoing. His mandatory urine specimen collected following the accident tested positive for cocaine use. Reconfirmation by an independent laboratory using GC/MS revealed 19,000 ng BE/mL and 600 ng cocaine/mL urine, unusually high concentrations considering his calm demeanor at the time of the accident. When given the ultimatum of either admitting cocaine use and entering drug abuse treatment voluntarily or denying cocaine use and losing his job, he protested. His union attorney, James Reif, of Brooklyn, New York, obtained a hair sample from Mr. Jioia and submitted it for analysis.

RIA revealed no detectable amounts of cocaine, cocaine metabolite, or immunologically similar materials in washings of the exterior portion of the hair or in the ethanolic extracts from the pulverized hair. Given studies showing some correlation between drug consumption and BE equivalents/mg hair (6,7), the finding of no detectable amount of BE or immunologically similar substances in the large hair sample (500 mg) supported the opinion that there was no evidence that Mr. Jioia had consumed cocaine during the past 6 months of hair growth tested.

The New York City Transit Authority attorney challenged the sensitivity of the test and questioned the amount of cocaine Mr. Jioia would have had to consume to cause a detectable result. Based on research with cocaine and other drugs (6,7,11), an

* Abuscreen radioimmunoassay for cocaine metabolite, Roche Diagnostics System, Nutley, NJ.

amount of approximately 500 mg of cocaine consumed would probably result in a detectable quantity in the large hair sample tested. Therefore, the test results showing no evidence of cocaine use were not consistent with the high concentrations of such a highly addictive, repeat-use drug as cocaine detected in Mr. Jioia's urine. Other aspects casting doubt on the results from this urine specimen included chain-of-custody questions, unaccounted-for urine volume discrepancies, and the initial laboratory's insistence that a laboratory error was absolutely impossible.

The arbiter in this dispute awarded Mr. Jioia lost wages, lost overtime compensation, and a return to his job.

C. Case 3

In August 1988, hair samples allegedly originating from Mrs. Ninni Burgel were collected surreptitiously by Mr. Burgel and submitted for analysis by attorney Peter Bronstein of New York City, who represented Mr. Burgel in the couple's child custody dispute. Mrs. Burgel freely admitted previous cocaine use but denied any recent, continuing use.

RIA analysis of the hair extract revealed amounts of BE or immunologically similar substances that exceeded the highest concentration standard. Mrs. Burgel declined to submit a second hair sample for GC/MS confirmation. Attorney Bronstein obtained a court order requiring that she appear before a licensed medical practitioner and permit head hair samples be cut for further analysis. Mrs. Burgel's attorneys, with assistance from forensic toxicologists, argued that the hair test proposed was not reliable and not generally accepted. The trial judge ruled in favor of the required hair test. Although Mrs. Burgel's attorneys appealed this ruling, the New York State Appeals Court upheld the trial court judge's option to order the hair test (8).

Within 6 days of the appeals court ruling, the parties reached an out-of-court settlement (9).

IV. DISCUSSION

Immunoassays, particularly the RIAs used in these examples, provide high specificity and sensitivity comparable to GC/MS. When performed properly, they provide excellent results. Given the human error potential of mass screening by technicians plus the advent of widely available GC/MS instruments, the practice of confirming positive initial results by GC/MS has solid scientific basis.

While external contamination has been raised as an issue in urine testing, hair can be washed and rinsed in various solutions to remove external contamination of hair. Possible sources of external contamination include touching hair with hands contaminated with traces of drug, perspiration, glandular secretion, and soaking hair in chemical solutions containing high concentrations of drugs (10). BE cutoff concentrations of 0.85 ng/10 mg hair (11) or 1.25 ng/5-61 mg hair (12) may prevent low-level contamination errors by the parent substance, cocaine. The presence of the metabolite BE further implicates cocaine ingestion. In addition, the presence of BE diminishes the likelihood of external contamination.

In each of these examples, cocaine substances were extracted using a previously published method (4,6). Many approaches (acids, bases, organic solvents, and enzymes) (10,11,13) may dissolve hair for drug removal. With cocaine it is important to avoid high pH because alkaline solutions will hydrolyze the ester bonds, converting cocaine (observed to exceed BE concentrations in hair) (11,14,15) to BE and, ultimately, to ecgonine.

One way to address questions about the reliability of hair test results is to collect sufficient sample (500 mg) to permit GC/MS confirmation, retest, and an additional retest by an independent laboratory, if requested.

V. SUMMARY

It would be interesting to know more about the presence of drugs in hair. For example, the mechanism(s) of drug incorporation, dose/response relationship, and effects of bleaching, dying, perming, age, race, and sex on detectability of drugs and their metabolites in hair would provide useful topics for research. None the less, as stated by James E. Starrs, professor of law and forensic sciences at George Washington University, "we don't need to understand the mechanism by which a drug gets into hair to know it's there" (9).

REFERENCES

1. W. Arnold. The determination of drugs and their substitutes in human hairs. *Forensic Sci. Int.*, *46*:17 (1990).
2. M. R. Moeller, Institute of Legal Medicine, Hamburg University, Germany, personal communication (1990).
3. Majdic trial continues with chemist's testimony. *Kodiak Daily Mirror*, 42(87):1 (1982).
4. F. P. Smith and R. H. Liu. Detection of cocaine metabolite in perspiration stain, menstrual bloodstain, and hair. *J. Forensic Sci.*, *31*:1269 (1986).
5. L. Buder, Court allows testing of hair for cocaine in a custody battle. *The New York Times*, October 28, p. 1 (1988).
6. A. M. Baumgartner, P. F. Jones, W. A. Baumgartner, and C. T. Black. Radioimmunoassay of hair for determining opiate-abuse histories. *J. Nuclear Med.*, *20*(7):748 (1979).
7. Y. Nakahara, K. Takahashi, Y. Takeda, K. Konuma, S. Fukui, and T. Tokui. Hair analysis for drug abuse, part ii. Hair analysis for monitoring of methamphetamine abuse by isotope dilution gas chromatography/mass spectrometry. *Forensic Sci. Int.*, *46*:243 (1990).
8. *Burgel v. Burgel*. New York State 2nd Appellate Div., 553:735 (1988).
9. Novel hair analysis said to prove drug use. *BNA Criminal Practice Manual Trial Practice Series*, 4(12):270 (1990).
10. D. A. Kidwell. Hair analysis: Techniques and problems. Presented at the 2nd International Congress of Therapeutic Drug Monitoring and Toxicology, Barcelona, Spain (1990).
11. W. A. Baumgartner, V. A. Hill, and W. H. Blahd. Hair analysis for drugs of abuse. *J. Forensic Sci.*, *34*:1433 (1989).
12. S. A. Reuschel and F. P. Smith. Benzoylecgonine detection in hair samples of jail detainees using RIA and GC/MS. *J. Forensic Sci.*, in press.
13. C. Offidani, A. Carnevale, and M. Chiarotti. Drugs in hair: A new extraction procedure. *Forensic Sci. Int.*, *41*:35 (1989).
14. R. M. Martz. The identification of cocaine in hair by GC/MS and MS/MS. *Crime Lab. Digest.*, *15*(3):67 (1988).
15. D. A. Kidwell. Analysis of drugs of abuse in hair by tandem mass spectrometry. Proc. 36th American Society of Mass Spectrometry Conference on Mass Spectrometry and Allied Topics, San Francisco (1989).

83

Hair Test to Verify Gestational Cocaine Exposure

Gideon Koren,[†] Julia Klein, Karen Graham, and Rachel Forman *The Hospital for Sick Children, Toronto, Ontario, Canada*

I. INTRODUCTION

During the last decade there has been increasing use of cocaine by young adults in North America. According to the Addiction Research Foundation in Toronto, about 10% of men and women between 18 and 28 years of age have consumed cocaine during the late 1980s. A variety of indicators, such as an increase in cocaine-related crimes or deaths and in drug seizures by police, coupled with a decreasing cost of this compound, indicate that these numbers are increasing (1).

The potential effects of cocaine on the fetus have induced high levels of public concern. To date, however, no homogeneous pattern of perinatal complications of malformations attributable to cocaine has reached a scientific consensus. Such children tend to be smaller, more likely to be premature, with small head circumference, and their mothers tend more often to suffer from abruptio placenta (2). However, because these women very often use other illicit drugs, alcohol, cigarettes, have sexually transmitted disease, poor nutrition, and negligible prenatal care, it is extremely difficult to assign a cause-and-effect role to cocaine in inducing fetal or neonatal complications.

One of the major stumbling blocks in assessing the reproductive risk of cocaine is the immense difficulties in documenting the pattern of maternal use of this compound. Traditionally, the evidence of illicit drug exposure comes from self-report or detection of the compound in urine or blood. However, there is ample evidence that drug users tend to be unreliable in their reports.

During gestation, the pregnant mother is ever less likely to yield an accurate report; in some parts of the United States her confession may incriminate her.

The validity of blood and urine tests is dependent on the elimination half-life of the compound in question. In the case of cocaine, which has a very short $t_{1/2}$; the drug and its metabolites are not likely to be detected for more than a few days in either

*This study is supported in part by Physician Services Incorporated, Toronto, & Health and Welfare Canada.
[†]Gideon Koren is a Career Scientist of Ontario Ministry of Health.

blood or urine. Even the newborn, which has a limited ability to metabolize and eliminate the drug, will not retain it for more than a week (3).

If cocaine exposure is suspected at birth, meconium is a very sensitive source of material for detection the drug, as cocaine swallowed with the amniotic fluid and potentially also reaching the gut by enterohepatic circulation can be detected in high concentrations (4).

Most studies evaluating the reproductive risks of cocaine compare outcome in mothers who are believed to have used the drug to those who presumably have not. Many of them have relied on maternal reports only, while others have based their assessment on urine tests (5). Because as many as 50% of women using cocaine may deny such habit and have negative urine tests, most studies to date are impaired in their most important element, namely, the exposure to the independent variable. This reality stresses an urgent need for a biological marker which does not lose its sensitivity within a few days after the end of exposure, and which may yield not only a dichotomous answer ("yes" or "no" exposure) but rather a cumulative reflection of the exposure.

II. THE USE OF HAIR IN MEDICAL TOXICOLOGY

A variety of compounds have been shown to be incorporated into adult hair, including trace metals, barbiturates, amphetamines, opiates, phencyclidine, nicotine, cocaine, etc. Although the contact of the hair with the blood is well established, the mechanisms leading to the absorption of xenobiotics into growing hair have not been elucidated (6). The fact that hair grows continuously at an average rate of 1 cm a month (7) gives a potential time dimension to the accumulation of drugs and their metabolites in the shaft. Should there be a proven correlation between ingested dose and hair concentration, then the hair may produce concentration-time account of systemic exposure to the xenobiotic in question.

While hair has been used to prove antemortem exposure to illicit and medicinal drugs in forensic medicine, the use of this material as evidence of intrauterine exposure is relatively new. The hair covering the head of the neonate is grown mainly during the last 3 months before birth and falls off within the first 4 to 5 months of life. Hence exposure of the pregnant woman to drugs only during the first trimester, for example, is not likely to be reflected in hair.

The best-documented use of hair accumulation of a xenobiotic to provide evidence of cumulative exposure is the case of methyl mercury. This compound is preferentially accumulated in the brain, causing neurotoxicity to both adults and fetuses. By measuring maternal hair concentrations of methyl mercury, Clarkson and his team can predict serious neurological damage to the newborn.

III. ANIMAL EXPERIMENTS WITH GESTATIONAL COCAINE EXPOSURE

For these experiments we have used the guinea pig, as it has abundant fur at birth. We could document accumulation of cocaine in adult fur following chronic subcutaneous administration of the drug (10 mg/kg for the last trimester of the pregnancy).

Theoretically, measurement of cocaine in neonatal hair could reflect contact of the hair with amniotic fluids containing the drug. To rule out this possibility, we exposed hair to an aqueous solution of benzylecgonine at the concentration range found in the amniotic fluid (375 ng/mL). Following washing of the hair there was no measurable benzylecgonine, suggesting that, similar to adult hair, the drug is detected after its growing into the neonatal hair.

IV. MEASUREMENT OF COCAINE IN HUMAN HAIR

Of 40 health professional volunteers, with negative history of cocaine use and no detected cocaine in urine, there was not a single case of a false positive. As important, out of 13 cocaine addicts submitted for treatment, hair was positive for cocaine in all instances.

To rule out the possibility that hair may show cocaine due to passive exposure to crack smoke in a 25-m^3 room, we exposed hair of three volunteers to smoke produced by 100 mg of crack and could not detect any benzylecgonine. We performed an in-vitro exposure of hair to smoke from incremental amounts of crack: 0.1 mg, 1 mg, 10 mg, and 100 mg in a 4-L closed jar. For every case, the passively accumulated benzylecgonine was washed out completely after 3 × 10 min; ethanol washes at 37°C.

In nine instances we received neonatal hair, where the mother admitted and/or was tested positive for urinary cocaine in late pregnancy. In all instances neonatal hair revealed measurable amounts of the drug (Figure 1). Similar tests done on toddlers or small children revealed decreasing sensitivity of the assay, probably due to shedding of the neonatal hair containing cocaine.

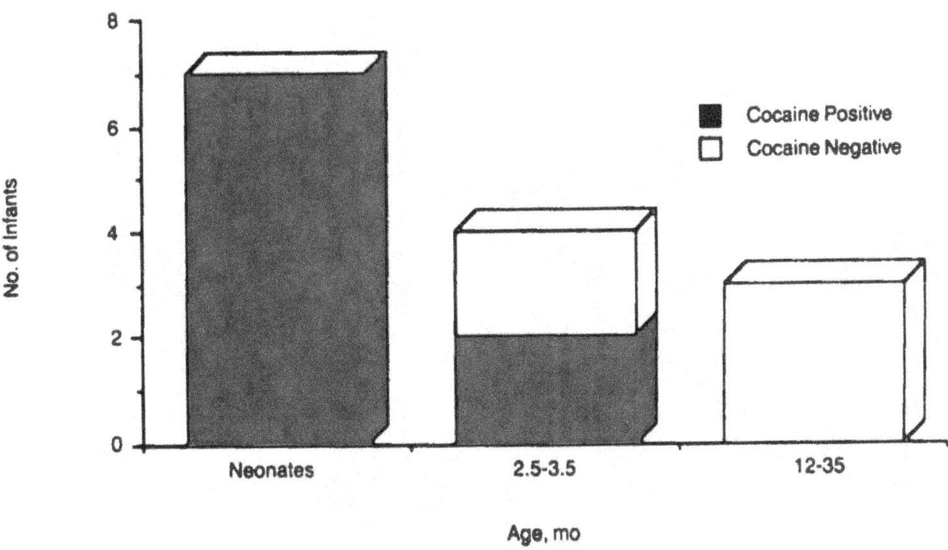

Figure 1 Results of analysis of hair from babies and young children with known in utero exposure to cocaine, as determined by maternal history.

V. HAIR ANALYSIS

Cocaine was extracted from hair according to a modification of the procedure described by Baumgartner and Berka (6). Benzylecgonine was measured by a commercially available radioimmunoassay (RIA). Positive results were invariably confirmed by gas chromatography/mass spectrometry (GS/MS).

Cocaine is merely one example of a drug where hair may become a better biological marker for intrauterine exposure. The next case may unmask the powerful potential of this method.

We were asked to assess whether a 6-month-old infant dying from sudden infant death syndrome was exposed in utero to cocaine. With maternal consent we analyzed the hair of the twin brother of the deceased baby and have found large amounts of cocaine; this was followed by the mother admitting cocaine use but denying exposure to any other illicit drug. However, GC/MS analysis of maternal hair revealed detectable concentrations of heroin, not previously suspected. In this case the hair test revealed the alarming fact that the surviving baby is cared for by a heroin user, a fact that may have a major impact on her ability to provide an acceptable level of care to the child.

VI. SUMMARY

Many questions are still unanswered in the use of hair test for cocaine in general and in neonates in particular. However, because cocaine is not a naturally occurring compound, there is no doubt that its detection in neonatal hair reflects fetal exposure to the agent.

Establishing a dose-versus-hair concentration curve for cocaine will be hampered by the poor quality of report by addicts on the dose used by them. However, we have preliminary evidence that heavy users (e.g., addicts) have substantially higher hair concentrations than social, nonaddict users (Figure 2).

Establishing the relationship between cumulative maternal cocaine dose and neonatal hair concentrations may use the maternal hair levels as an intermediate, after the maternal dose response curve has been established.

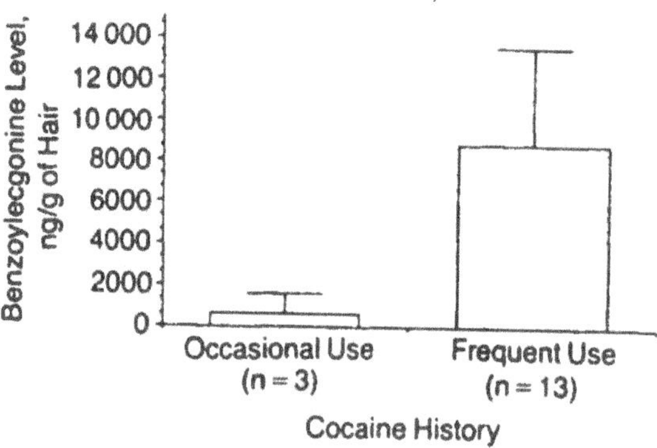

Figure 2 Detection of benzoylecgonine in the hair of 3 occasional and 13 frequent users (95% confidence interval shown).

The potential of this new method is huge; with carefully controlled studies we may gain improved biological markers for cumulative intrauterine exposure to xenobiotics.

REFERENCES

1. R. G. Smart. "Crack" cocaine use in Canada: A new epidemic? *Am. J. Epidemiol., 127*:1315–1317 (1988).
2. S. D. Dixon and A. Oro. Cocaine and amphetamine exposure in neonates: Perinatal consequences (abstract). *Ped. Res., 21*:359A (1987).
3. R. T. Jones. Psychopharmacology of cocaine. In *Cocaine: A Clinician's Handbook.* (A. M. Washton and M. S. Gold, eds.). John Wiley, Toronto (1987).
4. E. M. Ostrea, P. Parks, and M. Brady. The detection of heroin, cocaine and cannabinoid metabolites in meconium of infants of drug dependent mothers (IDDMs): Clinical significance. *Ped. Res., 25*:225A (1989).
5. D. A. Frank, B. S. Zuckerman, H. Amaro, et al. Cocaine use during pregnancy: Prevalence and correlates. *Pediatrics, 82*:888–895 (1988).
6. W. A. Baumgartner and C. Berka. Hair analysis for drugs of abuse. American Association for Clinical Chemistry, Inservice Training Program, TOX-TDM, *10*:7–21 (1989).
7. M. Saitoh, M. Uzuka, M. Sakamoto, and T. Kobori. Rate of hair growth. In *Advances of Biology of Skin,* Vol. 9 (W. Montagna and R. L. Dobson, eds.). Profman, Oxford, pp. 183–201 (1969).

84

Doping Control and Olympic Games

Jordi Segura *Institut Municipal d'Investigació Mèdica, Barcelona, Spain*

The approach of the next Olympic Games, to be held in 1992 in Barcelona, and the recent experiences of known doping cases in Seoul in 1988 turns the control of drug misuse in sport into an area of remarkable interest. Testing performance during Olympic Games is complicated by the necessity to process a large number of samples and the short time required for developing results. A potent and also flexible structure of personnel and instruments is needed to assure maximum reliability at the required high analysis performance rate.

The antidoping measures in the Games involve a complex program. However, the experience acquired at previous events shows that the very complexity of the problem allows an overall approach, which makes for a greater success rate than that to be expected in a situation of lesser significance. Suitable planning allows the laboratory staff to take adequate measures such as providing information for athletes and doctors, the use of reliable procedures, and in-depth discussion of the results, which in turn allows the correct decisions to be made.

With respect to information, there must be close contact with the teams' medical officers, both before and during the Games. Various helpful items are available. A Pharmacological Guide is issued before the Games, listing the medicines which will be available at the Official Pharmacy as well as those substances banned by the International Olympic Committee. Once approved, it is distributed to the National Olympic Committees. Information also has to be provided about alternative treatments for any illness. This is known as the "positive list," in the sense that the provision of authorized treatment can be considered a positive input. Although this list is not exclusive, it is supplied to team medical officers and has proved a useful instrument on numerous occasions. Another important item is detailed information regarding the procedures for the collection and analysis of samples. These procedures are approved and published beforehand, but at the beginning of the Games they are explained and commented on in detail. Finally, there is always the chance to consult the Medical Commission of the International Olympic Committee through its representative at the Olympic Village Medical Center.

Mention also should be made of an issue which is vitally important: the reliability of the procedures used in both the transport and analysis of samples. The means of procuring reliable samples of athletes' urine must be guaranteed by the existence of a permanent collection team with adequate information at each competition site. This team will include the officer in charge and a sufficient number of technicians, interpreters, and auxiliary personnel. The waiting and collecting rooms must be of sufficient size, and access to them has to be restricted and controlled. The collection procedure itself must be validated by the presence of a member of the federation involved and a member of the International Olympic Committee. The transport, which can be organized in various ways, must at all times guarantee that the samples reach the laboratory quickly and without any chance of tampering.

A maximum level of reliability has been always a must in doping analysis. In this regard, the confirmation of results by mass spectrometry is mandatory for all presumptive positive tests. Recent developments in this technique and its coupling to chromatographic methods (i.e., liquid chromatography) are being introduced into the laboratory's routine methodology. The need for approaches to detect and confirm newly banned drug classes such as natural or recombinant hormones and blood doping will assure interesting research areas for the future. Top-level working standards at the laboratory bench are assured by the strict fulfillment of Good Laboratory Practice regulations and quality assurance strategies. The program for harmonization and accreditation of antidoping laboratories by the Medical Commission of the International Olympic Committee is one of the relevant international approaches to guarantee the quality of drug analysis.

The third aspect, which is crucial for maintaining faith and credibility in the system, is the long, detailed discussion of any positive results by the IOC Medical Commission. This includes an audience with the athlete and the National Olympic Committee involved. Any disciplinary action is taken only after lengthy debate and close evaluation of the information available.

Finally, the data involved in collecting and analyzing more than 1600 samples during the course of the Games can provide answers to questions that are not directly related to the simple detection of known banned products. Subsequent analysis of the computerized data can be used in the interest of improving sport medical practice, both through the investigation of the variations in certain parameters of physiological interest in top-level athletes and also in the attempted early detection of the abuse of hitherto unknown substances.

85

Hair Analysis for Drugs of Abuse: Decontamination Issues

Werner A. Baumgartner[1,2,3] **and Virginia A. Hill**[2] [1]*Psychemedics Corporation, Santa Monica, California* [2]*West Los Angeles Veterans Administration Medical Center, Wadsworth Division, Los Angeles, California* [3]*University of Southern California School of Pharmacy, Los Angeles, California*

I. INTRODUCTION

The analysis of drugs of abuse in human hair, first reported in 1978 (1), is increasingly being recognized as a useful test to complement urine and blood analysis (2). This work was subsequently extended to therapeutic agents (3).

Current thinking suggests that hair complements urine by providing long-term information (from month to years) of an individual's drug use. In addition, hair analysis, in contrast to urinalysis, can also provide a measure of the amount and pattern of drug use.

The feasibility of obtaining measures of the amount of use from the drug levels in hair was investigated by studying the correlations between self-reports of illegal drug use (2) or known administered doses of therapeutic drug (2,3) and the amount of drug found in hair. With illicit drugs our research suggests that we were able to differentiate only between heavy, intermediate, and light use. More accurate estimates of use were precluded by unreliable self-reports and the variabilities of biochemical individuality.

A much higher degree of accuracy can be achieved when only the pattern of drug use is investigated. Here only relative changes in the drug content of hair are measured; that is, the patient serves as his own control, thereby allowing one to bypass the problem of biochemical individuality.

The measurement of relative changes in the drug content of hair is particularly useful for monitoring compliance in various clinical settings. For example, in addiction recovery programs, the drug content in hair segments which grew during rehabilitation could be compared to that which grew prior to entering the program. When compliance of medication intake is monitored, segmental hair analysis could compare the drug content in hair segments which grew while the patient was unsupervised with segments which grew during periods when medication was administered under supervision. Other clinical uses of hair analysis include diagnosis of drug abuse, toxic psychosis, and prenatal drug exposures.

However, the most extensive use of hair analysis to this date has been in nonclinical situations: the workplace, the criminal justice system, and the courts. The fact that hair analysis is difficult to evade, and its ability to provide long-term histories of drug use appear to be the main reasons for its popularity for nonclinical applications.

The most crucial issue facing any drug testing program, regardless of whether it involves urine, blood, saliva, or hair, is the avoidance of technical and evidentiary false positives. Technical false positives are caused by errors in the collection, processing, and analysis of specimens and the reporting of results, while evidentiary false positives are caused by passive exposure to the drug. Historically, the greatest attention had been given to preventing technical false positives by instituting appropriate blind quality control programs. Unfortunately, such programs are essentially ineffective against evidentiary false positives.

In the case of urinalysis, the evidentiary false positive problem has been addressed for passive marijuana smoke inhalation and poppy seed ingestion by adjustments of cutoffs levels and/or examination of the individual by a medical review officer. However, less attention has been given to the possibility of positive urinalysis results from passive exposure to other drugs, e.g., cocaine, PCP, and methamphetamines.

In the case of hair analysis, our laboratory focused on the passive exposure problem from the start (2,4), mainly because of the numerous approaches that were afforded by hair analysis: the ease of performing different types of passive exposure (contamination) experiments; the evaluation of a multitude of wash procedures along with kinetic analysis of wash results for decontaminating the hair specimen; testing the physical status of the hair by staining procedures; and measurement of metabolites. Not all of these experimental approaches are available to urinalysis.

The present report focuses extensively on the decontamination of hair and describes the approaches for preventing evidentiary false positives due to external contamination of the hair specimen. We do this not only to demonstrate the reliability of hair testing procedures but also to show that hair analysis can be a useful adjunct for solving some of the evidentiary false positive problems of urinalysis. Our methods for preventing evidentiary false positives have been evaluated with over 50,000 hair samples involving over 250,000 analyses.

II. METHODS

Two types of procedures have been developed for discriminating between active drug use and passive exposure to drugs; these are designated as the individualized forensic procedure and the mass production forensic procedure. These two methods, although distinct, are related by the fact that certain wash kinetic criteria can trigger the reexamination of a mass production sample by the individualized forensic procedure. These criteria, and their experimental and theoretical basis, will be discussed in Section III. In this section, we describe the sequence and methodology of the individualized forensic procedure and how this relates to mass forensic production. The methods for different types of contamination experiments are also described.

A. Methylene Blue Staining Procedure

As a first step in any individualized forensic test, it is necessary to evaluate the physical properties of a hair specimen, since this can greatly affect the interpretation of results. Whether a hair has been damaged by perming, dyeing, relaxing, or natural causes can be determined by staining it with methylene blue. Damaged hair takes up the methylene blue stain, but undamaged hair does not. The stain result determines which solvent is subsequently applied in the decontamination process: Ethanol is used for damaged hair, and water-based procedures are used for undamaged hair. The ethanol washes are "milder" than aqueous washes, in the sense that ethanol removes less of the blood-derived drug, i.e., drugs which entered the hair via the circulatory system during the synthesis of the hair fiber. Interpretation of the wash kinetics and the significance of the quantitative results are also determined by the results of the hair stain (see below).

The staining procedure consists of washing 5 to 10 hair strands in Prell shampoo (5% of commercial liquid preparation in water) for 20 min by shaking them at 100 cycles/min at room temperature in order to remove dirt, gels, or creams which can interfere with the staining. Prior to washing, the hair strands are tied together with a thin strip (1 to 2 mm) of heavy aluminum foil. After washing, the strands are rinsed well with water and then immersed in 0.5% w/v solution of methylene blue in water for 5 min at room temperature. Subsequently, the hair is rinsed repeatedly in about 100 mL of distilled water at room temperature until no further dye appears in the wash solution. The hair strands are then examined longitudinally under a microscope. It is more difficult to evaluate the uptake of the dark blue dye than in other colored hair. Consequently, it is advisable to compare black hair which has been exposed to the dye to a few strands of the same specimen which have not been treated with the dye. A microscope with strong illumination is also very helpful for evaluating the uptake of the dye in the case of dark hair.

From the microscopic examination of the hair specimen, one can readily determine which portion of the hair is damaged/treated and which is not. It is advisable for the subsequent analysis and interpretation of the data to cut the hair specimen into sections such that the untreated hair is separated from the treated portions.

In the case of mass production procedures, the hair is not stained as the first step in the analysis but only if the results of the first drug analysis demonstrate abnormal wash kinetics.

B. Wash Procedures

Hair is extensively washed to remove any external drug contamination. For mass production forensic testing, 15 to 20 mg of hair are washed in test tubes at 37°C with shaking at 100 cycles/min. The first wash is performed with 2 mL of ethanol for 15 min in order to remove greasy residues and loosely adhering surface contamination. This wash is then followed by three 30-min, 2-mL washes with 0.01 M pH 6 phosphate buffer at 37°C with shaking at 100 cycles/min. At this point the washing of mass production samples is stopped. For those samples with positive results in the subsequently applied hair analysis, the kinetics of the wash curve is determined by radioimmunoassay (RIA) analysis of the washes.

In the case of individual forensic samples, the wash is continued for appropriate time intervals until the wash kinetics approach a plateau. In the case of hair which was determined to be damaged by the staining procedure, the above-described wash procedures are performed with 2 mL of ethanol instead of water. As most RIA assays are adversely affected by ethanol, it is necessary to remove this under a stream of air prior to analyzing an aliquot of the wash. When measuring the cocaine and methamphetamine content of these washes, care must be taken not to lose these volatile analytes during the evaporation step. This can be avoided by evaporating the solvent at room temperature or below. To avoid the possibility of contaminating the washed hair sample in the laboratory, the specimen is kept in its stoppered wash tube for the subsequently applied digestion process.

C. Analysis of Drugs in Washed Hair

The most effective method for measuring the amount of drug remaining in washed hair involves dissolving the hair specimen so that one can be certain all the material entrapped in the protein matrix has been released. If solvent extraction procedures are used, complete extraction of analyte cannot be guaranteed, since extraction efficiency greatly depends on the physical properties of the hair, i.e., whether it is thick and intact or thin and damaged. Of course, the dissolution procedure must be extremely mild in order not to cause decomposition of analyte.

For mass spectrometric analyses of five drugs (cocaine/benzoylecgonine, PCP, morphine, methamphetamine, marijuana), we chose a mild enzymatic digestion procedure at a pH of 6.2, i.e., at a pH at which, according to our experience, cocaine does not hydrolyze to benzoylecgonine. No special precautions need to be taken to measure the other metabolites in hair, e.g., Δ-9-carboxy THC, monoacetylmorphine, and amphetamine. The digestion of the hair specimen requires the following mixture: 10 mL of 0.5 M Tris buffer, pH 6.2, to which is added 60 mg of dithiothreitol, 200 mg of sodium dodecyl sulfate, and 20 units of proteinase K. Add 2 mL of this solution per 20 mg of washed hair, the amount typically used for hair analysis. Place the mixture in a 37°C shaking water bath and shake at 80 to 100 oscillations/min overnight (16 to 18 h). Remove from bath water, mix, centrifuge, and remove the supernatant (leaving behind melanin pellet) for subsequent extraction and preparation for gas chromatography/mass spectrometry (GC/MS) procedures.

The resulting protein solution, minus the melanin fractions, may be analyzed by sensitive negative or positive chemical ionization mass spectrometric techniques in current use for blood analysis (5–8). In the case of marijuana, the method of Foltz et al. (9) had to be adapted to an ultrasensitive GC/MS/MS procedure which was developed to measure the low levels of Δ-9-carboxy THC in hair.

Mass spectrometric confirmations can be performed conveniently and reproducibly at the following drug concentrations: cocaine, 6 ng/10 mg hair; opiates, 1 ng/10 mg hair; PCP, 1 ng/10 mg hair; methamphetamine, 2 ng/10 mg hair; Δ-9-carboxy THC, 1 pg/100 mg hair. To facilitate effective wash kinetic studies, higher GC/MS cutoff levels were chosen for drug screening purposes: opiates, 5 ng/10 mg hair; PCP, 3 ng/10 mg hair; methamphetamine, 5 ng/10 mg hair.

As the amount of hair specimen used for the GC/MS analysis (i.e., 5 mg of hair), is considerably smaller than that used for urine analysis (i.e., 5 to 10 mL), it is necessary to use ultrasensitive mass spectrometric procedures. Of course, there is no correlation between urinalysis and hair analysis cutoff levels.

It should be recognized that the chemical conditions which lead to the dissolution of one of the most resilient protein structures, hair, are highly deleterious to the most sensitive of proteins, antibodies. Consequently, the above-described digest used for mass spectrometry is quite unsuitable for RIA analysis. In its place, Psychemedics uses its proprietary (patent pending) hair analysis procedure, which requires 0.5 to 1.0 mg of hair per analyzed drug. For those wishing to do preliminary RIA screens on hair specimens, various published solvent-based extraction procedures exist (1,10–12).

D. Contamination of Hair

The susceptibility of hair to contamination by exposure to drugs in solutions or vaporous states was extensively investigated by us. In the case of smoke, we performed experiments consisting simply of suspending locks of different types of hair (fine hair, strong asiatic hair, porous hair) in a closed container. In the case of cocaine, this involved evaporating 0.5 g of cocaine freebase in a 1-L glass flask. This was an extreme form of exposure, judging by the dense fumes that were generated. Exposure time to the lingering vapors was also extreme, lasting for several hours in the case of cocaine and overnight for l-methamphetamine. For the 1-methamphetamine study, the contents of one Vicks nasal inhaler were heated in the 1-L flask. In the case of PCP and marijuana, the exposures were equally severe. Here, smoke from a burning marijuana cigarette (1% THC) or a PCP-impregnated cigarette (10 mg/cigarette) was repeatedly passed over hair in a narrow tube (20 mL volume) using a suction device.

III. RESULTS

A. Contamination Experiments

There are several reasons for designing experiments for the introduction of drugs into hair from the exterior rather than through naturally occurring in-vivo processes, e.g., through the blood supply.

1. At one extreme there is the need to develop hair control material which is essentially indistinguishable from the hair of drug users. This can be accomplished with a range of extreme conditions of drug exposure, e.g., exposing hair at elevated temperatures and for extended times to drugs dissolved in keratin-penetrating solvents such as dimethylsulfoxide.
2. At the other extreme there is the need to develop realistic models for external contamination. Here the procedures used should simulate naturally occurring situations and span a relatively mild range of conditions.
3. Between the extreme conditions of 1 and 2 is the zone in which the conditions of drug exposure are of intermediate severity. This zone does not represent the worst-case scenario of external contamination, but may be used for estimating the worst case by extrapolating to less severe conditions. Similarly, extrapolation to more severe conditions can shed light on what is needed for the development of controls. In short, the exposure conditions in this intermediate zone lend themselves to evaluating the permeability of hair to externally applied drugs.

Our initial experiments investigating the permeability of hair to externally applied drugs and their removal by washing were performed under intermediate conditions of drug exposure (i.e., condition 3 above). They were by far more severe than the worst-case passive exposure scenarios, as we effectively placed the hair specimens in the bowl of a cocaine pipe or in a cigarette holder of a marijuana or PCP joint. These experimental conditions, however, were imposed on us by what we now recognize to be the rather remarkable impermeability of hair to externally applied drugs. Only by using such drastic exposure conditions were we able to get the necessary levels of drugs into or onto the hair for the desired wash kinetic studies.

The permeability of hair to cocaine freebase vapors was investigated with several different hair specimens whose physical properties ranged from strong and undamaged to weak and damaged. The contaminated hair specimens were then decontaminated by washing with phosphate buffer wash solutions for minimum periods of 30 min at 37°C with vigorous shaking. Longer washes were used near the extraction plateau. The minimum exposure period of 30 min is important, since this allows for the swelling of the hair fiber, thereby facilitating the more effective decontamination of hair.

Figure 1 shows the extraordinary resistance of hair to the penetration of cocaine vapor. This is evident from the sharply rising wash kinetics, the rapid attainment of a plateau indicating the near completion of the decontamination process, and by the small amount of residual cocaine found in the subsequently analyzed dissolved hair specimen. The residual drug content is expressed as a percentage of the total removed from the hair surface. In essence, these results suggest a piling up of cocaine smoke at the laminar surface of the cuticle—the protective layer of the hair.

As the drug concentrations in the first wash solutions were extremely high—at least two to three orders of magnitude above what is normally found in the hair of the highest drug users—we were concerned that the extended contact with the first wash solution could in itself be a cause for some penetration into deeper layers of the hair fiber. To avoid such wash solution-caused contamination of hair, we investigated the effectiveness of treating vapor-exposed hair to many short washes of no more than 1 min duration. Using the same hair, we also evaluated the effectiveness of different solvents with this short wash approach: phosphate buffer (pH 6, 0.01 M), ethanol, and Press shampoo.

Consistent with our expectations, we attained the wash kinetic plateau with short washes even more quickly than with the 30-min washes (Figure 2). The residual amount of drug in hair was also less by an order of magnitude. Buffer, ethanol, and shampoo proved to be equally effective for decontamination purposes.

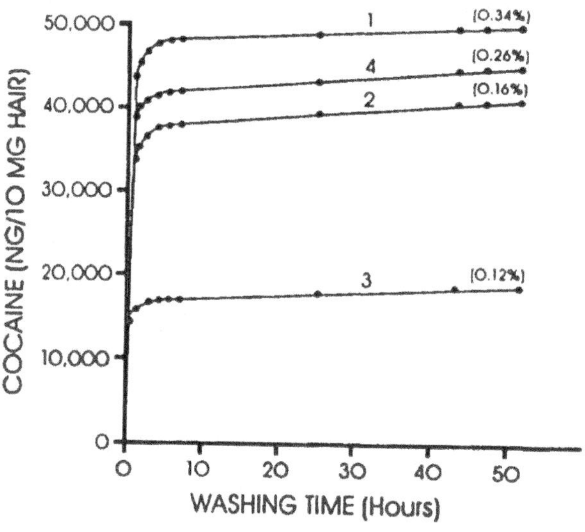

Figure 1 Cleaning of different types of hair with multiple phosphate buffer washes (pH 6) at 37°C subsequent to contamination of hair specimens with cocaine vapors. Residual amount of cocaine in hair after washing is expressed as a percentage of the total removed.

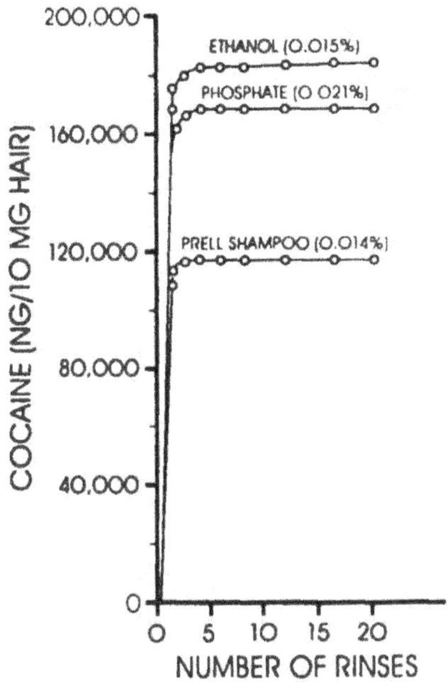

Figure 2 Cleaning of hair with multiple washes of short duration (1 to 2 min) at 37°C subsequent to contamination with cocaine vapors. Washes were performed with different solvents: Prell shampoo, phosphate buffer (pH 6), and ethanol. Residual amount of cocaine in hair after washing is expressed as a percentage of the total removed.

Although the short wash approach under conditions of severe surface contamination was more effective than the 30-min wash procedure, it would be incorrect to conclude that short washes are more effective than 30-min washes for the decontamination of hair exposed to drugs under more normal conditions, i.e., the decontamination of hair of drug users (see below). However, the short wash approach, because of insufficient time for penetration of the wash solution, simulates the conditions for decontamination by normal hygienic practices.

In Figure 3 we see the decontamination kinetics of different types of hair which were exposed to cocaine hydrochloride solutions (5 μg/mL) for 3.5 h at 37°C and subsequently dried and washed with either phosphate buffer or ethanol. The conditions of these experiments are, of course, once again far more extreme than natural contamination, e.g., the possibility of surface contamination being carried into deeper layers of the hair by sweat. Three types of hair were used for this experiment: porous permed hair (A), strong Asiatic hair (C), and Caucasian hair of intermediate qualities (B).

It is interesting that the amounts of drug taken up by the different hair types are in direct proportion to their porosities. But no major differences were observed between the decontamination kinetics for the different solvents. The residual amounts of drugs (if any) which were found in the dissolved hair were once again insignificant in relation to the total initial contamination.

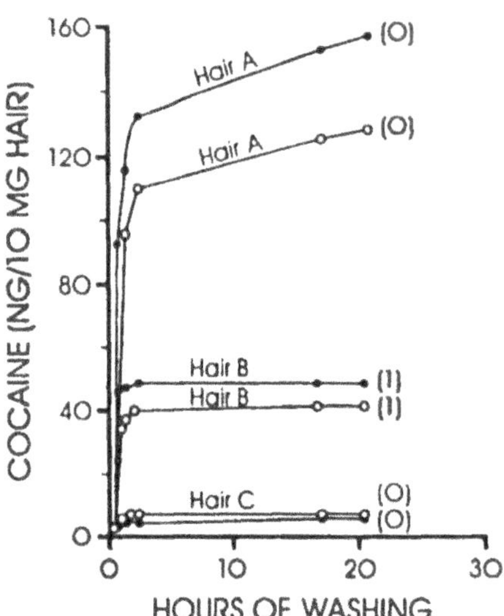

Figure 3 Cleaning of different types of hair with different solvents subsequent to contamination of hair by soaking specimens in cocaine hydrochloride solution (5 μg/ml) for 3.5 h at 37°C. Types of hair: A, porous treated hair; B, adult untreated h air; C, juvenile untreated hair. Solvents used: ●, phosphate buffer (pH 6); O, ethanol. Residual amount of cocaine in hair after washing is expressed as a percentage of the total removed.

Similarly satisfactory wash kinetics indicative of highly effective decontamination procedures and hair's resistance to penetration by smoke were obtained for PCP [with phosphate buffer or ethanol as solvent (Figure 4)], for marijuana [with ethanol as solvent (Figure 5)], and for methamphetamine [with phosphate buffer as solvent (Figure 6)].

B. Wash Studies with Hair from Drug Users

We performed a study of the wash kinetics of 700 positive hair specimens from cocaine users. A smaller number of hair samples from opiate, PCP, and methamphetamine users was also studied. However, since these showed essentially identical wash kinetics as the cocaine-positive samples, we will not distinguish these in our discussion. Marijuana-positive hair was not included in this study for reasons that will be discussed below.

Although many of the hair samples were subjected to protracted wash procedures until a plateau was obtained, we will focus here only on the results of the truncated wash kinetics approach, i.e., the method used in our mass production operations. This, it will be recalled, involves an initial 15-min wash with ethanol followed by three 30-min phosphate washes at 37°C with vigorous shaking.

The results of these experiments are summarized in Figure 7. Essentially, the wash results fell into one of the following five kinetic categories:

1. No drugs were found in the wash solution, but the amount found in the hair digest was large. The material found in the hair digest is considered to have entered the

hair by in-vivo processes, e.g., via the circulatory system. These kinetics occur with uncontaminated, strong hair from drug users (not shown in Figure 7).

2. Only the first alcohol wash, but not the phosphate buffer washes, contain detectible amounts of drugs, and the amount found in the digest is relatively large (curve D). Such plateaus in the wash kinetics occur in the case of strong but externally contaminated hair; i.e., no blood-derived material is removed by the phosphate buffer washes.
3. The drug content in the phosphate buffer washes is considerably greater than in the first alcohol wash, and a plateau or a near plateau is attained (curve C). The amounts found in the hair digest are considerably greater than those removed by washing. These wash kinetics are obtained with weaker hair, from which small amounts of blood-derived drugs along with surface contamination is being removed.
4. The wash kinetics do not attain a plateau (curve B). These kinetics are typical of hair damaged by hair treatments (perming, dyeing). Considerable quantities of blood-derived drugs are removed by the phosphate buffer, and consequently the amount of drug found in the digest is correspondingly smaller. Such hair are best washed by the ethanol procedure (see below).
5. Kinetics typical of contamination but no use exhibit a sharply rising curve, with most of the drug present in the first alcohol wash accompanied by the rapid attainment of a plateau and an insignificant amount of drug in the digest (curve A).

Hairs which, on the basis of the methylene blue stain, were extensively damaged were washed with ethanol instead of phosphate buffer. This was done because the

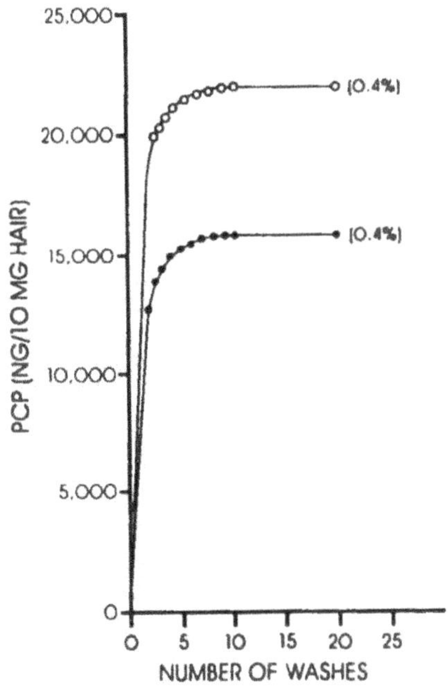

Figure 4 Cleaning of hair with multiple washes of short duration (1 to 2 min) subsequent to contamination of hair with PCP vapors. Washes were performed with different solvents: ●, phosphate buffer (0.01 M, pH 6); ○, ethanol. Residual amount of PCP in hair after washing is expressed as a percentage of the total removed.

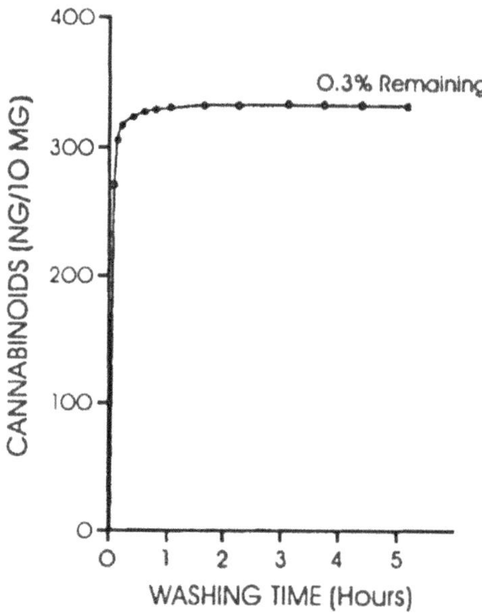

Figure 5 Cleaning of hair with multiple ethanol washes at 37°C subsequent to contamination of hair with marijuana smoke. Residual amount of cannabinoids in hair after washing is expressed as a percentage of the total removed.

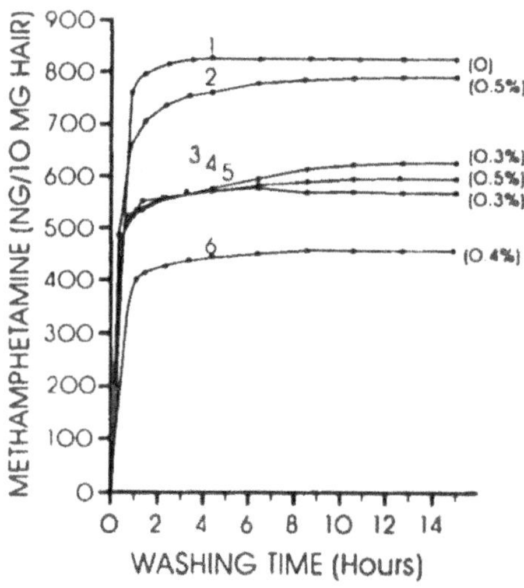

Figure 6 Cleaning of different types of hair with multiple phosphate buffer washes at 37°C (0.01 M, pH 6) subsequent to contamination of hair with methamphetamine vapors. Residual amount of methamphetamine in hair after washing is expressed as a percentage of the total removed.

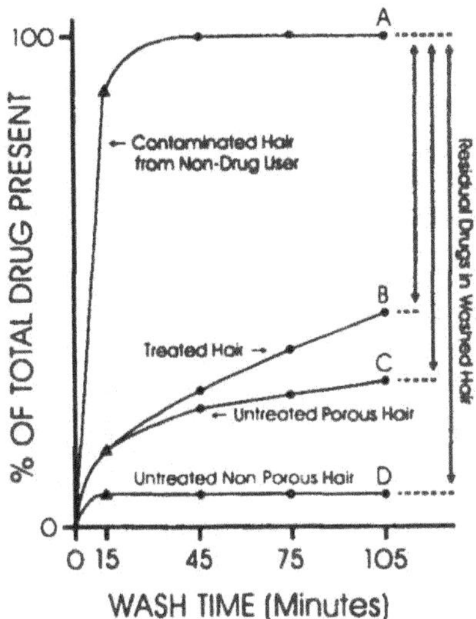

Figure 7 Wash kinetics of different types of hair from drug users and non-drug users. Δ, ethanol wash; ●, buffer wash.

hair-swelling properties of phosphate buffer frequently remove too much of the blood-derived drug. The milder properties of ethanol relative to phosphate buffer are illustrated in Figure 8. We see there that the extraction of drugs by ethanol tends to a plateau within 2 h of washing, but that relatively large amounts of blood-derived material can be removed by the subsequently applied phosphate buffer washes.

C. Wash Kinetic Ratios

The data from our decontamination and permeability studies can be interpreted by the model depicted in Figure 9. This model describes the effects of different wash procedures on the concentration of internally or externally derived drugs as a function of distance from the hair surface.

Newly grown hair of drug users will contain drugs essentially uniformly distributed across the hair fiber. This region (D_1-D) is designated as the "virgin blood zone." The curve B-D_6, on the other hand, represents the concentration of drug in hair as a function of distance from the hair surface, deposited there under the very extreme external contamination conditions of our permeability experiments. The short-term, repeated daily washing of hair, by allowing only limited penetration of aqueous solvent, eventually (after repeated applications) depletes all accessible drugs deposited from the bloodstream in the narrow region near the hair surface (D_1-D_2, hygiene zone). Such washing, however, removes extreme cases of contamination only to level C. The "optimum wash zone," i.e., the one defined by the attainment of a plateau in the wash kinetics after extended washing, is designated by region D_1-D_5. Most of the naturally occurring external contamination falls within this zone and is therefore completely removed along with all accessible blood-derived material in that region.

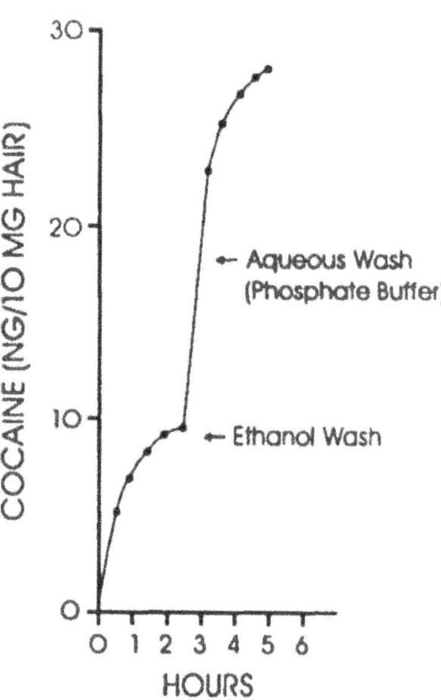

Figure 8 An example of the action of ethanol and water on cocaine-positive, treated (porous) hair.

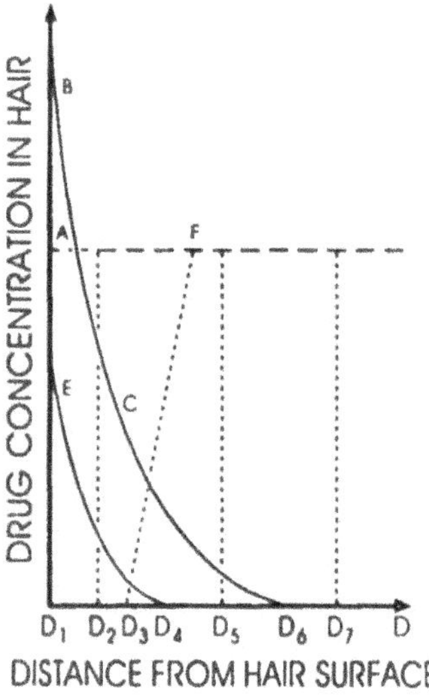

Figure 9 The effects of different wash procedures on the concentration of externally derived drug as a function of distance from the hair surface.

Now we postulate, for the sake of testing safety, that extreme but rarely occurring contamination scenarios can cause the penetration of drugs into region D_5-D_6, i.e., to a distance beyond the optimum wash zone. To guard against such a possibility, we have defined a "safety zone" (D_5-D_7) between the blood zone (D_7-D) and the optimum wash zone (D_1-D_5).

Recognition of this possibility has led us to introduce the "safety zone ratio" criterion for discriminating between use and external contamination. This ratio is defined by the amount of drug per 10 mg of hair in the hair digest divided by the amount of drug per 10 mg of hair in all four washes combined. On empirical grounds, i.e., on the basis of our initial experience with the wash kinetics with over 700 positive hair specimens, we have determined that this ratio should be greater than 0.33 for an interpretation of drug use rather than external contamination.

The limited number of washes of our mass production procedure cannot, of course, guarantee the removal of all external contamination. The limited effectiveness of the truncated wash kinetic approach is designated in Figure 9 by the sloping line D_3-F and region D_1-D_3. Diagrammatically, this shows that a limited number of washes may in many instances remove all external contamination, i.e., all of that which has penetrated no further than distance D_3, but only partially the contamination that has penetrated beyond that point. To compensate for this limitation which the practicalities of mass production impose, we have introduced the second wash kinetic criterion, that of the "extended wash ratio." This ratio is defined as the amount of drug per 10 mg of hair in the hair digest divided by the amount of drug per 10 mg of hair in the last wash. Again, on the basis of our experience, we have empirically determined that this value should be greater than 10 before a positive hair result can be attributed to drug use rather than external contamination.

In effect, what this ratio does is to theoretically extend washing for a minimum of 10 additional 30-min washes. This is a minimum estimate, since it assumes that the wash kinetics continue linearly. Of course, this is highly unlikely, as wash kinetics tend toward a plateau; i.e., the contents of each additional wash beyond the third phosphate wash tend to decrease rather than remain constant as projected by the extended wash ratio. Consequently, the benefit bestowed by this extended wash ratio generally goes well beyond the minimum of 10 additional washes.

Finally, a third wash kinetic criterion was defined, the "curvature ratio" of the wash kinetics. This ratio is defined as the total amount of drug in the three phosphate washes per 10 mg of hair divided by three times the drug content per 10 mg of hair in the last wash. This ratio provides a measure of the curvature of the wash kinetics and as such also a measure of the extent of depletion of a drug pool in hair accessible to a particular solvent. A value of 1 would indicate no evidence of curvature and little depletion of a particular drug pool. In cases where a plateau is attained and the pool is completely depleted, the value of the ratio approaches infinity.

On its own, the curvature ratio is used mainly for interpretive purposes. When low curvature is found in combination with a high drug content in the combined wash fraction, a more extensive investigation by the individualized forensic wash procedure is triggered.

The percentage distribution of values for the three wash kinetic ratios in our experiment population of 700 positive cocaine samples is given in Tables 1, 2, and 3. It should be noted that only a small fraction of the ratio values are near their decision cutoff values. Obviously, the certainty of an interpretation of a positive result as being

Table 1 Percentage Distribution of Extended Wash Ratios in 700 Hair Samples Positive for Cocaine

Extended wash ratio	Percentage of total samples
10–20	17.7
20–40	26.8
40–100	33.3
100–200	9.5
>200	12.6

Table 2 Percentage Distribution of Safety Zone Ratios in 700 Hair Samples Positive for Cocaine

Safety zone ratio	Percentage of total samples
0.33–1.0	7.1
1.0–5.0	36.8
5.0–10.0	25.6
>10	30.5

Table 3 Percentage Distribution of Curvature Ratios in 700 Hair Samples Positive for Cocaine

Curvature ratio	Percentage of total samples
1.3–1.5	13.4
1.5–2.0	19.2
2.0–5.0	38.8
5.0–10.0	9.4
>10	19.2

due to drug use increases the farther removed the ratio values are from their cutoff levels. All samples meeting the wash kinetic criteria are confirmed by GC/MS when nonclinical testing is performed. Samples which do not meet the ratio criteria and which take up methylene blue are reanalyzed by the individualized forensic method using all-ethanol washes.

A corollary to our wash kinetic analysis is the identification of criteria for realistic contamination models. One obvious criterion is that the drug exposure conditions must not be excessively severe. To this we now add that any partial decontamination of the hair should not be excessively harsh; i.e., it should, if applied, simulate only hygienic practices and not the presently described wash procedures.

In our initial trial with the 700 positive cocaine cases, the results of hair analysis were compared to self-reports, and in many cases also with concurrently performed urine tests. The few initial discrepancies between laboratory results and self-reports were investigated with the individualized forensic wash procedures on an newly collected sample (the "safety net") and with the application of an even more definitive test, i.e., the measurement of the metabolite, benzoylecgonine, in hair (see below). In all cases where a second hair specimen was donated, the original hair analysis results were confirmed by these additional investigations.

The wash ratio study was subsequently extended to a larger population, yielding approximately 2500 positive cocaine results. This involved mainly job applicants. Even with applicants, very few challenges to the original results were received. The few who mounted a challenge either refused to provide a second hair specimen (the safety net specimen), or they abandoned their challenge when the first result was confirmed by the more detailed analysis of the second (safety net) specimen. It should be emphasized that all industrial clients were strongly urged by Psychemedics in the interest of their own and the applicants' safety to insist on the donation of a second "safety net" specimen in cases where the results of the first analysis were challenged. This policy contributes not only to the safe operation of hair analysis, but most important, it greatly enhances the comfort level of the tested individual.

It is noteworthy that many employers did not merely wait for challenges but actively interviewed all individuals with positive test results. Some clients even initiated formal studies in which hair analysis results were compared to those of urinalysis. Surprisingly, in these studies applicants readily confirmed a positive hair result with a positive self-report. In one particular study, this occurred even in a majority of cases where drug use was low (once or twice a month), i.e., where the hair analysis results were close to the cutoff level of 5 ng/10 mg hair.

D. Marijuana Decontamination

Although marijuana smoke can be effectively removed from hair (Figure 5), we found that our wash kinetic criteria, although highly effective for cocaine, opiates, amphetamines, and PCP, did not provide the same margin of safety for this drug. This we found was caused largely by an unfavorable signal-to-noise ratio; i.e., small amounts of residual contamination were not always insignificant in comparison to the small amounts of blood-derived material. The latter was generally 2 to 3 orders of magnitude below the levels of the other drugs. It was therefore necessary to develop another approach for distinguishing between drug use and external contamination.

We investigated the possibility of making the distinction by measuring the metabolite Δ-9-carboxy THC in hair. We developed sensitive GC/MS/MS techniques for this purpose. The levels of Δ-9-carboxy THC in hair were in the range of 1 to 100 pg/10 mg hair. Extensive experiments with severely contaminated hair of the type depicted in Figure 5 showed no evidence of the presence of Δ-9-carboxy THC, even in unwashed hair.

E. Measurement of Cocaine and Opiate Metabolites in Hair

In spite of the success of our wash kinetic procedures, we currently also apply the measurement of metabolites to the other drugs. Initially, we focused on the cocaine metabolite, benzoylecgonine, the most prevalent positive hair analysis result with workplace testing. We have also extended this approach to opiates with the successful identification in hair of the heroin metabolite, monoacetylmorphine. A study investigating the incidence of monoacetylmorphine in the hair of heroin users is currently in progress.

There are several reasons for pursuing the measurement of metabolites in hair. Most important, it provides us with an independent and theoretical simple validation of our wash kinetic procedures. Ultimately it may even allow us to dispense with the more tedious wash procedures altogether. In addition, we were interested in establishing whether the benzoylecgonine-to-cocaine ratio could be used to differentiate between the use of freebase or cocaine salt.

The greatest difficulty for us was to develop an enzymatic digestion procedure that was both fast but also harmless to the cocaine. With respect to the latter, this meant that the digestion had to occur at a pH of 6.2 or below to prevent hydrolysis of cocaine to benzoylecgonine. Unfortunately, near this pH most digestive enzymes lose their activity.

The initial results in Table 4 show no obvious correlation between benzoylecgonine:cocaine ratios and the use of freebase or cocaine salt. Further work on this question is in progress with more clearly defined clinical populations. Ongoing field studies with cocaine-positive hair samples show good agreement between wash kinetic criteria indicative of use and the presence of benzoylecgonine.

F. Segmental Analysis

The question has been raised whether or not external contamination could interfere with the information provided by segmental analysis. Actually, the question is more complicated than that, for one must also consider the effects of dormant hair.

We have investigated these questions with a clinical population from the Schick-Shadel organization, where cessation of drug use subsequent to entry into the rehabilitation program was documented by various objective measures. Hair specimens were obtained from these patients at the time of entry into the program and after 6 to 8 weeks of aversion therapy. A hair specimen corresponding to 4 weeks prior to entry was analyzed and compared to a section corresponding to the last 2 weeks of therapy.

The results in Table 5 show that dormant hair effects do not interfere with demonstration of cessation of drug use. Clearly, an essentially 100% drop in the drug content in hair is indicated. Furthermore, as clients also frequently washed their hair while participating in the study, these results also show that the drugs present in the hair shaft are not "smeared out" over the segment which grew during the drug-free period.

Table 4 Benzoylecgonine:Cocaine Ratios in Hair from Known Cocaine Users[a]

Case no.	BE[b]	COC[c]	BE/COC	History
1	73	250	0.29	$35/week crack use
2	16	89	0.18	1–5 g/day, i.v.
3	<5	31	—	Light irregular user
4	31	200	0.16	$100/week crack use
5	26	65	0.40	$60/week, i.v.
6	16	130	0.12	Admits only 4 uses, i.v.
7	29	220	0.13	$140/week, i.v.
8	57	320	0.18	$300/week crack use
9	78	130	0.21	Admits only one sniff
10	664	500	0.16	Claims 5 g/day, i.v.
11	84	2337	0.28	$420/week crack use
12		138	0.61	$75/week snorting

[a]These samples were washed 24 h in phosphate buffer before analysis.
[b]BE = benzoylecgonine (ng/10 mg hair).
[c]COC = cocaine (ng/10 mg hair).

Table 5 Dormant Hair Effects in Hair Analysis for Determining Cocaine Use

Client no.	First sample[a] (ng BE/10 mg hair)	Second sample[b] (ng BE/10 mg hair)	Time between first and second (months)	Dormant hair effect (%) (2nd/1st × 100)
2037	32.5	0	1	0
2085	264	2.5	2	0.95
2349	56	0	2	0
3505	57	0	2	0
4259	88	0	2.5	0
4268	317	3.3	2	1.0
4309	101	0	2	0
5290	113	0	1.5	0
6456	115	0	2	0
6782	193	2	2	1.0
7520	41	0	2	0
4327	286	8.3	1	2.9
4303	168	3.8	2	2.3

[a]First sample was a 1.3-cm section.
[b]Second sample was a 0.6-cm section.

These findings agree with several hundred other segmental studies that we have performed where objective parameters (e.g., detainment in a drug-free environment) along with self-reports of drug use showed excellent correlation with segmental analysis results (13,14).

Even more convincing, of course, are segmental analysis studies with medications such as haloperidol, where the pattern of use is accurately known (3). Finally, if more fine-grained information on the cessation of drug use is needed, then the small contribution from dormant hair effects to segmental analysis can be readily calculated on the basis that approximately 15% of hair is in the dormant state (15).

IV. DISCUSSION

The main objective of this report is to describe methodologies and the underlying rationale for preventing evidentiary false positives due to external contamination of hair by drugs. As a preliminary to the development and field testing of wash procedures for forensic and clinical purposes, we tested the resistance of hair to penetration by drugs under severe conditions of drug exposure. Our results showed that hair was highly resistant to the penetration of air- or solution-borne drugs. This observation is consistent with numerous studies in the cosmetics industry (16,17).

Resistance to penetration was defined by us operationally in terms of wash kinetics. Consequently, our definition of impenetrable does not necessarily exclude the existence of accessible domains in hair (e.g., those causing swelling of hair as a result of the ready absorption of water), but rather that, in addition to these possible accessible domains, there must also exist highly inaccessible domains, e.g., the microfibrils, into which externally applied drugs cannot penetrate readily. The existence of such inaccessible domains provides the demarcation criterion for the observed difference between externally deposited drugs and drugs deposited by in-vivo processes.

Of course, if drugs dissolved in water are able to enter the regions accessible to the solvent, then our data show that whatever goes in easily is also readily removed by water. This, of course, is possible only if the drugs tested by us do not bind strongly to the keratin structures of hair. Our data show this to be the case in spite of the cationic nature of many of the drugs studied. It is not clear whether this lack of binding to the hair surface or the interior of the hair is an innate property of the drugs studied or the result of displacement effects caused by the myriad of cationic substances present in high concentrations in all cosmetic agents. The low concentration of these drugs, of course, also contributes to reduced binding.

Wash procedures were developed on the basis of these initial contamination experiments. Truncated wash kinetics were rationalized in terms of a five-zone model: the hygiene zone, the truncated wash zone, the optimum wash zone, the safety zone, and the blood zone. This model yields three kinetic parameters: the extended wash ratio, the safety zone ratio, and the curvature ratio. Extensive experience with over 50,000 hair samples (over 250,000 analyses) and over 3000 positive samples has demonstrated the efficacy of our wash kinetic approach.

Another laboratory (18) has recently attempted to apply the extended wash ratio criterion, but not the safety zone and curvature ratios, to the decontamination of a hair specimen that had been soaked in a drug-containing solution. It was reported that the extended wash ratio approach did not work for the decontamination of the particular sample which they studied. Putting aside the issue of whether this soaking experiment was a realistic model for perspiration-induced contamination, we must note that the wash procedures differed from ours in several respects. The most important difference was that their washes were of very short duration; apparently, hair was washed without agitation on a filter paper at room temperature (19). Consequently, their failure with one out of three of our wash criteria is not surprising.

The same laboratory also cited the case of a patient who had taken 125 mg of dihydrocodeine for a short period of time, but where the drug was found in regions of hair that did not correspond to the time of drug use (18).

This data is not in conflict with our position, as we do not deny the possibility of sweat-induced contamination or the partial effectiveness of decontamination of such hair by normal hygienic practices. What we do maintain is that our wash procedures and kinetic criteria are able to differentiate between use and contamination. It should be noted that the laboratory citing this case relied on the efficacy of normal hygienic practices—i.e., the investigators did not wash the hair prior to analysis, nor did they compare the rather low value, (0.4 ng/10 mg hair) to the presumably much higher values in the section of hair which corresponded to the patient's drug use.

The possibility of racial bias due to differences in the melanin content of hair is another issue that has been raised against hair analysis (18). The data cited, however, have involved only five pairs of subjects with black and brown hair along with dubious self-reports of drug use and this over a relatively small dose range. All these parameters showed considerable statistical scatter and therefore are not likely to be statistically significant. Furthermore, it appears that the data were culled from a study performed by our own laboratory (19). If this is the case, then it is certain that melanin did not bias the results, as this is removed prior to the analysis of the hair digest.

We agree, however, with the suggestion that certain drugs, e.g., haloperidol, may accumulate preferentially in melanin granules. A recent Japanese study with white and black mice showed major differences in the haloperidol content between black and white mouse hair. This report, however, does not clarify whether this difference is caused by genetic (metabolic) factors or by the absence or presence of melanin. However, we agree with the suggestion that methods which do not involve the dissolution of hair and removal of the melanin fraction are potentially subject to a melanin bias with certain drugs.

This bias also extends, of course, to solvent penetration effects. For instance, strong black Asiatic hair is more difficult to extract than black African hair. It is for all of the above reasons that we have chosen not to use solvent-based extraction procedures, but rather to dissolve hair under mild conditions, i.e., conditions which do not release analytes entrapped in the melanin fractions. The latter can be extracted only with appropriate solvents at elevated temperatures.

We found that the effectiveness of discriminating between drug use and contamination can be further enhanced by the application of several other measures. The most important one here is the identification of metabolites, since these arise as a result of ingestion. The ability to collect a second hair sample to guard against specimen mixup or for the investigation (by special wash procedures, etc.) or any special circumstances that may have affected the first result is another valuable means for preventing evidentiary false positives.

The many safety features of hair analysis cannot be matched by urinalysis. Of particular concern are a number of scenarios for evidentiary false positives in the case of urinalysis, against which no quality assurance programs can protect. For example, with the exception of marijuana, insufficient attention appears to have been given to the possibility of generating evidentiary false positives through passive exposure. For example, the study of Baselt and Chang (20) has shown that oral ingestion of as little as 25 mg of cocaine can cause peak urine values to be as high as 7940 ng/mL. The period over which positive values were obtained (4000 to 5000 ng/mL) was longer than 24 h. Consequently, be extrapolation, the passive exposure to only a few milligrams of cocaine is likely to produce positive urinalysis results. Furthermore, the lung and oral routes, unlike hair, do not offer resistance to penetration, nor are washing and the use of wash criteria possible. And then there is the problem of poppy seed ingestion and the possibility (or at least the claim) of subversive activities via spiked food or drink.

True, the body is constantly cleansed of drugs by urine excretion. In the case of hair this is achieved by regular washing. But such natural flushing is not particularly helpful, for example, to the spouse of a drug user, who suffers chronic environmental exposure to low amounts of drugs and consequently is at considerable risk of producing a positive urinalysis result. With hair analysis, the normally occurring decontamination processes are augmented by vigorously applied wash procedures, wash kinetic analyses, and measurement of metabolites.

One possible approach for avoiding the problems of evidentiary false positives is for urinalysis to raise its cutoff levels. Alternatively, or in addition, urinalysis testing programs could use hair analysis in case of a challenged positive result. Undoubtedly, a "safety net" program with hair analysis could greatly alleviate the apprehensions of those being tested by urinalysis, thereby contributing to improved employer/employee relations.

ACKNOWLEDGMENTS

The authors gratefully acknowledge the assistance of the staff of Psychemedics Corporation with this research: Susan Freedy, Gloria Ameigeiras, Jacob Edem, Gene Hayes, Henry Scholtz, and Thomas Donahue. We are particularly indebted to William H. Blahd, Chief of the Nuclear Medicine and Ultrasound Service, West Los Angeles V.A. Medical Center, Wadsworth Division, for supporting this research from its inception. The work was in part supported by a grant from the National Institutes of Justice (Grant #86-IJ-CX-0029) and by V.A. Medical Research Funds.

We thank David Kidwell (Naval Research Laboratory) and David Blank (Naval Military Personnel Command) for providing us with a republication copy of their Congress Proceedings manuscript and for the opportunity of discussing several aspects of their data.

REFERENCES

1. A. M. Baumgartner, P. F. Jones, W. A. Baumgartner, and C. T. Black. Radioimmunoassay of hair for determining opiate abuse histories. *J. Nuclear Med.*, *20*:749–752 (1979).
2. W. A. Baumgartner, V. Hill, and W. H. Blahd. Hair analysis for drugs of abuse. *J. Forensic Sci.*, *34*:1433–1453 (1989).
3. H. Matsuno, T. Uematsu, and M. Nakashima. The measurement of haloperidol and reduced haloperidol in hair as an index of dosage history. *Br. J. Clin. Pharmacol.*, *29*:187–194 (1990).
4. W. A. Baumgartner and V. A. Hill. Hair analysis for drugs of abuse: Forensic and policy issues. Conference on Hair Analysis for Drugs of Abuse, National Institute on Drug Abuse, May 1990 (Proceedings, in press).

5. R. L. Foltz, A. F. Fentiman, and R. B. Foltz. Methamphetamine. In *GC/MS Assays for Abused Drugs in Body Fluids*, NIDA Research Monograph 32. National Institute on Drug Abuse, Maryland (August 1980).
6. R. L. Foltz, A. F. Fentiman, Jr., and R. B. Foltz. Phencyclidine. In *GC/MS Assays for Abused Drugs in Body Fluids*, NIDA Research Monograph 32, National Institute on Drug Abuse, Maryland (August 1980).
7. J. M. Moore. Morphine determination. *J. Chromatogr.*, *147*:327 (1978).
8. R. W. Taylor, N. C. Jain, and M. P. George. Simultaneous identification of cocaine and benzoylecgonine using solid-phase extraction and gas chromatography/mass spectrometry. *J. Anal. Toxicol.*, *11*:233–234 (1987).
9. R. L. Foltz, K. M. McGinnis, and D. M. Chinn. Quantitative measurement of Δ-9-tetrahydrocannabinol and two major metabolites in physiological specimens using capillary column gas chromatography negative ion chemical ionization mass spectrometry. *Biomed. Mass Spectrom.*, *10*:316–331 (1983).
10. D. Valente, M. Cassini, M. Pigliaphchi, and G. Vincent. Hair as the sample in assessing morphine and cocaine addiction. *Clin. Chem.*, *27*:1952–1953 (1981).
11. A. M. Baumgartner, P. F. Jones, and C. T. Black. Detection of phencyclidine in hair. *J. Forensic Sci.*, *26*:576–581 (1981).
12. W. A. Baumgartner, P. F. Jones, C. T. Black, and W. H. Blahd. Radioimmunoassay of cocaine in hair. *J. Nuclear Med.*, *23*:790–792 (1982).
13. Hair analysis for cocaine and marijuana: Research report. Navy Contracts 53-5202-9167 and 53-5202-2129.
14. R. Siegel. Paper presented at Conference on Hair Analysis for Drugs of Abuse, National Institutes of Justice, Washington, DC, January 1989.
15. W. A. Baumgartner, J. Baer, and V. H. Hill. Hair analysis for the detection of drug use in pretrial, probation, and parole populations. Final Research Report, Grant 86-IJ-CX0029.
16. C. R. Robbins. *Chemical and Physical Behavior of Human Hair*. Van Nostrand Reinhold, New York, pp. 1–16 (1979).
17. C. R. Robbins. Private communication.
18. D. A. Kidwell. Conferences on hair analysis for drugs of abuse. National Institutes on Drug Abuse, May 1990 (Proceedings, in press).
19. D. A. Kidwell. Private communication.
20. R. S. Baselt and R. J. Chang. Urinary excretion of cocaine and benzoylecgonine following oral ingestion in a single subject. *J. Anal. Toxicol.*, *11*:81–82 (1987).

86

Analysis of Anabolic Steroids in Food and Biological Materials

Carlos H. Van Peteghem *State University of Ghent, Ghent, Belgium*

I. INTRODUCTION

Within the European Economic Community (EECC), the use of certain substances having a hormonal action and of any substances having a thyrostatic action is prohibited by Council Directive 85/358 of July 16, 1985 (1), supplementing Council Directive 81/602 of July 31, 1981 (2).

Another Council Directive, no. 86/469 of September 16, 1986, lays down the rules for the examination of animals and fresh meat for the presence of residues (3). A list of the substances which have to be traced on the basis of an approved sampling program in each of the member states is given in Table 1. The number of samples to be analyzed is proportional to the number of animals slaughtered and differs from species to species.

The methods which are authorized for detecting residues of the substances mentioned have been laid down in Commission Decision 87/410 of July 14, 1987 (4). They include immunoassays (IA), thin-layer chromatography (TLC), high-performance liquid chromatography (HPLC), gas chromatography (GC), mass spectrometry (MS), and spectrometry (SP). Each approved analytical procedure should be characterized by its own criteria for specificity, accuracy, precision, limit of detection, sensitivity, practicability, applicability, and, when required, limit of decision and limit of quantification.

II. FEATURES OF ANABOLIC STEROID RESIDUE ANALYSIS

The bioanalysis of anabolic steroid residues is characterized by the following.

1. *Low analyte concentrations.* Except for injection sites, the residues occur in the parts-per-billion (ppb) or even sub-parts-per-billion range. This implies in most cases a tedious extraction and cleanup of the sample, which makes the procedure slow and expensive. Methods using immunoassays, similar to those used for the screening of urine samples, may be the very rare exception.
2. *Pressure of work.* When samples are taken from the animal upon slaughtering, the carcass is retained in the slaughterhouse until the analytical results become available. Its quality and thus its commercial value rapidly lowers during the 4 to 5 days required to complete the analysis. When urine samples are taken from living

Table 1 Groups of Residues to Be Traced, According to Council Directive 86/469/EEC of September 16, 1986

A. Groups for all member states

Group I
1. Stilbenes, stilbene derivatives, salts and esters of them
2. Thyrostatics
3. Other substances with estrogenic, androgenic, or gestagenic action, except those of group II

Group II
Substances permitted by virtue of Article 4 of Directive 81/602/EEC and Article 2 of Directive 85/649/EEC

Group III
1. Inhibiting compounds: antibiotics, sulphonamides, and other antimicrobial agents
2. Chloramphenicol

B. Specific groups

Group I—other drugs
1. Substances against endo- and ectoparasites
2. Tranquilizers and beta-blockers
3. Other veterinary drugs

Group II—other residues
1. Contaminants in animal feed
2. Environmental contaminants
3. Other substances

animals, these may be neither slaughtered nor traded before the result of the analysis is known.

3. *The yes-or-no answer type*. This requires the highest obtainable certainty or reliability must be obtained and that, as a direct consequence of this, the analytical procedures may be arranged in a certain order. It also means that some techniques may not be at all suitable for certain sample types or matrices.
4. *The heavy economic and social consequences*. Positive carcasses must be destroyed. Positive living animals must be kept at the farm until a second analysis proves that they have become negative. If certain substances such as stilbenes are found and confirmed, the animal must be killed and the carcass destroyed. The owner of the animal may be prosecuted for violation of national or EEC legislation.

III. ANALYTICAL APPROACHES

The analyses are performed on samples which are taken either from living animals at the farm or from slaughtered animals in the abattoir. In the first situation either urine or feces, very seldom saliva or plasma, are taken, while in the latter case a much larger variety of samples is available: urine, injection site, fat, muscle, liver, bile, and kidney.

A general outline which indicates the particular analytical procedures actually being used in Belgium for a number of sample types are shown in Table 2. More detailed outlines of the analytical procedures for urine samples and muscle tissue samples, as they are currently used in our laboratory, are shown in Tables 3 and 4 (5–7). Details about the HPLC fractionation which is the keystone of the procedures which use MS detection can be read in detail in the publications concerned (6,7).

As a valuable alternative to HPLC fractionation, more and more use is being made of immunoaffinity chromatography. This procedure utilizes a column filled with an inert material to which antibodies are covalently bound. In passing the crude extract through such a column, the analytes are retained by an antigen-antibody interaction while the bulk of interferences can be removed by rinsing the column. The analytes are finally eluted by denaturation of the antibody with an alcohol solution, after which the column can be regenerated by means of buffer solutions (8,9).

IV. QUALITY ASSURANCE—QUALITY CONTROL

In view of what has been stated as to the required reliability of the results, a number of measures must be taken to exclude as efficiently as possible the occurrence of false positive results as well as of false negatives. Four different approaches may be valid.

Table 2 General Outline of Analytical Procedures Used in Belgium for the Detection of Anabolic Steroid Residues for a Number of Sample Types

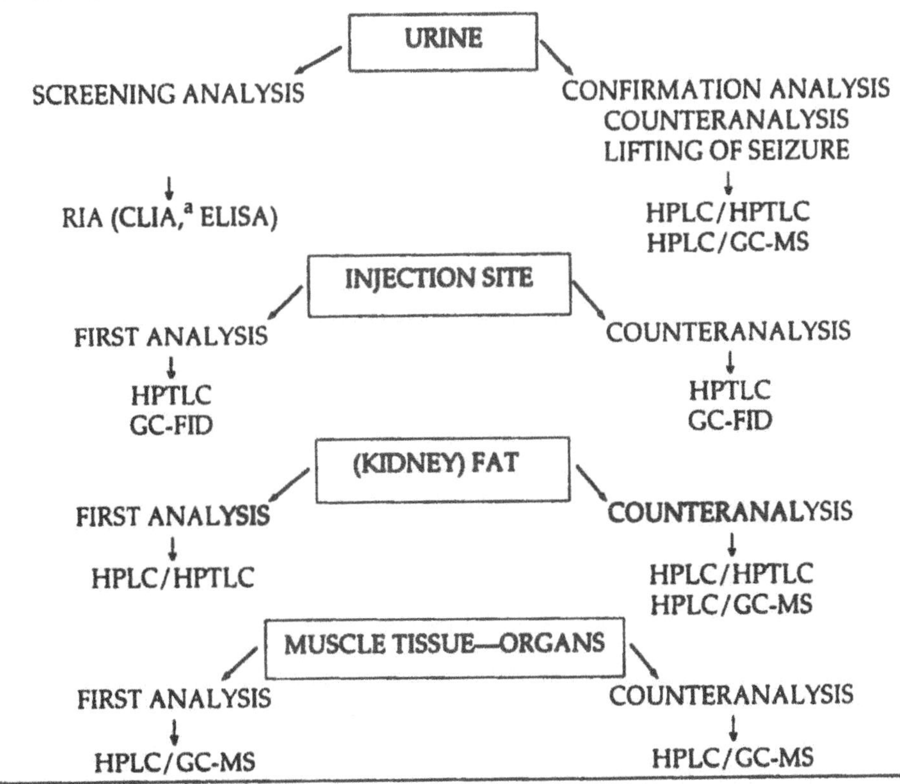

[a]CLIA = chemiluminescence immunoassay.

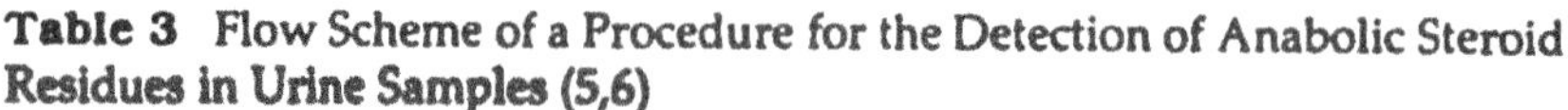

Table 3 Flow Scheme of a Procedure for the Detection of Anabolic Steroid Residues in Urine Samples (5,6)

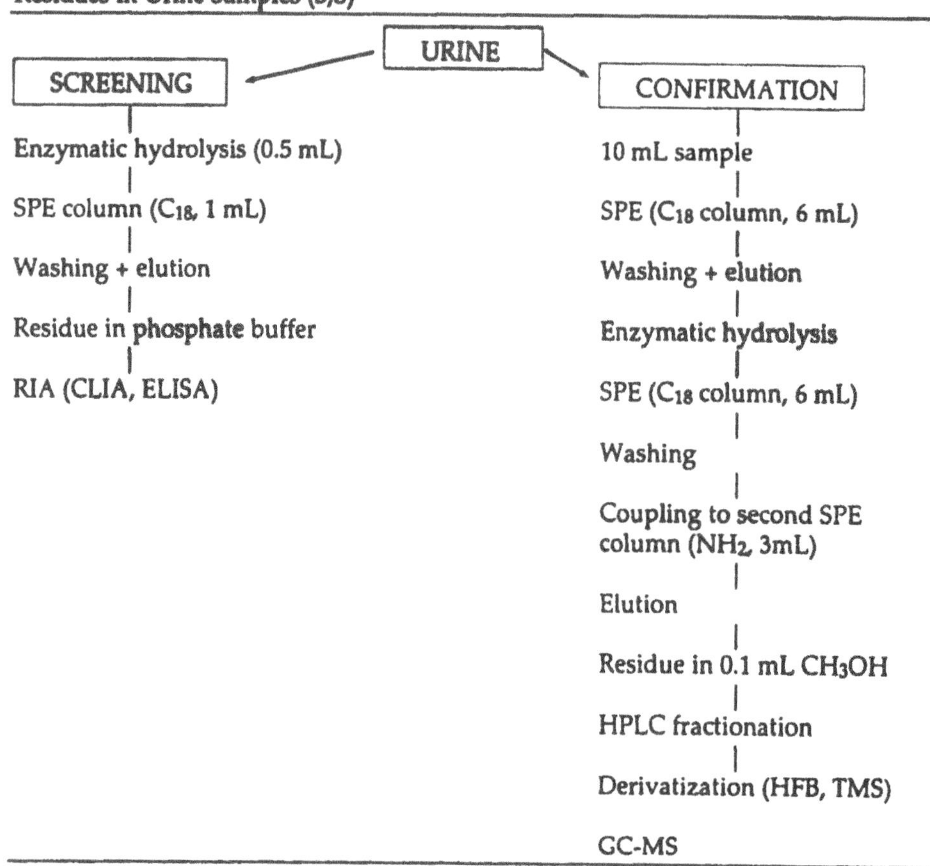

1. *Criteria for positive identification*, as laid down in EEC Commission Decision 67/410 (4). These criteria are intended mainly for the prevention of false positives. It is our experience that some of those criteria have been expressed so strongly that they may produce false negatives. A revision, after careful practical checking, has become desirable.
2. *Availability of reference materials*, as well analytes as "grown" or spiked materials. The Community Bureau of Reference (BCR) of the EEC is making big efforts to produce, in the decade to come, a number of reference materials. The first material, lyophilized urine, containing residues of stilbenes (diethylstilbestrol, hexestrol, and dienestrol), is about to be certified and will be the first product of what is expected to become a large series.
3. *Labeled compound*. Radioactively labeled homologs are very suitable for recovery controls, while deuterium-labeled molecules in mass spectrometric methods serve for overall control: compensation of losses during sample preparation, derivatization efficiency, and instrument check. Unfortunately, at present only a limited number of labeled compounds have become available.
4. *Collaborative studies*. The performances of analytical laboratories involved in the routine control for anabolic steroids abuse should be controlled on a regular basis

Table 4 Flow Scheme of a Procedure for the Detection of Anabolic Steroid Residues in Muscle Tissue

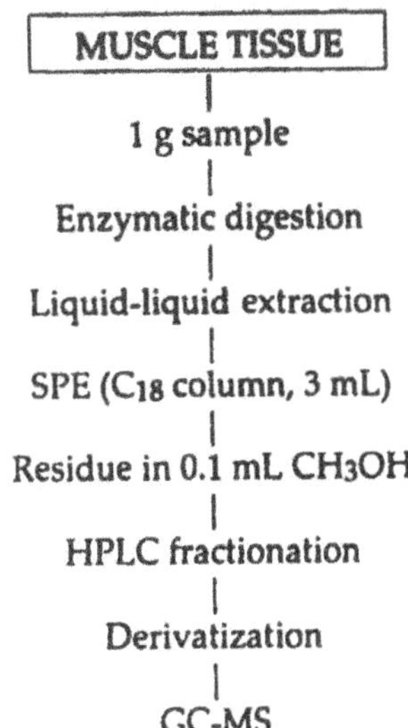

via international collaborative studies with follow-up discussions and eventually workshops. The Benelux Working Party on Hormones and Antihormones is making great efforts in that direction. The preliminary results prove the need of that kind of analytical refinement and refresher training.

V. STATE OF THE ART AND CONCLUSIONS

1. Anabolic steroid residue analysis currently can be performed with a sufficiently high reliability for various (not all) compounds in a number of matrices, provided that the most suitable techniques are used.
2. These techniques remain tedious and expensive and require experienced staff members and skilled technicians.
3. The interests of the laboratory people, the (political) health authorities, and the cattle breeders diverge greatly.

REFERENCES

1. *Official Journal of the European Communities*, pp. L 191/46-49, July 23, 1985.
2. *Official Journal of the European Communities*, pp. L 222/32-33, August 7, 1981.
3. *Official Journal of the European Communities*, pp. L 275/36-45, September 26, 1986.
4. *Official Journal of the European Communities*, pp. L 223/18-36, August 11, 1987.
5. P. Evrard, P. Gaspar, and G. Maghuin-Rogister. A specific radioimmunoassay for the detection of 19-nortestosterone residues in urine and plasma of cattle, *J. Immunoassay*, 7:353 (1986).
6. N. A. Schmidt, H. J. Borburgh, T. J. Penders, and C. W. Weykamp. Steroid profiling—An update. *Clin. Chem.*, *31*:637 (1985).
7. E. Daeseleire, A. De Guesquière, and C. Van Peteghem. Derivatization and GC/MS detection of anabolic steroid residues isolated from edible muscle tissues. *J. Chromatogr.*, 562:673 (1991).

8. L. A. Van Ginkel, R. W. Stephany, H. J. Van Rossum, H. M. Steinbuch, G. Zomer, E. Van de Heeft, and A. P. J. M. De Jong. Multiimmunoaffinity chromatography: A simple and highly selective clean-up method for multi-anabolic residue analysis of meat. *J. Chromatogr.*, *489*:111 (1989).
9. W. Haasnoot, R. Schilt, A. R. M. Hamers, and F. A. Huf. Determination of 17β-19-nortestosterone and its metabolite 17α-19-nortestosterone in biological samples at the sub parts per billion level by high-performance liquid chromatography with on-line immunoaffinity sample pretreatment. *J. Chromatogr., 489:157 (1988).*

87

Pharmacogenetics and Interpretation of Toxicological Results

Rafael de la Torre *Institut Municipal d'Investigació Médica, Barcelona, Spain*

I. INTRODUCTION

Genetic polymorphism related to certain metabolic enzymes can be the dominant cause of interindividual variations in the elimination of several drugs (1). This could lead to clinically significant differences in the pharmacological responses of some patients (2). Nevertheless, inheritance and, more specifically, genetic polymorphism in drug metabolism are among many other biological factors that influence the concentrations of xenobiotics in body fluids and their further interpretation (3).

II. POLYMORPHIC VARIATIONS IN DRUG METABOLISM

The major pathways showing genetic polymorphism in drug metabolism phase I reactions are acetylation (4) and oxidation (5). Reports on genetic polymorphism on phase II reactions (glucuronidation and sulphatation) are in general inconclusive, and further studies are needed with adequate probe drugs (6). An apparent polymorphic variation in glucuronidation has been observed for drugs with a polymorphic drug oxidation (7).

Some other metabolic polymorphisms due to enzyme deficiencies are acetaldehyde dehydrogenase (8), serum cholinesterase (9), glucose 6-phosphate dehydrogenase (10), and paraoxonase (11).

The identification of polymorphic differences in drug metabolism is of interest because of the clinical manifestations arising from these differences. The relevance of side effects and/or toxicity in the poor/slow metabolizer condition is highly dependent on the toxic effects derived from the accumulation in the body of the parent compound or the relevance of the pathway in its metabolism. Dextromethorphan and flunitrazepam show that not always can a higher incidence of toxic effects be subsequently derived, because there is a metabolic pathway under genetic control. Dextromethorphan is used in therapeutics as an antitussive, and it has some interest in pharmacogenetics because it has been proposed as a probe drug for population phenotyping of debrisoquine polymorphism (12). Dextromethorphan is mainly O-demethylated to dextrorphan, and poor metabolizers of the debrisoquine phenotype have this metabolic pathway impaired. Because of the safety of this drug, no major side effects due to its accumulation in the body are expected. Another drug of interest in Europe, related to the use of drugs of abuse, is the benzodiazepine flunitrazepam. One of its main metabolites, 7-aminoflunitrazepam, is subsequently acetylated. Because acetylation is a secondary pathway of this drug, it is unlikely that the appearance of side effects is related to this polymorphism.

Examples will be given of some common drugs involved in toxicological overdoses or in drugs of abuse testing where it is known that some metabolic pathways are under control of different oxidative polymorphisms such as codeine (debrisoquine phenotype) and diazepam (S-mephenytoin phenotype). Other genetic factors regulating nonoxidative enzymes important in the metabolism of drugs, as would be the case for pseudocholinesterase (cocaine), will be discussed. In addition, we will consider the case of some therapeutic agents such as d-propoxyphene, which are responsible for some pharmacological interactions of relevance when substrates with polymorphic metabolizing enzymes are involved.

III. METABOLISM OF SOME DRUGS REGULATED BY POLYMORPHIC METABOLIZING ENZYMES

A. Codeine (Debrisoquine Hydroxylase)

There is a monogenic deficiency of the capacity for debrisoquine hydroxylation in 5 to 10% of different Caucasian populations. Studies in Oriental populations show a lower incidence (2.5 to 0.5%) of this phenotype (13). The isozyme of the cytochrome P450 enzyme family involved in this polymorphism is cytochrome P450 II D6, and mutations on the gene encoding this proteins, CYP2D6, can explain the poor-metabolizer phenotype (14). There are a number of important drugs regulated by this polymorphism, including some belonging to the groups of tricyclic antidepressants, beta-blockers, antiarrhythmics, neuroleptics, and opiates.

Some opiates are under the same polymorphic genetic control as debrisoquine, e.g., dextromethorphan (12), and codeine (15,16). Figure 1 shows the plasma concentration-time curve of dextrorphan, the main metabolite of dextromethorphan, in a study in healthy volunteers (n = 12) receiving a dose of 30 mg of dextromethorphan hydrobromide by the oral route (17). One of the volunteers has this metabolic pathway impaired and was later phenotyped as a poor metabolizer of debrisoquine. This same poor metabolizer was subjected to an excretion study with codeine (30 mg of codeine phosphate given by the oral route). When comparing its metabolic excretion profile with that of an extensive metabolizer of debrisoquine, the absence of morphine, one of the main metabolites of codeine, and normorphine can be observed (Figure 2, from a chromatogram obtained by gas chromatography coupled to mass spectrometry). The concentration of free codeine in both phenotypes is the same, but those of norcodeine are higher in the poor metabolizer (Figure 3). The rate of codeine and norcodeine excreted in the free and conjugated fraction of urine in the poor metabolizer is almost constant during the time collection period but higher than in the extensive metabolizer. Then, after submitting the urine to an enzymatic hydrolysis, the total recovery of codeine is much higher in the poor metabolizer than in the extensive one.

No acute toxic effects are expected from the accumulation of codeine if it has been prescribed and ingested as a cough remedy. Of greater concern is the detection of

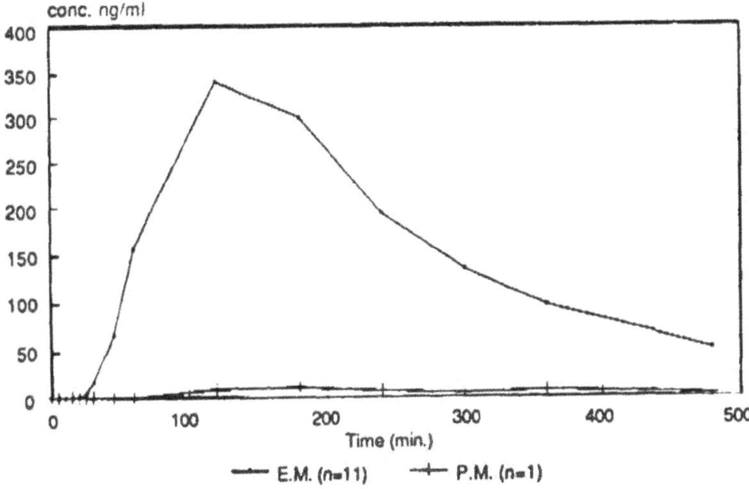

Figure 1 Plasma concentration vs. time curve of dextrorphan. E.M.: extensive metabolizers, P.M.: poor metabolizers.

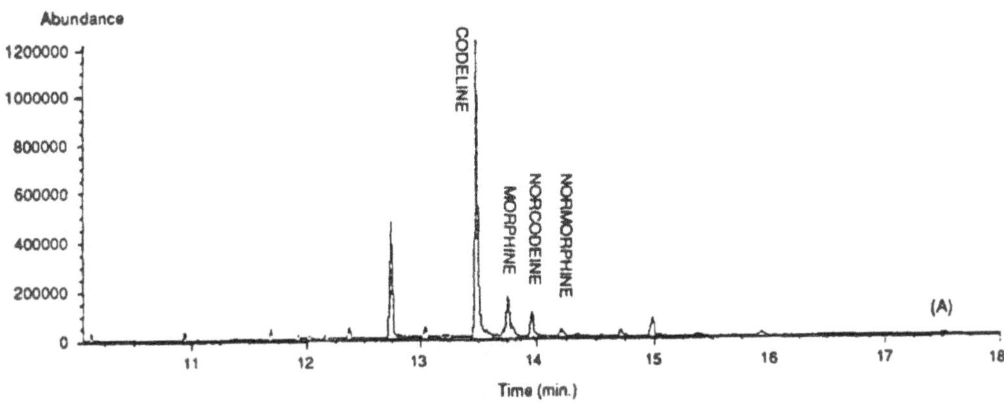

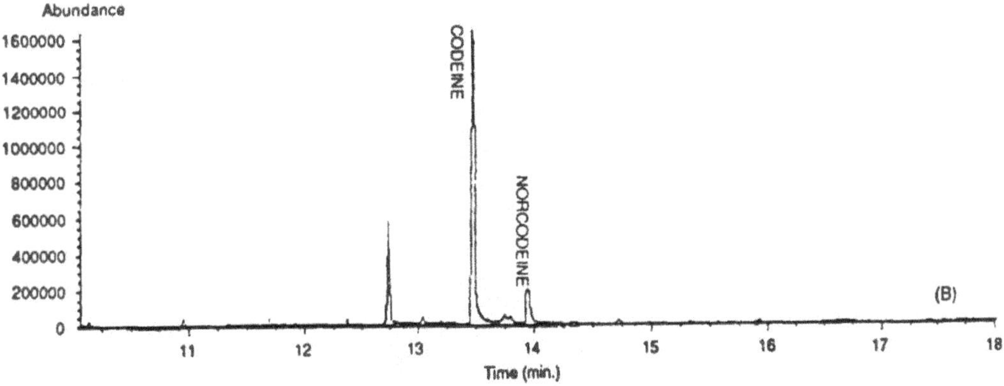

Figure 2 Urinary profile of an extensive metabolizer (A) and a poor metabolizer (B) of codeine.

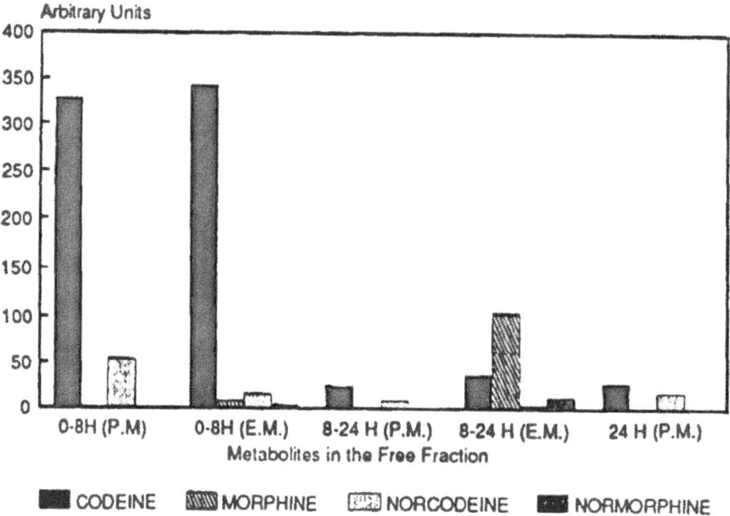

Figure 3 Relative proportions of codeine and its main metabolites in the free fraction or urine at different time collection periods. E.M.: extensive metabolizers, P.M.: poor metabolizers.

codeine during drug testing or in doping control. When opiates are detected in drug testing, quite often patients in any kind of treatment program claim the ingestion of codeine. Two questions arise:

Has codeine been ingested as a cough remedy?
Has codeine been abused because of a shortage of heroin?

Taking into account the above data, it is very difficult to answer these questions if the laboratory or the physician are dealing with a poor metabolizer of debrisoquine.

In doping control, codeine is a banned substance belonging to the group of narcotic analgesics in the International Olympic Committee List of Doping Classes and Methods. Because of the use of this drug as an antitussive or analgesic in association with paracetamol or aspirin, recently a cutoff level of 1 μg/mL has been suggested for reporting results to the sports medical commissions for evaluation. The pharmacogenetics of codeine and the interpretation of the whole metabolic profile must be taken into account in this evaluation of results. If the results after the evaluation are inconclusive, it is worthwhile to do a controlled excretion study in the individual involved with some of the probe drugs proposed for debrisoquine phenotyping.

B. Diazepam (S-Mephenytoin Hydroxylase)

In the metabolism of mephenytoin, it has been shown that the S- but not the R-enantiomer is hydroxylated in a polymorphic manner (18). The isozymes involved in this polymorphism belongs to the cytochrome P450 II C group. The number of substrates for which metabolism is regulated by this polymorphism is quite low and includes desmethylmephenytoin, mephobarbital, propranolol, diazepam, nordiazepam, and, to a certain extent (50%), the N-demethylation of imipramine. The incidence of poor metabolizers is less than 5% in the Caucasian population (19).

Among those drugs showing this polymorphism, benzodiazepines are of interest because they are among the main drugs involved in drug overdoses (20).

The oxidation of diazepam to nordiazepam and the oxidation of nordiazepam to oxazepam are impaired in poor metabolizers of mephenytoin. Pharmacokinetic studies in healthy volunteers after the administration of 10 mg of both substances show a reduction by one-half in the clearance of the drug and a two-times-longer half-life in poor metabolizers of mephenytoin (21).

C. Cocaine (Pseudocholinesterase)

Human plasma cholinesterase hydrolyzes a number of drugs, some of them related to drugs of abuse, e.g., cocaine (22) and heroin (23). The enzyme is defined by two genetic loci, E_1 and E_2. E_1 is primarily responsible for the clinically relevant enzymatic activity. One form of this gene, the E_1a, is responsible for the enzymatic activity of the atypical cholinesterase or pseudocholinesterase. The frequency of the abnormal enzyme in the Caucasian population is around 1/2500, and lower incidences have been shown in other populations (24).

Cocaine, a substrate of cholinesterase, is rapidly eliminated from the body by minimal urinary excretion but mostly by metabolism. While one of the main metabolites, benzoylecgonine, is formed nonenzymatically, ecgoninemethylester is formed by cholinesterase in liver and serum (25). It has been suggested that a deficiency in cholinesterase activity might be responsible in some cases for the acute toxicity of cocaine in drug abusers (26).

D. Propoxyphene

The increasing knowledge of specific substrates and inhibitors of the different polymorphic metabolic pathways allows precise predictions of drug-drug interactions. Some common drugs such as quinidine or d-propoxyphene are potent inhibitors of substrates cosegregating with the debrisoquine phenotype. In the case of d-propoxyphene, a polymorphic regulation in its metabolism has not been demonstrated, but its effect as inhibitor is potent enough to induce a misphenotyping. That would be the case of individuals submitted to a phenotyping test with a probe drug if d-propoxyphene is co-administered (27). Then the population of poor metabolizers of debrisoquine can be artifactually increased.

From a pharmacodynamic point of view, the main concern with these drugs is that they can lead to clinically relevant metabolic interactions when they are co-administered or when they are associated with other drugs that co-segregate with this polymorphism.

IV. CONCLUSIONS

Inheritance is one factor among others influencing the interpretation of toxicological results. Nevertheless, it is becoming more relevant because of increasing knowledge of the number of drugs of importance in therapeutics that show a metabolic polymorphism and a rise in the number of individuals subjected to drug testing in different fields as in sports, therapeutic drug monitoring, and at the workplace. Then, even in the case of metabolic polymorphisms with a low incidence among the population, toxicologists are more often confronted with cases where genetics can help to interpret toxicological results.

REFERENCES

1. L. P. Balant, U. Gundert-Remy, A. R. Boobis, and Ch. von Bahr. Relevance of genetic polymorphism in drug metabolism in the development of new drugs. *Eur. J. Clin. Pharmacol., 36*:551–554 (1989).
2. K. Brosen and L. F. Gram. Clinical significance of the sparteine/debrisoquine oxidation polymorphism. *Eur. J. Clin. Pharmacol., 36*:537–547 (1989).
3. A. Deom. L'intoxication d'origine inconnue, l'apport du laboratoire pour le clinicien. *Therapeutische Umschau, 43*:259–268 (1986).
3. W. Kalow, H. W. Goedde, and D. P. Agarwal, eds. Ethnic differences in reactions to drugs and xenobiotics. In *Progress in Clinical and Biological Research,* Vol. 214. Alan R. Liss, New York (1986).
4. W. W. Webber and D. W. Hein. N-acetylation pharmacogenetics. *Pharmacol. Rev., 37*:26–79 (1985).
5. M. Eichelbaum and A. S. Gross. The genetic polymorphism of debrisoquine/sparteine metabolism. Clinical aspects. *Pharm. Ther., 46*:377–394 (1990).
6. G. Alvan, L. P. Balant, P. R. Bechtel, and A. R. Boobis, Eds. European Consensus on Pharmacogenetics. COST B1 Medicine. Commission European Community, Brussels (1990).
7. Q. Y. Yue, J.-O. Svensson, C. Alm, F. Sjoqvist, and J. Sawe. Interindividual and interethnic differences in the demethylation and glucuronidation of codeine. *Br. J. Clin. Pharm. 28*: 629–637 (1989).
8. H. W. Goedde, and D. P. Agarwal. Pharmacogenetics of aldehyde dehydrogenase (ALDH). *Pharm. Ther., 45*:345–371 (1990).
9. W. Kalow, H. W. Goedde, and D. P. Agarwal, Eds. Ethnic differences in reactions to drugs and xenobiotics. In *Progress in Clinical and Biological Research,* Vol. 214. Alan R. Liss, New York (1986).
10. L. Luzzatto. Glucose-6-phosphate dehydrogenase and other genetic factors interacting with drugs. In *Ethnic Differences in Reactions to Drugs and Xenobiotics*. W. Kalow, H. W. Goedde, and D. P. Agarwal, Eds. Alan R. Liss, New York, pp. 385–399 (1986).
11. J. R. Playfer, L. C. Eze, M. F. Bullen, and D. A. P. Evans. Genetic polymorphism and interethnic variability of plasma paraoxonase activity. *J. Med. Gen., 13*:337–342 (1976).
12. B. Schmid, J. Bircher, R. Preisig, and A. Kupfer. Polymorphic dextrometorphan metabolism: Co-segregation of oxidative O-demethylation with debrisoquine hydroxylation. *Clin. Pharmacol. Ther., 38*:618–624 (1985).
13. W. Kalow. Genetic variation in the human cytochrome P-450 system. *Eur. J. Clin. Pharmacol., 31*:633–641 (1987).
14. M. Kagimoto, M. Heim, K. Kagimoto, Zeugin, T., and U. S. Meyer. Debrisoquine/sparteine polymorphism: Characterization of the mutations of the CYP2D6 causing deficient P450IID6 protein. In *Drug Metabolizing Enzymes: Genetics, Regulation and Toxicology. VIIIth International Symposium on Microsomes and Drug Oxidation* (J.-A. Gustafsson and S. Orrenius, Eds.), Stockholms Projektgrupp, Stockholm, p. 46 (1990).
15. O. Y. Yue, J.-O. Svensson, C. Alm, Sjoqvist, F., and J. Sawe. Codeine O-demethylation co-segregates with polymorphic debrisoquine hydroxylation. *Br. J. Clin. Pharm., 28*:639–645 (1989).
16. O. Mortimer, K. Persson, M. G. Ladona, D. Spalding, U. M. Zanger, U. A. Meyer, and A. Rane. Polymorphic formation of morphine from codeine in poor and extensive metabolizers of dextrometorphan: Relationship to the presence of immunoidentified cytochrome P450IIDI. *Clin. Pharmacol. Ther., 47*:27–35 (1990).

17. M. C. Caturla, R. de la Torre, M. Congost, and J. Segura. Plasma dextrometorphan metabolites and relation to urinary phenotyping: Pilot study. *Eur. J. Clin. Pharmacol., 36*:A149 (1989).
18. A. Kupfer and R. Preisig. Pharmacogenetics of mephenytoin: A new drug hydroxylation polymorphism in man. *Eur. J. Clin. Pharmacol., 26*:753–759 (1984).
19. E. J. Sanz, T. Villen, C. Alm, and L. Bertilsson. S-mephenytoin hydroxylation phenotypes in a Swedish population determined after co-administration with debrisoquine. *Clin. Pharmacol. Ther., 45*:495–499 (1989).
20. A. S. Brett. Implications of discordance between clinical impression and toxicology analysis in drug overdose. *Arch. Intern. Med., 148*:437–441 (1988).
21. L. Bertilsson, T. K. Henthorn, E. Sanz, G. Tybring, J. Sawe, and T. Villen. Importance of genetic factors in the regulation of diazepam metabolism: Relationship to S-mephenytoin, but not debrisoquine, hydroxylation polymorphism. *Clin. Pharm. Ther., 45*:348–354 (1989).
22. D. J. Stewart, T. Inaba, M. Lucassen, and W. Kalow. Cocaine metabolism: Cocaine and norcocaine hydrolysis by liver and serum esterases. *Clin. Pharmacol. Ther., 25*:464–468 (1979).
23. O. Lockridge, N. Mottershaw-Jackson, H. W. Eckerson, and B. N. LaDu. Hydrolysis of diacetylmorphine (heroin) by human serum cholinesterase. *J. Pharmacol. Exp. Ther., 215*:1–8 (1980).
24. W. Kalow. Ethnic differences in drug metabolism. *Clin. Pharmacokinet., 7*:373–400 (1982).
25. T. Inaba, D. J. Stewart, and W. Kalow. Metabolism of cocaine in man. *Clin. Pharmacol. Ther., 23*:547–552 (1978).
26. P. Devenyi. Cocaine complications and pseudocholinesterase. *Ann. Intern. Med., 110*: 167–168 (1989).
27. E. Sanz and L. Bertilsson. D-propoxyphene is a potent inhibitor of debrisoquine, but not S-mephenytoin 4-hydroxylation in vivo. *Ther. Drug Monitor.*, in press (1990).

Part VI

Inorganics

88

Selenium: An Overview from the Laboratory Perspective on Its Role in Infant Health

J. A. Cocho, J. R. Cervilla, and J. M. Fraga *Hospital General de Galicia, Santiago de Compostela, Spain*

I. SELENIUM ANALYSIS

One of the critical points in the study of selenium is the improvement in the quality of its analysis (1). Nowadays this is more easily achieved by (a) interlaboratory comparisons, (b) improvement in analytical methodology and instrumentation, and (c) use of controls and reference materials (2). All these allow an adequate standardization of the analytical methodology and a continued control of the laboratory results (2).

Results of interlaboratory comparisons (IC) reveal the wide variability in reported selenium results. The coefficients of variations (CVs) reported (3,4) ranged from 15% for a sample of 14 μmol/L to 55% at 0.4 μmol/L, and for the normal human blood selenium level, the CV is 19%. These CVs are >100% when only aqueous standards are used (4). It is alarming to note in the IC that the range of results among "experienced laboratories applying their preferred routine or research method to analysis of the pools" (1) is extended from reference range to altered values.

Fluorimetry and atomic absorption spectrometry with hydride generation or electrothermal atomization (GFAAS) are the most commonly used techniques for the analysis of selenium.

Recent developments in atomizer technology have helped minimize interferences and improve the accuracy and precision of GFAAS (5). In addition, the use of flow injection in combination with hydride-generation AAS (6) offers several advantages, particularly the reproducible use of small sample volumes and the ability to achieve a rapid sample throughput.

The little sample preparation, the simplicity of operation, the accuracy and precision, the potential for direct analysis, and the accessibility of equipment to many clinical laboratories have made GFAAS the technique of choice for selenium analysis.

The major analytical problems with this method are preatomization losses, spectral interferences (Fe and P), chemical interferences, and formation of carbonaceous deposits inside the tube.

Selenium may be stabilized at ashing temperatures as high as 1200°C by the addition of matrix modifiers. Good results in various systems have been reported with Ni alone, or with Ni plus Cu, Ag, Mg, or Pd (1,5). However, some of these modifiers should be used with caution, because of the inability to stabilize Se(IV) (5). Other modifiers also used are Cu + Mg (5), Pt (7), and Rh (8). Since 1987, however, Pd has been the most widely used modifier, either alone or reduced with ascorbic acid (9,10), in combination with reducing gas (5% H_2/95% Ar) (10), Pd + Cu, Pd + Mg (11), and Pd + Ni. The combination Pd + Mg has a special interest as a universal modifier for volatile elements. Our experience in a laboratory where many other elements are analyzed is that elements such as Ni or Cu cannot be used at elevated concentrations without later contamination problems.

The use of deuterium versus Zeeman background correction has been controversial (12), and Zeeman correction is essential for selenium analysis in blood and urine. Deuterium correction produces precise and accurate results with serum (10).

Moreover, the new matrix modifiers plus STPF conditions and the new spectrometers, which can achieve 200 abs. lect./min are capable of doing time resolution of background and selenium signals. These approaches permit the use of deuterium background corrector for other biological samples.

Finally, every laboratory must evaluate its proposed method using certified standards and controls. Standard reference materials are available for human serum from the U.S. National Bureau of Standards (SRM 909) and from Nyegaard & Co., Oslo (Seronorm 109 and 112); for urine SRM 2670 and Seronorm 108; for tissues, as bovine liver, SRM1577a; and for other biological matrices. A second generation of human serum control is available from the laboratory of Dr. Versieck (Belgium). Furthermore, frequently reviews of new reference materials are published (13). Between-run precision may be monitored using Levy-Jennings control charts (2).

II. SELENIUM: INFLUENCE ON INFANT HEALTH

Interest in the possible role of selenium in human nutrition has been increasing over the last 10 years (14). As a result, nutrition pathologies and biochemical troubles related to the deficit of Se in the human diet have been studied. These Se-related problems have been found in some areas of Se-deficients soils and food (continental China) and in severe digestive or surgical pathologies which require long-term parenteral total nutrition (15,16).

Selenium is also an indirect regulator of metabolism of the essential fatty acids through the role that glutathione peroxidase (GPx) plays in the oxidation of arachidonic acid in the formation of prostaglandins, leucotrienes, and thromboxans. As an antioxidant, GPx reduces the organic and inorganic peroxides in the intracellular medium, and it needs four atoms of Se per molecule.

In Galicia (northwest Spain), the soil has a low Se level (0.03 ppm) and consequently the amount of Se in the diet is also low (91 μg/day) (17). These values are lower than in other Western countries, the United States, and Japan (17,18).

There is a great interest in the estimation of the content of Se in milk, because infants are fed exclusively on milk, and they are one of the groups at a higher risk of suffering from nutritional Se deficits because of their fast growth rate. There is a "nutritional interest" in the level of Se in milk in order to know the Se intake in the first months of life.

On the other hand, the Se content of the different types of milk, including human milk (HM), as well as the needs of Se for the infants, are not known in our area, and there are no precise Recommended Dietary Allowances by the International Committee of Pediatrics Nutrition (19).

III. MATERIALS AND METHODS

We have measured the Se content in different types of milk: human milk, formula milk (FM), and cow milk (CM). From these quantities we have calculated the daily intake in healthy infants of 1 to 4 months who are receiving HM or FM without supplements of Se.

Se was measured by fluorimetry (17).

IV. RESULTS AND DISCUSSION

The Se content in HM and FM, with and without supplement of Se and CM, are shown in Table 1. The content in HM doubles the content in FM without supplement. The Se content varies according to the type of milk and the geographic origin.

Our results are concordant with others published about Barcelona (northeast Spain) (Table 2), and these results from Spain are among the lowest in the world. These are similar to the levels in Northern Europe and New Zealand, both with low levels of Se in soils and food (20–22).

Nutritional habits, food availability according to the geographic area, and other not well known factors have an influence on the content of Se in HM, CM, and FM from a particular geographic area. The chemical form of Se in food, and its influence on its bioavailability, is another important regulating factor (19,23).

Heating processes which milk undergoes to be conditioned and made suitable for human use reduce the content of Se because it is a volatile element (17).

There is a perfect adaptation of milk composition to the needs of growth and nutrition for every mammalian species. Therefore, it can be deduced that a higher level of Se in HM is due to an evolutional adaptation process that improves the

Table 1 Selenium Content (μg/L) of Several Types of Milk

	N	X ±	SD	Range
HM mature	37	11.7	4.2[a]	4–25
CM (Galicia)				
Coastal	8	12.0	4.0	6–16
Inland	15	5.5	1.3[a]	3–8
Mixed	49	8.5	3.2[a]	1–23
CM commercial[b]	60	9.3	2.5[a]	2–20
FM[c]				
With Se suppl.	3	15.3	4.5	7–21
Without Suppl. 114[d]	114[d]	5.0	1.6[a]	2–9

[a]$p < 0.001$
[b]Several Spanish areas, including Galicia.
[c]Reconstruction at 13%
[d]Number of brands 11.

Table 2 Median Selenium Content: Mature Human Milk of Several Countries

Area	Se (μg/L)
Germany	28.3
Spain	
Barcelona	11.4 (9–18)
Galicia	11.7 (4–25)[a]
Finland	10.0–11.8
Italy	13.3 (1–50)
Japan	17–18
New Zealand	9–13
Sweden	11.9
United States	
17 states	13.3
Illinois	15.3
Evansville	23.0
Salt Lake City	25.0

[a]This work.

antioxidant defense of human infants in the first stages of extrauterine life, against the sharp increase in blood oxygen pressure after birth.

The average intake of Se is higher in infants feeding on HM in comparison to those feeding on FM during the first 4 months of life, 6.5 versus 4.0 μ/day ($p < 0.001$). These intakes are among the lowest recorded, similar only to the Se form Se-deficient countries (22), and do not reach the minimum intakes recommended by the National Academy of Nutrition of the United States (NRC). These Recommended Dietary Allowances (RDA, 1989) (24) have recently been reduced to 10 μg/day for the first 6 months of life and 15 μg/day until the end of the first year.

We concluded that although we have not found any pathology in this short-term study related to any deficit in Se in our suckling babies, an insufficient intake during a longer period may cause disorders in the long term (18,20,25).

For this reason, in our area and others of similar characteristics, it is perhaps better to recommend the use of FM with Se supplement in those cases where human milk feeding is not possible and so to reach a minimum and sufficient intake (19,20,24).

REFERENCES

1. G. Lockitch. Selenium: Clinical significance and analytical concepts. *Crit. Rev. Clin. Lab. Sci.*, 27:483 (1989).
2. C. Bradley and F. Y. Leung. Quality control procedures for monitoring whole blood and plasma trace metal levels by atomic absorption spectrophotometry. *Sci. Total Environ.*, 89:353 (1989).
3. T. S. Koth. Interlaboratory study of blood Se determinations. *J. Assoc. Offic. Anal. Chem.*, 70:664 (1987).
4. T. S. Koth. Effects of blood standards on interlaboratory variation in the assay of blood selenium. *Anal. Chem.*, 59:597 (1987).
5. K. S. Subramanian. Determination of trace metals in blood by GFAAS: Recent studies. *Atomic Spectrosc.*, 9:169 (1988).

6. K. McLaughlin, D. Dadgar, M. R. Smyth, and D. McMaster. Determination of selenium in blood plasma and serum by flow injection hydride generation AAS. *Analyst.*, *115*:275 (1990).
7. K. Saeed. Direct electrothermal atomisation AAS determination of selenium in whole blood and serum. *J. Anal. Atomic Spectrom.*, 2:151 (1987).
8. D. Wagley, G. Schiemedel, E. Mainka, and H. J. Ache. Direct determination of some essential and toxic elements in milk and milk powder by graphite furnace atomic absorption spectrometry. *Atomic Spectrosc.*, *10*:106 (1989).
9. M. B. Knowles and K. G. Brodie. Determination of selenium in blood by Zeeman graphite furnace AAS using a palladium-ascorbic acid chemical modifier. *J. Anal. Atomic Spectrom.*, 3:511 (1988); 4:305 (1989).
10. B. E. Jacobson and G. Lockitch. Direct determination of selenium in serum by graphite furnace ASS with deuterium background correction and a reduced palladium modifier: Age-specific reference ranges. *Clin. Chem.*, *34*:709 (1988).
11. B. Welz, G. Schlemmer, and J. R. Mudakavi. Palladium nitrate-magnesium nitrate modifier for graphite furnace AAS. *J. Anal. Atomic Spectrom.*, 3:695 (1988).
12. V. A. Letorneau, B. M. Joshi, and L. C. Butler. Comparison between Zeeman and continuum background correction for graphite furnace AAS on environmental samples. *Atomic Spectrosc.*, *8*:145 (1987).
13. I. Roelandts. Additional biological reference materials. *Spectrochim. Acta*, *44B*:985 (1989).
14. S. S. Baker, R. R. Lerman, S. H. Kery, and K. S. Crocker. Selenium deficiency with total parenteral nutrition: Reversal of biochemical and functional abnormalities by Se Suppl.: A case report. *Am. J. Clin. Nutr.*, *38*:769 (1983).
15. D. Kelly, et al. Symptomatic selenium deficiency in a child on home parenteral nutrition. *Pediatr. Res.*, *21*:22 (1987).
16. A. M. Van Rij and C. D. Thomson, Selenium deficiency in total parenteral nutrition. *Am. J. Clin. Nutr.*, *32*:2076 (1979).
17. J. A. Cocho, C. Parrado, J. R. Cervilla, J. R. Alonso, and J. M. Fraga. Determinación fluorimétrica de selenio. Estudios en un grupo de boblación. *Quim. Clin.*, 3:19 (1984).
18. J. R. Cervilla, J. R. Fdez-Lorenzo, J. M. Fraga, J. A. Cocho, and J. I. Ramos. Glutathione peroxidase activity in erithrocytes of newborns fed maternal or formula milk. In *Selenium in Medicine and Biology*, Walter de Gruyter, Berlin, p. 215 (1988).
19. Committee on Nutrition, Selenium. Trace elements. In *Pediatric Nutrition Handbook*, 2nd ed. (G. B. Forbes and C. W. Wooddruff, Eds.), American Academy of Pediatrics, p. 129 (1985).
20. I. Lombeck, K. Kasperek, H. D. Harbisch, L. E. Feinendengen, and H. J. Bremer. Selenium content of human milk, cow's milk and infant formulas. *Eur. J. Pediatr.*, *129*:139 (1978).
21. R. E. Litov, V. S. Sickles, G. M. Chan, I. R. Hargett, and A. Cordano. Selenium status in term infants fed human milk or infant formula with or without added selenium. *Nutr. Res.*, *9*:585 (1989).
22. R. Walivaara, L. Jansson, and B. Akesson. Contenido de selenio en leche humana en muestras obtenidas en 1979 y 1983 en Suecia. *Acta Pediatr. Scand. (Span. Ed.)*, 3:259 (1986).
23. J. M. Fraga, J. R. Cervilla, J. V. Iglesias, J. Peña, and J. A. Cocho. Selenium levels and selenium intake in newborns. In *Selenium in Medicine and Biology*, Walter de Gruyter, Berlin, p. 211 (1988).
24. *Recommended Dietary Allowances*, 10th ed. National Academy Press, Washington, DC (1989).
25. J. R. Cervilla, J. Varela-Iglesias, J. Peña, and J. M. Fraga. Selenium intake and serum selenium in newborns infants under total parenteral nutrition. *C.I.C. Edizioni Internazionali, s.r.l.*, p. 207 (1989).

89

Serum Magnesium Concentrations After Three Different Maximal Exercises

Alfredo Cordova and Valentin del Villar *University of Valladolid-CUS, Soria, Spain* **Luis Rabadan** *Hospital Insalud, Soria, Spain*

I. INTRODUCTION

Magnesium (Mg) is the second most common intracellular cation in humans. It has important functions in numerous metabolic processes. It is reasonable to surmise that Mg status may influence exercise capacity and, conversely, that exercise may affect Mg status (1). For example, Mg is required for the activity of several enzymes that participate in energy metabolism, including those that utilize or form adenosine triphosphate. Mg also facilitates the delivery of oxygen to working muscles by inducing production of 2,3-diphosphoglycerate in erythrocytes (1–6).

There is a considerable body of data that demonstrates an effect of exercise on magnesium metabolism. Serum magnesium concentrations decreased immediately following a marathon (7). The lowering of serum Mg following intense exercise has been verified by numerous investigators (2,8,9). An increased magnesium content in exercising muscle during prolonged work was paralleled by a decline in plasma Mg (10). This observation suggests that the reduction in serum magnesium observed during exercise may be a function of both sweat losses and redistribution of serum Mg into the working muscle, into red blood cells, and into adipocytes (6,8,10). Others (11,12) have observed increases in serum magnesium concentrations after maximal exercise, both in humans and in rats.

As described above, it is evident that magnesium metabolism can be affected by exercise. Inconsistencies may be related to differences in experimental designs, work intensity, and duration, which might modify exercise-induced changes in Mg metabolism. This study investigated the effects of different maximal exercises (one of endurance exercise) on Mg metabolism in humans.

II. SUBJECTS AND METHODS

Eleven healthy male university students, 20 to 23 years old, and all volunteers, were studied. All were moderately fit (training 2 to 3 h/week). Physical characteristics

Table 1 Physical Characteristics[a]

Age (years)	21.5 ± 0.8
Height (cm)	177.8 ± 3.8
Weight (kg)	70.6 ± 4.0
VO_2 max (mL/kg/min)[b]	41.7 ± 3.7
Time training (h/week)	2.6 ± 0.8

[a]Values are means ± SD; n = 12.
[b]VO_2 max = maximum O_2 uptake.

(means ±SD) were obtained on all subjects (Table 1). The subjects completed a questionnaire, which revealed that their diet was balanced, they did not smoke, and they did not consume any alcohol. Also, they did not take any medication. All subjects had normal cardiopulmonary and electrocardiographic function. The test were performed at 9 a.m. (one each week). The maximal aerobic capacity (VO_2 max) of each subject was determined using a Jeager Ergopneumotest.

The exercise programs were performed using a mechanically braked Monark cycle ergometer. They consisted of: (a) triangular progressive test (TPT), increasing the load 30W/3 min, until maximal power was obtained (recognized when the last step was maintained for at least 2 min); (b) rectangular sustained test (RST), maintaining the maximal load obtained in TPT for at least 7 min, and (c) interval endurance test (IET) with peaks of 1 min at maximal power, and regulated every 4 min, for 45 min (13). The subjects were required to perform the three tests, one each week.

Each subject had a catheter inserted in an antecubital vein, and blood samples were drawn after about 45 min of rest and at the end of exercise. The serum from venous blood was separated by low-speed centrifugation. Before and immediately after finishing each exercise session, hematocrit (Hct) was measured in duplicate by the microcentrifuge method, and hemoglobin (Hb) was measured by the cyanomethemoglobin method, (Boehringer, Mannheim). Magnesium in serum was measured with a Perkin-Elmer atomic absorption spectrophotometer (model 272). Protein total (PT) was determined by nephelometry using the Beckman immunochemical system, and albumin (Alb) was determined by electrophoresis. The percent change in plasma volume (% CPV) was calculated using pre- and postexercise Hb and Hct (14).

Results are represented as means ± SD. Data were analyzed for significance of means differences by using the paired t-test. Only when the ANOVA indicated significance ($p < 0.05$) was the t-test applied to paired observations. Statistical significance was assumed to be $p < 0.05$.

III. RESULTS

Physical characteristics (mean ± SD), age, weight (WT), height (Ht), VO_2 max, and time training are showed in Table 1. The Ht and Hb increased with exercise (TPT, RST, and IET) (Table 2); PT also increased after exercise. The change of plasma volume was similar after three exercises (5–8%).

The changes in serum Mg concentrations are presented in Figure 1. Mg increased both after maximal subtained test (RST) and after interval endurance test (IET) with

Table 2 Variations of Hemoglobin (Hb), Hematocrit (Hct), and Protein Total (PT), at Rest and Immediately After Exercise

		TPT		RST		IET	
	RST	Before	After	Before	After	Before	After
Hb (g/dL)	15.7 ± 1.0	15.6 ± 1.1*	16.9 ± 1.0†	15.0 ± 0.9*	16.5 ± 0.9†	15.4 ± 1.1*	16.4 ± 1.1†
Hct (%)	45.4 ± 6.0	45.1 ± 5.3*	49.4 ± 5.6†	45.3 ± 5.8*	49.1 ± 7.0†	45.2 ± 5.5*	48.6 ± 7.1†
PT (g/dL)	7.6 ± 0.3	7.7 ± 0.4*	8.8 ± 0.3†	7.4 ± 0.3*	8.3 ± 0.3†	7.5 ± 0.3*	8.4 ± 0.4†

[a]Significant differences $p < 0.05$, * with respect to rest just before of exercise and †after exercise. Data are expressed as means ± SD.

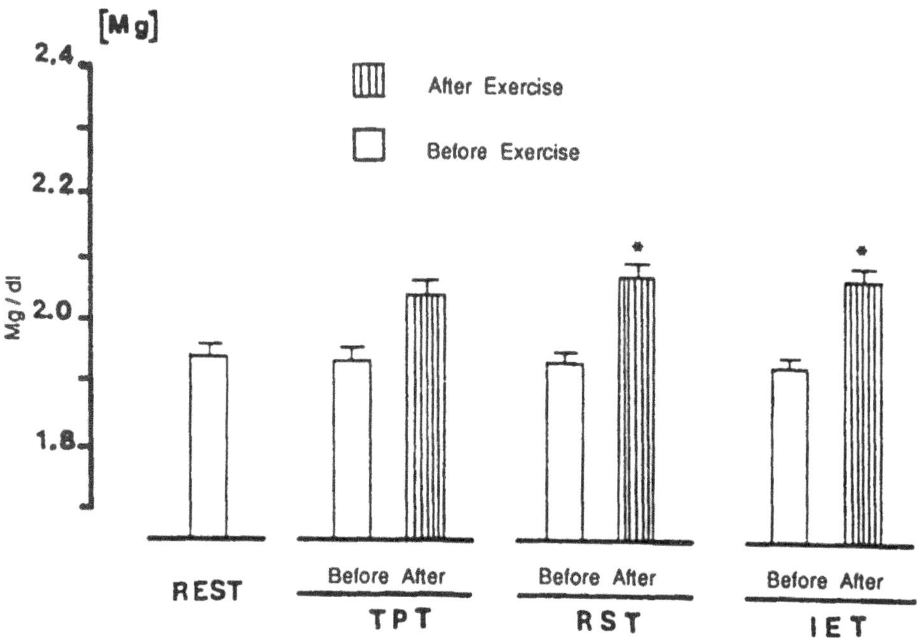

Figure 1 Serum Mg concentrations at rest and immediately after exercise: progressive (TPT), sustained (RST), and endurance test (IET). Significant differences with respect to rest are represented as $p < 0.05$.

respect to those obtained at rest. Immediately after the progressive test (TPT) it was also increased, but not significantly, the augmentation of serum Mg concentration was of the same order as the increase of Hb, Hct, and PT.

IV. DISCUSSION

The data showed that serum magnesium concentrations increase after sustained maximal exercise (RST) and after interval endurance test (IET). After progressive exercise (TPT), serum Mg also increases, but not significantly.

The effects of exercise on changes in plasma Mg have been investigated on several occasions (2–12). In general, a decline in plasma Mg concentration has been noted after prolonged submaximal exercise (2,3,8,15). Both chronic and acute exercise induces a redistribution of Mg (11,16). High-intensity intermittent exercise of an anaerobic nature induces transient compartmental Mg shifts in the blood (11). Cells with high metabolic activity have a higher Mg concentration than cells with lower activity (17).

On the other hand when the exercise lasts only 20 to 30 min (18), the serum Mg concentration may be unchanged or transiently increased. More recently, other authors (16) reported no changes in plasma concentrations but a significant reduction in erythrocyte Mg content following vigorous bicycle ergometry. We think that this phenomenon may be responsible for the increase in serum Mg concentrations. A recent report of an exercise-induced increase in plasma Mg concentration (4) is in agreement with data obtained in rats (12). Changes in plasma Mg during exercise essentially depend on work duration. Physical activity from 10 to 40 min leads to a significant increase (2), whereas exertion of a longer duration of any type is accompanied by a significant fall.

The changes in plasma volume (CPV) indicated by increases in Hb, Hct, and PT may be a result of decreased blood volume. One explanation of the elevated Hb levels might be that the stress of exercise produces shifts in the blood. These effects may explain the increases in serum magnesium concentrations.

V. SUMMARY

The purpose of this study was to investigate the effects of different maximal exercises (one of endurance exercise) on Mg metabolism in men. Eleven male healthy university students, volunteers between 20 and 23 years old, were studied. All were moderately fit (training 2 to 3 h/week). The exercise programs used a mechanically braked Monark cycle ergometer to measure: (a) triangular progressive test (TPT) by increasing the load 30W/3 min, until maximal power was obtained; (b) rectangular sustained test (RST), which required maintaining the maximal load obtained in TPT for at least 7 min; and (c) interval endurance test (IET) with peaks of 1 min at maximal power, and repeated every 4 min, for 45 min. Before and immediately after finishing each exercise session, Mg, PT, Hct, and Hb were measured. Mg increased, both after maximal subtained test (RST) and after interval endurance test (IET), relative to resting values. The augmentation of serum Mg concentration of the same order of increase as that of Hb, Hct, and PT. From these results it seems that the variations of serum magnesium concentrations after exercise are not time dependent, at least when the time of exercise is less than 45 min.

REFERENCES

1. A. S. Prasad. Magnesium. In *Trace Elements and Iron in Human Metabolism* (A. S. Prasad, ed.), Plenum Medical, New York, pp. 159–189 (1978).
2. G. Haralambie. Changes in electrolytes and trace elements during long-lasting exercise. In *Metabolic Adaptation to Prolonged Physical Exercise* (H. Howald and J. R. Poortmans, Eds.) Birkhäuser, Basel, pp. 340–351 (1975).
3. H. E. Refsum, B. Tveit, H. D. Meen, and S. B. Stromme. Serum electrolyte, fluid and acid-base balance after prolonged heavy exercise at low environmental temperature. *Scand. J. Clin. Lab. Invest.*, 32:117–122 (1973).
4. A. E. Olha, V. Klissouras, J. D. Sullivan, and S. C. Skoryna. Effect of exercise on concentration of elements in the serum. *J. Sports Med.*, 22:414–425 (1982).
5. R. McDonald and C. L. Keen. Iron, zinc and magnesium nutrition and athletic performance. *Sports Med.*, 5:171–184 (1988).
6. H. C. Luisaski, W. W. Bolonchuk, L. M. Klevay, D. B. Milne, and H. H. Sandstead. Maximal oxygen consumption as related to magnesium, copper, and zinc nutriture. *Am. J. Clin. Nutr.*, 37:407–415 (1983).
7. L. L. Rose, D. R. Carrou, S. L. Lowe, E. W. Peterson, and K. M. Cooper. Serum electrolyte changes after marathon running. *J. Appl. Physiol.*, 29:449–451 (1970).
8. R. B. Franz, H. Ruddel, G. L. Todd, T. A. Dormein, J. C. Buell, and R. S. Elliot. Physiologic changes during a marathon with special references to magnesium. *J. Am. Col. Nutr.*, 4:187–194 (1985).
9. G. Haralambie. Serum zinc in athletes in training. *Int. J. Sports Med.*, 2:135–138 (1981).
10. D. L. Costill, R. Cote, and W. Fink. Muscle water and electrolytes following varied levels of dehydration in man. *J. Appl. Physiol.*, 40:6–11 (1976).

11. P. A. Deuster, E. Dolev, S. B. Kyl, R. A. Anderson, and E. K. Schoomaker. Magnesium homeostasis during high-intensity anaerobic exercise in men. *J. Appl. Physiol.*, *62*:545–550 (1987).
12. A. Cordova, M. Gimenez, and J. F. Escanero. Effect of swimming to exhaustion at low temperatures, on serum and copper in rats. *Physiol. Behav.* (in press).
13. M. Gimenez, E. Servera, and W. Salinas. Square-wave endurance exercise test (SWEET) for training and assessment in trained and untrained subjects. I. Description and cardio-respiratory responses. *Eur. J. Appl. Physiol.*, *49*:359–368 (1982).
14. D. D. Dill and D. L. Costill. Calculation of percentage changes in volumes of blood, plasma, and red cells in dehydration. *J. Appl. Physiol.*, *37*:247–248 (1974).
15. S. B. Stromme, I. L. Stenwold, H. D. Meen, and H. E. Refsum. Magnesium metabolism during prolonged heavy exercise. In *Metabolic Adaptation to Prolonged Physical Exercise* (H. Howald and J. R. Poortmans, Eds.), Birkhäser, Basel, pp. 361–366 (1975).
16. S. W. Golf, L. Happel, and Graef. Plasma aldosterone, cortisol and electrolyte concentrations in physical exercise after magnesium supplementation. *J. Clin. Chem. Clin. Biochem.*, *22*:717–721 (1984).
17. H. E. Refsum, H. D. Meew, and S. B. Stromme. Whole blood, serum and erythrocyte magnesium concentrations after repeated heavy exercise of long duration. *Scand. J. Clin. Lab. Invest.*, *32*:123–127 (1973).
18. G. Haralambie and J. Keul. Der Einfluss von Muskerlarbeit auf den magnesiumspiegel und die neuromuskulare Erregbarkeit beim Menschen. *Med. Klin.*, *65*:1445–1449 (1970).

90

Relationship of Zinc to Growth and Development

J. Ignacio Monreal, Angel J. Monreal, RMC Da Cunha Ferreira, and Haidée Arias *Clínica Universitaria de Navarra, Pamplona, Spain*

I. PHYSIOLOGICAL LEVEL

Thirty years ago, Prasad et al. (1) reported, from observation on Iranian, Egyptian, and Turkish populations in the Mideast, a syndrome characterized by growth delay, anemia, and gonadal disfunction, which could be corrected by zinc supplementation. Zinc supplementation led to a height increment of 15 cm in 1 year.

This amazing picture became the starting point for several investigations about the effect of zinc deficiency on experimental animals. Zinc-deficient rats' growth is less compared with pair-fed or "ad libitum"-fed rats (Table 1).

In pregnant females, zinc deficiency not only has the same effect on growth, it is also teratogenic (2). Zinc-deficient terminal fetuses are smaller in comparison with pair-fed or "ad libitum"-fed ones. Besides, zinc-deficient fetuses have a lower weight at birth, diminished viability, and carential malformations. Some of these skeletal malformations were described: micrognathia, meningoencephalocele, fused vertebrae, hemivertebrae, syndactyly, oligodactyly, dysplasia in the bones of extremities, absent ribs, and curly tail.

Weight gain registered under experimental conditions is in relation to diet and other unknown factors. The first aspect leads to a decrease in food intake due to a lack of appetite and to a cyclic pattern of intake with a period of 3 or 4 days, as previously reported (3,4).

However, this is not the only reason for the lower weight gain, as the efficacy of food is lower in zinc-deficient animals than in pair-fed or controls (Table 1).

The first stage in the study of zinc deficiency was characterized by the observation of the external manifestations of total zinc deprivation on development. These studies evolved to a second stage in which the enzymatic level became predominant.

II. ENZYMATIC AND METABOLIC LEVEL

Some of the properties of the zinc ion must be known to explain its relationship with enzymes. The zinc ion is often necessary for the catalytic activity of an enzyme. In its absence the enzymes becomes inactive. In other instances Zn may play a structural

Table 1 Weight Gain, Food Efficiency, and Plasma Insulin of Male Rats with a Zinc-Deficient Diet (<0.5 μg/g) (ZD) During 3 Weeks, Rats with a Control Diet (75 μg/g) "ad Libitum" (C), and Rats with Control Diet in Restricted Amounts to the ZD Group (PF)

	Weight gain (g)	Food efficiency	Insulin (μg/L)
ZD	69 ± 11	0.244 ± 0.025	1.10 ± 0.29
PF	82 ± 17	0.288 ± 0.037	1.44 ± 1.05
C	132 ± 21	0.357 ± 0.039	3.09 ± 2.22

role in the enzyme or in other types of proteins. Finally, zinc functions in the positive and negative modulation of enzyme activity.

The zinc ion can accept pairs of electrons from atoms of oxygen, nitrogen, and sulfur, which facilitates its participation in hydrolysis reactions, similar to the action of phosphatases for phosphates or esterases in esters.

Another property of the zinc ion is its stability in the formation of complexes with coordination number 4,5, or 6, a circumstance that permits it to change ligands rapidly in a reaction and accommodate itself to a new disposition (5). Therefore, in the course of the catalytic action of a zinc-dependent enzyme, a molecule of water coordinated with the ion can be substituted with other ligands. Among the most common are the imidazole from the histidine group, the sulphydryl radical of cysteine, and the carboxyls of glutamate and aspartate.

Vallee and Galdes (6) reported a series of more than 200 zinc-dependent enzymes. Table 2 indicates enzymes metabolic pathways. They are enzymes of the gluconeogenesis, proteases, nucleic acid synthesis, etc.

Nutritional and tissue studies have been developed in order to study the biochemical tissue changes, mainly enzymatic and structural, due to zinc deficiency. The disposition of zinc in the organism is not absolute. There are several factors that regulate or interfere with the capture of the ion from the diet. A fiber-rich diet limits the intestinal absorption of zinc. The same happens with phytic acid or inositol hexaphosphate, which are abundant in cereals and vegetables. They form complexes with divalent ions in the intestinal lumen and prevents their absorption (7).

Table 2 Zinc-Dependent Enzymes

Enzyme	Pathway
DNA polymerase	Replication
RNA polymerase	Transription
Thymidine kinase	DNA synthesis
Carbonic anhydrase	Acid-base balance
Alcohol dehydrogenase	Alcohol metabolism
Carboxypeptidase A	Protease
Alkaline phosphatase	Bone synthesis
Fructose bisphosphatase	Gluconeogenesis
δ-Aminolevulinate dehydratase	Heme synthesis
Angiotensin-converting enzyme	Mineralocorticoids

In the intestine, as in other tissues, the zinc ion can be displaced by other cations such as calcium, iron, and cooper; also, a hypoproteic diet reduces zinc retention in the organism. Any of these factors, or the coincidence of two or more of them, can lead to subclinical states of Zn deficiency, such as those described in marginal populations of developed countries. Under these conditions, skin and mucosa alterations begin to appear, as well as alterations in sensorial organs, sexual function, and growth retardation.

On the experimental level, it has been possible to prove all these observations repeatedly. For example, regarding bone growth, femur and tibia length and bone mass are less in zinc-deficient animals (Figure 1).

If we analyze their composition, the changes do not affect the elements of organic matrix, uronic acids, hexosamines and hydroxyproline, but the mineral component as calcium and phosphorus.

Among tissue enzymes, decreased activity is observed more in those with osteoblastic function, such as alkaline phosphatase and pyrophosphatase (Figure 2), than in ones with osteoclastic function. Zinc-dependent enzymes are enzymatic cofactors and thus influence bone alteration.

On the other hand, different metabolic pathways change their activity, and this contributes to a lesser growth. In serum, a 3-week zinc deficiency reduces basal glucose and increases proteins, besides the typical changes of denutrition such as the fall of triglycerides and fatty acids (Figure 3). What is more, zinc deficiency leads to a relative intolerance to oral glucose overload. Altogether, the offer of energetic substrates to different tissues modifies and limits cellular development.

Growth also depends on the release and activity of some hormones. In zinc deficiency, food restriction does not trigger a stress response with increased levels of cortisol, as it does in pair-fed animals (Figure 4).

However, insulin, without modifying its storage in zinc-deficient adults or fetuses, suffer an altered release with a low concentration in plasma (Table 1). In young males, zinc deficiency markedly reduces insulin binding to hepatocite receptors, when compared with pair-fed males in which hormone binding capacity is remarkably increased (Figure 5).

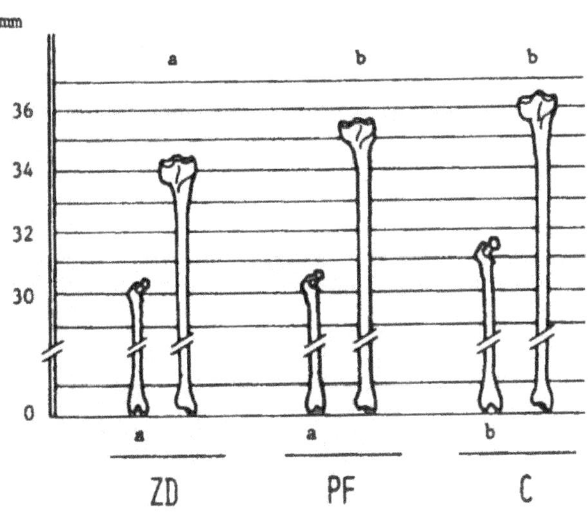

Figure 1 Femur and tibia length in the rats described in Table 1.

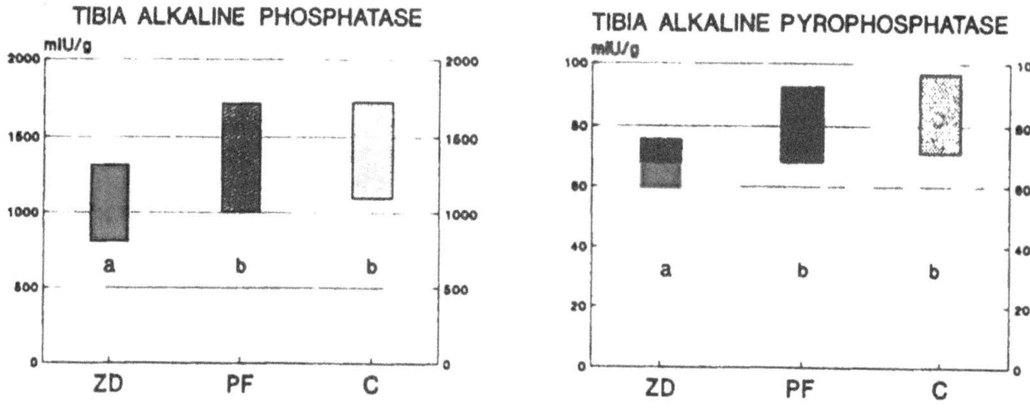

Figure 2 Tibia enzymes in the rats described in Table 1.

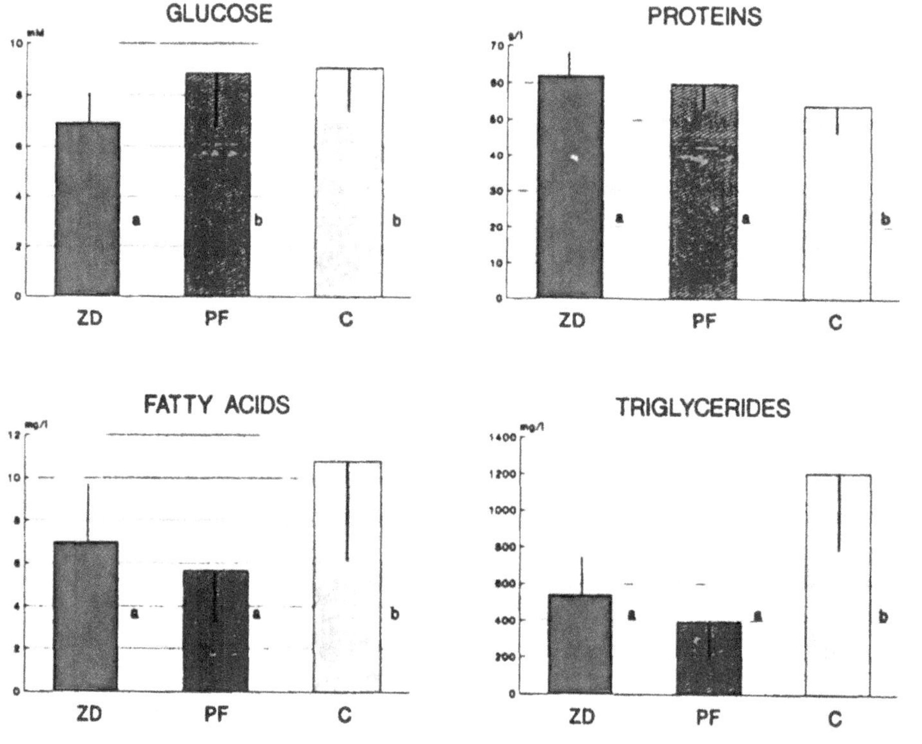

Figure 3 Serum glucose, proteins, fatty acids, and triglycerides in the rats described in Table 1.

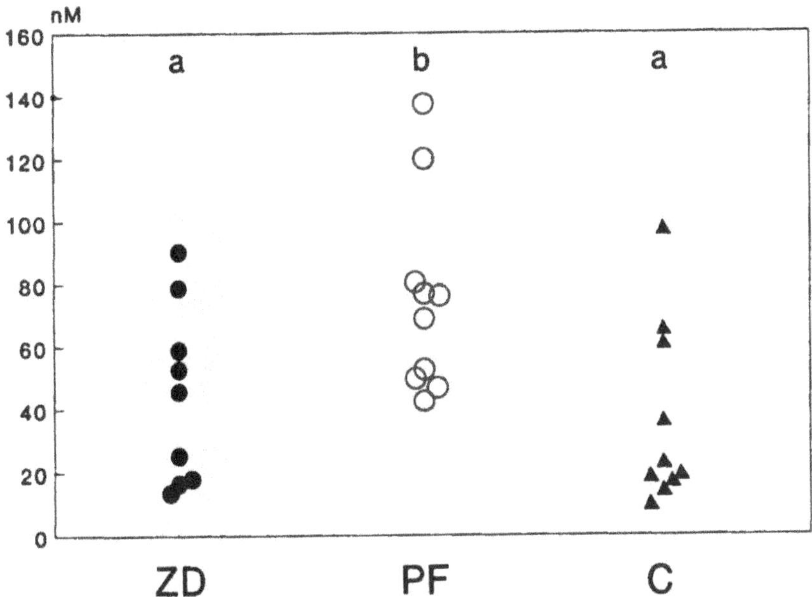

Figure 4 Serum cortisol in the rats described in Table 1.

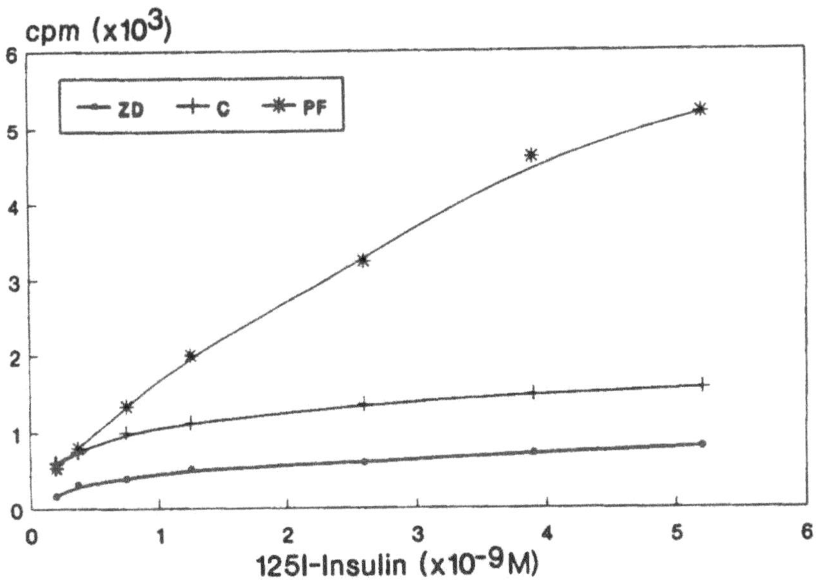

Figure 5 Insulin-hepatic receptor interaction in the rats described in Table 1. Membrane receptors were obtained by Cuatrecasas' method (12).

III. MOLECULAR AND GENETIC LEVEL

During the last few years, zinc participation in molecular processes on growth and development has been studied intensely. Protein kinase C is a key enzyme in the traduction of the hormonal signal to the cell. Protein phosphorylation in serine and threonine regulates metabolic pathways. Diacylglycerol ligation to its regulatory site, released by the action of phospholipase C on membrane phosphatidylinositol-4,5-biphosphate (PIP2), increases its calcium affinity and activates this enzyme, which plays an important role in cellular proliferation and division (8). Protein kinase C is found to be inactive in cytosol and active in membrane. Zinc modulates membrane binding, catalytic activity, and the regulatory effect on the enzyme of phorbol esters, derivatives of polycyclic alcohols which can promote tumors (8). Protein kinase C has cysteine-rich sequences fundamental for its regulatory site, although the fact of being the binding points of the cation has not been demonstrated.

Replication and transcription in zinc deficiency might be another source of disorders due to a catalytic role of the cation on DNA and RNA polymerases. Messenger RNA synthesis by gene transcription depends on the regulatory function of DNA promoting and suppressing sequences capable of ligating specific proteins which recognize DNA nucleotide sequences. This recognition takes place by means of structural elements which frequently repeat among proteins (9). One of the DNA binding structures already described includes zinc, from which it takes its name: zinc fingers.

Previously identified in the transcription factor TFIIIA of *Xenopus laevis*, its primary structure includes nine sequences rich in cysteine and histidine in which zinc forms a tetrahydric complex of coordination with the four extreme amino acids and constitutes a binding loop for DNA (9,10).

Structurally related to this complex is another one which is repeatedly observed in some DNA binding proteins: sequences of four to six cysteine residues and pairs of cysteines separated by two or four amino acids, to which zinc binds, one or more atoms, by tetrahydric coordination. This structure has been found in yeast regulating proteins and glucocorticoid receptors. Other steroid receptors show similar features (11).

IV. CONCLUSIONS

Some of the present aspects about the biochemical role of zinc have been presented. From the initial reports, in which somatic manifestations, malformations, and growth delay were the most striking findings, our knowledge has progressed to zinc-dependent metabolic pathways, nutritional aspects, absorption, transport, ionic competence, etc. At present, we also know how zinc takes part in the hormonal mechanisms of growth control and in the genetic and molecular level of cellular proliferation.

REFERENCES

1. A. S. Prasad, J. A. Halsted, and M. Nadini. Syndrome of iron deficiency, anemia, hepatosplenomegaly, hypogonadism, dwarfism and geophagia. *Am. J. Med.*, 31:532–546 (1961).
2. R. M. C. Da Cunha Ferreira, J. I. Monreal, and I. Villa. Teratogenicity of zinc deficiency in the rat: Study of the fetal skeleton. *Teratology*, 39:181–194 (1989).
3. R. B. Williams and J. K. Chesters. The effects of early zinc deficiency on DNA and protein synthesis in the rat. *Br. J. Nutr.*, 24:1053–1059 (1970).

4. J. K. Chesters and J. Quaterman. Effects of zinc deficiency on food intake and feeding patterns of rats. *Br. J. Nutr.*, 24:1061–1969 (1970).
5. M. N. Hughes. *The Inorganic Chemistry of Biological Processes*. John Wiley, New York (1972).
6. B. L. Vallee and A. Galdes. The metalobiochemistry of zinc enzymes. *Adv. Enzymol.*, 56:283–430 (1984).
7. N. R. Reddy, S. K. Sathe, and D. K. Salunnkhe. Phytates in legumes and cereals. *Adv. Food Technol.*, 28:1–92 (1982).
8. I. J. Forbes, P. D. Zalewshi, C. Giannakis, H. S. Petkoff, and P. A. Cowled. Interaction between protein kinase C and regulatory ligand is enhanced by a chelatable pool of cellular zinc. *Biochim. Biophys. Acta*, 1053:113–117 (1990).
9. R. M. Evans and S. M. Hollenberg. Zinc fingers: Guilt by association. *Cell*, 52:1–3 (1988).
10. P. F. Johnson and S. L. McNight. Eukaryotic transcriptional regulatory proteins. *Ann. Rev. Biochem.*, 58:799–839 (1989).
11. P. Davies and Rushmere. The structure and function of steroid receptors. *Sci. Prog. Oxford*, 72:563–578 (1988).
12. P. Cuatrecasas. Isolation of the insulin receptor of liver and fat cell membranes. *Proc. Natl. Acad. Sci. USA*, 69:318–322 (1972).

91

Clinical Usefulness of Magnesium(II) Lymphocytes Content Determination in Dilated Myocardial Disease

M. J. Merino Plaza *Hospital Princeps d'Espanya, Barcelona, Spain*

I. INTRODUCTION

Magnesium(II) is the second most abundant intracellular cation and has several critically important roles in the organism.

Over the last few years, numerous papers emphasizing the importance of magnesium(II) in cardiovascular pathophysiology have been published. Magnesium(II) deficiency has been implicated in the pathogenesis of ventricular arrhythmias, coronary spasm, hypertension, and atherosclerosis (1–3), and intracellular magnesium(II) is emerging as an important factor in the study of cardiac diseases.

Since less than 1% of total body magnesium(II) is extracellular, the serum magnesium(II) concentration may not be an accurate measure of total body magnesium(II) content (2,4–7).

Erythrocytes magnesium(II) content occasionally has been used to assess the magnesium(II) status of a patient, but these determinations have been shown to be poor predictors of intracellular magnesium(II) content.

Recently, it has been reported that magnesium(II) lymphocytes content determination may provide valuable information about intracellular magnesium(II) content, and this determination provides a better assessment of total body magnesium(II) status than does measurement of magnesium(II) concentration in plasma or erythrocytes (2,6–8).

II. MATERIALS AND METHODS

Serum magnesium(II) concentration and lymphocytes magnesium(II) content were measured in 20 patients with dilated myocardial disease (13 men and 7 women, ages 27 to 72), diagnosed with clinical and hemodynamic problems and treated with similar diuretic and digitalis therapy. None of them had a noncardiac disease that could predispose to magnesium(II) depletion, and 13 of them suffered arrhythmias or auricular fibrillation. Twenty healthy voluntaries served as reference group (6 men and 14 women, ages 26 to 52).

The magnesium(II) concentration was determined by atomic absorption spectrometry, and the cell lysate also assayed for protein by the method of Lowry.

Peripheral venous blood (20 mL) was obtained from all subjects by venepuncture (using heparin as anticoagulant); we used 15 mL for magnesium(II) lymphocyte content determination and 5 mL for the rest of the biochemistry determinations.

Lymphocytes were obtained using a modification of the method of Elin (6) and lysed by sonication.

Data were analyzed using the Mann-Whitney U test and the Spearman rank correlation nonparametric test.

III. RESULTS

Serum magnesium(II) concentration did not correlate with lymphocytes magnesium(II) content (Table 1). Serum magnesium(II) concentration in the patients group ($x = 0.62$ mmol/L; $s = -.06$ mmol/L) showed no significant difference from the reference group ($x = 0.65$ mmol/L; 0.04 mmol/L) (Tables 2 and 3).

Lymphocytes magnesium(II) content in the patients group ($x = 0.029$ mol/kg; $s = 0.010$ mol/kg) was significantly lower than in the reference group ($x = 0.04$ mol/kg; $s = 0.005$ mol/kg) (Tables 2 and 3).

Lymphocytes magnesium(II) content correlated with eyection fraction ($r = 0.517$; $p < 0.05$) in the patients group (Table 1), but serum magnesium(II) concentration did not correlate with eyection fraction in this group (Table 1).

In the group of 13 patients with adverse prognoses there was significant correlation between lymphocytes magnesium(II) content and telediastolic volume index ($r = 0.512$; $p < 0.05$) (Table 1).

Table 1 Correlations Between Lymphocytes Magnesium(II) Content and Other Values Studied[a]

	Patients		Patients with adverse prognoses	
	r	p	r	p
Mgs-MgL	0.1578	NS	0.186	NS
MgL-IVTD	0.1413	NS	0.512	p < 0.05
MgL-IVTS	0.013	NS	0.160	NS
MgL-FE	0.517	p < 0.05	0.714	p < 0.05
Mgs-FE	0.024	NS		
MgL-IC	0.081	NS	0.230	NS
IVTD-IVTS	0.8947	p < 0.01	0.916	p < 0.01
IVTD-FE	−0.11	NS	−0.04	NS
IVTD-IC	0.015	NS	0.216	NS
IVTS-FE	−0.428	p < 0.05	−0.506	p < 0.05
IVTS-IC	−0.4818	NS	−0.028	NS
FE-IC	0.4818	p < 0.05	0.320	NS

[a]Mgs, serum magnesium(II) concentration; MgL, lymphocytes magnesium(II) concentration; IVTS, telesistolic index; IVTD, telediastolic index; IC, cardiothoracic index; FE, eyection fraction; NS, no significance.

Table 2 Lymphocytes Magnesium(II) Content (Mg_L) and Serum Magnesium(II) (Mg_S) Concentration in the Different Groups Studied

	Mg_S (mmol/L)			Mg_L (mol/kg)		
	$\overline{X}$	S	N	$\overline{X}$	S	N
Reference group	0.65	0.04	20	0.040	0.005	20
Patients (total group)	0.62	0.06	20	0.029	0.010	20
Patients (adverse prognoses)	0.60	0.05	13	0.028	0.006	13
Patients (better prognoses)	0.64	0.07	7	0.033	0.002	7

IV. DISCUSSION AND CONCLUSIONS

There is experimental evidence of the vital role of magnesium(II) in maintaining cardiovascular integrity and normal function. Results of clinical, laboratory, and epidemiologic studies indicate an association between magnesium(II) deficiency and an increase in cardiac arrhythmias (1,2,5), and it appears to be a reasonable prophylactic measure to monitor closely magnesium(II) status in patients at risk.

Our data show low magnesium(II) lymphocytes content in patients with dilated myocardial disease compared with the reference group. As serum magnesium(II) concentration is similar in the two groups, the lymphocytes magnesium(II) content determination provides better information about intracellular magnesium(II) content in cardiac muscle cells. In clinical settings, this determination has advantages over serum magnesium(II) concentration determinations.

Furthermore, lymphocytes magnesium(II) content correlates with eyection fraction, especially in patients with adverse prognoses, and this determination may be of prognostic value in patients with dilated myocardial diseases.

These results are in agreement with the data found in the reviewed bibliography (2,7).

Table 3 Mean Comparison Between Lymphocytes Magnesium(II) Content and Serum Magnesium(II) Concentration in the Different Groups Studied

	Serum Mg significant differences	Lymphocytes Mg significant differences
Reference Group—patients	No	$p < 0.001$
Patients with adverse prognoses—patients with better prognoses	No	No
Patients with adverse prognoses—reference group	$p < 0.05$	$p < 0.001$
Patients with better prognoses—reference group	No	$p < 0.001$

REFERENCES

1. J. L. Seeling. Cardiovascular consequences of magnesium deficiency and loss: Pathogenesis, prevalence and manifestations. Magnesium and chloride loss in refractory potassium repletion. *Am. J. Cardiol.*, *63*:4G–21G (1989).
2. E. Ryzen, U. Elkayam, and R. K. Rude. Low blood mononuclear cell magnesium in intensive cardiac care unit. *Am. Heart J.*, *111*:475–480 (1986).
3. D. P. Lauler. Magnesium coming of age. *Am. J. Cardiol.*, *63*:1G–3G (1989).
4. R. K. Rude. Physiology of magnesium metabolism and the important role of magnesium in potassium deficiency. *Am. J. Cardiol.*, *63*:31G–34G (1989).
5. S. R. Whang, E. Elink, T. Dyckner, P. D. Wester, J. K. Aikawa, and M. P. Ryan. Magnesium depletion as a cause of refractory potassium repletion. *Arch. Intern. Med.*, *145*:1686–1689 (1985).
6. R. J. Elin and J. M. Hosseini. Magnesium content of mononuclear blood cells. *Clin. Chem.*, *31*:377–380 (1985).
7. R. A. Reinhart, J. J. Marx, R. G. Haas, and N. A. Desbiens. Intracellular magnesium of mononuclear cells from venous blood of clinically healthy subjects. *Clin. Chim. Acta*, *167*:187–195 (1987).
8. J. M. Queraltó, C. Hernandez, M. Tura, J. Garcia Picart, and J. M. Dominguez de Rozas. Mononuclear blood cell magnesium content. In *Biologie prospective comptes rendus du 7 colloque de Pont—Mousson* (M. M. Galteau, G. Siest, and J. Henny, Eds.), John Libbey Eurotext, Paris, 63:1G–3G (1989).

92

Aluminum Bone Content in Patients Undergoing Renal Hemodialysis

M. D. Fernandez, A. L. M. de Francisco, J. C. Garrido, F. J. Aguayo, and C. Alvarez *Hospital Universitario Marques de Valdecilla, Santander, Spain*

I. INTRODUCTION

From a biological point of view, aluminum is considered a nonessential trace element, and it was implicated a few years ago as a toxic metal. Its toxicity has been recognized primarily with chronic renal failure (1,2), although adverse effects in the presence of normal renal function remain to be clearly defined.

Long-term hemodialysis patients develop increased serum aluminum concentrations (3). Clinic sequelae associated with hyperaluminemia include encephalopathy, metabolic bone disease, anemia, and myopathy. The prevalent clinical manifestations probably defend on the extent of aluminum accumulation in the different tissues. In relation to bone tissue, aluminum stores induce osteomalacia with progressive skeletal pain, proximal muscle weakness, and fractures (4).

Since serum aluminum levels do not always provide a clear reflection of the patient's total burden of aluminum (5,6), we studied bone aluminum concentrations in chronic renal patients under dialysis for at least 2 years and compared them to patients with normal renal function (control group).

II. MATERIALS AND METHODS

We studied a group of 19 uremic patients under regular hemodialysis, at home or in hospital, for more than 24 months.

Bone biopsies were obtained from the iliac crest during surgical renal transplantation. Controls proceeded from necropsies of 39 patients without previous renal failure. Both samples were collected in ethanol containing less aluminum than 10 µg/L, and consisted of trabecular bone.

Samples were removed from ethanol, freezed-dried, and stored in screw-capped disposable 2-mL polypropylene tubes at −40°C until analyzed.

All tissue samples had a dry weight between 10 and 20 mg.

A wet digestion method was used (7). Bone samples were introduced to Teflon screw-capped 7-mL vessels and decomposed with 200 μL of concentrated nitric acid (Suprapur, Merck no. 441), allowed to stand at room temperature for 1 h, and heated at 85°C for 5 h. This resulted in a clear solution. Finally, the digest was diluted to 0.35 mL with reverse osmosis-treated water. Ten microliters of this solution were diluted again with 200 μL of the same deionized water.

Aluminum was measured by electrothermal atomic absorption spectrometry (Zeiss model FMD3) using a graphite furnace (Perkin-Elmer model HGA 76), and 20 μL were injected into the autosampler (Perkin-Elmer model AS-1) (8,9).

Step	Temp (°C)	Ramp time (min)	Hold time (min)	Argon flow (ml/min)
1	100	5	20	300
2	250	5	20	300
3	1.450	1	20	300
4	2.620	1	5	Stop

The heating and gas flow program was as follows:

The standard-addition method was used to calibrate and measure aluminum concentration in the diluted digest (10).

Precision was established using animal bone H-4 from the International Atomic Energy Agency (Austria) as control (11), yielding the following results: A1 mean value = 4.4 μg/g dry wt; SD = 0.2; CV% = 5.4; n = 15.

III. RESULTS

The chi-square test showed a nonparametric distribution for both patient and control groups. The 95% confidence range was established as the interval between the 2.5 and 97.5 percentiles. These results are summarized in Table 1.

Figure 1 shows a graphic representation of these results. The statistical comparison between the groups using the Wilcoxon test showed significant differences ($p < 0.001$).

IV. CONCLUSIONS

1. Distribution of aluminum content in both patient and control groups is nonparametric.

Table 1 Analysis of Bone Aluminum

	Al (μg/g dry wt)	
	Control	Patients
n	39	19
Median	7.2	39.6
2.5–97.5 percentiles	3.2–10.6	20.9–112.5

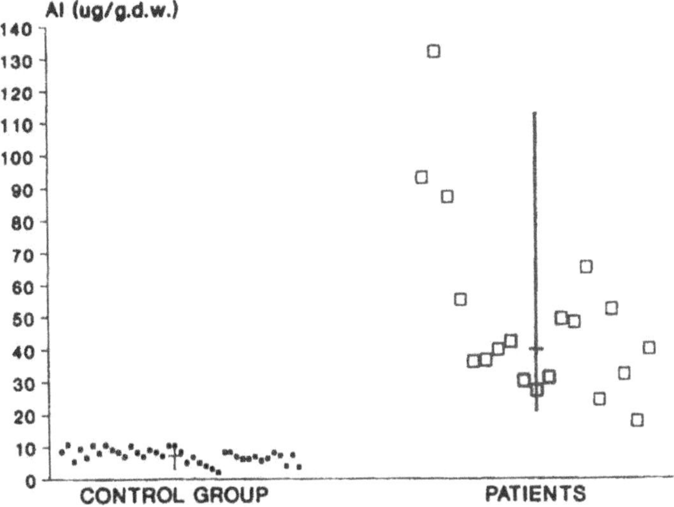

Figure 1 Aluminum bone content.

2. Aluminum values obtained from bone patients under hemodialysis are clearly higher than those from the control group.
3. Bone tissue seems to be a proper sample in which to evaluate aluminum overload in uremic patients.
4. Further investigations will be necessary to find out if there is any relationship between aluminum serum levels and bone content in patients undergoing chronic renal failure.

REFERENCES

1. International Conference on Aluminium in Renal Disease, University of Surrey, Guilford, U.K., 1985. *Trace Elem. Med.*, 2:103 (1985).
2. International Symposium on Aluminium Toxicity and Its Effects on Bone Metabolism, University of Oviedo, Oviedo, Spain, 1986. *Nefrologia, VI*:26 (1986).
3. M. R. Wills and J. Savory. Aluminium homeostasis: In *Aluminium and Other Trace Elements in Renal Disease* (A. Taylor, Ed.), London, Baillière Tindall, London, pp. 24–36 (1986).
4. G. D. Smith and R. J. Winney. Aluminium related osteomalacia—Response to reverse osmosis water treatment. In *Aluminium and Other Trace Elements in Renal Disease* (A. Taylor, Ed.), Baillière Tindall, London, pp. 98–105 (1986).
5. S. M. Channon, O. M. Osman, I. Boddy, R. C. S. Rodger, T. H. F. Goodship, P. A. Smith, R. Wilkinson, and M. K. Ward. Serum aluminium measure—A false sense of security? In *Aluminium and Other Trace Elements in Renal Diseases* (A. Taylor, Ed.), Baillière Tindall, London, pp. 171–176 (1986).
6. M. E. De Broe, F. L. Van de Viver, F. J. E. Silva, P. C. D'Haese, and A. H. Verbueken. Measuring aluminium in serum and tissues: Overview and perspectives. *Nefrologia, VI*: 51–45 (1986).
7. G. Knapp. New developments in the decomposition of biological materials. In *Trace Element Analytical Chemistry in Medicine and Biology*, Vol. 5 (P. Brätter and P. Schramel, Eds.), Walter de Gruyter, New York, pp. 64–71 (19xx).
8. D. H. Fromen and A. C. Alfrey. Aluminium. In *Quantitative Trace Analysis of Biological Materials* (H. A. Mckenzie and L. E. Smythe, Eds.), Amsterdam, pp. 633–647 (1988).

9. H. T. Delves, B. Suchak, and C. S. Fellows. The determination of aluminium in foods and biological materials. In *Aluminium in Foods and the Environment* (C. R. Massey and D. Taylor, Eds.), Royal Society of Chemistry, London, pp. 52–67 (1988).
10. L. Ximenez. *Espectroscopia de Absorcion Atomica*. Publicaciones Analiticas, Madrid (1982).
11. Y. F. Leung and R. A. Henderson. Assessment of quality control sera suitable for aluminium determination by absorption spectrometry. In *Chemical Toxicology and Clinical Chemistry of Metals* (S. Brown and J. Savory, Eds.), Academic Press, London, pp. 69–72 (1983).

93

Lead—Unsafe at Any Level?

Petrie M. Rainey *Yale University School of Medicine, New Haven, Connecticut*

"Lead is the oldest of the industrial poisons except carbon monoxide, which must have begun to take its toll soon after Prometheus made the gift of fire to man (1)." Hippocrates is credited with the first case report of industrial lead poisoning (2). The association of childhood lead poisoning with the ingestion of lead in paint chips was first described in Australia a century ago (3). Although symptomatic lead poisoning is still seen regularly in both industrially exposed workers and in children ingesting lead paint, great strides have been made in the past 50 years to reduce the incidence of such poisonings. The measurement of blood lead levels in these populations at risk has greatly facilitated this effort.

Awareness of the harmful effects of lead at levels below the symptomatic threshold is relatively recent, and still evolving. The whole-blood lead concentration considered to be safe is regularly being revised downward. The finding of undesirable effects at lead levels which are currently quite prevalent is perhaps not surprising. Although typical rates of lead intake have been substantially reduced by the elimination of leaded gasoline, they are still orders of magnitude greater than the intake levels and associated body burdens under which human beings evolved (4). Recently there have been studies in a number of countries which have shown deleterious effects of lead on childhood development at levels formerly considered free of risk. These have been recently reviewed (5,6). Although similar studies have reported no significant adverse effects, meta-analysis of the conjoint studies strongly supports a small but real effect (7). Moreover, these effects appear to have persistent sequelae (8). Although the impairment is small, the expected prevalence is high. Since some effects may persist for years, the cumulative impact has been estimated to result in an annual cost to society of billions of dollars (9). Demonstration of low-level lead effects has not been limited to children. Increases in blood pressure in adults have been correlated with blood lead concentration in the 0- to 10-μg/dL range (6,10). As a result of this growing of evidence, there must be another downward revision in lead levels considered to be safe, at least for children.

The U.S. Centers for Disease Control (CDC) have convened a panel of experts on the effects of low-level lead burdens to determine a new standard for judging pediatric lead exposure. Although a final statement is not expected until 1991, the panel has already met once, and from the minutes of their meeting (11) it may be inferred that no positive blood lead concentration will be considered to be "safe." A level of 5 μg/dL will be considered the threshold for observation of minimal effects.

Intervention to abate community exposure may be recommended at levels as low as 10 μg/dL. Abatement at the level of the individual may be recommended at levels greater than 15 μg/dL. There may be no changes in recommendations for chelation therapy, despite the availability of dimercaptosuccinic acid. This is an effective oral chelating agent with little demonstrated toxicity which is available in many countries and will soon be available in the United States. It offers a considerable improvement over chelation with ethylenediaminetetracetic acid. Nonetheless, at present there is a paucity of data to assess the benefits of chelation at low elevations of blood lead.

Although the guidelines will be only recommendations, it is expected that they will affect the practice of many pediatricians. In addition, they will be in a form which can readily be incorporated into regulations. Perhaps most important, the CDC is taking measures to facilitate the implementation of the anticipated guidelines (12). This includes efforts to make readily available accurate low-level calibrators and reference standards, as well as support for the development of improved instrumentation.

The new recommendations will affect the practice of many laboratories which provide blood lead levels for pediatric screening purposes. The currently acceptable standard of accuracy of ±6 μg/dL at low lead levels will be inadequate (11). Additionally, zinc protoporphyrin and free erythrocyte protoporphyrin measurements will no longer be tenable as screening techniques, since these parameters show significant elevations only when blood lead levels are 25 μg/dL or more (13). Measurement of blood lead as the initial test will be necessary.

In pediatric patients, obtaining venous blood requires more time and skill than obtaining capillary blood. However, many patients find a skillful venipuncture less traumatic than a fingerstick. Moreover, the use of capillary blood as a screening tool will require scrupulous attention to prevent contamination of the specimen by lead from the skin or environment. The choice of whether to use capillary or venous blood for screening purposes can best be decided using situational considerations. Positive screening results should be confirmed in all cases with a venous blood level, independent of whether capillary or venous blood was used in the initial measurement.

Two strategies are likely to emerge for screening. One will involve sending specimens to large central laboratories. This will allow the economies of volume, the purchase of high-capital-cost instrumentation, and the development of a high level of expertise. However, turnaround time may be measured in days (or even weeks), and a number of patients may be lost to follow-up. The other strategy will involve rapid on-site testing with identification of patients requiring confirmatory testing at the time of the initial visit. Under such circumstances, a lower level of precision may be acceptable for screening than for confirmatory testing. However, the new action levels are located relatively centrally in the current distributions of blood lead levels, rather than in the tail of the distribution, as were previous levels. Thus, the price for lower screening precision will be a considerable increase in the number of false positives or negatives, depending on the level at which the screening cutoff is set.

For example, if analytical imprecision was up to 6 μg/dL, it would be necessary to set the cutoff for a positive screen at 9 μg/dL in order to assure that the screen identified as positive all those patients with actual levels of 15 μg/dL or higher. If contamination contributed up to 4 μg/dL, patients with actual levels of 5 μg/dL would regularly be identified as positive. A few patients with actual levels as low as 1 μg/mL could test positive through the combined effects of contaminants and analytic variability. Given the current distribution of blood lead levels, as many as two-thirds of the screening tests might be positive under these conditions. Such a test would have little usefulness as a screen.

The volume of testing may increase substantially. Not only will there be the increase resulting from the screening blood samples, there will also be a substantial increase in the number of patients requiring confirmatory blood levels.

Accurate classification may require precision limits as low as ±1–2 μg/dL. The minimum detection limit of a method sets a theoretical threshold on achievable precision, since no proportional imprecision is included. Based on current detection limits, graphite-furnace atomic absorption and inductively coupled plasma mass spectrometry should be capable of achieving the required precision, although the performance currently achieved in actual practice is considerably less than the theoretical capability. Requisite precision should be achievable through greater attention to calibration and quality control. Both methods require instruments which are technically complex and have high initial costs, but allow substantial automation and provide high throughput. These systems are most appropriate for centralized testing and for confirmatory testing.

Lead can be measured by anodic stripping voltammetry using instruments which have relatively low initial costs and are also rugged and portable. Their simplicity of operation and rapid turnaround time (less than 5 min) could allow their placement at screening sites, providing a quick determination of those patients for whom a venous sample would be necessary. The disadvantages of anodic stripping voltammetry include the high cost of reagents necessary for specimen preparation (i.e., breakdown of lead complexes in blood) and relatively poor precision at low lead concentrations. Part of the precision problem may relate to a biphasic response curve, the slope of which is up to 25% steeper below 30 μg/dL than it is above this level (14). It may be possible to improve the low-level precision of currently available instrumentation simply by calibrating the instrument specifically for the 0- to 30-μg/dL range. This will require only the provision of the appropriate calibrators. Redesign of the electrode may allow for further improvements in low-level precision. With the incorporation of modifications to improve precision, anodic stripping voltammetry may be suitable for on-site screening as well as confirmatory testing.

An unresolved issue is the nature of interventions which will be cost effective in reducing the body burden of lead associated with these lower levels. This is particularly important given the large increase expected in the number of children who will be identified as requiring some intervention. An approach which targets multiple potential sources of exposure will undoubtedly be the most effective. Development of instrumentation and methodology which allows precise measurement of low levels of blood lead is a necessary first step, since accurate feedback will be necessary to document effective interventions.

REFERENCES

1. A. Hamilton. *Exploring the Dangerous Trades*. Little-Brown, Boston (1943).
2. R. P. Wedeen. *Poison in the Pot: The Legacy of Lead*. Southern Illinois University Press, Carbondale, IL (1984).
3. J. L. Gibson. Notes on lead poisoning as observed among children in Brisbane. *Proc. Intercolonial Med. Congr. Austr.*, 3:76–83 (1982).
4. D. M. Settle and C. C. Patterson. Lead in albacore: Guide to lead pollution in Americans. *Science*, 207:1167–1176 (1980).

5. P. Muschak, J. M. Davis, A. F. Crocetti, and L. D. Grant. *Environ. Res.*, *50*:11–36 (1989).
6. M. Lippman. Lead and human health: Background and recent findings. *Environ. Res.*, *51*:1–24 (1990).
7. H. L. Needleman and C. A. Gatsonis. Low level lead exposure and the IQ of children: A meta-analysis of modern studies. *J. Am. Med. Assoc.*, *263*:673–678 (1990).
8. H. L. Needleman et al. The long-term effects of exposure to low doses of lead in childhood: An 11 year follow-up report. *N. Engl. J. Med.*, *322*:83–88 (1990).
9. H. L. Needleman. The persistent threat of lead: Medical and sociological issues. *Curr. Probl. Pediatr.*, *18*:697–744 (1988).
10. J. Schwartz. The relationship between blood lead and blood pressure in the NHANES II survey. *Environ. Health Perspect.*, *78*:15–22 (1988).
11. Centers for Disease Control. *Advisory Committee on Childhood Lead Poisoning Prevention: Minutes of the First Meeting*. U.S. Dept. of Health and Human Services, Atlanta, GA (1990).
12. Centers for Disease Control. *Proposed CDC Blood Lead Laboratory Reference System*. U.S. Dept. Health and Human Services, Atlanta, GA (1990).
13. S. Piomelli et al. Threshold for lead damage to heme synthesis in urban children. *Proc. Natl. Acad. Sci. USA*, *79*:3335–3339 (1982).
14. J. D. Osterloh, D. S. Sharp, and D. Hata. Quality control data for low blood lead concentrations by three methods used in clinical studies. *J. Anal. Toxicol.*, *14*:8–11 (1990).

Part VII

Miscellaneous

94

State of the Art of the Metrological Quality of Analytical Methods for Drug Assay

A. Blanco-Font and X. Fuentes-Arderiu *Hospital Princeps d'Espanya, L'Hospitalet de Llobregat, Barcelona, Spain*

I. INTRODUCTION

Quantities measured for therapeutic drug monitoring (TDM) purposes have particular characteristics with respect to the rest of clinical chemistry analyses. These characteristics are due principally to two factors: Drugs are not physiologically present in the human body and, when present, their plasma concentrations vary with the time interval between doses and the elimination of half-life (both depending on patient idiosyncracy and clinical state). Taking into account these particularities, Fraser (1) has developed a theoretical strategy for the objective setting of a goal for between-day imprecision of analytical methods used in the measurement of TDM quantities.

The present study tries to investigate the state of the art, on an international scale, of between-day imprecision and total error for the analytical methods of the following drugs: N-acetylprocainamide, amikacin, carbamazepine, digoxin, ethosuximide, gentamicin, lidocaine, paracetamol, phenobarbital, phenytoin, primidone, procainamide, quinidine, salicylate, theophylline, tobramycin, valproate, and vancomycin. The state-of-the-art data have been assessed by comparison with the allowable between-day imprecision and allowable total error (2) using for their establishment the goal for between-day imprecision proposed by Fraser (1) for TDM analytical methods.

II. MATERIALS AND METHODS

The information used belongs to the external quality assessment scheme Stratus TDM (Baxter Healthcare Corporation, Miami, Florida). The control materials used in the scheme were three lyophilized sera with different concentrations of drugs (TDM control; and lots TDM-119, 219, and 319). The data used in this study were selected according to the following criteria:

Results obtained within 1989
Only those drugs with a ≥20 participant laboratories
For each drug, only those laboratories returning ≥20 results during the year
For each drug, only data from the control material whose reported concentrations were within, or near, the therapeutic range

For each laboratory and for each drug, the performance indicators considered have been: the standard deviation (s), the coefficient of variation (CV), the relative inaccuracy with respect to the consensual mean (Δ_r) and the total error (TE). Consensual means, standard deviations, and coefficients of variation have been taken directly from the scheme reports. Relative inaccuracy and total error have been calculated as follows:

$$\Delta_r = \frac{\bar{x}_{observed} - \bar{x}_{consensual}}{\bar{x}_{consensual}}$$

$$TE = |\bar{x}_{observed} - \bar{x}_{consensual}| + 2s$$

The consensual mean was considered the conventional true value because for the control materials used, concentration values obtained by definitive or reference methods were not available.

The allowable between-day imprecision for the analytical methods of each drug was calculated using the formula proposed by Fraser (1) for analytical goal setting for between-day imprecision in TDM:

$$CV_{allowable}\,(\%) \le 25 \left(\frac{2^{T/t} - 1}{2^{T/t} + 1}\right)$$

where T is the time interval between doses and t is the drug elimination half-life. Since T as well as t may be different for different patients, those that produce the lowest allowable between-day imprecision were chosen. This formula was not applied to lidocaine because it is usually administered by continuous perfusion and the time interval between doses has no meaning in this case.

The allowable total error was calculated taking into account that the goal for inaccuracy is always zero (3), and using the above goal for between-day imprecision, so that

$$TE_{allowable} \le \Delta_{allowable} + 2s_{allowable} \le 2\,\frac{CV_{allowable} + \bar{x}_{observed}}{100}$$

$$\le 0.5 \left(\frac{2^{T/t} - 1}{2^{T/t} + 1}\right)$$

III. RESULTS AND DISCUSSION

For each drug, the mean and the 0.10, 0.25, 0.50, 0.75, and 0.90 percentiles of the between-day imprecision and relative inaccuracy are shown in Tables 1 and 2, respectively.

Table 1 Between-Day Imprecision of the Analytical Methods for Drugs Assay

Drug	Number of laboratories	Mean	Percentile				
			0.10	0.25	0.50	0.75	0.90
N-Acetylprocainamide	99	5.3	3.0	3.7	4.9	6.1	8.0
Amikacin	53	7.9	4.4	5.2	6.5	10.1	13.8
Carbamazepine	127	5.4	3.4	4.0	4.9	6.2	7.7
Digoxin	185	9.6	6.3	7.4	8.7	10.7	13.9
Ethosuximide	13	7.1	4.9	5.8	7.1	8.3	10.0
Gentamycin	169	6.9	4.2	5.1	6.3	8.2	10.1
Lidocaine	41	5.7	3.5	4.4	5.3	6.2	8.8
Paracetamol	96	7.2	4.9	5.5	6.7	8.0	9.8
Phenobarbital	171	5.2	2.7	3.6	4.7	6.1	8.1
Phenytoin	226	4.9	3.0	3.6	4.4	5.7	7.8
Primidone	41	5.3	3.0	4.0	5.1	6.6	7.4
Procainamide	96	6.6	4.4	5.1	6.1	7.4	9.6
Quinidine	137	8.4	3.8	5.0	6.9	11.0	14.1
Salicylate	34	4.4	2.0	2.6	3.8	6.0	7.1
Theophylline	254	5.1	3.3	3.8	4.6	5.9	8.0
Tobramycin	115	6.9	4.1	4.2	6.3	8.1	9.8
Valproate	56	6.5	3.7	4.1	5.6	6.7	8.6
Vancomycin	27	5.3	3.8	4.0	4.6	5.4	7.2

Table 2 Relative Inaccuracy of Analytical Methods for Drugs Assay

Drug	Number of laboratories	Mean	Percentile				
			0.10	0.25	0.50	0.75	0.90
N-Acetylprocainamide	99	0.031	0.004	0.009	0.023	0.038	0.057
Amikacin	53	0.044	0.005	0.013	0.029	0.058	0.094
Carbamazepine	127	0.039	0.011	0.019	0.030	0.052	0.077
Digoxin	185	0.066	0.016	0.033	0.060	0.085	0.123
Ethosuximide	13	0.031	0.004	0.005	0.022	0.050	0.076
Gentamycin	169	0.057	0.007	0.019	0.044	0.069	0.114
Lidocaine	41	0.023	0.006	0.013	0.018	0.032	0.043
Paracetamol	96	0.035	0.004	0.012	0.024	0.045	0.078
Phenobarbital	171	0.037	0.005	0.013	0.028	0.047	0.074
Phenytoin	226	0.027	0.002	0.007	0.017	0.033	0.058
Primidone	41	0.053	0.022	0.032	0.050	0.070	0.083
Procainamide	96	0.034	0.002	0.010	0.027	0.046	0.071
Quinidine	137	0.049	0.004	0.017	0.035	0.078	0.109
Salicylate	34	0.056	0.018	0.004	0.058	0.071	0.090
Theophylline	254	0.043	0.006	0.011	0.025	0.045	0.077
Tobramycin	115	0.038	0.005	0.012	0.030	0.057	0.085
Valproate	56	0.067	0.008	0.019	0.070	0.093	0.105
Vancomycin	27	0.039	0.007	0.022	0.039	0.059	0.065

Table 3 Proportion (P) of Laboratories for Which Between-Day Imprecision is Lower Than, or Equal to, the Allowable Between-Day Imprecision ($CV_{allowable}$)

Drug	$CV_{allowable}$	P
Amikacin	15.00	0.94
Procainamide	8.33	0.85
Paracetamol	8.33	0.77
Phenytoin	5.12	0.64
Quinidine	6.36	0.46
Theophylline	3.82	0.26
Vancomycin	3.97	0.22
N-Acetylprocainamide	3.44	0.18
Valproate	3.44	0.07
Salicylate	1.15	0.03
Carbamazepine	2.07	0.02
Tobramycin	3.44	0.02
Gentamycin	3.44	0.02
Primidone	2.47	0.02
Phenobarbital	1.73	0.01
Digoxin	1.18	0.00
Ethosuximide	0.87	0.00

Table 4 Proportion (P) of Laboratories for Which Total Error is Lower Than, or Equal to, the Allowable Total Error ($TE_{allowable}$)

Drug	$TE_{allowable}$	P
Theophylline (µmol/L)	5.9	0.90
Tobramycin (µmol/L)	0.7	0.84
Amikacin (µmol/L)	2.8	0.83
Procainamide (µmol/L)	3.7	0.60
Paracetamol (µmol/L)	32.1	0.53
Phenytoin (µmol/L)	5.7	0.46
Quinidine (µmol/L)	2.4	0.29
Vancomycin (µmol/L)	1.3	0.07
N-Acetylprocainamide (µmol/L)	3.0	0.05
Carbamazepine (µmol/L)	1.8	0.01
Digoxin (nmol/L)	0.04	0.01
Gentamycin (µmol/L)	0.8	0.01
Phenobarbital (µmol/L)	3.2	0.01
Salicyclate (µmol/L)	2.7	0.00
Ethosuximide (µmol/L)	7.4	0.00
Valproate (µmol/L)	36.2	0.00
Primidone (µmol/L)	1.4	0.00

Table 3 shows the allowable imprecision for the analytical methods of the drugs considered and the proportion of laboratories whose between-day imprecision was lower than, or equal to, the allowable one. The allowable total error and the proportion of laboratories for which total error was lower than, or equal to, the allowable one is presented in Table 4.

In general, the international state of the art of quantitation observed in this study is not good enough. With regard to the between-day imprecision, only in the cases of amikacin, paracetamol, phenytoin, and procainamide, did one-half of the laboratories participating in the scheme have CV values lower than, or equal to, the $CV_{allowable}$. The same situation in the case of total error is observed only for amikacin, paracetamol, procainamide, theophylline, and tobramycin.

From this study, it may be concluded that the between-day precision of the analytical methods for TDM must be improved, principally for drugs situated in the last position of Table 3. Finally, in order to achieve a more rigorous assessment of the state of the art of inaccuracy and total error, the control materials used in external quality assessment schemes should be assayed by definitive or reference methods.

REFERENCES

1. C. G. Fraser. Desirable standards of performance for therapeutic drug monitoring. *Clin. Chem.*, 33:387–389 (1987).
2. J. O. Westgard, R. N. Carey, and S. Wold. Criteria for judging precision and accuracy in method development and evaluation. *Clin. Chem.*, 20:825–833 (1974).
3. Proceedings of the Subcommittee on Analytical Goals in Clinical Chemistry, World Association of Societies of Pathology, Ciba Foundation, London, England (U.K.), April 25–28, 1978. Analytical goals in clinical chemistry: Their relationship to medical care. *Am. J. Clin. Pathol.*, 71:624–630 (1979).

95

Documentation of European Drugs with Standardized TLC System

Fritz Degel and Norbert Paulus *Institut für Klinische Chemie, Klinikum Nürnberg, Germany*

I. INTRODUCTION

Our institute is in charge of the toxicological analyses requested by the hospital's big poison center. As a rule, analysis in our laboratory are performed for patient care, especially in emergency cases, and not to answer forensic or occupational-medical questions.

Those concerned with clinical-toxicological analysis must compromise between quickness and a maximum of reliability. There is a demand for quick and reliable analytical methods, which should be available around the clock.

A. Why We Chose a Standardized TLC System (Toxi-Lab)

One of our analytical tools in toxicological analysis is the standardized thin-layer chromatographic (TLC) system Toxi-Lab (Toxi-Lab, Inc., Irvine, California), which we began to use about 9 years ago. The reasons why we chose such a standardized system in clinical toxicological analysis are the following:

1. High speed and simplicity of handling. Test performance is easy to learn and can be carried out by normally trained chemical and medical technical personnel in a short time (chromatographic run: 15 min). Interpretation, however, should be done by an experienced analyst.
2. Standardization of TLC procedures. Many TLC procedures have been simplified and standardized (e.g., extraction, sample and reference standard application, etc.).
3. Good reproducibility of R_f values. TLC plates are controlled by the manufacturer to ensure constant R_f values. There is a lot-specific recommendation for ammonia dosage to meet this criterion.
4. Principle of "differentiating detection." Substance identification is achieved not only by determination of R_f values alone, but also be use of specific color development in different detection stages. This is done by dipping, not by spraying.

5. Good documentation (metabolite patterns included) (facilitated by use of printed worksheets whose interpretation is in turn facilitated by use of a reference standard compendium).
6. Good quality control. Twenty-six co-analyzed reference standards (in each run) give good control of chromatography and color development conditions. By use of them, a correction of R_f values is possible, too. External quality controls are available.
7. Good distribution of R_f values over the whole range. This makes the system well suited for drug screening procedures, especially in "general unknown" analyses.
8. Effective extraction procedure. The recovery rate of the extraction procedure is high for most of the substances in question. It produces clean extracts, which are well suited for confirmatory analyses, such as gas chromatography (GC) or even gas chromatography/mass spectrometry (GC/MS).
9. Low solvent consumption.

B. Drawbacks of the Toxi-Lab System

However, some drawbacks of the Toxi-Lab system became apparent in terms of its application to European drugs.

1. The system is built for the American market. The existing drug compendium has proven to be insufficient for the comparably bigger European or especially our German market.
2. Some substance groups are only poorly recognized, or produce only a nonspecific color reaction, using the standard procedure. Among these are the benzodiazepines, the salicylates, and the pyrazolones (antipyrine derivatives).
3. Some European substance groups need special procedures for discrimination (examines: β-blockers, benzodiazepines, antirheumatics).

To overcome those drawbacks, we proceeded to build up our own data base for drugs of the European market by photodocumentation and started to develop special procedures for certain drug classes. Our data base at this point contains about 450 entries of drugs and urine metabolite patterns, and it is still growing. The main part of our data is included in the latest Toxi-Search update (Toxi-Lab, Inc., Irvine, California) (a computer-aided toxicological search program), which was introduced at the LuxTox meeting in Luxembourg in May 1990. At this time we are preparing a compendium called *European Drug Monograph* (1) containing drug information and photographic documentation for more than 200 drugs (and urine metabolite patterns) of the German/European market. This work will be available at the beginning of 1991 from Toxi-Lab, Inc.

II. MATERIALS AND METHODS

For the Toxi-Lab system, Toxi-Grams A glass fiber plates, impregnated with silica and ammonium vanadate, and Toxi-Grams C8 were applied. Unless otherwise stated, the solvent systems and color reagents used were those described by Toxi-Lab (2). The color reactions are based on a sequence of the following detections: stage I, combination of Marquis/Mandelins reagents; stage II, water; stage III, UV light; stage IV, modified Dragendorff's reagent.

For system modifications and special procedures, the following reagents were used:

Benzodiazepine special procedure:
Extraction solvent: *n*-heptane/dichloromethane (60/40, v/v)
Solvent system I: toluene
Solvent system II: toluene/acetone (90/10, v/v)
Sodium nitrite, aqueous solution (20 g/100 mL)
Bratton-Marshall reagent: N-(1-naphthyl)-ethylenediaminedihydrochloride, methanolic solution (0.5 g/100 mL)

Pyrazolone/salicylate detection:
Ferric chloride, aqueous solution (5 g/100 mL)
Potassium ferricyanide, aqueous solution (1 g/100 mL)

All TLC analyses were performed after liquid/liquid extraction of urine and gastric content samples from patients suspected of acute poisoning, or buffered aqueous samples, spiked with parent drug compounds.

All R_f data are reported as corrected R_f values. In most of the cases with urine metabolite patterns we performed a confirmation of the identity of the metabolite spots by GC/MS. This was done by punching out of the spot, extracting it with methanolic ammonia, derivatization (if necessary), followed by GC/MS analysis.

III. RESULTS AND DISCUSSION

A. Objectives of Study

The *European Drug Monograph* will contain documentation on the following drug classes of the European market:

1. Classes of frequently abused drugs: psychopharmaca/neuroleptics, benzodiazepines and their urine metabolites, phenothiazines (partially with urine metabolite patterns), sympathomimetics, analgesics, β-blockers
2. Other drug classes: calcium antagonists, nonsteroidal antirheumatics, antihistaminics, xanthine derivatives
3. Miscellaneous drugs of toxicological relevance

B. Examples from Drug Classes Studied

1. *Benzodiazepines and Their Urine Metabolites*

Analysis of the benzodiazepine drugs is done by performing an acid hydrolysis of the urine samples, followed by chromatography using toluene or toluene/acetone (90/100) as mobile phases, and detection with Bratton-Marshal reagent after photolytic desalkylation of the resulting benzophenone metabolites.

2. *Pyrazolones (Antipyrine derivatives) and Salicylates*

Pyrazolone derivatives are still very common analgesics in Europe. They appear only as specific brown spots in color detection stage 4 using the standard Toxi-Lab A procedure. We confirm their presence using the standard chromatographic procedure and detection with ferric chloride reagent. Further confirmation and an enhancement of sensitivity can be achieved after drying of the plate and spraying (or dipping) with potassium ferricyanide reagent, resulting in big blue spots.

Salicylate derivatives (salicylamide, ethenzamide) react the same way, resulting in a violet spot in ferric chloride and blue in ferricyanide detection. Acidic salicylate derivatives (salicylic acid, acetylsalicylate) can only be found using the acid or neutral extraction (Toxi-Lab B procedure). Extraction of salicylamide in alkaline solution can result in cleavage of the acid-amide function and, hence, nonextractability of the drug using the A procedure.

3. *Phenothiazines*

Phenothiazine drugs usually result in a similar and complex urine metabolite pattern. As gastric contents are the first available material in acute clinical toxicology, we documented the parent drugs of European phenothiazines as well as some of their metabolite patterns. Some European species have other color characteristics of their urine metabolites (e.g., levomepromazine). The appearance of the parent compound is also dependent on the amount of drug ingested, and on the time of sampling.

4. *European β-Blockers*

β-Blockers are often abused substances in doping cases and also in cases of acute poisoning (accidental or suicide attempts). European species result in characteristic color reactions using the standard color detection sequence; differentiation by R_f value, however, is very poor using silica gel as a stationary phase. Due to their close structural relationship, enhancement of the "discriminating power" (3) cannot be achieved by variation of the mobile-phase system, but only by variation of the separation mechanism using reversed-phase chromatographic material (C_8). Even in cases of acute poisoning, parent drugs were predominant in most of the urine samples analyzed.

5. *Xanthine-Derivatives (Broncholytics)*

This class of drugs shows characteristic color development (gray spots) in color detection stage 4, which is normally not a discriminating stage. Sensitivity of detection is very different in this class; some members, such as theophylline and pentifyllin, can be detected only in high concentrations. As described in the Toxi-Lab compendium, Toxi-Tips (4), sensitivity can be greatly enhanced by omitting the stage 1 and 2 dippings (just saturate with formaldehyde and dip into Dragendorff's reagent).

6. *Nonsteroidal Antirheumatics*

Members of this class of drugs normally have acidic character; in spite of that, some of them can be extracted in good yield using the alkaline Toxi-Lab A procedure. Due to their acidic nature, migration in the standard solvent system containing ammonia is very low. In this class of drugs, however, change over to the standard C_8 procedure could not solve this problem, as in this system those acidic compounds migrate with the solvent front. A way to differentiate them is to omit the ammonia in solvent system A; the shape of the spots, however, appears streaky. A method for better discrimination of these drugs is under development.

C. Application of Toxi-Search to European Drugs

As mentioned above, most of our data are included in the latest update of the Toxi-Search program. For most of the drugs, we created entries for different concentration ranges: low, medium, and high. Color codings are always based on the "predominant" color of a spot. Some drugs show characteristic "center/halo" phenomena (e.g., the amphetamines and phenothiazines). As a rule, the color of the halo is the same as that appearing with low concentrations; that of the center is the "high" concentration color. All inputs were made according to this principle. Using

Table 1 Effect of Combinations of Systems and Color Reactions on Discriminating Power: Data Base for European Drugs (n = 451)

System	Discriminating power		Error window
	Maximum	Results	
TLA	0.810	0.785	10
TLA + colors	0.999	0.992	10/1/1/1/1

Toxi-Search for computer-assisted search of the general unknown, this program can only give assistance in sorting of bigger data bases, as it is nearly impossible to keep all these data and color appearances in mind, even after extensive TLC experience. The program never results in an "absolute" answer; the proposals should be confirmed by comparing with the actual drug appearance, either via photograms in the drug compendium, or by co-analyzing the suspected of drugs, or by confirmation using other techniques.

D. Use of C_8 Chromatograms for Confirmation of Clinical Samples

One possibility for confirmation of a suspected drug is the use of another TLC separation mechanism, such as the reversed-phase system C_8. As this technique is rapid and easy to perform using the extracts already prepared for the standard TLC system, it is especially suited for confirmation in acute cases of clinical toxicological analysis. We always evaporate the extracts of alkaline extraction under a stream of nitrogen, redissolve in 100 µL of toluene/methanol (90/10, v/v), and concentrate the main part of it into two Toxi-Disks, without any loss in sensitivity. The second disk can be used for confirmation purposes (e.g., C_8), the rest of the extracts for gas chromatography or GC/MS; all these systems can be run simultaneously in a minimum of time (less than 1 h) in cases of acute poisoning where a quick answer is crucial.

Tables 1 and 2 show system evaluations using the mathematical parameter "discriminating power" (DP), introduced by Moffat in 1973 (3) as a measure for the identification efficiency of TLC systems. This technique evaluates the separation power of a TLC system by use of the R_f distribution and reproducibility. The better the system, the higher should be the DP value (depending on the "error window," based on the threefold standard deviation of a series of R_f values). As can be seen, integra-

Table 2 Effect of Combinations of Systems and Color Reactions on Discriminating Power: Toxi-Lab Drug Compendium (n = 154)

System	Discriminating power		Error window
	Maximum	Results	
TLA	0.810	0.782	10
TLA + C_8	0.964	0.947	10/10
TLA + colors	0.999	0.984	10/1/1/1/1
TLA + C_8 + colors	0.999	0.996	10/10/1/1/1/1

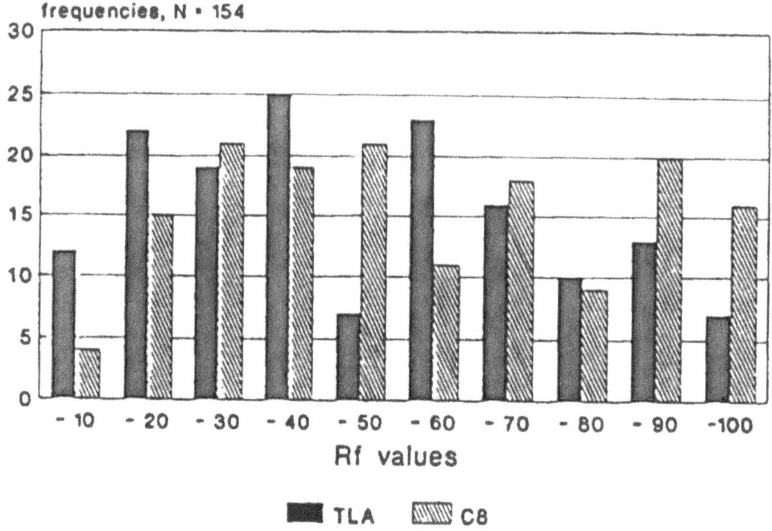

Figure 1 Distribution of R_f values: Toxi-Lab Drug Compendium (n = 154).

tion of different separation mechanisms and specific color reactions results in a nearly theoretical value for this parameter).

Figures 1 and 2 show the R_f distribution, Figures 3 and 4 the distribution of colors, appearing in the standard procedure for the drugs represented in the Toxi-Lab drug compendium and in our own data base for European drugs.

Table 3 shows a system comparison with capillary GLC, a chromatographic technique, known for its high separation efficiency, which is used in parallel in our laboratory, for confirmatory purposes. As can be seen, this system alone results in a

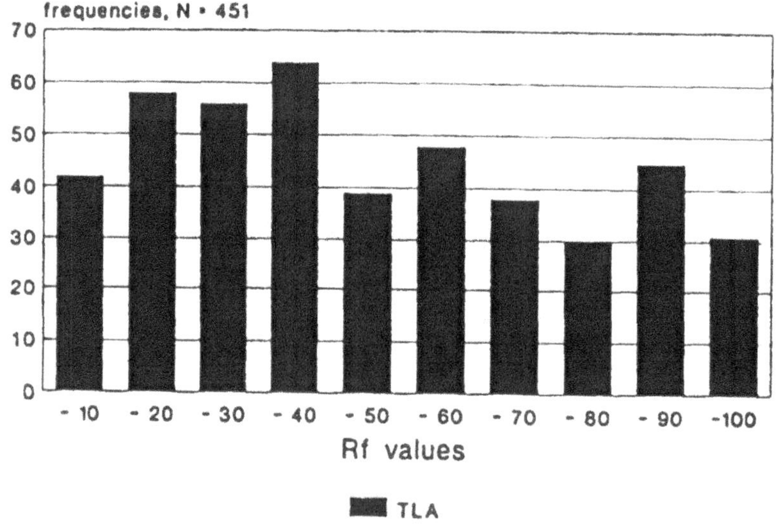

Figure 2 Distribution of R_f values: Data base for European drugs (n = 451).

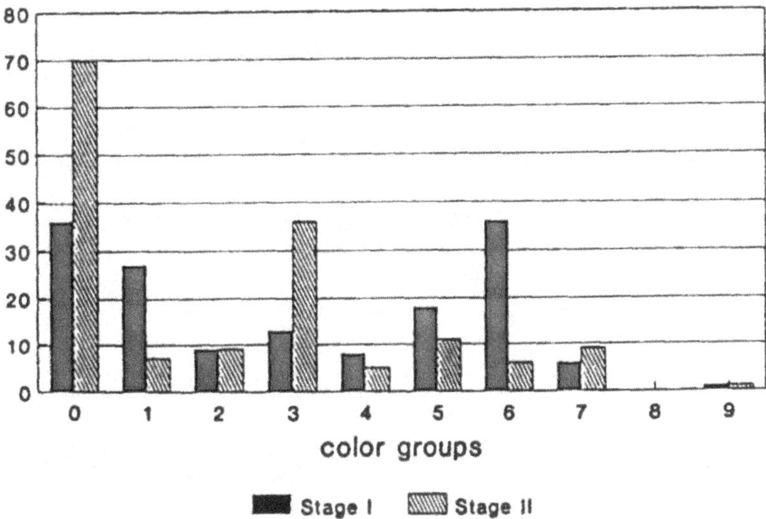

Figure 3 Distribution of colors in detection stages I and II: Toxi-Lab Drug Compendium (n = 154). Color codings according to the Toxi-Lab color wheel. (Color group 8: not defined.)

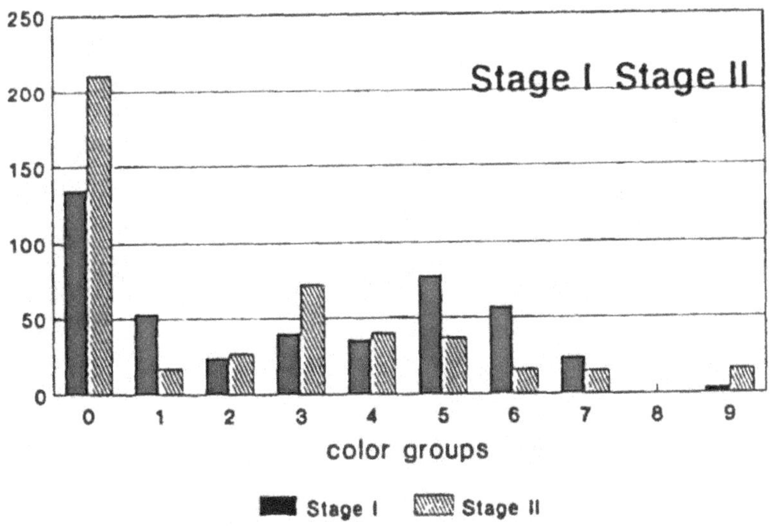

Figure 4 Distribution of colors in detection stages I and II: Data base for European drugs (n = 451). Color codings according to the Toxi-Lab color wheel (Color group 8: not defined.)

Table 3 Effect of Combinations of Systems and Color Reactions on Discriminating Power: Comparison of TLC and Color Reactions with GLC (n = 210)

System	Discriminating power		Error window
	Maximum	Results	
TLA	0.810	0.767	10
C_8	0.810	0.764	10
GLC (SE 54,capill.)	0.975	0.906	0.05
TLA + C_8	0.964	0.938	10/10
TLA + GLC	0.995	0.976	10/0.05
TLA + C_8 + GLC	0.999	0.993	10/10/0.05
TLA + colors	0.999	0.982	10/1/1/1/1
TLA + C_8 + colors	0.999	0.995	10/10/1/1/1/1
TLA + colors + GLC	0.999	0.998	10/1/1/1/1/0.05

DP that is unequaled by the TLC techniques. Combination of different separation mechanisms, however, and still more the integration of color reactions in TLC yields a markedly higher DP as compared with GLC alone. Of course, a combination of all these techniques results in a nearly theoretical value.

IV. CONCLUSIONS

As has been shown in this investigation, the identification possibilities of a highly efficient and sophisticated technique such as capillary GLC can be fairly overranged by an integration of different separation mechanisms and specific color reactions in such simple and easy-to-perform techniques at TLC.

REFERENCES

1. F. Degel and N. Paulus. *European Drug Monograph*, Toxi-Lab, Inc., Irvine, CA (in press).
2. *Toxi-Lab Instruction Manual*. Toxi-Lab, Inc., Irvine, CA.
3. A. C. Moffat and K. W. Smalldon. The calculation of discriminating power for a series of correlated attributes. *J. Forensic Sci. Soc., 13*:291–295 (1973).
4. *Toxi-Lab Drug Compendium*. Toxi-Lab, Inc., Irvine, CA.

96

Eurotox Proficiency Study 1990

I. C. Dijkhuis,* D.R.A. Uges,† J. J. Eerland,* and A. R. Harteveld* **The Hague Hospitals Pharmacy, The Hague, The Netherlands, †Academic Hospital Department of Pharmacy, Groningen, The Netherlands*

I. INTRODUCTION

In a few years there will be a United Europe, and the European Community will gradually replace the old national regulations. Many problems should be solved before 1992, the year in which European harmonization should be a fact.

This European harmonization includes agreement about qualities of products—not only in foods and drugs, but also in drug assays. The cost of an assay will become the European market price, which often will be in conflict with the quality of the results. Consequently, there should be a European control of laboratory performance. Perhaps there will be a European certification for laboratories to ensure the quality of their performance.

The European harmonization will probably not be limited to Western Europe: Eastern European countries will try to join the Western market. In the first 2 years they will have a hard time, because of their low budgets and consequently, poor instrumentation. But in 5 years their output might be similar to that of the West, and so, a European quality control should pay attention to Eastern Europe now, and not 5 years from now.

The existing associations on External Quality Assessment (EOA) in Western Europe should first of all coordinate their efforts in order to obtain European harmonization in therapeutic drug monitoring (TDM) and Toxicology—in other words, harmonization of qualities and quality criteria. These associations also should make contacts to EQA organizations in Eastern Europe or, when these do not exist, give a hand in setting up such programs.

The Dutch Association for Quality Assessment in TDM and Toxicology (abbreviated as KKGT) has tried a first step, using the Eurotox Proficiency Study 1990.

II. KKGT ASSOCIATION

The KKGT organization was set up in 1975 and is still located in the central hospital pharmacy in The Hague. There are approximately 100 participants in the KKGT programs, mainly Dutch laboratories.

KKGT has five TDM programs, based on quantitative assays in serum. The annual frequencies of tests and the number of participants when the various programs started

Table 1 External Quality Assessment in TDM/Toxicology: KKGT 1990

Program (Components)		Start	Frequency	Laboratories
Antiepileptics	(9)	1974	8	98
Antibiotics	(2)	1982		
Cardiacs	(9)	1977	6	81
Antidepressants	(4+4)	1980	5	55
Benzodiazepines	(4+4)	1980	5	49
"Experimental"	(2–4)	1988	4	54
Toxicology	(1–5)	1977	6	61

are given in Table 1. The KKGT programs use serum specimens to which drugs are added. Presently, newborn calf serum is used. These specimens are lyophilized to prevent decomposition of the drug added. The drug concentrations are assigned values, calculated by weighed-in quantities and corrected for volume increase after reconstruction.

All KKGT serum specimens are prepared in the hospital pharmacy in The Hague, which has large facilities for freeze-drying. Approximately 60 different batches of serum specimens are lyophilized each year.

The *routine* programs (Table 2) include drugs such as antiepileptics and cardiacs; most of them can be analyzed today even by laboratory technicians, using the enzymatic methods from Abbott or Syva. For example, laboratories will have no problems in assaying phenytoin (Fig. 1). Laboratory numbers are coded and the results are plotted on the diagram. Method statistics are given (outliers outside x ± 2.5 SD). Laboratory performance is also mentioned as excellent, good, moderate, or poor. The criteria depend on the grade of difficulty of the drug tested and its concentration. Criteria are often also dependent on the instrumentation that is available in the country. For phenytoin levels > 5 ml/L, a laboratory result outside 100 ± 20% of the assigned value is called "poor." Consequently, values between 80% and 90% and between 110% and 120% are "moderate" values, those between 90% and 95% or 105% and 110% are "good," and values between 95% and 105% are "excellent." The laboratories receive the KKGT survey within 3 weeks after deadline. In case of a "very poor" laboratory result, the laboratory is informed by return post. "Very poor" is outside 2x poor, so in case of this example <60 or >140%. Laboratory performance per test yields more or less credit points: excellent = 10, good = 7, moderate = 4, poor = 1 point. The annual score is the total amount of points, calculated for 10 tests; so the

Table 2 KKGT Routine Assays 1990

Phenobarbital	Phenytoin	Ethosuximide
Carbamazepine +m	Clonazepam	Quinidine
Valproic acid	Primidone	Lidocaine
Procainamide +m	Lithium	Gentamycin
Disopyramide +m	Digoxin	Tobramycin
Theophyllinle	Caffeine	

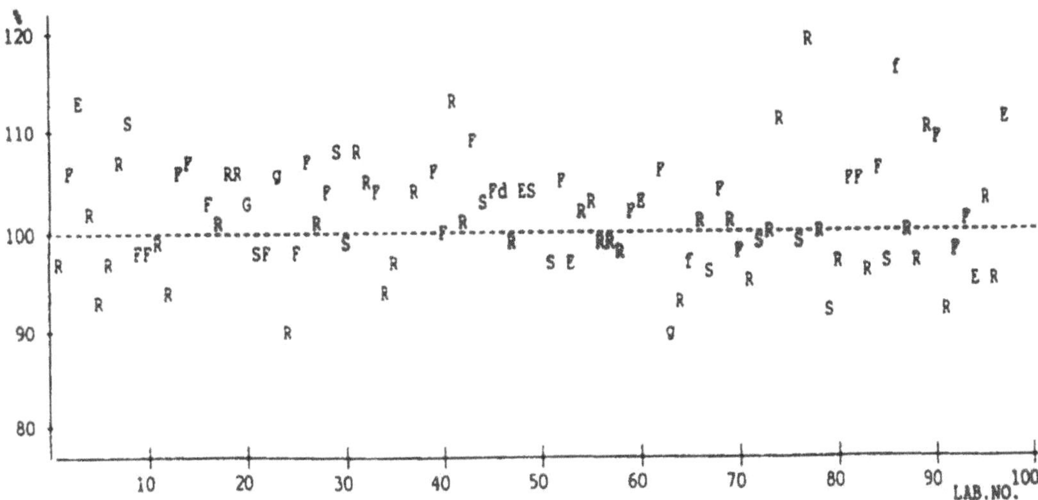

Figure 1 Phenytoin test II, 1989, tt mg/L.

minimum is 10, the maximum is 100 points. At least half + 1 of the total number of tests per year should be reported in order to get a year performance score (Figure 2).

According to KKGT criteria, a laboratory should have more than 55 points for a KKGT certification. Obviously, this performance score is subjective and made for the Dutch situation only.

The antidepressants and benzodiazepines may be called "*subroutine*," since occasionally instruction about methods is given (Table 3). The "*experimental*" program is a typical training program: in 1988–1990, 24 different drugs were assayed, and selected methods were reported in detail to all participants. In this program, attention is also paid to kinetics and dosage advice (Table 4).

The *clinical toxicology* program is based on real case histories with drug-spiked serum specimens and, occasionally, urine samples. In this report, identities and quantities as well as interpretation of these results should be reported. During the period 1975–1990, over 130 different components were investigated and discussed.

III. EUROTOX 1990

Our objective in this study was to investigate if an annual international proficiency testing program on clinical toxicology would be worthwhile—e.g., once yearly a study with three unknown cases and three or four known quantitative assays: After some years, the program should be evaluated.

In January 1990, some 300 laboratories in and outside Europe were invited to take part in this study. Approximately 200 of them answered positively, and to those laboratories serum specimens were sent. Finally, 140 laboratories from 25 different countries reported their results. In this study, the KKGT toxicology program served as a model to obtain information on laboratory performance and instrumentation used in the various countries. Drugs which might only be assayed by using automatic assays such as TDx or EMIT, were not used in the present study.

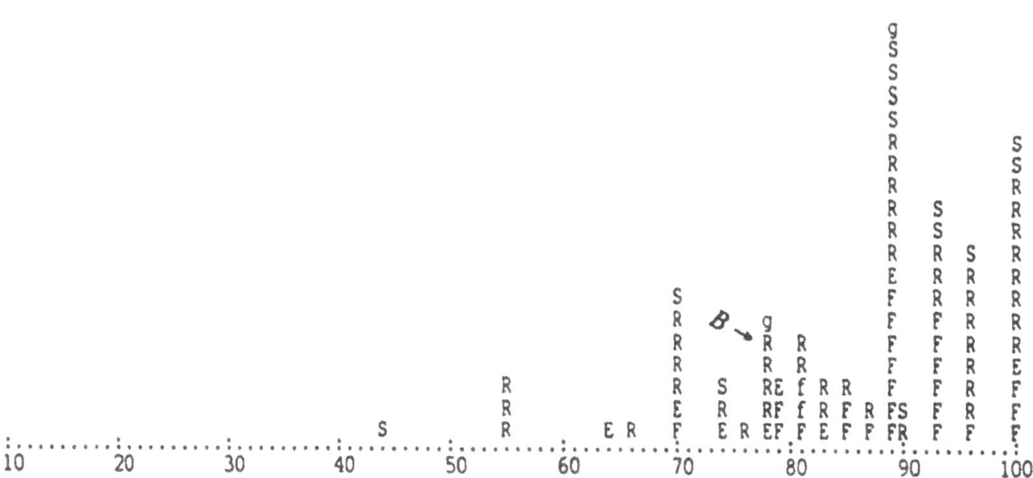

Figure 2 Phenytoin performance 1989. KKGT criterion for phenytoin (eight tests) is that at least five tests were reported and that more than 55 points were obtained. Laboratory A made a good performance score but reported eight tests, obtained 78 points, and received a certificate.

To illustrate our ideas, three unknown cases (serum specimens and case histories) were distributed.

Case 1: "A regular methadone and flunitrazepam user, locked up overnight in a police cell, was found comatose in the morning and transferred to hospital. In the clinic, he did not react to naloxone and flumazenil." Cause of coma was secobarbital, 25 mg/L in the serum.

Case 2: "A laboratory technician, found on the floor of his room by friends. He told them he had prepared rat poison that night and had tasted it. Strychnine might be used. No spasms were observed, only a high CPK value." For details see Section III.C.

Case 3: "A woman walked into the clinic, claiming she had taken malaria pills the day before. Two days later she died suddenly." The cause of death was chloroquine, 6.8 mg/L in the serum.

In order to investigate possible matrix effect on drug assays, some drugs were assayed both in serum and in blood: paracetamol, dextropropoxyphene, ethyl alcohol, lorazepam. The results for paracetamol and lorazepam are given below.

Table 3 KKGT Subroutine Assays 1990

Diazepam	+m	Amitriptyline	+m
Nitrazepam		Imipramine	+m
Oxazepam		Clomipramine	+m
Temazepam		Doxepin	+m
Flunitrazepam	+m	Maprotiline	+m
Desalkylflurazepam			

Table 4 KKGT "Experimental" Assays 1988-1990

Acenocoumarol	Flecainide	Propranolol
Amikacin	Flucytosine	Sulphamethoxazole
Amiodarone +m	Methotrexate	Thiopental +m
Atenolol	Metoprolol	Trimethoprim
Cimetidine +m	Mexiletine	Vancomycin
Clobazam	Mianserin +m	Warfarin
Cyclosporin	Phenoprocoumon	

A. Paracetamol

Paracetamol (acetaminophen for the United States) was determined in the past by spectrophotometry using the Glynn-Kendal method. All methods produce reliable data; laboratory performance was good as well (Figure 3). The all-participants mean value (101%) is in agreement with the weighed-in value. Liquid blood instead of lyophilized serum produces a matrix effect of 10%, not only in TDx and spectrophotometry but also for HPLC, which we do not understand. It has been proven that this was not due to decomposition of paracetamol.

B. Lorazepam

Lorazepam is rather difficult to assay. There is, at this moment, no enzymatic "black-box" method, so here you meet the real analytical professionals! Figure 4 shows that HPLC gives better results than GC, particularly when the participants are trained in this technique, as the Dutch group are. The all-participants mean (119% in serum) is *not* in agreement with the weighed-in value, due to the great number of poor results.

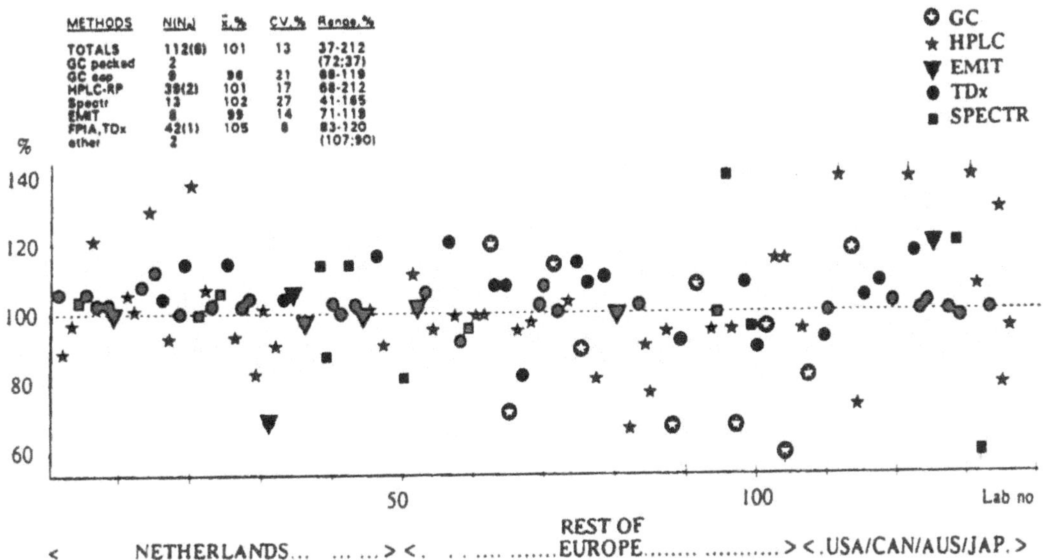

Figure 3 Paracetamol, 73.5 mg/L.

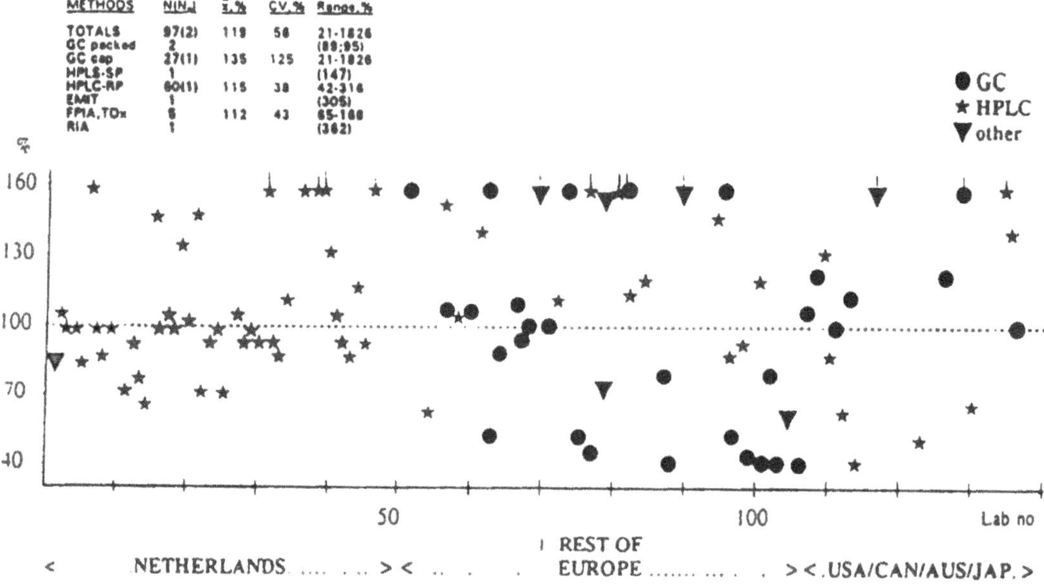

Figure 4 Lorazepam, 0.19 mg/L, serum.

In liquid blood, HPLC results were even better than in serum; GC results, however, were worse, perhaps because of the capillary systems. There is no matrix effect from blood, but during transport there will be a decomposition of lorazepam in liquid blood versus lyophilized material, approximately 8% during 2 weeks transport at +25°C. Decomposition may also occur when, after reconstitution of serum, the sample is not kept in the dark. There is a relatively high decomposition in serum versus blood because of light which comes to approximately in 2 weeks.

C. Rat Poison Case

This case was a real case, a very interesting one from an analytical point of view. *Case History*: "A laboratory technician was found by his friends in the morning, lying on the floor in his room. He murmured that he had tasted his home-made rat poison and said he probably also had put strychnine in it. At hospital, no spasms—characteristic of strychnine poisoning—were observed. The patient had a shoulder luxation. He was kept in the hospital for observation, and blood was sent to the chemistry lab, which reported a 1000-fold increase of CPK level. Then, blood taken on admission to hospital was sent to the toxicology lab."

The clinicians did not believe that the man had ingested strychnine, since he had no symptoms of poisoning. He only needed treatment for his shoulder. Only after the CPK message from the clinical chemistry laboratory did they become suspicious about strychnine, and the toxicology laboratory was then asked to do a screening. The toxicology laboratory found strychnine to be present: 0.4 mg/L, which is sublethal. At this concentration, the man must have had serious spasms during the night with strong muscle activities, explaining the high CPK value.

Table 5 Eurotox 1990–1992: Survey of Results

Participants	Netherlands	Europe	Others	Totals
Responded	34	52	28	114
Screened strychnine	25	36	13	74
Identified strychnine	23	25	10	58
Quantitated S < ±30%	14	13	6	33
Quantitated S > ±60%	1	9	—	10

Our objective in this case was to investigate the analytical procedures in strychnine analysis, since this poison was mentioned in the case history. First of all, the laboratory should find out whether strychnine was present or absent. Table 4 gives a survey of all results. About 65% of the laboratories looked for strychnine, 50% found this poison, 30% assayed strychnine within 30% of the correct concentration.

In Table 6, the participants results in identification are given in more detail. Strychnine can easily be attracted from serum with chloroform of dichloromethane (but *not* with ether). After concentration, strychnine can be identified using a simple TLC procedure: Mandelin spray gives purple spots changing into orange, which is both specific and sensitive (detection limit less than 0.1 μg). For GC, a high column temperature is needed, otherwise strychnine may not be seen (note the great number of false-negative results, even by GC-MS).

HPLC is, in our view, the method of choice, both in identification and in quantification.

IV. SUMMARY

In this study, organized by the Dutch Association for Quality Assurance in TDM and Clinical Toxicology (KKGT), 135 laboratories from 25 different countries participated. The objective was to investigate the usefulness and the feasibility of a regular—e.g., once yearly—exchange of knowledge and ideas within the European countries by using proficiency testing programs. In the present proficiency study, three toxicology cases were solved and four quantitative assays in two different matrices were made. Some results have been discussed in this report.

Table 6 Eurotox 1990–1992: Identification of Strychnine

Method	Positive	Negative	Negative, %
TLC	20	8	40
GC-packed	3	—	—
GC-cap + NPD/FID	19	3	16
GC-cap + MS	9	5	55
HPLC-SP	6	—	—
HPLC-RP + UV	9	1	9
HPLC + DAD	2	—	—

97

Statistical Approaches to Accuracy in Drug Screening

Vina Spiehler *Diagnostic Products Corporation, Los Angeles, California*

I. INTRODUCTION

Accuracy in drug screening can be expressed as the probability that a positive (or negative) result is correct. This probability can be calculated from Bayes' theorem if the performance characteristics of the test and the prevalence of the drug in the subject population are known (1). The probability that the test result is correct is called the predictive value of the test (2). The predictive value associated with a test result can be increased by sequential independent tests. To meet the requirements for independence, these tests must be performed on a second aliquot of the sample and must be based on a different physical or chemical property of the drug. The predictive value of a positive result from a single immunoassay screen on random urine specimens from a general workplace population has been reported to range from 0.10 to 0.70 (3). This is often not sufficient evidence for forensic applications. After confirmation with gas chromatography/mass spectrometry (GC/MS), the predictive value can be increased to the range from 0.90 to 0.99 or better. This provides the appropriate level of confidence that the result is truely positive.

II. BAYES' THEOREM

Bayes' theorem states that the probability that hypothesis H is true, given experimental results E, is the quotient of the probability of the result E when the hypothesis was known to be true, divided by the sum total of the occurrence of the result E in all experiments, whether or not the hypothesis was true:

$$P(H:E) = \frac{P(E:H)}{P(E:H) + P(E:H)}$$

In drug screening the hypothesis is that the urine specimen truly contains the drug or drug metabolite. The experimental evidence comes from the drug screening tests performed on the specimen. The numerator, the probability that the test will be positive when the specimen does not contain the drug, is equal to the sensitivity of the test times the prevalence of the drug in the population:

P(E:H) = (prevalence)(sensitivity)

when sensitivity is defined as the percent of all the actually positive specimens which give a positive test result (see Glossary, Table 1). Prevalence is the fraction of the specimens which actually contain the drug. This is commonly estimated from epidemiological studies of the test population.

The denominator is the sum total of all the positive test results in all the specimens:

P(E:H) + P(E:H) = (prevalence)(sensitivity) + (1 - prevalence)(1 - specificity)

when specificity (see Glossary) is defined as the fraction of the negative specimens which give a negative test result.

The probability that a positive test result is correct is called the predictive value of the test (2). Therefore, according to Bayes' theorem, the predictive value (PV) is equal to

$$\text{PV} = \frac{(\text{prevalence})(\text{sensitivity})}{(\text{prevalence})(\text{sensitivity}) + (1 - \text{prevalence})(1 - \text{specificity})}$$

In the study of known specimens, an equal number of positive and negative specimens are selected. In this case the prevalence equals 0.50, 1 - prevalence = 0.50, and the prevalence term can be factored out.

An example calculation of predictive value for a test with 96% sensitivity and 93% specificity is shown in Table 2. The prevalence of the drug in the population being screened in Table 2A is estimated to be 10%.

III. SENSITIVITY AND SPECIFICITY

The sensitivity and specificity for a test are calculated from a tabulation of the test performance on known specimens. In drug testing, gas chromatography/mass spectrometry is often accepted as establishing the true state of the specimen. The test results can be classified as true positives, false positives, false negatives, and true negatives by comparing the answer to the GC/MS result (Table 1). The sensitivity of the screening test is equal to {TP/(TP + FN)} or the true positive rate. The specificity of the screening test is equal to {TN/(TN + FP)}.

The performance characteristics of the test, sensitivity and specificity, change with the cutoff chosen to differentiate positive from negative results. This is shown for a drug screening test in Figure 1. Each cutoff has a different percentage of false positive and false negative results. Use of a high cutoff threshold results in many false negatives. The specificity of the test is improved at the expense of the sensitivity. However,

Table 1 Calculation of Sensitivity and Specificity[a]

	Test positive	Test negative	Total
Drug present by GC/MS	92	4	96
Drug absent by GC/MS	7	97	104

[a]Sensitivity = TP(TP + FN) = 92/96 = 0.96 or 96%; specificity = TN/(TN + FP) = 97/104 = 0.93 or 93%.

Table 2A Calculation of Predictive Value for a Test with 96% Sensitivity, 93% Specificity, and a Population Prevalence of 10%[a]

	Prevalence	Probability of positive test	Probability of negative test
Drug present	0.10	0.96 ± 0.10 = 0.096	0.04 ± 0.10 = 0.004
Drug absent	0.90	0.07 ± 0.90 = 0.063	0.93 ± 0.90 = 0.837
Total		0.159	0.841

[a]Predictive value of a positive result = 0.096/0.159 = 0.603 or 60%; predictive value of a negative result = 0.837/0.841 = 0.995 or 99.5%.

Table 2B Calculation of Predictive Value for Positive Result After Confirmation Test with 99% Sensitivity and Specificity[a]

	Prevalence	Probability of positive test	Probability of negative test
Drug present	0.60	0.99 ± 0.60 = 0.594	0.04 ± 0.10 = 0.004
Drug absent	0.40	0.01 ± 0.40 = 0.004	0.99 ± 0.40 = 0.396
Total		0.598	0.400

[a]Predictive value of a confirmed positive = 0.594/0.598 = 0.993 or 99%.

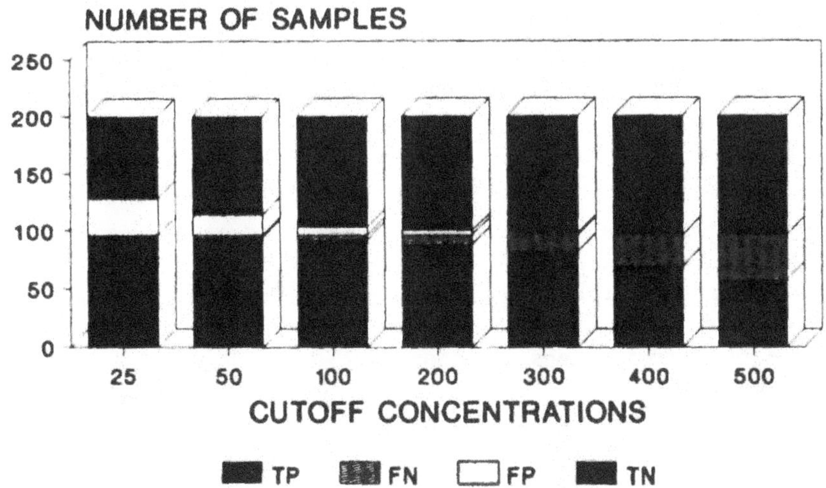

Figure 1 Drug screening immunoassay results.

Table 3 Effect of Prevalence on Predictive Value When Sensitivity is 99% and Specificity is 98%

Prevalence	Predictive value of a positive test
0.1%	9%
0.5%	33%
1.0%	50%
5.0%	84%
10.0%	92%
25.0%	97%
50.0%	99%

as can be seen from the data in Figure 1, choosing a low threshold for the cutoff concentration improves the sensitivity at the expense of the specificity. An optimum choice of cutoff might be at a concentration which minimizes the probability of either a false positive or a false negative. That sensitivity and specificity for the threshold chosen ad the cutoff concentration are used for calculation of the predictive value.

IV. PREVALENCE

In addition to the performance characteristics of the screening test, the predictive value is a function of the prevalence of drugs in the population being screened. In toxicology screening the prevalence can vary quite a bit. Among persons arrested for a crime in the United States, the prevalence of drugs in the body ranges from 54% to 83% (4). But the prevalence can vary by drug or by location. For example, the prevalence of methamphetamine in arrestees is 36% in San Diego but only 1% in Philadelphia (4). The prevalence of cocaine in persons admitted to an emergency room in 1989 in the United States with drug-related morbidity was 40% (5), but cocaine prevalence is less than 3% in employees (3). As a result, the predictive value of a drug screen for these different populations would be greatest for populations with a high prevalence and much lower for populations with a low prevalence (Table 3).

V. CONFIRMATION

As shown in the example calculation in Table 1, the predictive value for a positive result after a single screening test may not be high enough for reporting a result which may be used in forensic proceedings. In a population with a low prevalence, this is true even for very sensitive and specific tests. Additional experimental information from subsequent independent tests can increase the probability that the drug is present. The amount of increase can be calculated by again applying Bayes' theorem. The sensitivity and specificity of the new test are used. The probability after the first test is used as the prevalence, since this is the prevalence of true positives in all the positive screen results. For a confirmation method with 99% sensitivity and specificity, the predictive value after confirmation for the example in Table 2 would be 99.3%.

The predictive value of a negative result in the screening test in the example is 99.5%. Since the probability that this result is correct is better than 0.99, it is not necessary to continue to test this specimen.

VI. CONCLUSION

The accuracy of a qualitative test such as a drug screening test can be expressed in terms of clinical sensitivity and specificity calculated from the test performance with known specimens. By using Bayes' theorem, the sensitivity and specificity of the test can be combined with the prevalence of the drug in the population to estimate the probability that a positive test result is a true result. In general populations in which drug prevalence is low, the probability that a positive screen test is a true positive is usually less than 75%. That is, one out of four positive results may falsely indicate drug in a specimen. By confirming positive screen results by a second independent test based on a different chemical or physical property of the drug, the probability that the result is true can be increased to a level of certainty (0.90% or better) which allows the laboratory to report the result with confidence.

GLOSSARY

Bayes' theorem: a formula for combination of probabilities to reflect increasing or decreasing confidence in a hypothesis with additional experimental results.

False negative: a negative test result for a specimen in which the drug was actually present.

False positive: a positive test result for a specimen in which the drug was not present.

Predictive value: the probability that the test result is correct.

Prevalence: the proportion of positive cases per 100,000 population.

Sensitivity: the proportion of all the known positive specimens which test positive

$$\text{sensitivity} = \frac{\text{true positives}}{\text{true positives} + \text{false positives}}$$

Specificity: the proportion of all the known negative specimens which test negative.

$$\text{specificity} = \frac{\text{true negative}}{\text{true negatives} + \text{false positives}}$$

True negative: a negative test result for a specimen which does not contain the drug.

True positive: a positive test result for a specimen which does contain the drug.

REFERENCES

1. V. R. Spiehler, C. M. O'Donnell, and D. V. Gokhale. Confirmation and certainty in toxicology screening. *Clin. Chem.*, 33:1535–1539 (1988).
2. V. E. Wells, W. Halprin, and M. Thun. The estimated predictive value of screening for illicit drugs in the workplace. *Am. J. Public Health*, *78*(7):817–819 (1988).

3. R. S. Galen and S. R. Gambino. *Beyond Normality: The Predictive Value and Efficiency of Medical Diagnoses*. John Wiley, New York (1975).
4. National Institute of Justice. *1988 Drug Use Forecasting Research Update*, p. 3 (December 1989).
5. National Institute on Drug Abuse. *Drug Abuse Warning Network (DAWN)*, ser. G, no. 24 (1989).

98

Synthetic Calibrator Matrices in Immunoassays

Anand Akerkar *Creative Scientific Technology, Inc., Great Neck, New York*

Human or mammalian blood products have been traditionally used as matrices for in-vitro calibrators and reference controls. Two or more calibrators containing specific levels of analyte are typically used in an immunoassay for the purpose of constructing a dose-response curve (standard curve) by means of which the signal responses of reference controls and unknown specimens can be calculated in terms of concentration or mass units of a given analyte. Similar reference controls or calibrators are also utilized for verifying assay performance in other clinical chemistry procedures.

In view of the compositional variability of serum or plasma and the other disadvantages enumerated in Table 1, one may reasonably raise the question under what conditions it would be feasible to substitute protein-free matrices in immunoassays analogous to the current trend toward low-protein matrices in tissue culture and monoclonal antibody production (1,2).

The use of such "synthetic" matrices would eliminate most of the disadvantages listed in Table 1 and would clearly be advantageous from the manufacturers' as well as the users' perspectives. This report addresses the conditions necessary to achieve the full benefit and wider applicability of synthetic calibrator matrices in immunoassays.

The ideal calibrator matrix for use in immunoassays would have a well-defined composition and would exactly match the composition of the unknown specimen except for the unknown analyte concentration. These conditions are rarely attainable in practice due to the wide diversity of normal and abnormal physiological blood or urine specimens.

Next best would be a homologous serum-based matrix that approximately matches the typical specimen composition. However, some techniques utilized for removal of endogenous analytes from serum, such as "stripping" with charcoal or ion-exchange resins, frequently cause major alterations in the lipid and protein compositions. Affinity chromatography, in contrast, can selectively remove endogenous analytes without significant alteration of the matrix. The high cost of the requisite antibodies and their low binding capacities make this procedure generally practical only for removing antigens present in extremely low concentrations.

Less optimal would be the use of calibrator matrices containing heterologous sera or albumin fractions in appropriate buffer formulations. Such protein-based matrices

Table 1 Disadvantages of Protein-Based Calibrators

1. Lot-to-lot variability.
2. Inherent instability of proteins or protein solutions requiring storage at freezer temperatures.
3. Potential contamination of protein matrices with unacceptably high levels of endogenous analytes, necessitating in-house processing or purchase of more costly "stripped" protein matrices.
4. Potential contamination with, and the need for testing for, pathogenic or nonpathogenic microorganisms and viruses, including hepatitis virus and HIV.
5. Presence of viruses and infectious agents in some approved lots due to the limitations of current testing procedures (3).
6. Protein matrices are excellent growth media requiring aseptic manipulation and/or use of effective antimicrobial agents.
7. The variable and frequently diminishing endogenous enzyme levels and their unpredictable effect on calibrator performance.
8. The relatively high cost of human and animal products, particularly of purified blood fractions.
9. Large and expensive inventories must be maintained to minimize screening and validation expenses.
10. Potential shortages in the supply of human blood products due to fewer donors, high donor rejection rates, and improved whole blood utilization resulting in fewer outdated blood units becoming available for in vitro use.

are widely used in commercially produced calibrators. The undefined nature of serum and the lack of homogeneity of protein fractions require time-consuming screening of raw materials to match the performance of prior lots of calibrators. These and other shortcomings of protein-based calibrators are enumerated in Table 1.

Total elimination of proteins from calibrator matrices would solve most of these problems, but is rarely applicable due to the potential for severe specimen-calibrator mismatch in analyte kinetics. Some attempts have been made to overcome these drawbacks by replacing proteins with defined synthetic materials. Synthetic matrices are commercially available (4). Several manufacturers also have incorporated synthetic matrices into immunoassay kits. However, little information on the comparable performance of such matrices is available.

Calibrator matrices, whether natural or synthetic, must meet the following criteria:

1. Precision
   Intraassay
   Interassay
2. Accuracy
   Analyte potencies
   Binding kinetics of analyte
   Matched nonspecified binding in calibrators and specimens

Since no mixing of calibrators and specimens occurs in the assay except for possible use of the zero calibrator as a specimen diluent for out-of-range specimens, the calibrators need not reflect absolute or true analyte levels but could represent nominal or assay-specific potencies. Nominal calibrators are frequently used in so-called free analyte assays.

As a consequence of kinetic control rather than equilibrium attainment in most commercial assays of short incubation duration, temperature control is desirable in both synthetic matrices as well as with many suboptimally matched protein matrices. Incubations of such assays at room temperature are clearly not conducive to good interassay control of assay parameters and accuracy. Controlled incubations in a water bath at 25°C instead of at room temperature would also improve the correlations in interlaboratory comparisons of immunoassay results, particularly when comparing data generated in European countries where laboratory temperatures may be more than 5°C lower than in the United States. Furthermore, it is known that antigen-antibody interactions are affected by temperature changes and that the affinity constants of most antigen-antibody interactions increase with a decrease in temperature (5).

The need for matching the nonspecific binding of both analyte and labeled analyte in synthetic matrices and specimens presents a major obstacle to the wider use of synthetic matrices. The occurrence of variable protein-dependent nonspecific interactions in specimens and the generally lower but constant interactions in calibrators may result in significant bias in assay results. In total analyte systems, bound analytes may be liberated from low-affinity binding sites by means of appropriate competitive blocking agents. The substitution of synthetic matrices for serum matrices clearly requires not only effective control of assay conditions, but also more demanding optimization of the matrix composition to eliminate potential assay bias. Complete elimination of bias may not be achievable with certain analytes, even with nominal calibrators.

Although the temperature dependence in immunoassay is normally not as high as in enzyme assays, immunoassay kit manufacturers and end users could potentially enhance their assay precision and accuracy by tighter control over incubation temperatures and by closer matching of calibrator and specimen matrices. These recommendations for assay optimization apply equally to synthetic and protein-based calibrator matrices.

Under the trade names SeraSub (synthetic serum) and UriSub (synthetic urine), Creative Scientific Technology, Inc., has developed these two products, which are ready for the preparation of calibrators and controls.

SeraSub is a nonbiological, organic polymer material which can be used as a base for standards and controls. Both SeraSub and UriSub are stable at ambient temperature for at least 30 months.

We undertook extensive study to determine the usefulness of SeraSub and UriSub and their stability after the particular analytes were spiked in. The study protocol was as follows.

The standard calibrators were prepared using SeraSub and UriSub. Each bottle was exposed to 45°C, 37°C, ambient temperature (20–30°C), +4°C, and −20°C, for 1 week (to simulate shipping conditions) and then stored at ambient temperature for the entire study period. The analysis was performed periodically. In all cases, the study indicated no differences in the calibration curve. The data are shown in Tables 2 through 5.

Table 2

Kit	Number of samples	Slope	Correlation coefficient
Tobramicin	163	0.98	0.99
Theophylline	126	0.97	0.99
Digoxin	137	0.99	0.96
Gentamicin	138	0.97	0.98
Phenobarbital	137	0.98	0.99

Table 3

Product	Number of samples	Correlation coefficient
THC	178	0.93
Benzodiazepam	93	0.91
Cocaine	133	0.89

Table 4 Theophylline EIA

	Level recovered					
Level spiked (ng/mL)	1 day	7 days	1 month	6 months	12 months	18 months
0	0	0	0	0	0	0
5	5	5	5.3	4.8	5.6	5.3
10	11	11	11.3	9.8	9.7	9.7
15	14.8	15.6	15.8	14.6	15.2	16.0
20	21.3	21.7	21.6	19.2	19.1	19.35
40	39.0	43.0	38.6	37.9	39	40.2

Table 5 Digoxin RIA

	Level recovered					
Level spiked (ng/mL)	1 day	7 days	1 month	6 months	12 months	16 months
0	0	0	0	0	0	0
0.5	0.5	0.6	0.4	0.5	0.5	0.5
1.0	1.0	1.0	1.1	1.01	1.8	1.1
2.0	2.0	2.0	1.9	1.86	1.92	1.98
5.0	5.0	5.3	5.1	5.3	5.0	4.8

In conclusion, SeraSub and UriSub offer the following features:

1. *Consistent properties.* Because the material is not derived from a living host, it can be produced in large quantities with consistent properties, therefore eliminating lot-to-lot variations commonly associated with biological matrices. Benefits to the manufacturer include higher yield, lower cost to produce, less inspection required, and the fact that it takes only 1/2 hour to produce a homogeneous solution. The principal benefit to the laboratorian is continuity of quality control due to consistency between lots.
2. *Stability without lyophilization.* Because it is nonbiological, the material is not susceptible to degradation due to elevated temperature exposure during typical handling. This produces the following benefits.
   For the manufacturer:
      No lyophilization process required
      No special packaging
      Stable in liquid form
      Storable at room temperature
      Shippable at ambient temperature
      Large single lots possible
      Reduced cost of manufacture storage, packaging, quality control, and shipping
   For the laboratorian:
      Ready to use without reconstitution, reducing errors and improving ease of use
      Easy to store
      Reduces lot change problems
3. *No endogenous analytes.* Because the material is nonbiological, it has no endogenous interfering substances. Hence, the matrix has the following attributes:
      No "stripping of unwanted substances" required
      Easy to make "0" standard
      Very low (NSB) nonspecific binding
      Benefits for the manufacturer are reduced quality control efforts, improved yield, and reduced cost. The benefit to the laboratorian is simplified user calculations or adjustments.
4. *Multianalyte controls and standards.* Because the material has no endogenous substances, it is possible to produce multianalyte controls and standards and simple "labeled subsets" as needed without concern about interferences. Therefore, one multianalyte lot will suffice for a complete range of "subset" standards (i.e., B_{12}/folate, thyroid panel, or complete panel as required), and the same matrix can be used for either standard or control without stripping.
   Benefits for the manufacturer are:
      Single stock, various labels
      Interchangeable material for standard and controls
      Ease of manufacture of full line
      Lower cost of manufacture and quality control
   Benefits for the laboratorian are:
      Both subset and complete standards or controls can be the same
      Reduced cost
      Reduced variability
5. *Safety.* Because the material is nonbiological, it does not have the HIV, hepatitis, and other infective possibilities of HSA and other materials. Benefits for the manufacturer are safety, reduced testing and handling cost, and reduced liability. Benefits for the laboratorian are safety and reduced handling cost.

REFERENCES

1. W. L. Cleveland, I. Wood, and B. F. Erlanger. Routine large-scale production of monoclonal antibodies in a protein-free culture medium. *J. Immunol. Methods*, 56:221–234 (1983).
2. M. C. Glassy, J. P. Thanakan, and P. C. Chan. Serum-free media in hybridoma culture and monoclonal antibody production. *Biotech. Bioeng.*, 32:1015–1028 (1988).
3. S. J. Stramer, J. S. Heller, R. W. Coombs, J. V. Parry, D. D. Ho, and J. P . Allain. Markers of HIV infection prior to IgG antibody seropositivity. *J. Am. Med. Assoc.*, 262:64–69 (1989).
4. SeraSub. Creative Scientific Technology, Inc., Great Neck, NY.
5. P. M. Keane, W. H. C. Walker, J. Gauldie, and G. E. Abraham. Thermodynamic aspects of some radioassays. *Clin. Chem.*, 22:70–73 (1976).

99

New Delivery Methods for Intravenous Drugs

Milap C. Nahata *The Ohio State University and Children's Hospital, Columbus, Ohio*

I. INTRODUCTION

The use of intravenous (i.v.) drugs is extremely important in the management of hospitalized patients. It has been estimated that over 65% of patients in hospitals require some form of i.v. infusion for drugs, nutrition, fluids, and electrolytes (1). Intravenous drugs are also being used in the ambulatory setting to contain health care costs; the market for home health care infusion products is growing at a rate of 35% per year. The fastest-growing market, at a rate of about 60% per year, is patient-controlled analgesia (PCA) (1).

An i.v. route for drug administration is used for various reasons, e.g., when a rapid pharmacologic effect is desired or when other routes of administration cannot be used due to the condition of the patient or low bioavailability of drugs. It has been shown that the method of i.v. drug delivery can markedly influence its serum concentration in patients. An enormous body of literature demonstrates a relationship between serum concentrations of certain drugs and their pharmacological activity (efficacy and/or adverse effects). Some of these i.v. drugs may include aminoglycosides, cyclosporine, digoxin, methotrexate, phenytoin, and theophylline (2). The serum concentrations of these drugs are routinely monitored to optimize therapy. Thus, it can be readily understood that such monitoring will be beneficial only when an i.v. drug is administered by an optimal method.

The objective of this report are (a) to describe desirable features of an infusion system; (b) to discuss various types of drug delivery methods; and (c) to underscore the importance of factors affecting drug delivery in both pediatric and adult patients.

I. DESIRABLE FEATURES OF AN INFUSION SYSTEM

The primary goal of using an infusion system is to deliver an i.v. drug most effectively, safely, conveniently, and economically. Several factors should be considered to attain this goal.

A. Accuracy and Precision

An infusion system should deliver each dose of the drug accurately and precisely. The entire dose should be delivered within a desired time to attain efficacy.

B. Diversity and Adaptability

Both the patient population and dosage regimens of drugs can vary markedly in most institutions. Thus, an infusion system should be capable of delivering different drugs at low and high flow rates, using a variety of commercially available syringes or i.v. tubing. It should be programmable to deliver multiple doses and solutions at different rates.

C. Ease and Convenience of Use

The infusion system should be very easy to use by nurses, staff, patients, and their family members. It should be lightweight, particularly for use in ambulatory patients. The procedures for operating and using the device, and making changes in dose requirements, should be easily understood to minimize potential errors.

D. Safety

An infusion device must have an adequate alarm system that will function when (a) the desired flow rate is not maintained; (b) air is introduced into the system; (c) occlusion occurs; (d) a power source is required; or (e) the infusion period is completed. The individuals operating the system should be well trained to minimize any potential for misuse.

E. Cost

There is a tremendous need to acquire and maintain the least expensive infusion devices. Since these are compensable by a fixed amount of money received from the patient or an insurance/governmental agency for the treatment of a particular disease, infusion therapy must be provided at the lowest possible cost without compromising patient care. The cost calculations should incorporate the prices of the device, i.v. set and other accessories, nursing time for drug administration and infusion monitoring, and pharmacy time for drug preparation.

Thus, an optimal infusion device should be accurate, precise, flexible, versatile, easy to use, safe, and economical. It should be usable in a wide variety of patients and for administering a variety of i.v. drugs.

III. TYPES OF DRUG DELIVERY METHODS

The infusion devices can be classified in various ways. One classification divides devices into controllers and pumps. The controllers use only the pressure of gravity, so height of the solution container and the resulting head pressure are important for flow regulation. The infusion pumps, on the other hand, use positive pressure to (a) overcome minor occlusions, which are generally associated with viscous solutions and i.v. systems; (b) overcome back-pressure created by arterial lines; and (c) accurately administer drugs at very low flow rates.

Another method of classifying infusion pumps is based on their mechanism of operation. *Peristaltic systems* uses rotary cams (rotary peristaltic) or fingerlike projections (linear peristaltic). The fluid is propelled when these cams or projections press on the fluid-filled tubing. There is a tendency for the tubing to be stretched, limiting the accuracy to ±5% to 10%. IVAC 530 would be an example of such a system.

Cassette systems operate in two cycles—a filling cycle to fill the chamber with a precise volume and a delivery cycle to infuse this volume. The IMED 900 series for large-volume infusions and the Parker Micropump for ambulatory use are examples of this system.

The *elastomeric reservoir* is the simplest pump mechanism available currently. The elasticity of the balloonlike reservoir exerts a constant pressure, and the drug solution in the balloon is forced through a flow restrictor to control the rate. External factors such as temperature and fluid viscosity may affect delivery, as the internal pressure within the reservoir is low. The Baxter PCA and Intermate Chemomate are examples of this system.

Syringe pumps have become the most widely used devices due to their low cost and the ability to use standard syringes of different sizes. The motor-driven lead screw or a gear (pulsatile) mechanism moves the syringe plunger, and this speed determines the flow rate. One device, the 3M/AVI Medifuse system, uses a spring (nonelectronic) mechanism to apply a constant force on the plungers. These pumps provide an accuracy of ±2% to 5%, but the largest syringe size which can be accommodated is 60 ml.

IV. NEW TYPES OF DRUG DELIVERY DEVICES

The progress in technology has resulted in significant changes in methods for i.v. drug delivery. It is now possible to administer a drug or solution at multiple rates, and multiple solutions at different rates. Similarly, syringe pumps and pumps for PCA therapy are suitable for both institutional and ambulatory settings.

Although infusion pumps generally deliver one solution at one rate, multiple-rate programming, available in newer pumps, eliminates the need for manual changes in rates, e.g., for cycling TPN solutions. Table 1 describes examples of such pumps, which can be programmed for up to nine different rates and volumes for a single solution, or those which automatically ramp up or down to a specified rate without requiring programming of rates to achieve ramping.

Multiple solutions can be infused by one pump at independent rates (Table 2). This can minimize the need for multiple infusion pumps for the same patient. Different systems use one or more i.v. sets, a combination of continuous and intermittent rates, and intermittent flushing of tubing.

Recently, the Multiplex Series 100 Fluid Management System has been marketed by Baxter. This system is designed to deliver up to 10 medications through two separate venous sites, with the capability to print a data base for drug compatibility with other drugs, electrolytes, and i.v. fluids, and patient's vital signs.

Syringe pumps can produce cost savings, since syringes are much less expensive than i.v. bags or volumetric chambers (e.g., Volutrol or Buretrol); however, the need to use a specific i.v. set with the pump can reduce the savings to some degree. Less nursing time for drug administration and pharmacy time for drug preparation may also produce cost savings (Table 3).

Table 1 Infusion Pumps with Multiple-Rate Programming Capabilities[a]

Pump (maker)	Cost	Weight	Rates (mL/h)	Special features
Quest 521 Intelligent Pump (Quest, Kendall McGaw)	$2895	12 lb	1–999	9 rates
Provider One infusion system (Pancretec)	$2995	14 oz	1–400	Up/down rates
Micropump 2100 (Parker)	$2700	70 g	0.1–4.5	PCA
Minimed 504S (Minimed)	$3195	3.7 oz	1–72 μL/h	4 rates/day
Minimed 404SP (Minimed)	$3495	3.7 oz	2–720 μL/h	6 rates/day

[a]Modified from Ref. 3.

Table 2 Infusion Pumps with Multiple-Solultion Programming Capabilities[a]

Pump (maker)	Cost	Weight	Rates (mL/h)	Special features
Gemini PC 2 (IMED)	$4250	18.50 lb	1–999	2 fluids
Life Care 5000 Plum (Abbott)	$3500	13.00 lb	0.1–999	2 fluids
Intelliject (Intellimed)	$4900	3.40 lb	0.3–40.5	4 fluids; PCA
Omni Flow 4000 (Omni Flow)	$4995	13.25 lb	1.4–800	4 fluids; auto air elimination
Minimed III (Minimed)	$3395	13.00 lb	1–800	3 fluids; PCA
Omni Flow Therapist (Omni Flow)				4 fluids; air elimination; prints report

[a]Modified from Ref. 3.

Table 3 Syringe Pumps

Pump (maker)	Cost	Weight	Special features
Bard Mini-Infuser Model 150 XL (Bard)	$695	2.7 lb	Infusion duration depends on syringe size and solution volume
Becton-Dickinson 360 Infuser (Becton Dickinson)	$695	4 lb	Programmed for varying rate or duration
AutoSyringe AS20A (AutoSyringe)	$1795	27 oz	Multiple doses from single syringe
Bard 400 (Bard)	$1195	2.7 lb	Multiple doses from single syringe
Intelliject (Intellimed)			Programmed to deliver multiple doses from four syringes
Med Fusion (Med Fusion)	$1695	5 lb	Suitable for neonates and pediatrics

PCA devices are being used with increasing frequency. A specific dose (e.g., mL or mg), maximum number of doses over a certain period, and a lockout interval can be programmed by a health professional. The devices can also produce records such as cumulative dose, number of doses, and number of attempts by patients (Table 4).

Several devices can be used for both institutional and ambulatory (or home-care) settings. These include the Micropump model 2004, Single-Day Infusor, Infumed 300, CADD-PLUS, Intermate Chemomate, Daymate, Provider Plus 2000, Provider Plus 4000, and Provider 5000. The flow rate can be as low as 0.1 mL/h or as high as 250 mL/h.

A. Devices Under Development

Although much progress has been made in infusion therapy, there are several types of devices under investigation. These comprise implantable pumps (Infusaid, Medtronics SynchroMed) for delivering drugs including antineoplastics, morphine, and heparin; osmotic pressure devices (Alza) for delivering hormones and cisplatin; and pumps generating electromagnetic force (Flowsmart) to drive the drug across the membrane. Membrane technology (MICROS) has also been used for delivering gentamicin by gravity flow (4). Finally, the future of closed-loop systems appears promising, although the biological sensors need to be developed further to provide feedback for automatic adjustment of dosing based on physiological response. An example of this system is the Ames Biostator for insulin administration, which uses an external glucose electrode to monitor blood glucose concentrations, a microcomputer to read

Table 4 PCA Pumps[a]

Pump (maker)	Cost	Weight	Rates (mL/h)	Special features
Harvard PCA (Bard)	\$3550	10 lb	0.1–99.9 (continuous, 150)	Connected to external printer; deliveries; 0.1- to 9.9-mL dosage volume; lockout interval: 3 to 60 min
Lifecare PCA Classic (Abbott)	\$3450	14 lb	1 mL/14 s, continuous	Delivers 0.1 to 5 mL dosage volume; programmable in mg; lockout interval 5 to 99 min
Lifecare PCA Plus 4100 (Abbott)	\$3750	14 lb	1 mL/14 s, continuous	Can be connected to an external printer and has features of Lifecare PCA Classic; lockout interval 5 to 99 min
BD PCA Infuser (Becton Dickinson)	\$3210	6 lb	1–60	Delivers dosage volumes of 0 to 10 mL; programmable in mg
Stratofuse PCA Infuser (Strato Medical, Baxter)	\$3900	3.8 lb	0.1 mL/h to 2.3 mL/min	Internal printer; delivers 0.1 to 5 mL; lockout interval 5 to 60 min
Graseby PCAS (Graseby Medical)	\$3495	6 lb	0–9.9	Programmable in mg; can be connected to an external printer
CADD PCA (Pharmacia Deltec)	\$2995	15 oz	0.05–20	Delivers 0 to 3.95 m L; lockout interval 5 to 100 min; lock levels for programming and lockable pole clamp
Baxter PCA (Baxter)	\$37[b]	1 oz	2 or 5	Delivers up to 0.5 mL; lockout intervals 6 or 15 min
Bard Ambulatory PCA (Bard)	\$3195	11 oz	0–20, continuous	Delivers volumes of 0 to 9.9 mL; lockout interval 3 to 240 min
Bard PCA I (Bard)	\$3195	4.2 oz	0–9.9, continuous	Delivers 0 to 6 mL; lock-out intervals 3 to 60 min; lockable pole clamp

[a]Adapted from Ref. 3.
[b]Cost of infuser and control module.

the concentration and determine the insulin dose, using patient weight and blood glucose results, and a pump to deliver the drug. An IMED 929 has been interfaced with a microprocessor to adjusts nitroprusside dose based on the patient's blood pressure (5). The IVAC 560i is also being evaluated for nitroprusside administration.

B. Clinical Implications

Since the serum concentrations of certain i.v. drugs have been correlated with efficacy and/or adverse effects (2,6,7), an attempt is made to carefully monitor these concentrations in patients. Several studies have demonstrated that factors including flow rate, injection site, diameter of tubing, and type of infusion device can markedly influence the serum concentrations and pharmacokinetics of drugs in neonates, infants, children, and adult patients (8–14).

Although it is not possible to recommend an ideal infusion system for drug delivery in all institutions, it is important to evaluate the specific needs within each institution before selecting one or two devices for use. We have selected one volumetric infusion pump and one syringe pump for delivering drugs to infants and children at our hospital. A multidisciplinary committee including i.v. therapists, nurses, pharmacists, and physicians should evaluate and monitor the use of infusion devices. We have used such an approach and have also developed guidelines for the administration of i.v. drugs, and collection of blood samples for therapeutic drug monitoring. Physicians issuing the directions for drug doses and fluid requirements and requesting serum concentration measurements, pharmacists preparing drugs for infusion and interpreting pharmacokinetic data, nurses administering drugs and recording times for drug infusion, phlebotomists drawing samples of biological fluids, and laboratory personnel measuring drug concentrations in biological fluids must all work together to develop the best scheme for drug delivery and pharmacokinetic monitoring at each institution.

Finally, a multidisciplinary approach should also be used to critically evaluate the cost-benefit analysis of various infusion devices to maximize therapeutic benefits and minimize cost of therapy.

V. SUMMARY

Optimal delivery of intravenous drugs is extremely important in the management of seriously ill patients. Both efficacy and adverse effects have been correlated with serum concentrations of drugs. Although numerous methods are used for drug delivery, important features of the devices include accuracy, predictability, diversity, adaptability, ease of use, safety and cost. Mechanical devices may use peristaltic, cassette, or elastomeric reservoir mechanisms for drug delivery. New technology allows multiple-rate programming, multiple-solution infusions, single or combined functions of a pump and controller, a syringe pump for intermittent infusions, patient-controlled analgesia pumps, and devices for hospital or ambulatory care. Recent developments include site-specific drug delivery, improved delivery of drugs by gravity flow, implantable pumps, osmotic pressure devices, and pumps to deliver drugs based on the patient's physiological status. Intravenous flow rates, injection site within the intravenous tubing, diameter of the tubing, volume of the fluid and drug to be infused, and type of infusion device used are among the factors which have been shown to influence serum concentrations of drugs in patients. A collaborative effort among medical, pharmacy, nursing, and laboratory personnel has been successful in developing and utilizing specific guidelines at our institution to administer each intravenous drug. These guidelines are used to optimize therapeutic drug monitoring in our patients.

REFERENCES

1. J. R. Talley. Infusion technology. *J. Pharm. Pract.*, *1*:128–130 (1988).
2. W. E. Evans, J. J. Schentag, and W. J. Jusko, Eds. *Applied Pharmacokinetics: Principles of Therapeutic Drug Monitoring*, 2nd ed. Applied Therapeutics, Spokane, WA (1986).
3. J. W. Kwan. High technology IV infusion devices. *Am. J. Hosp. Pharm.*, *46*:320–335 (1989).
4. M. C. Nahata, M. Miller, and D. Durrell. Evaluation of a controlled release membrane infusion device for delivery of gentamicin in healthy adults. *Chemotherapy*, *36*:8–12 (1990).
5. D. R. Porter, J. T. B. Moyle, R. J. Lester, et al. Closed loop control of vasoactive drug infusion. *Anesthesia*, *39*:670–677 (1984).
6. R. D. Moore, C. R. Smith, and P. S. Lietman. Association of aminoglycoside plasma levels with therapeutic outcome in gram-negative pneumonia. *Am. J. Med.*, *77*:657–662 (1984).
7. J. L. Bootman, A. J. Wertheimer, D. E. Zaske, et al. Individualizing gentamicin dosage regimens in burn patients with gram negative septicemia: A cost-benefit analysis. *J. Pharm. Sci.*, *68*:267–271 (1979).
8. M. C. Nahata, D. A. Powell, J. P. Glazer, and M. D. Hilty. Effect of intravenous flow rate and injection site on in vitro delivery of chloramphenicol succinate and in vivo kinetics. *J. Pediatr.*, *99*:463 (1981).
9. M. C. Nahata, D. A. Powell, D. E. Durrell, M. A. Miller, and J. P. Clazer. Effect of infusion methods on tobramycin serum concentrations in newborn infants. *J. Pediatr.*, *104*:136–138 (1984).
10. M. C. Nahata. Influence of infusion systems on pharmacokinetic parameters of tobramycin in newborn infants. *Chemotherapy*, *34*:361–366 (1988).
11. J. A. Armitstead and M. C. Nahata. Effect of variables associated with intermittent gentamicin infusions on pharmacokinetic predictions. *Clin. Pharm.*, *2*:153–156 (1983).
12. M. C. Nahata, D. E. Durrell, and M. A. Miller. Tobramycin delivery using a controller with a volumetric chamber and syringe. *Am. J. Hosp. Pharm.*, *43*:2239–2241 (1986).
13. R. A. Pleasants, W. T. Sawyer, D. M. Williams, W. R. McKenna, and J. R. Powell. Effect of four intravenous infusion methods on tobramycin pharmacokinetics. *Clin. Pharm.*, *7*: 374–379 (1988).
14. D. M. Munoz, E. R. Green, M. M. Chrymko, J. M. Mylotte, and J. G. Kitrenos. Delivery of gentamicin by a controlled-release infusion system versus a minibag system. *Am. J. Hosp. Pharm.*, *7*:303–307 (1988).

100

Unified Liquid Chromatographic and Gas Chromatographic Assays for Therapeutic Drug Monitoring and Toxicology

D. A. Svinarov *Medical Academy, Sofia, Bulgaria*

I. INTRODUCTION

Analysis of biological material for bronchodilators, anticonvulsants, antiarrhythmics, antifolates, chloramphenicol, sedatives, analgesics, and tranquilizers accounts for the bulk of therapeutic drug monitoring (TDM) and toxicology (TOX) workload in many laboratories. Both immunoassays and chromatographic techniques, with their typical benefits and shortcomings, have found a place in the analytical arena for these drug groups (1–4). Gas chromatography (GC) is more broadly applied in TOX screen programs, and liquid chromatography (LC) is the main analytical alternative to immunoassays for TDM. The principle advantage of chromatographic techniques is that they simultaneously resolve parent compounds, their metabolites, and chemically and therapeutically unrelated drugs. They have several common disadvantages, including laborious and time-consuming sample preparation, different pretreatment prior to GC and LC, and separate chromatographic conditions for the different drug groups. Although important efforts have been made recently to circumvent these shortcomings (5–8), the concept of unified GC and LC assays for the routine TDM/TOX laboratory still remains to be developed. It is highly desirable to use identical, fast, and easy sample pretreatment and a versatile chromatographic system of similar working conditions for most LC and GC analyses. This goal will always stimulate the chromatographer to introduce a better analytical basis for more clinically relevant TDM/TOX service (9).

Recently we described a LC method for the simultaneous determination of some bronchodilators, anticonvulsants, chloramphenicol, and hypnotic agents, with a micro Chromosorb P column used for sample preparation, which can be applied in TDM and TOX mode (8). This method has been slightly modified and extended to a unified analytical system for determination of some bronchodilators, anticonvulsants, antiarrhythmics, antifolates, analgesics, tranquilizers, and chloramphenicol in biological material. The system combines identifical sample preparation with

Chromosorb P microcolumns, and three chromatographic modes. Chromatographic mode I is the above-mentioned LC technique, modified into three methods: method TDM-1 for fast determination of theophylline and caffeine; method TDM-2 for theophylline and caffeine, nine anticonvulsants and metabolites, and chloramphenicol; and the TOX-1 screening method for 16 TOX analytes. Chromatographic mode II [C-8 analytical column and acetonitrile (ACN)/phosphate buffer as mobile phase] consists of four methods: method TDM-3 for the determination of procainamide and N-acetylprocainamide; method TDM-4 for the determination of trimethoprim, sulfamethoxazole, and N-acetylsulfamethoxazole; method TDM-5 for the determination of lidocaine, mexiletine, disopyramide, and quinidine; and method TOX-2 for the screening of salicylate, acetaminophen, chlordiazepoxide, nitrazepam, clonazepam, medazepam, and diazepam. Chromatographic mode III is a GC screening for salicylate, meprobamate, barbital, methyprylon, butabarbital, amobarbital, pentobarbital, secobarbital, caffeine, glutethimide, mephobarbital, phenobarbital, methaqualone, phenytoin, diazepam, nitrazepam, and clonazepam, separated on a megabore HP-5 column with flame ionization detection.

II. MATERIALS AND METHODS

A. Equipment

Chromatographic mode I uses a modular liquid chromatograph consisting of a series M-510 pump, WISP-710B automatic injector, Lambda Max 481 spectrophotometer, and 730-Data Module (all from Millipore-Waters, Milford, Massachusetts), a 150-mm × 4.6-mm (I.D.) reversed-phase column, packed with Supelcosil-LC-8, 5-μm particle size (Supelco, Gland, Switzerland), and a mobile phase composed of 400 mL of ACN and 1600 mL of double-distilled water.

Chromatographic mode II uses a modular liquid chromatograph consisting of a series M-45 pump, a U6K manual injector, and M-441 absorbance detector and a single-pen recorder (all from Millipore-Waters, Milford, Massachusetts), a 50-mm × 4.6-mm (I.D.) reversed-phase column packed with Supelcosil-LC-8, 5-μm particle size (Supelco, Gland, Switzerland). The mobile phase for this chromatographic mode consisted of ACN (variable proportions)/0.02 M H_3PO_4, pH 3.3, plus 0.01% dibutylamine: 8% ACN to resolve procainamide and N-acetylprocainamide (TDM-3); 11% ACN for the determination of trimethoprim, sulfamethoxazole, and N-acetylsulfamethoxazole (TDM-4); 15% ACN for lidocaine, mexiletine, disopyramide, and quinidine (TDM-5); 30% for the screening of salicylate, acetaminophen, clonazepam, nitrazepam, diazepam, medazepam, and chlordiazepoxide.

Chromatographic mode III uses a HP 5890/series II gas chromatograph equipped with 5% phenylmethylsilicone 10 m × 0.53 mm × 2.65 μ (film thickness) analytical column, flame ionization detector, and integrator HP-3396 (Hewlett Packard, Avondale, Pennsylvania).

B. Reagents and Standards

HPLC-grade ACN, methanol, dichlormethane, isopropanol, and Chromosorb P/NAW, 8-100 mesh (diatomite support), were purchased from Merck (Darmstadt, FRG). All other chemicals were analytical grade, from standard chemical suppliers.

Theophylline, caffeine, isobutylmethylxanthine, phenobarbital, phenytoin, procainamide, N-acetylprocainamide, N-propionylprocainamide, lidocaine, disopyramide, quinidine, and methylphenylphenylhydantoin were purchased from Sigma Chemical Co. (St. Louis, Missouri); butabarbital, pentobarbital, amobarbital, mephobarbital, secobarbital, methyprylon, glutethimide, methaqualone, mexiletine, methylmexiletine, acetaminophen, chlordiazepoxide, meprobamate, trimethoprim, sulfamethoxazole, N-acetylsulfamethoxazole, and sulfisoxazole were gifts from C. E. Pippenger (Cleveland Clinic Foundation, Cleveland, Ohio); barbital, medazepam, clonazepam, diazepam, nitrazepam, and acetylsalicylic acid were obtained from the National Drug Control Institute, Sofia, Bulgaria.

Stock standards (1.0 to 10.0 g/L in methanol) were used to prepare combined mixtures in methanol, corresponding to the respective TDM and TOX methods of the system described. These combined standard mixtures were added to drug-free sera to prepare the working standards and controls. Internal standard working solutions were as follows: TDM-1, isobutylmethylxanthine, 20 mg/L in 0.2 M HCl; TDM-2, TOX-1 as published (8); TDM-3, N-propionylprocainamide, 20 mg/L in 0.2 M NaOH; TDM-4, sulfisoxazole, 40 mg/L in 0.2 M HCl; TDM-5, methylmexiletine, 25 mg/L in 0.2 M NaOH; TOX-2, methylphenylphenylhydantoin, 250 mg/L in 0.2 M HCl; TOX-3, isobutylmethylxanthine, 150 mg/L in 1.0 M HCl.

C. Microcolumn for Sample Preparation

The Chromosorb P microcolumn has been described previously (8). Briefly, plug a 1-mL plastic pipette tip with a small amount of glass wool or cotton, fill the low conical part with about 150 mg of Chromosorb P/NAW, and place the prepared column on a strand, above an appropriate-size glass collection tube.

D. Procedure

Apply 50 µL (TDM-1, TDM-2, TOX-1), 100 µL (TDM-3, TDM-4, TDM-5, TOX-2), or 300 µL of plasma to the microcolumn for sample preparation, then add 50 µL of respective internal standard. After about 2 min, apply 1.0 mL (TDM-1, TDM-2, TOX-1), 2 × 1.0 mL (TDM-3, TDM-4, TDM-5, TOX-2) of dichlormethane/isopropanol (9/1 by vol), or 3 × 1.0 mL of dichlormethane (TOX-3), collect the eluate, and evaporate it under nitrogen in a water bath at 30°C. Reconstitute the residue with 50 to 100 µL of the respective mobile phase (LC), or with 30 µL of methanol (GC). Inject 1 to 2 µL (GC) or 10 to 20 µL (LC) and chromatograph as follows.

LC working conditions: TDM-1, flow rate 3.0 mL/min (pressure 16 MPa), detector 273 nm, sensitivity 0.02 A full scale; TDM-2 and TOX-1, as described previously (8); TDM-3, flow rate 1.5 mL/min (pressure 8 to 10 MPa), detector 254 nm, sensitivity 0.05 A full scale; TDM-4, flow rate 3.0 mL/min (pressure 10 to 14 MPa), detector 254 nm, sensitivity 0.05 (0 to 3 min) to 0.2 A full scale (3 to 9 min); TDM-5, flow rate 3.0 mL/min (pressure 9 to 11 MPa), detector 214 nm, sensitivity 0.05 A full scale; TOX-2, flow rate 1.5 mL/min (pressure 7 to 10 MPa), detector 254 nm, sensitivity 0.1 A full scale. All LC assays are performed at ambient temperature.

GC working conditions: injector temperature 270°C, detector temperature 300°C; oven temperature profile, initial 150°C, initial hold time 4 min, linear program rate 13°C, final temperature 250°C, final hold time 3 min; carrier (argon) flow 15 mL/min, hydrogen flow 30 mL/min, air flow 300 mL/min.

For all TDM methods, construct a calibration curve in "internal standard" mode and calculate unknowns from peak height or peak area ratios. For TOX methods, identify unknowns from respective relative retention times (retention time of a compound divided by the retention time of the internal standard), report as "positive" or "negative," and confirm the results by an alternative method (GC versus LC). Quantitation of TOX analytes is performed (if required), after confirmation, utilizing "internal standard" calculation.

III. RESULTS AND DISCUSSION

Figure 1 is a schematic illustration of the analytical stages of our unified approach for TDM and TOX assays. Sample preparation is performed in exactly the same way for each LC and GC method included in the system. Separate internal standard solutions, and specific volumes of sample or extractant, are the only differences at this stage of work. Unification concerns the extractant as well: Dichlormethane/isopropanol (9/1 by volume) is used in all LC methods, and dichlormethane alone is the extractant for the GC analysis. Only three pipettings, and a fast evaporation step, are required prior to chromatography. The sample is extracted, purified, and filtered as it passes through the Chromosorb P column bed. Pretreatment of a single specimen takes about 10 min; a batch of 20 samples can be prepared for chromatography in 30 min. Unification at the stage of chromatographic separation is achieved by the application of an identical stationary phase for all LC assays, usage of ACN as the organic part of the mobile phases, isocratic runs, and close chromatographic conditions for all TDM and TOX assays performed by LC. This unified approach provided considerable versatility, simplicity, and convenience for our laboratory.

Figure 2 shows a chromatogram for serum standard of the TDM-1 analyses. Retention times and capacity factors are listed in Table 1. Theophylline and internal standard peaks are resolved in less than 3 min, which results in an output of 20 specimens per hour. Despite the possibility of measuring theophylline and caffeine by the TDM-2 method, we introduced the above modification with the aim of performing bronchodilator analyses faster and more simply; with TDM-1 we often report up to 40 results in an hour, thus challenging the output of some analyzers based on immunological principles. The sample volume for theophylline and caffeine assay can easily be scaled down to 20 μL, which greatly facilitates TDM in pediatric intensive-care units.

TDM-2 and TOX-1 assays are performed in exactly the same way as published, with the exception of a different extractant: dichlormethane/isopropanol (9/1 by volume) instead of chloroform/isopropanol (6/1 by volume). This change of extractant is made for the sake of unity with chromatographic mode II, and it does not affect the performance characteristics or chromatogram appearance of TDM-2 and TOX-2 analyses.

Figure 3 presents typical chromatograms for the methods included in chromatographic mode II. Retention times and capacity factors are listed in Table 1. TDM-3 separation is complete within 3 min, and the run time of 8 to 10 min for TDM-4, TDM-5, and TOX-2 assays allows for analysis of 6 to 8 specimens per hour. Use of various proportions of ACN and various flow rates of the mobile phases is the main difference among the methods comprising chromatographic mode II. Isocratic separation at specific wavelengths is preferred to an eventual gradient run, combining all methods into one, because, with the instrumentation used in our laboratory, it provided faster and more simple analyses for batches of samples, and optimal UV detection for each drug group.

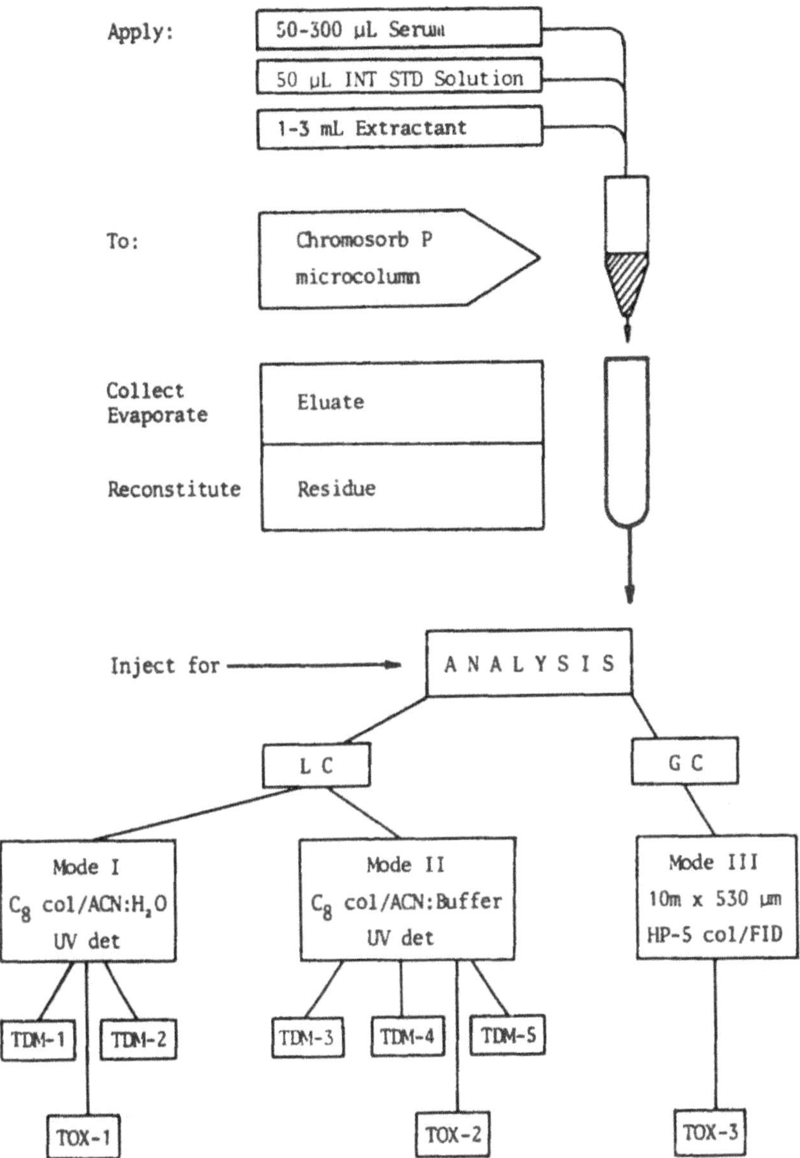

Figure 1 Flow diagram of the analytical stages by this system.

Figure 4 illustrates a chromatogram of a serum-based standard, containing all TOX analytes, screened by chromatographic mode III (TOX-3). Retention times and relative retention times of respective peaks are presented in Table 2. "X" peaks are unidentified endogenous compounds, which did not interfere with drugs, measured by this assay. The run is complete within 14 min, and a single-step oven temperature program was sufficient to separate all peaks of interest.

No significant interference was observed, from endogenous compounds and eventually co-administered drugs, for all TDM and TOX methods at the chromatographic

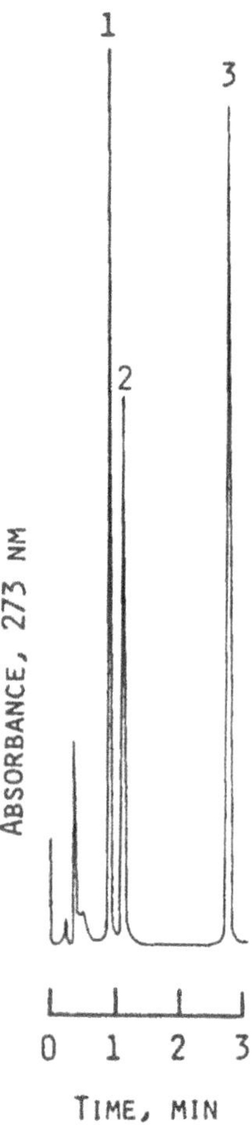

Figure 2 Chromatogram of a serum calibration standard for TDM-1, chromatographic mode I: theophylline 15 mg/L (1), caffeine 11.25 mg/L (2), isobutylmethylxanthine 20 mg/L (3).

conditions used, with three exceptions: the above-mentioned endogenous peaks for TOX-3(a), the co-elution of pentobarbital and mephobarbital in TOX-1(b), and the overlap between theophylline and phenobarbital in TOX-3(c). In our TOX-3 method theophylline was omitted because it gave a very weak FID response, with significant peak tailing, in contrast to the sharp and gaussian phenobarbital peak. In case of analytical uncertainty the result is not reported before its confirmation by an alternative procedure (GC versus LC, or chromatography versus immunological technique).

All analytes gave linear responses with the concentration studied: theophilline 0 to 60 mg/L, caffeine 0 to 45 mg/L, procainamide 0 to 45 mg/L, N-acethylprocainamide 0 to 45 mg/L, trimetroprim 0 to 16 mg/L, sulfamethoxazole 0 to 300 mg/L, N-acetylsulfamethoxazole 0 to 150 mg/L, lidocaine 0 to 20 mg/L, mexiletine 0 to 10 mg/L, disopyramide 0 to 20 mg/L, quinidine 0 to 20 mg/L, acetaminophen 0 to 300 mg/L, salicylate 0 to 500 mg/L, chlordiazepoxide 0 to 30 mg/L, nitrazepam 0 to 30 mg/L,

Table 1 Retention Times (R_t), Capacity (K′), and Performance Characteristics for the Drugs and Metabolites by All LC Assays of This System[a]

Drug	R_t (min)	K′	Between-run CV (%) (n = 20)	Accuracy (%)	Extraction efficiency (%)	Detection limit (mg/L)
Theophylline	0.96	0.92	<5.7	±3.4	90	0.05
Caffeine	1.19	1.38	<5.1	±5.1	93	0.05
Isobutylmethylxanthine[b]	2.71	4.42				
Procainamide	0.59	1.26	<5.8	±4.9	93	0.10
N-acetylprocainamide	1.06	3.08	<4.9	±4.8	94	0.10
N-propionylprocainamide[b]	2.39	7.85				
Trimethoprim	2.21	4.02	<5.5	±4.7	97	0.20
Sulfamethoxazole	3.94	7.95	<3.3	±2.8	101	0.05
Sulfisoxazole[b]	5.60	11.73				
N-acetylsulfamethoxazole	8.06	17.32	<2.9	±4.2	99	0.05
Lidocaine	1.26	2.15	<7.7	±5.9	88	0.10
Mexiletine	2.80	6.00	<3.8	±1.6	95	0.20
Disopyramide	3.58	7.95	<5.8	±4.4	89	0.50
Quinidine	4.63	10.58	<8.1	±6.8	92	0.20
Methylmexiletine[b]	7.21	17.02				
Acetaminophen	0.60	0.30	<7.1	±7.7	92	0.50
Salicylate	1.14	2.96	<5.9	±6.9	102	2.00
Chlordiazepoxide	2.34	4.09	<6.9	±4.6	98	0.50
Nitrazepam	3.34	6.26	<4.1	±3.9	94	0.10
Clonazepam	4.10	7.91	<3.9	±4.4	93	0.20
MPPH[b]	4.92	9.70				
Medazepam	7.00	14.22	<5.8	±7.5	90	0.50
Diazepam	9.20	19.00	<4.2	±5.8	97	0.50

[a]TDM-2 and TOX-1 analytes not included.
[b]Internal standard.

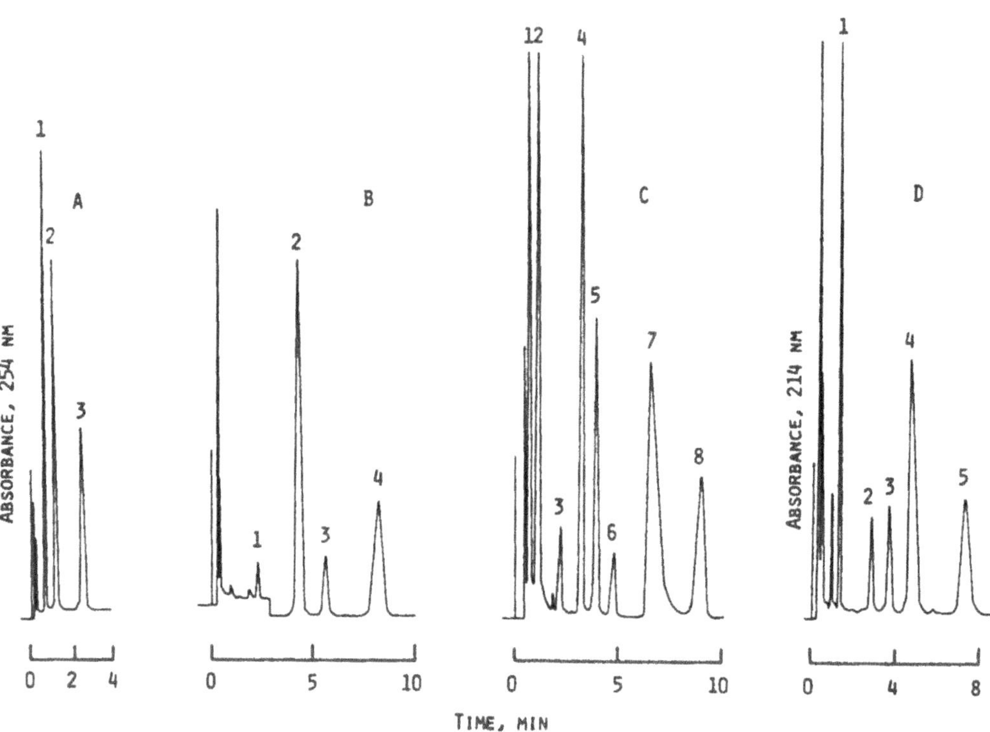

Figure 3 Chromatograms of serum calibration standards, derived according to chromatographic mode II: A, TDM-3, procainamide 10 mg/L (1), N-acetylprocainamide 10 mg/L (2), N-propionylprocainamide 10 mg/L (3); B, TDM-4, trimethoprim 1.0 mg/L (1), sulfamethoxazole 50 mg/L (2), sulfisoxazole 20 mg/L (3), N-acetylsulfamethoxazole 25 mg/L (4); C, TOX-2, acetaminophen 50 mg/L (1), salicylate 300 mg/L (2), chlordiazepoxide 10 mg/L (3), nitazepam 10 mg/L (4), clonazepam 10 mg/L (5), methylphenylphenylhydantoin 250 mg/L (6), medazepam 10 mg/L (7), diazepam 10 mg/L (8); D, TDM-5, lidocaine 5.0 mg/L (1), mexiletine 2.5 mg/L (2), disopyramide 5.0 mg/L (3), quinidine 5.0 mg/L (4), methylmexiletine 12.5 mg/L (5).

clonazepam 0 to 30 mg/L, medazepam 0 to 30 mg/L, diazepam 0 to 30 mg/L, meprobamate 0 to 200 mg/L, barbital 0 to 75 mg/L, butabarbital 0 to 75 mg/L, amobarbital 0 to 50 mg/L, pentobarbital 0 to 50 mg/L, secobarbital 0 to 30 mg/L, glutethimide 0 to 10 mg/L, mephobarbital 0 to 75 mg/L, phenobarbital 0 to 100 mg/L, methaqualone 0 to 30 mg/L, phenytoin 0 to 100 mg/L. The correlation coefficients for plots of peak area (or height) ratios of drugs to internal standard versus expected concentrations were all >0.99. The slopes of these plots ranged from 0.903 to 1.19; the intersepts varied between –0.09 and +0.08.

Tables 1 and 2 list performance characteristics of the system. The extraction efficiency was calculated to measure the loss during sample pretreatment as follows: peak area (or height) was divided by the peak area (or height) of the same concentration of the drug in mobile phase (methanol, GC), directly injected for chromatography. High extraction efficiency is one of the benefits of our microcolumn technology

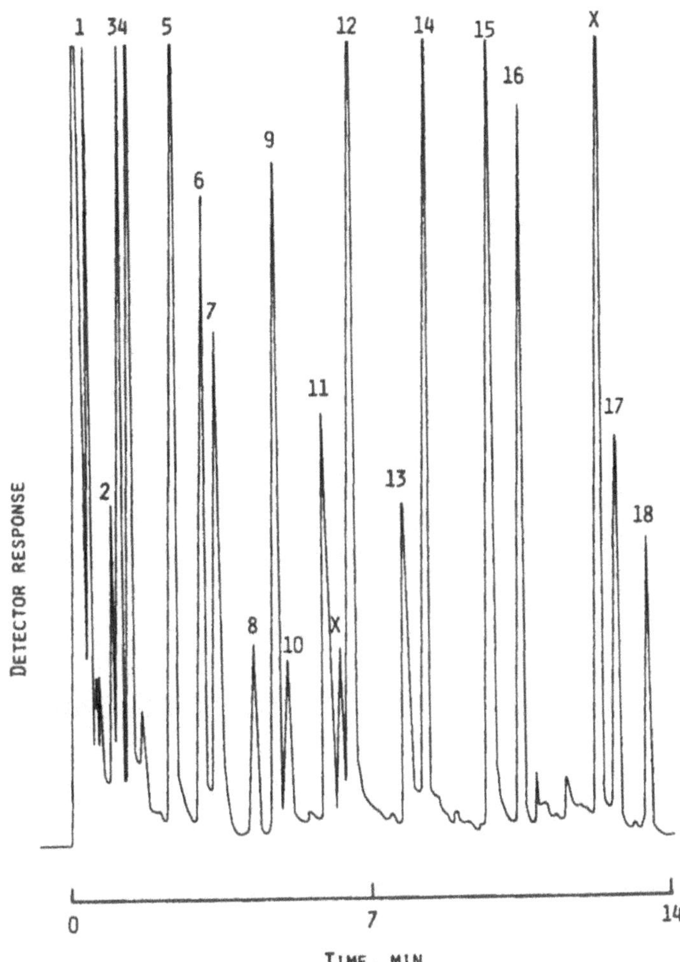

Figure 4 Chromatogram of a serum calibration standard for TOX-3, chromatographic mode III: salicylate 250 mg/L (1), meprobamate 30 mg/L (2), barbital 30 mg/L (3), methyprylon mg/L (4), butabarbital 30 mg/L (5), amobarbital 20 mg/L (6), pentobarbital 20 mg/L (7), secobarbital 10 mg/L (8), caffeine 50 mg/L (9), glutethimide 5.0 mg/L (10), mephobarbital 30 mg/L (11), phenobarbital 60 mg/L (12), isobutylmethylxanthine 25 mg/L (13), methaqualone 20 mg/L (14), phenytoin 30 mg/L (15), diazepam 15 mg/L (16), nitrazepam 15 mg/L (17), clonazepam 15 mg/L (18); X, unidentified endogenous peaks.

for sample preparation. The accuracy was calculated according to the formula [(expected concentration—found concentration)/expected concentration] × 100, and found to be between ±8.1%. The precision data were obtained by replicate measurements of drug-supplemented sera and commercial therapeutic drug controls at two or three concentration levels. The lower limits of detection were calculated at concentrations that gave a signal-to-noise ratio of 5/1.

The system has been used successfully in our laboratory for more than 1 year. It is applicable to the analysis of serum, plasma, plasma modification of sample pretreat-

Table 2 Retention Times (R_t), Relative Retention Times (RR_t), and Performance Characteristics for the Drugs by the GC Assay of This System

Drug	R_t (min)	RR_t	Between-run CV (%) (n = 20)	Accuracy (%)	Extraction efficiency (%)	Detection limit (mg/L)
Salicylate	0.45	0.055	<7.9	±8.1	89	5.0
Meprobamate	0.92	0.114	<7.8	±6.2	97	5.0
Barbital	1.12	0.139	<6.2	±5.9	92	2.0
Methyprylon	1.39	0.172	<4.9	±6.3	99	1.0
Butabarbital	2.39	0.297	<6.2	±7.1	91	1.0
Amobarbital	3.14	0.390	<8.1	±5.2	104	1.5
Pentobarbital	3.45	0.429	<6.9	±6.1	100	1.5
Secobarbital	4.33	0.538	<6.6	±6.6	88	2.0
Caffeine	4.78	0.594	<4.4	±4.2	95	1.5
Glutethimide	5.13	0.638	<3.8	±4.4	95	1.0
Mephobarbital	5.87	0.730	<6.1	±3.9	98	2.0
Phenobarbital	6.58	0.818	<5.3	±4.1	97	3.0
Isobutylmethylxanthine[a]	8.04					
Methaqualone	8.28	1.029	<2.9	±6.5	89	1.0
Phenytoin	9.77	1.215	<4.4	±6.3	94	1.5
Diazepam	10.45	1.299	<5.2	±4.1	96	0.5
Nitrazepam	12.75	1.585	<6.3	±5.1	92	1.0
Clonazepam	13.50	1.679	<4.9	±6.4	93	1.5

[a] Internal standard.

ment or chromatography. The overall performance characteristics well meet the requirements for routine TDM and TOX assays.

We conclude that the approach for development of unified LC and GC assays for TDM and toxicology provides several advantages and benefits: The ability to determine most of the main anticonvulsants, antiarrhythmics, bronchodilators, antifolates, sedatives, analgesics, tranquilizers, and chloramphenicol in an integrated chromatographic system is achieved; identical sample preparation with Chromosorb P microcolumn is simple, cost-effective, and provides rapid and excellent purification; application of the same chromatographic mode both for TDM and TOX analysis allows for routine as well as stat samples be assayed with simplicity and convenience.

IV. SUMMARY

Identical sample preparation and three chromatographic modes were combined in an analytical system for the measurement of some bronchodilators, anticonvulsants, antiarrhythmics, antifolates, chloramphenicol, sedatives, analgesics, and tranquilizers in biological material. Aliquots of 50 to 300 μL of plasma or other biologic fluid, and 50 μL of acidic or basic internal standard solution, are applied to a micro Chromosorb P column. Drugs are then eluted with 1 to 3 mL of extractant, evaporated to dryness under nitrogen, reconstituted with 30 to 100 μL of the respective mobile phase (LC), or with methanol (GC), and 2 to 20 μL are injected for analysis. Chromatographic mode I is a slight modification of a previously published LC technique (8). Chromatographic mode II (LC) utilizes a Supelco C-8 5-μm 5-SM column and a mobile phase of acetonitrile (ACN, variable proportions)/0.02 M H_3PO_4, pH 3.3, plus 0.01% dibutylamine and UV detection at 254 nm (30% ACN for the TOX screen of salicylate, acetaminophen, and 5-benzodiazepines; 11% ACN to separate trimethoprim, sulfamethoxazole, and N-acetylsulfamethoxazole; 8% ACN for procainamide and N-acetylprocainamide) or UV detection at 214 nm (15% ACN for the quantitation of lidocaine, mexiletine, disopyramide, and quinidine). Chromatographic mode III is a GC screening for salicylate, caffeine, four tranquilizers, and 11 sedatives, separated on a 10-m × 0.53-mm megabore HP-5 column with flame ionization detection. Lower limits of detection varied from 0.05 to 0.5 mg/L (TDM) to 0.1 to 5.0 mg/L (TOX); analytical recovery was from 88% to 104%; CVs were <8.1 between runs.

REFERENCES

1. J. A. F. De Silva. Analytical strategies in therapeutic monitoring of drugs in biological fluids. *J. Chromatogr.*, *340*:3–30 (1985).
2. D. W. Holt. Therapeutic monitoring of antiarrhythmic drugs. In *Therapeutic Drug Monitoring* (B. Widdop, ed.), Churchill Livingstone, London, pp. 154–181 (1985).
3. M. Ahnoff, M. Ervik, P. O. Langerstrom, B. A. Persson, and J. Vessman. Drug level monitoring: Cardiovascular drugs. *J. Chromatogr.*, *340*:73–138 (1985).
4. R. N. Gupta. Drug level monitoring: Sedative hypnotics. *J. Chromatogr.*, *340*:139–172 (1985).
5. R. Meatherall and D. Ford. Isocratic liquid chromatographic determination of theophylline, acetaminophen, chloramphenicol, caffeine, anticonvulsants and barbiturates in serum. *Ther. Drug. Monitor.*, *10*:101–115 (1988).
6. R. Meatherall. High-performance liquid chromatographic determination of trimethoprim in serum. *Ther. Drug Monitor.*, *11*:79–83 (1989).

7. S. C. Laizure, C. L. Holden, and R. C. Stevens. Ion-paired high-performance liquid chromatographic separation of trimethoprim, sulfamethoxazole and N^4-acetylsulfamethoxazole with solid phase extraction. *J. Chromatogr.*, *528*:235–242 (1990).
8. D. A. Svinarov and D. C. Dotchev. Simultaneous liquid-chromatographic determination of some bronchodilators, anticonvulsants, chloramphenicol, and hypnotic agents, with Chromosorb P columns used for sample preparation. *Clin. Chem.*, *35*:1615–1618 (1989).
9. C. E. Pippenger. Personal communication.

101

Biological Indicators of Environmental Cancer Risk

B. Sinués, M. L. Bernal, M. A. Sáenz, J. Lanuza, and M. Bartolomé *University of Zaragoza, Zaragoza, Spain*

I. INTRODUCTION

In the following paragraphs, general aspects about biological indicators of environmental cancer risk are considered. The most commonly used methods and their significance are also reviewed in relation to some studies from our group. When drugs are suspected, therapeutic drug monitoring and toxicological analysis help greatly to assess the risk determinants more accurately.

The theory about the environmental origin of cancer derives from the fact that migrants from one area to another tend to adopt the cancer incidence pattern of the country to which they migrate. Keeping in mind that over 80% of human cancers are environmentally induced by physical, chemical, and biological agents, they can, in principle, be preventable.

The initial and essential step in induced carcinogenesis is assumed to be the initiation, which seems to be a mutational event, derived from the reaction between genotoxic compounds with nucleophiles in DNA. Promotion and progression follow, leading to development of cancer.

The extent of DNA damage depends on the dose of xenobiotic. In addition, there are two different kinds of processes that modulate the effect (Figure 1). The initiator concentration at DNA depends on absorption, excretion, and, in particular, metabolism, in which activation or inactivation can be produced. On the other hand, DNA repair rates and the type of repair will determine the final DNA damage. Therefore, enzyme proteins are importantly involved in the initiation step. These proteins are genetically codified, but the environment modifies protein synthesis by inducing or inhibiting it. So humans represent a unique biological entity in which environmental and genetic factors interact to produce the pharmacological response, both therapeutic and toxic.

In carcinogenesis, the main consequences are (a) the interindividual difference in susceptibility for cancer development, and (b) the possibility of genetically determined individuals who are at higher risk.

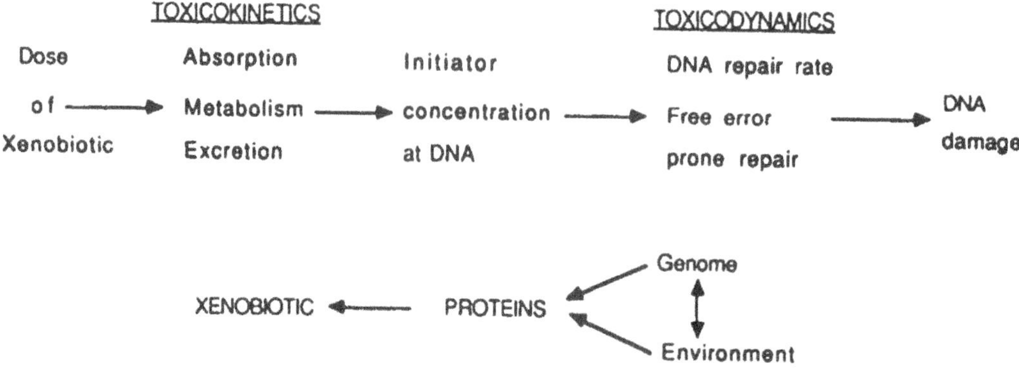

Figure 1 Influence of proteins on processes affecting DNA lesions.

II. BIOLOGICAL MONITORING OF EXPOSURE

The ultimate aim of biomonitoring in humans exposed to carcinogens should be the best estimation of risk. Biological monitoring can be used (a) for improving the evaluation of the external dose, (b) for the detection of reversible biological effects, and thus prevention of irreversible ones, (c) to identify the cell events occurring in the genesis of neoplasia, (d) to delimit groups at risk, and (e) to identify the mechanism of toxic action.

Epidemiological studies give important information about the response of humans exposed, but only detect fairly large increases in relative risk. Moreover, they often cannot provide direct proof of a cause-and-effect relationship. Therefore, exposure biomonitoring is necessary without excluding external dosimetry and epidemiology.

It is necessary to integrate the results of biological monitoring with the epidemiological data as well as with that emerging from experimental carcinogenicity. Biological monitoring of exposure can overcome many of the limitations of these kinds of studies for the assessment of human risk.

To establish the association between biological monitoring and exposure dose is of particular interest. This depends not only on the external dose (e.g., ppm in the air, number of cigarettes per day, radiation dose, etc.), but also on the duration of the exposure. In the case of drug intake, therapeutic drug monitoring and toxicological analysis can help determine the imputability and estimate more accurately the magnitude of human risk.

The methods currently available for biomonitoring of exposure can measure (a) early biological effects, (b) internal dose, or (c) biologically effective dose (19).

Early biological effects are evaluated by short-term cytogenetic tests on cultured blood lymphocytes, such as chromosome aberrations (CA), sister chromatid exchange frequency (SCE), and occurrence of micronuclei (MN). *Internal dosimetry* can be performed mainly by measuring the concentration of agents or their metabolites in body fluids or by the analysis of the mutagenicity in excreta. Highly specific methods, measuring *biologically effective dose*, such as adults to DNA, RNA, and proteins, are suitable only when a single and known agent is involved.

A. Early Biological Effects

Chromosome aberration analysis is the most extensively employed method to assess the early effects of exposures to genotoxic agents. This determination is a direct test to measure gross changes occurring in DNA. Structural aberrations in chromosomes result from breakage and rearrangement of whole chromosomes into abnormal forms. They are induced most efficiently by substances which break the backbone of DNA or distort the DNA helix.

Chromosome aberrations can be divided into: chromosome-type and chromatid-type aberrations. It has been observed that chromosome-type aberrations are more frequently induced by physical agents (e.g., dicentric rings, etc.), while the chromatid type are more effectively produced by chemicals. The latter is due to the fact that the DNA lesions have to pass through an S phase in order to be transformed into aberrations. In the induction of SCE, these S-dependent agents are also extremely efficient (4).

There may exist many data suggesting that increased CA indicate exposure to factors which increase the risk of cancer illness. Increased occurrence of chromosome aberrations is associated with (a) some hereditary diseases with higher incidence of cancer (ataxia telangiectasia, Fanconi's anemia, Bloom's syndrome), (b) populations exposed to physical and chemical genotoxic agents, and (c) chemical agents that have been evaluated as carcinogens in humans. In addition, many types of cancers are associated with specific or nonspecific chromosomal aberrations (21).

Sister chromatid exchanges are the cytological manifestation of breakage and rejoining at homologous sites on the two chromatids of a single chromosome. On the addition of 5-bromodeoxyuridine, we can observe SCE in any cell which has completed two replication cycles. Some compounds, which are capable of forming covalent adducts to the DNA metabolism or repair, give rise to the production of SCE (5). Ever though the molecular nature of SCE is unknown, DNA synthesis may be involved, as this process is absolutely necessary for the formulation of SCE (14). It has been demonstrated on a population basis that there is an increase in SCE due to the exposure of known mutagens and carcinogens.

Micronuclei are formed during cell division from acentric fragments or whole chromosomes that are left behind during anaphase movements, as a result of the previous existence of chromosome aberrations. They are easily detected in interphase cells as free intracytoplasmic bodies in cytokinesis-blocked cells by cytochalasin B (6). Micronuclei occurrence has been used as an effective biological dosimeter in populations exposed to genotoxic physical and chemical agents (1,17).

To identify and eliminate factors that may confound the cytogenetic analysis in biomonitoring studies, it is important to obtain relevant information about age, radiation exposure, smoking history, viral infection, personal and family history, nutrition, etc. It is important also to have a matched control population, in order to establish any conclusion. In any case, the ideal situation would be to do prospective studies whenever possible.

Table 1 presents data from our studies on populations exposed to genotoxic agents, either suspected or known, and those on patients with chronic illnesses associated with an increased incidence of cancer. Reference is given to the application, size, and results.

As an example of prospective cytogenetic study, we shall now comment on some results of the biological monitoring of exposure to ionizing radiation plus contrast

Table 1 Cytogenetic Studies on Human Populations Exposed to Suspect or Known Genotoxic Agents (Endpoints: SCE, AC, MN)

	Application	n	Data[a]
1. Physical agents			
Ionizing radiation	Abdominal scanning	20	(–)
Ionizing radiation plus contrast agents	Excretory urography	20	+
Diatrizoate			
Ioxaglate		20	+
2. Chemical agents			
Personal habits			
Tobacco smoking	Heavy smokers	28	+
	Mild smokers	25	(–)
Heroin	Heroin abusers	18	(+)
Drugs			
Phenytoin	Epileptics	54	(–)
Carbamazepine	Epileptics	23	(+)
	Trigeminal neuralgia	10	(+)
Halothane	Gynecological surgery	25	(–)
Theophylline	Asthmatics	30	(+)
Insulin	Type I diabetics	31	(–)
Occupational exposure			
Vinyl chloride	Plastics	52	+
Aromatic amines	Textiles	56	(+)
Polycyclic hydrocarbons	Asphalts	51	+
3. Chronic illnesses			
Renal failure	Anuria	37	+
Diabetes	Type I diabetics	33	(+)
4. Biological agents			
Human immunodeficiency virus	Men at risk for AIDS	31	+
Hepatitis B virus	Seropositives	21	(+)

[a]+: Positive data in more than one test; (+): Positive data in one test (–): Negative data.

agents used in urography (11). The human material consisted of two groups of 20 patients undergoing excretory urography. Group I was treated with diatrizoate as contrast agent and group II with ioxaglate. Three samples from each patient were taken: before exploration (A), immediately after exploration (B), and 1 week after urography (C).

We could see a similar effect on chromosome aberrations and on SCE (Figure 2). After exploration, SCE increased in both groups, being more remarkable in the diatrizoate group, in which SCE frequency continued to be higher 7 days later with respect to baseline values, while in the ioxaglate group there was no increase at this time. Radiation dose was similar in all patients, with no statistically significant differences between groups. On the other hand, the dose of contrast medium was the same for each patient (in terms of iodine, it was 0.32 gr/kg of weight).

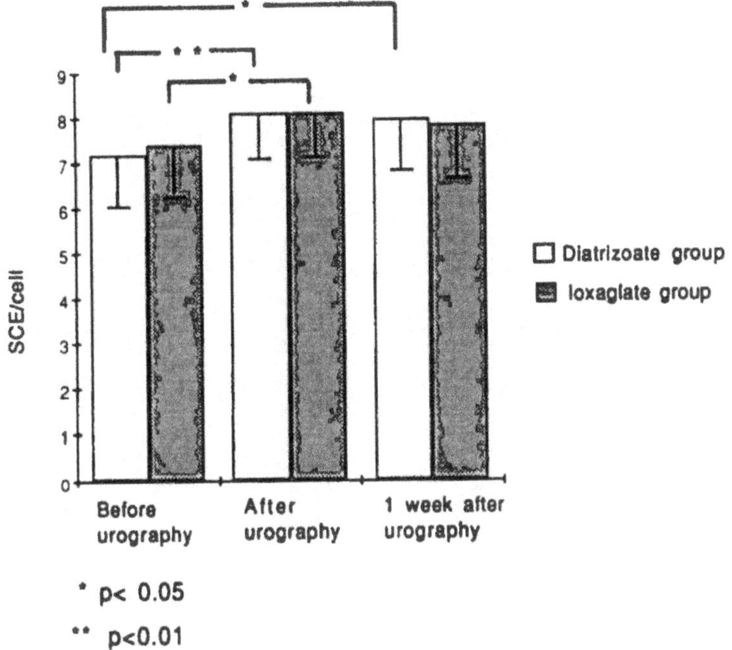

Figure 2 SCE frequency in patient undergoing urography.

Our work does not clarify whether contrast media produce these effects independent of radiation, but some data emphasize the relevance of contrast agents in the production of cytogenetic effects. So the differences between groups leads us to this conclusion, since these cannot be attributed to the different radiation dose or to the different photoelectric effect of iodine. Therefore, factors linked to the chemical nature of the contrast must probably be implicated in the extent and duration of the biological effects.

B. Internal Dosimetry

Internal dosimetry is usually performed by the analysis of (a) urinary thioether excretion, and (b) mutagenicity in urine.

First we shall refer to *thioether determination.* Phase I enzymes and, in particular, monooxygenases are implicated in the activation of parent compounds to intermediate electrophylic reactives, whose covalent binding to DNA, RNA, and cellular macromolecules produce chemical lesions, such as necrosis immunoreactivity, degeneration, and/or mutation. Gluthathione conjugation of electrophyles originating from xenobiotics results in the formation of thioethers, which are ultimately excreted mainly as mercapturic acids in urine. So conjugation is generally considered a detoxication mechanism, as it can prevent covalent binding of these agents to cellular macromolecules.

In our laboratory we have used a nonspecific method (20) based on the Ellman reaction in acidified urine samples extracted with ethyl acetate before alkaline

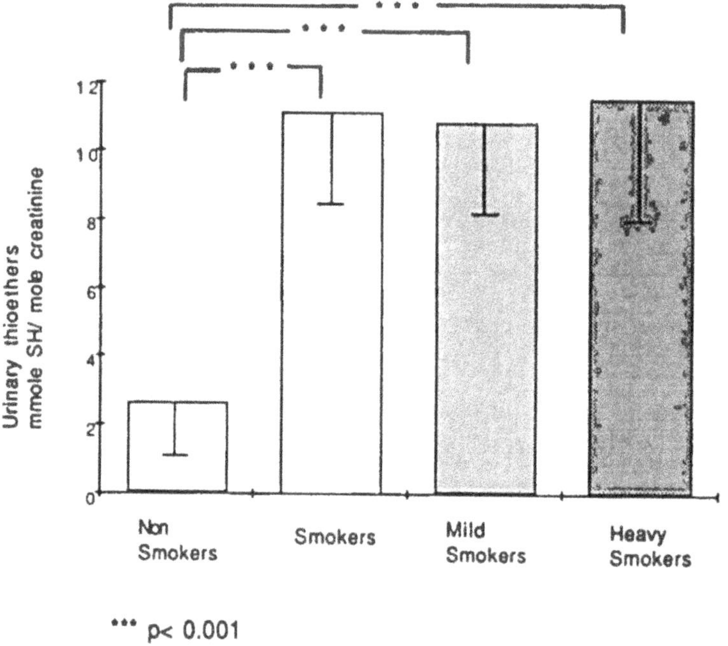

Figure 3 Thioether excretion as a biological indicator of exposure to tobacco smoke.

hydrolysis. This method partly reduces the high background values and the large interindividual variations. It is a useful tool to monitor the internal dose of mixtures of chemicals, such as tobacco smoke. In the case of smokers (16), it has been a good indicator (Figure 3).

It has been suggested that the application of this procedure to the biological monitoring of exposure in population groups is limited by the lack of chemical specificity. However, this method was able to distinguish two very similar analytes. These products were halothane and isofluorane, both of them halogenated hydrocarbons administered to women as anesthetics. These doses did not have any statistically significant differences. Thioether elimination resulted in a significant increase after administration (10) only in the halothane group, but not in the isoflurane group (Figure 4). So, in spite of the poor chemical specificity, the method was able to differentiate two very closely chemically related products, one of them implicated in hepatotoxicity—halothane—and the other not. Other exposures, in which thioether excretion has been shown to be a good indicator, through our studies, include: vinyl chloride (15), paracetamol (3), carbamazepine (7), and asphalts (12).

Finally, the analysis of the *mutagenicity in urine* is based on Ames test (2). We evaluate the mutagenic index and the premutagenic index (the latter with mycrosomal activation), and express both the number of reverted strains of *Salmonella typhimurium*, TA 98, per millimoles of creatinine. For the extraction procedure, urine is concentrated with XAD-2 columns and these are eluted with acetone. The eluted materials, redissolved in dimethylsulfoxide (DMSO), are tested for mutagenic activity.

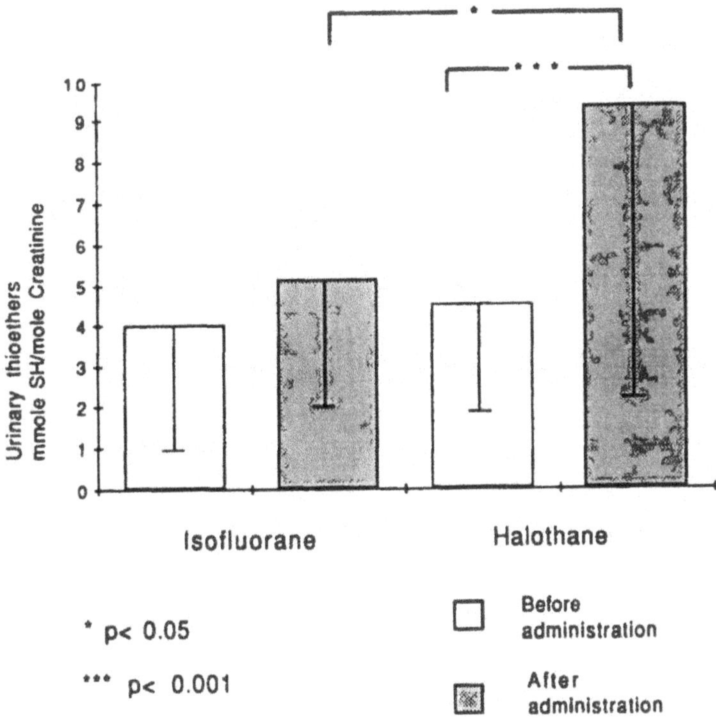

Figure 4 Thioether excretion before and after anesthetic administration.

With regard to these indicators, we found that the mutagenic index was not higher in smokers than in nonsmokers. However, the premutagenic index resulted to be a valid indicator of tobacco smoke exposure, given that the smokers, as a group, showed a much higher index than the nonsmokers (12) (Figure 5).

To conclude, the case of aromatic amines well illustrates the importance of the interaction between genetic and environmental factors.

Some aromatic amines have been identified as human bladder carcinogens (8). Biological monitoring of exposure is of particular interest, since genetic factors are highly important in their metabolism. So susceptible individuals can be identified only through biomonitoring. Moreover, epidemiological studies have shown a strong association between low acetyltransferase activity and risk of bladder cancer.

We consider very suggestive Lower's (9) hypothesis on the metabolism of aryl amines and the possible mechanism of bladder carcinogenesis.

Based on this hypothesis, we determined the mutagenic index, premutagenic index, and the mutagenic index in urine treated with β-glucuronidase in 70 workers from the textile industry who were exposed to aryl amines (15). The N-acetylator phenotype was assessed with isoniazide. Neither mutagenic nor premutagenic index turned out to be good indicators of exposure. In contrast, the mutagenic index with β-glucuronidase was clearly higher in workers than in controls, the increase being higher in the group of high exposure.

With regard to the influence of the N-acetylation pharmacogenetics, the slow acetylators showed an increase in relation to the fast acetylators, independent of

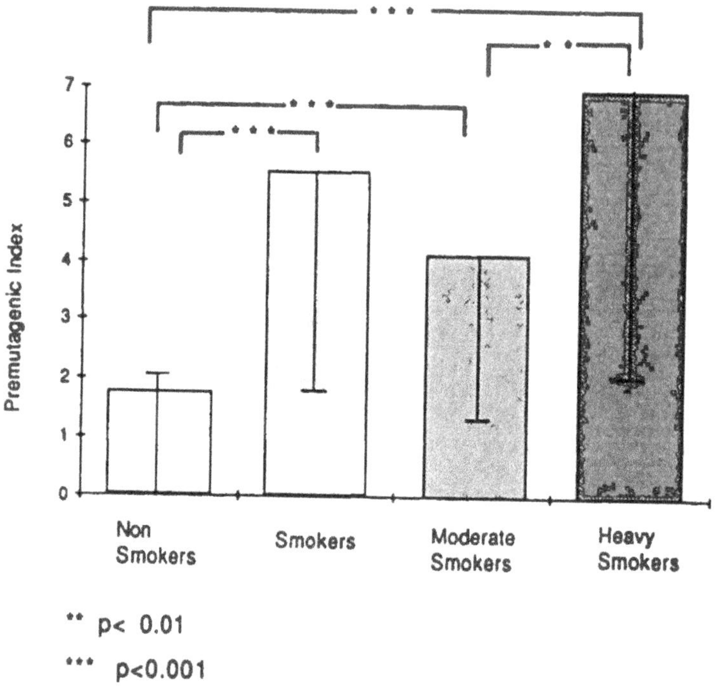

Figure 5 Premutagenic index in mine from nonsmokers and smokers.

whether they were nonsmokers or smokers. The main conclusion may be that this indicator seems to be able to delimit groups at higher risk, both genetically and environmentally determined.

REFERENCES

1. Z. Almássy, A. Krepinsky, C. Bianco, and G. J. Köteles. The present state and perspectives of micronucleus assay in radiation protection. A Review. *Appl. Radiat. Isot., 38*(4):241–249 (1987).
2. B. N. Ames, F. D. Lee, and W. E. Durston. An improved bacterial test system for the detection and classification of mitogens and carcinogens. *Proc. Natl. Acad. Sci. USA, 70*: 782–786 (1973).
3. M. L. Bernal, M. Bartolomé, and B. Sinués. Treatment with acetaminophen on children and its influence on erythrocytary glutathione and urinary thioethers. *2nd International Congress of Therapeutic Drug Monitoring and Toxicology*, Barcelona, 1990, Abstracts.
4. A. V. Carrano and A. T. Natarajan. Considerations for population monitoring using cytogenetic techniques. *Mutat. Res., 204*:379–406 (1988).
5. A. V. Carrano and L. H. Thompson. Sister chromatid exchange and gene mutation. *Cytogenet. Cell Genet., 33*:57–61 (1982).
6. M. Fenech and A. Morley. Measurement of micronuclei in lymphocytes. *Mutat. Res., 147*:29–36 (1985).
7. J. Gazulla, M. L. Bernal, and B. Sinués. Cytogenetic study and internal dosimetry in patients undergoing treatment with carbamazepine. *2nd International Congress of Therapeutic Drug Monitoring and Toxicology*, Barcelona, 1990, Abstracts.
8. International Agency for Research on Cancer. Chemicals, Industrial Processes and Industries Associated with Cancer in Humans (IARC monographs 1–29). (IARC Monographs on

the Evaluation of the Carcinogenic Risk of Chemicals to Humans, Suppl. 4), pp. 31–34, 37–38, 166–167, Lyon (1982).

9. G. M. Lower, T. Nilsson, C. E. Nelson, H. Wolf, T. E. Gamsky, and G. T. Byran. N-Acetyltransferase phenotype and risk in urinary bladder cancer: Approaches in molecular epidemiology: Preliminary results in Sweden and Denmark. *Environ. Health Perspect.*, 29:71–79 (1979).
10. A. Navarro. Biomonitorización de la exposición a halotano. Tesis doctoral, Zaragoza University, Zaragoza, Spain (1990).
11. M. E. Nünez and B. Sinués. Cytogenetic effects of diatrizoate and loxaglate on patients undergoing excretory urography. *Invest. Radiol.*, 25(60):692–697 (1990).
12. J. O. Orden. Biomonitoriazción de la exposición a los asfaltos en trabajadores de las brigadas rodantes. Tesis doctoral. Zaragoza University, Zaragoza, Spain (1990).
13. M. A. Sáenz, P. Rueda, M. L. Bernal, A. Alcalá, and B. Sinués. Acetylator phenotype and cytogenetic effects in biomonitoring of exposure to tobacco smoke. *2nd International Congress of Therapeutic Drug Monitoring and Toxicology*, Barcelona, 1990, Abstracts.
14. D. A. Shafer. Alternate replication bypass mechanisms for sister chromatid exchange formation. *Prog. Topics Cytogenet.*, 2:67–98 (1982).
15. B. Sinués, M. L. Bernal, M. A. Sáenz, and A. Alcaló. Mutagenicity in urine and acetylator phenotype in workers exposed to arylamines. *2nd International Congress of Therapeutic Drug Monitoring and Toxicology*, Barcelona, 1990, Abstracts.
16. B. Sinués, M. Izquierdo, and J. Pérez Viguera. Chromosome aberrations and urinary thioethers in smokers. *Mutat. Res.*, 240:289–293 (1990).
17. B. Sinués, A. Sáenz, M. L. Bernal, A. Tres, J. Lanuza, C. Ceballos, and M. S. Sáenz. Sister chromatid exchanges, proliferating rate index and micronuclei in biomonitoring of internal exposure to vinyl chloride monomer in plastic industry workers. *Toxicol. Appl. Pharmacol.* (in press).
18. B. Sinués, M. Izquierdo, A. Castillo, J. Pérez, and J. Gazulla. Biological dosimetry of exposure to vinyl chloride in plastic workers. *Cytotechnology, Suppl. 81* (1988).
19. H. Vainio. Current trends in the biological monitoring of exposure to carcinogens. *Scand. J. Work Environ. Health*, *11*:1–6 (1985).
20. R. Van Doorn, R. P. Leijdekkers, R. M. E. Bos, Brouns, P. IH. Henderson. Detection of human exposure to electrophylic compounds by assay of thioether detoxication products in urine. *Ann. Occup. Hyg.*, 24(1):77–92 (1981).
21. J. J. Yunis. The chromosomal basis of human neoplasia. *Science*. 221:227–236 (1983).

102

Lipid and Lipoproteins After Maximal Exercise

Alfredo Cordova,* M. Rosario Casado, Lucía Perez-Gallardo,* and Luis Rabadan**** **University of Valladolid, Soria, Spain, **Hospital Insalud, Soria, Spain*

I. INTRODUCTION

Lipoprotein abnormalities constitute a major risk for development of cardiovascular disease. The increased popularity of physical activity in recent years has been created, in part, by the conviction that regular exercise helps to prevent coronary heart disease (CHD). Exercise has been reported to reduce coronary mortality (1). Physical exertion may decrease coronary mortality by its effects on lipoproteins, or by changing other lipid-modifying co-variables, such as body composition (2). The changes in serum lipid levels caused by continuous long-term exertion are due mainly to increased energy needs in working muscles (3,4). The immediate effects from one to several bouts of physical activity seem to influence lipoprotein levels. A reduction in triglycerides has been shown after physical exertion (4,5). It is generally accepted that the longer the duration and the lower the intensity of exercise, the higher is the relative share of lipids as energy sources (4,5).

Initial studies of exercise training's effects on total cholesterol did not differentiate changes in HDL and LDL cholesterol. Although HDL cholesterol has been shown to increase in certain studies, the response has been variable in other investigations (6–9). However, since exercise is accompanied by many co-variables which also favorably alter these levels, it is difficult to determine the direct effect of regular physical activity. This work investigated the influence of a session of maximal exercise on plasma lipids and lipoprotein concentrations in male subjects.

II. SUBJECTS AND METHODS

Eleven male, normolipidemic, healthy, volunteer university students, were studied before and after maximal exercise. All were moderately fit (training 2 to 3 h/week). Physical characteristics (means ±SD) were obtained on all subjects. The subjects completed a questionnaire, which revealed that their diet was balanced, they did not smoke, and they did not consume any alcohol. Also, they were not taking any medication. All subjects, after cardiopulmonary and electrocardiographic examination, were normal. The tests were performed at 9 a.m. each week.

Two types of maximal exercise were performed on a mechanically braked Monark cycloergometer: (a) triangular progressive test (TPT), increasing the load 30 W/3 min, until obtaining the maximum power (considered when the last step was maintained for at least 2 min); (b) rectangular sustained test (RST), maintaining the maximum load obtained in TPT during 7 min.

Each subject had a catheter inserted in an antecubital vein, and blood samples were drawn at rest and at the end of exercise. The serum from venous blood was separated by low-speed centrifugation at 4°C. Total cholesterol (TC) and (TGs) were determined by enzymatic methods (Boehringer, Mannheim). HDL-O was determined by $MgCl_2$-Na phosphotungstate precipitation (Boehringer, Mannheim). LDL-O levels were calculated according to Friedwald et al. (10). Hematocrit (Hct) was measured in duplicate by the microcentrifuge method, and hemoglobin (Hb) was measured by the cyanomethemoglobin method (Boehringer, Mannheim).

Results are represented as means ±SD. Data were analyzed for significance of mean differences by using the paired t-test (and linear regression analysis). Only when the ANOVA indicated significance ($p < 0.05$) was the t-test applied to paired observations. Statistical significance was assumed when $p < 0.05$.

III. RESULTS

The physical characteristics and exercise time data are presented in Table 1.

In Table 2, the serum lipid (TC, TG) and lipoprotein (HDL-C, LDL-C) profiles are shown at rest and at the end of exercise. There was a significant increase in TC levels after TPT (14%) and RST (9.5%) with respect to TC at rest. TG levels also increased, but not significantly, after exercise (TPT 17%, RST 5.7%). Lipoproteins (HDL-C, LDL-C) levels rose significantly only after TPT exercise. Changes in plasma volume were judged by hematocrit (Hct) and hemoglobin (Hb) measurements.

IV. DISCUSSION

Triglycerides (TGs) levels in our study were not significantly different in the pre- and postexercises (TPT,RST). Although most studies (11–13) have demonstrated reduced levels, controversy still exists over the effects of prolonged training on triglycerides. Some prospective studies have demonstrated a relationship between exercise conditioning and triglycerides. The exercise produced a reduction of serum (TGs) among adult normolipidemic men (12); however, when the reductions were investigated (13), it was found that they were relatively transient (13). Other data (14) have not found this favorable relationship between exercise and (TGs) concentration. Studies of short-term physical training programs have not demonstrated a reduction in triglyceride levels (14,15). Increases were found (16) in serum triglycerides after volleyball

Table 1 Physical Characteristics[a]

Age (yr)	21.5 ± 0.8
Weight (kg)	70.6 ± 4.0
Height (cm)	178.0 ± 3.8
V_{O_2} max (mL/kg/min)	41.7 ± 3.7
Time training (h/wk)	2.6 ± 0.2

[a]Values are means ± SD; n = 12.

Table 2 Variations of Lipids and Lipoprotein Concentrations (mg/dL) and Hematologic Values (Hct %, Hb g/dL), at Rest and Immediately After Maximal Exercise (TPT, RST)[a]

	Rest	TPT	RST
TC	157 ± 32.7	179 ± 35.4*	172 ± 36.6*
TGs	122 ± 37.9	143 ± 57.2	129 ± 37.1
HDL-C	49 ± 7.1	55 ± 6.3*	53 ± 9.3
LDL-C	83 ± 26.8	95 ± 30.6*	92 ± 31.2
Hct	45.4 ± 2.4	49.4 ± 2.4*	49.2 ± 3.5*
Hb	15.7 ± 1.0	16.9 ± 1.0*	16.8 ± 0.9*

[a]Values are means ± SD, n = 12.
*Significant differences $p < 0.05$.

playing, which were explained as being due to an increased release of catecholamines following the exercise. There is general acceptance that a moderate workload during ordinary human activity does not alter plasma TGs levels, while long-distance skiing, considered to be prolonged heavy exercise, decreases the level (3,13,17). However, though plasma (TGs) tends to decrease in prolonged exercise and during training, its contribution to the total energy expenditure is relatively small (3).

The serum total cholesterol (TC) concentrations were increased significantly following the end of exercises (TPT, RST). This is in agreement with other findings (9,18,19). During the period of long-lasting exercise, plasma cholesterol levels may or may not change, while cholesterol oxidation increases. Under normal conditions a considerable amount of cholesterol is present in the tissues, and under pathologic conditions it can be stored in abundance (3). The exercise state may affect cholesterol turnover rate without changing its plasma concentration and, at the same time, lower the tissue cholesterol pools (3).

The lipoprotein (HDL-C, LDL-C) levels were increased following TPT. The influence of physical activity on lipoprotein lipase activity correlating with increased serum HDL-C is known (20). The effects of endurance training on lipoprotein cholesterol may reflect an enhancement of the lipoprotein-lipase system. The decreased hepatic lipase found in trained subjects could contribute to the increased HDL-C fraction after training (21). This may be induced by an improved receptor sensitivity for catecholamines released during physical exercise. These hormones, directly or indirectly, are involved in mobilization and utilization of substrates or in the process of deposition of substrate as fat and glycogen in muscle cells (2,3,6,17). The increased HDL-C observed may be attributed to an increased triglyceride turnover with a consequent increase in plasma VLDL, the catabolism of which leads to an increase in the mass of the HDL in the circulation (16,22). With respect to LDL-C, our results show that a pattern of increase similar to that high LDL levels are associated with a low coronary risk and that a negative correlation exists between HDL-C and CHD, independent of triglycerides, LDL-C values, and other risk factors.

V. SUMMARY

Eleven male, normolipidemic, healthy, volunteer university students were studied before and after maximal exercise. All were moderately fit (training 2 to 3 h/week).

Two types of maximal exercise were performed on a mechanically braked Monark cycloergometer: (a) triangular progressive test (TPT), increasing the load 30 W/3 min, until obtaining the maximum power (considered when the last step was maintained for at least 2 min); (b) rectangular sustained test(RST), maintaining the maximal load obtained in TPT for 7 min. Total cholesterol (TC) and TGs, HDL-C, LDL-C, hematocrit (Hct), and hemoglobin (Hb) were measured. The serum lipid (TC, TG) and lipoprotein (HDL-C, LDL-C) profiles were determined at rest and at the end of the exercise test. There was a significant increase in TC levels after TPT (14%) and RST (9.5%) with respect to TC at rest. TG levels also increased after exercise, but not significantly (TPT 17%, RST 5.7%). Lipoprotein (HDL-C, LDL-C) levels rose significantly only after TPT exercise. In conclusion, maximal progressive exercise causes more variations on serum lipids and lipoproteins than a maximal sustained test. However, the present data and the concepts discussed encourage us to investigate the lipid patterns in relation to type of sport and amount of training.

REFERENCES

1. R. S. Paffenbarger, R. T. Hyde, A. L. Wing, and C. C. Asieh. Physical activity, all cause mortality, and longevity of college alumni. *N. Engl. J. Med., 314*:605–613 (1986).
2. L. Golberg and D. L. Elliot. The effects of exercise on lipid metabolism in men and women. *Sports Med., 4*:307–321 (1987).
3. P. Paul. Effects of long lasting physical exercise and training on lipid metabolism. In *Metabolic Adaptation to Prolonged Physical Exercise* (H. Howald and J. R. Poortmans, eds.), Birkhauser, Basel, pp. 156–193 (1974).
4. B. Essen. Intramuscular substrate utilization during prolonged exercise. In *The Marathon: Physiological, Medical, Epidemiological and Psychological Studies*, (P. Milvy, ed.), *Ann. N.Y. Acad. Sci., 301*:30–44 (1977).
5. P. D. Gollnik. Free fatty acid turnover and the availability of substrates as a limiting factor in prolonged exercise. In *The Marathon: Physiological, Medical, Epidemiological, and Psychological Studies* (P. Milvy, ed.). *Ann. N.Y. Acad. Sci., 301*:64–76 (1977).
6. B. Dufaux, G. Assmann, and W. Hollman. Plasma lipoproteins and physical activity: A review. *Int. J. Sports Med., 3*:123–136 (1982).
7. C. Tsopanakis, D. Kotsarellis, and A. D. Tsopanakis. Lipoprotein and lipid profiles of elite athletes in Olympic sports. *Int. J. Sports Med., 7*:316–321 (1986).
8. P. A. Vodak, P. D. Wood, W. L. Haskell, and P. T. Williams. HDL-cholesterol and other plasma lipid and lipoprotein concentrations in middle-aged male and female tennis players. *Metabolism, 29*:745–752 (1980).
9. E. R. Skinner, C. Watt, and R. I. Maughan. The acute effect of marathon running on plasma lipoproteins in female subjects. *Eur. J. Appl. Physiol., 56*:451–456 (1987).
10. W. T. Friedwald, R. K. Levy, and D. S. Fredrickson. Estimation of the concentration of low-density lipoprotein cholesterol in plasma without use of the preparative ultracentrifuge. *Clin. Chem., 18*:499–502 (1972).
11. R. P. Martin, W. L. Haskell, and P. D. Wood. Blood chemistry and lipid profiles of elite distance runners. *Ann. N.Y. Acad. Sci., 301*:346–360 (1977).
12. S. A. Lopez, R. Vial, L. Ballart, and G. Arroyave. Effect of exercise and physical fitness on serum lipids and lipoproteins. *Atherosclerosis, 20*:1–9 (1974).
13. P. D. Thompson, E. Culliname, L. O. Henderson, and P. N. Herbert. Acute effects of prolonged exercise on serum lipids. *Metabolism, 29*:662–665 (1980).
14. K. D. Brownell, P. S. Bachorik, and R. S. Ayerle. Changes in plasma lipid and lipoprotein levels in men and women after a program of moderate exercise. *Circulation, 65*:477–484 (1982).
15. E. B. Alterkruse and J. H. Wilmore. Changes in blood chemistries following a controlled exercise program. *J. Occup. Med., 15*:110–113 (1973).

16. A. Bonetti, A. Catapano, A. Novarini, R. Pascale, P. Zeppilli, and U. Zuliani. Changes in lipid metabolism induced by volley ball playing. *J. Sports Med. Phys. Fitness, 28*:40–44 (1988).
17. L. Golberg and D. J. Elliot. Effect of physical activity on lipid and lipoprotein levels. *Med. Clin. N. Am., 69*:41–55 (1985).
18. A. D. Tsopanakis, E. P. Sgouraki, K. N. Pavlou, E. R. Nadel, and S. R. Bussolar. Lipids and lipoprotein profiles in a 4-h endurance test on a recumbent cycloergometer. *Am. J. Clin. Nutr., 49*:980–984 (1989).
19. N. Nagao, Y. Imai, J. Arie, Y. Sawada, and K. Karatsu. Comparison of serum apolipoproteins and lipoproteins in active and inactive males. *J. Sports Med. Phys. Fitness., 28*:67–73 (1988).
20. E. A. Nikkila, M. R. Taskinen, S. Rehunen, and M. Harkonen. Lipoprotein lipase activity in adipose tissue and skeletal muscle of runners: Relation to serum lipoproteins. *Metabolism, 27*:1661–1671 (1978).
21. J. Marniemi, P. Peltonen, I. Vuori, and E. Hietanen. Lipoprotein-lipase of human post-heparin plasma and adipose-tissue in relation to physical training. *Acta Physiol. Scand., 110*:131–135 (1980).
22. A. Berg, J. Johns, M. Baumstark, W. Kreutz, and J. Keul. Changes in HDL subfractions after a single, extended episode of physical exercise. *Atherosclerosis, 47*:231–240 (1983).

103

Lipoproteins in Rats with Uranyl Acetate-Induced Acute Renal Failure

Jordi Camps, Arturo Ortega, Elisabet Vilella, Cristina Constantí, Josep M. Simó, and Jorge Joven *Hospital de Sant Joan and Unitat de Toxicologia, Facultat de Medicina, Reus (Tarragona), Spain*

I. INTRODUCTION

It is well known that hyperlipoproteinemia is often related to coronary and cerebrovascular diseases. These alterations are invariably present in patients and in experimental animals affected by renal disturbances such as nephrotic syndrome (1,2) and chronic renal failure (3). However, lipoprotein disturbances in acute renal failure (ARF) have not been studied in humans or in animals. Such studies are extremely difficult in humans, since hyperlipidemia already accompanies many of the causes of ARF. On the other hand, animal intoxication with uranyl nitrate has been used as an experimental model for ARF. Several morphological and functional disturbances which occur in the kidneys of uranyl acetate-treated animals have been described previously (4–7). Hence, the aim of the present study was to ascertain the presence or absence of lipoprotein disturbances in rats with ARF.

II. MATERIALS AND METHODS

The study was performed on 30 male Sprague-Dawley rats weighing 250 to 275 g (Panlab, Barcelona, Spain). On day 0, 24-h urine samples were collected from 6 of 30 animals (control group). Urinary volume and osmolality were measured, and 2-mL aliquots were stored at –20°C. Immediately afterward, these animals were killed and exsanguinated through the posterior vena cava. Two 1-mL aliquots of plasma were stored at –20°C, and an additional 2 mL was immediately subjected to sequential preparative ultracentrifugation.

The remaining 24 animals were given a single intravenous (i.v.) injection of 10 μg/g of uranyl acetate dihydrate (analytical grade, E. Merck, Darmstadt, FRG), dissolved in 0.9% saline at a concentration of 25 g/L. Rats were randomly separated into four groups of six animals, on which the same study as for the control group was performed on days 2, 5, 13, and 20 after uranium administration.

Urinary excretion of sodium and potassium were determined in a Corning 614 analyzer (Ciba Corning, Essex, UK). Urine osmolality was measured by depression of the freezing point (Advanced Instruments osmometer model 3MO, Needham Heights, Massachusetts). Urinary excretion of N-acetyl-β-d-glucosaminidase (NAG), a sensitive marker of acute tubular necrosis (8), was determined by the method of Maruhn (9). Cholesterol and triglyceride in plasma and the isolated lipoprotein subfractions were measured enzymatically in a Monarch 2000 automatic analyzer (Instrumentation Laboratories SpA, Milan, Italy). Creatinine in plasma and urine was determined by standard techniques (10).

An aliquot (2 mL) of plasma was subjected to sequential preparative ultracentrifugation to separate very-low-density lipoproteins (VLDL, d < 1.006 kg/L) and low-density lipoproteins (LDL, d = 1.006 to 1.063 kg/L) (11) in a Kontron TFT 45-6 rotor (Kontron Instruments, Zurich, Switzerland). The fractions were quantitatively isolated using tube slicing. Measurements of the infranate of the LDL isolation stage were considered representative of high-density lipoproteins (HDL).

III. RESULTS

The effects of uranyl acetate on biochemical parameters are shown in Table 1. This compound, when injected intravenously as a single dose, produced marked renal tubular necrosis, as was demonstrated by the increase in the urinary NAG excretion. An increase in the urinary volume and a decrease in the urine osmolality with a considerable reduction in creatinine clearance were also noted.

Cholesterol and triglyceride levels, both in plasma and in lipoproteins, were not significantly modified by the i.v. administration of uranyl acetate (Table 2).

Table 1 Urinary Volume, Osmolality, Creatinine Clearance, and Urinary Parameters (expressed as units per day and 100 g body weight)

		Days after injection			
	Control rats	2	5	13	20
Volume excreted (mL)	10 ± 5.8	23 ± 8.3*	21 ± 8.3*	18 ± 10.2	14 ± 5.1
Osmolality (mOsm/kg)	1629 ± 290	598 ± 174***	598 ± 174***	1157 ± 291*	1582 ± 386
NAG (μkat/mol creatinine)	28 ± 12	2235 ± 455***	2235 ± 455***	413 ± 212**	74 ± 52
Creatinine clearance (μL/s)	8.3 ± 4.2	0.9 ± 0.5**	0.9 ± 0.5**	2.6 ± 0.6**	3.7 ± 0.3**
Proteins excreted (g)	4 ± 1.2	32 ± 6.9**	32 ± 6.9**	32 ± 30.4	7 ± 07
Sodium excreted (mmol)	0.22 ± 0.11	0.22 ± 0.14	0.22 ± 0.14	0.37 ± 0.24	0.21 ± 0.14
Potassium excreted(mmol)	0.32 ± 0.10	0.49 ± 0.17	0.49 ± 0.17	0.76 ± 0.41	0.74 ± 0.17
Creatinine excreted(μmol)	23.0 ± 4.3	13.2 ± 3.37**	13.2 ± 3.37**	17.1 ± 3.53*	17.4 ± 1.31*

Results are presented as means ± SD.
*, **, ***Significantly different from control ($p < 0.05$, $p < 0.01$, $p < 0.001$, respectively).

Table 2 Cholesterol and Triglyceride Levels in Plasma and Isolated Lipoprotein Fractions

		Days after injection			
	Control rats	2	5	13	20
Cholesterol (mmol/L)					
Total	1.43 ± 0.49	1.88 ± 0.40	1.56 ± 0.44	2.04 ± 0.24	1.70 ± 0.24
VLDL	0.32 ± 0.38	0.34 ± 0.39	0.30 ± 0.41	0.31 ± 0.20	0.36 ± 0.09
LDL	0.25 ± 0.32	0.23 ± 0.28	0.21 ± 0.25	0.56 ± 0.36	0.23 ± 0.15
HDL	0.86 ± 0.52	1.31 ± 0.34	1.05 ± 0.37	1.17 ± 0.70	1.11 ± 0.17
Triglyceride (mmol/L)					
Total	0.88 ± 0.47	0.54 ± 0.23	0.55 ± 0.17	0.82 ± 0.24	0.91 ± 0.21
VLDL	0.75 ± 0.51	0.42 ± 0.20	0.53 ± 0.12	0.69 ± 0.27	0.81 ± 0.20
LDL	0.07 ± 0.06	0.07 ± 0.05	0.06 ± 0.08	0.09 ± 0.06	0.07 ± 0.06
HDL	0.06 ± 0.05	0.05 ± 0.03	0.06 ± 0.04	0.03 ± 0.02	0.03 ± 0.02

Results are presented as means ± SD.

IV. DISCUSSION

Several authors have studied the existence of lipoprotein disturbances in subjects with chronic renal failure and nephrotic syndrome. However, the effect of ARF on plasma lipoprotein levels remains unknown, probably because of the difficulty of studying acutely affected patients. Hypertriglyceridemia and low HDL levels are the most striking features in patients affected with chronic renal failure undergoing conservative treatment or receiving hemodialysis (12,13). On the other hand, the existence of hyperlipoproteinemia in patients with nephrotic syndrome is well established (1).

The i.v. administration of 10 μg/g of uranyl acetate into rats has shown to be an appropriate experimental model with morphological and functional modifications consistent with severe and transient renal impairment. Our results show that cholesterol and triglyceride composition in VLDL, LDL, and HDL remained unchanged during both necrosis and recovery periods following drug administration, thereby providing experimental evidence against the hypothesis that ARF could affect lipoprotein metabolism.

REFERENCES

1. J. S. Chopra, N. P. Mallick, and M. C. Stone, Hyperlipoproteinemia in nephrotic syndrome. *Lancet*, *1*:317 (1971).
2. J. Joven, L. Masana, C. Villabona, E. Vilella, T. Bargalló, M. Trias, M. Figueras, and P. R. Turner. Low density lipoprotein metabolism in rats with puromycin aminonucleoside-induced nephrotic syndrome. *Metabolism*, *38*:491 (1989).
3. P. Rubiés-Prat, E. Espinel, J. Joven, M. R. Ras, and L. Pira. High-density lipoprotein cholesterol subfractions in chronic uremia. *Am. J. Kidney Dis.*, *9*:60 (1987).
4. K. Nomiyama and E. C. Foulkes. Some effects of uranyl acetate on proximal tubular function in rabbit kidney. *Toxicol. Appl. Pharmacol.*, *13*:89 (1968).
5. M. J. Giglio, C. E. Bozzini, J. A. Barcat, and E. Arrizurieta. Relationship between severity of renal damage and erythropoietin production in uranyl nitrate-induced acute renal failure. *Exp. Hematol.*, *14*:257 (1986).

6. J. L. Domingo, A. Ortega, J. L. Paternain, and J. Corbella. Evaluation of the perinatal and postnatal effects of uranium in mice upon oral administration. *Arch. Environ. Health,* 44:395 (1989).
7. J. L.Domingo, A. Ortega, J. M. Llobet, and J. Corbella. Effectiveness of chelation therapy with time after acute uranium intoxication. *Fund. Appl. Toxicol.,* 14:88 (1990).
8. C. Cojocel, N. Dociu, E. Ceacmacudis, and K. Baumann. Nephrotoxic effects of aminoglycoside treatment of renal protein reabsorption and accumulation. *Nephron.,* 37:113 (1984).
9. D. Maruhn. Rapid colorimetric assay of β-galactosidase and N-acetyl-β-glucosaminidase in human urine. *Clin. Chim. Acta,* 73:453 (1976).
10. J. M. Simó, N. Bertran, M. Juanpere, J. Camps, and J. Joven. Evaluación preliminar de un analizador automático Monarch 2000. *Quim. Clín.,* 8:329 (1989).
11. R. J. Havel, H. A. Eder, and J. M. Bragdon. The distribution and chemical composition of ultracentrifugally separated lipoproteins in human serum. *J. Clin. Invest.,* 34:1345 (1955).
12. R. C. Blantz. The mechanism of acute renal failure after uranyl nitrate. *J. Clin. Invest.,* 55:621 (1975).
13. J. Camps, X. Sola, A. Rimola, et al., Comparative study of aminoglycoside nephrotoxicity in normal rats and rats with experimental cirrhosis. *Hepatology,* 4:837 (1988).
14. R. G. Meeks. The rat. In *The Clinical Chemistry of Laboratory Animals* (W. F. Loeb and F. W. Quimby, eds.), Pergamon Press, New York, p. 19 (1989).
15. J. D. Bagdade and J. J. Albers. Plasma high-density lipoprotein concentration in chronic-hemodialysis and renal-transplant patients. *N. Engl. J. Med.,* 296:1436 (1977).
16. J. D. Brunzell, J. J. Albers, L. B. Haas, et al. Prevalence of serum lipid abnormalities in chronic hemodialysis. *Metabolism,* 26:903 (1977).

104

Effect of Sodium Metavanadate on Plasma Proteins and Fructosamine in Streptozotocin-Induced Diabetic Rats

C. Constantí, J. Camps, A. Ortega, J. Llobet, J. M. Simó, and J. Joven *Hospital de Sant Joan and Unitat de Toxicologia, Facultat de Medicina, Reus (Tarragona), Spain*

I. INTRODUCTION

Vanadium salts have been shown to have insulinlike effects on glucose levels in streptozotocin (STZ)-induced diabetic rats (1,2). However, toxic effects on various tissues and on the serum concentration of different analytes have been reported (3). The present investigation studied the influence of sodium vanadate on serum protein and fructosamine concentration in STZ-diabetic rats.

II. MATERIALS AND METHODS

Twenty-six male Sprague-Dawley rats (190 to 210 g) were obtained from Panlab (Barcelona, Spain). After 8 days of adaptation to plastic metabolic cages (day 0), diabetes was induced in all the animals by two s.c. injections (given on day 0 and day 3 at 12.00 h) of STZ (40 mg/kg) in cold 0.1 M citrate buffer. Sodium metavanadate (20 mg/kg/day, 25 days) was given i.p. to 16 of the 26 diabetic rats (VAN-rats). The remaining 10 animals did not receive any additional treatment (control group). Blood samples were taken on days 0, 7, 14, 21, and 25 by retroorbital puncture and centrifuged. The sera were stored at −20°C until biochemical measurements were performed. Fructosamine was measured by the method of nitroblue tetrazolium (Uni-Kit II Roche, Hoffmann-La Roche & Co., Ltd., Diagnostica, Basle, Switzerland) adapted to a Monarch 2000 automatic analyzer (Instrumentation Laboratori Spa, Milan, Italy). Glucose and total proteins were determined by standard techniques (4).

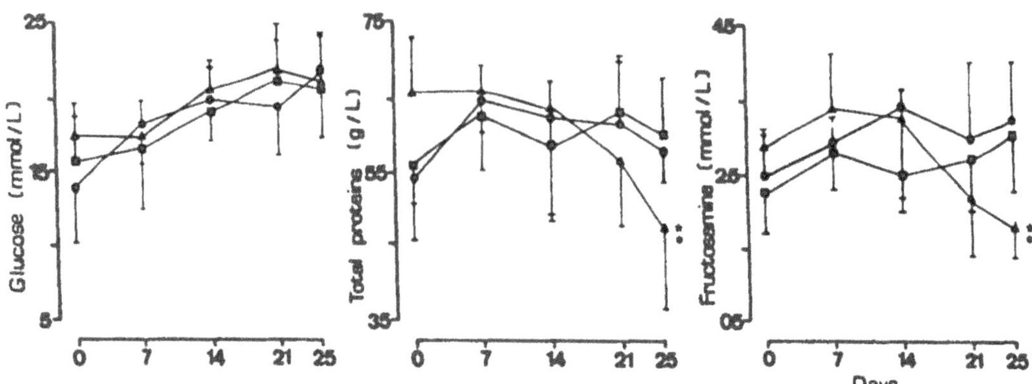

Figure 1 Glucose, total proteins, and fructosamine concentrations in control group (●) VAN-1 (▲) and VAN-2 rats (■). (★): $p < 0.01$ with respect to day 0; (✱) $p < 0.01$ with respect to VAN-2 rats.

III. RESULTS AND DISCUSSION

Results are shown in Figure 1. In the control group, glucose progressively increased after STZ administration, as did fructosamine. Serum protein concentration did not change significantly during the experiment. Serum glucose was decreased in VAN-rats with respect to that in the controls. Six of the 16 VAN-rats (VAN-1 rats) presented with a decrease of serum total protein from day 21 and on serum fructosamine on day 25. Both parameters remained steady in the remaining VAN-rats (VAN-2 rats). Glycemia was not significantly different in VAN-1 and VAN-2 rats. In the whole group of VAN-rats there was a significant direct correlation between total protein and fructosamine on days 21 to 25 ($r = 0.61$; $p < 0.01$). These results show that in STZ-diabetic rats, as in humans, serum fructosamine is related not only to glycemia but also to serum protein concentration. Clinical and in-vitro studies have previously reported a relationship between fructosamine and serum proteins (5–10) which was first described (5,6) for fructosamine and albumin in humans. Serum immunoglobulins may also influence fructosamine concentration (7). There are direct correlations between in-vitro glycated albumin and fructosamine and between glycated immunoglobulin G and fructosamine (9). However, at present no in-vivo model to further investigate serum protein-fructosamine relationship has been described. Results of the present study indicate that serum fructosamine concentration is not an appropriate index for glycemic control in STZ-diabetic rats treated with sodium vanadate but, on the other hand, this experimental model may be useful to study the protein-fructosamine relationship.

REFERENCES

1. C. E. Heyliger, A. G. Tahiliani, and J. H. McNeil. Effects of vanadate on elevated blood glucose and depressed cardiac performance of diabetic rats. *Science*, 227:1474 (1985).
2. J. Meyerovitch, Z. Farfel, J. Sack, and Y. Schecter. Oral administration of vanadate normalizes blood glucose levels in streptozotocin-related rats. *J. Biol. Chem.*, 262:6658 (1987).

3. J. Domingo, J. Llobet, M. Gomez, J. Corbella, and C. Keen. Effects of oral vanadium administration in streptozotocin diabetic rats. *Proc. 1st Int. Symp. Metal Ions in Biology and Medicine*, Reims France, pp. 312–314.
4. J. M. Simó, N. Bertrán, M. Juanpere, J. Camps, and J. Joven. Evaluación preliminar del analizador automático Monarch 2000. *Quím. Clín.*, *8*:329 (1989).
5. J. R. Baker, J. P. O'Connor, P. A. Metcalf, M. R. Lawson, and R. N. Johnson. Clinical usefulness of estimation of serum fructosamine concentration as a screening test for diabetes mellitus. *Br. Med. J.*, *287*:863 (1983).
6. M. P. Van Dieijen-Visser, C. Seynaeve, and P. J. Brombacher. Influence of variations in albumin or total-protein concentration on serum fructosamine concentration. *Clin. Chem.*, *32*:1610 (1986).
7. S. Rodríguez-Segade, S. Lojo, M. F. Camiña, J. M. Paz, and R. Del Río. Effects of various serum proteins on quantification of fructosamine. *Clin. Chem.*, *35*:134 (1989).
8. M. H. Dominiczak, J. M. Orrell, and W. E. I. Finlay. The effect of hypoalbuminaemia, hyperbilirubinaemia and renal failure on serum fructosamine concentration in non-diabetic individuals. *Clin. Chim. Acta.*, *182*:123 (1989).
9. A. Mosca, A. Carenini, F. Zoppi, A. Carpinelli, G. Banfi, F. Ceriotti, P. Bonini, and G. Pozza. Plasma protein glycation as measured by fructosamine assay. *Clin. Chem.*, *33*:1141 (1987).
10. J. De Schepper, M. P. Derde, P. Goubert, and F. Gorus. Reference values for fructosamine concentrations in children's sera: Influence of protein concentration, age and sex. *Clin. Chem.*, *34*:2444 (1988).

105

Binding of Cadmium to Albumin and Its Inhibition by Zinc, Copper, EDTA, and Penicillamine

F. J. Isern,* L. Pérez-Gallardo,* and J. F. Escanero** *Universidad de Valladolid, Valladolid, Spain, **Universidad de Zaragoza, Zaragoza, Spain*

I. INTRODUCTION

Cadmium (Cd) is considered to be a nonessential and toxic element for living beings (14), although humans continue to use it for various purposes (7). The average rates of Cd that are absorbed by humans are 30% by inhalation and 8% ingested in the diet (20). In both cases Cd enters into the bloodstream, where most of it binds to plasma proteins (7,12), particularly to serum albumin, the most abundant plasma protein. It is thereby distributed to different body organs and tissues (1,13). Cd also binds to metallothioneins, proteins induced by metals, which intervene in the detoxication of Cd and other ions (3). Nevertheless, serum albumin binds mainly divalent ions in both specific and nonspecific sites. The quaternary structure of albumin allows this protein to bind to almost all ligands, with high affinity in some few places (primary sites) and in many others with less affinity (secondary sites). Cd, like other ions, is included in the second group. With respect to the association between Cd and serum albumin, it is necessary to determine the Cd content in fatty acids (FA), since these influence the binding of any ligand, as well as other factors such as pH, temperature, and the type and concentration of salts in the solution, which may alter albumin.

The physiological functions of plasma albumin can be summed up in the following: maintenance of the colloid-osmotic pressure; nourishment, once it has been degraded; and detoxification, when it acts as a store for noxious products which are bound to this protein such as heavy metals, toxic chemical compounds, and other compounds which have completed their physiological function. In the molecule of serum albumin, at least six regions of binding to several compounds have been described. Positive inorganic ions associate reversely to the so-called IV region of binding. This bond is made in a well-defined site, but there also exist other binding sites in the molecule (SH, COO^-, imidazol, etc.) (15). The plasma proteins distribute 95% of Cd to the body organs and tissues. Nevertheless, with time, Cd accumulates mainly in tissues, primarily in the liver and in the kidneys (7,8,20).

Interferences exist between toxic and essential elements in different physiological processes (4–6,16), such as intestinal absorption and renal and blood transport (4,16). Thus, it is important to know about these interferences in order to prevent deficiencies of the essential elements or to avoid contamination by the toxic ones (3). One of the less studied processes is transport, which represents a crucial point in the study of contamination by toxic elements.

On the other hand, several chelators of Cd, such as L-cysteine, sodium tripolyphosphate (STPP), diethylenetriaminetetracetic acid (DPTA), deferoxamine, dithiocarbamates, dimercaptosuccinic acid, dimercaprol (2,3-dimercaptopropanol or BAL), disodium dicalcium edetate (EDTA diNa-Ca), and d- or l-penicilamine, when bound to Cd, prevent its binding to proteins and thus its accumulation in various body organs and tissues (2,9,10,12). Chelates are complexes constituted by the association between a donor of electrons and a metal, forming a heterocyclic ring. When this ring includes five or six atoms, the chelate is very stable, and thus the chelating agent has a high affinity for the metal, so that the former can remove the latter from organic tissues, especially in the case of toxic heavy metals (12).

The aim of this study was to establish the kinetics of the binding of Cd to serum albumin and to try to modify it in vitro through the addition of Zn and Cu and the chelating agents EDTA and l-penicillamine.

II. MATERIALS AND METHODS

For this study a solution in distilled water of 100 μmol/L lyophilized bovine serum albumin (BSA), at physiological pH (7.35 to 7.40) and room temperature (20 to 22°C), was used. With this solution, to which the elements and molecules for the study were added, membrane cones (CF50A Centriflo, Amicon) were filled to perform the ultrafiltration. This was done by centrifugation for 15 min at 3000 rpm (Hettich Universal II). Afterwards, the absence of proteins in the ultrafiltrate was verified. The concentration of ions was determined by atomic absorption spectrophotometry, of EDTA diNa-Ca by spectrophotometry, and of L-penicillamine by the ninhydrin test. These measurements were performed before (E_{total}) and after ($E_{ultrafiltrate}$) the ultrafiltration, and the element bound to BSA (E_{bound}) was estimated by the difference between them.

A. Study of the Cd-BSA Binding

The studies of the Cd-BSA binding were performed by determining the intrinsic association constant (K_A) and the number of binding sites (n). This type of study requires the analysis of a great range of values of Cd bound to BSA (from 28.0 to 3633.0 μmol/L), analyzed with Scatchard's equation (17): $V_E/E_{UF} = n{\cdot}K_A{-}KV_{Cd}$, where V_E is the concentration of the bound cation per mole of albumin, E_{UF}, is the concentration of the cation in the ultrafiltrate, K_A is the intrinsic constant of association, and n is the number of binding sites.

B. Study of the Inhibition of the Cd-BSA Binding

The studies of the inhibition of the Cd-BSA binding were also performed with the physiological metallic ligands Zn and Cu and with the organic nonphysiological ones EDTA diNa-Ca and L-penicillamine. Cd was added to the original solution to reach a concentration of 200 μmol/L, together with increasing concentrations of the supposed inhibitors, to obtain concentrations of ultrafiltrated Cd above 90% in all cases. The final concentrations added were the following:

Zn: 250, 400, 850, 1850, and 2700 μmol/L
Cu: 90, 225, 400, 650, 2000, 4700, 6100, and 12300 μmol/L
EDTA diNa-Ca: 190, 250, 550, 1100, 1900, and 3200 μmol/L
L-Penicillamine: 250, 350, 650, 1300, and 3600 μmol/L

C. Statistical Treatment

The significance of the regression lines of the binding of Cd to albumin as well as the study of its inhibition by the presence of other ligands was analyzed by the Student-Fischer t-test.

III. RESULTS

A. Bound Cd-BSA

The data obtained after the application of the Scatchard's equation show two types of binding sites, saturable and nonsaturable. In the case of the saturable sites, the values were 163 ± 6.263 L/mol for the intrinsic constant of association (K_A) and 2.285 binding sites (n) per mole of BSA. For the nonsaturable sites an affinity coefficient, $N_{A2}K_{A2}$, of 6.969 L/mol was obtained.

The percentage of binding ranged from 97.24 to 84.98% for Cd concentrations between 29 and 236 μM, respectively. The results obtained with this technique show a high correlation ($r = 0.991$, $p < 0.001$). In Figure 1 the regression lines of the saturable and nonsaturable sites for the bound Cd-BSA are represented.

B. Inhibition of the Bound Cd-BSA

The results of the inhibition of the bound Cd-BSA by the physiological elements Zn and Cu and by the chelating agents EDTAdiNa-Ca and L-penicillamine are presented in Table 1.

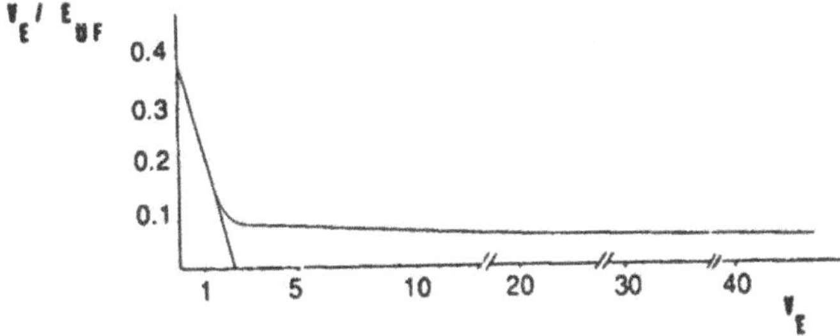

Figure 1 Binding of cadmium to serum albumin. Scatchard plot of successive binding points: saturable (correlation: 0.991) and non-saturable (correlation: 0.889). Each point is the mean of eighth determinations. For explanation of abbreviations, see Section II.

Table 1 Percentages of Ultrafiltrable (UF) Cd by the Addition of Increasing Amounts of Zn and Cu, EDTA, and L-Penicillamine[a]

Zn added μmol/L	0	250	400	850	1850	2700			
% Cd Ult.	10.70	11.60	12.00	67.15	72.73	90.70			
Cu added μmol/L	0	90	225	400	650	2000	4700	6100	12300
% Cd Ult.	10.70	17.14	37.78	39.13	42.31	50.98	84.82	85.71	92.86
EDTA diNa-Ca added μmol/L	0	190	250	550	1100	1900	3200		
% Cd Ult.	10.70	42.11	57.14	85.84	85.00	89.47	95.44		
L-penicillamine added μmol/L	0	250	350	650	1300	3600			
% Cd Ult.	10.70	55.81	56.82	60.00	78.61	97.74			

[a]Each value is the mean of eight determinations.
[*]Significant differences ($p < 0.001$) in relation to remaining percentages.

In the case of the Zn and Cu, a different behavior was observed. By adding 2.70 mmol/L of Zn, the percentage of ultrafiltrated Cd increased from 10.7% to 90.70%. This value represents a 89.56% inhibition. On the other hand, to achieve similar percentages (92.%) with Cu, 12.30 mmol/L were needed, indicating that Zn inhibits the binding of Cd to albumin more efficiently than does Cu. With regard to the chelating agents, if EDTA di-Na-Ca is not added, the Cd ultrafiltrated is 10.70%, passing to 94.89% by adding 3.20 mmol/L. L-penicillamine produces similar inhibition values (97.47%) by adding 3.6 mmol/L.

IV. DISCUSSION

In relation to the kinetics of the bound cadmium serum albumin, the results obtained in this work for the saturable site are: association constant, K_A, 1.63×10^2 L/mol and n, number of binding sites, 2.285 sites/mol. These results are notably different from the higher reported values of $K_A = 1.30 \times 10^3$ L/mol and n = 17 sites/mol (19). They were probably due to the use of a different pH, 5.95. Although this difference would not seem to be enough to explain such disagreement in the values obtained, the lack of other studies on the same subject makes it difficult to interpret the parameters of the Cd-BSA binding. On the other hand, our work presents the existence of not only saturable sites for that binding, but a region of nonsaturable sites, with an apparent constant of association, $N_{A2}K_{A2}$, of 6.959 L/mol.

With regard to inhibition, it was reported (15) that the first binding site of Cu is the amino-terminal end of serum albumin, forming a chelate with the last three amino acids Asp-Ala-His in human serum albumin and with Asp-Ala-His in the bovine one. The binding of this ion to the protein could be established through bonds between the imidazol group and nitrogen and oxygen or, instead, another nitrogen. Copper binds to this site with more strength, followed by zinc and then by cadmium (15). This region is also reported to be the binding site of metallic ions (11), using two atoms of nitrogen, one in the imidazol group of the histidine and the other in the amino group of the aspartic acid. After various experiments, the first two amino acids (Asp and Ala in humans and Asp and Thr in bovines) were shown not to be indispensable, because they are interchangeable with others, whereas the substitution of histidine revokes the bond (11). Thus, this amino acid in the third position is a key point for the binding. Zinc, on the other hand, binds to the imidazol groups of the histidine and to the carboxyl group of the aspartic acid, rather than to the amino-terminal group. The missing coordination sites for this ion would be occupied by molecules of the solvent.

With the results obtained in this work we can conclude that copper is strongly bound to the amino-terminal end, which is highly specific for this ion, and its first binding site is to the serum albumin. However, zinc binds to the amino acidic groups of histidine all along the albumin molecule, in a similar way to cadmium; thus both elements inhibit at lower concentrations.

The chelating compounds, EDTA diNa-Ca and L-penicillamine, have shown a similar efficiency in the inhibition of the binding of cadmium to serum albumin. The percentages of ultrafiltrated Cd both by EDTA diNa-Ca (18), 95.44% by adding 3.20 mmol/L of this compound, and by L-penicilamine, 97.10% by adding 3.60% mmol/L, can be considered to be in the same range. The possible mechanism of this inhibition might be its competition with the cation for the albumin, due to its chelating action.

REFERENCES

1. O. Andersen. Oral cadmium toxicology. *Acta Pharmacol. Toxicol.* (Copenhagen), *Suppl. 7*:44–47 (1984).
2. O. Andersen, J. Nielsen, and P. Svendsen. Oral cadmium chloride intoxication in mice: Effects of chelation. *Toxicology, 52*:65–79 (1988).
3. V. Burgat-Scaze, et al. Toxicité a long term du cadmium administré a tres faible dose chez la rat: Repartition du cadmium, du zinc et du cuivre. *Toxicol. Appl. Pharmacol., 46*:781–791 (1978).
4. B. A. Chowdhury and R. K. Chandra. Biological and health implications of toxic heavy metal and essential trace element interactions. *Proc. Food Nutr. Sci., 11*:55–113 (1987).
5. F. C. Foulkes and C. Voner. Effects of Zn status, bile and other endogenous factors on jejunal Cd absorption. *Toxicology, 22*:115–122 (1981).
6. E. González Fernández. Toxicocinética y Evaluación de riesgos para la salud producidos por la exposición a cadmio. *Med. Segur. Trab., 35*:3–17 (1988).
7. G. A. Goodman, et al., eds. *Las Bases Farmacológicas de la Terapéutica*. Medica-Panamericana, pp. 1531–1538 (1986).
8. W. H. Hallenbeck. Human health effects of exposure to cadmium. *Experientia, 40*:136–142 (1984).
9. S. Kojima, K. Kazuyoshi, M. Kiyozumi, and T. Honda. Comparative effects of three chelating agents on distribution and excretion of cadmium in rats. *Toxicol. Appl. Pharmacol., 83*:516–524 (1986).
10. S. G. Jones et al. Dependence on chelating agent properties of nephrotoxicity and testicular damage in male mice during cadmium decorporation. *Toxicology, 53*:135–146 (1988).
11. V. Kragh-Hansen. Molecular aspects of ligand binding to serum albumin. *Pharm. Rev., 33*:17–52 (1981).
12. M. Litter, ed. *Farmacología Experimental y Clinica*. El Ateneo, pp. 1709–1793 (1988).
13. A. Lugea, A. Barber, and F. Ponz. Inhibition by cadmium of d-galactose and l-phenylalanine transport by rat intestine "in vitro." *Rev. Esp. Fisiol., 44*:121–126 (1988).
14. J. M. Ortem and O. W. Neuhaus, eds. *Bioquímica Humana*. Panamericana, pp. 730–757 (1984).
15. M. S. N. Rao and H. Lal. Metal protein interactions in buffer solutions. Part III. Interaction of Cu^{II}, Zn^{II}, Cd^{II}, Co^{II}, and (Ni^{II}) with native and modified serum albumin. *J. Am. Chem. Soc., 80*:3226–3335 (1958).
16. H. H. Sandstead. Interactions of toxic elements with essential elements: Introduction. Micronutrient interactions: Vitamins, minerals and hazardous elements. *Ann. N.Y. Acad. Sci., 355*:131–365 (1980).
17. G. Scatchard. The attraction of proteins for small molecules and ions. *Ann. N.Y. Acad. Sci., 51*:660–672 (1949).
18. F. Vázquez and Vasák. Comparative Cd-N.M.R. studies on rabbit $^{113}Cd_7$-, (Zn_1,Cd_6) and partially metal depleted $^{113}Cd_6$-metallothionein-2n. *Biochem. J., 253*:611–614 (1988).
19. H. Waldmann-Meyer. Thermodinamic proton-, cadmium-, and zinc-binding constants of serum albumin determined by zone electrophoresis. *J. Biol. Chem.*, 235:3337–3341 (1960).
20. J. K. Yost. Cadmium, the environment and human health: An overview. *Experientia*, 157–164 (1984).

106

Random Sequential Mechanism of Human Placental Glutathione S-Transferase

M. T. Serafini and A. Romeu *Barcelona University, Tarragona, Spain*

I. INTRODUCTION

Glutathione S-transferases (GST) (EC 2.5.1.18) are a family of isoenzymes involved in the binding, transport, and metabolism of xenobiotics and endogenous substances (1). Compounds with an electrophilic center conjugate readily with glutathione (γ-Glu-CysH-Gly) (GSH) (2). GSH S-conjugates are typically converted to mercapturic acids (3); the γ-Glu moiety of the conjugate is removed, and the resulting cysteinylglycine conjugate is converted to the cysteine conjugate, which is N-acetylated and then excreted (4). The pathway varies with different compounds and species. Several organs and the intestinal flora may also be involved (5).

Certain GSTs were previously known as ligands (which bind bilirubin, steroids, azo dyes, and certain carcinogens) and were thought to function in the transport of these compounds from blood plasma into liver cells (6). GSTs from liver are dimeric enzymes that can contain four types of subunits (7). The GSTs account for about 10% of the soluble protein (8). The GSTs are transcriptionally activated by xenobiotics during chemical carcinogenesis (9). During investigation of changes in GST isoenzymes in rat hepatocarcinogenesis, the neutral form of GST from rat placenta (GSTp) is increased in the hyperplastic nodule (HN)-bearing rat liver (10). By single radial immunodiffusion using anti-GSTp antibody and, later, by immunohistochemical methods, it was found that the placental form is present at a very low level in normal rat liver, but it is present in significant quantities in adult rat kidney, lung, pancreas, and spleen, and also in brain (11). Thus, the GSTp form seems to be almost universally albeit weakly expressed in various rat tissues. It is now grouped into the class "pi" of the species- independent classification of GST (12).

The steady-state kinetic mechanisms of several GSTs have been studied extensively, and several mechanisms, including random, Ping-Pong, and sequential, have been proposed (13). At this time, the mechanism postulated for the GST is a primitive one that depends only on the close juxtaposition of the interacting substrates, i.e., a proximity effect (14). The active site of the enzyme may be viewed as consisting of a specific binding area for GSH that is close to a second site which is characterized by its lipophilic character. Thus, when a sufficiently reactive electrophile is present at the site adjacent to the glutathione thiolate ion, reaction will occur (14). Spacious lipophilic surroundings are necessary for accommodating the vast array of substrates, ranging in size and shape from iodomethane to the steroids and polynuclear aromatic hydrocarbons that have some hydrophilic aspect and are bound (14). The results of kinetic mechanism studies of the transferase-catalyzed reaction are controversial. Nevertheless, it is clear that under physiological conditions, a sequential mechanism obtains which requires that both substrates be on the enzyme before any product is released (15).

II. EXPERIMENTAL PROCEDURE

The steady-state kinetics has been studied for the conjugation of GSH with 1-chloro-2,4-dinitrobenzene (CDNB). GSH, CDNB, and purified human placental GST were obtained from Sigma Chemical. The other compounds were readily available commercial products. GST enzymatic activity was measured by the method given by Habig et al. (16). A unit of activity is defined as the amount of enzyme catalyzing the formation of 1 μmol of product per minute under the conditions of the specific assay. The GST reaction system contained a suitable amount of enzyme (diluted in 1% albumin and buffer), varying concentrations of GSH (up to 1 mM), as well as DCNB (up to 1 mM) and 0.1 M Na-phosphate buffer, pH 6.5, in 1 mL of assay mixture. The initial rates were determined by monitoring changes in absorbance at 340 nm with a Hitachi U2000 double-beam spectrophotometer. Data represent triplicate determinations in a simple experiment. Model fitting was performed using the BMDP biomedical computer program (University of California, Berkeley) (17).

III. RESULTS AND DISCUSSION

Consider a transferase reaction of the form AX + B = A + BX in which X is transferred from A to B. Such a scheme can represent a large number of enzyme-catalyzed reactions. In the presence of an enzyme there are a number of possible pathways by which the reaction may proceed, and a general scheme illustrating the most common pathways is as follows:

```
      EAX  ——  EX  ——  EBX
     /    \           /    \
   E       EAXB —— EABX      E
     \    /           \    /
      EB                EA
```

The upper pathway is a scheme in which the two substrates never meet as such on the surface of the enzyme, the reaction proceeding by way of a complex of X with the enzyme and by complexes of the enzyme with only one substrate or product molecule. Such complexes are known as binary complexes. In the remainder of the scheme there are, in addition to binary complexes, complexes containing the two substrate molecules bound to the enzyme. In a mechanism involving ternary complexes, it is obviously possible that the substrates can bind in any order, as illustrated in the scheme. This is known as a random-order-of-addition mechanism, or more commonly a random-order mechanism (18).

The initial velocity data were fitted to rate equation describing a steady-state, ordered, bi-bi kinetic system:

$$v = \frac{V}{1 + K_{M,A}/[A] + K_{M,B}/[B] + K_{S,A}K_{M,B}[A][B]}$$

Where v and V are the initial and maximum velocities, respectively; [A] and [B] are the mean GSH and CDNB concentrations, respectively; $K_{M,A}$ and $K_{M,B}$ are the Michaelis constants of GSH and CDNB; and $K_{S,A}$ is the dissociation constant of the binary complex GST-GSH. At any fixed concentration of one of the substrates, variation of the other gives a Michaelis curve. At very large concentrations of B the equation reduces to the Michaelis-Menten equation for a single-substrate reaction, and hence $K_{M,A}$ can be defined as the concentration of A which, at saturation concentrations of B, will give half-maximum velocity, and $K_{M,A}$ can be defined in a similar fashion; in other words, these constants are true Michaelis constants. Furthermore, the individual kinetic constants can be obtained by a simple graphical procedure, taking reciprocals from the above equation (19). A reciprocal plot of 1/v against 1/[A] at a series of B concentrations (primary plot) gives a series of straight lines which intersect above the −1/[A] axis as shown in the figures; lines intersect the 1/v axis at $(1 + K_{M,B}/[B])/V$. Thus the individual $K_{M,B}$ were calculated from the linear relation that exists between the algebraic values of these intercepts on the 1/v axis and the inverse of the fixed B concentrations (secondary plot). Figure 1 shows the double-reciprocal plots of individual velocity data over a range of GSH concentrations from 0.1 to 1 mM at concentrations of CDNB ranging from 0.1 to 1 mM. Data of the intercepts on the 1/v axis obtained at fixed concentrations of CDNB plotted against 1/[CDNB] gave a straight line (y = 0.00155 + 0.00212x), and the true Michaelis constant for CDNB was 1.845 mM. Figure 2 shows the double-reciprocal plots of

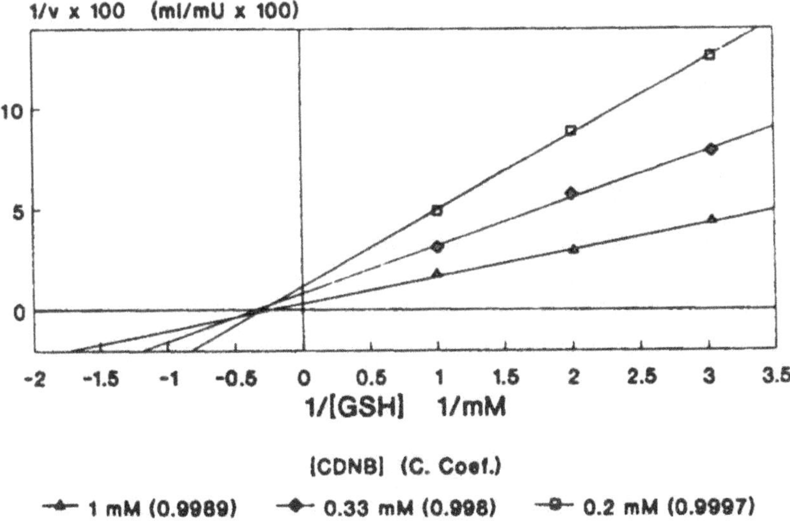

Figure 1 Double-reciprocal plots of individual velocity data over a range of GSH concentrations at concentrations of CDNB ranging from 0.1 to 1 mM. Data represent triplicate determinations in a simple experiment.

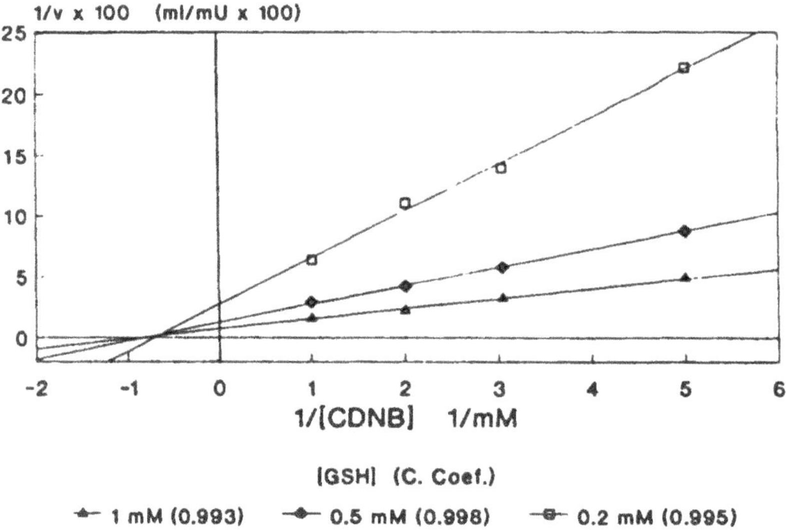

Figure 2 Double-reciprocal plots of individual velocity data over a range of CDNB concentrations at concentrations of GSH ranging from 0.1 to 1 mM. Data represent triplicate determinations in a simple experiment.

individual velocity data over a range of CDNB concentrations from 0.1 to 1 mM at concentrations of GSH ranging f rom 0.1 to 1 mM. Algebraic values of these intercepts on the 1/v axis obtained at fixed concentrations of GSH plotted against 1/[GSH] also gave a straight line (y = 0.00242 + 0.00496x), and the true Michaelis constant for GSH was 2.048 mM.

The slope of the Lineweaver-Burk reciprocal plot (1/v versus 1/a) is ($K_{M,A}$ + $K_{S,A}K_{M,B}$/[B])/V. Thus the dissociation constants of the binary complex enzyme-substrate were also calculated from the secondary plots of these slopes versus 1/[B]. Thus, when the slopes of the Lineweaver-Burk reciprocal plot obtained at fixed concentrations of CDNB were plotted against 1/[CDNB], a straight line was obtained (y = 0.00635 + 0.00616x), and the dissociation constant of the binary complex GST-GSH was 2.880 mM. On the other hand, from the linear relationship between slopes at fixed concentrations of GSH and 1/[GSH] (y = 0.00017 + 0.00765x), the dissociation constant of the binary complex GST-CDNB was 1.543 mM. All correlation coefficients (R^2) were greater than 0.99.

It should be emphasized that the K_S obtained in this way does not necessarily represent the true binding constant for the reaction E + A = EA, since if the EA complex undergoes any subsequent transformation before the binding of B, all steps that occur before the binding of B will be included in the apparent dissociation constant (20). Data summarized in the present work agree with the rapid equilibrium, random sequential, bi-bi mechanism for human placental glutathione S-transferase (21). However, the present results are consistent with the mathematical condition ($K_{S,A}$) ($K_{M,B}$) = ($K_{M,A}$) ($K_{S,B}$) that obeys the random-order equilibrium pathways of enzymatic reactions involving two substrates, where each substrate combines with its own specific site and there are few effects of one substrate on the affinity for the other.

REFERENCES

1. C. B. Pickett and A. Y. H. Lu. *Ann. Rev. Biochem.*, *58*:743–764 (1988).
2. D. J. Reed. *Biochem. Pharmacol.*, *35*:7–13 (1986).
3. H. Sies and A. Wendel, eds. *Glutathione, Functions in liver and kidney*. Springer Verlag, Berlin (1978).
4. E. Boyland and L. F. Chasseaud. *Adv. Enzymol.*, *32*:173–219 (1969).
5. A. J. Meister. *J. Biol. Chem.*, *263*:17205–17208 (1988).
6. I. M. Arias and W. B. Jacoby, eds. *Glutathione, Metabolism and Function*. Raven Press, New York (1976).
7. B. Mannervik and U. H. Danielson. *Crit. Rev. Biochem.*, *23*:283–337 (1988).
8. A. Meister and M. E. Anderson. *Ann. Rev. Biochem.*, *52*:711–760 (1983).
9. N. W. I. Chow, J. Wang-Peng, Ch. Kao-Shan, M. F. Tam, H. J. Lai, and P. Tu Ch. *J. Biol. Chem.*, *263*:12797–12800 (1988).
10. K. Sato. *Jpn. J. Cancer Res* (Gann), *79*:556–572 (1988).
11. K. Sato, K. Satoh, I. Hatayama, S. Tsuchida, Y. Soma, Y. Shiratori, N. Tateoka, Y. Inada, and A. Kitara. In *Glutathione S-Transferase and Carcinogenesis* (T. J. Mantle, C. B. Pickett, and J. D. Hayes, eds.). Taylor & Francis, London, pp. 127–137 (1987).
12. T. D. Boyer. *Hepatology*, *9*:48–496 (1989).
13. B. Nannervik. *Adv. Enzymol.*, *57*:357–417 (1985).
14. W. B. Jakoby and W. H. Habig. In *Enzymatic Basis of Detoxication* (W. B. Jacoby, ed.). Academic Press, New York, pp. 63–94 (1980).
15. M. J. Pabst, W. H. Habig, and W. B. Jakoby. *J. Biol. Chem.*, *247*:7140–7148 (1974).
16. W. H. Habig, M. J. Pabst, and W. B. Jakoby. *J. Biol. Chem.*, *249*:7130–7139 (1974).
17. W. J. Dixon and M. P. Brown. *BMDP Biomedical Computer Program*. University of California Press, Berkeley (1977).
18. I. H. Segel. *Enzyme Kinetics*. John Wiley, New York (1975).
19. J. R. Florini and C. S. Vestling. *Biochem. Biophys. Acta*, *251*:575–583 (1957).
20. W. J. Dixon and E. C. Webb. *Enzymes*. Longman Group, London (1979).
21. K. M. Ivanetich and R. D. Goold. *Biochem. Biophys. Acta*, *998*:1–13 (1989).

107

Imipramine Receptors: Protein Influence and Membrane Versus Whole Platelet Binding

Belén Arranz, Pilar Rosel, and Francisco Perez-Arnau *Hospital de Bellvitge, Barcelona, Spain*

I. INTRODUCTION

Although the existence of a decreased number of platelet ^{3}H-imipramine binding sites in various affective disorders has been widely described (1–4), several authors have also reported no difference in, or even higher levels of, these binding sites between depressed patients and normal controls (5,6).

These discrepancies could be due to several factors, i.e., ^{3}H-imipramine specific activity, ^{3}H-imipramine and desipramine concentrations, protein concentrations used in the assay, etc.

On the other hand, although most authors work with platelet membranes, using whole platelets could facilitate the assay by reducing the pellet isolation step.

Our aim has been to assess the influence of the protein concentration on ^{3}H-imipramine binding and to compare membrane and whole-platelet binding sites (B_{max}) and affinity (K_d).

Specific ^{3}H-imipramine binding was calculated as the difference between total and nonspecific binding (10 μL desipramine 8 mmol/L). B_{max} and K_d values were obtained from Scatchard analysis (EBDA program). Results were expressed as fmol/mg protein and nmol/L, respectively. Whole-platelet binding results were expressed as number of receptors/cell. Statistical evaluation was performed by Student's, Mann Whitney's, and Wilcoxon's tests.

II. MATERIALS AND METHODS

Protein influence was studied from 60 supposedly healthy volunteers (30 men and 30 women) aged 19 to 62 years. Membrane versus whole platelet binding was determined from 14 subjects (7 men and 7 women) aged 25 to 42 years. All samples were collected at 9.00 a.m. from March to May 1990. Subjects with current or past history of

Table 1 Mean Protein Concentration (mg/mL)

	$(\bar{x} \pm s)$	n	P
Men	0.81 ± 0.33	30	N.S.
Women	0.81 ± 0.32	30	N.S.
Whole sample	0.90 ± 0.34	60	N.S.

N.S. = not significant.

psychiatric or medical disorders were excluded from the study. All of the volunteers studied had Hamilton depression scores of less than 7.

^{3}H-Imipramine binding sites were determined following the method of Langer et al. (7) with slight modifications. Membranes were obtained from lysis and homogenization of platelet-rich plasma in a low-ionic-strength buffer. Lysis was not performed for the whole platelet binding. Membrane and whole-platelet protein concentrations were determined by the biuret method.

For the ^{3}H-imipramine assay, membrane aliquots (100 μL) were incubated with ^{3}H-imipramine (40.4 Ci/mmol) at six concentrations ranging between 0.5 and 5 nmol/L in a final volume of 500 μL for 60 min at 4°C. After incubation, membranes were rapidly filtered, under vacuum, through Whatman glass fiber filters. Radioactivity was then measured in a beta counter.

III. RESULTS

A. Protein Influence

Table 1 reflects the mean protein concentrations obtained for both sexes and in the whole group. Protein concentrations for men did not differ statistically from those obtained for women. When we considered a protein concentration cutoff value of 0.90 mg/mL (the mean concentration obtained in our group), the population was divided as shown in Table 2. We did not obtain statistically significant differences for B_{max} or K_d values between the two groups thus defined; neither did we find differences when we considered each sex independently (Figure 1).

Table 2 B_{max} and K_d Values for a Protein Concentration Cutoff Value of 0.90 (mg/mL)

	Protein Concentration		
	< 0.90 mg/mL	> 0.90 mg.mL	P
B_{max} (fmol/mg prot)	2672 ± 1069	2300 ± 1068	N.S.
K_d (nmol/L)	2.87 ± 1.61	2.93 ± 1.50	N.S.
Number of cases	32	28	

N.S. = not significant.

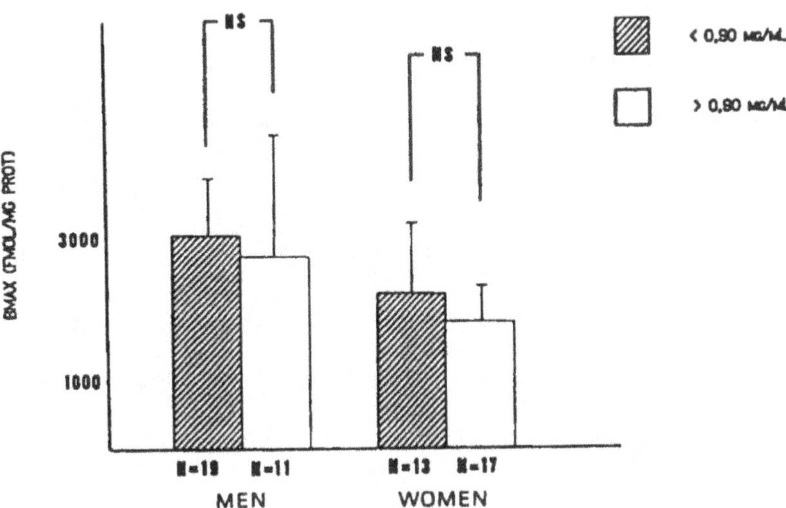

Figure 1 B_{max} values in both men and women for a protein concentration cutoff value of 0.90 mg/mL. NS = not significant.

B. Membrane Versus Whole Platelet Binding

As shown in Table 3, we found statistical differences for B_{max} and K_d values between membranes and whole platelets. The mean number of receptors per cell was 1691 ± 837.

IV. DISCUSSION AND CONCLUSIONS

The results shown above indicate that B_{max} and K_d values are not influenced by a protein concentration within the range studied by our group (0.56 to 1.24 mg/mL). Thus, the discrepancies observed by several authors regarding the ^{3}H-imipramine binding in depressed patients cannot be attributed to the different protein concentrations used in the assay. Other putative factors should be taken into account in order to explain these discrepancies. Moreover, the interferences inherent in the method for protein determination (Lowry method) used by most authors could explain the wide diversity of results.

On the other hand, and despite its longer isolation step, we highly recommend the utilization of membrane platelets, as described by most authors (7), due to their better affinity and higher binding sites. These results should be subjected to a deeper interpretation process, as the differences obtained between membrane and whole

Table 3 B_{max} and K_d Values for Membrane and Whole Platelet

	Membrane	Whole platelet	P
B_{max} (fmol/mg prot)	2915 ± 1049	2015 ± 663	$p < 0.05$
K_d (nmol/L)	2.11 ± 0.49	2.85 ± 0.71	$p < 0.05$
Number of cases	14	14	

platelet could be the result only of the receptor configuration state and not of the true presence of a higher number of binding sites in membrane platelets.

REFERENCES

1. K. B. Asarch, J. C. Shih, and A. Kulesar. Decreased ^{3}H-imipramine binding in depressed males and females. *Commun. Psychopharmacol.*, *4*:425–442 (1980).
2. R. Raisman, D. Sechter, M. S. Briley, E. Zarifian, and S. Z. Langer. High affinity binding in platelets from untreated and treated depressed patients compared to healthy volunteers. *Psychopharmacol.* *75*:368–371 (1981).
3. R. Raisman, M. S. Briley, F. Bouchami, D. Sechter, E. Zarifian, and S. Z. Langer. ^{3}H-Imipramine binding and serotonin uptake in platelets from untreated depressed patients and control volunteers. *Psychopharmacol.* *77*:332–335 (1982).
4. S. Z. Langer, and R. Raisman. Binding of ^{3}H-imipramine and ^{3}H-desipramine as biochemical tools for studies in depression. *Neuropharmacol.* 22:407–413 (1983).
5. P. M. Whittaker, J. J. Warsh, H. C. Stancer, E. Persad, and C. K. Vint. Seasonal variations in platelet ^{3}H-imipramine binding: Comparable values in control and depressed populations. *Psychiatr. Res.*, *11*:127–131 (1984).
6. M. Baron, A. Barkai, R. Gruen, S. Kowalik, and F. Quitkin. ^{3}H-imipramine platelet binding sites in unipolar depression. *Biol. Psychiatr.*, *18*:1403–1409 (1983).
7. S. Z. Langer, M. S. Briley, and R. Raisman. Specific ^{3}H-imipramine binding in human platelets: Influence of age and sex. *Naunyn-Schmiedeberg's Arch. Pharmacol.*, *313*:189–194 (1980).

108

Chemical Structure and Hepatotoxicity of Nonsteroidal Antiinflammatory Drugs in Isolated Liver Cells

M. P. Míguez Santiyan and J. D. Pedrera Zamorano *Facultad de Veterinaria, Cáceres, Spain*

I. INTRODUCTION

Nonsteroidal antiinflammatory drugs (NSAIDs) are a heterogeneous group of compounds, often chemically unrelated, which nevertheless share many therapeutic actions and side effects and have, as their major therapeutic use, the management of several types of rheumatic diseases. The side effects are of varying degrees of severity, including gastrointestinal irritation, skin hypersensitivity reactions, blood dyscrasias, and renal and hepatic impairment (1). A number of clinical studies have reported the hepatotoxicity of several drugs of this group in patients with no previous history of hepatic diseases (2–7). Recently, five cases of hepatitis associated with a commonly used NSAID, diclofenac, have been published (8). This had led to the withdrawal of some of these drugs (ibufenac, fenclozic acid, benoxaprofen) from clinical use. Hepatic alterations have also been reported with naproxen (9–11) and piroxicam (12).

In view of this well-documented NSAID-induced hepatotoxicity, a careful evaluation of potential hepatotoxicity of current and newly developed drugs is appropriate. In recent years isolated hepatocytes have proven to be a particularly useful in-vitro system for assessing the mechanism of the toxic effects of xenobiotics.

In the present work we have included two commonly used NSAIDs which are chemically unrelated: naproxen, which contains a carboxylic group, and piroxicam, which belongs to the oxicam group. Since in-vivo and in-vitro studies have demonstrated that NSAIDs can alter hepatic glucose metabolism (13,14), the aim of the present work was to demonstrate the toxicity in vitro of naproxen and piroxicam on glucose production using freshly isolated rat hepatocytes and to examine the possibility that the toxic action mechanism may be related to the different chemical structure of these NSAIDs.

II. MATERIALS AND METHODS

A. Chemicals

Collagenase, NAD, NADH, NADP, and NADPH were obtained from Boehringer Mannheim, lactate dehydrogenase and 3-hydroxybutyrate dehydrogenase from Sigma Chemical Co. (St. Louis); all other reagents used were of the highest analytical purity available from commercial sources.

Naproxen was a generous gift of Elmu, S. A. (Madrid), and piroxicam was a gift of Laboratorios Robert, S. A. (Barcelona). These NSAIDs were dissolved on the day of use in 100 mM Na_2CO_3 and the solution adjusted to pH 7.5-8.5 with 1.2 N HCl. Studies using comparable concentrations of Na_2CO_3/HCl showed no effect on any of processes measured.

B. Hepatocyte Isolation and Incubation

Hepatocytes were isolated from male, 24- to 48-h-starved Wistar rats 9250 to 300 g by a modification of the method of Berry and Friend (15) as previously described (16). Hepatocyte preparations used in this study were 85% to 90% viable on the basis of the trypan blue exclusion test. Cells (150 to 200 mg wet wt) in a final volume of 4 mL of Krebs-Henseleit bicarbonate were incubated at 37°C in 25-mL Erlenmeyer flasks in a shaking bath for 30 min under an atmosphere of 95% O_2/5%CO_2. Cells were exposed to several concentrations of the drugs, below and above the therapeutic plasma concentration (TCP) (naproxen, 0.25 mM; piroxicam, 0.025 mM). The reactions were stopped with perchloric acid (2% w/v final concentration), and the precipitate was removed by centrifugation. Neutralized supernatant was used for further analysis.

C. Gluconeogenesis and Redox State

Gluconeogenesis by hepatocytes from 48-h-starved rats was determined as the rate of glucose production in the incubations in the presence of gluconeogenic substrates (lactate, pyruvate, glycerol, fructose, or galactose). Glucose concentration in the incubation medium was measured after 0, 10, 20, and 30 min by a glucose oxidase method. The increase in glucose concentration was linear with time over the incubation period.

Lactate, pyruvate, 3-hydroxybutyrate, and acetoacetate were measured enzymatically in the cells from 24-h-starved rats, by the following methods: lactate (17), pyruvate (18), 3-hydroxybutyrate (19), and acetoacetate (20).

D. Statistical Analysis

The data obtained from hepatocyte studies were analyzed utilizing the ANOVA (analysis of variance) program. Results are expressed as the mean ± standard deviation of 5 to 7 experiments.

III. RESULTS

The effect of two structurally distinct nonsteroidal antiinflammatory drugs (naproxen and piroxicam) on gluconeogenesis in isolated rat hepatocytes has been investigated. Addition of these NSAIDs, at a range of concentrations equivalent to 1 to 20 times the therapeutic plasma concentration (TPC), to a suspension of freshly isolated hepatocytes did not affect viability, as judged by the trypan blue exclusion test. Figure 1 shows as a dose-response curve for the inhibition of glucose by naproxen and piroxicam. It can be seen that even at a concentration equivalent to 20 times the TPC, the greatest inhibition of piroxicam is not achieved.

Tables 1 and 2 show how the two NSAIDs produced a clear dose-dependent inhibition of glucose production from lactate, pyruvate, and glycerol, although the response was different for each drug tested, and they had no effect on gluconeogenesis from fructose and galactose. The highest level of inhibition found with the carboxylic acid NSAID (naproxen) was observed from lactate, while gluconeogenesis from pyruvate was the least affected. However, although the highest level of inhibition caused by piroxicam was also observed from lactate, in this case the least level of inhibition was found with glycerol as substrate.

Naproxen was more active at impairing gluconeogenesis from lactate and glycerol than was piroxicam. In the incorporation of this substrates to the gluconeogenic pathway, two enzymes play an important role, lactate dehydrogenase and glycerol phosphate dehydrogenase. These enzymes require NAD^+ for their action, therefore a modification in the supply of this co-factor could affect the gluconeogenic flux. The

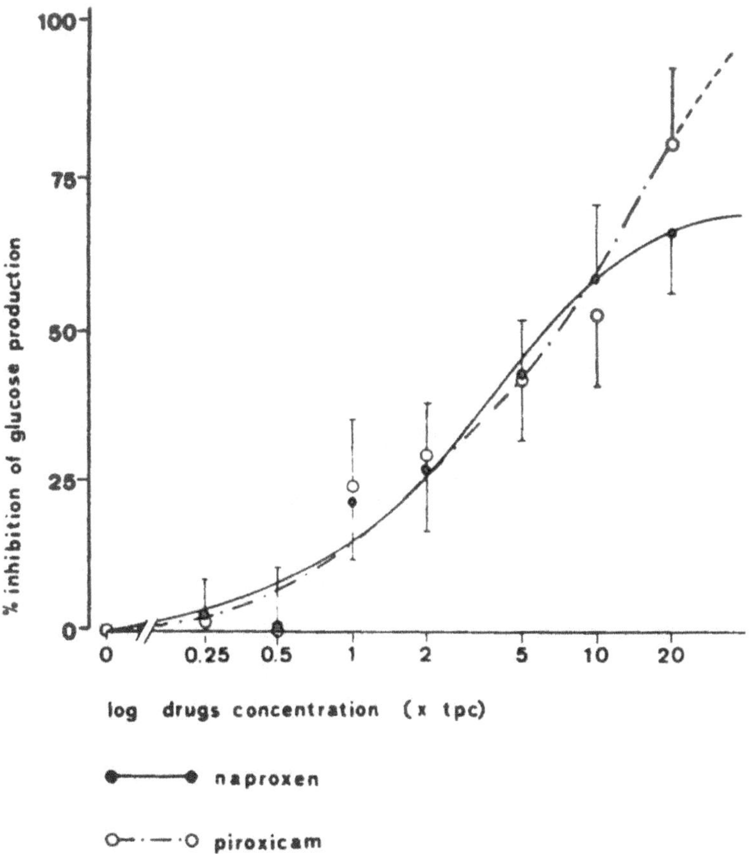

Figure 1 Dose-response curve of the effects of naproxen and piroxicam on gluconeogenesis from pyruvate in isolated hepatocytes.

Table 1 Effects of Naproxen on Gluconeogenesis for Different Substrates in Isolated Hepatocytes

Naproxen (x TPC)	Lactate	Pyruvate	Glycercol	Fructose	Galactose
	(n mol/min/g wet wt)				
0	390 ± 130[a]	516 ± 189[a]	575 ± 192[a]	1795 ± 139[a]	520 ± 169[a]
1	228 ± 97[b]	408 ± 100[b]	404 ± 103[b]	1582 ± 231[a]	545 ± 127[a]
2	181 ± 79[b]	380 ± 96[b]	350 ± 91[b]	1855 ± 355[a]	474 ± 91[a]
5	979 ± 39[c]	301 ± 98[c]	288 ± 116[c]	1895 ± 247[a]	476 ± 92[a]

Isolated hepatocytes obtained from 48-h-fasted rats were incubated at 37°C for 30 min with the indicated substrates in the presence and absence of naproxen. Final concentration of substrates was 10 mM. Glucose production was determined as the difference between glucose level in hepatocytes at zero time and after 30 min of incubation and is expressed as nmol glucose/min/g wet weight. The results are the mean ± standard deviation of 5 to 7 duplicate experiments.
[a]Control group or with a response similar to control.
[b]Group whose response was homogeneous but different with regard to group a or group c.
[c]Group whose response was homogeneous but different with regard to group a or group b.

redox state of the NADH/NAD$^+$ in the cytosol has been indirectly assessed by measuring the lactate/pyruvate concentration ratio. In hepatocytes incubated without exogenous substrates, the lactate/pyruvate ratio was 4.94 (Figure 2). Addition of naproxen caused an elevation of this ratio at the TPC, reflecting a more reduced cytosolic state. Piroxicam had no effect on the lactate/pyruvate ratio.

The redox state of the NADH/NAD$^+$ coupling in the mitochondria has been assessed by measuring the 3-hydroxybutyrate/acetoacetate concentration ratio. In control cells incubated without exogenous substrates, the 3-hydroxybutyrate/acetoacetate ratio was 0.33 (Figure 3). Naproxen produced a decrease in this ratio that was significant only at 10 × TPC and is indicative of a more oxidized redox state. Piroxicam, however, seems to have a biphasic effect. At low concentrations, piroxicam

Table 2 Effects of Piroxicam on Gluconeogenesis for Different Substrates in Isolated Hepatocytes

Piroxicam (x TPC)	actate	Pyruvate	Glycercol	Fructose	Galactose
	(n mol/min/g wet wt)				
0	390 ± 130[a]	516 ± 189[a]	575 ± 192[a]	1795 ± 139[a]	520 ± 169[a]
1	286 ± 105[b]	386 ± 109[b]	464 ± 143[b]	1806 ± 404[a]	497 ± 102[a]
2	244 ± 103[b]	368 ± 135[b]	451 ± 120[b]	1743 ± 430[a]	469 ± 164[a]
5	149 ± 64[c]	304 ± 108[c]	397 ± 116[b]	1386 ± 308[b]	450 ± 150[a]

Isolated hepatocytes obtained from 48-h-fasted rats were incubated at 37°C for 30 min with the indicated substrates in the presence and absence of naproxen. Final concentration of substrates was 10 mM. Glucose production was determined as the difference between glucose level in hepatocytes at zero time and after 30 min of incubation and is expressed as nmol glucose/min/g wet weight. The results are the mean ± standard deviation of 5 to 7 duplicate experiments.
[a]Control group or with a response similar to control.
[b]Group whose response was homogeneous but different with regard to group a or group c.
[c]Group whose response was homogeneous but different with regard to group a or group b.

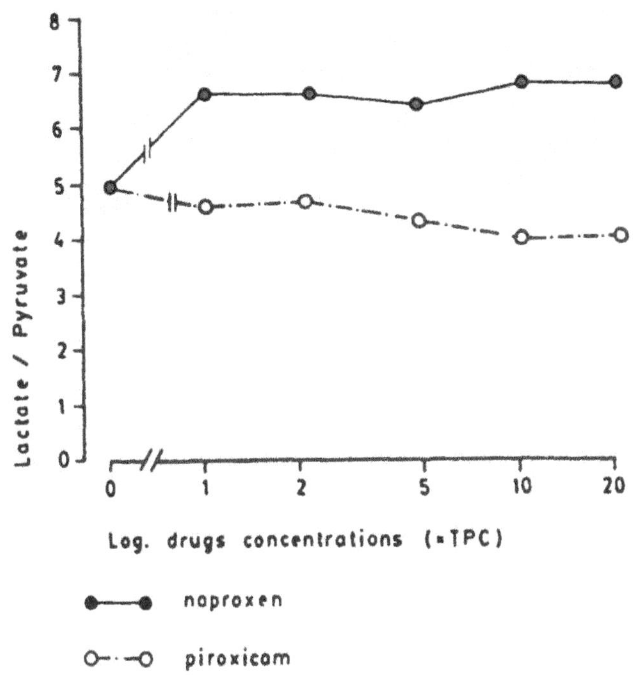

Figure 2 Effects of naproxen and piroxicam on the cytosolic redox state.

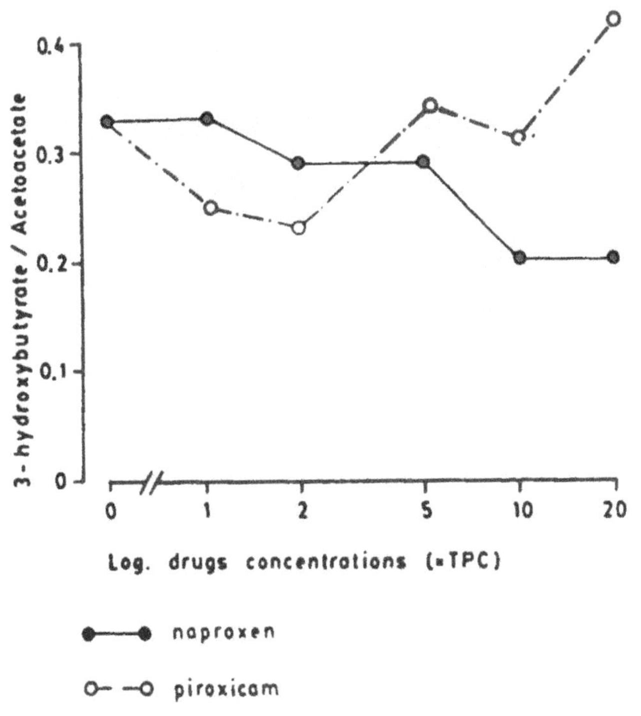

Figure 3 Effects of naproxen and piroxicam on the mitochondrial redox state.

caused a decrease of the 3-hydroxybutyrate/acetoacetate ratio; and the highest concentration produced an increase in this ratio.

IV. DISCUSSION

The present study demonstrates that naproxen and piroxicam inhibit gluconeogenesis from several common hepatic glucose precursors in isolated liver cells derived from 48-h-fasted rats. The results suggest that the inhibition of glucose formation may occur mainly before the triosephosphate level, since the glucose production from fructose and galactose was not modified by these drugs. A possible effect of these NSAIDs at the level of either pyruvate carboxylase or phosphoenol-pyruvate carboxykinase cannot be rejected. When the effects of each NSAID were compared, naproxen was more active at impairing glucose production from lactate and glycerol than piroxicam. The NSAID-inhibited gluconeogenesis from glycerol may also be at some of the steps between the glycerol and the triosephosphate level.

Since lactate dehydrogenase and glycerolphosphate dehydrogenase are two key enzymes necessary to the incorporation of lactate and glycerol in the gluconeogenic route, a modification of these enzymes or in some of their co-factors could be suggested.

Lactate/pyruvate ratio indicative of the cytosolic NADH/NAD^+ ratio was determined in hepatocytes exposed to several concentrations of the drugs. The increase in this parameter produced by naproxen could suggest a minor availability of NAD^+ for these enzymes, due to a more reduced cytosolic space. However, the 3-hydroxybutyrate/acetoacetate ratio in hepatocytes was lowered by naproxen, reflecting a more oxidized redox state in the mitochondria. This more oxidized redox state could be due to: (a) increased transfer of reducing equivalents to the cytosol; or (b) increased oxidation of NADH by the electron-transport chain (21). Both explanations could occur with naproxen. On the one hand, more reducing equivalents have been found in the cytosol as judged by the increase lactate/pyruvate ratio observed, which could stem from the mitochondrial compartment. On the other hand, the NSAIDs have been widely reported (16,22,23) as uncoupling agents, increasing oxidation of NADH by the electron-transport chain.

The redox state of the cytosolic and mitochondrial NAD system showed by naproxen in our experiments reflect this uncoupling effect. In a manner similar to livers perfused with classical uncouplers (24), naproxen reduced the cytosolic NAD system while it was oxidized simultaneously in the mitochondria compartment, suggesting that rates of cytosolic NADH production via glucolysis, and mitochondrial consumption, via respiration occur at an increased rate during uncoupling.

Nevertheless, piroxicam acted in a different manner on cellular redox state. It did not affect the lactate/pyruvate ratio and showed a biphasic behavior on the 3-hydroxybutyrate/acetoacetate ratio. First, at low concentration (1 to 2 × TPC), it diminished the 3-hydroxybutyrate/acetoacetate ratio; second, at 5 to 10 × TPC, the ratio remain unchanged; and finally, at the highest assayed concentration (20 × TPC), the ratio showed an elevation. These results are difficult to explain in relation to the piroxicam-inhibited gluconeogenesis. Further investigations are necessary to elucidate these effects.

Gluconeogenesis also depends on ionic calcium release from intracellular stores and, more basically, depends on mitochondrial metabolism (25).

These different effects caused by the carboxylic acid NSAID (naproxen) and the other NSAID, which is not a carboxylic acid (piroxicam), are consistent with those reported recently (26), demonstrating that NSAIDs which are carboxylic acids interfere with intracellular sequestration of calcium and membrane binding of calcium in liver tissue. These effects would tend to increase cytoplasmic calcium concentrations, an effect known to stimulate glucogenolysis (27). In fact, this occurs with NSAIDs which are carboxylic acids. However, piroxicam did not stimulate glycogenolysis (14), nor did it interfere with hepatic calcium sequestration (26). It has also been suggested that mitochondrial sequestration is dependent on the mitochondrial redox state (28).

However, further investigations are required to establish if a relationship exists between NSAID-inhibited gluconeogenesis and the cellular calcium level. The nonsteroidal antiinflammatory drugs are generally administered for long periods and could cause subclinical hepatotoxicity in certain metabolic disorders or in susceptible individuals.

Although some of the effects described in this work are produced at high drug concentrations, the results would indicate that it is necessary to control the drug plasma level in NSAID-treated patients.

V. SUMMARY

The effects of two chemically unrelated nonsteroidal antiinflammatory drugs (NSAIDs), naproxen and piroxicam, on isolated rat hepatocytes were examined. Addition of these NSAIDs at a range of concentrations equivalent to 1 to 20 times the therapeutic plasma concentration to a suspension of freshly isolated cells did not affect cell viability, as judged by the trypan blue exclusion test. Naproxen, which contains a carboxylic acid group, and piroxicam, which is an oxicam, both showed a markedly dose-dependent inhibition of gluconeogenesis from lactate, pyruvate, or glycerol starting from the therapeutic plasma concentration, but the effects were dependent on the gluconeogenic substrate used. Neither of the two affected the glucose production from fructose or galactose.

On the other hand, this NSAID, which is a carboxylic acid, increased the lactate/pyruvate ratio and decreased the 3-hydroxybutyrate/acetoacetate ratio. Nevertheless, piroxicam, which is not a carboxylic acid, did not affect the lactate/pyruvate ratio and showed biphasic behavior on the 3-hydroxybutyrate/acetoacetate ratio. These observations are consistent with the hypothesis that some early functional changes may be mediated by altered ion homeostasis, since it has been demonstrated that carboxylic acid NSAIDs interfere with intracellular sequestration of calcium while piroxicam does not show this effect.

A different toxic mechanism therefore seems to exist for chemically different NSAIDs.

REFERENCES

1. K. Rainsford. Side-effects of anti-inflammatory/analgesic drugs: Renal, hepatic and other systems. *Trends Pharmacol. Sci.*, 205–209 (1984).
2. M. G. Cuthbert. Adverse reactions to non-steroidal anti-inflammatory drug. *Curr. Med. Res. Doin.*, 2:600–610 (1974).
3. P. Gaspareto. Un caso di ittero epatocellulare in corso di tratamento con ibuprofen. *Minerva Pediatrica*, 26:531–533 (1974).

4. D. A. Stempel and J. J. Miller. Lymphopenia and hepatic toxicity with ibuprofen. *J. Pediatr., 90*:657 (1977).
5. D. J. Stennett, W. Simonson, and A. C. Hall. Fenoprofen induced hepatotoxicity. *Ann. J. Hosp. Pharm., 35*:901 (1978).
6. P. Sternlieb and R. M. Robinson. Stevens-Johnson syndrome plus toxic hepatitis due to ibuprofen. *N.Y. State J. Med., 78*:1239 (1978).
7. H. M. C. A. Taggart and J. N. Alderice. Fatal cholestatic jaundice in elderly patients taking benoxaprofen. *Br. Med. J., 284*:1372 (1982).
8. T. J. Iveson, N. G. Ryley, P. M. Kelly, J. M. Trowell, J. O. McGee, and R. W. Chapman. Diclofenac associated hepatitis. *J. Hepatol., 10*:85–89 (1990).
9. L. Giarelli, G. Falconieri, and M. Delendi. Fulminant hepatitis following naproxen administration. *Human Pathol., 17*:10 (1986).
10. B. H. Bass. Jaundice associated with naproxen. *Lancet, 1*:998 (1974).
11. I. P. Law and H. Knight. Jaundice associated with naproxen. *N. Engl. J. Med., 295*:1–201 (1976).
12. M. Rahman and Col. *Clin. Pharmacol. Ther., 25*:243 (1979), cited in *Martindale, The Extra Pharmacopeia*, (James E. F. Reynolds, ed.), The Pharmaceutical Press, London, p. 275 (1982).
13. S. Ganguli, M. A. Sperling, C. Frame, and R. Christensen. Inhibition of glucagon-induced hepatic glucose production by indomethacin. *Am. J. Physiol., 236*:E258–E365 (1979).
14. E. P. Brass and M. J. Garrity. Effect of non-steroidal anti-inflammatory drugs on glycogenolysis in isolated hepatocytes. *Br. J. Pharmacol., 86*:491–496 (1985).
15. M. N. Berry and D. S. Friend. High yield preparation of isolated rat liver parenchymal cells. *J. Cell. Biol., 43*:506–520 (1969).
16. M. P. Miguez, A. Jorda, and J. Cabo. Inhibition of ureogenesis in isolated rat liver cells by a non-steroidal anti-inflammatory drug (Butibufen). *Biochem. Pharmacol., 35*:2145–2148 (1986).
17. I. Gutmann and A. W. Wahlefeld. Determination of L-(+)-lactate with lactate dehydrogenase and NAD. In *Methods of Enzymatic Analysis* (H. U. Bergmeyer, ed.). Academic Press, New York, vol. 4, pp. 1464–1468 (1974).
18. R. Czok and W. Lamprecht. Determination of pyruvate. In *Methods of Enzymatic Analysis*, 2nd ed. (H. U. Bergmeyer, ed.), Academic Press, New York, vol. 4, pp. 1446–1451 (1974).
19. D. H. Williamson and J. Mellamby. Determination of D-(–)-3-hydroxybutyrate with 3-HBDH. In *Methods of Enzymatic Analysis* (H. U. Bergmeller, ed.), Academic Press, vol. 4, pp. 1836–1839 (1974).
20. J. Mellamby and D. H. Williamson. Determination of acetoacetate with 3-HBDH. In *Methods of Enzymatic Analysis* (H. U. Bergmeyer, ed.), Academic Press, New York, vol. 4, pp. 1840–1843 (1974).
21. L. Agius, M. H. Chowdhury, and G. M. M. Alberti. Regulation of ketogenesis, gluconeogenesis and the mitochondrial redox state by dexamethasone in hepatocytes monolayer cultures. *Biochem. J., 239*:593–601 (1986).
22. J. P. Famaey and J. Mockel. Importance of sulfhydryl groups for the uncoupling activity of non-steroidal acidic anti-inflammatory drugs and valinomycin in oxidative phosphorylation. *Biochem. Pharmacol., 22*:1487–1498 (1973).
23. Y. Tokumitsu, S. Lee, and M. Ui. "In vitro" effects of non-steroidal anti-inflammatory drugs on oxidative phosphorylation in rat liver mitochondria. Biochem. Pharmacol., 26:2101–2106 (1977).
24. R. Elbers, S. Soboll, and H. G. Kampffmeyer. Alterations in cellular intermediary metabolism by 4-dimethylaminophenol in the isolated perfused rat liver and the implications for 4-dimethylaminophenol toxicity. *Biochem. Pharmacol., 29*:1747–1753 (1980).
25. N. Kraus-Friedmann. Hormonal regulation of hepatic gluconeogenesis. *Physiol. Rev., 64*:170–259 (1984).
26. R. M. Burch, W. C. Wise, and P. V. Halushka. Prostaglandin-independent inhibition of calcium transport by non-steroidal anti-inflammatory drugs: Differential effects of carboxylic acids and piroxicam. *J. Pharmacol. Ther., 227*:84–91 (1983).
27. J. H. Exton. Regulation of carbohydrate metabolism by cyclic nucleotides. In *Cyclic Nucleotides II, Handbook of Experimental Pharmacology* (J. H. Kehabian and J. A. Nathanson, eds.), Springer-Verlag, Berlin and Heidelberg, vol. 58, pp. 3–88 (1982).
28. S. Orrenius, S. A. Jewell, G. Bellomo, H. Thor, D. P. Jones, and M. T. Smith. Regulation of calcium compartementation in the hepatocyte. A critical role of glutathione, In *Functions of Glutathione: Biochemical, Physiological, Toxicological and Clinical Aspects* (A. Larsson, S. Orrenius, A. Holmgren, and B. Mannervick, eds.), Raven Press, New York, p. 261 (1983).

109

DNA Synthesis Stimulation by Aluminum Ions in Human Fibroblasts Cultured in Vitro

Carmen Domínguez,* Antonio Moreno, and Angel Ballabriga*** *Hospital Infantil Vall d'Hebrón, Barcelona, Spain, and **Hospital Son Dureta, Palma de Mallorca, Spain*

I. INTRODUCTION

Aluminum (Al) is present in very small amounts in living organisms but is abundant in the environment. Although it is not considered an essential element in the diet and has no known physiological function, it exerts certain biological effects, and, during recent years, an increasing number of toxic effects have been attributed to it.

It has become generally accepted that Al may cause intoxication and disease in some rather artificial situations, e.g., chronic hemodialysis (1,2) and heavy occupational exposure to Al fumes or dust of soluble Al compounds (3). Due to its unquestionably neurotoxic effects, Al has been implicated in the etiology of several human brain diseases. These conditions include Alzheimer's disease, parkinsonism-dementia of Guam, and amyotrophic lateral sclerosis (4,5). Increased amounts of Al have been found in the brain tissue of these patients and have been identified at the core of senile plaques of Alzheimer's patients (6). Other studies also reveal that Al binds to chromatin of the cell nucleus and has a high affinity to DNA, probably to the phosphate groups, as well as to RNA and many nucleotides (7,8). The most documented effect of Al is probably hexokinase inhibition, which is caused by the tendency of ATP to form stronger complexes with Al than with Mg (9). Other enzymes inhibited by Al are adenylate cyclase (10,11), alkaline phosphatase (12), and 3',5-cyclic nucleotide phosphodiesterase, probably due to high-affinity interaction between Al and calmodulin, the normal activator of the enzyme (13).

Cultured cells permit study of the biological actions of Al at both natural and toxicological concentrations. The importance of trace metal ions in the stimulation and maintenance of growth of mammal cells cultured in vitro has been demonstrated with Zn, Se, Cu, and Va, which increase clonal growth in some cell types. Al is found in the environment in greater amounts than any of these elements, and its ability to influence cellular metabolism has been partially demonstrated.

The stimulatory effect of Al ions on DNA synthesis and cellular mitosis has been studied in Nakano mouse lens epithelial (NMLE) cells and Swiss 3T3K cells (14,15).

Previously, the mitogenic capacity of Al was shown (15) to be greatly potentiated by insulin or cholera toxin, and that the maximal stimulation of DNA synthesis was 400 to 550 μg/L in cultures of 3T3 and 3T6 cells.

The aim of this study was to evaluate the effect of Al on DNA synthesis and cellular division in human fibroblasts cultured in vitro, using "high" and "low" concentrations of Al and different periods of incubation.

II. MATERIALS AND METHODS

A. Cell Culture

Human fibroblasts from skin biopsies were cultured by standard methods. Minimal Essential Medium (MEM), supplemented by 10% fetal calf serum (FCS) was used for cell growth and until the cells were assayed. When cells reached confluency, they were routinely trypsinized with 0.25% trypsin. Fibroblasts used in these experiments were not used until the fifth passage, when they had attained cellular homogeneity.

B. Aluminum Incorporation

Cells were trypsinized and plated in 96-well culture plates with Dulbecco's medium (DM) containing 10% FCS and 10 mM Hepes buffer, pH 7.4. Cells were plated at 10,000 cells/well; after the 24 h necessary for cell adhesion to the culture surface, cells were washed with serum-free medium. Subsequently, serum-free Dulbecco's Medium was added; cultures were kept with this medium for 1 or 2 days until cell growth was stopped (quiescent cultures). Cell exposure to different concentrations of Al was carried out with a standard aluminum nitrate (BDH) solution diluted in DM with 2% FCS; these Al solutions were incorporated into cell cultures.

C. Thymidine Incorporation

After the incorporation periods established for each experiment, ^{3}H-thymidine (Amersham) (1 μCi/mL) was added to the wells for an additional 24-h incubation period. TCA-soluble material was then extracted by two washings with Hank's isotonic solution, followed by a further two washings with ice-cold 10% TCA. A fixing solution of ethanol:ether (3:1) was later added, with which cultures were kept for an hour. Finally, cultures were digested with 1 N NaOH, and radioactivity was counted by means of scintillation cocktail.

D. Cell Counts

Fibroblast cultures used for cell-count experiments were managed basically in the same way as those described above, and were carried out parallel to ^{3}H-thymidine incorporation. Twenty-four-well plates with a culture surface of 1.9 cm^2 were used; 100,000 cells were plated in each well. At the end of the incubation periods with Al, cells were washed with 0.9% saline serum, trypsinized, and finally counted by microscope with a Neubauer chamber.

E. Statistical Methods

Repeated measure analyses of variance with Scheffe's test were used to compare data. All values were mean ± SEM. A p value of 0.05 were used to judge significance.

F. Aluminum Determination

On a regular basis we measured the aluminum content of culture media to check the aluminum concentration actually incorporated into the assays.

Aluminum analysis was made with a model 3030 B atomic absorption spectrometer (Perkin-Elmer), a HGA-400 graphite furnace, and an AS 40 autosampler.

Instrumental conditions were as follows: wavelengths, 309 nm; slit, 0.7 nm; signal, peak area; lamp current, 25 mA; tube type, graphite-coated with L'vov platform.

Graphite furnace parameters were as follows:

Step 1: temp., 110°C; ramp time, 5 s; hold time, 30 s
Step 2: temp., 250°C; ramp time, 15 s, hold time, 30 s
Step 3: temp., 1500°C; ramp time, 1 s; hold time, 45 s
Step 4: temp., 2500°C; ramp time, 0 s; hold time, 3 s; miniflow
Step 5: temp., 2600°C; ramp time, 1 s; hold time, 3 s
Step 6: temp., 20°C; ramp time, 10 s; hold time, 5 s

III. RESULTS

In this study we assessed the stimulatory capacity of Al ions on ^{3}H-thymidine incorporation into DNA in cultured quiescent human fibroblasts, using two groups of Al concentrations. In the first group of experiments, Al concentrations used were 50, 100, 150, and 200 μg/L and incubation periods were 1, 2, 3, and 5 days (Figure 1). In the second group, the concentrations used were 250, 500, 1000, 1500, and 2000 μg/L, and incubation periods were 1, 2, 4, and 5 days (Figure 2).

When the incubation period is 1 day, Al does not stimulate ^{3}H-thymidine incorporation at any of the concentrations used (Figure 1A). The stimulatory capacity of Al begins to be significant only when Al concentrations of 100 μg/L are used in the culture medium, and occurs when incubation periods are of 2, 3, and 4 days (Figure 1B, 1C, and 1D). Aluminum induces significant increases in DNA synthesis which rise progressively as incubation periods are prolonged and also when concentrations are increased (Figure 1). Thus, Al ions add to the stimulatory capacity of 2% FCS on DNA synthesis, which is quantified in a range starting from 31% above the control when 100 μg/L in culture medium are used after 2 days of incubation and reaches 202% after 5 days of incubation with 200 μg/L of Al. Figure 2 shows the effect of Al on DNA synthesis in the second group of experiments with higher Al concentrations in a range of 250 to 2000 μg/L.

The addition of increasing amounts of Al to the culture medium produces greater DNA synthesis stimulation, which also increases significantly when incubation periods are prolonged from 1 to 2, 4, and 5 days. In this way, the Al concentration of 500 μg/L and 2 days of incubation increases stimulation by 1.3-fold above the control, 2.6-fold after 4 days, and 3.4-fold after 5 days. The most notable effect of Al can be seen in the 5-day incubation period, in which Al significantly and progressively increases the stimulatory ability of 2% FCS in quiescent human fibroblasts from the concentration of 250 μg/L (252% above the control) up to 2000 μg/L (422% above the control).

In contrast, we assessed cell growth of cultured human fibroblasts in Al concentrations ranging from 250 to 2000 μg/L and with incubation periods of 1 to 8 days, and found slight but not significant stimulation at 5 and 8 days, but much lower than the percentage of stimulation on DNA synthesis—3% above the control with an Al

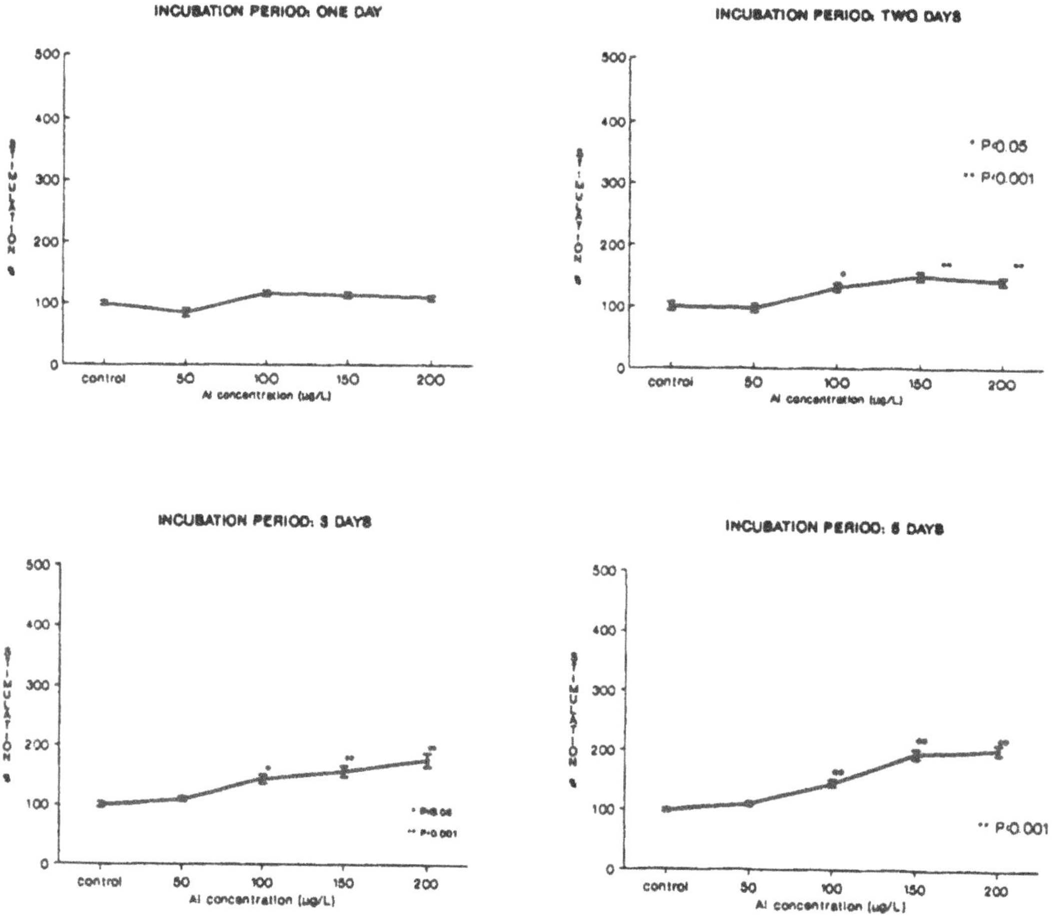

Figure 1 Effect of aluminum on ^{3}H-thymidine incorporation in human fibroblasts in culture. Each point represents the mean of 16 to 32 wells. Vertical bars represent standard error of the means. The percentages of stimulation are calculated in each experiment above the mean of controls (quiescent cells incubated with DM + 2% FCS).

concentration of 500 μg/L at 5 days' incubation—compared with 241% stimulation of DNA synthesis under the same conditions (Figure 3).

IV. DISCUSSION

Results of the present in-vitro study show clearly that, in cultured human fibroblasts, Al alone produces increases in ^{3}H-thymidine incorporation into DNA which begin to be significant only after 2 days of incubation. The presence of Al in the culture medium at concentrations varying from 50 to 2,000 μg/L induces dose-dependent increases in DNA synthesis. Minimum Al requirements for stimulating ^{3}H-thymidine incorporation in vitro are 100 μg/L; this concentration is much higher than that considered normal in serum, which should be lower than 10 μg/L. The former concentration is similar to that which produces toxic effects in Al-intoxicated uremic

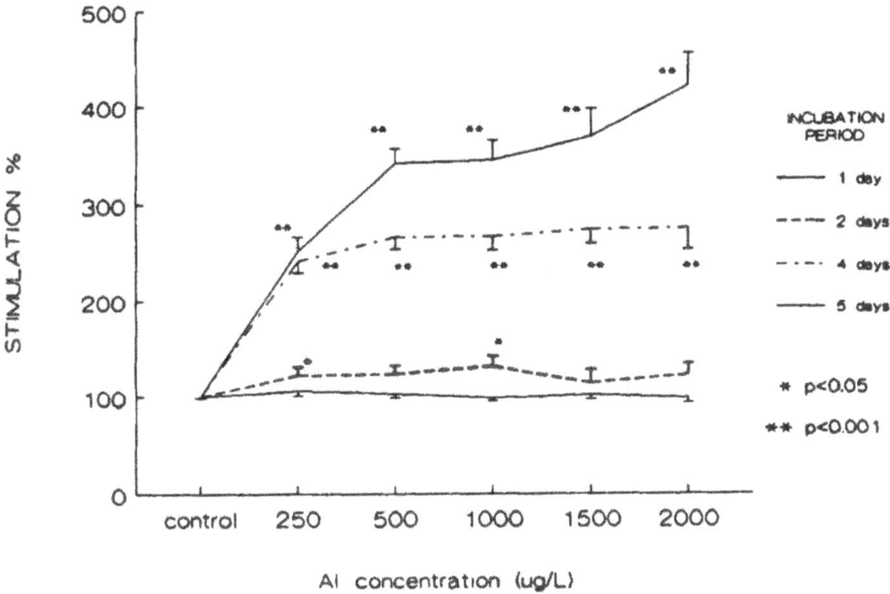

Figure 2 Effect of Al ions on ^{3}H-thymidine incorporation in human fibroblasts in culture. Each point represents the mean of 14 to 112 wells. Vertical bars represent standard error of the means. The percentages of stimulation are calculated above the controls (quiescent cells incubated with DM + 2% FCS).

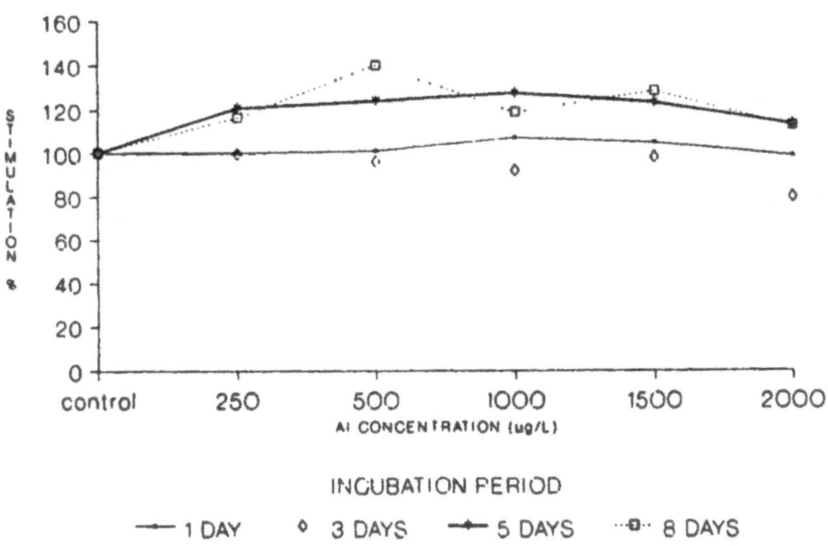

Figure 3 Effect of aluminum on cell number in human fibroblasts in culture. Each point represents the mean of 10 to 12 wells counted, calculated as percentage above the controls (quiescent cells incubated with DM + 2% FCS).

patients (serum Al of 100 μg/L). However, serum Al concentrations may not reflect Al levels in those tissues in which its toxicity has been demonstrated.

Under our experimental conditions, the maximum response to Al of cultured human fibroblasts occurs when a concentration of 2000 μg/L is used; this concentration was tested by evaluating Al by atomic absorption spectroscopy at the beginning and end of the incubation period, although it causes no harm to cellular morphology or cell adhesion to the culture surface (observations made with a phase-contrast microscope). Smith (15) found that Al induces a twofold increase in nontransformed Swiss 3T3 cells and a fivefold increase in transformed 3T6 cells on ^{3}H-thymidine incorporation. His results in nontransformed cells are comparable to ours with human fibroblasts for short incubation periods, although our data more clearly show the action of Al ions at longer incubation periods and their dose-dependent stimulatory action.

Using different cell lines, Al produces approximately sevenfold increases in 3T3K cells and up to 21-fold in Nakano mouse lens epithelial cells (14). Comparison of data obtained with several cell types indicates that different cell-type sensitivity exists; thus, there are varying degrees of response of cell lines to Al.

In our experiments with cultured human fibroblasts, we have failed to demonstrate the mitogenic capacity of Al ions alone as previously described (14,15) in other cell types with greater division capacity. Further studies are required to ascertain whether the synergistic action of other mitogenic agents is necessary for stimulation of cell division in human fibroblasts.

V. CONCLUSIONS

Our results indicate that Al already has an effect on DNA synthesis at the lowest concentration used (50 μg/L), although this is significant only after 5 days of incubation ($p = 0.001$). The action of Al is clearly evident, and stimulates DNA synthesis from the concentration of 100 μg/L in culture medium, with an incubation period of 2 days: 24% stimulation above the control; $p < 0.001$. This stimulation capacity becomes progressively greater when the Al concentration is increased—322% above the control with Al concentrations of 2000 μg/L, compared with 151% with 250 μg/L at 5 days of incubation—and when the incubation period is prolonged: 86% above the control with Al concentration of 200 μg/L and 5 days of incubation, compared with 31% at 2 days of incubation.

In this study, we also observed that Al ions do not significantly increase cellular division in cultured human fibroblasts in any of the concentrations or periods of incubation experimented.

In summary, we found Al to possess a remarkable capacity for stimulating DNA synthesis, which contrasts sharply with its poor mitogenic activity on human fibroblasts. Our results indicate that for Al to stimulate cell division, the presence of a further mitotic agent would be required.

REFERENCES

1. A. C. Alfrey, A. Hegg, and P. Craswell. Metabolism and toxicity of aluminum in renal failure. *Am. J. Clin. Nutr.*, 33:1509 (1980).
2. M. R. Wills and J. Savory. Aluminum poisoning: Dialysis encephalopathy, osteomalacia and anaemia. *Lancet*, *ii*:29 (1983).

3. I. Norseth. Aluminum. In *Handbook on the Toxicology of Metals* (L. Friberg, G. H. Nordberg, and V. B. Vouk, eds.), Elsevier/North-Holland. Biomedical Press, New York, p. 275 (1979).
4. D. Perl, D. C. Gajdusek, R. M. Garruto, R. T. Yanagihara, and C. J. Gibbs, Jr. Intraneuronal aluminum accumulation in amyotrophic lateral sclerosis and Parkinsonism-dementia of Guam. *Science, 217*:1053 (1982).
5. R. M. Garruto, R. Fukatsu, R. Yanagihara, D. C. Gajdusek, G. Hook, and C. E. Fiori. Imaging of calcium and aluminum in neurofibrillary tangle-bearing neurons in parkinsonism-dementia of Guam. *Proc. Natl. Acad. Sci. USA, 81*:1875 (1984).
6. D. P. Perl and A. R. Brody. Alzheimer's disease: x-ray spectrometric evidence of aluminum accumulation in neurofibrillary tanglle-bearing neurons. *Science, 208*:297 (1980).
7. S. J. Karlik, G. L. Eichhorn, and D. R. Crapper. Molecular interactions of aluminum with DNA. *Neurotoxicology, 1*:83 (1980).
8. C. A. Miller and E. M. Levine. Effects of aluminum salts on cultured neuroblastoma cells. *J. Neurochem., 22*:751 (1974).
9. J. C. K. Lai and J. P. Blass. Inhibition of brain glycolysis by aluminum. *J. Neurochem., 42*:438 (1984).
10. P. C. Sternweis and A. G. Gilman. Aluminum. A requirement for activation of the regulatory component of adenylate cyclase by fluoride. *Proc. Natl. Acad. Sci. USA, 79*:4888 (1982).
11. J. M. Mansour, A. Ehrlich, and T. E. Mansour. The dual effect of aluminum as activator and inhibitor of adenylate cyclase in the liver fluke *Fasciola hepatica. Biochem. Biophys. Res. Commun., 112*:911 (1983).
12. R. Rej and J. P. Bretaudiere. Effects of metal ions on the measurement of alkaline phosphatase activity. *Clin. Chem., 26*:423 (1980).
13. N. Siegel and A. Haug. Aluminum interaction with calmodulin. Evidence for altered structure and function from optical and enzymatic studies. *Biochem. Biophys. Acta, 744*:36 (1983).
14. T. R. Jones, D. L. Antonetti, and T. W. Reid. Aluminum ions mitosis in murine cells in tissue culture. *J. Cell. Biochem., 30*:31 (1986).
15. J. B. Smith. Aluminum ions stimulate DNA synthesis in quiescent cultures of Swiss 3T3 and 3T6 cells. *J. Cell. Physiol., 118*:298 (1984).

110

Modulation of Human Microsomal Aminopyrine N-Demethylase Activity by Cimetidine, Caffeine, and Demethylated Metabolites

José Augusto García-Agúndez, Antonio Luengo, and Julio Benítez *University of Extremadura, Badajoz, Spain*

I. INTRODUCTION

The ability to metabolize xenobiotics is a determining factor in their disposition and therefore in the monitoring of pharmacological treatments. The evaluation of the cytochrome P-450 mixed oxidase function is commonly used for this purpose. Human microsomal cytochrome P-450-dependent activities involved in drug metabolism, and particularly the mechanisms by which these activities are modulated, has been the focus of a considerable number of pharmacological studies in the last years. Several forms of cytochrome P-450 activities have been isolated and characterized in human liver microsomes, but human microsomal aminopyrine N-demethylase activity has only recently been purified (1), and little is known about mechanisms that could modulate its activity. We have previously shown that interindividual variability in microsomal N-demethylase activity is probably not due to the presence of different isozymes (since kinetic properties are similar in individuals with very different rates) but rather to different amounts of the same isozyme, or to the existence of an as yet unknown regulatory mechanism (2).

This report describes the study of modulation of aminopyrine N-demethylase activity by cimetidine [a known inhibitor of a number of human cytochrome P-450s (3)], caffeine [which in human liver microsomes is partly metabolized by demethylation (4–6)], and their metabolites produced in vitro [theophylline, theobromine, paraxanthine, and 1,3,7-trimethyluric acid (1,3,7-TMU)]. Aminopyrine N-demethylation could be somewhat related to caffeine demethylation (both may be metabolized by the same or by closely related enzymes), as our previous correlation studies suggest (6). Since we have found a strong inhibition on aminopyrine N-demethylase activity by 1,3,7-TMU, but not by caffeine, which is structurally very similar, we have checked to see if 1,3,7-TMU is directly responsible of the inhibition, or whether any

1,3,7-TMU metabolite produced in the microsomal preparation, or present as a contaminant in commercial preparations, could be responsible. For this we have studied whether 1,3,7-TMU is metabolized in human liver microsomes and, in addition, tried to ascertain the occurrence of 1,3,7-TMU-derived contaminants in commercial preparations.

II. METHODS

A liver sample (15 g) from an organ donor (female, 35 years) obtained 20 min after circulatory arrest was used in this study. Microsomal preparation, as well as determination of aminopyrine N-demethylase activity, NADPH reductase activity, cytochrome P-450 content, and cytochrome b_5 content were performed as described previously (2).

A method suitable for the study of microsomal caffeine metabolism (7) was used to study the occurrence of 1,3,7-TMU derivatives in the commercial preparations and after microsomal 1,3,7-TMU metabolism. The reaction mixture contained 5 mM 1,3,7-TMU, 5 mM $MgCl_2$, 0.5 mM NADPH, 50 mM glucose 6-phosphate, four enzyme units of glucose 6-phosphate dehydrogenase, and 50 mM Tris-HCl buffer (pH 7.5) in 1 ml. The reaction was started, after 5 min preincubation at 37°C, by addition of 1.5 mg of microsomal protein. Samples were incubated at 37°C for 45 min, and the reaction was stopped by the addition of ammonium sulfate at a final saturation of 100%. Then, 2 ml of chloroform:isopropanol (85:15) were added and samples were shaken for 2 h at room temperature. Controls were carried out by stopping the reaction just as it started. The organic phase was removed, dried under a nitrogen stream at 40°C, and resuspended in 500 μL of HPLC mobile phase. The equipment used for the high-performance liquid chromatography (HPLC) analysis consisted of a Beckman model 331 isocratic liquid chromatograph equipped with a model 160 (280 nm) detector and a Hewlett-Packard 3390-A integrator. The column was a Beckman Ultrasphere ODS, 5-μm particle size (25 cm × 4.6 mm), and the mobile phase was tetrahydrofuran/acetic acid/acetonitrile/water (6.66/2.74/62/928.6), at a flow of 1 mL/min.

III. RESULTS

Aminopyrine N-demethylase activity of the microsomal preparation, measured with 10 mM aminopyrine, was 1.1 nmol/min/mg protein, cytochrome P-450 was 0.73 nmol/mg, cytochrome b_5 was 0.46 nmol/mg, and NADPH-reductase activity was 16.2 nmol/min/mg protein. These values, as well as kinetic properties (K_m = 7 mM and V_{max} = 1.21 nmol/min/mg), are in the same range as those described previously in human liver microsomes frozen, immediately after circulatory arrest, at −80°C (2). These results suggest that warm ischemia (see Section II) did not seem to affect significantly these parameters in our sample, being therefore appropriate for this study.

Due to the direct correlation between caffeine demethylase activity and aminopyrine N-demethylase activity in human liver microsomes (6), and the fact that, in human liver microsomes, only primary (demethylated) metabolites of caffeine and 1,3,7-TMU can be detected as described (4,5), aminopyrine N-demethylase activity was assayed in the presence of high concentrations of caffeine and demethylated metabolites (theophylline, theobromine and paraxanthine), as well as 1,3,7-TMU and cimetidine. The results are shown in Table 1. Caffeine, theophylline, paraxanthine,

Table 1

Inhibitor	Concentration (mM)	Activity (%)
None	—	100
Cimetidine	5	25 ± 5.0
Caffeine	5	84 ± 2.2
Theophylline	5	70 ± 9.5
Theobromine	2.5	84 ± 2.2
Paraxanthine	5	90 ± 7.0
Trimethyluric acid	5	0.5 ± 0.8

Activity is expressed as the percentage of the N-demethylase activity. Results are means ± SD of five independent experiments. Due to its poor solubility, theobromine could not be assayed at concentrations higher than 2.5 mM.

and theobromine appear to cause a slight inhibition of the N-demethylase activity. In contrast, cimetidine produced a marked inhibition, and 1,3,7-TMU produced full inhibition at the concentration (5 mM) tested.

Since caffeine and its derivatives are widely used in beverages and as drugs (8), this study focused on the inhibitor effect of 1,3,7-TMU. Figure 1 shows the inhibition of aminopyrine N-demethylase activity with different concentrations of 1,3,7-TMU. Full inhibition was reached at concentrations over 500 μM, being IC_{50} values of 45 ± 20 μM (mean ± SE of four independent experiments). A Hill plot (not shown) of the data corresponding to the experiment shown in Figure 1 gave a Hill's coefficient of 2.4 with a correlation coefficient (r) = 0.987. This suggests that there is positive cooperation in the inhibition process. These results can be explained by the existence of more than one interaction site between cytochrome P-450 and 1,3,7-TMU.

When N-demethylase activity kinetics for aminopyrine are studied in the presence and in the absence of 100 μM 1,3,7-TMU, the results suggest that the mechanism of the

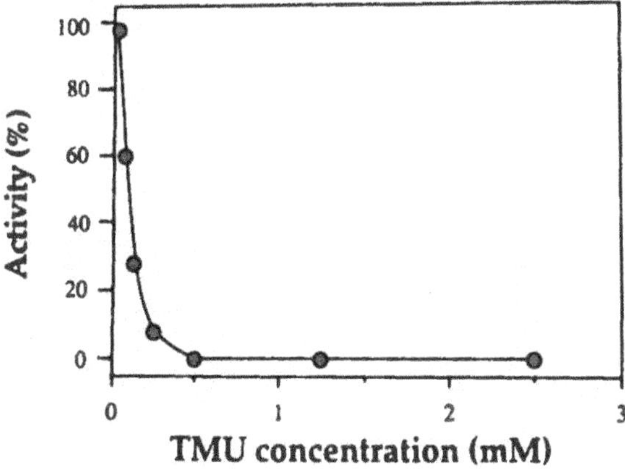

Figure 1 Inhibition of aminopyrine N-demethylase activity in the presence of different concentrations of trimethyluric acid, referred to the activity in the absence of trimethyluric acid (100%).

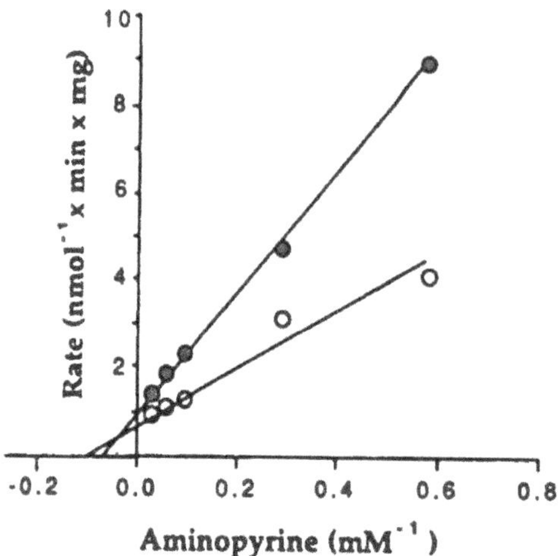

Figure 2 Double-reciprocal plots of aminopyrine N-demethylase activity in the presence of different concentrations of aminopyrine, in the absence (O) and in the presence of (●) of 100 μM trimethyluric acid.

inhibition follows a mixed pattern (see Figure 2) because the point of the intersection of the lines does not appear to be in 1/velocity nor 1/concentrations axis. V_{max} values were 1.21 nmol/min/mg without, and 1.01 nmol/min/mg with inhibitor, the K_m value was 7 mM, and K_i was 13.5 mM, respectively, in the presence or in the absence of 1,3,7-TMU.

The results obtained suggest that 1,3,7-TMU and/or some metabolite produced in the mixture during aminopyrine N-demethylase activity assay inhibits N-demethylase activity. The low values of IC_{50} suggest that 1,3,7-TMU and not any metabolite produced in small amounts, or the presence of some 1,3,7-TMU derivative on the commercial trimethyluric acid preparation, is responsible for the inhibition. This is confirmed by the results obtained in the study of purity of commercial trimethyluric acid and trimethyluric acid metabolism in human liver microsomes. Purity of commercial 1,3,7-TMU was checked in two different flasks (both Sigma lot 101F-0236; both produced inhibition of N-demethylase activity). In all the cases, no contaminants such as 1,3-dimethyluric acid (1,3-DMU), 1,7-dimethyluric acid (1,7-DMU), 3,7-dimethyluric acid (3,7-DMU), 1-methyluric acid (1-MU), 3-methyluric acid (3-MU), or 7-methyluric acid (7-MU) were detected in a range under 1/10,000. Although this result does not rule out the possibility that some of these 1,3,7-TMU derivatives could be present and be responsible for the inhibition, this possibility becomes highly improbable because if IC_{50} for 1,3,7-TMU is under 10^{-4}M, IC_{50} value of any undetected 1,3,7-TMU derived should be under 10^{-9}M.

Human microsomal metabolism of 1,3,7-TMU acid was also studied in the same experimental conditions used for the study of inhibition of N-demethylase activity. We have selected a concentration of 1,3,7-TMU of 5 mM (10 times the concentration that gives full inhibition of N-demethylase activity, as shown in Figure 1). No metabolites of 1,3,7-TMU were detected in these conditions, neither in the presence of 10 mM aminopyrine nor microsomes, being the recovery of 1,3,7-TMU of 85 ± 12%. All these results taken together confirm that trimethyluric acid, which in our experimental conditions is not metabolized in human liver microsomes, inhibits microsomal aminopyrine N-demethylase activity.

IV. DISCUSSION

Despite the large studies of human aminopyrine N-demethylase activity using the aminopyrine breath test, which is commonly used method to evaluate the capacity in the biodisposition of xenobiotics (9–11), liver microsomal aminopyrine N-demethylase activity is one of the less known cytochrome P-450-dependent activities. Very wide interindividual differences in this activity have been found, with usual values between 0.4 and 4.5 nmol/min/mg protein are usual (2,12,13). The cause of this interindividual variability and the mechanisms of modulation of this activity remain unknown. In rats, we have shown sex- and age-related differences in aminopyrine N-demethylase activity that suggest some kind of hormonal regulation mechanism. However, ovariectomy does not induce changes in sex-related differences, suggesting that, if hormonal regulation occurs, estrogens are not involved (14). In contrast, in human liver microsomes no sex-related differences have been found, and interindividual differences appear to be due to unspecific quantitative differences rather than specific qualitative differences (2).

The correlation between aminopyrine and caffeine N-demethylase activities described in human liver microsomes (6) agree with the results shown in this report. Though caffeine, theophylline, paraxanthine, and theobromine produced a slight (10–30%) inhibition at concentrations of 2.5 (theobromine) or 5 mM (caffeine, paraxanthine, and theophylline), it is remarkable that full inhibition is produced with 5 to 10 times smaller concentrations of 1,3,7-TMU. Moreover, considering that in adult human liver microsomes caffeine is biotransformed, producing 1,3,7-TMU at a rate that can be estimated to be about 20 to 30 pmol/min/mg protein (4), it is difficult to be sure that it is caffeine and not the 1,3,7-TMU produced (considering that this rate could reach a final concentration in our assay samples of 1 to 2 μM) that is responsible for the slight inhibition observed in the presence of caffeine. Our own unpublished results have shown that, with the same experimental conditions used for the experiments shown in Table 1, the concentration of 1,3,7-TMU produced as a result of microsomal caffeine metabolism is under 10^{-7} M, which is a concentration too small to produce significant inhibition of N-demethylase activity.

The difference between the effects of caffeine and trimethyluric acid, which are structurally very similar, on aminopyrine N-demethylase activity suggest that C_8-oxidation should be essential in the inhibitor effect. That could also explain the lack of significant inhibitor effect of theophylline, theobromine, and paraxanthine, which, as well as caffeine, possess methyl groups but not the oxidized C_8.

Many questions remain unclear in this inhibitor effect and will be the object of further studies. Nevertheless, it should be pointed out that, since caffeine-containing beverages intake is very common, and 1,3,7-TMU is one of the main metabolites of caffeine, this compound could be present in amounts sufficient to produce inhibition of N-demethylase activity. Therefore, results obtained in the evaluation of aminopyrine N-demethylase activity should be interpreted cautiously.

ACKNOWLEDGMENT

Financial support for this work was provided by CICYT-FAR 89-0081.

REFERENCES

1. M. Komori, T. Hashizume, H. Ohi, T. Miura, M. Kitada, K. Nagashima, and T. Kamataki. Cytochrome P-450 in human liver microsomes: High-performance liquid chromatographic isolation of three forms and their characterization. *J. Biochem.*, *43*:912 (1988).
2. J. A. García-Agúndez, A. Luengo, and J. Benítez. Aminopyrine N-demethylase activity in human liver microsomes. *Clin. Pharmacol. Ther.* (in press).
3. M. Murray. Mechanisms of the inhibition of cytochrome P-450-mediated drug oxidation by therapeutic agents. *Drug Metab. Rev.*, *18*:55 (1987).
4. F. Berthou, D. Ratanasavanh, C. Riche, D. Picart, T. Voirin, and A. Guillouzo. Comparison of caffeine metabolism by slices, microsomes and hepatocyte cultures from adult human liver. *Xenobiotica*, *19*:401 (1989).
5. M. Campbell, D. Grant, T. Inaba, and W. Kalow. Biotransformation of caffeine, paraxanthine, theophylline and theobromine by polycyclic aromatic hydrocarbon-inducible cytochrome(s) P-450 in human liver microsomes. *Drug Metab. Dispos.*, *15*:237 (1987).
6. J. Benítez, J. A. García-Agúndez, and A. Luengo. Human microsomal caffeine metabolism. Correlation with aminopyrine metabolism. *Proc. VIIIth Int. Symp. on Microsomes and Drug Oxidations*, Stockholm, Sweden, p. 193 (1990).
7. J. A. García-Agúndez and J. Benítez. A simplified isocratic HPLC method to determine microsomal caffeine metabolism. Application to D.A. rats. *Proc. VIIIth Int. Symp. on Microsomes and Drug Oxidations*, Stockholm, Sweden, p. 196 (1990).
8. J. J. Barone and H. Roberts. Human consumption of caffeine. In *Caffeine: Perspectives from Recent Research* (P. B. Dews, ed.), Springer-Verlag, Berlin, p. 59 (1984).
9. J. F. Schneider, D. A. Schoeller, B. Nemchansky, J. L. Boyer, and P. Klein. Validation of the $^{13}CO_2$ breath analysis as a measurement of demethylation of stable isotope labeled aminopyrine in man. *Clin. Chim. Acta*, *84*:153 (1978).
10. A. N. Kotake, B. D. Schreider, and J. R. Latts. The *in vivo* measurement of expired $^{14}CO_2$ derived from N-demethylation of aminopyrine as a reflection of the *in vitro* hepatic cytochrome P-450 drug metabolism activity in rats. *Drug. Metab. Dispos.*, *10*:251 (1982).
11. G. W. Hepner and E. S. Vessell. Assessment of aminopyrine metabolism in man by breath analysis after oral administration of ^{14}C-aminopyrine. *N. Engl. J. Med.*, *291*:1384 (1974).
12. H. Souhaili-el Amri, A. M. Batt, and G. Siest. Comparison of cytochrome P-450 content and activities in liver microsomes of seven animal species, including man. *Xenobiotica*, *16*:351 (1986).
13. A. R. Boobis and D. S. Davies. Human cytochromes P-450. *Xenobiotica*, *14*:151 (1984).
14. J. Benitez and J. A. García-Agúndez. Sex differences in aminopyrine N-demethylase activity in D.A. rat liver microsomes. *Eur. J. Clin. Pharmacol.*, *36*:202 (1989).

111

Drug Metabolism as an Index of Liver Function

M. Oellerich, M. Burdelski, K. Goeschen, H. U. Lautz, and H. Wedeking *Medizinische Hochschule Hannover, Hannover, Germany*

I. INTRODUCTION

Liver disease or dysfunction may influence the disposition of drugs (1). Therefore it has been proposed to use the capacity of the liver to metabolize certain drugs as a measure of hepatic function (2). In the past few years the search for novel approaches to assess hepatic performance was greatly stimulated by the rapid expansion of organ transplantation. This development has confronted us with new diagnostic and prognostic problems related to the assessment of donor liver quality or the evaluation of graft dysfunction and pretransplant prognosis. Clinically useful information is often not obtainable by conventional tests in these situations.

The objective of this report is to describe some of the recent advances that have been made in using drug metabolism as an index of liver function.

II. DYNAMIC LIVER FUNCTION TESTS

For the assessment of liver function, static or dynamic tests can be used (Figure 1). The traditional static tests are only an indirect measure of hepatic function. These tests involve one measurement of the concentration of an endogenous substance (e.g., metabolite, enzyme) at a single point in time. Dynamic tests, on the other hand, are a measure of actual performance and therefore reflect the functional state of the liver. The usual approach is to determine drug clearance or metabolite formation kinetics.

The best-studied dynamic liver function tests based on drug disposition are listed in Figure 2. In the first group of tests, the clearance or half-life or indocyanine green, caffeine, or antipyrine is used as a measure of liver function (1–3). A further related approach involves galactose elimination capacity (4). Tests based on the rate of metabolite formation are of particular interest. In the well-known ^{14}C-aminopyrine breath test (5,6) and a recently developed ^{14}C-erythromycin breath test (7), $^{14}CO_2$ exhalation gives a measure of hepatic oxidative function. These breath tests are relatively complex, however, and they have not been widely adopted in clinical practice.

In a modification of this approach, lidocaine metabolite formation has been proposed as an index of hepatic function (8,9). Lidocaine is rapidly converted to its primary metabolite, monoethylglycinexylidine (MEGX), by the hepatic cytochrome

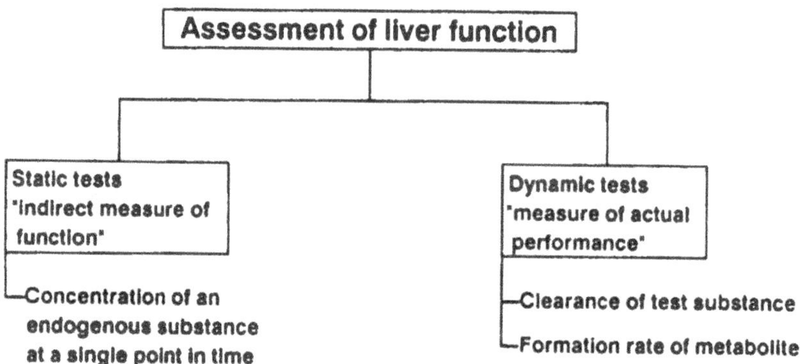

Figure 1 Static and dynamic liver function tests.

P-450 system. In studies with human liver microsomes, cytochrome P-450 III A 4 was identified as one of the enzymes involved in MEGX formation (10). A destruction of cytochrome P-450 structures of major changes of hepatic blood flow—for instance, due to protosystemic shunting—result in a decrease of MEGX formation.

This concept led to the development of the MEGX test, which is rapid and easy to perform (9). Blood specimens are taken before and 15 or 30 min after an i.v. bolus injection of a small lidocaine test dose (1 mg/kg). MEGX is determined in serum by use of fluorescence polarization immunoassay (Abbott Laboratories, Chicago) within about 20 min. The pharmacokinetic basis of this test has recently been described (11).

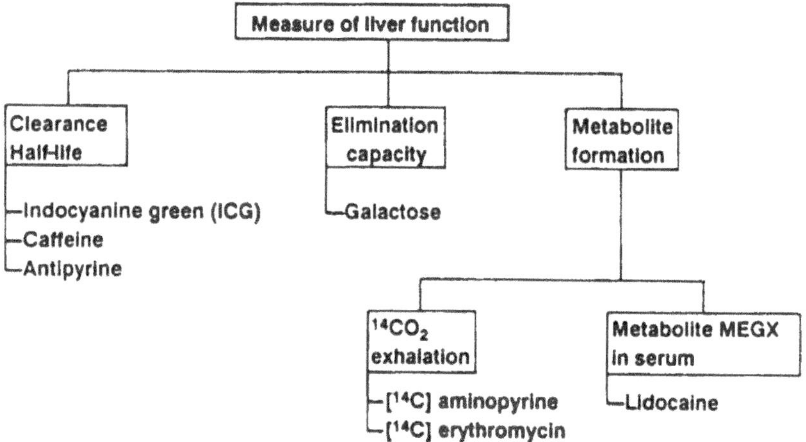

Figure 2 Dynamic liver function tests based on drug disposition.

III. SITUATIONS WHERE DYNAMIC LIVER FUNCTION TESTS ARE APPLICABLE

There are a number of instances when the use of dynamic liver function test should be considered. Such tests could be helpful in clinical transplantation to improve the selection criteria for donor organs or transplant candidates. Appropriate tests may also be useful for the evaluation of acute or chronic rejection episodes after transplantation. A further area of application encompasses patients with liver resection or with acute or chronic liver diseases of metabolic or infectious origin. In various prospective studies, the clinical utility of certain dynamic liver function tests such as MEGX formation has been demonstrated (11–16).

A. Donor Rating

Previously published data indicate that the determination of MEGX formation in liver donors yielded useful prognostic information with regard to graft survival in the corresponding recipients (12,14,15). The probability of graft survival over 120 days was significantly higher for livers from donors with MEGX test results above 90 μg/L than for those from donors with MEGX values of 90 μg/L or lower (12,14). Other liver function tests (bilirubin, prothrombin time, activity of aminotransferases, glutamate dehydrogenase, and cholinesterase, indocyanine green clearance, and galactose elimination capacity) were inefficient at predicting early outcome of transplantation (12). These studies suggest that the MEGX test provides useful information on donor organ quality and therefore may be useful to identify donor organs which show a high risk for early complications.

B. Pretransplant Prognosis of Cirrhotics

Furthermore, it could be demonstrated in prospective studies (11,13) that the MEGX and ICG tests are useful for the assessment of short-term pretransplant prognosis in patients with advanced cirrhosis. It was found that, in patients with cirrhosis, caffeine and antipyrine clearances were less favorable prognostic indicators than ICG elimination or MEGX formation (11,17). The available data favor the primarily flow-dependent ICG and MEGX test over the capacity-dependent caffeine and antipyrine tests as predictors of short-term survival in cirrhotics. These findings seem to support the contention that the implications of intra- and extrahepatic shunting are of great prognostic importance in transplant candidates with cirrhosis.

In an extension of our prospective studies (11,13), we evaluated the prognostic value of the MEGX and ICG tests in 141 adults with advanced biliary or postnecrotic cirrhosis. The follow-up period was 120 days after testing liver function. Nineteen of 141 patients were lost to follow-up: 18 patients underwent liver transplantation within the observation period, and one patient died in acute cardiac failure. The remaining 122 patients were enrolled in the prospective study (18). Twenty patients died from their liver disease within 120 days.

The patients were classified at inclusion according to the Pugh modification of the Child-Turcotte criteria: 32 patients were Child class A, 53 were class B, and 37 were class C. In addition to MEGX formation and ICG half-life, various liver function tests were done in each patient: the serum concentrations of bilirubin (BIL) and albumin (ALB), the serum catalytic concentrations of alkaline phosphatase (AP) and cholinesterase (CHE), and the prothrombin time (PT). Moreover, the following clinical features were recorded at the time of inclusion: ascites (ASC) and encephalopathy (ENC).

In order to find out which of the parameters studied were the best prognostic indicators, a stepwise survival analysis was performed by means of the Cox proportional hazards model (BMDP program 2L). Figure 3 shows the influence of transplant candidate variables as co-variates on the hazard function in the Cox model. At step zero, ICG and MEGX showed the highest chi-square values to enter the stepwise analysis. At the final step, the ICG and MEGX test results had entered the model and could not be removed. Ascites was the only additional variable that entered the model. However, the resulting improvement of our predictive ability was relatively low ($p = 0.087$) when findings for this clinical feature were added to ICG and MEGX test results. The data suggest that, of the parameters studied, the ICG and MEGX tests were the best short-term prognostic indicators.

There was a severe impairment of MEGX formation and ICG elimination in patients who died during the follow-up period of 120 days. In accord with this finding, Gremse et al. (16) observed strongly decreased MEGX test results in children with cirrhosis belonging to a high-mortality-risk group. Using data from an extended prospective study (19) in pediatric transplant candidates, Kaplan-Meier survival curves were calculated for patients with MEGX values ≥ 10 μg/L and for those with MEGX values < 10 μgL (Figure 4). The probability of surviving without transplantation at least 1 year after MEGX testing was significantly higher for patients with favorable MEGX values ($p < 0.0005$). The currently available data suggest that the MEGX and the ICG test could be promising tools for the improvement of our decision-making process with respect to the selection of transplant candidates.

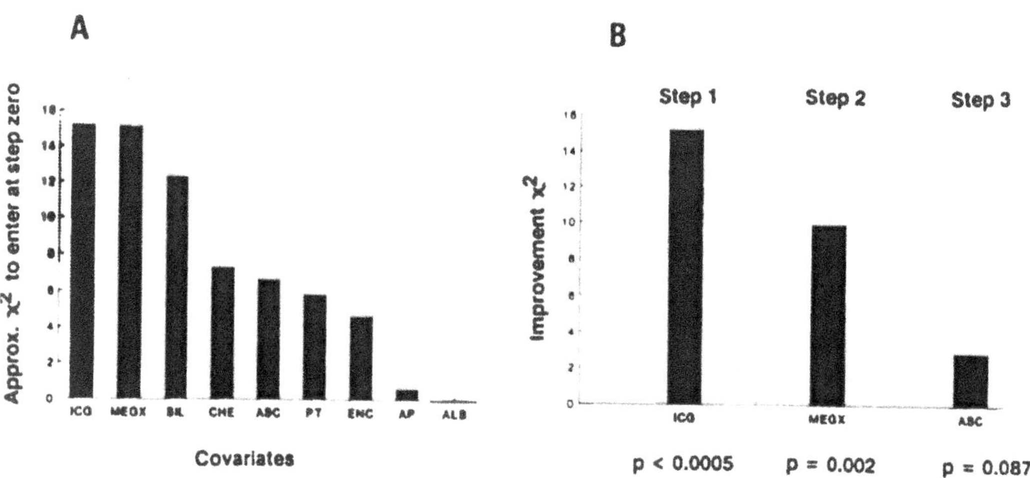

Figure 3 Relationship between transplant candidate variables and pretransplant survival. (A) Influence of transplant candidate variables as co-variates on the hazard function in the Cox model (adults: 20 dead, 102 censored). (B) Significant predictors of 120-day pretransplant survival predictors of 120-day pretransplant survival (summary of stepwise results).

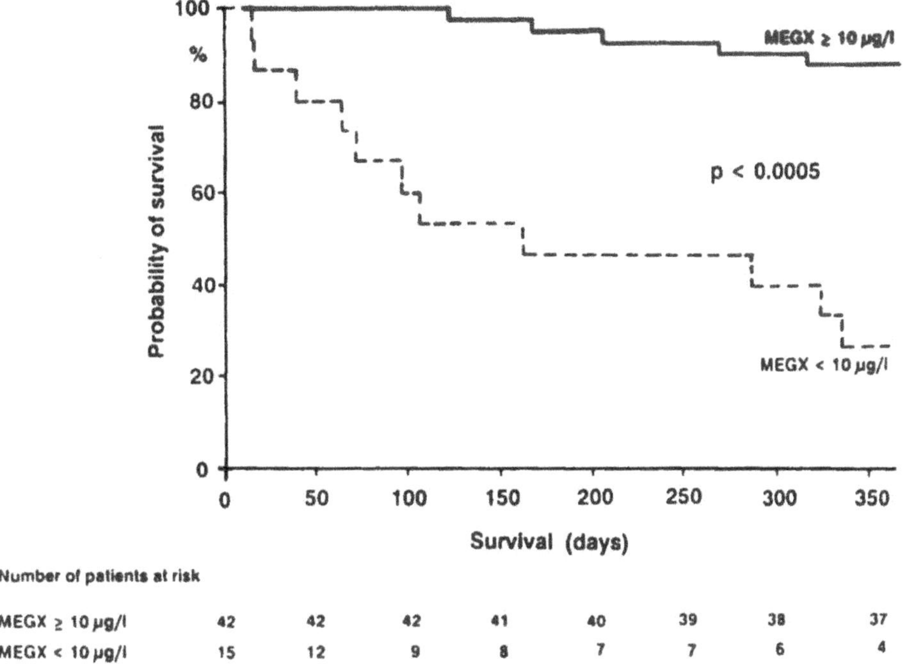

Figure 4 Life table showing probability of pretransplant patient survival in relation to MEGX test results.

C. Pregnancy

The usefulness of the MEGX test as a measure of liver function has also been evaluated in risk patients with hepatic impairment during the last trimester of pregnancy (20). In patients with pregnancy-induced hypertension, the HELLP syndrome is a serious complication. MEGX test results were significantly lower in patients with HELLP syndrome or pregnancy-induced hypertension than in women with normal pregnancy (Figure 5). In pregnant women and in liver-transplanted women with uncomplicated pregnancy, the MEGX values were in the range observed in nonpregnant females.

The dramatic decrease of MEGX values in patients with HELLP syndrome points to a severe impairment of hepatic function. These findings suggest that the flow-dependent MEGX test could be useful to monitor high-risk patients with HELLP syndrome or pregnancy-induced hypertension.

IV. LEVELS OF APPLICATION FOR DYNAMIC LIVER FUNCTION TESTS

The diagnostic use of dynamic liver function tests is directed at the evaluation of organ-specific partial functions (Figure 6). Appropriate tests can provide clinically useful information on the actual function state of the liver. The MEGX test, for example, indicates major disturbances of hepatic blood flow and changes of the oxidative metabolizing capacity. The prognostic use of these tests involves the evaluation of risk in connection with therapeutic decisions. For example, certain tests can be

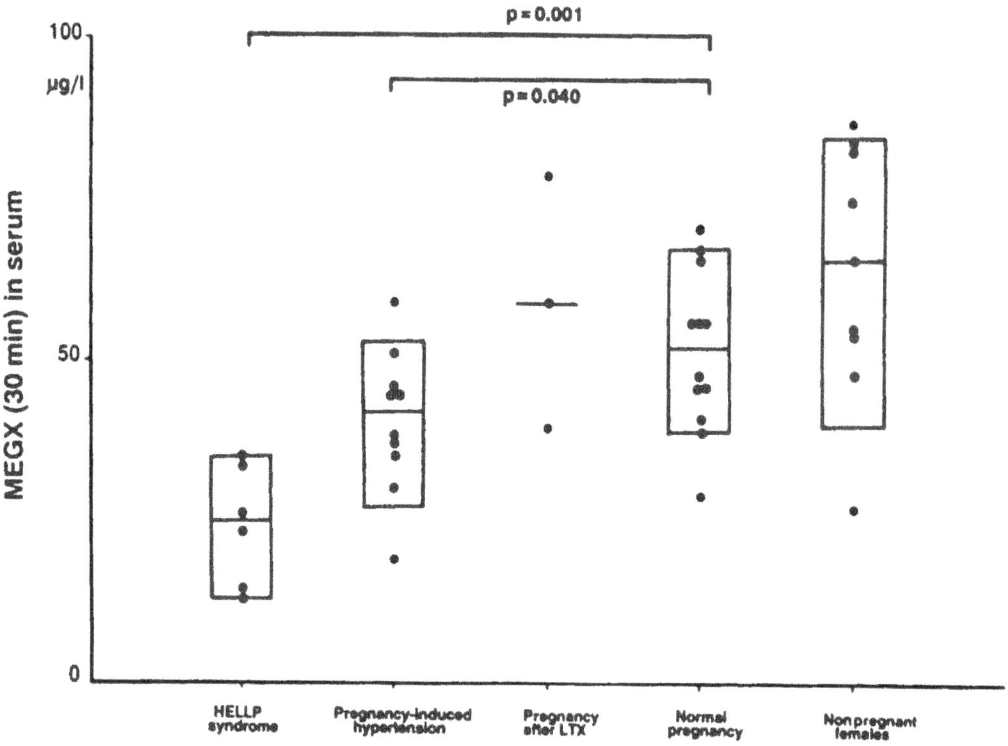

Figure 5 MEGX formation in pregnancy. Boxes delimit the 16th to 84th percentiles. The horizontal bar indicates the median.

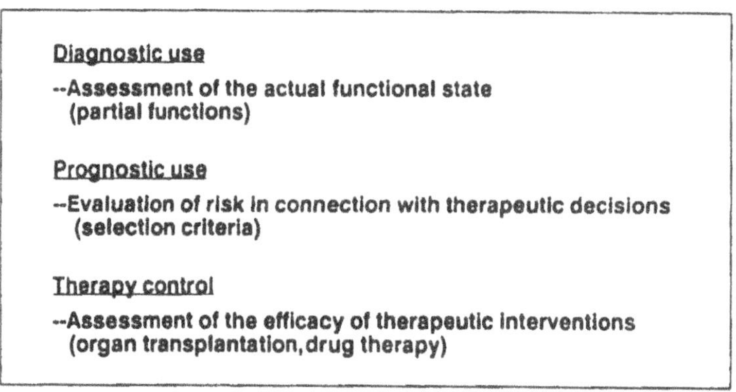

Figure 6 Levels of application for dynamic liver function tests.

used to improve selection criteria for donor organs or transplant candidates. Dynamic liver function tests may also play a role in therapy control. Such tests are helpful to assess the efficacy of therapeutic interventions, particularly in clinical transplantation.

It is to be expected that the use of drugs as indicators of organ function will open up new perspectives for the field of therapeutic drug monitoring.

REFERENCES

1. C. W. Howden, G. G. Birnie, and M. J. Brodie. Drug metabolism in liver disease. *Pharmacol. Ther., 40*:439–474 (1989).
2. R. A. Branch. Drugs as indicators of hepatic function. *Hepatology, 2*:97–105 (1982).
3. E. Renner, H. Wietholtz, P. Huguenin, M. J. Arnold, and R. Preising. Caffeine: A model compound for measuring liver function. *Hepatology, 4*:38–46 (1984).
4. N. Tygstrup. Determination of the hepatic elimination capacity (Lm) of galactose by single injection. *Scand. J. Clin. Lab. Invest., 18*(Suppl. 92):118–125 (1966).
5. B. Lauterberg and J. Bircher. Hepatic microsomal drug metabolizing capacity measured in vivo by breath analysis. *Gastroenterology, 65*:556 (1973).
6. G. W. Hepner and E. S. Vesell. Assessment of aminopyrine metabolism in man by breath analysis after oral administration of C-aminopyrine. *N. Engl. J. Med., 291*:1384–1388 (1974).
7. P. B. Watkins, S. A. Murray, L. G. Winkelman, D. M. Heuman, S. A. Wrighton, and P. S. Guzelian. Erythromycin breath test as an assay of glucocorticoid inducible liver cytochromes P-450: Studies in rats and patients. *J. Clin. Invest., 83*:688–697 (1989).
8. M. Oellerich. Clearancewerte und Metabolitenkinetik in der Leberdiagnostik. In *Funktion und Funktionsdiagnostik der Leber, Merck-Symposium 1985* (D. Seidel and H. Lang, eds.), Springer, Heidelberg, pp. 53–55 (1987).
9. M. Oellerich, E. Raude, M. Burdelski, et al. Monoethylglycinexylidide formation kinetics: A novel approach to assessment of liver function. *J. Clin. Chem. Clin. Biochem., 25*:845–853 (1987).
10. M. J. Bargetzi, T. Aoyama, F. J. Gonzalez, and U. A. Meyer. Lidocaine metabolism in human liver microsomes by cytochrome P450 III A4. *Clin. Pharmacol. Ther., 46*:521–527 (1989).
11. M. Oellerich, M. Burdelski, H. U. Lautz, M. Schultz, F. W. Schmidt, and H. Herrmann. Lidocaine metabolite formation as a measure of liver function in patients with cirrhosis. *Ther. Drug. Monitor., 12*:219–226 (1990).
12. M. Oellerich, M. Burdelski, B. Ringe, et al. Lignocaine metabolite formation as a measure of pretransplant liver function. *Lancet 1*:640–642 (1989).
13. M. Oellerich, M. Burdelski, H. U. Lautz, et al. Assessment of pre-transplant prognosis in patients with cirrhosis. *Transplantation* (in press).
14. M. Oellerich, M. Burdelski, B. Ringe, et al. Functional state of the donor liver and early outcome of transplantation. *Transplant Proc.* (in press).
15. T. J. Schroeder, D. A. Gremse, M. E. Mansour, et al. Lidocaine metabolism as an index of liver function in hepatic transplant donors and recipients. *Transplant Proc., 21*:2299–2301 (1989).
16. D. A. Gremse, H. H. A-Kader, T. J. Schroeder, and W. F. Balistreri. Assessment of lidocaine metabolite formation as a quantitative liver function test in children. *Hepatology* (in press).
17. G. G. Birnie, G. G. Thomson, A. Cooke, and M. J. Brodie. Antipyrine and indocyanine green kinetics in the prediction of the natural history of liver disease. *Br. J . Clin. Pharmacol.* 27:615P–616P (1987).
18. H. U. Lautz, M. Oellerich, M. Burdelski, et al. Assessment of short-term prognosis in transplant candidates with postnecrotic or biliary cirrhosis. *Transplant. Proc.* (in press).
19. M. Burdelski, M. Oellerich, J. Düwel, B. Rodeck, and J. Brodehl. Predictors of one-year survival in pediatric liver cirrhosis. *Hepatology* (in press).
20. M. Oellerich, K. Goeschen, H. Wedeking, D. Elling, and M. Burdelski. Lidocaine metabolite formation as a measure of liver function in pregnancy. *Hepatology* (in press).

112

Transitory Effect of Antiepileptic Drugs on Carnitine Homeostasis in Mouse Serum and Tissues

M. F. Camiña, I. Rozas, J. M. Paz, C. Alonso, and S. Rodríguez-Segade *University of Santiago de Compostela, Santiago de Compostela, Spain*

I. INTRODUCTION

The chief antiepileptic drugs, including valproic acid (VPA), carbamazepine (CBZ), phenytoin (PHT), and phenobarbital (PHB), cause numerous side effects. In particular, almost all are to some extent hepatotoxic (though fatal hepatic reactions are rare) (1). Since most epileptics receive long-term treatment with these drugs, it is important to watch for signs of chronic intoxication. The observation that, in many cases of anticonvulsant-induced chronic hepatic poisoning, two or more anticonvulsants are being taken simultaneously, with the apparent implication that the drug interactions are in many cases largely responsible for the toxic effects, has encouraged the prescription of single-drug treatment (2). Unfortunately, monotherapy does not always suffice to control attacks.

No significant deficiency was found between serum carnitine concentrations of healthy subjects and those of 11 epileptics treated with antiepileptics other than VPA (3). We recently found that 21.5% of a group of 149 such patients suffered hypocarnitinaemia (4). Though this figure is less than a third of the corresponding percentage for VPA, 76.5%, the number of patients taking non-VPA antiepileptics is four or five times greater than the number of patients treated with VPA.

In the present work we aimed to throw some light on the mechanism by which VPA, CBZ, PHT, and PHB administration alter carnitine homeostasis. For this purpose, we determined the effects of these drugs on free and acyl-carnitine concentrations in plasma, liver, kidney, heart, and skeletal muscle, and tried to ascertain if the effect is reversible after several hours.

II. METHODS

Male adults mice weighing 30 to 35 g were used. Therapeutic doses of antiepileptics (50 mg/kg of VPA, 20 mg/kg of CBZ, 10 mg/kg of PHT, or 6 mg/kg of PHB or isoosmotic sodium chloride solution for control groups) were administered by a gastric catheter, and all were slaughtered after 8 h fasting and, depending on the group, 30 or 480 min post-administration (p.a.).

Blood was collected from several neck vessels, and plasma was obtained. Immediately after decapitation, liver, kidney, heart, and muscle were removed and frozen in liquid nitrogen pending analysis. They were homogenized in a Potter-Elverjhem with four times their fresh weight of 50 mM Hepes buffer/10 mM EDTA (pH 7.5).

Free carnitine concentrations in plasma and tissue homogenates were determined using the method of Rodríguez-Segade et al. (5). Total carnitine was measured by the same procedure after hydrolysis with 1 M KOH (6). Acyl carnitine concentrations were calculed by substracting free carnitine concentrations from total carnitine.

Drugs levels were measured by fluorescence polarization on TD_x. Ammonium levels in blood were determined by a colorimetric method on slides, and the color changes were measured by a reflectanter meter (Blood Ammonia Checker System, Kyoto Daiichikagaku Company).

III. RESULTS

A. Effects of Antiepileptic Drugs on Serum Carnitine Concentrations

Figure 1 shows total, free, and acyl-carnitine concentrations in mouse plasma 0.5 and 8 h after oral administration of VPA, CBZ, PHT, PHB, or isoosmotic sodium chloride. In all groups, total carnitine concentration was significantly decreased 0.5 h p.a.: 65.8% for VPA, 34.8% for PHB, 18.5% for CBZ, and 15% for PHT. In all groups, the deficit in total carnitine involved a large deficit of acyl-carnitine (between 41.7% for PHT and 81.6% for VPA), whereas only VPA reduced free carnitine concentration (by 49.4%).

The effects on acyl-carnitine concentrations of VPA, CBZ, PHT, and PHB had disappeared at 8 h p.a., when they had reached control levels, whereas free carnitine concentration exceeded that of controls by 32.7% (CBZ), 34.6% (PHT), and 24.2% (PHB).

B. Antiepileptic Drug Concentrations in Serum

Figure 2 shows the concentrations of VPA, CBZ, PHT, and PHB in mouse serum 0.5 and 8 h p.a. VPA concentration was maximum at 0.5 h p.a. (45.5 ± 11.9 μg/mL) and had practically disappeared 8 h p.a. (0.3 ± 0.2 μg/mL). CBZ concentrations had fallen from 4.2 ± 0.3 μg/mL to near zero (0.6 ± 0.2 μg/mL) at 8 h p.a. PHT concentration was 3.1 ± 1.1 μg/mL at 0.5 h p.a. and was still high at 8 h p.a. (4.9 ± 0.9 μg/mL). PHB decreased from 7.6 ± 1.1 μg mL at 0.5 h p.a. to 3.5 ± 1.5 μg mL at 8 h p.a.

C. Effects of Antiepileptic Drugs on Blood Ammonia Levels

None of the treated groups exhibited hyperammoniaemia at any time after anticonvulsant administration.

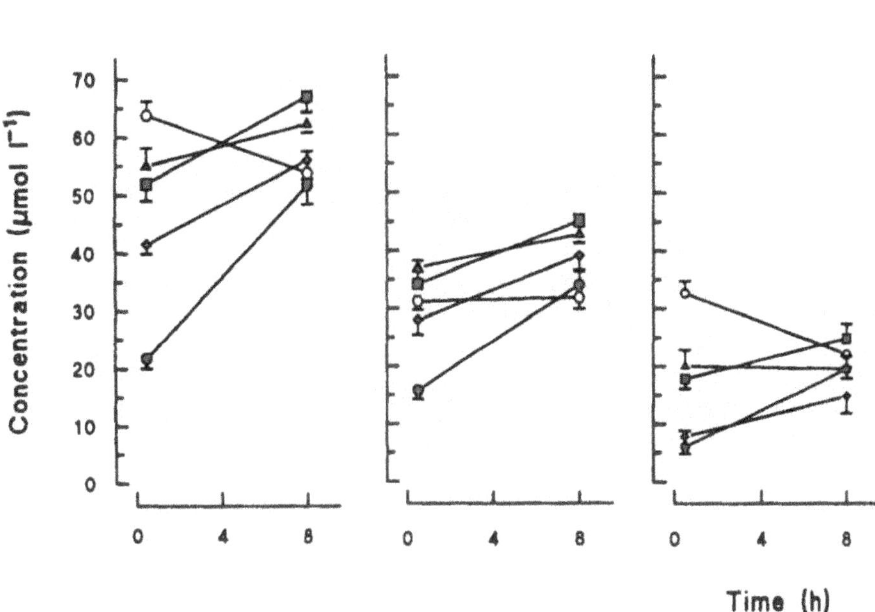

Figure 1 Total, free and acyl-carnitine in mouse serum, 0.5 and 8 h after administration of saline or therapeutic drug doses. Results are mean ± SEM.

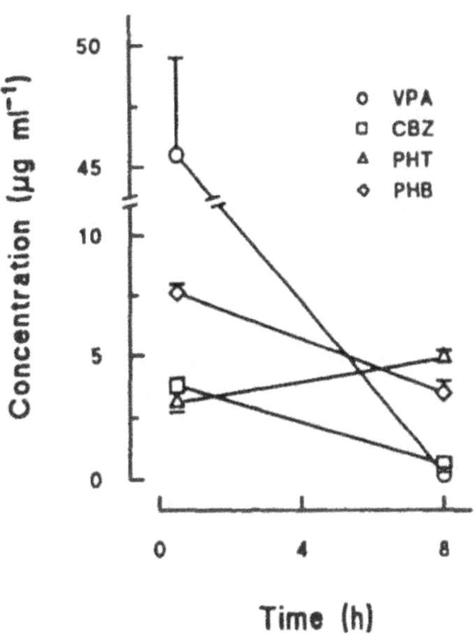

Figure 2 Drug levels in mice at 0.5 and 8 h post-administration of therapeutic dose. Results are mean ± SEM.

D. Effects of Antiepileptic Drugs on Carnitine Concentrations in Tissues

Table 1 lists the total carnitine as well as the free and acyl-carnitine concentrations in mouse liver, kidney, heart, and muscle at 0.5 and 8 h p.a.

The greatest effect in liver was caused by CBZ, with which free and acyl-carnitine concentrations were 21.5% and 114.8% (respectively) greater than in controls at 0.5 h p.a. No significant effects remained in any of the treated groups 8 h p.a.

Free carnitine concentrations in kidney were significantly affected by neither VPA nor CBZ, but were reduced by PHT (43.8% at 0.5 p.a. and 9.6% at 8 h p .a. (which was still statistically significant) and by PHB (15.5% at 8 h p.a.). Acyl-carnitine was also reduced: VPA cause a 69.1% deficit 0.5 h p.a.; PHT caused 66.7% deficits at the same time. These effects had almost disappeared at 8 h p.a.

In muscle, VPA had a significant effect of 19.6% deficit on free carnitine 0.5 h p.a., when VPA concentration in serum was greatest. CBZ caused free and acyl-carnitine deficits (47.7% and 35.6%), but control levels were recovered 8 h p.a. PHB raised free carnitine (32.2%) at 0.5 h p.a., but greatly depressed acyl-carnitine concentrations at the same time (66.1%) as PHT, which caused a deficit (53.4%). Both these drugs restored, or exceeded (34.8% for PHB), control levels.

In heart, VPA caused a significant deficit of acyl-carnitine 50.7% at 8 h p.a., whereas it had no effect at 0.5 h p.a. CBZ mainly affected acyl-carnitine concentrations, causing large deficits (between 75.8% at 0.5 p.a. and 45.4% at 8 p.a.). With PHT there was free carnitine deficit only at 0.5 h (21.5%), and PHB behaved similarly, with significant free carnitine deficits (27.4% and 22.4% for 0.5 and 8 h p.a., respectively).

IV. DISCUSSION

The existence of carnitine deficiency secondary to VPA treatment is well established (3,4,7,8), but until a recent study of ours (4) this side effect had not been reported for other antiepileptic drugs. In this report we found that VPA caused a marked deficit of free carnitine, while CBZ, PHT, and PHB did not reduce it (indeed, they produced a slight excess 8 h p.a.). Acyl-carnitine was reduced by all drugs at 0.5 h p.a. but control levels were reached after 8 h p.a.

The above findings show that, in mice, the effects of the drugs studied are immediate but transitory, all serum carnitine deficits having disappeared 8 h p.a. The persistent hypocarnitinaemia observed in patients treated with VPA (3,4,7,8) or other anticonvulsants (4) nevertheless suggests that these drugs have very different pharmacokinetics in humans.

Moreover, the fact that in this study the non-VPA antiepileptics affected only the ester of serum carnitine appears to explain the absence of hypocarnitinaemia in most patients treated with these drugs, since the A/F ratio of 0.13 observed in human controls (6.2 μM of acyl-carnitine for 47.1 μM of free carnitine) implies that in humans even the 80% reductions in serum acyl-carnitine levels observed in our mice (A/F ratio 1.04) would hardly affect the total serum carnitine concentrations (4).

Nishida et al. (9) found normal levels of free carnitine in the livers of rats treated with VPA. In the present study the only significant differences between treated mice and controls with regard to hepatic free carnitine and acyl-carnitine levels were positive. Thus anticonvulsant-induced hypocarnitinemia is not a result of generalized hepatic dysfunction, a conclusion likewise supported by the absence of significant

Table 1 Total, Free, and Acyl-Carnitine Concentrations (nmol/g) in Mouse Liver, Kidney, Heart, and Muscle at 0.5 and 8 h Post-Administration of Saline or Therapeutic Drugs Doses[a]

	Liver		Kidney		Heart		Muscle	
	0.5 h	8 h	0.5 h	8 h	0.5 h	8 h	0.5 h	8 h
Total carnitine								
Control	357 ± 46	297 ± 24	825 ± 85	834 ± 74	1423 ± 154	1432 ± 52	317 ± 52	335 ± 51
VPA	393 ± 25	310 ± 50	727 ± 59[c]	818 ± 64	1411 ± 150	1382 ± 100	270 ± 29[c]	361 ± 46
CBZ	461 ± 45[a]	304 ± 19	815 ± 75	887 ± 11	1168 ± 35[a]	1218 ± 113[a]	180 ± 28[a]	372 ± 70
PHT	328 ± 48	315 ± 55	445 ± 13[a]	750 ± 42	1122 ± 121[a]	1227 ± 170[b]	237 ± 29[a]	369 ± 51
PHB	319 ± 30	320 ± 46	727 ± 59[c]	705 ± 90	1018 ± 94[a]	1130 ± 65[c]	317 ± 33	397 ± 50[c]
Free carnitine								
Control	331 ± 44	276 ± 23	744 ± 89	774 ± 59	1274 ± 124	1280 ± 94	199 ± 49	245 ± 41
VPA	364 ± 29	287 ± 50	702 ± 59	760 ± 72	1281 ± 124	1308 ± 114	160 ± 27[c]	260 ± 37
CBZ	402 ± 25[a]	284 ± 17	753 ± 75	813 ± 31	1132 ± 28[b]	1135 ± 103[b]	104 ± 28[a]	249 ± 66
PHT	300 ± 49	305 ± 54	418 ± 9[a]	700 ± 46	1004 ± 117[a]	1125 ± 188[b]	198 ± 32	261 ± 33
PHB	307 ± 32	301 ± 46	671 ± 94	654 ± 84	925 ± 74[a]	993 ± 60[b]	263 ± 33[b]	277 ± 50
Acyl-carnitine								
Control	27 ± 12	21 ± 16	81 ± 25	60 ± 29	149 ± 76	152 ± 87	118 ± 16	89 ± 24
VPA	29 ± 18	23 ± 10	25 ± 11[a]	58 ± 13	129 ± 49	75 ± 21[c]	110 ± 17	101 ± 40
CBZ	58 ± 39[c]	20 ± 4	63 ± 42	75 ± 22	36 ± 18[b]	83 ± 28[c]	76 ± 24[a]	123 ± 55
PHT	28 ± 22	10 ± 3	27 ± 5[a]	50 ± 8	118 ± 34	102 ± 56	40 ± 16[a]	108 ± 35
PHB	12 ± 8[b]	19 ± 7	56 ± 31	51 ± 13	93 ± 30	137 ± 39	55 ± 25[c]	120 ± 32[c]

[a]Results are mean ± SD. Significantly different from controls: (a) $p < 0.001$; (b) $p < 0.01$; (c) $p < 0.05$.

differences in transaminase activities between VPA-treated patients and healthy controls (3,10).

The high serum A/F ratio in VPA-treated patients (4.8) has suggested that carnitine deficiency might be due to increased conversion of free carnitine to acyl-carnitine. In this study, however, the VPA-induced deficit in serum free carnitine 0.5 h p.a. was balanced by the deficit of acyl-carnitine (which was observed with all drugs); in no case was the A/F ratio at any time above normal. The deficits in serum free and acyl-carnitine, moreover, were not offset by any significant excess in any of the tissues examined, and cannot be explained in terms of increased uptake of carnitine by tissues.

The effects observed in the various tissues examined varied with both drug and tissue. In general, control levels of both free and acyl-carnitine recovered with time. The only effects on free carnitine concentrations in liver were the slight excesses produced by CBZ and PHT. PHT and PHB caused significant deficits in heart and kidney, and CBZ in muscle. Most tissues have carnitine concentrations over 10 times serum levels, so active carnitine uptake by these tissues must take place. The rate of uptake varies widely, however; the turnover times in rat kidney, liver, heart, and skeletal muscle are, respectively, 0.4, 1.3, 21 and 105 h (11). If similar figures hold for mouse tissues, the differences in carnitine concentration found in this study cannot be due solely to the differences in turnover time. This is particularly so with regard to free carnitine concentration, which in serum was reduced only by VPA but in kidney was markedly reduced by PHT and PHB, and in muscle by CBZ.

REFERENCES

1. F. E. Dreifuss and D. H. Langer. Hepatic considerations in the use of antiepileptic drugs. *Epilepsia, 28*:523–529.
2. R. J. Porter. How to use antiepileptic drugs. In *Antiepileptic Drugs* (R. H. Levy, F. E. Dreifuss, R. H. Mattson, B. S. Meldrum, and J. K. Penry, eds.), Raven Press, New York, pp. 117–132 (1989).
3. Y. Ohtani, F. Endo, and I. Matsuda. Carnitine deficiency and hyperammonemia associated with valproic acid therapy. *J. Pediatr., 101*:782–785.
4. S. Rodriguez, C. Alonso de la Peña, J. C. Tutor, J. M. Paz, M. D. Fernandez, I. Rozas, and R. del Rio. Carnitine deficiency associated with anticonvulsant therapy. *Clin. Chem., 181*:175–182 (1989).
5. S. Rodriguez-Segade, C. Alonso de la Peña, J. M. Paz, and R. del Rios. Determination of L-carnitine in serum, and implementation on the ABA-100 and CentrifiChem 600. *Clin. Chem., 31*:754–757 (1985).
6. J. D. McGarry and D. W. Foster. An improved and simplified radioisotopic assay for the determination of free and esterified carnitine. *J. Lipid. Res., 17*:277–281 (1976).
7. D. L. Coulter. Carnitine deficiency: A possible mechanism for valproate hepatotoxicity. *Lancet, I*:689 (1984).
8. I. Matsuda, Y. Ohtani, and N. Ninomiya. Renal handling of carnitine in children with carnitine deficiency and hyperammonemia associated with valproate therapy. *J. Pediatr., 109*:131–134 (1986).
9. N. Nishida, R. Sugimoto, A. Araki, M. Woo, Y. Sakane, and Y. Kbayaski. Carnitine metabolism in valproate-treated rats: The effect of L-carnitine supplementation. *Pediatr. Res.*, 22:500–503 (1987).
10. M. Castro-Gago, S. Otero, I. Novo, E. Rodrigo, I. Rozas, and S. Rodriguez-Segade. Deficiencia de carnitana asociada a hiperamoniemia en niños a tratamiento con ácido valproico. *Rev. Esp. Epilep., 3*:169–172 (1988).
11. D. E. Brooks and J. E. A. McIntosh. Turnover of carnitine by rat tissues. *Biochem. J., 148*:439–445 (1975).

113

Influence of the Inhalation of High Concentrations of Oxygen on Infections by *Aeroginosas peudomonas*: Experimental Study

B. Aparicio, D. Ludeña, C. Garcia, M. Capurro, A. Sanchez, and A. Bullon *Hospital Clinico Universitario de Salamanca, Salamanca, Spain*

I. INTRODUCTION

Patients with acute lung damage usually develop secondary intrahospital infections. Clinically, this is particularly clear after receiving high concentrations of oxygen in the intensive-care unit (ICU). However, many different parameters are possibly involved in this problem; both the intubation system and the depression of the immune defense system could be implicated as well as the oxygen therapy in this kind of superinfection.

Previous results (1,2) suggest that hyperoxia is strongly involved in this problem.

In Spain, hospital infections by *Pseudomonas aeruginosa* (PA) are statistically predominant (31%) (3).

In this study we attempted to establish experimentally whether hyperoxia as an isolated factor might or might not be involved in the susceptibility of the lung to secondary infections, and whether previous hyperoxia due to lung alterations facilitates this infection.

II. MATERIALS AND METHODS

Wistar rats, weighing 150 to 200 g, were exposed to oxygen in a 1.5 M^3 Plexiglas chamber. Oxygen, air, and relative humidity (40% to 60%) were monitored daily in the chamber, and oxygen flow was adjusted to maintain the oxygen concentration higher than 85% to 95% at 1 atm over 4 days. The temperature in the chamber ranged from 22 to 24°C.

A clinically isolated *Pseudomonas aeruginosa* (PA) was used in this study at a doses of 6×10^3 cfu/mL, each suspended in 0.25 mL of PBS. For the PA instillation we introduced PA directly into the trachea after anesthesia with intraperitoneal ketamine.

For this experiment we applied the following experimental protocol (Table 1).

Group A, B, and C are different types of controls with respect to our final purpose, using 20 animals in each group.

Group A had PA instillation breathing ambient air. This group was designed to elucidate the effect of PA instillation under an ambient air atmosphere.

Group B was formed of rats of the some origin as the other groups, breathing ambient air without PA instillation. The aim here was to discover the individual susceptibility of these animals to the infection, living under the same conditions as the other groups.

Group C was treated with 4 days of hyperoxia without PA instillation in order to test only the hyperoxia lesions.

Groups D and E were exposed to hyperoxia for 4 days. These groups were instilled first at the beginning and second at the end of the hyperoxia period. The purpose of these groups was to learn with certainty whether the infection is related to a well-established lesion on the lung parenchyma or whether it is dependent only on the oxygen therapy combination.

All the groups breathed ambient air for 7 days after the treatment.

After deep anesthesia we took specimens from blood, bronchoalveolar lavage (BAL) of the lung, liver, kidney, and spleen for microbiological analysis and after a complete necropsy we processed the lung for light and electron microscopy and the liver, kidney, and spleen only for light microscopy.

For microbiological analysis each specimen was weighed and then individually homogenized. Serial 10-fold dilutions were made and cultured in Mueller Hinton agar and Trypticase soy agar plates containing 25 mg/mL gentamicin and incubated at 37°C. The colonies of PA were evaluated at 24 to 48 h.

For light microscopy the specimens were fixed for 24 h in 10% formalin, dehydrated in graded alcohols, and embedded in paraffin. Multiple blocks were taken from each one of the five lobes of the lung and sectioned at 4 μ.

Table 1 Experimental Protocol Mortality. Ambient air except for □ area which represents hyperoxia 85–95%, 1 atmosphere.

GROUP A				★				Ω (1)			¥
GROUP B											¥
GROUP C		HYPEROXIA									¥
GROUP D	★		Ω (1)								¥
GROUP E				★				Ω (3)	Ω (1)		¥
DAYS	1	2	3	4	5	6	7	8	9	10	11

(20 Rats each Group)

Ω Dead before the end of the experiment

★ Instillations of Clinical Isolated *PseudAuroginosa*

¥ Sacrifice - End of the experiment

Breathing ambient Air

Hyperoxia 85 - 95 % 1 Atmosphere

For electron microscopy the specimens from all lobes of the lung were fixed for 2 h in 2.5% glutaraldehyde, postfixed in osmium tetroxide for 2 h, and then dehydrated in alcohol and embedded in araldite. The specimens selected in 1-μm cuts were cut on an LKB-III ultramicrotome and examined with a Philips 201 electron microscope.

III. RESULTS

A. Mortality

The mortality in this experiment (Table 1) demonstrates a positive effect of the hyperoxia-instillation combination with respect to the survival period of the rats. Only in group E was there a clear increase in dead rats with respect to the other groups.

Fortunately, a mortality of only four rats permitted us to obtain valid conclusions for this group.

In our study we excluded the rats that died before the end of the experiment, that is, on the eleventh day, from microbiological analysis, although all of them were studied by light microscopy.

B. Morphological Results

Group A: The animals died before the end of the experiment (on day 8), and one animal, sacrificed at the end of the experiment, had pneumonia. In both animals the pneumonia was necrotizing and hemorrhagic with mononuclear cells and polymorphonuclear leucocytes (Figure 1).

Group B (controls): No changes were found in any organs of the animals studied.

Group C: Only a lung hyperoxia lesion was found: Increases were observed in the alveolar wall (Figure 2), and occasionally there was intraalveolar edema.

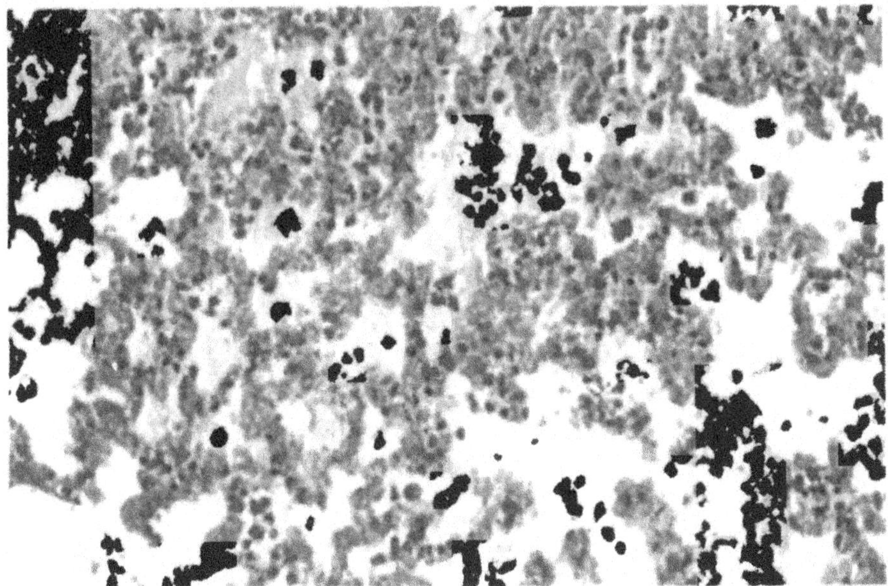

Figure 1 Influence of the inhalation of high concentrations of oxygen in the infections by *Aeroginosas pseudomona*: Experimental study.

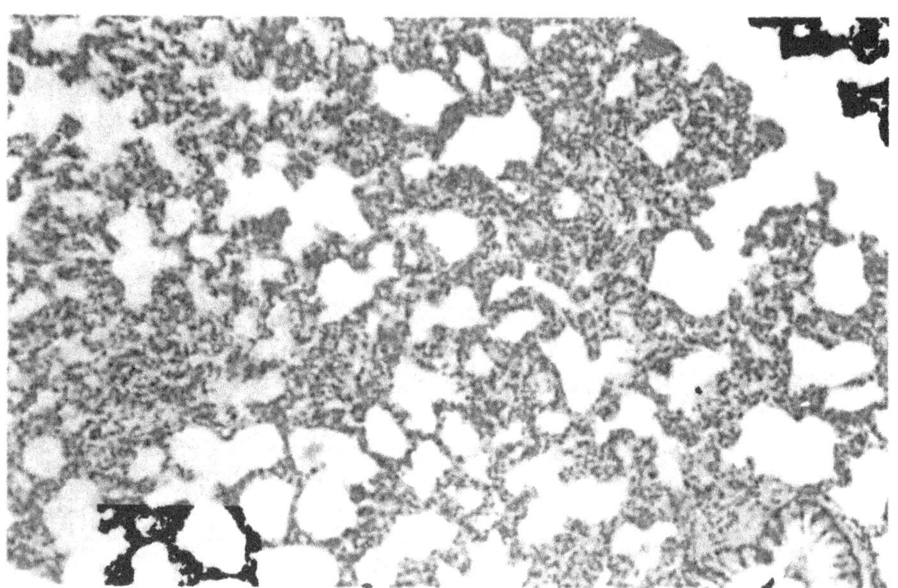

Figure 2 Influence of the inhalation of high concentrations of oxygen in the infections by *Aeroginosas pseudomona*: Experimental study.

Ultrastructurally, the most significant oxygen lesion in this group was a vascular lesion (Figure 3) with alterations in the endothelial cells. Also, it was common to observe interstitial edema and increases in interstitial cells.

Group D: All animals had lung hyperoxia lesions, and two animals (one dead before the end of the experiment and one sacrificed at the end of the experiment) had the same characteristics as did group A.

Group E: All animals had lung hyperoxia lesions and 16 had pneumonia (4 died before the end of the experiment and 12 were sacrificed at the end of the experiment). The pneumonia in this group was similar to that observed in the other group, but in three animals we observed the formations of abscesses (Figures 4 and 5).

There were no kidney, liver, and spleen lesions in any of the groups.

C. Microbiological Results

Group A: The microbiological cultures of BAL fluid, lung, and blood in one animal was viable for PA (200 cuf/mL in BAL, 250 cuf/mL in lung, and 100 cuf/mL in blood).

Groups B and C: All microbiological cultures were negative to PA.

Group D: In BAL, two animals had microbiological cultures positive to PA (300 cuf/mL), and one of these animals also showed positivity in the lung culture (250 cuf/mL).

Group E: In 11 animals, the microbiological cultures were positive to PA (between 150 and 350 cuf/mL) to BAL; 12 animals (between 200 and 350 cuf/mL) were positive to PA in the lung, and 4 of these animals were positive to PA in the blood (150 cfu/mL).

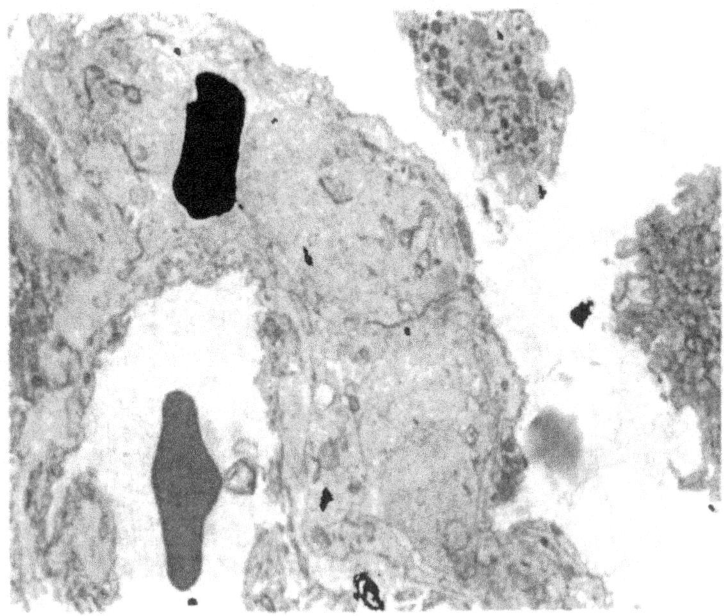

Figure 3 Influence of the inhalation of high concentrations of oxygen in the infections by *Aeroginosas pseudomona*: Experimental study.

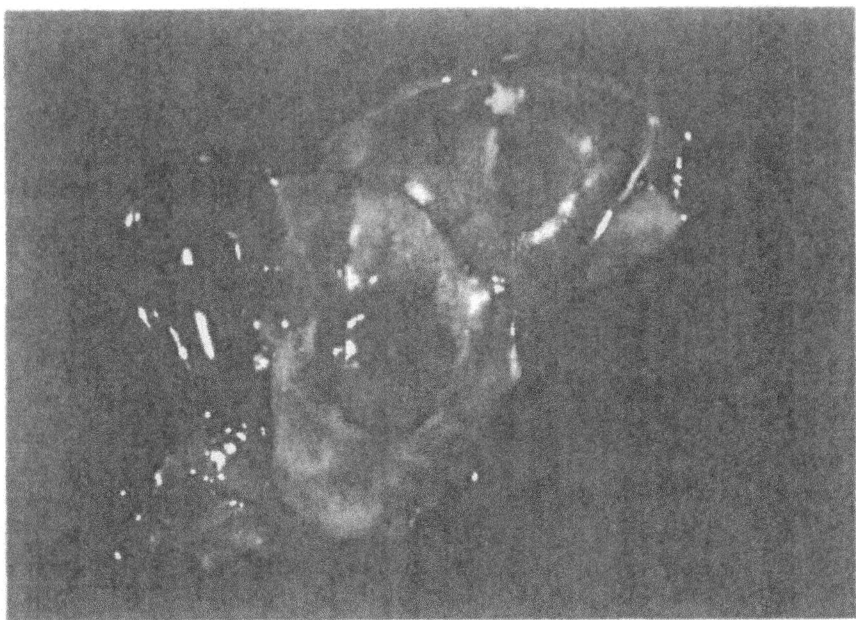

Figure 4 Influence of the inhalation of high concentrations of oxygen in the infecetions by *Aeroginosas pseudomona*: Experimental study.

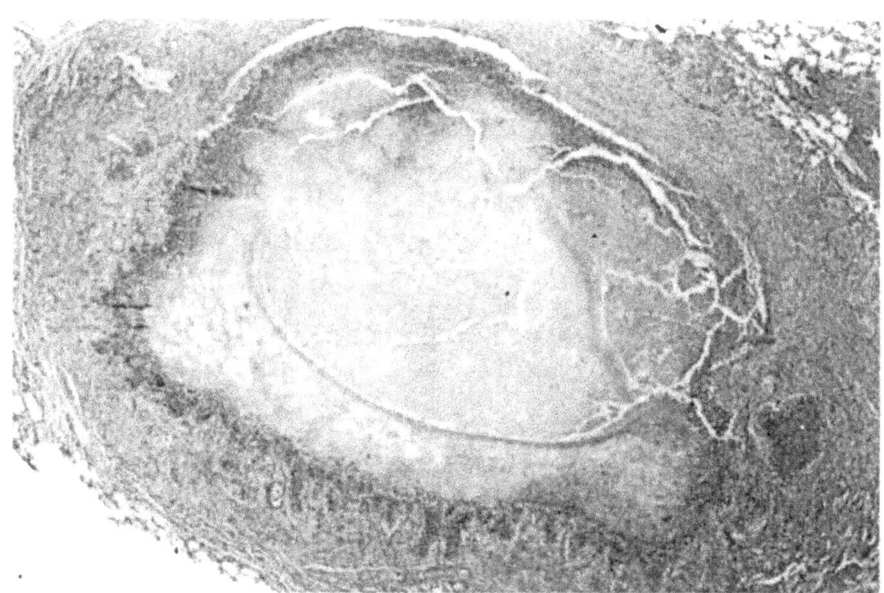

Figure 5 Influence of the inhalation of high concentrations of oxygen in the infections by *Aeroginosas pseudomona*: Experimental study.

Table 2 Morphological and Microbiological Results

	PA positive	Number of rats	Type of lesion	Number of rats	Total number of rats studied
Group A (PA inst. and amb. air)	BAL	1	Pneumonia	1	19
	Lung	1			
	Blood	1			
Group D (PA inst. and hyperoxia)	BAL	2	Hyperoxia lesions	19	19
	Lung	1	Pneumonia	1	
	Blood	—			
Group E (Hyperoxia and PA inst.)	BAL	11	Hyperoxia lesions	16	16
	Lung	12	Pneumonia	12	
	Blood	4			

Table 3 Rats with Pneumonia Lesions

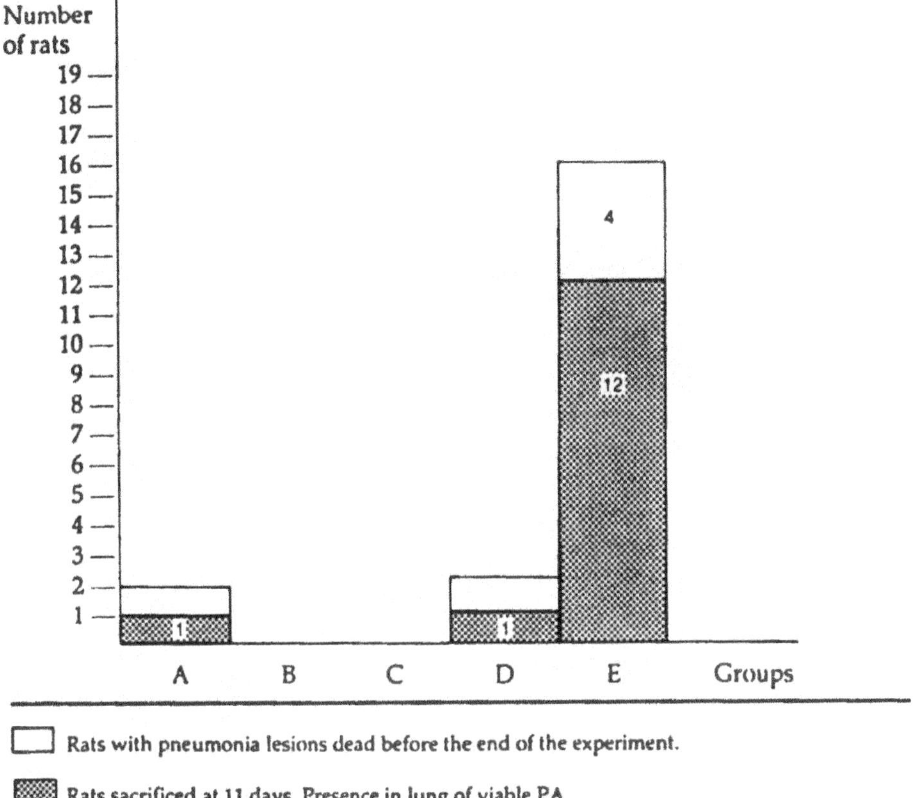

Rats with pneumonia lesions dead before the end of the experiment.

Rats sacrificed at 11 days. Presence in lung of viable PA.

The morphological lesions and the PA-positive microbiological study are summarized in Table 2.

However, what was essential for confirming our initial hypothesis can be seen in Table 3. With respect to the distribution of the pneumonia lesions in the different groups, group E had many more cases of pneumonia: 12 with lesions and microbiological control and 4 dead before the end of the experiment, without microbiological control but with clear morphological lesions.

IV. DISCUSSION

In this study we observed a noteworthy difference among the different groups.

In groups A and D we observed the same mortality and the same numbers of pneumonias. Only two difference were seen in these groups: in group A, one animal had bacteremia (the same animal had a positive culture to PA in BAL and lung); and in group B, two animals had positive cultures to PA in BAL and lung (Table 3).

However, in group E (with PA instillation after the hyperoxia treatment), we observed pneumonias in 80% of the animals studied (of 16 animals, 4 died before the end of the experiment) (Table 3) and 12 of these animals had positive cultures to PA in lung—11 of them in BAL and 4 in blood (Table 2).

What is important about these experiments is that two combinations of hyperoxia and well-established lesions with PA instillation produced an increase in pneumonia lung lesions and a much higher degree of dissemination of the infection with bacteremia in rats as compared with the other groups.

Possible mechanisms of impaired pulmonary bacterial clearance in this setting include the presence of pulmonary edema (4,5), abnormalities of the alveolar macrophages (6–8), or PMN numbers or function (2,9). In previous results, our team (10) reported the presence of intraalveolar edema during this period of intoxication and also alterations in macrophage viability (11). With respect to the influence and importance of PMN cells, many differences have been described. Dunn and Smith (2), suggesting that hyperoxia impairs the pulmonary clearance of PA by decreasing the influx of PMNs, because their results suggest that the reduction in the number of PMNs following hyperoxic exposure plays a major role in impairing the clearance. In our previous results (10) and also in this study we observed an increase of PMN cells, and these cells were predominant in the pneumonia lesions by PA. However, it is possible that hyperoxia might impair PMN function. In our opinion the most important feature of these secondary infections (PA infections) is the previous hyperoxia due to lung alterations, thus confirming our initial hypothesis.

REFERENCES

1. W. G. Johanson, Jr., J. J. Higuchi, D. E. Woods, P. Gomez, and J. J. Coalson. Dissemination of *Peudomona aeruginosa* during lung infection in hamsters. Role of oxygen-induced lung injury. *Am. Rev. Respir. Dis.*, *132*:385–361.
2. M. M. Dunn and L. J. Smith. The effect of hyperoxia on pulmonary clearance of *Pseudomona aeruginosa*. *J. Infect. Dis.*, *153*:676–681 (1986).
3. J. A. Garcia, A. Jimenez, A. Esteban, S. Ruiz Santana, L. Guerra, B. Alvarez, S. Corcia, J. Gudin, A. Martinez, E. Quintana, S. Armengol, A. Araenzana, J. Gregori, P. Merino, and A. San Martin. Microbiologia de las neumonias nosocomiales en seis unidades de medicina intensiva. Estudio multicentrico. *Med. Intensiva*, *12*:404–406 (1988).
4. J. F. Mullane, F. M. LaForce, and G. L. Huber. Variations in lung water and pulmonary host defense mechanism. *Am. Surg.*, *39*:630–636 (1973).
5. F. M. LaForce, J. F. Mullane, R. F. Boehme, W. J. Kelly, and G. L. Huber. The effect of pulmonary edema on antibacterial defenses of the lung. *J. Lab. Clin. Med.*, *83*:634–648 (1973).
6. S. A. Murphy, J. S. Hyams, A. B. Fisher, and R. K. Root. Effect of oxygen exposure on in vitro function of pulmonary alveolar macrophages. *J. Clin. Invest.*, *56*:503–511 (1975).
7. N. Suttorp and L. M. Simon. Decrease bactericidal function and impaired respiratory burst in lung macrophages after sustained in vitro hyperoxia. *Am. Rev. Respir. Dis.*, *128*:486–490 (1983).
8. R. H. Demling. The pathogenesis of respiratory failure after trauma and sepsis. *Surg. Clin. N. A m.*, *60*:1373–1390 (1980).
9. B. Dubaybo and R. W. Carlson, Postinfections ARDS: Mechanism of lung injury and repair. *Crit. Care Clinics*, *4*:229–243 (1988).
10. M. D. Ludeña, F. Sanchez, M. A. Merchan, and A. Bullon. Anatomia patologica subcelular de las lesiones pulmonares precoces inducidas por la respiracion en atmosfera con altas concentraciones de oxigeno. *Morf. Norm. Patol.(B).*, *7*:153–160 (1983).
11. A. Ramos. Estudio experimental sobre la prevencion de las lesiones pulmonares producidas por hiperoxia normobarica. Tesis Doctoral, Universidad de Salamanca, Salamanca, Spain (1986).

Index

Milton Keynes UK
Ingram Content Group UK Ltd.
UKHW052031071024
449327UK00027B/2520

9 780367 402839